INTERNATIONAL SERIES OF MONOGRAPHS ON PHYSICS

INTERNATIONAL SERIES OF MONOGRAPHS ON PHYSICS

163. B.J. Dalton, J. Jeffers, S.M. Barnett: *Phase space methods for degenerate quantum gases*
162. W.D. McComb: *Homogeneous, isotropic turbulence – phenomenology, renormalization and statistical closures*
161. V.Z. Kresin, H. Morawitz, S.A. Wolf: *Superconducting state – mechanisms and properties*
160. C. Barrabès, P.A. Hogan: *Advanced general relativity – gravity waves, spinning particles, and black holes*
159. W. Barford: *Electronic and optical properties of conjugated polymers, Second edition*
158. F. Strocchi: *An introduction to non-perturbative foundations of quantum field theory*
157. K.H. Bennemann, J.B. Ketterson: *Novel superfluids, Volume 2*
156. K.H. Bennemann, J.B. Ketterson: *Novel superfluids, Volume 1*
155. C. Kiefer: *Quantum gravity, Third edition*
154. L. Mestel: *Stellar magnetism, Second edition*
153. R.A. Klemm: *Layered superconductors, Volume 1*
152. E.L. Wolf: *Principles of electron tunneling spectroscopy, Second edition*
151. R. Blinc: *Advanced ferroelectricity*
150. L. Berthier, G. Biroli, J.-P. Bouchaud, W. van Saarloos, L. Cipelletti: *Dynamical heterogeneities in glasses, colloids, and granular media*
149. J. Wesson: *Tokamaks, Fourth edition*
148. H. Asada, T. Futamase, P. Hogan: *Equations of motion in general relativity*
147. A. Yaouanc, P. Dalmas de Réotier: *Muon spin rotation, relaxation, and resonance*
146. B. McCoy: *Advanced statistical mechanics*
145. M. Bordag, G.L. Klimchitskaya, U. Mohideen, V.M. Mostepanenko: *Advances in the Casimir effect*
144. T.R. Field: *Electromagnetic scattering from random media*
143. W. Götze: *Complex dynamics of glass-forming liquids – a mode-coupling theory*
142. V.M. Agranovich: *Excitations in organic solids*
141. W.T. Grandy: *Entropy and the time evolution of macroscopic systems*
140. M. Alcubierre: *Introduction to 3 + 1 numerical relativity*
139. A.L. Ivanov, S.G. Tikhodeev: *Problems of condensed matter physics – quantum coherence phenomena in electron-hole and coupled matter-light systems*
138. I.M. Vardavas, F.W. Taylor: *Radiation and climate*
137. A.F. Borghesani: *Ions and electrons in liquid helium*
135. V. Fortov, I. Iakubov, A. Khrapak: *Physics of strongly coupled plasma*
134. G. Fredrickson: *The equilibrium theory of inhomogeneous polymers*
133. H. Suhl: *Relaxation processes in micromagnetics*
132. J. Terning: *Modern supersymmetry*
131. M. Mariño: *Chern-Simons theory, matrix models, and topological strings*
130. V. Gantmakher: *Electrons and disorder in solids*
129. W. Barford: *Electronic and optical properties of conjugated polymers*
128. R.E. Raab, O.L. de Lange: *Multipole theory in electromagnetism*
127. A. Larkin, A. Varlamov: *Theory of fluctuations in superconductors*
126. P. Goldbart, N. Goldenfeld, D. Sherrington: *Stealing the gold*
125. S. Atzeni, J. Meyer-ter-Vehn: *The physics of inertial fusion*

123. T. Fujimoto: *Plasma spectroscopy*
122. K. Fujikawa, H. Suzuki: *Path integrals and quantum anomalies*
121. T. Giamarchi: *Quantum physics in one dimension*
120. M. Warner, E. Terentjev: *Liquid crystal elastomers*
119. L. Jacak, P. Sitko, K. Wieczorek, A. Wojs: *Quantum Hall systems*
117. G. Volovik: *The Universe in a helium droplet*
116. L. Pitaevskii, S. Stringari: *Bose–Einstein condensation*
115. G. Dissertori, I.G. Knowles, M. Schmelling: *Quantum chromodynamics*
114. B. DeWitt: *The global approach to quantum field theory*
113. J. Zinn-Justin: *Quantum field theory and critical phenomena, Fourth edition*
112. R.M. Mazo: *Brownian motion – fluctuations, dynamics, and applications*
111. H. Nishimori: *Statistical physics of spin glasses and information processing – an introduction*
110. N.B. Kopnin: *Theory of nonequilibrium superconductivity*
109. A. Aharoni: *Introduction to the theory of ferromagnetism, Second edition*
108. R. Dobbs: *Helium three*
107. R. Wigmans: *Calorimetry*
106. J. Kübler: *Theory of itinerant electron magnetism*
105. Y. Kuramoto, Y. Kitaoka: *Dynamics of heavy electrons*
104. D. Bardin, G. Passarino: *The Standard Model in the making*
103. G.C. Branco, L. Lavoura, J.P. Silva: *CP violation*
102. T.C. Choy: *Effective medium theory*
101. H. Araki: *Mathematical theory of quantum fields*
100. L.M. Pismen: *Vortices in nonlinear fields*
99. L. Mestel: *Stellar magnetism*
98. K.H. Bennemann: *Nonlinear optics in metals*
94. S. Chikazumi: *Physics of ferromagnetism*
91. R.A. Bertlmann: *Anomalies in quantum field theory*
90. P.K. Gosh: *Ion traps*
87. P.S. Joshi: *Global aspects in gravitation and cosmology*
86. E.R. Pike, S. Sarkar: *The quantum theory of radiation*
83. P.G. de Gennes, J. Prost: *The physics of liquid crystals*
73. M. Doi, S.F. Edwards: *The theory of polymer dynamics*
69. S. Chandrasekhar: *The mathematical theory of black holes*
51. C. Møller: *The theory of relativity*
46. H.E. Stanley: *Introduction to phase transitions and critical phenomena*
32. A. Abragam: *Principles of nuclear magnetism*
27. P.A.M. Dirac: *Principles of quantum mechanics*
23. R.E. Peierls: *Quantum theory of solids*

Phase Space Methods for Degenerate Quantum Gases

Bryan J. Dalton
Centre for Quantum and Optical Science, Swinburne University of Technology, Melbourne, Victoria, Australia

John Jeffers
Department of Physics, University of Strathclyde, Glasgow, UK

Stephen M. Barnett
School of Physics and Astronomy, University of Glasgow, Glasgow, UK

OXFORD
UNIVERSITY PRESS

OXFORD
UNIVERSITY PRESS

Great Clarendon Street, Oxford, OX2 6DP,
United Kingdom

Oxford University Press is a department of the University of Oxford.
It furthers the University's objective of excellence in research, scholarship,
and education by publishing worldwide. Oxford is a registered trade mark of
Oxford University Press in the UK and in certain other countries

First Edition published in 2015

Published in the United States of America by Oxford University Press
198 Madison Avenue, New York, NY 10016, United States of America

British Library Cataloguing in Publication Data
Data available

Library of Congress Control Number: 2014939573

ISBN 978–0–19–956274–9

Preface

The aim of this book is to present a comprehensive theoretical description of phase space methods for both bosonic and fermionic systems in order to provide a useful textbook for postgraduate students, as well as a reference book for researchers in the newly emerging field of quantum atom optics. Phase space distribution function methods involving phase space variables suitable for systems where small numbers of modes are involved are complemented by phase space distribution functional methods involving field functions for the study of systems with large mode numbers, such as when macroscopic numbers of bosons or fermions are present. The approach for bosonic systems involves c-number quantities, whilst that for fermionic cases involves Grassmann quantities. The book covers both the Fokker–Planck-type equations that determine the distribution functions or functionals, and Langevin-type equations which govern stochastic forms of the variables or fields.

The approach taken to treat bosonic and fermionic systems can be regarded as complementary to approaches taken in other branches of physics, notably quantum field theory, particle physics and statistical physics. In those disciplines, path integrals and Feynman diagrams rather than Fokker–Planck-type equations are the method of choice.

Representative applications to physical systems are presented as examples of the methods, but no attempt is made to review the content of the broad subject of quantum atom optics itself. There are other books and reviews that do this.

The book provides proofs of important results, with detail presented in the Appendices. Each chapter contains a number of problems for students to solve. As Grassmann algebra and calculus will generally be unfamiliar to students and to researchers in quantum atom optics, the main points of this topic are given appropriate coverage.

Chapters dealing with the following topics are included:

- states and operators in bosonic and fermionic systems;
- complex numbers and Grassmann numbers;
- Grassmann calculus;
- fermion and boson coherent states;
- canonical transformations and their applications;
- phase space distributions for fermions and bosons;
- Fokker–Planck equations;
- Langevin equations;
- application to few-mode systems;

- functional calculus for c-number and Grassmann fields;
- distribution functionals in quantum-atom optics;
- functional Fokker–Planck equations;
- Langevin field equations;
- application to multi-mode systems;
- further developments.

Acknowledgements

This book would not have been written without helpful discussions with and comments from colleagues on key theoretical issues over the past several years. In particular, we wish to acknowledge M. Babiker, R. Ballagh, T. Busch, J. Corney, J. Cresser, P. Deuar, P. Drummond, A. Filinov, M. Fromhold, B. Garraway, C. Gilson, M. Olsen, L. Plimak, J. Ruostekowski, K. Rzazewski and R. Walser. This book would not have been completed without the patience and continued support of OUP. Work on this book was supported by the Australian Research Council via the Centre of Excellence for Quantum-Atom Optics (2003–2010). BJD thanks E. Hinds and S. Maricic for the hospitality of the Centre for Cold Matter, Imperial College, London during the writing of this book.

The authors are grateful to Maureen, Hazel and Claire for their patience during the writing of this book.

Contents

1
Introduction

1.1 Bosons and Fermions, Commuting and Anticommuting Numbers

Quantum physics allows for two fundamentally different classes of particles, bosons and fermions. The former are identified by the fact that they carry integer spin (or helicity), while the latter have half-integer spin. This single difference is simply related to the symmetry properties of multi-particle states; it leads, in particular, to the law that while many bosons can occupy the same state, it is forbidden for more than one fermion to occupy a single state. Laser light and Bose–Einstein condensates of ultracold atoms are macroscopic examples of many bosons occupying the same single-particle state. The different chemical properties of the elements, by way of contrast, derive from the fundamental property of the constituent fermionic electrons.

The properties of multi-particle states are most naturally encapsulated in the forms of the creation and annihilation operators, the actions of which add or remove a single particle. The boson creation and annihilation operators, which we denote $\hat{a}_i^\dagger$, $\hat{a}_i$, satisfy the *commutation* relations

$$\begin{aligned}\left[\hat{a}_i, \hat{a}_j^\dagger\right] &\equiv \hat{a}_i\hat{a}_j^\dagger - \hat{a}_j^\dagger\hat{a}_i = \delta_{ij},\\ \left[\hat{a}_i, \hat{a}_j\right] &= 0 = \left[\hat{a}_i^\dagger, \hat{a}_j^\dagger\right],\end{aligned} \tag{1.1}$$

where the indices i, j label orthogonal particle states or field modes. The fermion creation and annihilation operators, which we denote $\hat{c}_i^\dagger$, $\hat{c}_i$, satisfy instead the anticommutation relations

$$\begin{aligned}\left\{\hat{c}_i, \hat{c}_j^\dagger\right\} &\equiv \hat{c}_i\hat{c}_j^\dagger + \hat{c}_j^\dagger\hat{c}_i = \delta_{ij},\\ \left\{\hat{c}_i, \hat{c}_j\right\} &= 0 = \left\{\hat{c}_i^\dagger, \hat{c}_j^\dagger\right\}.\end{aligned} \tag{1.2}$$

For reasons of notational simplicity the mode indices i, j are assumed *discrete*, but a straightforward generalisation deals with cases where the indices are *continuous* (or part discrete, part continuous). Phase space methods involve mapping quantum states, operators and dynamics onto a complex space. For bosons this is a space of complex numbers, or c-numbers, so that, in particular, the operators $\hat{a}_i^\dagger$ and $\hat{a}_i$ are associated with a pair of complex numbers α_i^+ and α_i. For fermions, however, the properties of the operators are in the form of anticommutation relations and the natural representation of them is in terms of anticommuting numbers, known as *Grassmann numbers*, or

Phase Space Methods for Degenerate Quantum Gases. First Edition. Bryan J. Dalton, John Jeffers, and Stephen M. Barnett.
Published in 2015 by Oxford University Press.

g-numbers. The fermionic operators $\hat{c}_i^\dagger$ and $\hat{c}_i$ are associated with the g-numbers g_i^+ and g_i. Theoretical treatments involving both complex numbers and Grassmann numbers are needed for a complete theory. While the former are well known to most physicists, the latter are less familiar, so some brief historical background is warranted.

Complex numbers were first discussed in the eighteenth century by de Moivre and Euler and developed further in the nineteenth century by Gauss and Cauchy. These satisfy the familiar associative, distributive and commutative rules for multiplication and addition that apply to real numbers. Applications of complex numbers in physics and other sciences are wide-ranging – it is hard to think of an area where they are not used. In the field of electromagnetism, the association of currents, charges and fields with c-number quantities, with the amplitude and phase of the physical quantity represented by the magnitude and phase of the c-number involved, is a well-known method. The field of quantum physics is inextricably connected to c-numbers; for example, the transition amplitude for a quantum process is a c-number, whose modulus squared gives the transition probability.

The development of phase space methods for bosons, especially for photons, was one of the first applications of the coherent states introduced by Glauber [1, 2]. Indeed, such methods have played a fundamental role in the analysis of quantum optical models for more than forty years and have been described in numerous texts [3–5].

Grassmann numbers were first discussed in the nineteenth century by the German mathematician of that name in a study in the field of abstract algebra investigating the consequences of modifying the standard rules of multiplication and addition that applied to c-numbers. Applications of Grassmann numbers have been relatively limited, the main applications to date having been in physics in the areas of particle physics and condensed matter physics, where these quantities have proved to be very useful in treating fermion systems [6–8]. The textbook by Berezin [9] gives a comprehensive account of Grassmann numbers for applications in physics.

In 1999 Cahill and Glauber [10] presented the first phase space approach for fermions based on Grassmann variables applicable to quantum optics, following similar lines to their more well-known work for bosons published thirty years earlier [11, 12]. More recently, the field of quantum atom optics has involved treating spatially extended systems containing fermionic and bosonic atoms as well as the electromagnetic field. Now Grassmann fields are starting to be used in the treatment of the fermionic cases. The methods employed are analogous to the treatment of bosonic fields via c-number fields such as in the work of Gatti et al. on spatially squeezed electromagnetic fields [13]. Plimak et al. [14, 15], and Shresta et al. [16], Anastopoulos and Hu [17], Tyc et al. [18], Tempere and Devreese [19] are among the still small number of authors applying or suggesting the use of Grassmann variables in quantum atom optics. Mathematical studies for stochastic analysis using Grassmann variables have also been carried out [20–22]. With the recent rapid development of this field it is timely to attempt a unified presentation of such methods, for both bosons and fermions. This is our purpose.

1.2 Quantum Correlation and Phase Space Distribution Functions

Many applications in quantum atom optics require the determination of expectation values of normally ordered products of creation and annihilation operators, a type of quantum correlation function. If the state of the system is represented by a density operator $\hat{\rho}$, then the quantum correlation functions are given by

$$\begin{aligned} G_{\mathrm{b}}(i, j, \cdots, k, l, \cdots) &= \left\langle \hat{a}_i^\dagger \hat{a}_j^\dagger \cdots \hat{a}_k \hat{a}_l \cdots \right\rangle \\ &= \mathrm{Tr}\left(\hat{\rho}\hat{a}_i^\dagger \hat{a}_j^\dagger \cdots \hat{a}_k \hat{a}_l \cdots\right) \\ &= \mathrm{Tr}\left(\hat{a}_k \hat{a}_l \cdots \hat{\rho}\hat{a}_i^\dagger \hat{a}_j^\dagger \cdots\right) \end{aligned} \tag{1.3}$$

for bosons and

$$\begin{aligned} G_{\mathrm{f}}(i, j, \cdots, k, l, \cdots) &= \left\langle \hat{c}_i^\dagger \hat{c}_j^\dagger \cdots \hat{c}_k \hat{c}_l \cdots \right\rangle \\ &= \mathrm{Tr}\left(\hat{\rho}\hat{c}_i^\dagger \hat{c}_j^\dagger \cdots \hat{c}_k \hat{c}_l \cdots\right) \\ &= \mathrm{Tr}\left(\hat{c}_k \hat{c}_l \cdots \hat{\rho}\hat{c}_i^\dagger \hat{c}_j^\dagger \cdots\right) \end{aligned} \tag{1.4}$$

for fermions. The quantities $i, j, \cdots, k, l, \cdots$ denote the modes of the quantum system, and $\hat{a}_i^\dagger$, $\hat{a}_i$ and $\hat{c}_i^\dagger$, $\hat{c}_i$ are creation and annihilation operators for the boson and fermion cases, respectively. Normal ordering means that creation operators appear on the left of annihilation operators when the density operator is placed on either side of these operators. The cyclic property of the trace means that the density operator can be moved to the middle and, as explained in the next section, this expression for the quantum correlation function is the basis of our phase space approach. A normally ordered product of operators containing at least one annihilation operator therefore gives a zero expectation value if the mode of the system corresponding to that particular operator is in its vacuum or zero-particle state. The theory of photodetection in quantum optics [23] provides a well-known example of where quantum correlation functions appear.

Other operator orderings are possible, and indeed some applications require quantum correlation functions in which the operators are antinormally ordered or symmetrically ordered. These correlation functions can be determined from the normally ordered quantum correlation functions by applying the commutation or anticommutation rules, but, alternatively, phase space methods similar to those presented here for the normally ordered case can also be formulated.

If the number of modes that need to be considered is not too large, then the correlation functions can be determined from distribution functions based on phase space methods. In such methods the boson creation and annihilation operators $\hat{a}_i^\dagger, \hat{a}_i$ are associated with the commuting c-number variables α_i^+, α_i, but the fermion creation and annihilation operators $\hat{c}_i^\dagger, \hat{c}_i$ are associated with the (anticommuting) Grassmann variables g_i^+, g_i. The density operator $\hat{\rho}$ is represented by a phase space distribution function, which for bosons is a c-number function of α_i^+, α_i and their complex conjugates written as $P_{\mathrm{b}}(\alpha_i, \alpha_i^+, \alpha_i^*, \alpha_i^{+*})$, and which in the fermion case is a g-number function of Grassmann variables g_i, g_i^+, written as $P_{\mathrm{f}}(g_i, g_i^+)$.

The phase space methodology involves considering first the so-called characteristic function. The characteristic function for normally ordered quantum correlation functions is the trace of the density operator multiplied on either side by exponential factors involving the annihilation and creation operators. The characteristic function is determined from the distribution function via a phase space integral. For bosons the phase space integral involves standard c-number integration, and for fermions Grassmann integrals are required. Starting from expressions in terms of the characteristic function, the correlation functions can be written as a phase space integral involving the distribution function, with the creation and annihilation operators for each quantum mode replaced by their associated phase space variables. The expressions are analogous to averages of products of the phase space variables, with the distribution function acting as a probability distribution for the phase space variables. However, in the fermion case the distribution function is a Grassmann function with phase space variables that are g-numbers, and even in the boson case, where it is a real c-number function, the phase space variables are c-numbers and certain types of distribution function (such as the Glauber–Sudarshan P function) can be negative in parts of phase space. In the present treatment the bosonic case involves the positive P distribution in a double phase space, which can be chosen to be real and positive. However, it is not unique. In general, the phase space variables cannot be interpreted as possible measured values for physical quantities as they are not real numbers, nor does the distribution function have the required features of reality, positivity and uniqueness to be interpreted as a probability distribution. Indeed, distribution functions are not unique in general and cannot be interpreted as true probability distributions for the variables. Their role, rather, is to aid calculation of the correlation functions. Hence the positive P distribution is often referred to as a quasi-distribution function. The existence of the distribution function is established via canonical forms of the density operator, which is expressed in terms of phase space integrals involving normalised projector operators based on bosonic or fermionic coherent states. These states are eigenstates of the annihilation operators with the phase variables as the eigenvalues. In addition, unnormalised distribution functions $B_{\mathrm{b}}(\alpha_i, \alpha_i^+, \alpha_i^*, \alpha_i^{+*})$ and $B_{\mathrm{f}}(g_i, g_i^+)$ are introduced, based on unnormalised projectors using bosonic or fermionic Bargmann coherent states. These give simple phase space integral expressions for Fock state populations and coherences.

The time dependence for the density operator is determined from either a Liouville–von Neumann equation or a master equation, and from these equations the distribution function can be shown to satisfy a *Fokker–Planck* equation [24]. This step is accomplished by applying the so-called *correspondence rules*. These relate the effect of annihilation and creation operators on the density operator to combinations of differentiations and multiplications with respect to phase space variables acting on the distribution function in the Fokker–Planck equation. For bosons, standard c-number differentiation is involved; for fermionic systems, left and right Grassmann differentiation are required (as we shall see, Grassmann differentiation depends on the direction from which it acts). Conventional Fokker–Planck equations involve only first- and second-order derivatives with respect to the phase variables; however, as will be seen below, generalised Fokker–Planck equations involving higher-order derivatives can occur for some distribution functions [24–26] – Wigner functions being one such case. Fokker–Planck equations are also established for the unnormalised distribution functions B_{b} and B_{f}.

In a further development, the phase space variables are treated as stochastic quantities satisfying *Ito stochastic* equations (Langevin equations), which provide equivalent evolution to the Fokker–Planck equation, assuming the latter involves only first- and second-order differentiation. The Langevin equations contain both deterministic and random noise contributions, whose form is determined from the Fokker–Planck equation. The phase space integral expressions for the quantum correlation functions can then be replaced by stochastic averages. Determination of the phase space distribution function over enough of the phase space may involve a significant amount of calculation. The stochastic approach is often more convenient for numerical work in that sufficiently accurate results for the quantum correlation functions can often be obtained from a relatively small sample of stochastic trajectories. This feature is important when large numbers of bosons or fermions are involved. For Fock state populations and coherences the unnormalised phase space distribution functions provide simpler expressions, so the Langevin equations for the stochastic phase space variables are also determined. This is of particular importance for fermions [14] as the Langevin equations are linear in the stochastic phase space variables, which enables stochastic averages of phase space variables to be obtained from stochastic averages of c-number stochastic functions and stochastic averages of phase space variables at an initial time. The latter quantities are determined from initial conditions for the Fock state populations and coherences. This enables fermion cases to be treated numerically, since only c-number quantities need to be represented on the computer.

The treatment just described is designed to determine normally ordered correlation functions using a distribution function of the Glauber–Sudarshan P type, as generalised to the Drummond–Gardiner positive P version. If symmetrically ordered or antinormally ordered correlation functions are required, then characteristic and distribution functions of the Wigner (W) or Husimi (Q) type can be introduced. However, the present textbook is focused on positive P representations. For systems involving Hamiltonians containing one- or two-body interactions and with linear couplings to the environment, the Fokker–Planck equation will involve only first- and second-order differentiation in the case of the positive P representation, thereby enabling Langevin equations to be developed – an outcome not always possible with the other distributions. For these other distribution functions, the correspondence rules and hence the Fokker–Planck equations will be different. These also may involve only first- and second-order differentiation (for example if approximations are possible), in which case Ito stochastic equations can be obtained that are related to the Fokker–Planck equation as in the positive P case. Other distributions will be briefly treated for completeness.

1.3 Field Operators

In systems which contain a large number of modes, the distribution function treatment becomes unwieldy and a switch to a treatment avoiding a consideration of separate modes is highly desirable. The system is then described in terms of *field operators* $\hat{\Psi}^{\dagger}(\mathbf{r}), \hat{\Psi}(\mathbf{r})$, where $\mathbf{r}$ is the particle position. As we will see, these operators are associated with the creation and destruction of bosonic or fermionic particles at particular positions. For these operators our fundamental discrete mode commutation and anticommutation relations are replaced by

$$\begin{aligned} \left[\hat{\Psi}_{\mathrm{b}}(\mathbf{r}), \hat{\Psi}_{\mathrm{b}}^{\dagger}(\mathbf{r}')\right] &= \delta(\mathbf{r}-\mathbf{r}'), \\ \left[\hat{\Psi}_{\mathrm{b}}(\mathbf{r}), \hat{\Psi}_{\mathrm{b}}(\mathbf{r}')\right] &= 0 = \left[\hat{\Psi}_{\mathrm{b}}^{\dagger}(\mathbf{r}), \hat{\Psi}_{\mathrm{b}}^{\dagger}(\mathbf{r}')\right] \end{aligned} \tag{1.5}$$

for bosons and

$$\begin{aligned} \left\{\hat{\Psi}_{\mathrm{f}}(\mathbf{r}), \hat{\Psi}_{\mathrm{f}}^{\dagger}(\mathbf{r}')\right\} &= \delta(\mathbf{r}-\mathbf{r}'), \\ \left\{\hat{\Psi}_{\mathrm{f}}(\mathbf{r}), \hat{\Psi}_{\mathrm{f}}(\mathbf{r}')\right\} &= 0 = \left\{\hat{\Psi}_{\mathrm{f}}^{\dagger}(\mathbf{r}), \hat{\Psi}_{\mathrm{f}}^{\dagger}(\mathbf{r}')\right\} \end{aligned} \tag{1.6}$$

for fermions. This change, in turn, requires a generalisation of the phase space method involving functionals. Distribution functional methods are required for cases involving large numbers of occupied spatial modes, such as in models of spatial squeezing [13]. For systems with large numbers of fermions, the Pauli exclusion principle limits the occupation number to one and so, automatically, a large number of modes are occupied. Thus functional methods become increasingly important.

The normally ordered quantum correlation functions of interest now involve the field creation and annihilation operators $\hat{\Psi}^{\dagger}(\mathbf{r}), \hat{\Psi}(\mathbf{r})$ and are of the form

$$\begin{aligned} G^{(p,q)}(\mathbf{r}_1, \cdots, \mathbf{r}_p; \mathbf{s}_q, \cdots, \mathbf{s}_1) &= \left\langle \hat{\Psi}^{\dagger}(\mathbf{r}_1)\cdots\hat{\Psi}^{\dagger}(\mathbf{r}_p)\hat{\Psi}(\mathbf{s}_q)\cdots\hat{\Psi}(\mathbf{s}_1)\right\rangle \\ &= \mathrm{Tr}\left(\hat{\Psi}(\mathbf{s}_q)\cdots\hat{\Psi}(\mathbf{s}_1)\hat{\rho}(t)\hat{\Psi}^{\dagger}(\mathbf{r}_1)\cdots\hat{\Psi}^{\dagger}(\mathbf{r}_p)\right), \end{aligned} \tag{1.7}$$

where for an N-particle system we require $p, q \leq N$ to give a non-zero result. In the usual case, $p = q$; the correlation function with $p = q = 1$ is referred to as the first-order correlation function, with $p = q = 2$ it is referred to as the second-order correlation function and so on. In the last equation, we have used the cyclic properties of the trace to place the field annihilation (or creation) operators to the left (or right) of the density operator. In this form, the quantum correlation function is closely related to its appearance in a quantum measurement theory of multiple-particle detection for the quantum state $\hat{\rho}(t)$, the detection process being one in which the state after each successive detection reflects the previous detections and where the system may finally end up in any state. Thus, if the system were in a pure state $|\Psi_i\rangle$, then the probability of finding it in state $|\Psi_f\rangle$ after some measurement process is given by

$$P_{fi} = \langle\Psi_f|\hat{\Omega}|\Psi_i\rangle\langle\Psi_i|\hat{\Omega}^{\dagger}|\Psi_f\rangle, \tag{1.8}$$

where $\hat{\Omega}$ is an operator that reflects the nature of the measurement process, depending for example on a set of positions $\mathbf{r}_1, \mathbf{r}_2, \cdots, \mathbf{r}_n$ where particles are detected. If the initial state was mixed and described by a density operator $\hat{\rho} = \sum_i p_i|\Psi_i\rangle\langle\Psi_i|$ and the final state the system was found in was not recorded, then the overall measurement probability becomes

$$\begin{aligned} P &= \sum_{i,f} p_i\langle\Psi_f|\hat{\Omega}|\Psi_i\rangle\langle\Psi_i|\hat{\Omega}^{\dagger}|\Psi_f\rangle \\ &= \mathrm{Tr}(\hat{\Omega}\hat{\rho}\hat{\Omega}^{\dagger}). \end{aligned} \tag{1.9}$$

It is this form for the quantum correlation function, with the density operator in the centre, which will form the basis of our phase space approach. Of course, owing to the cyclic properties, we also have $P = \mathrm{Tr}(\hat{\rho}\hat{\Omega}^{\dagger}\hat{\Omega})$, the form that is more familiar.

In the phase space functional method, the field operators are associated with field functions rather than the separate mode creation and annihilation operators being represented via phase space variables. The density operator is represented by a distribution functional of the field functions. For boson systems, the field operators obey commutation rules under multiplication and thus it is natural to associate these operators with boson fields involving c-numbers, which commute. For fermion systems, the field operators used obey anticommutation rules under multiplication and thus it is natural to associate these operators with Grassmann fields, involving g-numbers and which anticommute.

The methodology involves considering first the so-called characteristic functional, which is determined from the distribution functional via a functional integral. For the bosonic case, standard c-number functional integration is involved; for the fermionic case, a Grassmann functional integral is required. By using the characteristic functional, the correlation functions can be written as a functional integral involving the distribution functional and with the field operators replaced by field functions.

From either a Liouville–von Neumann equation or a master equation for the quantum density operator, the distribution functional can be shown to satisfy a functional Fokker–Planck equation. This equation involves functional differentiation of the distribution functional with respect to the field functions. For bosons, standard c-number functional differentiation is involved; for fermions, left and right Grassmann functional differentiation is used.

As with the discrete modes, the phase space fields may be treated as stochastic field quantities satisfying Ito stochastic field equations, which are equivalent to the functional Fokker–Planck equation. These contain both deterministic and random noise contributions, determined from the functional Fokker–Planck equation. The phase space functional integral expressions for the quantum correlation functions are then replaced by stochastic averages of the fields. The comment regarding numerical work about the need for stochastic equations in the case of systems with small mode numbers is even more relevant for systems with large mode and particle numbers. Indeed, it may be impractical to solve the functional Fokker–Planck equations numerically, whereas numerical solutions of the stochastic field equations may still be feasible by obtaining enough trajectories to determine the quantum correlation functions accurately.

It should be noted that whilst we have focused on field creation and annihilation operators associated with creating and destroying particles at particular positions, we can alternatively consider field creation and annihilation operators $\hat{\Phi}^{\dagger}(\mathbf{p}), \hat{\Phi}(\mathbf{p})$ which create and destroy particles with a particular *momentum* $\mathbf{p}$. The two sets of field operators are interrelated via Fourier transforms. Our emphasis on the former is because most quantum-atom optics measurements determine particle positions.

2
States and Operators

The physics of systems of identical bosons and fermions involves dynamical processes between basis states describing these systems when the interactions between the particles are absent. The dynamical interactions cause transitions to occur between these states. Basis states are obtained by applying either a symmetrising (bosons) or an antisymmetrising (fermions) operator to products of single-particle states (or modes) occupied by the particles. In the fermion case each mode is occupied by at most one fermion, whereas more than one boson can occupy a particle state.

Physical quantities for such systems involve sums of identical terms, usually involving either one particle at a time or pairs of particles. It is convenient to replace the conventional quantum physics notation by the occupation number notation, in which all single-particle states are listed along with the numbers of identical particles occupying each single-particle state. We can then introduce annihilation and creation operators for each mode, defined by their effect in turning a general basis state into a related basis state – as the names suggest, the new basis state will contain one particle fewer in the mode for the annihilation operator case and one more for the creation operator case. The basis states (now referred to as Fock states) can then be obtained via operating the relevant number of creation operators on a special state with no particles occupying any mode – the vacuum state.

Furthermore, the operators describing physical quantities can be written in terms of these annihilation and creation operators – one-particle terms involving a creation and an annihilation operator for each pair of modes, and two-particle terms involving two creation operators and two annihilation operators for each quartet of modes. From the annihilation and creation operators and their associated spatial mode functions we can then define spatially dependent field operators. These operators are simply annihilation or creation operators for single-particle states with modes given by Dirac delta functions, so need not be regarded as anything fundamentally different. Physical quantities can then be expressed in terms of spatial integrals involving the field operators. This whole approach is often referred to as second quantisation to distinguish it from the so-called first quantisation approach, where creation and annihilation operators are not used, and the aim of this chapter is to describe this second quantisation approach. The distinction is because annihilation and creation operators connect basis states in which the total number of identical particles changes by one, so that a Hilbert space of vectors with all possible total particle numbers is involved, whereas the first quantisation approach restricts this to one fixed value.

2.1 Physical States

As we have seen, fundamental particles and even composite systems such as atoms or nuclei fall into two distinct categories – bosons, which have integer spin, and fermions, with half-integer spin. The state vector for a system of identical particles must satisfy the symmetrisation principle, namely that the state vector is either symmetric (bosons) or antisymmetric (fermions) when any pair of particles is interchanged.

The physical states for identical-particle systems can be developed from a set of basis states in which the identical particles occupy single-particle states, or quantum modes. These single-particle states $|m_1\rangle, |m_2\rangle, \cdots, |m_i\rangle, \cdots, |m_n\rangle$ may be eigenstates for some suitable single-particle Hamiltonian, but other ways of defining them (such as being the eigenvectors for one-particle quantum correlation functions) also exist and the choice made is usually dictated by the physics. We list all members of this set of states as $m = \{m_1, m_2, \cdots, m_n\}$, where we choose $m_1 < m_2 < \cdots < m_n$. For simplicity, we assume there are a finite number n of such single-particle states, though this restriction is not necessary. Single-particle states are chosen to be orthonormal:

$$\langle m_i | m_j \rangle = \delta_{ij}. \tag{2.1}$$

The symbols m_i used to designate a single-particle state may involve energy and spin quantum numbers as well as those associated with spatial motion, such as orbital-angular-momentum quantum numbers. The simplest boson case involves bosons with spin 0. Then the modes are distinguished only via their different spatial functions. The simplest fermion case, however, involves fermions with spin $\frac{1}{2}$. For this case single-fermion states with two distinct magnetic quantum numbers $-\frac{1}{2}$ and $+\frac{1}{2}$ occur, though the associated spatial functions can be the same for the two different magnetic sub-states. Bosons with non-zero spin and all fermion systems require a spin magnetic quantum number to designate the single-particle state. In general the subscripts i, j will specify the magnetic spin state as well as the spatial function – sometimes we may make this explicit by introducing a pair of quantum numbers such as α_i, β_j.

The treatment that follows sometimes requires us to consider subsets of the complete set of single-particle states. For example, we may consider just the subset of quantum modes that are occupied by identical particles. We introduce the convention that the symbol $m = \{m_1, m_2, \cdots, m_n\}$ will always designate the entire set of n single-particle states and symbols such as $l = \{l_1, l_2, \cdots, l_p\}$, $k = \{k_1, k_2, \cdots, k_q\}$ etc. will be used to designate subsets with p, q etc. ($\leq n$) members. However, ordering conventions $l_1 \leq l_2 \leq \cdots \leq l_p$, $k_1 \leq k_2 \leq \cdots \leq k_q$ etc. will still apply, the equality sign allowing for cases where a particular single-particle state appears more than once.

Suppose there are N identical particles. The state in which particle μ_1 occupies state $|l_1\rangle$, particle μ_2 occupies state $|l_2\rangle, \cdots$ and particle μ_N occupies state $|l_N\rangle$ will be designated via the tensor product state

$$|l_1(\mu_1)\rangle |l_2(\mu_2)\rangle \cdots |l_N(\mu_N)\rangle, \tag{2.2}$$

where now the occupied modes are the set $l = \{l_1, l_2, \cdots, l_N\}$, which is a subset of m, with $m_1 \leq l_1 \leq l_2 \leq \cdots \leq l_N \leq m_n$ – which allows for the possibility that a particular mode contains more than one identical particle.

These product states do not satisfy the symmetrisation principle required by states of identical quantum particles. The physical states are constructed by applying either a symmetrising operator $\hat{\mathcal{S}}$ or an antisymmetrising operator $\hat{\mathcal{A}}$, which are defined in terms of permutation operators $\hat{P}(\mu_1, \mu_2, \cdots, \mu_N)$ – the action of which replaces particle 1 by μ_1, 2 by $\mu_2, \cdots, N$ by μ_N – where the quantity $\epsilon(\mu_1, \mu_2, \cdots, \mu_N) = +1, -1$ if $\hat{P}(\mu_1, \mu_2, \cdots, \mu_N)$ is an even or odd permutation:

$$\begin{aligned} \hat{\mathcal{S}} &= \mathcal{N}_\mathrm{b} \sum_P \hat{P}(\mu_1, \mu_2, \cdots, \mu_N), \\ \hat{\mathcal{A}} &= \mathcal{N}_\mathrm{f} \sum_P \epsilon(\mu_1, \mu_2, \cdots, \mu_N) \hat{P}(\mu_1, \mu_2, \cdots, \mu_N), \end{aligned} \tag{2.3}$$

and $\mathcal{N}$ is a normalising factor.

The corresponding physical states for N identical particles satisfying the symmetrisation principle are

$$\begin{aligned} |l_1, l_2, \cdots, l_N\rangle &= \hat{\mathcal{S}}|l_1(1)\rangle|l_2(2)\rangle \cdots |l_N(N)\rangle \qquad \text{bosons,} \\ &= \hat{\mathcal{A}}|l_1(1)\rangle|l_2(2)\rangle \cdots |l_N(N)\rangle \qquad \text{fermions,} \end{aligned} \tag{2.4}$$

and as the physical state vector no longer distinguishes which identical particle occupies which mode, the state can be specified by stating only which single-particle states are occupied. These states can be shown to be orthonormal:

$$\langle l_1, l_2, \cdots, l_N | k_1, k_2, \cdots, k_N\rangle = \delta_{l_1 k_1} \cdots \delta_{l_N k_N}. \tag{2.5}$$

For fermions, the Pauli exclusion principle that no two fermions can occupy the same mode follows directly. For if two modes l_i and l_j were the same, then the antisymmetrising operator would produce zero when acting on $|l_1(1)\rangle \cdots |l(i)\rangle \cdots |l(j)\rangle \cdots |l_N(N)\rangle$. The physical consequence of this is that fermions tend to fill up the single-particle states one at a time from the lowest in energy until the last fermion is accommodated. For electrons in a metal, this results in high-energy states being occupied even at very low temperatures, with the electrons filling states up to a Fermi surface in momentum space. Applying the Pauli principle to electrons in an atom explains the periodic table of the elements, as chemical properties are associated with the angular-momentum quantum numbers of electrons in the spatially outermost states. As the maximum occupancy of states with orbital angular momentum l is $2(2l + 1)$, there is a periodic repetition of these outer-electron angular momenta, as states with different principal quantum number n are filled up.

For bosons, there is no restriction on the occupancy of any mode. The symmetrising operator even acts on the product vector $|l(1)\rangle \cdots |l(N)\rangle$, where all the modes are the same, and does not give zero. At low temperatures bosons tend to fill up the lowest-energy mode, leading to the formation of Bose–Einstein condensates. This macroscopic occupancy of a single mode leads to many interesting coherence effects.

A better notation would be to list *all* the single-particle states m_i and then specify the *occupation number* ν_i, which gives how many identical particles occupy each mode m_i. For bosons this number ν_i can be $0, 1, 2, \cdots$, but for fermions it is $0, 1$ only.

The normalised basis states may now be designated as $|m_1, \nu_1; m_2, \nu_2; \cdots, m_n, \nu_n\rangle$ and the general physical state $|\Psi\rangle$ for N identical particles will be written as a quantum superposition of the basis states. However, in order to avoid too many symbols we shall generally use a more abbreviated notation. As by convention all the modes m_i are included, it is only necessary to specify the occupation numbers ν_i. Thus the normalised basis state $|m_1, \nu_1; \cdots, m_n, \nu_n\rangle$ is designated $|\nu_1, \cdots, \nu_n\rangle$ or $|\nu\rangle$ for short, where by the symbol $\nu = \{\nu_1, \cdots, \nu_n\}$ we denote the whole set of $\nu_1, \cdots, \nu_n$ and we have

$$|\Psi\rangle_N = \sum_{\nu_1, \cdots, \nu_n} B_N(\nu)|\nu\rangle, \tag{2.6}$$

$$\langle\nu|\xi\rangle = \delta_{\nu_1\xi_1} \cdots \delta_{\nu_n\xi_n}, \tag{2.7}$$

$$\sum_{\nu} |B_N(\nu)|^2 = 1, \tag{2.8}$$

where the sum runs over the occupancy numbers $\nu_1, \cdots, \nu_n$, the $B_N(\nu)$ are complex coefficients and $\sum_{i=1}^{n} \nu_i = N$. Note that the sum is only over occupation numbers, subject to the constraint that the total occupancy is N. The last two equations make sure that the states are orthonormal. However, for the physical states of identical-particle systems we are not restricted to pure states of the form (2.6).

For closed systems (in which the number of particles N is prescribed), there are mixed states described by a quantum density operator $\hat{\rho}_N$ of the form

$$\hat{\rho}_N = \sum_{\nu} \sum_{\xi} \rho_N(\nu, \xi)|\nu\rangle\langle\xi|. \tag{2.9}$$

The complex coefficients $\rho_N(\nu, \xi)$ are the density matrix elements. The requirements that the density operator is Hermitian, i.e. $\hat{\rho}_N = \hat{\rho}_N^\dagger$, has unit trace $\mathrm{Tr}(\hat{\rho}_N) = 1$ and for a mixed state satisfies the condition $\mathrm{Tr}(\hat{\rho}_N^2) < 1$ lead to well-known constraints on the density matrix elements. For pure states, $\hat{\rho}_N = |\Psi\rangle_N\langle\Psi|$ and $\mathrm{Tr}\ \hat{\rho}_N^2 = 1$, and in this case the density operator can be written with $\rho_N(\nu, \xi) = B_N(\nu)B_N^*(\xi)$.

If we consider all the states $|\nu\rangle$ arranged in order of total occupancy $\sum_i \nu_i = 1, 2, 3, \cdots$, a hierarchy of basis states for identical-particle systems with various total particle numbers $N = 1, 2, \cdots$ can be listed. In addition to these states, we can introduce the vacuum state $|0\rangle$ for the system of zero identical particles. It is then convenient to mathematically define a Hilbert space containing all superpositions of the form

$$|\Psi\rangle = \sum_N C_N|\Psi\rangle_N, \tag{2.10}$$

where the C_N are complex coefficients. Such a state is not physical, as super-selection rules do not allow superpositions of states with differing total particle numbers (apart, it seems, from the special case of photons). However, states of this form are useful mathematically, even though the pure physical states are restricted to subspaces of this Hilbert space.

For open systems (which particles can enter or leave), physical states can be prepared in which the number of identical particles is not prescribed. A generalisation of the system concept is required to incorporate such cases where the system composition is indefinite. Such states are mixed and can only be described by a density operator $\hat{\rho}$, not a state vector. However, super-selection rules require that the density operator can only be of the form

$$\begin{aligned}\hat{\rho} &= \sum_N f_N |\Psi\rangle_N \langle\Psi| \\ &= \sum_N f_N \hat{\rho}_N,\end{aligned} \tag{2.11}$$

where $\sum_N f_N = 1, f_N \geq 0$, which only involves component density operators $\hat{\rho}_N$ for systems with specified total particle numbers N (again, except for photons), each weighted by real, positive f_N. The density operator satisfies the earlier requirements for mixed states. Pure states have only one $f_N = 1$ and all others vanish. More general density operators such as

$$\hat{\rho} = \sum_N \sum_M \rho_{N,M} |\Psi\rangle_{NM} \langle\Psi| \tag{2.12}$$

with non-zero $\rho_{N,M}$ do not represent physical states, even though they are mathematical operators in the general mathematical Hilbert space describing identical-particle systems with all possible total particle numbers N.

The basic justification of the super-selection rule is that in the non-relativistic quantum physics of many-atom systems that we are considering, the Hamiltonian commutes with the total particle number, and hence if $|\Psi\rangle$ is a solution of the Schrödinger equation it follows that the $|C_N|^2$ are time-independent. Hence the quantum superposition (if it exists as a physical state) would need to have been prepared at the initial time. Such a superposition is of states whose energies differ by relativistic energies of order of the rest mass energy mc^2 – the single-atom and atom–atom interaction energies are non-relativistic and there are no non-relativistic processes available that would create such a state. For example, processes such as the dissociation of up to N diatomic molecules into pairs of atoms can be described via entangled molecule–atom states. However, the state for the atoms alone will be a mixed state described by a (reduced) density operator with atom numbers ranging from 0 up to $2N$. Furthermore, even if such a superposition as in (2.10) existed, there is no observable that could measure the phase difference between states with differing total atom numbers – this phase difference would be associated with a relativistic frequency. The case of the photon is of course different – its rest mass energy is zero and photon modes with different frequencies have only non-relativistic energy differences. Only measurements of the total atom number are feasible – which would be proportional to $|C_N|^2$ for a pure state.

In spite of these considerations, it has been asserted that superpositions of states with differing total particle numbers, such as the boson coherent states described in Chapter 5, are needed to describe coherence and interference effects in systems such as Bose–Einstein condensates. However, as Leggett [27] has pointed out (see also Bach and Rzazewskii [28] and Dalton and Ghanbari [29]), a highly occupied number state for a single mode with N bosons has coherence properties of high order n, as long as $n \ll N$. The introduction of a coherent state is *not* required to account for coherence effects.

However, for systems containing both bosons and fermions, or two different types of either, the total number of particles need not remain unchanged. For example, the bosons may be molecules formed from pairs of fermionic atoms. A bosonic molecule can be split into its constituent fermionic atoms or created from their recombination. In such processes the number of bosons decreases (or increases) by one and the number of fermions increases (or decreases) by two. This constraint of changing fermion numbers by even integers is, as far as we know, a fundamental selection rule. Such situations can be treated, but N would then refer to $\sum_i \nu_i^{\mathrm{B}} + 2\sum_j \xi_j^{\mathrm{f}}$, where ν_i^{B} and ξ_j^{f} would be the occupancies of the boson and fermion modes, respectively. More general basis states are needed, but the density operator (2.11) is still diagonal.

2.2 Annihilation and Creation Operators

We can now define creation and annihilation operators associated with each mode by their effect on the basis states $|\nu\rangle$. The annihilation operator is defined to reduce the number of particles occupying a single-particle state by one, and the creation operator increases it by one. These operators therefore change N-particle states into $N-1$- and $N+1$-particle states, respectively.

For *fermions*, the annihilation operator $\hat{c}_i$ for the mode m_i is defined by its effect on the two possible states, occupied and unoccupied:

$$\begin{aligned}\hat{c}_i|m_1,\nu_1;\cdots;m_i,0;\cdots,m_n,\nu_n\rangle &= 0,\\ \hat{c}_i|m_1,\nu_1;\cdots;m_i,1;\cdots,m_n,\nu_n\rangle &= (-1)^{\eta_i}|m_1,\nu_1;\cdots;m_i,0;\cdots,m_n,\nu_n\rangle,\end{aligned} \tag{2.13}$$

where $(-1)^{\eta_i} = +1$ or -1 according to whether there is an even or odd number of modes listed preceding the mode m_i which are occupied ($\eta_i = \sum_{j<i}\nu_j$). This factor is required for consistency with the antisymmetry of fermion states. Orthogonality of the basis states means that the only non-vanishing matrix element of $\hat{c}_i$ is

$$\langle m_1,\nu_1;\cdots;m_i,0;\cdots,m_n,\nu_n|\hat{c}_i|m_1,\nu_1;\cdots;m_i,1;\cdots,m_n,\nu_n\rangle = (-1)^{\eta_i}. \tag{2.14}$$

By taking the complex conjugate, it is then not difficult to work out the non-vanishing matrix element for the creation operator $\hat{c}_i^\dagger$. We have

$$\langle m_1,\nu_1;\cdots;m_i,1;\cdots,m_n,\nu_n|\hat{c}_i^\dagger|m_1,\nu_1;\cdots;m_i,0;\cdots,m_n,\nu_n\rangle = (-1)^{\eta_i}, \tag{2.15}$$

and hence the effect of the creation operator is

$$\begin{aligned}\hat{c}_i^\dagger|m_1,\nu_1;\cdots;m_i,0;\cdots,m_n,\nu_n\rangle &= (-1)^{\eta_i}|m_1,\nu_1;\cdots;m_i,1;\cdots,m_n,\nu_n\rangle,\\ \hat{c}_i^\dagger|m_1,\nu_1;\cdots;m_i,1;\cdots,m_n,\nu_n\rangle &= 0.\end{aligned} \tag{2.16}$$

For fermions, the mode occupancy is restricted to be 0 or 1. This means that fermion annihilation and creation operators give zero when applied twice or more. The basic results for the fermion annihilation and creation operators can be summarised as

$$\begin{aligned}\hat{c}_i|\nu_1;\cdots;\nu_i;\cdots,\nu_n\rangle &= \nu_i(-1)^{\eta_i}|\nu_1;\cdots;1-\nu_i;\cdots,\nu_n\rangle,\\ \hat{c}_i^\dagger|\nu_1;\cdots;\nu_i;\cdots,\nu_n\rangle &= (1-\nu_i)(-1)^{\eta_i}|\nu_1;\cdots;1-\nu_i;\cdots,\nu_n\rangle.\end{aligned} \tag{2.17}$$

It is then straightforward to establish that the fermion annihilation and creation operators satisfy the anticommutation rules

$$\left\{\hat{c}_i, \hat{c}_j^\dagger\right\} = \hat{c}_i\hat{c}_j^\dagger + \hat{c}_j^\dagger\hat{c}_i = \delta_{ij},$$
$$\{\hat{c}_i, \hat{c}_j\} = 0 = \left\{\hat{c}_i^\dagger, \hat{c}_j^\dagger\right\}, \tag{2.18}$$

where the subscripts i, j refer to modes. The presence of the factor $(-1)^{\eta_i}$ is required for establishing these rules, which are proved in Appendix A. Note that (2.18) implies that fermion annihilation and creation operators give zero when applied twice:

$$(\hat{c}_i)^2 = (\hat{c}_i^\dagger)^2 = 0. \tag{2.19}$$

For *bosons*, the annihilation operator $\hat{a}_i$ for the mode m_i is defined by its effect on the state with occupancy ν_i. Square root factors are introduced so that the creation and annihilation operators will have the same commutation rules as those for a quantum harmonic oscillator, so

$$\hat{a}_i|\nu\rangle = \sqrt{\nu_i}|\nu_1;\cdots;\nu_i - 1;\cdots,\nu_n\rangle \tag{2.20}$$

and hence the non-zero matrix elements of $\hat{a}_i$ and $\hat{a}_i^\dagger$ are

$$\langle\nu_1;\cdots;\nu_i - 1;\cdots,\nu_n|\hat{a}_i|\nu_1;\cdots;\nu_i;\cdots,\nu_n\rangle = \sqrt{\nu_i}, \tag{2.21}$$
$$\langle\nu_1;\cdots;\nu_i - 1;\cdots,\nu_n|\hat{a}_i^\dagger|\nu_1;\cdots;\nu_i;\cdots,\nu_n\rangle = \sqrt{\nu_i + 1}, \tag{2.22}$$

where the non-zero matrix elements for the creation operator $\hat{a}_i^\dagger$ are obtained by taking the complex conjugate. We then have

$$\hat{a}_i^\dagger|\nu_1;\cdots;\nu_i;\cdots,\nu_n\rangle = \sqrt{\nu_i + 1}|\nu_1;\cdots;\nu_i + 1;\cdots,\nu_n\rangle. \tag{2.23}$$

Thus for bosons the mode occupancy is unrestricted. We can establish further that the boson annihilation and creation operators satisfy the commutation rules

$$\left[\hat{a}_i, \hat{a}_j^\dagger\right] = \hat{a}_i\hat{a}_j^\dagger - \hat{a}_j^\dagger\hat{a}_i = \delta_{ij},$$
$$[\hat{a}_i, \hat{a}_j] = 0 = \left[\hat{a}_i^\dagger, \hat{a}_j^\dagger\right]. \tag{2.24}$$

2.3 Fock States

The basis states can now be determined in terms of the annihilation and creation operators. For physical states in *fermion* systems, the basis states involving a set of occupied modes l, with $m_1 \leq l_1 < l_2 < \cdots < l_p \leq m_n$, are given by

$$|l_1, 1; l_2, 1; \cdots, l_p, 1\rangle \equiv \hat{c}_{l_1}^\dagger\hat{c}_{l_2}^\dagger\cdots\hat{c}_{l_p}^\dagger|0\rangle. \tag{2.25}$$

The proof of this result is left as an exercise. The Hilbert space spanned by the full set of such basis states is the same mathematical linear vector space as we have previously encountered, *Fock space*. As emphasised previously, not all of its vectors represent physical states. The basis states themselves (2.25) are referred to as *Fock states*.

The Fock states will now be written in terms of the occupancy notation ν for all of the modes m, with mode m_i occupied by ν_i fermions, as

$$|\nu\rangle = \left(\hat{c}_{m_1}^\dagger\right)^{\nu_1} \left(\hat{c}_{m_2}^\dagger\right)^{\nu_2} \cdots \left(\hat{c}_{m_n}^\dagger\right)^{\nu_n} |0\rangle, \tag{2.26}$$

where the convention is that in the product the $\hat{c}_{m_i}^\dagger$ are arranged in order with $\hat{c}_{m_1}^\dagger$ on the left and $\hat{c}_{m_n}^\dagger$ on the right. The occupancies are $\nu_i = 0, 1$, and $(\hat{c}_{m_i}^\dagger)^0 = \hat{1}$ and $(\hat{c}_{m_i}^\dagger)^1 = \hat{c}_{m_i}^\dagger$. This expression for the basis states is consistent with the definition of the annihilation operators $\hat{c}_i$, and the state satisfies the required orthonormality conditions. The occupancy ν_i $(= 0, 1)$ is an eigenvalue of the mode number operator $\hat{n}_i = \hat{c}_i^\dagger \hat{c}_i$.

The general non-physical state in Fock space will be given by superpositions of the Fock states (2.25) with c-number coefficients $B(\nu_1, \nu_2, \cdots, \nu_n) \equiv B(\nu)$,

$$|\Phi\rangle_{\mathrm{f}} = \sum_{\nu} B(\nu)|\nu\rangle. \tag{2.27}$$

As there are n modes, each of which has two possible occupancies $\nu_i = 0, 1$, there must be a total of 2^n different sets of occupancies. As each set is associated with a c-number coefficient $B(\nu)$, a finite number 2^n of c-numbers are involved in specifying an arbitrary (non-physical) state in Fock space allowing for all possible fermion numbers.

If only physical states with N fermions are allowed, then the only non-zero coefficients in (2.27) are such that $\sum_i \nu_i = N$, because the Pauli principle requires that only one fermion occupies each mode and hence there must be N occupied modes. As there are ${}^nC_N$ different sets of N modes that can be chosen from the available n, there will be ${}^nC_N$ different basis vectors for states of the N-fermion system. To specify physical pure states of an N-fermion system only ${}^nC_N = n!/(N![n-N]!)$ c-number coefficients are therefore required, which is a much smaller number than 2^n. In fact, for $N \ll n$ we find that

$${}^nC_N \approx \sqrt{\frac{2}{\pi n}} 2^n \exp\left(-\frac{[n-2N]^2}{2n}\right), \tag{2.28}$$

which is exponentially smaller than 2^n. Of course, with n modes, each only able to contain one fermion, the maximum number of fermions whose states are in Fock space is $N = n$.

By specialising to a physical N-fermion state we find

$$|\Psi\rangle_{N\mathrm{f}} = \sum_{\nu} B_N(\nu)(\hat{c}_{m_1}^\dagger)^{\nu_1}(\hat{c}_{m_2}^\dagger)^{\nu_2} \cdots (\hat{c}_{m_n}^\dagger)^{\nu_n}|0\rangle, \tag{2.29}$$

where now only coefficients where $\sum_i \nu_i = N$ are non-zero.

For *bosonic* physical states involving a set of occupied modes l, with $m_1 \leq l_1 < l_2 < \cdots < l_p \leq m_n$ and with mode l_i occupied ξ_i times, the basis vectors are given by

$$|l_1, \xi_1, \cdots, l_p, \xi_p\rangle = \frac{\left(\hat{a}_{l_1}^\dagger\right)^{\xi_1}}{\sqrt{\xi_1!}} \cdots \frac{\left(\hat{a}_{l_p}^\dagger\right)^{\xi_p}}{\sqrt{\xi_p!}} |0\rangle. \tag{2.30}$$

The proof of this result is left as an exercise. The Hilbert space spanned by the full set of such basis states is the same mathematical linear vector space we have previously encountered and is again known as Fock space. As emphasised previously, not all of its vectors represent physical states. The basis states themselves (2.30) are also referred to as Fock states.

The Fock states will now be expressed in the occupancy notation for all of the modes, with mode m_i occupied by ν_i bosons, as

$$|\nu\rangle = \frac{\left(\hat{a}^\dagger_{m_1}\right)^{\nu_1}}{\sqrt{\nu_1!}} \cdots \frac{\left(\hat{a}^\dagger_{m_n}\right)^{\nu_n}}{\sqrt{\nu_n!}} |0\rangle. \tag{2.31}$$

The order of the creation operators in the product is immaterial for bosonic operators as they mutually commute. The occupancy ν_i is an eigenvalue of the mode number operator $\hat{n}_i = \hat{a}^\dagger_i \hat{a}_i$.

The general non-physical state in Fock space is formed from superpositions of the Fock states (2.31) with c-number coefficients $B(\nu)$,

$$|\Phi\rangle_\mathrm{b} = \sum_\nu B(\nu)|\nu\rangle. \tag{2.32}$$

As each mode can have any occupancy, even with a finite number of modes the total number of distinct basis vectors in the bosonic Fock space will be infinite. Thus an infinity of different c-number coefficients $B(\nu)$ is required to specify an arbitrary state in Fock space.

If only physical states with N bosons are allowed, then the only non-zero coefficients are such that $\sum_i \nu_i = N$. Any number of bosons can occupy each mode, so it follows that the number of occupied modes p can be any number from 1 to N. There are ${}^nC_p$ different sets of p modes that can be chosen from the n that are available, and for each of this set of p modes we denote $F(N,p)$ as the number of ways of choosing the occupancies such that $\sum_i \nu_i = N$. Hence there are $\sum_p F(N,p)^n C_p$ different basis vectors for the N-boson system, and this is the number of c-number coefficients required to specify any physical pure state. For n modes, each able to contain any number of bosons, there is no maximum number of bosons whose states are in Fock space, in contrast to the fermion case.

For the case of a physical N-boson state, we have

$$|\Psi\rangle_{N\mathrm{b}} = \sum_\nu B_N(\nu) \frac{\left(\hat{a}^\dagger_{m_1}\right)^{\nu_1}}{\sqrt{\nu_1!}} \cdots \frac{\left(\hat{a}^\dagger_{m_n}\right)^{\nu_n}}{\sqrt{\nu_n!}} |0\rangle, \tag{2.33}$$

where now only coefficients where $\sum_i \nu_i = N$ are non-zero. The Fock space treatment of mixed states for closed and open systems, both fermionic and bosonic, in terms of density operators is a straightforward development of the results in Section 2.1.

Two important Fock states for non-interacting particles are the ideal single-mode boson state and the fermion ground state. In both cases we consider N identical particles. The bosons all occupy the same mode, and if this mode has the lowest single-particle energy then this state is the standard single-mode *Bose–Einstein condensate.*

The fermions all have to occupy differing modes because of the Pauli exclusion principle, and if these have energies from the lowest up to an energy called the Fermi energy then the state is the *Fermi ground state.*

For the boson case the quantum state is

$$|\Psi\rangle_{\mathrm{BEC}} = \frac{\left(\hat{a}_1^\dagger\right)^n}{\sqrt{N!}}|0\rangle, \tag{2.34}$$

where the lowest-energy mode has creation operator $\hat{a}_1^\dagger$. The fermion quantum state is

$$|\Psi\rangle_{\mathrm{Fermi}} = \hat{c}_1^\dagger \hat{c}_2^\dagger \cdots \hat{c}_N^\dagger |0\rangle, \tag{2.35}$$

where the occupied modes have creation operators $\hat{c}_1^\dagger, \hat{c}_2^\dagger, \cdots, \hat{c}_N^\dagger$. Note that for non-interacting spin-1/2 fermions, modes with spin up and spin down would have the same energy. This state, in which all the fermions have occupied the lowest energy levels available, would also have the lowest total energy, because exciting a fermion into an unoccupied higher energy level would increase the total energy.

2.4 Two-Mode Systems

In order to further illustrate the above notation using interesting states, we need to treat cases where more than one mode is involved. As discussed above, if only one mode is excited, then for systems with N bosons the only possible quantum state involves occupying the single mode with all N bosons; that is, the quantum state must be $|m_1, N\rangle$ – the single-mode Bose–Einstein condensate. Superposition states of the form $\sum_N B(m_1, N)|m_1, N\rangle$ can be written down, but these are non-physical. For fermions, if only one mode is considered, the physical states are restricted to having $N = 0$ or $N = 1$ and are $|m_1, 0\rangle = |0\rangle$ (the vacuum state) or $|m_1, 1\rangle = (\hat{c}_1)^\dagger|0\rangle$ (the state with one fermion in mode m_1). Superposition states of the form $(B(m_1, 0)|m_1, 0\rangle + B(m_1, 1)|m_1, 1\rangle)$ can be written down, but again these are non-physical. However, states that still have physically interesting features can be obtained even if the number of modes is restricted to two, and we now consider this situation.

For the *boson two-mode case*, basis states for N-boson systems are of the form

$$\left|\frac{N}{2}, k\right\rangle \equiv \left|\frac{N}{2} - k, \frac{N}{2} + k\right\rangle = \frac{\left(\hat{a}_{m_1}^\dagger\right)^{(N/2-k)}}{\sqrt{(N/2-k)!}} \frac{\left(\hat{a}_{m_2}^\dagger\right)^{(N/2+k)}}{\sqrt{(N/2+k)!}}|0\rangle, \tag{2.36}$$

where $k = -N/2, -N/2+1, \cdots, N/2-1, N/2$. There are $N+1$ such basis states. These states have $\nu_1 = N/2 - k$ bosons in mode m_1 and $\nu_2 = N/2 + k$ bosons in mode m_2. As well as being eigenstates of the total boson number operator $\hat{n}_2 + \hat{n}_1$ with eigenvalue N, these states $|N/2, k\rangle$ are also *relative-number* eigenstates for the operator $\frac{1}{2}(\hat{n}_2 - \hat{n}_1)$ with eigenvalue k. The physical states for the N-boson system will be of the form

$$|\Psi\rangle_{N\mathrm{b}} = \sum_k B_N\left(\frac{N}{2}-k, \frac{N}{2}+k\right) \frac{\left(\hat{a}^\dagger_{m_1}\right)^{(N/2-k)}}{\sqrt{(N/2-k)!}} \frac{\left(\hat{a}^\dagger_{m_2}\right)^{(N/2+k)}}{\sqrt{(N/2+k)!}}|0\rangle$$
$$= \sum_k b\left(\frac{N}{2}, k\right)\left|\frac{N}{2}, k\right\rangle, \tag{2.37}$$

with the expansion coefficients satisfying normalisation, i.e. $\sum_k |b(N/2,k)|^2 = 1$. A wide variety of physical states can be described in this way. These range from states for unfragmented Bose–Einstein condensates, where all the bosons are either in mode m_1 or mode m_2 ($|N/2, -N/2\rangle$ or $|N/2, N/2\rangle$), to fragmented Bose–Einstein condensates, where half of the bosons are in mode m_1 and half in mode m_2 ($|N/2, 0\rangle$). States with a small variance in the relative boson number are called *number-squeezed* states.

When the amplitudes are given by $b(N/2, k) = \exp(ik\theta_p)/\sqrt{N+1}$, where θ_p is a relative phase given by $\theta_p = p \times 2\pi/(N+1)$ with $p = -N/2, -N/2+1, \cdots, +N/2$, an interesting special case occurs. Such states are *relative-phase* eigenstates [29, 30], denoted

$$\left|\frac{N}{2}, \theta_p\right\rangle = \frac{1}{\sqrt{N+1}} \sum_{Nk} \exp(ik\theta_p)\left|\frac{N}{2}, k\right\rangle, \tag{2.38}$$

and are useful in describing two-mode interferometry using Bose–Einstein condensates. They are also states with a maximum two-mode entanglement.

Non-physical states with a superposition of states with different total boson numbers can be written in the form

$$|\Phi\rangle_{\mathrm{b}} = \sum_{Nk} b\left(\frac{N}{2}, k\right)\left|\frac{N}{2}, k\right\rangle \equiv \sum_{\nu_1,\nu_2} B(\nu_1,\nu_2)\frac{\left(\hat{a}^\dagger_{m_1}\right)^{\nu_1}}{\sqrt{\nu_1!}}\frac{\left(\hat{a}^\dagger_{m_2}\right)^{\nu_2}}{\sqrt{\nu_2!}}|0\rangle.$$

For the *fermion two-mode case*, the basis states for a system with zero fermions are of the form $|m_1, 0; m_2, 0\rangle$, the basis states for one-fermion systems are $|m_1, 1; m_2, 0\rangle$ and $|m_1, 0; m_2, 1\rangle$, and the basis state for a two-fermion system is $|m_1, 1; m_2, 1\rangle$, where

$$|m_1, 0; m_2, 0\rangle = |0\rangle,$$
$$|m_1, 1; m_2, 0\rangle = (\hat{c}_1)^\dagger|0\rangle, \qquad |m_1, 0; m_2, 1\rangle = (\hat{c}_2)^\dagger|0\rangle,$$
$$|m_1, 1; m_2, 1\rangle = (\hat{c}_1)^\dagger(\hat{c}_2)^\dagger|0\rangle. \tag{2.39}$$

The physical states for the zero-, one- and two-fermion systems will be

$$|\Psi\rangle_{0\mathrm{f}} = |0\rangle,$$
$$|\Psi\rangle_{1\mathrm{f}} = B_1(1,0)(\hat{c}_1)^\dagger|0\rangle + B_1(0,1)(\hat{c}_2)^\dagger|0\rangle,$$
$$|\Psi\rangle_{2\mathrm{f}} = (\hat{c}_1)^\dagger(\hat{c}_2)^\dagger|0\rangle, \tag{2.40}$$

with only the one-fermion system allowing any variety. For the non-physical states we can write

$$|\Phi\rangle_{\mathrm{f}} = B_0(0,0)|0\rangle + B_1(1,0)(\hat{c}_1)^\dagger|0\rangle + B_1(0,1)(\hat{c}_2)^\dagger|0\rangle \\ +B_2(1,1)(\hat{c}_1)^\dagger(\hat{c}_2)^\dagger|0\rangle, \tag{2.41}$$

plus others in which the expansion coefficients are Grassmann numbers.

It can sometimes be useful to replace the original set of one-particle states $|m_1\rangle \cdots |m_n\rangle$ by a new set of one-particle states, which we might designate as $|p_1\rangle \cdots |p_n\rangle$. The two-mode case is useful to illustrate the issues that are involved. Let us introduce two new orthonormal one-particle states via the general relationship

$$|p_1\rangle = \cos\theta \exp\left(-i\frac{\chi}{2}\right)|m_1\rangle + \sin\theta \exp\left(+i\frac{\chi}{2}\right)|m_2\rangle, \\ |p_2\rangle = -\sin\theta \exp\left(-i\frac{\chi}{2}\right)|m_1\rangle + \cos\theta \exp\left(+i\frac{\chi}{2}\right)|m_2\rangle. \tag{2.42}$$

For all choices of the real parameters θ, χ, the new one-particle states are orthonormal. We also introduce new annihilation and creation operators via

$$\hat{b}_1 = \cos\theta \exp\left(+i\frac{\chi}{2}\right)\hat{a}_1 + \sin\theta \exp\left(-i\frac{\chi}{2}\right)\hat{a}_2, \\ \hat{b}_2 = -\sin\theta \exp\left(+i\frac{\chi}{2}\right)\hat{a}_1 + \cos\theta \exp\left(-i\frac{\chi}{2}\right)\hat{a}_2 \tag{2.43}$$

for bosons and

$$\hat{d}_1 = \cos\theta \exp\left(+i\frac{\chi}{2}\right)\hat{c}_1 + \sin\theta \exp\left(-i\frac{\chi}{2}\right)\hat{c}_2, \\ \hat{d}_2 = -\sin\theta \exp\left(+i\frac{\chi}{2}\right)\hat{c}_1 + \cos\theta \exp\left(-i\frac{\chi}{2}\right)\hat{c}_2 \tag{2.44}$$

for fermions, all of which satisfy the expected commutation or anticommutation rules. To provide an interpretation of these operators, we consider the effect of the creation operators on the vacuum state. For the bosonic case

$$\hat{b}_1^\dagger|0\rangle = \cos\theta \exp\left(-i\frac{\chi}{2}\right)\hat{a}_1^\dagger|0\rangle + \sin\theta \exp\left(+i\frac{\chi}{2}\right)\hat{a}_2^\dagger|0\rangle \\ = |p_1, 1; p_2, 0\rangle, \\ \hat{b}_2^\dagger|0\rangle = -\sin\theta \exp\left(-i\frac{\chi}{2}\right)\hat{a}_1^\dagger|0\rangle + \cos\theta \exp\left(+i\frac{\chi}{2}\right)\hat{a}_2^\dagger|0\rangle \\ = |p_1, 0; p_2, 1\rangle, \tag{2.45}$$

so that $\hat{b}_1^\dagger$ really does create a boson in the new single-particle state p_1 and $\hat{b}_2^\dagger$ really does create a boson in the new single-particle state p_2. The analogous annihilation features for $\hat{b}_1$ and $\hat{b}_2$ are established similarly, and the effect of operating with powers of the annihilation or creation operators confirms the interpretation. Analogous interpretations for $\hat{d}_1^\dagger$, $\hat{d}_2^\dagger$, $\hat{d}_1$ and $\hat{d}_2$ apply in the fermion case.

For bosons, it is of interest to consider the physical state in which all N bosons occupy one of the new single-particle states, p_1 for example. This state is parameterised by θ, χ and is called the *binomial state*,

$$|\theta, \chi\rangle_{\mathrm{b}} = \frac{(\hat{b}_1^{\dagger})^n}{\sqrt{N!}}|0\rangle. \tag{2.46}$$

A straightforward substitution for $\hat{b}_1^{\dagger}$ in terms of $\hat{a}_1^{\dagger}$ and $\hat{a}_2^{\dagger}$ and evaluation of $(\hat{b}_1^{\dagger})^n$ using the binomial expansion show that this state can be written

$$|\theta, \chi\rangle_{\mathrm{b}} = \sum_k b_k(\theta, \chi)\left|\frac{N}{2}, k\right\rangle, \tag{2.47}$$

where

$$b_k(\theta, \chi) = \left(C^N_{N/2-k}\right)^{1/2}(\cos\theta)^{N/2-k}(\sin\theta)^{N/2+k}\exp(+ik\chi). \tag{2.48}$$

The particles are binomially distributed between the two modes, hence the name 'binomial state'. A simple special case is provided by the state

$$\left|\theta = \frac{\pi}{4}, \chi = 0\right\rangle_b = \frac{1}{\sqrt{N!}}\left(\frac{\hat{a}_1^{\dagger} + \hat{a}_2^{\dagger}}{\sqrt{2}}\right)^N |0\rangle, \tag{2.49}$$

in which all the bosons occupy the symmetric single-particle state $(1/\sqrt{2})(|m_1\rangle + |m_2\rangle)$.

2.5 Physical Quantities and Field Operators

Fermion and boson operators $\hat{c}_i, \hat{c}_j^{\dagger}$ and $\hat{a}_i, \hat{a}_j^{\dagger}$ can be combined in various orders, of which the simplest non-zero examples are:

- a single zero-order operator, the unit operator $\hat{1}$;
- first-order operators, the annihilation and creation operators;
- second-order operators, the products of creation and annihilation operators.

We may obtain Hilbert space *operators in second quantisation form* by taking linear combinations of the annihilation and creation operators with c-number coefficients. If the combinations are Hermitian, then such operators may represent physical quantities. For example, the Hamiltonian for a fermion or boson system involving one- and two-particle interactions may be written in terms of one-particle and two-particle operators $\hat{h}(a)$ and $\hat{V}(a,b) = \hat{V}(b,a)$ as

$$\hat{H} = \sum_a \hat{h}(a) + \frac{1}{2}\sum_{a,b}\hat{V}(a,b), \tag{2.50}$$

where the sums are over the N identical particles and where the expressions are invariant under any permutation of the identical particles. In terms of annihilation and creation operators, the Hamiltonian given in first quantisation in the form (2.50) can be written in the second quantisation form as

$$\hat{H}_{\rm f} = \sum_{i,j} h_{ij}\hat{c}_i^\dagger \hat{c}_j + \frac{1}{2}\sum_{i,j,k,l} v_{ijkl}\hat{c}_i^\dagger \hat{c}_j^\dagger \hat{c}_l \hat{c}_k, \tag{2.51}$$

$$\hat{H}_{\rm b} = \sum_{i,j} k_{ij}\hat{a}_i^\dagger \hat{a}_j + \frac{1}{2}\sum_{i,j,k,l} \xi_{ijkl}\hat{a}_i^\dagger \hat{a}_j^\dagger \hat{a}_l \hat{a}_k, \tag{2.52}$$

for fermions and bosons, respectively. The coefficients are given in terms of the single-particle states (modes) via one- or two-particle matrix elements as

$$h_{ij}, k_{ij} = \langle i(a)|\hat{h}(a)|j(a)\rangle, \tag{2.53}$$

$$v_{ijkl}, \xi_{ijkl} = \langle i(a)|\langle j(b)|\hat{V}(a,b)|k(a)\rangle|l(b)\rangle. \tag{2.54}$$

Hermiticity and symmetry guarantee that

$$\begin{aligned} h_{ij}, &= h_{ji}^*, \qquad & v_{ijkl} &= v_{klij}^*, \qquad & v_{ijkl} &= v_{jilk}, \\ k_{ij}, &= k_{ji}^*, \qquad & \xi_{ijkl} &= \xi_{klij}^*, \qquad & \xi_{ijkl} &= \xi_{jilk}. \end{aligned} \tag{2.55}$$

Since $\hat{c}_i^\dagger \hat{c}_i^\dagger = 0$ and $\hat{c}_l \hat{c}_l = 0$, it follows that for fermions only the v_{ijkl} with $i \neq j$ and $l \neq k$ need be included. Note that the same number of annihilation operators as creation operators occur in each term, which results in the Hamiltonian commuting with the total number operator $\hat{N}$:

$$\hat{N}_{\rm f} = \sum_{i=1}^{n} \hat{c}_i^\dagger \hat{c}_i = \sum_{i=1}^{n} \hat{n}_i, \tag{2.56}$$

$$\hat{N}_{\rm b} = \sum_{i=1}^{n} \hat{a}_i^\dagger \hat{a}_i = \sum_{i=1}^{n} \hat{n}_i. \tag{2.57}$$

This means that the total number of identical bosons or fermions is conserved in dynamical processes, so if there was a given number of such particles initially then the number will not change for isolated systems. This makes it impossible to generate a quantum superposition of states with differing numbers of bosons or fermions from such an initial state – hence the super-selection rule. Analogous results to the first and second terms in (2.51) and (2.52) apply for other sums over one- and two-body operators where the expressions are invariant under any permutation of the particles.

The Hamiltonians for mixed systems of both bosons and fermions involve terms such as the above for the purely bosonic and fermionic subsystems, but with additional interaction terms with both bosonic and fermionic operators. For example, interactions associated with splitting a bosonic molecule into two fermionic atoms are of the form

$$\hat{V}_{\rm bf} = \frac{1}{2}\sum_{i,k,l} (\eta_{ikl}\hat{a}_i^\dagger \hat{c}_l \hat{c}_k + \eta_{ikl}^* \hat{c}_k^\dagger \hat{c}_l^\dagger \hat{a}_i), \tag{2.58}$$

where the η_{ikl} are given by matrix elements involving the bosonic state i and the two fermionic states k, l. Note that the fermions are created or destroyed in *pairs*.

Another linear combination involving annihilation and creation operators gives the *field operators*. For example, the field operators associated with annihilating or creating a particle at position $\mathbf{r}$ can be written

$$\hat{\Psi}_{\mathrm{f}}(\mathbf{r}) = \sum_i \hat{c}_i \phi_i(\mathbf{r}), \quad \hat{\Psi}^\dagger_{\mathrm{f}}(\mathbf{r}) = \sum_i \hat{c}^\dagger_i \phi^*_i(\mathbf{r}), \tag{2.59}$$

$$\hat{\Psi}_{\mathrm{b}}(\mathbf{r}) = \sum_i \hat{a}_i \phi_i(\mathbf{r}), \quad \hat{\Psi}^\dagger_{\mathrm{b}}(\mathbf{r}) = \sum_i \hat{a}^\dagger_i \phi^*_i(\mathbf{r}), \tag{2.60}$$

for fermions and bosons, respectively. Hence we have introduced a complete set of orthonormal single-mode functions $\phi_i(\mathbf{r})$ satisfying

$$\int d\mathbf{r}\ \phi^*_i(\mathbf{r})\phi_j(\mathbf{r}) = \delta_{i;j}, \tag{2.61}$$

$$\sum_i \phi^*_i(\mathbf{r})\phi_i(\mathbf{r}') = \boldsymbol{\delta}(\mathbf{r}\text{-}\mathbf{r}'). \tag{2.62}$$

To show that the field creation operators do indeed create a particle at a particular position, we note that the single-particle state $|m_i\rangle$ can be expressed in terms of Dirac-delta-function-normalised *position eigenstates* $|\mathbf{r}\rangle$ via

$$|m_i\rangle = \int d\mathbf{r}\ |\mathbf{r}\rangle\,\langle\mathbf{r}\,|m_i\rangle = \int d\mathbf{r}\ |\mathbf{r}\rangle\,\phi_i(\mathbf{r})$$
$$\langle\mathbf{r}\,|\mathbf{r}'\rangle = \delta(\mathbf{r}-\mathbf{r}') \tag{2.63}$$

where the mode function $\phi_i(\mathbf{r})$ is the position representation of $|m_i\rangle$. Using completeness, it then follows that

$$|\mathbf{r}\rangle = \sum_i \phi^*_i(\mathbf{r})|m_i\rangle \tag{2.64}$$

gives the position eigenstates in terms of the single-particle states $|m_i\rangle$.

We now consider a set of distinct particle positions $\mathbf{r}_1, \mathbf{r}_2, \cdots, \mathbf{r}_N$. As in the case of the single-particle modes, an ordering convention is invoked to avoid duplication. Thus $\mathbf{r}_1 < \mathbf{r}_2 < \cdots < \mathbf{r}_N$. Now, in first quantisation the state

$$\begin{aligned}\sum_{l_1,\cdots,l_N} \phi^*_{l_1}(\mathbf{r}_1)\cdots\phi^*_{l_N}(\mathbf{r}_N)|l_1\cdots l_N\rangle &= \sum_{l_1,\cdots,l_N} \phi^*_{l_1}(\mathbf{r}_1)\cdots\phi^*_{l_N}(\mathbf{r}_N)(\mathcal{S}\ \text{ or }\mathcal{A})|l_1(1)\rangle\cdots|l_N(N)\rangle \\ &= (\mathcal{S}\ \text{ or }\mathcal{A})|\mathbf{r}_1(1)\rangle\cdots|\mathbf{r}_N(N)\rangle \\ &= |\mathbf{r}_1\cdots\mathbf{r}_N\rangle\end{aligned} \tag{2.65}$$

must be a symmetrised state $|\mathbf{r}_1\cdots\mathbf{r}_N\rangle$ in which there is one particle at $\mathbf{r}_1$, a second at $\mathbf{r}_2$, $\cdots$ and an Nth at $\mathbf{r}_N$. But in second quantisation

$$\begin{aligned}\sum_{l_1,\cdots,l_N} \phi^*_{l_1}(\mathbf{r}_1)\cdots\phi^*_{l_N}(\mathbf{r}_N)|l_1\cdots l_N\rangle &= \sum_{l_1,\cdots,l_N} \phi^*_{l_1}(\mathbf{r}_1)\cdots\phi^*_{l_N}(\mathbf{r}_N)\hat{c}^\dagger_{l_1}\cdots\hat{c}^\dagger_{l_N}|0\rangle \\ &= \sum_{l_1,\cdots,l_N} \phi^*_{l_1}(\mathbf{r}_1)\cdots\phi^*_{l_N}(\mathbf{r}_N)\hat{a}^\dagger_{l_1}\cdots\hat{a}^\dagger_{l_N}|0\rangle\end{aligned} \tag{2.66}$$

for fermions and bosons, respectively. Hence, on comparing this result with the expressions (2.59) and (2.60), we see that the symmetrised state $|\mathbf{r}_1 \cdots \mathbf{r}_N\rangle$ is given by

$$|\mathbf{r}_1 \cdots \mathbf{r}_N\rangle = \hat{\Psi}_{\mathrm{f}}^{\dagger}(\mathbf{r}_1) \cdots \hat{\Psi}_{\mathrm{f}}^{\dagger}(\mathbf{r}_N)|0\rangle \tag{2.67}$$

$$= \hat{\Psi}_{\mathrm{b}}^{\dagger}(\mathbf{r}_1) \cdots \hat{\Psi}_{\mathrm{b}}^{\dagger}(\mathbf{r}_N)|0\rangle \tag{2.68}$$

for fermions and bosons, respectively. Thus the field creation operators do in fact create particles at particular positions, ordered as $\mathbf{r}_1 < \mathbf{r}_2 < \cdots < \mathbf{r}_N$.

The normalisation condition for the position states is

$$\langle \mathbf{r}_1 \cdots \mathbf{r}_N | \mathbf{s}_1 \cdots \mathbf{s}_N\rangle = \delta(\mathbf{r}_1 - \mathbf{s}_1) \cdots \delta(\mathbf{r}_N - \mathbf{s}_N). \tag{2.69}$$

The proof is left as an exercise.

It is also useful to define momentum field operators via spatial Fourier transforms. For both the boson and the fermion cases, we write the position field creation operators $\hat{\Psi}_{\mathrm{b,f}}(\mathbf{r})$ as

$$\hat{\Psi}_{\mathrm{b,f}}(\mathbf{r}) = \int \frac{d\mathbf{k}}{(2\pi)^{3/2}} \exp(i\mathbf{k}\cdot\mathbf{r})\hat{\Phi}_{\mathrm{b,f}}(\mathbf{k}),$$

$$\hat{\Phi}_{\mathrm{b,f}}(\mathbf{k}) = \int \frac{d\mathbf{r}}{(2\pi)^{3/2}} \exp(-i\mathbf{k}\cdot\mathbf{r})\hat{\Psi}_{\mathrm{b,f}}(\mathbf{r}), \tag{2.70}$$

i.e. as spatial integrals involving momentum field operators $\hat{\Phi}_{\mathrm{b,f}}(\mathbf{k})$, the inverse relation also being stated. The interpretation of the momentum field operators is that

$$|\mathbf{k}_1 \cdots \mathbf{k}_N\rangle = \hat{\Phi}_{\mathrm{b,f}}^{\dagger}(\mathbf{k}_1) \cdots \hat{\Phi}_{\mathrm{b,f}}^{\dagger}(\mathbf{k}_N)|0\rangle \tag{2.71}$$

is a state with one particle having momentum $\hbar\mathbf{k}_1$, a second particle having momentum $\hbar\mathbf{k}_2$, $\cdots$ and the Nth particle having momentum $\hbar\mathbf{k}_N$. As with position eigenstates, the momentum Fock states are delta function normalised:

$$\langle \mathbf{k}_1 \cdots \mathbf{k}_N | \mathbf{l}_1 \cdots \mathbf{l}_N\rangle = \delta(\mathbf{k}_1 - \mathbf{l}_1) \cdots \delta(\mathbf{k}_N - \mathbf{l}_N). \tag{2.72}$$

The proofs of these properties of momentum field operators and momentum Fock states are left as an exercise.

Up to now, we have ignored the situation where the bosonic or fermionic particles have *internal* degrees of freedom in addition to their position (and momentum). The internal states of bosonic and fermionic atoms are associated with a variety of quantum numbers, of which the total angular momentum F and the total magnetic quantum number M_f (these defining the different *hyperfine states*) are the most important for experiments at ultralow temperatures. However, many experiments involve atoms in just one hyperfine state, so for simplicity in exposition we will generally ignore the internal quantum states. For completeness, cases where several internal states are involved can be described by introducing annihilation and creation field operators $\hat{\Psi}_F(\mathbf{r}),\hat{\Psi}_F^{\dagger}(\mathbf{r})$ for each internal state – these states being labelled by F. For both bosons and fermions, there are mode expansions for the field operators analogous to (2.59) and

(2.60) involving a complete set of orthonormal single-particle mode functions $\phi_{Fi}(\mathbf{r})$ satisfying equations analogous to (2.61) and (2.62). Thus we have

$$\hat{\Psi}_{\text{f}F}(\mathbf{r}) = \sum_i \hat{c}_{Fi}\phi_{Fi}(\mathbf{r}), \quad \hat{\Psi}^{\dagger}_{\text{f}F}(\mathbf{r}) = \sum_i \hat{c}^{\dagger}_{Fi}\phi^*_{Fi}(\mathbf{r}), \tag{2.73}$$

$$\hat{\Psi}_{\text{b}F}(\mathbf{r}) = \sum_i \hat{a}_{Fi}\phi_{Fi}(\mathbf{r}), \quad \hat{\Psi}^{\dagger}_{\text{b}F}(\mathbf{r}) = \sum_i \hat{a}^{\dagger}_{Fi}\phi^*_{Fi}(\mathbf{r}), \tag{2.74}$$

for fermions and bosons, respectively. Note that the mode functions $\phi_{Fi}(\mathbf{r})$ for different internal states F are *not* required to be mutually orthogonal – it is convenient in some cases to choose them to be the same. A similar interpretation to that in the case of (2.68) and (2.67) applies for particles created at positions $\mathbf{r}_1, \cdots, \mathbf{r}_N$ in specific internal states. Asking which identical boson or fermion is created at these positions is not a meaningful question. Orthonormalisation conditions analogous to those in (2.61) apply.

The anticommutation rules for fermionic annihilation and creation operators directly imply that the fermionic field operators satisfy the anticommutation rules

$$\begin{aligned} \left\{\hat{\Psi}_{\text{f}F}(\mathbf{r}), \hat{\Psi}^{\dagger}_{\text{f}G}(\mathbf{s})\right\} &= \delta_{FG}\delta(\mathbf{r}-\mathbf{s}), \\ \left\{\hat{\Psi}_{\text{f}F}(\mathbf{r}), \hat{\Psi}_{\text{f}G}(\mathbf{s})\right\} &= \left\{\hat{\Psi}^{\dagger}_{\text{f}}(\mathbf{r}), \hat{\Psi}^{\dagger}_{\text{f}G}(\mathbf{s})\right\} = 0. \end{aligned} \tag{2.75}$$

Similar considerations apply for boson systems. The bosonic field operators satisfy the following commutation rules:

$$\begin{aligned} \left[\hat{\Psi}_{\text{b}F}(\mathbf{r}), \hat{\Psi}^{\dagger}_{\text{b}G}(\mathbf{s})\right] &= \delta_{FG}\delta(\mathbf{r}-\mathbf{s}), \\ \left[\hat{\Psi}_{\text{b}F}(\mathbf{r}), \hat{\Psi}_{\text{b}G}(\mathbf{s})\right] &= \left[\hat{\Psi}^{\dagger}_{\text{b}F}(\mathbf{r}), \hat{\Psi}^{\dagger}_{\text{b}G}(\mathbf{s})\right] = 0. \end{aligned} \tag{2.76}$$

We have stated that the annihilation and creation operators $\hat{\Psi}_{\text{f}F}(\mathbf{r})$, $\hat{\Psi}^{\dagger}_{\text{f}F}(\mathbf{r})$ etc. act to remove or create a single particle with internal state F at position $\mathbf{r}$. To demonstrate this, it suffices to calculate the effect of $\hat{\Psi}_{\text{f}F}(\mathbf{r})$ or $\hat{\Psi}_{\text{b}F}(\mathbf{r})$ on a single particle prepared in the state $\hat{c}^{\dagger}_{Fi}|0\rangle$ or $\hat{a}^{\dagger}_{Fi}|0\rangle$. We find

$$\begin{aligned} \hat{\Psi}_{\text{f}F}(\mathbf{r})\hat{c}^{\dagger}_{Fi}|0\rangle &= \phi_{Fi}(\mathbf{r})|0\rangle, \\ \hat{\Psi}_{\text{b}F}(\mathbf{r})\hat{a}^{\dagger}_{Fi}|0\rangle &= \phi_{Fi}(\mathbf{r})|0\rangle, \end{aligned} \tag{2.77}$$

for fermions and bosons respectively. In each case the amplitude associated with destroying the particle is precisely that for the mode ϕ_{Fi} at the position $\mathbf{r}$. For multi-particle states, the commutation and anticommutation properties of the operators enforce automatically the required symmetry or antisymmetry properties of the state. This is the great virtue of using them.

Field annihilation and creation operators do not represent physical quantities, and have the effect of changing N-particle physical states into $N \pm 1$-particle physical states. Physical quantities can be constructed from the field operators, however, which always have the same number of creation and annihilation operators. For example, in terms of field operators, typical Hamiltonians for a boson system with spin-0 bosons of mass M and for a fermion system with spin-$\frac{1}{2}$ fermions of mass m are

$$\hat{H}_{\mathrm{b}} = \int d\mathbf{r} \left(\frac{\hbar^2}{2M} \nabla \hat{\Psi}(\mathbf{r})^\dagger \cdot \nabla \hat{\Psi}(\mathbf{r}) + \hat{\Psi}(\mathbf{r})^\dagger V_{\mathrm{b}} \hat{\Psi}(\mathbf{r}) + \frac{g_{\mathrm{b}}}{2} \hat{\Psi}(\mathbf{r})^\dagger \hat{\Psi}(\mathbf{r})^\dagger \hat{\Psi}(\mathbf{r}) \hat{\Psi}(\mathbf{r}) \right), \tag{2.78}$$

$$\hat{H}_{\mathrm{f}} = \int d\mathbf{r} \left(\sum_{\alpha} \frac{\hbar^2}{2m} \nabla \hat{\Psi}_\alpha(\mathbf{r})^\dagger \cdot \nabla \hat{\Psi}_\alpha(\mathbf{r}) + \sum_{\alpha} \hat{\Psi}_\alpha(\mathbf{r})^\dagger V_{\mathrm{f}} \hat{\Psi}_\alpha(\mathbf{r}) \right.$$
$$\left. + \frac{g_{\mathrm{f}}}{2} \sum_{\alpha} \hat{\Psi}_\alpha(\mathbf{r})^\dagger \hat{\Psi}_{-\alpha}(\mathbf{r})^\dagger \hat{\Psi}_{-\alpha}(\mathbf{r}) \hat{\Psi}_\alpha(\mathbf{r}) \right), \tag{2.79}$$

where $\alpha = -\frac{1}{2}, +\frac{1}{2}$ lists two internal states with $M_{\mathrm{f}} = -\frac{1}{2}, +\frac{1}{2}$, the zero-range approximation is used for interactions between the particles and V_{b}, V_{f} are trapping potentials. The fermion–fermion interaction is required to be between fermions of opposite spin because $\hat{\Psi}_\alpha^2(\mathbf{r}) = 0$. For bosons strong s-wave interactions occur, but for fermions only weak p-wave collisions occur unless the fermions have opposite spin, because the Pauli principle prevents two fermions with the same spin from being at the same position.

The Hamiltonians in terms of field operators are entirely equivalent to their forms in terms of mode annihilation and creation operators, as can be seen if the mode expansion forms are substituted for the field operators. For example, doing this for the bosonic Hamiltonian in (2.78) gives the Hamiltonian (2.52) with the following expressions for the coefficients:

$$k_{ij} = \int d\mathbf{r} \left(\frac{\hbar^2}{2M} \nabla \phi_i^*(\mathbf{r}) \cdot \nabla \phi_j(\mathbf{r}) + \phi_i^*(\mathbf{r}) V_b \phi_j(\mathbf{r}) \right),$$
$$\xi_{ijkl} = g_b \int d\mathbf{r} (\phi_i^*(\mathbf{r}) \phi_j^*(\mathbf{r}) \phi_l(\mathbf{r}) \phi_k(\mathbf{r})). \tag{2.80}$$

Similar results may be obtained in the fermion case.

2.6 Dynamical Processes

If an isolated system is prepared in an initial state, it evolves via either the *time-dependent Schrödinger equation* in the case of pure states or a *Liouville–von Neumann equation* if it is a mixed state. These equations involve only the Hamiltonian operator $\hat{H}$ for the system, and are given respectively by

$$i\hbar \frac{\partial}{\partial t} |\Psi\rangle_N = \hat{H} |\Psi\rangle_N, \tag{2.81}$$
$$\frac{\partial}{\partial t} \hat{\rho} = \frac{1}{i\hbar} [\hat{H}, \hat{\rho}]. \tag{2.82}$$

If the system is not isolated from the environment, then the system evolution may be described by a *master equation* [4, 31]. The evolution involves system operators relating to the coupling to the environment, as well as the system Hamiltonian $\hat{H}$

given by (2.52) and (2.51). Where reservoir relaxation times are very short, typical Markovian master equations describing weak environmental coupling have the form

$$\frac{\partial}{\partial t}\hat{\rho} = \frac{1}{i\hbar}[\hat{H}, \hat{\rho}] + \frac{1}{2}\sum_{ab}\Gamma_{ab}\left(2\hat{S}_b^\dagger\hat{\rho}\hat{S}_a - \hat{\rho}\hat{S}_a\hat{S}_b^\dagger - \hat{S}_a\hat{S}_b^\dagger\hat{\rho}\right), \tag{2.83}$$

where $a \equiv i, j$ and $b \equiv k, l$ each denote pairs of boson or fermion modes and

$$\hat{S}_a = \hat{c}_j^\dagger\hat{c}_i, \qquad \hat{S}_b = \hat{c}_l^\dagger\hat{c}_k \qquad \text{fermions}, \tag{2.84}$$

$$\hat{S}_a = \hat{a}_j^\dagger\hat{a}_i, \qquad \hat{S}_b = \hat{a}_l^\dagger\hat{a}_k \qquad \text{bosons}, \tag{2.85}$$

$$\Gamma_{ab} \equiv \Gamma_{ij;kl}, \qquad \Gamma_{ab} = \Gamma_{ba}^*, \qquad \Gamma_{ij;kl} = \Gamma_{kl;ij}^*. \tag{2.86}$$

Such a master equation could be used to describe a system of particles of either type weakly coupled to a reservoir of foreign atoms, where two-body collisions between a particle and a reservoir atom cause the boson/fermion to be removed from mode i and added to mode j. The $\hat{S}_a$ act as transition operators. The $\Gamma_{ij;kl}$ are relaxation coefficients. Detailed expressions for the master equation coefficients – which depend upon the reservoir parameters – can be found in several places (see [4, 5, 32] and Appendix B). The master equation is constrained to have the Lindblad form (2.83) by general physical requirements of complete positivity and conservation of probability, as expressed by the trace of $\hat{\rho}$ being unity. The master equation is number-conserving – particles are neither created nor destroyed. It is straightforward to show that

$$\frac{\partial}{\partial t}\langle\hat{N}_{\mathrm{b,f}}\rangle = \frac{\partial}{\partial t}\mathrm{Tr}(\hat{N}_{\mathrm{b,f}}\hat{\rho}) = 0, \tag{2.87}$$

where $\hat{N}_{\mathrm{b,f}}$ is the boson or fermion number operator.

The canonical thermal equilibrium state has the unnormalised density operator

$$\hat{\rho} = \exp\left(-\beta\{\hat{H} - \mu\hat{N}\}\right), \tag{2.88}$$

where the temperature is $T = 1/(k_{\mathrm{b}}\beta)$ and μ is the chemical potential. In this case the unnormalised density operator satisfies the *Matsubara equation* [33],

$$\frac{\partial}{\partial\beta}\hat{\rho} = -\frac{1}{2}\left\{(\hat{H} - \mu\hat{N}), \hat{\rho}\right\}, \tag{2.89}$$

which involves the anticommutator. This is similar to a dynamical equation with an imaginary time $i\beta$. Physical quantities such as the mean energy can be evaluated from expressions such as

$$\langle\hat{H}\rangle = \frac{\mathrm{Tr}(\hat{H}\hat{\rho})}{\mathrm{Tr}(\hat{\rho})}, \tag{2.90}$$

and phase space methods can be applied to determining the distribution function for $\hat{\rho}(\beta)$.

For the isolated-system case, the dynamical behaviour involves matrix elements of terms in the Hamiltonian between typical system states and it is useful to consider the most common cases. The first are the *one-particle processes*, where the term includes one annihilation operator and one creation operator, for example $\hat{c}_j^\dagger \hat{c}_i$ for fermions or $\hat{a}_j^\dagger \hat{a}_i$ for bosons. From the results (2.13), we see that for fermions and for $i \neq j$

$$\langle \mu | \hat{c}_j^\dagger \hat{c}_i | \nu \rangle = \delta_{\nu_i;1} (-1)^{\eta_i} \delta_{\mu_j;1} (-1)^{\eta_j} \prod_{l \neq i,j} \delta_{\nu_l, \mu_l}.$$

This matrix element vanishes unless $\nu_i = 1$, $\nu_j = 0$ and $\mu_i = 0$, $\mu_j = 1$. At most two modes change occupancy; the others do not. This means that the step caused by this term is only possible if the mode m_i is initially occupied by one fermion and the mode m_j is initially unoccupied. After the step, the mode m_i is unoccupied and the mode m_j is occupied. This feature, in which the occupancy of a mode m_j prevents a fermion process from occurring owing to the term $\hat{c}_j^\dagger \hat{c}_i$, is called the *Pauli blockade.*

From the results (2.20), we see that for the boson case and for $i \neq j$

$$\langle \mu | \hat{a}_j^\dagger \hat{a}_i | \nu \rangle = \sqrt{\nu_i} \sqrt{\mu_j} \delta_{\mu_i, \nu_i - 1} \delta_{\mu_j - 1, \nu_j} \prod_{l \neq i,j} \delta_{\mu_l, \nu_l}.$$

This matrix element vanishes if $\mu_i \neq \nu_i - 1$ and $\mu_j \neq \nu_j + 1$, so the step caused by this term is only possible if the mode m_i is initially occupied by at least one boson. Then the mode m_i is finally occupied by one boson fewer and the mode m_j is occupied by one more. The other modes have unchanged occupancies, so at most two modes change occupancy. The overall matrix element is $\sqrt{\nu_i(\nu_j + 1)}$, which increases with the initial occupancy of both of the modes m_i and m_j. This feature is called *Bose enhancement.*

The second most common cases are the two-particle processes in which two particles are transferred between modes. These appear, for example, in the interaction terms in the Hamiltonians (2.52) and (2.51) and require us to calculate the expectation values of a product of two creation and annihilation operators $\hat{c}_i^\dagger \hat{c}_j^\dagger \hat{c}_l \hat{c}_k$ for fermions or $\hat{a}_i^\dagger \hat{a}_j^\dagger \hat{a}_l \hat{a}_k$ for bosons. The calculation of these for any given state is straightforward using the commutation or anticommutation properties of the operators.

2.7 Normally Ordered Forms

So far, many expressions (such as the quantum correlation functions and the Hamiltonians) involving annihilation and creation operators have been written with these operators already placed in normal order, with creation operators on the left of the annihilation operators. However, not all expressions are normally ordered, and it is useful to define the so-called *normally ordered forms* of other expressions involving the annihilation and creation operators [34]. This applies for both mode and field operators, though only the former will be discussed here as the extension to field operators is straightforward. As the operator forms of most interest can be written as linear combinations of products of operators with c-number coefficients, the normally ordered forms of products will be defined first. Thus for two operators

$$\begin{aligned}
&\mathcal{N}(\hat{a}_i \hat{a}_j) = \hat{a}_i \hat{a}_j, \quad \mathcal{N}\left(\hat{a}_i^\dagger \hat{a}_j^\dagger\right) = \hat{a}_i^\dagger \hat{a}_j^\dagger, \quad \mathcal{N}\left(\hat{a}_i^\dagger \hat{a}_j\right) = \hat{a}_i^\dagger \hat{a}_j, \quad \mathcal{N}\left(\hat{a}_i \hat{a}_j^\dagger\right) = \hat{a}_j^\dagger \hat{a}_i \\
&\mathcal{N}(\hat{c}_i \hat{c}_j) = \hat{c}_i \hat{c}_j, \quad \mathcal{N}\left(\hat{c}_i^\dagger \hat{c}_j^\dagger\right) = \hat{c}_i^\dagger \hat{c}_j^\dagger, \quad \mathcal{N}\left(\hat{c}_i^\dagger \hat{c}_j\right) = \hat{c}_i^\dagger \hat{c}_j, \quad \mathcal{N}\left(\hat{c}_i \hat{c}_j^\dagger\right) = -\hat{c}_j^\dagger \hat{c}_i,
\end{aligned} \tag{2.91}$$

for boson and fermion operators, respectively. The normally ordered form of an operator $\hat{A}$ is typically denoted $:\hat{A}:$ or $\mathcal{N}(\hat{A})$. Note the last result for fermions – the minus sign is because a permutation of the operators is needed to place the creation operator on the left. In all cases, the expressions have all creation operators to the left of all annihilation operators. This property survives for products involving more than two operators, with the creation and annihilation operators separately in their original order. In the fermion case, the expression is multiplied by +1 or −1 depending on whether an odd or even permutation of the operators is needed to obtain the normally ordered form. To give a few examples involving four operators,

$$\begin{aligned}\mathcal{N}\left(\hat{a}_i\hat{a}_j^\dagger\hat{a}_k\hat{a}_l^\dagger\right) &= \hat{a}_j^\dagger\hat{a}_l^\dagger\hat{a}_i\hat{a}_k, \qquad \mathcal{N}\left(\hat{a}_i\hat{a}_j^\dagger\hat{a}_l^\dagger\hat{a}_k\right) = \hat{a}_j^\dagger\hat{a}_l^\dagger\hat{a}_i\hat{a}_k,\\ \mathcal{N}\left(\hat{c}_i\hat{c}_j^\dagger\hat{c}_k\hat{c}_l^\dagger\right) &= -\hat{c}_j^\dagger\hat{c}_l^\dagger\hat{c}_i\hat{c}_k, \qquad \mathcal{N}\left(\hat{c}_i\hat{c}_j^\dagger\hat{c}_l^\dagger\hat{c}_k\right) = +\hat{c}_j^\dagger\hat{c}_l^\dagger\hat{c}_i\hat{c}_k,\end{aligned} \tag{2.92}$$

where in the first fermion example an odd permutation is needed, and in the second the permutation is even. Note that the normally ordered form for fermion operators may vanish if two annihilation or two creation operators are the same. The normally ordered form of a linear combination of products $\hat{P}_i$ is defined as the corresponding linear combination of the normally ordered products. Thus

$$\mathcal{N}(\gamma_1\hat{P}_1 + \gamma_2\hat{P}_2) = \gamma_1\mathcal{N}(\hat{P}_1) + \gamma_2\mathcal{N}(\hat{P}_2) \tag{2.93}$$

for bosons or fermions, where the γ_i are c-numbers.

In the case where annihilation and creation operators for different modes are involved, the following useful theorem can be established:

$$\begin{aligned}\mathcal{N}\left(F\left(\hat{a}_i^\dagger\hat{a}_i\right)G\left(\hat{a}_j^\dagger\hat{a}_j\right)\right) &= \mathcal{N}\left(F\left(\hat{a}_i^\dagger\hat{a}_i\right)\right)\mathcal{N}\left(G\left(\hat{a}_j^\dagger\hat{a}_j\right)\right) \qquad (i\neq j),\\ \mathcal{N}\left(F\left(\hat{c}_i^\dagger\hat{c}_i\right)G\left(\hat{c}_j^\dagger\hat{c}_j\right)\right) &= \mathcal{N}\left(F\left(\hat{c}_i^\dagger\hat{c}_i\right)\right)\mathcal{N}\left(G\left(\hat{c}_j^\dagger\hat{c}_j\right)\right) \qquad (i\neq j),\end{aligned} \tag{2.94}$$

where F, G are arbitrary functions. This result depends only on the annihilation and creation operators commuting (or anticommuting) for $i \neq j$. The proof is left as an exercise.

The vacuum expectation value of any normally ordered form vanishes:

$$\langle 0|\,\mathcal{N}(\hat{F})\,|0\rangle = 0, \tag{2.95}$$

where $\hat{F}$ is a function of the annihilation and creation operators. Also, the commutation and anticommutation rules for the basic operators may be expressed in a standard form for both bosons and fermions as

$$\hat{a}_i\hat{a}_j^\dagger - \mathcal{N}\left(\hat{a}_i\hat{a}_j^\dagger\right) = \delta_{ij}, \qquad \hat{c}_i\hat{c}_j^\dagger - \mathcal{N}\left(\hat{c}_i\hat{c}_j^\dagger\right) = \delta_{ij}. \tag{2.96}$$

These last two results form the basis of *Wick's theorem* [6] for evaluating the vacuum expectation value of products of several annihilation and creation operators.

2.8 Vacuum Projector

The vacuum state $|0\rangle$, which contains no particles in any mode, is associated with the *vacuum projector* $|0\rangle\langle 0|$. This projector can be written using normally ordered forms (see Section 2.7) of products of exponential operators based on number operators. We have

$$|0\rangle\langle 0| = \mathcal{N}\left(\prod_i \exp(-\hat{a}_i^\dagger \hat{a}_i)\right) = \mathcal{N}\left[\exp\left(-\sum_i \hat{a}_i^\dagger \hat{a}_i\right)\right] \tag{2.97}$$

$$= \mathcal{N}\left(\prod_i \exp(-\hat{c}_i^\dagger \hat{c}_i)\right) = \mathcal{N}\left[\exp\left(-\sum_i \hat{c}_i^\dagger \hat{c}_i\right)\right] \tag{2.98}$$

for bosons and fermions, respectively. The result for bosons is given in [4, p. 45].

To prove the result, we note that if the projector operates on any state $|\Psi\rangle_{\mathrm{Perp}}$ orthogonal to the vacuum state the result is zero:

$$(|0\rangle\langle 0|)\,|\Psi\rangle_{\mathrm{Perp}} = 0, \qquad \langle\Psi|_{\mathrm{Perp}}\,(|0\rangle\langle 0|) = 0. \tag{2.99}$$

This is equivalent to

$$\begin{aligned} \hat{a}_i\,|0\rangle\langle 0| &= 0, & |0\rangle\langle 0|\,\hat{a}_i^\dagger &= 0, \\ \hat{c}_i\,|0\rangle\langle 0| &= 0, & |0\rangle\langle 0|\,\hat{c}_i^\dagger &= 0, \end{aligned} \tag{2.100}$$

as any such orthogonal state must involve basis states with at least one creation operator acting on the vacuum state. Hence, if it can be shown that the expressions in (2.97) and (2.98) satisfy these conditions, then the expressions must equal the vacuum projector, apart from a possible multiplying constant. As the vacuum state matrix element of the projector is unity, the multiplicative constant will also equal unity if the matrix element of the expressions in (2.97) and (2.98) is equal to unity.

For the fermion case,

$$\begin{aligned} \hat{c}_i\mathcal{N}\left(\prod_j \exp(-\hat{c}_j^\dagger \hat{c}_j)\right) &= \hat{c}_i\mathcal{N}\left(\exp(-\hat{c}_i^\dagger \hat{c}_i)\prod_{j\neq i} \exp(-\hat{c}_j^\dagger \hat{c}_j)\right) \\ &= \hat{c}_i\left(\left\{1 - \hat{c}_i^\dagger \hat{c}_i - \frac{1}{2}\hat{c}_i^\dagger\,\hat{c}_i^\dagger \hat{c}_i \hat{c}_i - \cdots\right\}\mathcal{N}\left\{\prod_{j\neq i} \exp(-\hat{c}_j^\dagger \hat{c}_j)\right\}\right) \\ &= \left(\left\{\hat{c}_i - \hat{c}_i\hat{c}_i^\dagger \hat{c}_i\right\}\mathcal{N}\left\{\prod_{j\neq i} \exp(-\hat{c}_j^\dagger \hat{c}_j)\right\}\right) \\ &= \left(\left\{\hat{c}_i - (1 - \hat{c}_i^\dagger \hat{c}_i)\hat{c}_i\right\}\mathcal{N}\left\{\prod_{j\neq i} \exp(-\hat{c}_j^\dagger \hat{c}_j)\right\}\right) \\ &= 0, \end{aligned} \tag{2.101}$$

where line two follows using the theorem in (2.94). Similarly,

$$
\begin{aligned}
\mathcal{N}\left(\prod_j \exp(-\hat{c}_j^\dagger \hat{c}_j)\right)\hat{c}_i^\dagger &= \mathcal{N}\left(\prod_{j\neq i} \exp(-\hat{c}_j^\dagger \hat{c}_j)\exp(-\hat{c}_i^\dagger \hat{c}_i)\right)\hat{c}_i^\dagger \\
&= \left(\mathcal{N}\left\{\prod_{j\neq i}\exp(-\hat{c}_j^\dagger \hat{c}_j)\right\}\left\{1-\hat{c}_i^\dagger\hat{c}_i-\frac{1}{2}\hat{c}_i^\dagger\hat{c}_i^\dagger\hat{c}_i\hat{c}_i-\cdots\right\}\right)\hat{c}_i^\dagger \\
&= \left(\mathcal{N}\left\{\prod_{j\neq i}\exp(-\hat{c}_j^\dagger \hat{c}_j)\right\}\left\{\hat{c}_i^\dagger-\hat{c}_i^\dagger\hat{c}_i\hat{c}_i^\dagger\right\}\right) \\
&= \left(\mathcal{N}\left\{\prod_{j\neq i}\exp(-\hat{c}_j^\dagger \hat{c}_j)\right\}\left\{\hat{c}_i^\dagger-\hat{c}_i^\dagger(1-\hat{c}_i^\dagger\hat{c}_i)\right\}\right) \\
&= 0.
\end{aligned}
\tag{2.102}
$$

Also, we can show that $\langle 0|\,(|0\rangle\,\langle 0|)\,|0\rangle = 1$ for the expression given for the vacuum state projector:

$$
\begin{aligned}
&\langle 0|\mathcal{N}\left(\prod_j \exp(-\hat{c}_j^\dagger \hat{c}_j)\right)|0\rangle \\
&= \langle 0|\left(\mathcal{N}\left\{\prod_{j\neq i}\exp(-\hat{c}_j^\dagger \hat{c}_j)\right\}\left\{1-\hat{c}_i^\dagger\hat{c}_i-\frac{1}{2}\hat{c}_i^\dagger\hat{c}_i^\dagger\hat{c}_i\hat{c}_i-\cdots\right\}\right)|0\rangle \\
&= \langle 0|\left(\mathcal{N}\left\{\prod_{j\neq i}\exp(-\hat{c}_j^\dagger \hat{c}_j)\right\}\right)|0\rangle \\
&= 1,
\end{aligned}
\tag{2.103}
$$

after successively considering the remaining modes $i = 2, \cdots$. The result for bosons is left as an exercise.

Finally, in term of field operators, the vacuum projector is given by

$$
|0\rangle\,\langle 0| = \mathcal{N}\left(\exp\left(-\int d\mathbf{r}\,\hat{\Psi}_{\mathrm{f}}^\dagger(\mathbf{r})\hat{\Psi}_f(\mathbf{r})\right)\right) \tag{2.104}
$$

$$
= \mathcal{N}\left(\exp\left(-\int d\mathbf{r}\,\hat{\Psi}_{\mathrm{b}}^\dagger(\,\mathbf{r})\hat{\Psi}_b(\mathbf{r})\right)\right) \tag{2.105}
$$

for fermions and bosons, respectively.

2.9 Position Measurements and Quantum Correlation Functions

Measurement results can often be expressed in terms of *quantum correlation functions*, which involve expectation values of products of field operators. For bosons, for example, a typical quantum correlation function is

$$G^n(\mathbf{r}_1,\cdots,\mathbf{r}_p;\mathbf{s}_q,\cdots,\mathbf{s}_1) = \left\langle \hat{\Psi}(\mathbf{r}_1)^\dagger \cdots \hat{\Psi}(\mathbf{r}_p)^\dagger \hat{\Psi}(\mathbf{s}_q) \cdots \hat{\Psi}(\mathbf{s}_1) \right\rangle$$
$$= \mathrm{Tr}\Big(\hat{\Psi}(\mathbf{s}_q) \cdots \hat{\Psi}(\mathbf{s}_1) \hat{\rho}(t) \hat{\Psi}(\mathbf{r}_1)^\dagger \cdots \hat{\Psi}(\mathbf{r}_p)^\dagger \Big), \tag{2.106}$$

where for an N-particle system we require $p, q \leq N$ to give a non-zero result. Various spatial-interference and coherence effects in Bose–Einstein condensates can be described via such correlation functions. The case for $p = q$ where $\mathbf{r}_i = \mathbf{s}_i$ for all i is proportional to the probability of simultaneously detecting bosons at $\mathbf{r}_1, \mathbf{r}_2, \cdots, \mathbf{r}_p$.

To show this important result, we see that the probability density $P(\mathbf{r}_1, \cdots, \mathbf{r}_N)$ of detecting all N bosons in volumes $\delta\mathbf{r}_1, \delta\mathbf{r}_2, \cdots, \delta\mathbf{r}_N$ around an ordered set of positions at $\mathbf{r}_1, \mathbf{r}_2, \cdots, \mathbf{r}_N$ for a state $\hat{\rho}(t)$ is given in terms of the projector onto the position state $|\mathbf{r}_1 \cdots \mathbf{r}_N\rangle$ as

$$P(\mathbf{r}_1,\cdots,\mathbf{r}_N)\delta\mathbf{r}_1 \cdots \delta\mathbf{r}_N = \mathrm{Tr}\left(|\mathbf{r}_1 \cdots \mathbf{r}_N\rangle \langle \mathbf{r}_1 \cdots \mathbf{r}_N| \, \hat{\rho}(t) \right) \delta\mathbf{r}_1 \cdots \delta\mathbf{r}_N.$$

Hence

$$P(\mathbf{r}_1,\cdots,\mathbf{r}_N) = \mathrm{Tr}\left(\hat{\Psi}_\mathrm{b}^\dagger(\mathbf{r}_1) \cdots \hat{\Psi}_\mathrm{b}^\dagger(\mathbf{r}_N) |0\rangle \langle 0| \, \hat{\Psi}_\mathrm{b}(\mathbf{r}_N) \cdots \hat{\Psi}_\mathrm{b}(\mathbf{r}_1) \hat{\rho}(t) \right)$$
$$= \mathrm{Tr}\left(|0\rangle\langle 0| \hat{\Psi}_\mathrm{b}(\mathbf{r}_N) \cdots \hat{\Psi}_\mathrm{b}(\mathbf{r}_1) \hat{\rho}(t) \hat{\Psi}_\mathrm{b}^\dagger(\mathbf{r}_1) \cdots \hat{\Psi}_\mathrm{b}^\dagger(\mathbf{r}_N) \right). \tag{2.107}$$

The last result is similar to a quantum correlation function apart from the presence of the vacuum projector $|0\rangle\langle 0|$. However, using (2.105) and replacing the integral by a sum over small regions $\delta\mathbf{s}_j$, we have

$$P(\mathbf{r}_1,\cdots,\mathbf{r}_N) = \mathrm{Tr}\Bigg[\mathcal{N}\left(\exp\left(-\sum_j \delta\mathbf{s}_j \hat{\Psi}_\mathrm{b}^\dagger(\mathbf{s}_j) \hat{\Psi}_\mathrm{b}(\mathbf{s}_j) \right) \right)$$
$$\times \hat{\Psi}_\mathrm{b}(\mathbf{r}_N) \cdots \hat{\Psi}_\mathrm{b}(\mathbf{r}_1) \hat{\rho}(t) \hat{\Psi}_\mathrm{b}^\dagger(\mathbf{r}_1) \cdots \hat{\Psi}_\mathrm{b}^\dagger(\mathbf{r}_N) \Bigg]$$
$$= \mathrm{Tr}\Bigg[\mathcal{N}\left(\prod_j \exp\left(-\delta\mathbf{s}_j \hat{\Psi}_\mathrm{b}^\dagger(\mathbf{s}_j) \hat{\Psi}_\mathrm{b}(\mathbf{s}_j) \right) \right)$$
$$\times \hat{\Psi}_\mathrm{b}(\mathbf{r}_N) \cdots \hat{\Psi}_\mathrm{b}(\mathbf{r}_1) \, \hat{\rho}(t) \, \hat{\Psi}_\mathrm{b}^\dagger(\mathbf{r}_1) \cdots \hat{\Psi}_\mathrm{b}^\dagger(\mathbf{r}_N) \Bigg]$$
$$= \mathrm{Tr}\Bigg[\mathcal{N}\left(\prod_j (1 - \delta\mathbf{s}_i \hat{\Psi}_\mathrm{b}^\dagger(\mathbf{s}_j) \hat{\Psi}_\mathrm{b}(\mathbf{s}_j) + \cdots) \right)$$
$$\times \left[\hat{\Psi}_\mathrm{b}(\mathbf{r}_N) \cdots \hat{\Psi}_\mathrm{b}(\mathbf{r}_1) \hat{\rho}(t) \, \hat{\Psi}_\mathrm{b}^\dagger(\mathbf{r}_1) \cdots \hat{\Psi}_\mathrm{b}^\dagger(\mathbf{r}_N) \right] \Bigg]. \tag{2.108}$$

The quantity in the large round brackets can be expanded as a power series involving the volume elements $\delta\mathbf{s}_j$. The density operator must, in general, contain pure states with at least N bosons in order for the quantity in square brackets in the last line to be non-zero, as the field operators $\hat{\Psi}_{\mathrm{b}}(\mathbf{r}_i)$ from the left and $\hat{\Psi}_{\mathrm{b}}^{\dagger}(\mathbf{r}_i)$ from the right destroy one boson in each operation. Suppose there are states with at most M bosons in $\hat{\rho}(t)$. Then the quantity $[\cdots]$ will vanish if the number of bosons to be detected is greater than M. Suppose we choose to detect exactly $N = M$ bosons. Then the quantity $[\cdots]$ is proportional to the vacuum state projector, so the field annihilation operators $\hat{\Psi}_{\mathrm{b}}(\mathbf{s}_j)$ in $(\cdots)$ will give zero, leaving only the contributions from the unity terms. Thus for $N = M$

$$\begin{aligned} P(\mathbf{r}_1,\cdots,\mathbf{r}_N) &= \mathrm{Tr}\big(\hat{\Psi}_{\mathrm{b}}(\mathbf{r}_N)\cdots\hat{\Psi}_{\mathrm{b}}(\mathbf{r}_1)\hat{\rho}(t)\hat{\Psi}_{\mathrm{b}}^{\dagger}(\mathbf{r}_1)\cdots\hat{\Psi}_{\mathrm{b}}^{\dagger}(\mathbf{r}_N)\big) \\ &= G^n(\mathbf{r}_N,\cdots,\mathbf{r}_1;\mathbf{r}_1,\cdots,\mathbf{r}_N), \end{aligned} \tag{2.109}$$

showing that the position probability density is indeed given by the diagonal case of the quantum correlation function. This result is also valid for the case $N < M$. In that situation the quantity $[\cdots]$ will consist of a number of terms including one proportional to the vacuum projector, but also terms proportional to projectors for states with up to $M - N$ bosons. For the terms involving projectors with one or more bosons, there will be one or more field annihilation operators $\hat{\Psi}_{\mathrm{b}}(\mathbf{s}_j)$ in $(\cdots)$ that could give a non-zero contribution. However, such terms are accompanied by volume elements $\delta\mathbf{s}_i$, leading to contributions that are at least linear in the volume elements. As the volume elements $\delta\mathbf{s}_i$ are allowed to become small, such terms are negligible compared with those from the unity factors in $(\cdots)$. Thus (2.109) gives the correct probability density for position measurements for all N, including $N > M$, where the probability density is zero.

Exercises

(2.1) Show that the boson annihilation and creation operators satisfy the commutation rules in (2.24).

(2.2) Show that the fermion Fock state $|l_1,1;l_2,1;\cdots l_p,1\rangle$ is given by (2.25) in terms of fermion creation operators $\hat{c}_i^{\dagger}$ and the vacuum state $|0\rangle$.

(2.3) Show that the boson Fock state $|l_1,\xi_1,\cdots,l_p,\xi_p\rangle$ is given by (2.30) in terms of boson creation operators $\hat{a}_i^{\dagger}$ and the vacuum state $|0\rangle$.

(2.4) Show that the two new modes in (2.42) are orthogonal.

(2.5) Show that the new mode annihilation and creation operators in (2.43) and (2.44) satisfy the expected commutation and anticommutation rules.

(2.6) Confirm that the two-boson state $(\hat{b}_1)^{\dagger}(\hat{b}_2)^{\dagger}|0\rangle$ is a state with one boson in each of the modes 1 and 2.

(2.7) Confirm that the two-fermion state $(\hat{d}_1)^{\dagger}(\hat{d}_2)^{\dagger}|0\rangle$ is a state with one fermion in each of the modes 1 and 2.

(2.8) Justify the expressions in (2.52) for one- and two-body terms in the general Hamiltonian (2.50). (*Hint:* Consider matrix elements between arbitrary Fock states in both first and second quantised forms.)

(2.9) Justify the expressions in (2.51) for one- and two-body terms in the general Hamiltonian (2.50). (*Hint:* Consider matrix elements between arbitrary Fock states in both first and second quantised forms.)

(2.10) Prove the orthonormalisation condition in (2.69) for many-body position states.

(2.11) Justify the interpretation of the momentum Fock states (2.71) of creating particles with momenta $\hbar\mathbf{k}_1, \hbar\mathbf{k}_2, \cdots, \hbar\mathbf{k}_N$.

(2.12) Prove the orthonormalisation condition in (2.72) for many-body momentum states.

(2.13) Describe the possible two-particle processes for the fermion and boson cases.

(2.14) Prove the result in (2.94) for the fermion case.

3

Complex Numbers and Grassmann Numbers

The algebra of the *real* numbers is the subject of many elementary courses and textbooks in mathematics; the rules of addition and multiplication applying to these quantities will certainly be familiar to the reader. At a higher level, we encounter quantities such as *complex* numbers (or c-numbers for short) and *matrices.* The c-numbers satisfy the same algebraic rules as the reals but there is the additional operation of complex conjugation. Matrices have rather different properties and, in particular, the product of any two will usually depend on the order in which they appear. It is perhaps not surprising, therefore, that further quantities can be introduced into mathematics which do not obey all of the familiar addition and multiplication rules as the real or complex numbers do. One simple example is the quaternions, introduced by Hamilton and used by Maxwell in his work on electricity and magnetism. A second, and for our purposes more important, example is the *Grassmann* numbers (or g-numbers), introduced by Hermann Grassmann in the middle of the nineteenth century. The aim of this chapter is to present the algebra of the g-numbers and to compare and contrast this with that of the more familiar complex numbers. We are familiar with a calculus based on the real numbers, and this can be extended to treat complex and Grassmann numbers. This will be the topic of the following chapter.

3.1 Algebra of Grassmann and Complex Numbers

We start with a formal definition of the Grassmann numbers and then explore the consequences of this definition and, in particular, the similarities and differences between them and the complex numbers. A *Grassmann algebra* over the c-numbers is an *associative* algebra constructed from the *unit* 1 and a set of *anticommuting generators* (also called *Grassmann numbers*) $\{g_1, g_2, \cdots, g_i, \cdots, g_n\}$ that satisfy the *multiplication rule*

$$g_i g_j + g_j g_i = \{g_i, g_j\} = 0 \qquad (i, j = 1, \cdots, n). \tag{3.1}$$

The corresponding multiplication rule for a set of c-numbers $\{\alpha_1, \cdots, \alpha_i, \cdots, \alpha_n\}$, of course, is

$$\alpha_i \alpha_j - \alpha_j \alpha_i = [\alpha_i, \alpha_j] = 0 \qquad (i, j = 1, \cdots, n). \tag{3.2}$$

Thus Grassmann numbers *anticommute* on multiplication, whereas complex numbers *commute.* It is the anticommutation rule that is the defining feature of the g-numbers

Phase Space Methods for Degenerate Quantum Gases. First Edition. Bryan J. Dalton, John Jeffers, and Stephen M. Barnett.
Published in 2015 by Oxford University Press.

and also hints at their significance for describing fermions. The standard notation is $[a, b]$ for the *commutator* and $\{g, h\}$ for the *anticommutator* for c-numbers and g-numbers, as already described for operators.

A complete algebra of both c-numbers and g-numbers is created by defining the two basic *operations* of *addition* and *multiplication*, in which there is both a *zero* number 0 and a *unit* number 1. These properties are listed below:

(a) *Addition*:

$$\begin{aligned} g_i + g_j &= g_j + g_i, \\ \alpha_i + \alpha_j &= \alpha_j + \alpha_i, \\ g_i + 0 &= 0 + g_i = g_i, \\ \alpha_i + 0 &= 0 + \alpha_i = \alpha_i. \end{aligned} \tag{3.3}$$

(b) *Multiplication*:

$$\begin{aligned} g_i g_j &= -g_j g_i, \\ \alpha_i \alpha_j &= \alpha_j \alpha_i, \\ 1 g_i &= g_i 1 = g_i, \qquad 0 g_i = g_i 0 = 0, \\ 1 \alpha_i &= \alpha_i 1 = \alpha_i, \qquad 0 \alpha_i = \alpha_i 0 = 0. \end{aligned} \tag{3.4}$$

To emphasise these features, we can say that addition and multiplication are *commutative* for c-numbers, but for g-numbers only addition is commutative, whereas multiplication is anticommutative. For both types of number, adding zero or multiplying by one leaves the number unchanged, and multiplying by zero gives the zero complex or Grassmann number. Note, however, that 1 is *not* a Grassmann number.

The rules of addition and multiplication are such that associative laws apply,

$$\begin{aligned} g_i + (g_j + g_k) &= (g_i + g_j) + g_k = g_i + g_j + g_k, \\ \alpha_i + (\alpha_j + \alpha_k) &= (\alpha_i + \alpha_j) + \alpha_k = \alpha_i + \alpha_j + \alpha_k, \\ g_i(g_j g_k) &= (g_i g_j) g_k = g_i g_j g_k, \\ \alpha_i(\alpha_j \alpha_k) &= (\alpha_i \alpha_j)\alpha_k = \alpha_i \alpha_j \alpha_k, \end{aligned} \tag{3.5}$$

as do the distributive laws,

$$\begin{aligned} g_i(g_j + g_k) &= g_i g_j + g_i g_k, \\ \alpha_i(\alpha_j + \alpha_k) &= \alpha_i \alpha_j + \alpha_i \alpha_k, \\ (g_j + g_k) g_i &= g_j g_i + g_k g_i, \\ (\alpha_j + \alpha_k)\alpha_i &= \alpha_j \alpha_i + \alpha_k \alpha_i. \end{aligned} \tag{3.6}$$

Thus the key difference between complex and Grassmann numbers is that products of Grassmann numbers exhibit the feature that changing the order of the factors produces a sign change,

$$g_i g_j = -g_j g_i, \tag{3.7}$$

whereas for c-numbers

$$\alpha_i \alpha_j = +\alpha_j \alpha_i \tag{3.8}$$

and there is no sign change.

An important consequence of the above properties is that Grassmann numbers exhibit a feature quite different from that for c-numbers: the square and any higher power of a g-number are *zero*, even though the Grassmann number itself is non-zero:

$$g_i^2 = g_i^3 = \cdots = 0. \tag{3.9}$$

For c-numbers, the only number whose square is zero is 0 itself. It follows that the only c-number that is also a Grassmann number is zero and, of course, that the only g-number that is also a c-number is zero. Although c-numbers such as unity can multiply Grassmann numbers, they are not themselves g-numbers, as the square of a non-zero c-number is itself non-zero.

It follows from the distributive laws (3.6) that the sum (or difference) of two Grassmann numbers anticommutes in multiplication with other g-numbers. Hence the sums and differences of Grassmann numbers are themselves Grassmann numbers. Products of two Grassmann numbers, however, commute rather than anticommute on multiplication:

$$\begin{aligned}[(g_i g_j), (g_k g_l)] &= (g_i g_j)(g_k g_l) - (g_k g_l)(g_i g_j) = g_i g_j g_k g_l - (-1)^4 g_i g_j g_k g_l \\ &= 0.\end{aligned} \tag{3.10}$$

Here the four factors of -1 arise from the four swapping operations needed to commute $g_i g_j$ through $g_k g_l$. Hence, unlike c-numbers, where the sums and products of c-numbers are also c-numbers, the product of two g-numbers is not itself a Grassmann number. A product of two g-numbers is not a c-number, however, as its square is zero. An immediate consequence of this is that there is no inverse for a Grassmann number (the inverse of a non-zero c-number α_i is, of course, simply $1/\alpha_i$). Suppose that there did exist a g-number, or indeed any other kind of number, $(g_i)^{-1}$ which, when multiplied by g_i gave the product 1. It would then follow that $(g_i)^{-1} g_i = 1$, but squaring each leads to the contradiction $0 = 1$. It follows, similarly, that there is no process corresponding to division of g-numbers. There is no Grassmann number (g_i/g_j) which when multiplied by the Grassmann number g_j will give g_i. If there were, then the product of two Grassmann numbers would be a Grassmann number.

Grassmann numbers and complex numbers can be combined by addition or by multiplication. A moment's thought will convince you that the sum of a g-number and a c-number is neither a Grassmann number nor a complex number. The product of a Grassmann number and a c-number obeys the usual commutative c-number rules:

$$c g_i = g_i c, \tag{3.11}$$

$$c(g_i + g_j) = c g_i + c g_j = (g_i + g_j) c, \tag{3.12}$$

where c is any c-number. It is straightforward to show that the product of a c-number and a g-number is itself a Grassmann number. In Section 3.3 we will consider so-called *Grassmann functions*, which are linear combinations of products of Grassmann numbers with c-number coefficients.

Grassmann numbers are essentially abstract quantities and cannot be represented in terms of c-numbers – just as the imaginary number i cannot be represented in terms of real numbers. However, we can find matrices that have the same properties as g-numbers. For example, the 2×2 matrix

$$[g] = \begin{bmatrix} 1 & -1 \\ 1 & -1 \end{bmatrix} \tag{3.13}$$

satisfies the equation $[g]^2 = 0$, just as for any g-number g. The two 4×4 matrices

$$[g_1] = \begin{bmatrix} 0 & 0 & 0 & 0 \\ 1 & 0 & 0 & 0 \\ 0 & 0 & 0 & 0 \\ 0 & 0 & 1 & 0 \end{bmatrix}, \qquad [g_2] = \begin{bmatrix} 0 & 0 & 0 & 0 \\ 0 & 0 & 0 & 0 \\ 1 & 0 & 0 & 0 \\ 0 & 1 & 0 & 0 \end{bmatrix} \tag{3.14}$$

satisfy the equations $[g_i]^2 = 0$ and $[g_1][g_2] + [g_2][g_1] = 0$, just as would a pair of g-numbers g_1, g_2. More generally, a set of n g-numbers can be represented by n $2^n \times 2^n$ square matrices. It is not usually helpful to think of g-numbers as matrices, however, but rather as abstract quantities defined by their algebraic properties.

3.2 Complex Conjugation

Complex conjugation is an important operation for complex numbers and has a fundamentally important role in quantum mechanics as the means by which positive probabilities are obtained from complex amplitudes. Any complex number α can be written in terms of two real numbers a and b and the imaginary quantity i in the form

$$\alpha = a + ib. \tag{3.15}$$

The complex conjugate of α is denoted α^* (although other notations such as $\bar{\alpha}$ are often used) and is obtained from α by replacing i with $-i$:

$$\alpha^* = a - ib. \tag{3.16}$$

More generally, for functions of one or more complex variables the complex conjugate is obtained by replacing i with $-i$ everywhere it occurs.

It is useful to introduce a notion of complex conjugation for Grassmann numbers, and we do this by simply defining the properties of g_i^*, the complex conjugate of g_i. This quantity is also a Grassmann number and is subject to the following rules:

$$\begin{aligned} (g_i^*)^* &= g_i, \\ (g_i + g_j)^* &= g_i^* + g_j^*, \\ (g_i g_j)^* &= g_j^* g_i^*, \end{aligned} \tag{3.17}$$

so that complex conjugation is reflexive (taking the complex conjugate twice gives us the original quantity). Although the conjugate of a *sum* is the sum of the conjugates, we define the conjugate of a product to be the product of the conjugates in the reverse order. This procedure is reminiscent of the rule for taking the Hermitian conjugate of a product of matrices. For comparison, we recall that the process of complex conjugation for c-numbers has similar features; if α_i is a complex number then the complex conjugate α_i^* is also a complex number subject to the following rules:

$$\begin{aligned} (\alpha_i^*)^* &= \alpha_i, \\ (\alpha_i + \alpha_j)^* &= \alpha_i^* + \alpha_j^*, \\ (\alpha_i \alpha_j)^* &= \alpha_j^* \alpha_i^* = \alpha_i^* \alpha_j^*. \end{aligned} \tag{3.18}$$

Complex conjugation for c-numbers is reflexive, the conjugate of a *sum* is the sum of the conjugates and the conjugate of a product is the product of the conjugates in the same order (or, because they commute, the reverse order). This distinction is important for algebraic calculations with g-numbers.

As with other g-numbers, the conjugate obeys all the previous rules of Grassmann algebra, including multiplication with a c-number c, so that

$$(cg_i)^* = c^* g_i^* = g_i^* c^* = (g_i c)^*. \tag{3.19}$$

It is straightforward to prove that the associative and distributive rules that apply to the original g-numbers also apply to the conjugates, which also anticommute for multiplication, i.e. $g_i^* g_j^* = -g_j^* g_i^*$. Again, the square and higher powers of g-number conjugates are also zero, i.e. $(g_i^*)^2 = 0$.

It is also possible to define the real and imaginary parts g_{R} and g_{I} for a Grassmann number in a manner analogous to that for c-numbers:

$$\begin{aligned} g &= g_{\mathrm{R}} + i g_{\mathrm{I}}, & g^* &= g_{\mathrm{R}} - i g_{\mathrm{I}}, \\ g_{\mathrm{R}} &= \frac{1}{2}(g + g^*), & g_{\mathrm{I}} &= \frac{1}{2i}(g - g^*), \end{aligned} \tag{3.20}$$

where $g_{\mathrm{R}} = g_{\mathrm{R}}^*$ and $g_{\mathrm{I}} = g_{\mathrm{I}}^*$. Thus the formal requirements of being equal to their complex conjugates is satisfied by g_{R} and g_{I}, and in this way they behave somewhat like the real and imaginary parts of a complex number.

3.3 Monomials and Grassmann Functions

Complex numbers allow the development of the idea of *functions*, in which each member of a set of the original c-numbers $\alpha = \{\alpha_1, \cdots, \alpha_i, \cdots, \alpha_n\}$ corresponds to a new quantity $f(\alpha) \equiv f(\alpha_1, \cdots, \alpha_i, \cdots \alpha_n)$, called a *complex function*. Similarly, we can take a set of Grassmann numbers $g = \{g_1, \cdots, g_i, \cdots, g_n\}$ and make it correspond to a new quantity $f(g) \equiv f(g_1, \cdots, g_i, \cdots, g_n)$, called a *Grassmann function*. There is, however, an important difference: in the c-number case the complex function is in general another c-number, whereas in the g-number case the Grassmann function is not in general a g-number. As we have already seen, for example, the quantity $f(g) = g_1 g_2$ is not a g-number, although it is a Grassmann function.

A natural way to start our discussion of Grassmann functions is with the *monomials* of various orders that can be formed from the g-numbers g_i via successive multiplication of different g-numbers. Thus

$$
\begin{array}{ll}
1 & \text{Order } 0 \\
g_1, g_2, \cdots, g_i, \cdots, g_n & \text{Order } 1 \\
g_1g_2, g_1g_3, \cdots, g_1g_n; g_2g_3, g_2g_4, \cdots, g_2g_n; \cdots; g_{n-1}g_n & \text{Order } 2 \\
\vdots & \\
g_1g_2 \cdots g_i \cdots g_n & \text{Order } n
\end{array}
\tag{3.21}
$$

is a list of all the distinct non-zero monomials that can be formed from the n g-numbers $g_1, \cdots, g_n$. As $g_2g_1 = -g_1g_2$, there is no need to list g_2g_1 separately, while the order-two function g_1^2 is automatically zero, as is any product in which a g-number appears more than once. Clearly, if there are n original g-numbers there can be no non-zero monomial of order greater than n. The number of distinct monomials of order p is given by ${}^nC_p$, so the total number of monomials of all orders is given by $\sum_{p=0}^{n} {}^nC_p = 2^n$.

The set of distinct monomials forms the basis of a *linear vector space* of dimension 2^n, with elements f defined by

$$
\begin{aligned}
f(g) \equiv f(g_1, \cdots, g_i, \cdots, g_n) &= f_0 + \sum_{k_1} f_1(k_1) g_{k_1} + \sum_{k_1 k_2} f_2(k_1, k_2) g_{k_1} g_{k_2} + \cdots \\
&\quad + \sum_{k_1 \cdots k_n} f_n(k_1, \cdots, k_n) g_{k_1} \cdots g_{k_n},
\end{aligned}
\tag{3.22}
$$

where the f_i are c-numbers. Each such expression is an element in the Grassmann algebra. The quantity f so defined is a function of the Grassmann generators $f(g)$ and is referred to as a *Grassmann function*. If each of the sums over $k_1, \cdots, k_n$ runs over *all* of the values $1, \cdots, n$ then the c-numbers f_i are not unique as all possible permutations of the g-numbers would then appear. If, however, we restrict the sums so that each non-zero monomial only appears *once*, then the c-numbers f_i will be unique. Alternatively, if the f_i are restricted to being *antisymmetric* with respect to permutations of the $k_1, \cdots, k_i$, then the f_i are unique. Naturally, if two of the k_i are equal, then the corresponding f_i can be set to zero as the associated product of Grassmann numbers is itself zero.

For a single Grassmann generator with $g = \{g_1\}$, the most general Grassmann function is simply

$$
f(g) = f_0 + f_1 g_1. \tag{3.23}
$$

For a pair of Grassmann generators with $g = \{g_1, g_2\}$, the most general Grassmann function includes a term proportional to the product g_1g_2:

$$
\begin{aligned}
f(g) &= f_0 + f_1(1)g_1 + f_1(2)g_2 + f_2(1,2)g_1g_2 \\
&= f_0 + f_1(1)g_1 + f_1(2)g_2 - f_2(1,2)g_2g_1.
\end{aligned}
\tag{3.24}
$$

More generally, a Grassmann function of n g-numbers can be expressed as linear combinations of monomials, so that to define each Grassmann function a total of 2^n c-numbers $f_0, f_1(k_1), \cdots, f_n(k_1 \cdots k_n)$ are required. This equivalence of Grassmann functions to a set of c-numbers is very useful for numerical calculations involving Grassmann functions. If we include the conjugates with the original g-numbers then we have the $2n$ g-numbers specified as the set $g, g^* = \{g_1, \cdots, g_n, g_1^*, \cdots, g_n^*\}$; we can then obtain monomials up to order 2^{2n}, and the Grassmann algebra contains elements expressed as a straightforward generalisation of (3.22). Each element requires 2^{2n} c-numbers to specify it.

The basic operations defined for the elements of the Grassmann algebra, or Grassmann functions, are:

(a) *Multiplication by c-numbers*:

$$\begin{aligned} cf &= cf_0 + \sum_{k_1} cf_1(k_1)g_{k_1} + \sum_{k_1k_2} cf_2(k_1,k_2)g_{k_1}g_{k_2} \\ &\quad + \cdots + \sum_{k_1k_2\cdots k_n} cf_n(k_1,k_2,\cdots,k_n)g_{k_1}g_{k_2}\cdots g_{k_n} \\ &= fc. \end{aligned} \tag{3.25}$$

(b) *Addition*:

$$\begin{aligned} e + f &= (e_0 + f_0) + \sum_{k_1}(e_1(k_1) + f_1(k_1))g_{k_1} \\ &\quad + \sum_{k_1k_2}(e_2(k_1,k_2) + f_2(k_1,k_2))g_{k_1}g_{k_2} + \cdots \\ &\quad + \sum_{k_1k_2\cdots k_n}(e_n(k_1,k_2,\cdots,k_n) + f_n(k_1,k_2,\cdots,k_n))g_{k_1}g_{k_2}\cdots g_{k_n} \\ &= f + e. \end{aligned} \tag{3.26}$$

(c) *Multiplication*:

$$ef \neq fe, -fe. \tag{3.27}$$

Products of two Grassmann elements may be evaluated by applying the multiplication and addition rules listed in the previous section and always result in a Grassmann element. Clearly, any linear combination of elements in the Grassmann algebra is also an element of the Grassmann algebra. Multiplication of elements is in general neither commutative nor anticommutative, as can be seen by multiplying two elements of order 1.

The elements of the Grassmann algebra can be obtained from two subsets, one being the even Grassmann functions f_{E} that include only monomials of even order and the other being the odd Grassmann functions f_{O} that include only monomials of odd order. In general, therefore, these will have the forms

$$f_{\mathrm{E}}(g) = f_0 + \sum_{k_1k_2} f_2(k_1,k_2)g_{k_1}g_{k_2} + \cdots, \tag{3.28}$$

$$f_O(g) = \sum_{k_1} f_1(k_1) g_{k_1} + \sum_{k_1 k_2 k_3} f_3(k_1, k_2, k_3) g_{k_1} g_{k_2} g_{k_3} + \cdots . \tag{3.29}$$

It is then self-evident that any Grassmann function can be expressed as the sum of an even and an odd Grassmann function:

$$f = f_E + f_O. \tag{3.30}$$

New Grassmann functions can be defined as functions of an original Grassmann function $f(g_1, \cdots, g_i, \cdots, g_n)$. Important examples include *powers* and *exponentials*, which are defined in the same way as for c-numbers:

$$\begin{aligned} f(g)^n &\equiv f f \cdots f \qquad n \text{ times}, \\ \exp\left[f(g)\right] &= 1 + f + f^2/2! + f^3/3! + \cdots . \end{aligned} \tag{3.31}$$

Given that the original Grassmann function terminates after the nth order, it is not surprising that even if a function is defined by an infinite series, the series will often terminate. A simple example is $\exp(g) = 1 + g$, as all the higher-order terms are zero.

Although Grassmann variables do not possess an inverse, the inverse of a Grassmann function often does exist. Any Grassmann function with $f_0 = 1$ can be written as $(1 + g_1 f(g_2, \cdots, g_n))$, where $f(g_2, \cdots, g_n)$ is a Grassmann function of the other g-numbers. As

$$\begin{aligned} (1 + g_1 f(g_2, \cdots, g_n))(1 - g_1 f(g_2, \cdots, g_n)) &= 1 \\ = (1 - g_1 f(g_2, \cdots, g_n))(1 + g_1 f(g_2, \cdots, g_n)), \end{aligned} \tag{3.32}$$

the left and right inverse of $(1 + g_1 f(g_2, \cdots, g_n))$ is $(1 - g_1 f(g_2, \cdots, g_n))$. More generally, the inverse of a function with $f_0 \neq 0$ can be found by multiplying appropriately by f_0.

As with the g-numbers themselves, Grassmann functions satisfy the standard associative and distributive laws familiar for c-numbers. For g-numbers, the commutative law for multiplication and the division or inverse process are missing. As a consequence, most of the rules and features applying to the algebra of quantum mechanical operators also apply to Grassmann functions. In particular, consider the commutator of two Grassmann functions f, h. If one or both of the two functions are even, then the commutator is zero. If both functions are odd, then the commutator is non-zero in general, but it will always be an even function. Thus

$$\begin{aligned} &[f_E, h_E] = 0, \qquad [f_E, h_O] = 0, \qquad [f_O, h_E] = 0, \\ &[f_O, h_O] = 2 f_O h_O, \end{aligned} \tag{3.33}$$

so that for an arbitrary pair of functions f, h as in (3.30) we have

$$[f, g] = 2 f_O h_O. \tag{3.34}$$

It follows immediately that the commutator of $[f, h]$ with either f or h, or indeed any other Grassmann function, will be zero:

$$[f, [f, h]] = [h, [f, h]] = [g, [f, h]] = 0. \tag{3.35}$$

Odd Grassmann functions do not commute; however, they do have the property of anticommuting:

$$\begin{aligned} \{f_O, h_O\} &= f_O h_O + h_O f_O = 0, \\ f_O^2 &= f_O^3 = \cdots = 0, \end{aligned} \tag{3.36}$$

and clearly, the square and any higher power of any odd Grassmann function are zero. Furthermore, the complex conjugate of the product of two odd Grassmann functions is equal to the product of the complex conjugates taken in reverse order:

$$(f_O h_O)^* = h_O^* f_O^*. \tag{3.37}$$

Taken together, these two features of odd Grassmann functions demonstrate an important feature of *odd* Grassmann functions, namely that they behave algebraically as if they were themselves Grassmann variables.

This interesting general feature of Grassmann functions is quite useful in evaluating the products of exponential functions of g-functions. The *Baker–Hausdorff theorem* applies to matrices, operators and any other non-commuting objects and states that

$$\exp(A)\exp(B) = \exp(A + B)\exp\left(\frac{1}{2}[A, B]\right) \tag{3.38}$$

if A and B both commute with $[A, B]$. It then follows that the Baker–Hausdorff theorem applies to Grassmann functions:

$$\exp(f) \times \exp(h) = \exp(f + h) \times \exp\left(\frac{1}{2}[f, h]\right). \tag{3.39}$$

If either f or g (or both) is even, then the normal c-number rule $\exp(f) \times \exp(h) = \exp(f + h)$ applies.

The exponential of a Grassmann function has an *inverse* given as the exponential of minus the Grassmann function,

$$\exp(f) \times \exp(-f) = \exp(-f) \times \exp(f) = 1, \tag{3.40}$$

as can readily be obtained from (3.39). We frequently encounter the exponential transformation law, which has the general form

$$\exp(A)B\exp(-A) = B + [A, B] + \frac{1}{2!}[A, [A, B]] + \frac{1}{3!}[A, [A, [A, B]]] + \cdots. \tag{3.41}$$

It then follows that for a pair of Grassmann functions,

$$\exp(f) \times h \times \exp(-f) = g + [f, h]. \tag{3.42}$$

The series in (3.41) terminates after the second term because $[f,[f,h]]=0$. The complex conjugate of the product of two Grassmann functions is the product of the two conjugates in reverse order:

$$(f_1(g)f_2(g))^* = f_2(g)^* f_1(g)^*. \tag{3.43}$$

This follows from the rules for complex conjugation of Grassmann numbers.

Furthermore, we may introduce transformations of a set of Grassmann variables $g=\{g_1,\cdots,g_i,\cdots,g_n\}$ to produce a new set of quantities $h=\{h_1,\cdots,h_i,\cdots,h_n\}$ defined by

$$h_i = \sum_j A_{ij} g_j, \tag{3.44}$$

where the A_{ij} are a set of matrix elements each of which is a Grassmann function. In general the h_i will only be Grassmann functions, but it is straightforward to show that if all the A_{ij} are even Grassmann functions, then the new quantities h_i are actually Grassmann variables, satisfying all the basic laws (3.3)–(3.6) and (3.17) as required. In particular, the A_{ij} may be all c-numbers.

Finally, in manipulating products of Grassmann variables there are several useful rules for: (a) re-ordering the terms in opposite order (see (3.45)), (b) separating products of pairs of different types of g-numbers into the product of terms of the same type (see (3.46)) and (c) transferring products of two different types of g-numbers into the opposite order (see (3.47)).

(a) *Reordering*

$$h_1h_2\cdots h_n = h_n\cdots h_2h_1(-1)^{n(n-1)/2} \tag{3.45}$$

(b) *Separating*

$$g_1h_1\,g_2h_2\cdots g_nh_n = g_1g_2\cdots g_n\,h_1h_2\cdots h_n(-1)^{n(n-1)/2} \tag{3.46}$$

(c) *Transferring*

$$h_1h_2\cdots h_n\,g_1g_2\cdots g_n = g_1g_2\cdots g_n\,h_1h_2\cdots h_n(-1)^{n^2} \tag{3.47}$$

The proof of these rules is left as an exercise.

Exercises

(3.1) If α_1,α_2 are any two c-numbers and g_1,g_2 are any two g-numbers, then what type of number is $\alpha_1g_1+\alpha_2g_2$?

(3.2) Evaluate the square of the product of g-numbers g_1g_2, thereby confirming that this product is not a c-number.

(3.3) Three-vectors $\{\vec{V},\vec{W},\cdots\}$ have something in common with g-numbers if multiplication is defined as the vector product. In particular, we have $\vec{V}\times\vec{W}=-\vec{W}\times\vec{V}$ and $\vec{V}\times\vec{V}=\vec{0}$. Are such vectors then examples of Grassmann numbers?

(3.4) Prove that the number of distinct Grassmann monomials of order p is given by ${}^nC_p$, where n is the number of g-numbers. Hence confirm that the total number of monomials of all orders is $\sum_{p=0}^{n} {}^nC_p = 2^n$.

(3.5) Prove the set of properties (3.17) using the fundamental properties of Grassmann numbers.

(3.6) Confirm that, for the Grassmann function (3.22), if each of the c-number functions f_i is antisymmetric with respect to permutations of the $k_1, k_2, \cdots, k_i$ then the f_i are unique.

(3.7) Prove the anticommutation rule (3.36) for odd Grassmann functions.

(3.8) Confirm that a Grassmann function has the same left and right inverse and calculate the inverse of $f_0 + g_1 f(g_2, \cdots, g_n)$. What happens if $f_0 = 0$?

(3.9) Do the square roots of the Grassmann function $f_0 + f_1 g_1$ exist within the algebra of c-numbers and g-numbers? If they do, what are they?

(3.10) Prove the Baker–Hausdorff theorem (3.38) and the exponential transformation rule (3.41). (*Hint:* You might try replacing A and B by λA and λB (with λ real) and differentiating with respect to λ.)

(3.11) Use the Baker–Hausdorff theorem for g-numbers (3.39) to show that $\exp(f)\exp(h) = \exp(f + h)$ if both $f_O = 0$ and $h_O = 0$.

(3.12) Prove the rule $(f_1 f_2)^* = (f_2^*)(f_1^*)$ for the product of two Grassmann functions.

(3.13) Show that if the transformation matrix elements A_{ij} are even Grassmann functions, then the new variables h_i related to an original set of Grassmann variables g_i via $h_i = \sum_j A_{ij} g_j$ are themselves Grassmann variables.

(3.14) Using induction, prove the rules in (3.45), (3.46) and (3.47) for reordering, separating and transferring products of Grassmann variables.

4 Grassmann Calculus

The ordinary calculus of functions involving c-numbers is well known. The basic processes of differentiation and integration, respectively, correspond to the gradient of the tangent to and the area under a function. Their mathematical structure is based on constructing limits which correspond to these quantities:

$$\frac{df}{dx} = \lim_{\delta x \to 0} \{f(x + \delta x) - f(x)\}/\delta x, \tag{4.1}$$

$$\int f(x)\, dx = \lim_{\delta x \to 0} \left\{ \sum_i f(x_i) \right\} \delta x, \tag{4.2}$$

where $\delta x = x_{i+1} - x_i$. The formal rules of c-number calculus are the rules by which these quantities transform.

In contrast, Grassmann calculus is introduced in a purely formal manner – a consequence of the fact that Grassmann numbers cannot be understood in terms of spatial coordinates: a Grassmann derivative cannot be interpreted as the gradient of a graph, and a Grassmann integral does not correspond to the area under a curve. Indeed, there is no concept of a definite Grassmann integral. Instead, the basic rules of Grassmann calculus are provided by analogy with those for c-numbers. This leads to some of the derived rules of Grassmann calculus being the same as those for c-number calculus. The anticommuting property of Grassmann numbers, however, means that other rules, such as the product rule for differentiation, are different. It also means that differentiation (or integration) acting from the left, in general, gives different results from differentiation acting from the right.

The approach to defining Grassmann differentiation and integration is founded on the simple feature that for a particular Grassmann variable g_i, the only non-zero powers are $g_i^0 = 1$ and $g_i^1 = g_i$. Higher powers are all zero, as $g_i^2 = 0$. Hence we only need to define derivatives and integrals for 1 and g_i. For derivatives and integrals involving Grassmann functions, where each g_i may appear in several terms, we simply prescribe that linearity rules apply and determine the derivatives and integrals from the basic rules for 1 and g_i. This simple procedure allows us to construct a complete Grassmann calculus, albeit an unfamiliar one. For example, the processes of differentiation and integration produce the same result when applied to a Grassmann function. This means that they are not mutually inverse processes, as they are in ordinary calculus. Many of the results for Grassmann functions also apply for Grassmann states and operators, as we shall see in Chapter 5.

In the next section we describe the form of c-number calculus which is normally used in phase-space-based formulations of quantum mechanics, and on which the Grassmann variable treatments described later are based. The following sections are devoted to Grassmann calculus itself, first differentiation and then integration.

4.1 C-number Calculus in Complex Phase Space

In mathematics, c-number calculus usually means the calculus of functions of a complex variable, normally denoted $z = x + iy$. The derivative of a complex function f can be defined at a particular point in the complex plane only if the function is analytic there. In order to satisfy this, the function must have a unique derivative, independent of the direction of the limit taken in the x–y Argand plane. Integrals are normally along lines in the complex plane (contours), and their value depends on the path taken. The c-number calculus of interest to us, however, is of a different character. We shall be concerned primarily with integrals of non-analytic functions over the entire complex plane.

The phase space associated with a one-component system in classical physics is defined by two axes, one corresponding to the system's position and the other to the momentum. A distribution $P(x, p)$ can be defined which gives the probability density that the system can be found at point x with momentum p. The evolution of the probability density is governed by a differential equation, the solution of which allows expectation values of any system variable to be found at any time. In quantum physics, however, the accuracy with which a system can be placed in such a space is limited by the Heisenberg uncertainty principle $\Delta x\ \Delta p \geq \hbar/2$. This means that a quantum phase-space probability distribution does not exist in this sense.

In place of the probability distribution, the primary quantity in quantum physics is the complex wave function, the squared modulus of which provides probability densities. The probability densities are not defined over phase space, owing to the limits imposed by the Heisenberg principle. Rather, they are probability densities over position or momentum space, depending on the representation used for defining the wave function, and they allow calculation of expectation values of observables. The evolution of the wave function, which in non-relativistic quantum mechanics is provided by the Hamiltonian, governs the evolution of the probability densities.

The concept of phase space densities is such a useful one that probability-density-like distributions have been defined for quantum systems. These so-called *quasiprobability* distributions are defined over a complex phase space and exhibit many of the features of their classical counterpart [4, 31]. Like the wave function, they form a representation of the state. The phase space is not normally based on x and p for bosonic systems, but more typically on the dimensionless complex variables $\alpha = (2\hbar\omega)^{-1/2}(\omega x + ip)$ and $\alpha^* = (2\hbar\omega)^{-1/2}(\omega x - ip)$, which are treated as independent variables (ω is the oscillation frequency of the system – the optical frequency in quantum optics). These complex variables are related to bosonic annihilation and creation operators, respectively. A number of different quasiprobability distributions have been defined, and three of the best known are the Glauber–Sudarshan P distribution, the Wigner W distribution and the Husimi Q distribution. Their properties and their applications in determining quantum correlation functions for normally, symmetrically and antinormally ordered products of bosonic creation and annihilation operators are given in many quantum optics textbooks [4]. By starting from the system Hamiltonian or a master equation, an evolution equation for the quasiprobability can be derived, normally called a Fokker–Planck equation. For quasi-distributions depending on α and α^*, this equation always contains terms which depend on partial derivatives with respect to the complex variables,

$$\frac{\partial}{\partial\alpha}, \frac{\partial}{\partial\alpha^*}, \tag{4.3}$$

where α and α^* are treated as independent variables. Solutions of this equation provide the quasiprobability as a function of time, from which expectation values can be calculated by integrating over phase space. Such integrals are over the whole *area* of phase space

$$\int d^2\alpha \equiv \int_{-\infty}^{+\infty} \int_{-\infty}^{+\infty} d\alpha_x \, d\alpha_y, \tag{4.4}$$

where $\alpha = \alpha_x + i\alpha_y$ defines the real and imaginary components.

We treat α and α^* as independent variables, which means that we need to be able to vary one without changing the other. Note that we will use the convention that $\alpha^* = \alpha_x^* + i\alpha_y^* = \alpha_x - i\alpha_y$. However, this does not in turn mean that a change $\delta\alpha_x$ in α will be equal to a change $\delta\alpha_x^*$ in α^*, as the former is the real part of an increment in α and the latter is the real part of a change in α^*. By definition, we have

$$\int d^2\alpha^* \equiv \int_{-\infty}^{+\infty} \int_{-\infty}^{+\infty} d\alpha_x^* \, d\alpha_y^*. \tag{4.5}$$

The transformation between $d^2\alpha$ and $d^2\alpha^*$ involves the Jacobian, as described by Courant [35]. Thus, for a function that may involve both α and α^*, $f(\alpha, \alpha^*) = f(\alpha_x + i\alpha_y, \alpha_x - i\alpha_y) = F(\alpha_x, \alpha_y)$, we have

$$\begin{aligned} \int d^2\alpha \, f(\alpha, \alpha^*) &= \int_{-\infty}^{+\infty} \int_{-\infty}^{+\infty} d\alpha_x \, d\alpha_y \, F(\alpha_x, \alpha_y) \\ &= \int_{-\infty}^{+\infty} \int_{+\infty}^{-\infty} d\alpha_x^* \, d\alpha_y^* \, \frac{\partial(\alpha_x, \alpha_y)}{\partial(\alpha_x^*, \alpha_y^*)} \, F(\alpha_x(\alpha_x^*, \alpha_y^*), \alpha_y(\alpha_x^*, \alpha_y^*)) \end{aligned} \tag{4.6}$$

$$\begin{aligned} &= \int_{-\infty}^{+\infty} \int_{-\infty}^{+\infty} d\alpha_x^* \, d\alpha_y^* \, f(\alpha_x^* - i\alpha_y^*, \alpha_x^* + i\alpha_y^*) \\ &= \int d^2\alpha^* \, f((\alpha^*)^*, (\alpha^*)) = \int d^2\alpha^* \, f(\alpha, \alpha^*). \end{aligned} \tag{4.7}$$

With this convention, we can just replace $\int d^2\alpha$ by $\int d^2\alpha^*$, leaving the function unaltered.

The theory of functions of a complex variable is covered in numerous textbooks [36, 37] and need not be reproduced here. It is worth recalling, however, that a function of a complex variable z is analytic within a region of the complex plane if it has a unique derivative in that region. A simple test for this is that a function $f(z)$ is analytic at $z = x + iy$ if it satisfies the Cauchy–Riemann equations

$$\frac{\partial u}{\partial x} = \frac{\partial v}{\partial y}, \qquad \frac{\partial v}{\partial x} = -\frac{\partial u}{\partial y}, \tag{4.8}$$

where u and v are the real and imaginary parts of f. Not all functions will be analytic and, indeed, *no function* of z^* (except for the trivial constant function) is an analytic function of z. A second important result is an identity for analytic functions $f(\alpha)$ that is not widely known, but which was proved by Glauber [1]. This states that

$$\begin{aligned} f(\alpha^*) &= \frac{1}{\pi}\int d^2\beta\, f(\beta^*)\, \exp\left(\alpha^*\beta - |\beta|^2\right), \\ f(\alpha) &= \frac{1}{\pi}\int d^2\beta\, f(\beta)\, \exp\left(\alpha\beta^* - |\beta|^2\right), \end{aligned} \tag{4.9}$$

and it is assumed that $f(\alpha)$ can be expanded in a convergent power series in the α plane. The second form of the result follows via (4.7).

Our main interest will be in non-analytic functions $f(\alpha, \alpha^*)$, and we need to be able to differentiate these with respect to both α and α^*. The natural way to do this is to treat α and α^* as independent variables so that, for example, if $f(\alpha, \alpha^*) = |\alpha|^2$ then $(\partial/\partial\alpha)f = \alpha^*$. There is a second and fully equivalent approach, which is to replace $f(\alpha, \alpha^*)$ by a function, $F(\alpha_x, \alpha_y)$, of two independent real variables, $\alpha_x = (\alpha + \alpha^*)/2$ and $\alpha_y = (\alpha - \alpha^*)/(2i)$. Applying standard calculus methods then gives the rules

$$\begin{aligned} \frac{\partial}{\partial\alpha} f(\alpha, \alpha^*) &= \frac{1}{2}\left(\frac{\partial}{\partial\alpha_x} - i\frac{\partial}{\partial\alpha_y}\right) F(\alpha_x, \alpha_y), \\ \frac{\partial}{\partial\alpha^*} f(\alpha, \alpha^*) &= \frac{1}{2}\left(\frac{\partial}{\partial\alpha_x} + i\frac{\partial}{\partial\alpha_y}\right) F(\alpha_x, \alpha_y). \end{aligned} \tag{4.10}$$

These lead to the most satisfactory result that if $f(\alpha)$ is an analytic function of α then the Cauchy–Riemann equations give $(\partial/\partial\alpha^*)f(\alpha) = 0$, which is consistent with treating α and α^* as independent variables.

In the present work, we will mainly be concerned with non-analytic quasiprobability distributions in double phase spaces α, α^+, so in this case we consider functions involving four complex variables $f(\alpha, \alpha^*, \alpha^+, \alpha^{+*})$ or, equivalently, functions of the four real and imaginary components $F(\alpha_x, \alpha_y, \alpha_x^+, \alpha_y^+)$. In the latter case, differentiations and integrations with respect to $\alpha_x, \alpha_y, \alpha_x^+, \alpha_y^+$ will be involved, rather than with respect to $\alpha, \alpha^*, \alpha^+, \alpha^{+*}$. Again, there are two alternative approaches that may be used for dealing with such non-analytic double-space probability distributions: one involves considering $f(\alpha, \alpha^*, \alpha^+, \alpha^{+*})$ with the four complex $\alpha, \alpha^*, \alpha^+, \alpha^{+*}$ considered as independent, and the second involves considering the equivalent function $F(\alpha_x, \alpha_y, \alpha_x^+, \alpha_y^+)$ of the four real variables. In the present work we will often use the latter approach. Note that we can apply (4.9) to functions $f(\alpha, \alpha^+)$ that are analytic in two variables α, α^+.

4.2 Grassmann Differentiation

4.2.1 Definition

Grassmann functions are sums of monomials in which each Grassmann variable appears at most once. Products of Grassmann numbers depend on the order in which they appear and, in a not dissimilar manner, derivatives with respect to Grassmann variables can give different results depending on whether the derivative is on the left acting to the right (left derivative) or is on the right acting to the left (right derivative). The left and right derivatives of a Grassmann function $f(g) \equiv f(g_1, \cdots, g_n)$ are defined via rules for each component monomial, which are based on the fundamental *rules for left and right differentiation* of the non-zero powers of a Grassmann variable. As we shall want to make a clear distinction between the two forms of differentiation, we shall use the arrow notation: $\overrightarrow{\partial}/\partial g_i$ for left differentiation and $\overleftarrow{\partial}/\partial g_i$ for right differentiation. The effects of these two types of derivatives have the form

$$\frac{\overrightarrow{\partial}}{\partial g_i}1 = 0 = 1\frac{\overleftarrow{\partial}}{\partial g_i}, \qquad \frac{\overrightarrow{\partial}}{\partial g_i}g_j = \delta_{ij} = g_j\frac{\overleftarrow{\partial}}{\partial g_i}. \tag{4.11}$$

In each monomial, a particular g_i can only occur once at most. The fundamental rules as defined above make all of the Grassmann derivatives of all monomials vanish apart from the derivatives with respect to those variables which are on the far left and right of the monomial. For a product of Grassmann variables which includes that being differentiated, we define the process of differentiation as follows: (a) move the differentiated variable g_i to become the left factor of the monomial (left differentiation) or to become the right factor (right differentiation) using the anticommuting property; then (b) remove the differentiated variable, as is consistent with the derivative of g_i with respect to itself being unity; and finally, (c) the other g_j are treated as if they are constants, as in ordinary differentiation. Thus the differentiation rules for a monomial are

$$\begin{aligned}
\frac{\overrightarrow{\partial}}{\partial g_i}g_{i_1}g_{i_2}\cdots g_{i_s} &= \delta_{ii_1}g_{i_2}\cdots g_{i_s} - \delta_{ii_2}g_{i_1}\cdots g_{i_s} + \cdots + (-1)^{s-1}\delta_{ii_s}g_{i_2}\cdots g_{i_{s-1}}, \\
g_{i_1}g_{i_2}\cdots g_{i_s}\frac{\overleftarrow{\partial}}{\partial g_i} &= \delta_{ii_s}g_{i_1}\cdots g_{i_{s-1}} - \delta_{ii_{s-1}}g_{i_1}\cdots g_{i_{s-1}}g_{i_s} + \cdots \\
&\quad +(-1)^{s-1}\delta_{ii_1}g_{i_2}\cdots g_{i_s}.
\end{aligned} \tag{4.12}$$

Only if a particular g_i does not occur is the derivative zero. A simple example is the differentiation of a product of two Grassmann variables:

$$\begin{aligned}
\frac{\overrightarrow{\partial}}{\partial g_i}g_kg_l &= \delta_{ik}g_l - \delta_{il}g_k, \\
g_kg_l\frac{\overleftarrow{\partial}}{\partial g_i} &= \delta_{il}g_k - \delta_{ik}g_l = -\frac{\overrightarrow{\partial}}{\partial g_i}g_kg_l,
\end{aligned} \tag{4.13}$$

which makes clear the difference between the results of left and right differentiation.

4.2.2 Differentiation Rules for Grassmann Functions

Differential calculus for c-numbers is aided by the existence of simple rules, such as the chain rule, the product rule and the quotient rule, each of which can be derived from the basic formulae. Here we describe the analogues for Grassmann functions. As discussed earlier, these are just sums of monomials weighted by c-numbers, and the usual linear rules of differentiation can be applied. Thus the left derivative of a Grassmann function $f(g)$ as in (3.22) is

$$\frac{\overrightarrow{\partial}}{\partial g_i} f(g) = \sum_{k_1} f_1(k_1) \frac{\overrightarrow{\partial}}{\partial g_i} g_{k_1} + \sum_{k_1 k_2} f_2(k_1, k_2) \frac{\overrightarrow{\partial}}{\partial g_i} g_{k_1} g_{k_2}$$

$$+ \cdots + \sum_{k_1 \cdots k_n} f_n(k_1, \cdots, k_n) \frac{\overrightarrow{\partial}}{\partial g_i} g_{k_1} \cdots g_{k_n}, \tag{4.14}$$

with a similar expression for the right derivative.

In order to formulate further a set of rules for formally differentiating or integrating a Grassmann function, we note that any Grassmann function involving g_i can always be written as the sum of two terms, the first, F_0^i, not involving g_i at all, and the second being a term in which g_i appears only once. For convenience, we use the notation $\overline{g_i}$ to denote the set $\overline{g_i} = \{g_1, \cdots, g_{i-1}, g_{i+1}, \cdots, g_n\}$, in which only g_i is absent. The second term can be written, using the anticommutation property of g-numbers, with g_i placed either on the left of the remaining factor $F_{\mathrm{L}1}^i$ or to the right of a different remaining factor $F_{\mathrm{R}1}^i$. Thus

$$f(g_1, \cdots, g_n) = F_0^i(\overline{g_i}) + g_i F_{\mathrm{L}1}^i(\overline{g_i}) \tag{4.15}$$

$$= F_0^i(\overline{g_i}) + F_{\mathrm{R}1}^i(\overline{g_i}) g_i, \tag{4.16}$$

where the Grassmann functions $F_0^i, F_{\mathrm{L}1}^i$ and $F_{\mathrm{R}1}^i$ are independent of g_i. Differentiation or integration with respect to g_i from the left just gives $F_{\mathrm{L}1}^i$, and the same processes from the right give $F_{\mathrm{R}1}^i$, the derivative of the F_0^i term giving zero. The results of the differentiation process will be a Grassmann function of the remaining Grassmann variables $g_1, g_2, \cdots, g_{i-1}, g_{i+1}, \cdots, g_n$ as follows:

$$\frac{\overrightarrow{\partial}}{\partial g_i} f(g_1, \cdots, g_n) = F_{\mathrm{L}1}^i(\overline{g_i}), \tag{4.17}$$

$$f(g_1, \cdots, g_n) \frac{\overleftarrow{\partial}}{\partial g_i} = F_{\mathrm{R}1}^i(\overline{g_i}). \tag{4.18}$$

In general, there is no relationship between the left and right derivatives of a Grassmann function. However, the left and right derivatives for *even and odd Grassmann functions* are simply related by the factors -1 and $+1$, respectively. To see this, consider how a typical term with n Grassmann numbers in which the Grassmann number g_i occurs in the pth place can be written in two ways:

$$\begin{aligned} & f_n(k_1, k_2, \cdots, k_n) g_{k_1} g_{k_2} \cdots g_{k_{p-1}} g_i g_{k_{p+1}} \cdots g_{k_n} \\ & = (-1)^{p-1} f_n(k_1, k_2, \cdots, k_n) g_i g_{k_1} g_{k_2} \cdots g_{k_{p-1}} g_{k_{p+1}} \cdots g_{k_n} \\ & = (-1)^{n-p} f_n(k_1, k_2, \cdots, k_n) g_{k_1} g_{k_2} \cdots g_{k_{p-1}} g_{k_{p+1}} \cdots g_{k_n} g_i, \end{aligned} \tag{4.19}$$

where g_i has been moved to the left in the first way of writing the term and to the right in the second. Hence differentiating this term from the two directions gives

$$\begin{aligned} \frac{\overrightarrow{\partial}}{\partial g_i} f_n g_{k_1} \cdots g_{k_{p-1}} g_i g_{k_{p+1}} \cdots g_{k_n} &= (-1)^{p-1} f_n g_{k_1} \cdots g_{k_{p-1}} g_{k_{p+1}} \cdots g_{k_n}, \\ f_n g_{k_1} \cdots g_{k_{p-1}} g_i g_{k_{p+1}} \cdots g_{k_n} \frac{\overleftarrow{\partial}}{\partial g_i} &= (-1)^{n-p} f_n g_{k_1} \cdots g_{k_{p-1}} g_{k_{p+1}} \cdots g_{k_n} \\ &= (-1)^{n-2p+1} \frac{\overrightarrow{\partial}}{\partial g_i} f_n g_{k_1} \cdots g_{k_{p-1}} g_i g_{k_{p+1}} \cdots g_{k_n}. \end{aligned} \tag{4.20}$$

For an even Grassmann function, each term in the monomial expansion will have an even number of g-numbers so that n is even. For odd Grassmann functions, however, n is odd. It follows that for even Grassmann functions, left and right differentiation differ by a factor -1, and for odd Grassmann functions left and right differentiation give the same result:

$$\frac{\overrightarrow{\partial}}{\partial g_i} f_{\mathrm{E}}(g) = (-1) f_{\mathrm{E}}(g) \frac{\overleftarrow{\partial}}{\partial g_i}, \tag{4.21}$$

$$\frac{\overrightarrow{\partial}}{\partial g_i} f_{\mathrm{O}}(g) = (+1) f_{\mathrm{O}}(g) \frac{\overleftarrow{\partial}}{\partial g_i}. \tag{4.22}$$

Higher-order derivatives are defined, as with c-number differentiation, by the successive application of differential operations. These are applied starting with the derivative nearest the Grassmann function and moving successively outwards so that

$$\frac{\overrightarrow{\partial}}{\partial g_N} \cdots \frac{\overrightarrow{\partial}}{\partial g_2} \frac{\overrightarrow{\partial}}{\partial g_1} f(g) = \frac{\overrightarrow{\partial}}{\partial g_N} \left(\cdots \left(\frac{\overrightarrow{\partial}}{\partial g_2} \left(\frac{\overrightarrow{\partial}}{\partial g_1} f(g) \right) \right) \right), \tag{4.23}$$

$$f(g) \frac{\overleftarrow{\partial}}{\partial g_1} \frac{\overleftarrow{\partial}}{\partial g_2} \cdots \frac{\overleftarrow{\partial}}{\partial g_N} = \left(\left(\left(f(g) \frac{\overleftarrow{\partial}}{\partial g_1} \right) \frac{\overleftarrow{\partial}}{\partial g_2} \right) \cdots \frac{\overleftarrow{\partial}}{\partial g_N} \right). \tag{4.24}$$

It is straightforward to show that Grassmann derivatives anticommute. To see this, we note that any Grassmann function can be written in the form

$$f(g) = F_0(\overline{g_{1,2}}) + g_1 F_1(\overline{g_{1,2}}) + g_2 F_2(\overline{g_{1,2}}) + g_1 g_2 F_{1,2}(\overline{g_{1,2}}),$$

where the $F(\overline{g_{1,2}})$ are functions of all the Grassmann variables except g_1 and g_2. It then follows that

$$\frac{\overrightarrow{\partial}}{\partial g_1}\frac{\overrightarrow{\partial}}{\partial g_2}f(g) = F_{1,2}(\overline{g_{1,2}}) = -\frac{\overrightarrow{\partial}}{\partial g_2}\frac{\overrightarrow{\partial}}{\partial g_1}f(g). \tag{4.25}$$

An analogous result applies for right differentiation. Mixed left and right differentiation can be applied in either order. To see this, we note that any Grassmann function can also be written in the form

$$f(g) = A_0(\overline{g_{1,2}}) + g_1 A_1(\overline{g_{1,2}}) + A_2(\overline{g_{1,2}})g_2 + g_1 A_{1,2}(\overline{g_{1,2}})g_2,$$

where the $A(\overline{g_{1,2}})$ are functions of all the Grassmann variables except g_1 and g_2. It then follows that

$$\left(\frac{\overrightarrow{\partial}}{\partial g_1}f(g)\right)\frac{\overleftarrow{\partial}}{\partial g_2} = A_{1,2}(\overline{g_{1,2}}) = \frac{\overrightarrow{\partial}}{\partial g_1}\left(f(g)\frac{\overleftarrow{\partial}}{\partial g_2}\right) = \frac{\overrightarrow{\partial}}{\partial g_1}f(g)\frac{\overleftarrow{\partial}}{\partial g_2}. \tag{4.26}$$

It may also be shown that

$$\frac{\overrightarrow{\partial}}{\partial g_1}f(g)\frac{\overleftarrow{\partial}}{\partial g_2} = -\frac{\overrightarrow{\partial}}{\partial g_2}f(g)\frac{\overleftarrow{\partial}}{\partial g_1} \tag{4.27}$$

for both even and odd Grassmann functions, showing that mixed left and right differentiation can also be swapped subject to a sign change.

Product rules for differentiation can be derived. These depend on whether the factors are even or odd Grassmann functions:

$$\begin{aligned}
\frac{\overrightarrow{\partial}}{\partial g_i}(f_1^{\mathrm{E}} f_2) &= \left(\frac{\overrightarrow{\partial}}{\partial g_i}f_1^{\mathrm{E}}\right) f_2 + f_1^{\mathrm{E}}\left(\frac{\overrightarrow{\partial}}{\partial g_i}f_2\right), \\
\frac{\overrightarrow{\partial}}{\partial g_i}(f_1^{\mathrm{O}} f_2) &= \left(\frac{\overrightarrow{\partial}}{\partial g_i}f_1^{\mathrm{O}}\right) f_2 - f_1^{\mathrm{O}}\left(\frac{\overrightarrow{\partial}}{\partial g_i}f_2\right), \\
(f_2 f_1^{\mathrm{E}})\frac{\overleftarrow{\partial}}{\partial g_i} &= f_2\left(f_1^{\mathrm{E}}\frac{\overleftarrow{\partial}}{\partial g_i}\right) + \left(f_2\frac{\overleftarrow{\partial}}{\partial g_i}\right) f_1^{\mathrm{E}}, \\
(f_2 f_1^{\mathrm{O}})\frac{\overleftarrow{\partial}}{\partial g_i} &= f_2\left(f_1^{\mathrm{O}}\frac{\overleftarrow{\partial}}{\partial g_i}\right) - \left(f_2\frac{\overleftarrow{\partial}}{\partial g_i}\right) f_1^{\mathrm{O}}.
\end{aligned} \tag{4.28}$$

Thus the product rule is different in general from that in ordinary calculus. For Grassmann functions that are neither even nor odd, the derivative of a product can be obtained from (4.28) after writing the function as the sum of its even and odd components. In cases where the Grassmann functions have inverses, quotient rules can be derived, and some cases are given as an exercise.

A chain rule for differentiation can also be obtained. The main case of interest is that involving linear transformations between Grassmann variables. Consider two sets of g-numbers g, h related by a c-number non-singular transformation matrix A_{ij}:

$$g_i = \sum_j A_{ij} h_j \,. \tag{4.29}$$

Under this transformation, the Grassmann function $f(g)$ becomes the new Grassmann function $e(h) \equiv f(g(h))$ and we have

$$\frac{\overrightarrow{\partial}}{\partial h_j} e(h) = \sum_i \left(\frac{\overrightarrow{\partial}}{\partial g_i} f \right) A_{ij}, \tag{4.30}$$

$$e(h) \frac{\overleftarrow{\partial}}{\partial h_j} = \sum_i \left(f \frac{\overleftarrow{\partial}}{\partial g_i} \right) A_{ij}, \tag{4.31}$$

as in standard calculus.

Grassmann variables g_i may also be considered as functions of a real parameter t via a t-dependent transformation matrix $A_{ij}(t)$. The *total derivative* with respect to t of the Grassmann function $f[g_1(t), g_2(t), \cdots, g_i(t), \cdots, g_n(t)]$ is

$$\begin{aligned} \frac{d}{dt} f(g(t)) &= \sum_i \frac{dg_i}{dt} \left(\frac{\overrightarrow{\partial}}{\partial g_i} f \right) \\ &= \sum_i \left(f \frac{\overleftarrow{\partial}}{\partial g_i} \right) \frac{dg_i}{dt}. \end{aligned} \tag{4.32}$$

Again, these rules are as in c-number calculus.

It is useful to introduce pairs of Grassmann variables related by complex conjugation, g_i, g_i^*. Indeed, this occurs in the analogue of the bosonic phase space spanned by independent c-numbers α and α^* for bosons. The rules above still apply, with the set of Grassmann variables now being augmented to include the conjugate set in a combined set $g, g^* = \{g_1, \cdots, g_n, g_1^*, g_2^*, \cdots, g_i^*, \cdots, g_n^*\}$. Grassmann functions of the full set of variables $f(g, g^*)$ can be defined. In other cases, depending on the type of quasiprobability used to represent the system, there may be two Grassmann variables g_i, g_i^+ for each quantum mode *not* related by complex conjugation; derivatives with respect to g_i^+ now also occur. The rules for differentiation derived above remain in force, with g_i and g_i^* or g_i and g_i^+ being independent Grassmann variables.

4.2.3 Taylor Series

The Grassmann numbers, unlike c-numbers, do not have a property of size or magnitude. To see this, we need only note that the square of any Grassmann number is zero. It is, nevertheless, useful to introduce the counterpart of a small variation of a Grassmann function by replacing each g-number g_k by $g_k + \delta g_k$, where δg_k is a further

Grassmann variable. More generally, δg_k can be any odd Grassmann function, but this subtlety will not be important for us.

Let us write a Grassmann function $f(g)$, making explicit the g_k dependence:

$$f(g) = f_0 + \sum_{k_1} g_{k_1} f_1(k_1) + \sum_{k_1,k_2} g_{k_1} g_{k_2} f_2(k_1, k_2) + \sum_{k_1,k_2,k_3} g_{k_1} g_{k_2} g_{k_3} f_3(k_1, k_2, k_3) + \cdots, \tag{4.33}$$

where the coefficients f_i are c-numbers and where there is a convention that these coefficients are zero unless $k_1 < k_2$ in $f_2(k_1, k_2)$, $k_1 < k_2 < k_3$ in $f_3(k_1, k_2, k_3)$ and so on. The summations are not restricted. Then we also have, with $g + \delta g = \{g_1 + \delta g_1, \cdots, g_k + \delta g_k, \cdots\}$,

$$\begin{aligned} f(g + \delta g) &= f_0 + \sum_{k_1} (g_{k_1} + \delta g_{k_1}) f_1(k_1) \\ &\quad + \sum_{k_1,k_2} (g_{k_1} + \delta g_{k_1})(g_{k_2} + \delta g_{k_2}) f_2(k_1, k_2) \\ &\quad + \sum_{k_1,k_2,k_3} (g_{k_1} + \delta g_{k_1})(g_{k_2} + \delta g_{k_2})(g_{k_3} + \delta g_{k_3}) f_3(k_1, k_2, k_3) + \cdots, \end{aligned} \tag{4.34}$$

which may be written as a series in the δg_{k_i}. It is immediately clear that we can write

$$f(g + \delta g) - f(g) = \sum_k \delta g_k \left\{ \frac{\overrightarrow{\partial}}{\partial g_k} f(g) \right\}, \tag{4.35}$$

correct to first order in the δg_k. Alternatively, we can use the right derivative to find

$$f(g + \delta g) - f(g) = \sum_k \left\{ f(g) \frac{\overleftarrow{\partial}}{\partial g_k} \right\} \delta g_k\,, \tag{4.36}$$

again correct to first order in the δg_k.

The treatment can be extended readily to second and higher orders. The results correct to second order, for example, are

$$f(g + \delta g) - f(g) = \sum_k \delta g_k \left\{ \frac{\overrightarrow{\partial}}{\partial g_k} f(g) \right\} + \frac{1}{2} \sum_{kl} \delta g_k\, \delta g_l \left\{ \frac{\overrightarrow{\partial}}{\partial g_l} \frac{\overrightarrow{\partial}}{\partial g_k} f(g) \right\} \tag{4.37}$$

$$= \sum_k \left\{ f(g) \frac{\overleftarrow{\partial}}{\partial g_k} \right\} \delta g_k + \frac{1}{2} \sum_{kl} \left\{ f(g) \frac{\overleftarrow{\partial}}{\partial g_k} \frac{\overleftarrow{\partial}}{\partial g_l} \right\} \delta g_l\, \delta g_k, \tag{4.38}$$

giving the left and right derivative forms, where there are no order restrictions on the double sums. Note that in the second-order terms the processes occur in the order *kllk*. The reader will have noticed the formal similarity between these and the first terms in a Taylor series for a c-number function.

4.3 Grassmann Integration

4.3.1 Definition

Just as we can define a differential calculus by analogy with c-number differential calculus, it is equally possible to define an integral calculus for Grassmann variables. We first need to introduce the concept of Grassmann differentials $dg_1, dg_2, \cdots, dg_n$, which satisfy all the standard rules for g-numbers,

$$dg_i\, dg_j + dg_j\, dg_i = 0 \qquad (i,j = 1, \cdots, n), \tag{4.39}$$

$$g_i\, dg_j + dg_j\, g_i = 0 \qquad (i,j = 1, \cdots, n), \tag{4.40}$$

and which also anticommute with fermion annihilation and creation operators, and commute with c-numbers and boson operators.

As with differentiation, two forms of integration can be defined: left and right integration. These are different in general and are distinguished by the position of the Grassmann differential. The left integral of a Grassmann function $f(g) \equiv f(g_1, \cdots, g_n)$ is defined via rules for each component monomial. This is based on the fundamental rules for left integrals of the two non-zero powers of g_j, namely $g_j^0 = 1, g_j^1 = g_j$:

$$\int dg_i\; 1 = 0, \qquad \int dg_i\, g_j = \delta_{ij}\,. \tag{4.41}$$

This situation is quite different from c-number integration, where results for both definite and indefinite integrals of all powers of α apply and which are obtained via reverse differentiation. Left integration is designed to link to left differentiation, and we see that the left integral and left derivative of g_i with respect to g_i both give unity. In that sense, integration and differentiation can be said to be the same. They are not reverse processes in any simple sense. The anticommutation features of Grassmann numbers mean that the basic rules for right integration are

$$\int 1\, dg_i = 0, \qquad \int g_j\, dg_i \;= -\delta_{ij}. \tag{4.42}$$

Thus, unlike differentiation, right and left integration differ by a sign change. Left integration is more useful for our purposes, and so from here on we assume all Grassmann integrals to be of this type unless stated otherwise explicitly.

In each monomial a particular g_i can only occur once at most, so the reordering process that was introduced for monomial differentials is used. The integration process involves first moving this g_i to become the left factor and then integrating it via the fundamental rules. Thus

$$\int dg_i\, g_{i_1} g_{i_2} \cdots g_{i_s} = \delta_{ii_1} g_{i_2} \cdots g_{i_s} - \delta_{ii_2} g_{i_1} \cdots g_{i_s} + \cdots + (-1)^{s-1} \delta_{ii_s} g_{i_2} \cdots g_{i_{s-1}}. \tag{4.43}$$

Thus if a particular g_i does not occur then the integral is zero, and if it does occur then the integration results in the deletion of that variable from the monomial. The -1 factors are just due to transporting the g_i to the left of the expression. The other g_j are treated by the integration as if they were constants, as occurs in ordinary integration, and they are left unchanged in the single integration over g_i.

The integration of a Grassmann function $f(g)$ is carried out using the usual linear rules of integration. Each monomial term which forms part of the Grassmann function is integrated independently. Thus the integral of a Grassmann function $f(g)$ as given by (3.22) is

$$\begin{aligned}\int dg_i\, f(g) &= f_0 \int dg_i\, 1 + \sum_{k_1} f_1(k_1) \int dg_i\, g_{k_1} + \sum_{k_1 k_2} f_2(k_1, k_2) \int dg_i\, g_{k_1} g_{k_2} \\ &+ \cdots + \sum_{k_1 \cdots k_n} f_n(k_1, \cdots, k_n) \int dg_i\, g_{k_1} \cdots g_{k_n}. \end{aligned} \tag{4.44}$$

If a given g_i does not appear in a factor, then the integral for that factor is zero.

Multiple integrals are understood as iterated single integrals carried out in succession starting from the right:

$$\int \cdots \int dg_s \cdots dg_j\, dg_i\, f(g) \equiv \int dg_s \left(\cdots \left(\int dg_j \left(\int dg_i\, f(g) \right) \right) \cdots \right). \tag{4.45}$$

Each basic step is carried out using the fundamental integration rules previously introduced.

As in ordinary calculus, there are more sophisticated integration rules for Grassmann functions which can be derived on the basis of the fundamental rules. The integral of a Grassmann function as in (3.22) with (unique) antisymmetric expansion coefficients over all Grassmann variables $g_1, \cdots, g_n$ is given by

$$\begin{aligned}\int d\mathbf{g}\, f(g) &\equiv \int \cdots \int dg_n \cdots dg_i \cdots dg_2\; dg_1 f(g_1, \cdots, g_n) \\ &= n! f_n(1, \cdots, n), \end{aligned} \tag{4.46}$$

which is a c-number. Here the arguments $1, \cdots, n$ correspond to the n Grassmann modes. When writing $d\mathbf{g}$, of course, we mean the product of Grassmann differentials in a particular order: $d\mathbf{g} = dg_n \cdots dg_i \cdots dg_2\; dg_1 = \prod_i dg_i$. The proof is left as an exercise. Note that this is the same result as is found for the nth-order derivative. An important result is that the full Grassmann *integral of the derivative* of a Grassmann function is zero:

$$\begin{aligned}\int d\mathbf{g}\, \frac{\overrightarrow{\partial}}{\partial g_i} f(g) &= 0, \\ \int d\mathbf{g}\; f(g) \frac{\overleftarrow{\partial}}{\partial g_i} &= 0. \end{aligned} \tag{4.47}$$

This is because differentiation removes the g_i variable, and so the fundamental rules require that the integral over dg_i must give zero. This is in contrast to the rule for normal calculus, which would give the function itself, evaluated on the boundaries if the integral is definite. Whether the final result is zero then depends on whether the function becomes zero on the boundary.

The case in which the original g_i are replaced by new h_j via the linear transformation defined in (4.29) requires us to consider how products of g-numbers such as $g_1 g_2 \cdots g_i \cdots g_n$ and products of differentials $dg_n \cdots dg_i \cdots dg_2\, dg_1$ transform. Let us write for non-singular matrices A, B

$$g_i = \sum_k A_{ik} h_k, \tag{4.48}$$

$$dg_i = \sum_k B_{ik}\, dh_k, \tag{4.49}$$

where we cannot just assume that the differential transforms in the same way as the g-number. It is straightforward to show that $h_k h_l + h_l h_k = 0$, $h_k\, dh_l + dh_l\, h_k = 0$ and $dh_k\, dh_l + dh_l\, dh_k = 0$ for any A, B, so the new quantities h_k and dh_l satisfy the same anticommutation rules as the old. The matrices A, B, however, are not independent, because the Grassmann integrals must yield the standard results

$$\begin{aligned} \int dh_k\ 1 &= 0, \\ \int dh_k\, h_l &= \delta_{kl}. \end{aligned} \tag{4.50}$$

The first result just follows from (4.49). For the second we require

$$\begin{aligned} \int dh_k\, h_l &= \int \sum_i B^{-1}_{ki} dg_i \sum_j A^{-1}_{lj} g_j \\ &= \sum_i B^{-1}_{ki} \sum_j A^{-1}_{lj} \delta_{ij} \\ &= \left(B^{-1}(A^T)^{-1}\right)_{kl} = \delta_{kl}, \end{aligned} \tag{4.51}$$

where T denotes the transpose. From this we obtain the conditions

$$B = \left(A^T\right)^{-1} = (A^{-1})^T, \tag{4.52}$$

$$dg_i = \sum_k (A^{-1})_{ki}\, dh_k, \tag{4.53}$$

so that the transformation rule for the differentials involves the transpose of the inverse of the matrix for transforming the g-numbers. For the c-number case, we would have found $B = A$.

We can now use these results to transform the product of all g-numbers and of all differentials:

$$\begin{aligned} g_1 \cdots g_n &= \sum_{k_1 \cdots k_n} A_{1k_1} h_{k_1} \cdots A_{1k_n} h_{k_n} \\ &= \sum_{k_1 \cdots k_n} A_{1k_1} \cdots A_{1k_n} \epsilon(k_1, \cdots, k_n) h_1 \cdots h_i \cdots h_n, \end{aligned} \tag{4.54}$$

where $\epsilon(k_1, \cdots, k_n) = +1$ or -1 if $k_1, \cdots, k_n$ is an even or odd permutation $P(1 \to k_1; 2 \to k_2; \cdots; n \to k_n)$, respectively, of the numbers $1, 2, \cdots, n$ and $\epsilon(k_1, \cdots, k_n) = 0$ if $k_1, \cdots, k_n$ is not a permutation (and thus contains two k_i that are the same). This result follows because if any pair of h_{k_i} are the same, then the right side is zero, and if the h_{k_i} are all different, then we can rearrange the order to be $h_1 h_2 \cdots h_n$ multiplied by the factor $\epsilon(k_1, \cdots, k_n)$. As the multiple summation generates the *determinant* of the matrix A, this shows the following important transformation results:

$$g_1 \cdots g_n = (\text{Det } A)\, h_1 h_2 \cdots h_i \cdots h_n, \tag{4.55}$$

$$dg_n \cdots dg_i \cdots dg_2\, dg_1 = (\text{Det } B)\, dh_n \cdots dh_i \cdots dh_2\, dh_1 \tag{4.56}$$

$$= (\text{Det } A)^{-1} dh_n \cdots dh_i \cdots dh_2\, dh_1 . \tag{4.57}$$

Consider the function $e(h) \equiv e(h_1, \cdots, h_n)$ obtained by substituting for each g_i in the function $f(g) \equiv f(g_1, \cdots, g_n)$, and thus $e(h) = f(g(h))$. The expansion of $e(h)$ in monomials will be

$$e(h) = f_0 + \sum_{k_1} \sum_{l_1} f_1(k_1) A_{k_1 l_1} h_{l_1} + \cdots + \sum_{k_1 \cdots k_n} \sum_{l_1 \cdots l_n} f_n(k_1, \cdots, k_n) A_{k_1 l_1} A_{k_2 l_2} \cdots A_{k_n l_n} h_{l_1} h_{l_2} \cdots h_{l_n}, \tag{4.58}$$

and hence

$$e_0 = f_0,$$
$$e_1(l_1) = \sum_{k_1} f_1(k_1) A_{k_1 l_1},$$
$$\vdots$$
$$e_n(l_1, l_2, \cdots, l_n) = \sum_{k_1 \cdots k_n} f_n(k_1, \cdots, k_n) A_{k_1 l_1} A_{k_2 l_2} \cdots A_{k_n l_n}, \tag{4.59}$$

where the e_i are also antisymmetric, as the f_i are antisymmetric and zero if any two l_i are equal. In particular, we can show that

$$e_n(l_1, \cdots, l_n) = \epsilon(l_1, \cdots, l_n) e_n(1, 2, \cdots, n),$$
$$f_n(k_1, \cdots, k_n) = \epsilon(k_1, \cdots, k_n) f_n(1, 2, \cdots, n), \tag{4.60}$$

and thus

$$e_n(1, 2, \cdots, n) = (\text{Det A}) f_n(1, 2, \cdots, n) . \tag{4.61}$$

Let us recall that the Grassmann integral of a function $f(g)$ is

$$\int \mathrm{dg}\ f(g) = \int \prod_i dg_i \sum_{k_1 \cdots k_n} f_n(k_1, \cdots, k_n) g_{k_1} \cdots g_{k_n} = n! f_n(1, 2, \cdots, n) . \tag{4.62}$$

Suppose, however, that we change the variable, use (4.55) and (4.56) and follow the same approach as in the derivation of (4.46). Using (4.55) and (4.56), after writing $g_{k_1} \cdots g_{k_n} = \epsilon(k_1, \cdots, k_n) g_1 \cdots g_n$, we find

$$\begin{aligned}\int d\mathbf{g}\, f(g) &= \int \prod_i dg_i \sum_{k_1 \cdots k_n} f_n(k_1, \cdots, k_n)\epsilon(k_1, \cdots, k_n) g_1 \cdots g_n \\ &= (\text{Det A})(\text{Det B}) \int \prod_i dh_i \sum_{k_1 \cdots k_n} f_n(1, \cdots, n)\epsilon(k_1, \cdots, k_n)^2 h_1 \cdots h_i \cdots h_n \\ &= (\text{Det A})(\text{Det B}) n! f_n(1, \cdots, n) \\ &= (\text{Det B}) n! e_n(1, \cdots, n). \end{aligned} \tag{4.63}$$

It must be true, however, that the Grassmann integral of $e(h)$ is

$$\int d\mathbf{h}\; e(h) = n! e_n(1, \cdots, n), \tag{4.64}$$

so, using $(\text{Det B}) = (\text{Det A})^{-1}$, we see that the transformation rule for the Grassmann integral is

$$\int d\mathbf{g}\; f(g) = (\text{Det A})^{-1} \int d\mathbf{h}\; e(h), \tag{4.65}$$

where $e(h) = f(g(h))$ under the linear transformation $g_i = \sum_j A_{ij} h_j$ with a non-singular matrix A. This differs from the usual c-number result involving the Jacobian, in which the multiplying factor would have been Det A rather than $(\text{Det A})^{-1}$.

Finally, we note that the process of taking the complex conjugate of a Grassmann integral changes left integration into right integration with the differentials complex conjugated and placed in reverse order. Thus we have

$$\begin{aligned}\left(\int dg_i\, f(g)\right)^* &= \int f(g)^*\, dg_i^*, \\ \left(\int dg_j\; dg_i\, f(g)\right)^* &= \int f(g)^*\, dg_i^*\; dg_j^*. \end{aligned} \tag{4.66}$$

4.3.2 Pairs of Grassmann Variables

In cases where there are pairs of Grassmann variables g_i, g_i^*, related via complex conjugation, $2n$-dimensional multiple integrals naturally occur of the form

$$\begin{aligned}&\int \cdots \int dg_n^*\; dg_n \cdots dg_i^*\; dg_i \cdots dg_1^*\; dg_1\; f(g_1, \cdots, g_n, g_1^*, \cdots, g_i^*, \cdots, g_n^*) \\ &\equiv \int d^2\mathbf{g}\; f(g, g^*). \end{aligned} \tag{4.67}$$

Here $d^2\mathbf{g}$ represents $d^2 g_n \cdots d^2 g_1$, with $d^2 g_i = dg_i^*\; dg_i$, and the Grassmann function now depends on the set of Grassmann variables $g, g^* = \{g_1, \cdots, g_n, g_1^*, \cdots, g_n^*\}$. In a natural notation, the result for this integral is $(2n)!f_{2n}(1,\cdots,n,1^*,\cdots,n^*)$. As with differentiation, the conjugate Grassmann variables are treated as independent variables. A simple example will illustrate the method:

$$\begin{aligned}\int d^2g_i \exp\left(-g_j^* g_k\right) &= \int d^*g_i\ dg_i(1-g_j^* g_k)\\ &= -\int d^*g_i\ dg_i\ g_j^* g_k\\ &= +\int d^*g_i\ g_j^* \int dg_i\ g_k\\ &= \delta_{ij}\delta_{ik}. \end{aligned} \tag{4.68}$$

Often we shall employ two Grassmann variables g_i, g_i^+ for each quantum mode, which are *not* related via complex conjugation. Different $2n$-dimensional multiple integrals occur of a form analogous to (4.67) with $d^2\mathbf{g} = \prod_i d^2g_i$, but now with $d^2g_i = dg_i^+ dg_i$ and the Grassmann function $f(g, g^+)$ now depending on the set of Grassmann variables $g, g^+ = \{g_1, \cdots, g_n, g_1^+, g_2^+, \cdots, g_n^+\}$. In an obvious notation, the result for the analogous integral is $(2n)!f_{2n}(1,2,\cdots,n,1^+,2^+,\cdots,n^+)$.

A simple example, one that is particularly useful, involves the integral of a Grassmann function $F(g, g^+)$ of two Grassmann variables:

$$\int dh^+ dh\, F(h, h^+)\,(h-g)(h^+-g^+) = F(g, g^+). \tag{4.69}$$

Note the order of the factors. If F also depends on other Grassmann variables, i.e. $F = F(g, g^+, e_1, e_2, \cdots)$, then a straightforward extension gives

$$\int dh^+ dh\, F(h, h^+, e_1, e_2, \cdots)(h-g)(h^+-g^+) = F(g, g^+, e_1, e_2, \cdots). \tag{4.70}$$

The proof is left as an exercise.

Another useful example involves an analogy to the Fourier theorem,

$$\int dk^+ dk\ F(k, k^+) \int dg^+ dg \exp(ig[h^+-k^+]) \exp(i[h-k]g^+) = F(h, h^+), \tag{4.71}$$

and its generalisation

$$\int d\mathbf{k}^+ d\mathbf{k}\ F(\mathbf{k}, \mathbf{k}^+) \int d\mathbf{g}^+ d\mathbf{g} \exp(i\mathbf{g}\cdot[\mathbf{h}^+-\mathbf{k}^+]) \exp(i[\mathbf{h}-\mathbf{k}]\cdot\mathbf{g}^+) = F(\mathbf{h}, \mathbf{h}^+). \tag{4.72}$$

The proof is given in Appendix C.

As before, the full Grassmann integral of a Grassmann derivative of a Grassmann function $f(g, g^+)$ is zero:

$$\int d^2\mathbf{g}\, \frac{\overrightarrow{\partial}}{\partial g_i} f(g, g^+) = \int d^2\mathbf{g}\, \frac{\overrightarrow{\partial}}{\partial g_i^+} f(g, g^+) = 0,$$

$$\int d^2\mathbf{g}\, f(g, g^+) \frac{\overleftarrow{\partial}}{\partial g_i} = \int d^2\mathbf{g}\, f(g, g^+) \frac{\overleftarrow{\partial}}{\partial g_i^+} = 0. \tag{4.73}$$

As before, this is because the differentiation from either the left or the right removes the g_i or g_i^+ variable from f, so the integral over each of these variables gives zero. A similar result applies when each mode is associated with a complex pair of Grassmann variables g_i and g_i^* and a Grassmann function $f(g, g^*)$ is differentiated with respect to g_i or g_i^*. In this case $d^2 g_i = dg_i^*\, dg_i$.

Formulae for integration by parts can be derived from the last equation by considering the product of two functions $e(g, g^+) f(g, g^+)$. Using the result (4.28) and (4.73), we find that, with $d^2 g_i = dg_i^+ dg_i$,

$$\int d^2\mathbf{g} \left(\frac{\overrightarrow{\partial}}{\partial g_i} e(g, g^+) \right) f(g, g^+) = -\,\sigma(e) \int d^2\mathbf{g}\, e(g, g^+) \left(\frac{\overrightarrow{\partial}}{\partial g_i} f(g, g^+) \right),$$

$$\int d^2\mathbf{g} \left(f(g, g^+) \frac{\overleftarrow{\partial}}{\partial g_i} \right) e(g, g^+) = -\,\sigma(e) \int d^2\mathbf{g}\, f(g, g^+) \left(e(g, g^+) \frac{\overleftarrow{\partial}}{\partial g_i} \right),$$

$$\int d^2\mathbf{g} \left(\frac{\overrightarrow{\partial}}{\partial g_i^+} e(g, g^+) \right) f(g, g^+) = -\,\sigma(e) \int d^2\mathbf{g}\, e(g, g^+) \left(\frac{\overrightarrow{\partial}}{\partial g_i^+} f(g, g^+) \right),$$

$$\int d^2\mathbf{g} \left(f(g, g^+) \frac{\overleftarrow{\partial}}{\partial g_i^+} \right) e(g, g^+) = -\,\sigma(e) \int d^2\mathbf{g}\, f(g, g^+) \left(e(g, g^+) \frac{\overleftarrow{\partial}}{\partial g_i^+} \right), \tag{4.74}$$

where $\sigma(e, f) = +1$ or -1 depending on whether e, f is even or odd. A similar result applies when each mode is associated with a complex pair of Grassmann variables g_i and g_i^* and the Grassmann functions $e(g, g^*)$ and $f(g, g^*)$ are differentiated with respect to g_i or g_i^* In this case $d^2 g_i = dg_i^*\, dg_i$.

Consider a function $F(g, h)$ of two sets of Grassmann variables $g = \{g_1, \cdots, g_n\}$ and $h = \{h_1, \cdots\cdots, h_n\}$ which is integrated over all of the g_i. The result, of course, will be a Grassmann function of the h_j. A rule for differentiating the integral with respect to the h_j can be established, showing that it is possible to obtain the result either by differentiating under the integral or after the integral has been performed. To establish this result, we first write the function as a linear combination of monomials with the g_i placed to the left of the h_j in the form

$$F(g, h) = \sum_{i_1 < i_2 < \cdots < i_p} \sum_{j_1 < j_2 < \cdots < j_q} F_{p+q}^{i_1 i_2 \cdots i_p; j_1 j_2 \cdots j_q} g_{i_1} g_{i_2} \cdots g_{i_p} h_{j_1} h_{j_2} \cdots h_{j_q},$$

with all the g_i placed in increasing index order to the left of the h_j, also similarly ordered. Then, with $d\mathbf{g} = dg_n \cdots dg_2\, dg_1$, we can establish the rules for differentiating the (left) integral from either the left or the right:

$$\frac{\overrightarrow{\partial}}{\partial h_j}\left(\int d\mathbf{g}\, F(g,h)\right) = (-1)^n \int d\mathbf{g}\left(\frac{\overrightarrow{\partial}}{\partial h_j}F(g,h)\right), \tag{4.75}$$

$$\left(\int d\mathbf{g}\, F(g,h)\right)\frac{\overleftarrow{\partial}}{\partial h_j} = \int d\mathbf{g}\left(F(g,h)\frac{\overleftarrow{\partial}}{\partial h_j}\right). \tag{4.76}$$

Thus we see that the usual c-number rule does not apply in the case of left differentiation – there is an extra factor of $(-1)^n$ involved. However, for right differentiation the usual rule does apply. The derivation of this result is in Appendix C.

If, however, we consider a function $F(g, g^+, h, h^+)$ of two sets of Grassmann variables g, g^+ and h, h^+, where there are *two variables* per mode g_i, g_i^+ and h_j, h_j^+, which is integrated over *all* the g_i and g_i^+, then differentiation of the integral does not involve a (± 1) factor. To show this for left differentiation, we express the function as the sum of two terms, one which is independent of h_j, h_j^+, and the other with h_j, h_j^+ on the left of a factor that is independent of them. Differentiation from the left after integration requires shifting h_j, h_j^+ to the left through $2n$ differentials. The factor $(-1)^{2n}$ is always unity, showing that for left differentiation the result is the same as when left differentiation is carried out before integration. Thus, with $d^2\mathbf{g}=\prod_i dg_i^+\, dg_i$, we have

$$\frac{\overrightarrow{\partial}}{\partial h_j}\int d^2\mathbf{g}\, F(g,g^+,h,h^+) = \int d^2\mathbf{g}\, \frac{\overrightarrow{\partial}}{\partial h_j}F(g,g^+,h,h^+), \tag{4.77}$$

$$\frac{\overrightarrow{\partial}}{\partial h_j^+}\int d^2\mathbf{g}\, F(g,g^+,h,h^+) = \int d^2\mathbf{g}\, \frac{\overrightarrow{\partial}}{\partial h_j^+}F(g,g^+,h,h^+). \tag{4.78}$$

Results analogous to these apply for right differentiation, but for brevity we will not include these formulae. Differentiating from the right after integration requires shifting h_j, h_j^+ to the right, showing that for right differentiation the result is the same as when right differentiation is carried out before integration. Similar results apply to the case of functions $F(g, g^*, h, h^*)$ involving two sets of Grassmann variables g, g^* and h, h^*.

Exercises

(4.1) Derive the Glauber identity (4.9).

(4.2) Derive the results for the left and right derivatives of $g_k g_m g_n$ with respect to g_i.

(4.3) Derive the following result for the nth-order derivative of the Grassmann function (3.22):

$$\frac{\overrightarrow{\partial}}{\partial g_n}\cdots\frac{\overrightarrow{\partial}}{\partial g_1}f(g_1\cdots g_n) = n!f_n(g_1, g_2, \cdot, g_n).$$

(4.4) Derive the quotient rules

$$\frac{\overrightarrow{\partial}}{\partial g_i}\left(\left(f_1^{\mathrm{O}}\right)^{-1} f_2\right) = +\left(f_1^{\mathrm{O}}\right)^{-1}\left(\frac{\overrightarrow{\partial}}{\partial g_i} f_1^{\mathrm{O}}\right)\left(f_1^{\mathrm{O}}\right)^{-1} f_2 - \left(f_1^{\mathrm{O}}\right)^{-1}\left(\frac{\overrightarrow{\partial}}{\partial g_i} f_2\right),$$

$$\left(f_2\left(f_1^{\mathrm{O}}\right)^{-1}\right)\frac{\overleftarrow{\partial}}{\partial g_i} = +f_2\left(f_1^{\mathrm{O}}\right)^{-1}\left(f_1^{\mathrm{O}}\frac{\overleftarrow{\partial}}{\partial g_i}\right)\left(f_1^{\mathrm{O}}\right)^{-1} - \left(f_2\frac{\overleftarrow{\partial}}{\partial g_i}\right)\left(f_1^{\mathrm{O}}\right)^{-1}.$$

(4.5) Prove the second-derivative results (4.25) and (4.27).

(4.6) Derive the second-order expansion rules (4.37) and (4.38) for Grassmann functions.

(4.7) Prove the result (4.46) for the total integration of a Grassmann function.

(4.8) Derive the result $\int d^2g_i \exp\{h_i^* g_i + g_i^* e_i + g_i g_i^*\} = \exp\{h_i^* e_i\}$.

(4.9) Derive the result (4.69).

(4.10) Verify the first and second results for integration by parts in (4.74).

(4.11) Prove the results (4.75) and (4.76) for differentiating under a Grassmann integral.

(4.12) Prove the results (4.66) for the complex conjugates of Grassmann integrals.

5

Coherent States

The phase space methods that we shall be studying require us to be able to map states onto a space parameterised by c-numbers for bosons or by Grassmann numbers for fermions. A necessary first step is to introduce the coherent states, both for bosons and for fermions. These states are superpositions of Fock states and so do not have a specified particle number. They are labelled by c-numbers or Grassmann numbers and are generated by a unitary transformation on the vacuum, or zero-particle, state. It is often convenient to write the coherent states as the product of an exponential factor containing complex conjugate variables α_i^* or g_i^* and Bargmann coherent states. These have the useful property that they depend only on the amplitudes α_i or g_i.

The coherent states are *overcomplete* in the sense that we can write any state as a superposition of coherent states, but they are not mutually orthogonal. There are, in a sense, many more coherent states than Fock states. This overcompleteness means that we can write the identity operator as a phase-space integral of mode variables, α_i and α_i^* for bosons or g_i and g_i^* for fermions. This completeness identity enables bosonic and fermionic operators such as the density operator to be expressed in canonical forms based on the positive P representation. These involve integrals over a pair of complex planes α_i, α_i^+ for the bosonic modes and a pair of Grassmann variables g_i, g_i^+ for the fermionic modes. The integrands are products of projectors based on the coherent states together with the canonical representation function of these variables specifying the particular operator. The latter is given in terms of matrix elements of the operator in terms of coherent states. In addition, the completeness relationships enable expressions for the traces of bosonic and fermionic operators to be obtained in terms of integrals involving their coherent-state matrix elements.

The reason why the states are referred to as coherent states will become apparent when we consider the quantum correlation functions for a system in a coherent state. These can be evaluated using the feature that the coherent states are eigenstates for the field operators. For both the bosonic and the fermionic coherent states, the higher-order quantum correlation functions factorise into a product of first-order quantum correlation functions. This feature indicates that there is coherence to all orders, hence the designation as coherent states.

5.1 Grassmann States and Grassmann Operators

For fermions, vectors can be defined in a generalised form of Hilbert space that involves linear combinations of basis vectors with Grassmann numbers as the coefficients. Such

Grassmann vectors are not to be taken to represent physical states (where the coefficients must be c-numbers) even if a fixed number of fermions are involved, but they have uses in mathematical manipulations. The fermion coherent states are Grassmann vectors.

Similarly, we may introduce generalised operators in this new Hilbert space by taking linear combinations of products of the fermion operators with Grassmann numbers as the coefficients. Such *Grassmann operators* do not represent physical quantities (which must involve c-numbers as coefficients) or symmetry operations, but again are useful mathematically. The unitary displacement operators defined below in (5.11) are Grassmann operators. Many of the results for Grassmann functions also apply to Grassmann operators and states, in which some of the Grassmann variables are replaced by fermion annihilation or creation operators. The anticommuting feature of the fermion operators with Grassmann variables enables the same proofs to be used.

For fermions, the treatment of coherent states involves products of annihilation and creation operators with Grassmann numbers, so the basic rules applying to such products must first be stated. The fermion and boson annihilation and creation operators can be multiplied by c-numbers in a standard manner. However, multiplication by Grassmann numbers requires different rules: g-numbers commute with boson operators and anticommute with fermion operators:

$$\begin{aligned} g\hat{a}_i &= \hat{a}_i g, & g\hat{a}_i^\dagger &= \hat{a}_i^\dagger g, \\ g\hat{c}_i &= -\hat{c}_i g, & g\hat{c}_i^\dagger &= -\hat{c}_i^\dagger g. \end{aligned} \tag{5.1}$$

Note that combinations such as $\hat{c}_i g_i$ and $g_j^* \hat{c}_j^\dagger$ commute for $i \neq j$:

$$\hat{c}_i g_i g_j^* \hat{c}_j^\dagger = (-1)^2 \hat{c}_i \hat{c}_j^\dagger g_i g_j^* = (-1)^4 \hat{c}_j^\dagger \hat{c}_i g_j^* g_i = (-1)^6 g_j^* \hat{c}_j^\dagger \hat{c}_i g_i = g_j^* \hat{c}_j^\dagger \hat{c}_i g_i\,, \tag{5.2}$$

where the proof involves six anticommutations.

As the Fock states are products of creation operators acting on the vacuum state $|0\rangle$, it is necessary to state the basic rules for multiplying the vacuum state by Grassmann numbers. The natural rule is that the Grassmann numbers commute with $|0\rangle$ or $\langle 0|$:

$$g|0\rangle = |0\rangle g, \qquad g\langle 0| = \langle 0|g. \tag{5.3}$$

Note that as a consequence

$$\begin{aligned} g|1_i\rangle &= g\left(\hat{c}_i^\dagger\right)|0\rangle = -\left(\hat{c}_i^\dagger\right)|0\rangle g = -|1_i\rangle g, \\ g|1_i 1_j\rangle &= g\left(\hat{c}_i^\dagger \hat{c}_j^\dagger\right)|0\rangle = +\left(\hat{c}_i^\dagger \hat{c}_j^\dagger\right)|0\rangle g = +|1_i 1_j\rangle g. \end{aligned} \tag{5.4}$$

In general, a Grassmann number anticommutes with an odd-fermion-number Fock state and commutes with an even-fermion-number state. More generally, the bra and ket vectors may be classified as even or odd vectors depending on whether they only contain terms with even or odd numbers of fermions. Thus any superposition of the zero-fermion state $|0\rangle$ and the two-fermion state $+|1_i 1_j\rangle$ is an even vector, whilst a superposition of a one-fermion state $|1_i\rangle$ and a three-fermion state $|1_i 1_j 1_k\rangle$ is odd. These concepts may be extended to include Grassmann vectors as well as ordinary state vectors. In this case all terms in even (or odd) vectors contain an even (or odd, respectively) number of fermion creation operators, annihilation operators *and* Grassmann numbers, and consequently commute (or anticommute, respectively) with any Grassmann number. Thus the Grassmann vectors $(\hat{1} + \hat{c}^\dagger h)|0\rangle = |0\rangle - h|1\rangle$ and $\langle 0|(\hat{1} + h^* \hat{c}) = \langle 0| - \langle 1|h^*$ are even vectors.

The process of taking the adjoint is covered by the following rules:

$$\begin{aligned} (g\hat{a}_i)^\dagger &= \hat{a}_i^\dagger g^*, \qquad & (g\hat{a}_i^\dagger)^\dagger &= \hat{a}_i g^*, \\ (g\hat{c}_i)^\dagger &= \hat{c}_i^\dagger g^*, \qquad & (g\hat{c}_i^\dagger)^\dagger &= \hat{c}_i g^*. \end{aligned} \tag{5.5}$$

These follow from the standard definition of an adjoint, $\langle \Phi|\hat{\Omega}|\Psi\rangle = \langle \Psi|(\hat{\Omega})^\dagger|\Phi\rangle^*$.

Some operators, and especially those that represent physical quantities, contain an even number of fermion creation and annihilation operators. These operators are called *even operators*, and these will commute with a Grassmann number. Other operators of interest, such as the fermion field operators, involve odd numbers of creation and annihilation operators; the operator $\hat{\Psi}_\mathrm{f}(\mathbf{r})$ in (2.59), for example, is a sum of terms with a single annihilation operator. These operators are called *odd operators*. Operators either commute or anticommute with a Grassmann number depending on whether they are even or odd. Hence

$$\begin{aligned} g\hat{H}_\mathrm{f} &= \hat{H}_\mathrm{f} g, \\ g\hat{\Psi}_\mathrm{f}(\mathbf{r}) &= -\hat{\Psi}_\mathrm{f}(\mathbf{r}) g \qquad g\hat{\Psi}_\mathrm{f}^\dagger(\mathbf{r}) = -\hat{\Psi}_\mathrm{f}^\dagger(\mathbf{r}) g, \end{aligned} \tag{5.6}$$

where $\hat{H}_\mathrm{f}$ is the fermion Hamiltonian. Arbitrary operators can always be expressed as the sum of an even operator and an odd operator. A Grassmann number will commute with the even component and anticommute with the odd component. These concepts may be extended to include Grassmann operators. In this case all terms in even (or odd) operators contain an even (or odd) number of fermion creation operators, annihilation operators *and* Grassmann numbers, and consequently commute (or anticommute, respectively) with a Grassmann number. Thus $(\hat{1} + h\hat{c})$ is an even operator, but $(\hat{c} + h\hat{c}^\dagger\hat{c})$ is odd. Operators in which there are no Grassmann variables involved are just a special case in which evenness or oddness depends only on the number of fermion creation and annihilation operators. Operators for bosonic systems need not be classified as even or odd, as both types commute with Grassmann numbers, so the distinction is unimportant.

5.2 Unitary Displacement Operators

The unitary *Glauber displacement operator* $\hat{D}_\mathrm{b}(\boldsymbol{\alpha}, \boldsymbol{\alpha}^*)$ for bosons is defined by

$$\hat{D}_\mathrm{b}(\boldsymbol{\alpha}, \boldsymbol{\alpha}^*) = \exp\left(\sum_{i=1}^{n} \left(\hat{a}_i^\dagger \alpha_i - \alpha_i^* \hat{a}_i\right)\right), \tag{5.7}$$

where here $\boldsymbol{\alpha} \equiv \{\alpha_1, \cdots, \alpha_i, \cdots, \alpha_n\}$ and $\boldsymbol{\alpha}^* \equiv \{\alpha_1^*, \cdots, \alpha_i^*, \cdots, \alpha_n^*\}$. Note that $\hat{D}_{\mathrm{b}}(\boldsymbol{\alpha}, \boldsymbol{\alpha}^*)$ depends on all the α_i and their complex conjugates α_i^*. We can write it as a product of single-mode unitary operators – the proof is left as an exercise –

$$\hat{D}_{\mathrm{b}}(\boldsymbol{\alpha}, \boldsymbol{\alpha}^*) = \prod_{i=1}^{n} \exp\left(\hat{a}_i^\dagger \alpha_i - \alpha_i^* \hat{a}_i\right). \tag{5.8}$$

The inverse operator is simply

$$\hat{D}_{\mathrm{b}}(\boldsymbol{\alpha}, \boldsymbol{\alpha}^*)^{-1} = \hat{D}_{\mathrm{b}}(\boldsymbol{\alpha}, \boldsymbol{\alpha}^*)^\dagger = \hat{D}_{\mathrm{b}}(-\boldsymbol{\alpha}, -\boldsymbol{\alpha}^*). \tag{5.9}$$

This plays a key role in defining the boson coherent states. A product of two displacement operators gives a further displacement operator, multiplied by a phase factor:

$$\hat{D}_{\mathrm{b}}(\boldsymbol{\alpha}, \boldsymbol{\alpha}^*)\hat{D}_{\mathrm{b}}(\boldsymbol{\beta}, \boldsymbol{\beta}^*) = \hat{D}_{\mathrm{b}}(\boldsymbol{\alpha} + \boldsymbol{\beta}, \boldsymbol{\alpha}^* + \boldsymbol{\beta}^*) \exp\left[\frac{1}{2}\sum_i (\beta_i^* \alpha_i - \alpha_i^* \beta_i)\right]. \tag{5.10}$$

The proof is left as an exercise.

The unitary *fermion displacement operator* is an example of a Grassmann operator; it is an operator function $\hat{D}_{\mathrm{f}}(\mathbf{g}, \mathbf{g}^*)$ of Grassmann numbers, where here $\mathbf{g} \equiv \{g_1, \cdots, g_i, \cdots, g_n\}$ and $\mathbf{g}^* \equiv \{g_1^*, \cdots, g_i^*, \cdots, g_n^*\}$. It is defined by

$$\hat{D}_{\mathrm{f}}(\mathbf{g}, \mathbf{g}^*) = \exp\left(\sum_{i=1}^{n} \left(\hat{c}_i^\dagger g_i - g_i^* \hat{c}_i\right)\right), \tag{5.11}$$

where each mode in the sum is associated with a pair of g-numbers g_i, g_i^*. Like its bosonic counterpart, $\hat{D}_{\mathrm{f}}(\mathbf{g}, \mathbf{g}^*)$ depends on all the g_i and their complex conjugates g_i^*. Equivalent forms of the fermion displacement operator are

$$\begin{aligned}\hat{D}_{\mathrm{f}}(\mathbf{g}, \mathbf{g}^*) &= \prod_{i=1}^{n} \exp\left(\hat{c}_i^\dagger g_i - g_i^* \hat{c}_i\right) \\ &= \prod_{i=1}^{n} \left[1 + \hat{c}_i^\dagger g_i - g_i^* \hat{c}_i + \left(\hat{c}_i^\dagger \hat{c}_i - \frac{1}{2}\right) g_i^* g_i\right].\end{aligned} \tag{5.12}$$

The proof is left as an exercise. The last form shows that the unitary displacement operator involves monomials of orders up to 2^{2n}. Note that although $\hat{D}(\mathbf{g}, \mathbf{g}^*)$ is a Grassmann operator, it involves fermion creation or annihilation operators and Grassmann numbers in pairs, and hence is an even operator which will commute with Grassmann numbers.

The adjoint of $\hat{D}_{\mathrm{f}}(\mathbf{g}, \mathbf{g}^*)$ is given by

$$\hat{D}_{\mathrm{f}}(\mathbf{g}, \mathbf{g}^*)^\dagger = \exp\left[\sum_{i=1}^{n} \left(g_i^* \hat{c}_i - \hat{c}_i^\dagger g_i\right)\right] = \hat{D}_{\mathrm{f}}(-\mathbf{g}, -\mathbf{g}^*), \tag{5.13}$$

which follows from (5.5). It then follows that $\hat{D}_{\mathrm{f}}(\mathbf{g},\mathbf{g}^*)^\dagger$ is the inverse of $\hat{D}_{\mathrm{f}}(\mathbf{g},\mathbf{g}^*)$:

$$\hat{D}_{\mathrm{f}}(\mathbf{g},\mathbf{g}^*)^\dagger\hat{D}_{\mathrm{f}}(\mathbf{g},\mathbf{g}^*) = \hat{D}_{\mathrm{f}}(-\mathbf{g},-\mathbf{g}^*)\hat{D}_{\mathrm{f}}(\mathbf{g},\mathbf{g}^*) = \hat{1}, \tag{5.14}$$

as required by unitarity.

If $\mathbf{g} \equiv \{g_1,\cdots,g_n\}, \mathbf{g}^* \equiv \{g_1^*,\cdots,g_n^*\}$ and $\mathbf{h} \equiv \{h_1,\cdots,h_n\}, \mathbf{h}^* \equiv \{h_1^*,\cdots,h_n^*\}$ are two sets of Grassmann numbers and their conjugates, then we can write the product of two fermion displacement operators as

$$\hat{D}_{\mathrm{f}}(\mathbf{g},\mathbf{g}^*)\hat{D}_{\mathrm{f}}(\mathbf{h},\mathbf{h}^*) = \hat{D}_{\mathrm{f}}(\mathbf{g}+\mathbf{h},\mathbf{g}^*+\mathbf{h}^*)\exp\left[\frac{1}{2}\sum_i(h_i^*g_i - g_i^*h_i)\right]. \tag{5.15}$$

It is straightforward to prove this (see the Exercises), mode by mode, using (5.12). The product formula (5.15) is called a *ray representation* of the additive group of Grassmann numbers. We will use this representation later in deriving fermionic-state properties.

The displacement operator as already defined is written in symmetric ordering. It is typically more useful to have specific normally and antinormally ordered forms of such operators, as applying these forms of the operators to specific states is easier. Normally and antinormally ordered unitary forms of the displacement operators can easily be found using the Baker–Hausdorff theorem, and express the symmetrically ordered displacement operator as the product of two exponential operators involving just creation or annihilation operators (the normally or antinormally ordered displacement operator) with an exponential of a scalar product. Thus

$$\begin{aligned}
\hat{D}_{\mathrm{b}}(\boldsymbol{\alpha},\boldsymbol{\alpha}^*) &= \exp\left(\sum_{i=1}^n \hat{a}_i^\dagger\alpha_i\right)\exp\left(\sum_{i=1}^n -\alpha_i^*\hat{a}_i\right)\exp\left(-\frac{1}{2}\boldsymbol{\alpha}^*\cdot\boldsymbol{\alpha}\right)\\
&= \exp\left(\sum_{i=1}^n -\alpha_i^*\hat{a}_i\right)\exp\left(\sum_{i=1}^n \hat{a}_i^\dagger\alpha_i\right)\exp\left(+\frac{1}{2}\boldsymbol{\alpha}^*\cdot\boldsymbol{\alpha}\right),\\
\hat{D}_{\mathrm{f}}(\mathbf{g},\mathbf{g}^*) &= \exp\left(\sum_{i=1}^n \hat{c}_i^\dagger g_i\right)\exp\left(\sum_{i=1}^n -g_i^*\hat{c}_i\right)\exp\left(-\frac{1}{2}\mathbf{g}^*\cdot\mathbf{g}\right)\\
&= \exp\left(\sum_{i=1}^n -g_i^*\hat{c}_i\right)\exp\left(\sum_{i=1}^n \hat{c}_i^\dagger g_i\right)\exp\left(+\frac{1}{2}\mathbf{g}^*\cdot\mathbf{g}\right),
\end{aligned} \tag{5.16}$$

where the general definitions of the scalar products are

$$\boldsymbol{\beta}\cdot\boldsymbol{\alpha} = \sum_i \beta_i\alpha_i \qquad \mathbf{h}\cdot\mathbf{g} = \sum_i h_i g_i. \tag{5.17}$$

The *transformation properties* of the boson operators $\hat{a}_i, \hat{a}_i^\dagger$ under the unitary displacement operator merely add a simple c-number displacement:

$$\begin{aligned}
\hat{D}_{\mathrm{b}}(\boldsymbol{\alpha},\boldsymbol{\alpha}^*)^\dagger\hat{a}_i\hat{D}_{\mathrm{b}}(\boldsymbol{\alpha},\boldsymbol{\alpha}^*) &= \hat{a}_i + \alpha_i,\\
\hat{D}_{\mathrm{b}}(\boldsymbol{\alpha},\boldsymbol{\alpha}^*)^\dagger\hat{a}_i^\dagger\hat{D}_{\mathrm{b}}(\boldsymbol{\alpha},\boldsymbol{\alpha}^*) &= \hat{a}_i^\dagger + \alpha_i^*.
\end{aligned} \tag{5.18}$$

The corresponding transformation properties of the fermion operators $\hat{c}_i, \hat{c}_i^\dagger$ are

$$\begin{aligned}\hat{D}_{\mathrm{f}}(\mathbf{g}, \mathbf{g}^*)^\dagger \hat{c}_i \hat{D}_{\mathrm{f}}(\mathbf{g}, \mathbf{g}^*) &= \hat{c}_i + g_i, \\ \hat{D}_{\mathrm{f}}(\mathbf{g}, \mathbf{g}^*)^\dagger \hat{c}_i^\dagger \hat{D}_{\mathrm{f}}(\mathbf{g}, \mathbf{g}^*) &= \hat{c}_i^\dagger + g_i^*. \end{aligned} \tag{5.19}$$

The effect of the displacement operator is to add a c-number or a Grassmann number to the annihilation and creation operators. The proof of the transformation rules is left as an exercise for the reader.

5.3 Boson and Fermion Coherent States

The *boson coherent states* are given by a c-number vector

$$|\boldsymbol{\alpha}, \boldsymbol{\alpha}^*\rangle = \hat{D}_{\mathrm{b}}(\boldsymbol{\alpha}, \boldsymbol{\alpha}^*)|0\rangle. \tag{5.20}$$

A boson coherent state involves all $2n$ c-numbers $\alpha_1, \cdots, \alpha_n, \alpha_1^*, \cdots, \alpha_n^*$. The boson coherent states may be written as products of coherent states, one for each of the modes:

$$|\boldsymbol{\alpha}, \boldsymbol{\alpha}^*\rangle = |\alpha_1, \alpha_1^*\rangle \cdots |\alpha_n, \alpha_n^*\rangle = \prod_i |\alpha_i, \alpha_i^*\rangle, \tag{5.21}$$

where

$$|\alpha_i, \alpha_i^*\rangle = \exp\left(\hat{a}_i^\dagger \alpha_i - \alpha_i^* \hat{a}_i\right) |0_i\rangle, \tag{5.22}$$

and where $|0_i\rangle$ is the vacuum state for the ith mode. Note that the vacuum state is itself a boson coherent state, with $\boldsymbol{\alpha} = \boldsymbol{\alpha}^* = 0$. The boson coherent states are right eigenvectors for the annihilation operators $\hat{a}_i$,

$$\hat{a}_i|\boldsymbol{\alpha}, \boldsymbol{\alpha}^*\rangle = \alpha_i|\boldsymbol{\alpha}, \boldsymbol{\alpha}^*\rangle, \tag{5.23}$$

and left eigenstates for the creation operators $\hat{a}_i^\dagger$,

$$\langle \boldsymbol{\alpha}, \boldsymbol{\alpha}^*|\hat{a}_i^\dagger = \langle \boldsymbol{\alpha}, \boldsymbol{\alpha}^*|\alpha_i^*, \tag{5.24}$$

where, in standard quantum notation,

$$\langle \boldsymbol{\alpha}, \boldsymbol{\alpha}^*| = (|\boldsymbol{\alpha}, \boldsymbol{\alpha}^*\rangle)^\dagger. \tag{5.25}$$

The coherent states are normalised, but not orthogonal:

$$\langle \boldsymbol{\alpha}, \boldsymbol{\alpha}^*|\boldsymbol{\alpha}, \boldsymbol{\alpha}^*\rangle = 1, \tag{5.26}$$

$$\langle \boldsymbol{\alpha}, \boldsymbol{\alpha}^*|\boldsymbol{\beta}, \boldsymbol{\beta}^*\rangle = \exp\left[\boldsymbol{\alpha}^* \cdot \boldsymbol{\beta} - \frac{1}{2}(\boldsymbol{\alpha}^* \cdot \boldsymbol{\alpha} + \boldsymbol{\beta}^* \cdot \boldsymbol{\beta})\right]. \tag{5.27}$$

They become approximately orthogonal if the difference in their c-number amplitudes is large. The proof of (5.27) is left as an exercise.

A *fermion coherent state* is a specific example of a Grassmann vector:

$$
\begin{aligned}
|\mathbf{g},\mathbf{g}^*\rangle &= \hat{D}_{\mathrm{f}}(g,g^*)|0\rangle \\
&= \prod_{i=1}^{n}\left[1+\hat{c}_i^\dagger g_i - g_i^*\hat{c}_i + \left(\hat{c}_i^\dagger\hat{c}_i - \frac{1}{2}\right)g_i^* g_i\right]|0\rangle.
\end{aligned}
\tag{5.28}
$$

A fermion coherent state involves all $2n$ Grassmann numbers $g_1,\cdots,g_n,g_1^*,\cdots,g_n^*$ and can be written as a product of coherent states $|g_i,g_i^*\rangle$ for the separate modes:

$$
|\mathbf{g},\mathbf{g}^*\rangle = |g_1,g_1^*\cdots|g_n,g_n^*\rangle = \prod_i |g_i,g_i^*\rangle,
\tag{5.29}
$$

where

$$
|g_i,g_i^*\rangle = \exp\left(\hat{c}_i^\dagger g_i - g_i^*\hat{c}_i\right)|0_i\rangle
\tag{5.30}
$$

and $|0_i\rangle$ is the vacuum state for the ith mode. The vacuum state is itself a fermion coherent state, with $\mathbf{g}=\mathbf{g}^*=0$. Note that the fermion coherent states are examples of even Grassmann vectors.

The significance of the fermion coherent states is similar to that for the case of bosons, namely that they are right eigenstates of the annihilation operators $\hat{c}_i$:

$$
\hat{c}_i|\mathbf{g},\mathbf{g}^*\rangle = g_i|\mathbf{g},\mathbf{g}^*\rangle.
\tag{5.31}
$$

The difference is that for the fermion coherent states the eigenvalues are Grassmann numbers, not c-numbers. The proof of this result is given in Appendix D. The adjoint fermion coherent state satisfies an eigenvalue equation for the creation operators $\hat{c}_i^\dagger$,

$$
\langle\mathbf{g},\mathbf{g}^*|\hat{c}_i^\dagger = \langle\mathbf{g},\mathbf{g}^*|g_i^*,
\tag{5.32}
$$

where, in the standard quantum notation,

$$
\langle\mathbf{g},\mathbf{g}^*| = (|\mathbf{g},\mathbf{g}^*\rangle)^\dagger.
\tag{5.33}
$$

Like their bosonic counterparts, these states are normalised but not orthogonal:

$$
\langle\mathbf{g},\mathbf{g}^*|\mathbf{g},\mathbf{g}^*\rangle = \langle 0|\hat{D}_{\mathrm{f}}(\mathbf{g},\mathbf{g}^*)^\dagger\hat{D}_{\mathrm{f}}(\mathbf{g},\mathbf{g}^*)|0\rangle = \langle 0|0\rangle = 1,
\tag{5.34}
$$

$$
\langle\mathbf{g},\mathbf{g}^*|\mathbf{h},\mathbf{h}^*\rangle = \exp\left[\mathbf{g}^*\cdot\mathbf{h} - \frac{1}{2}\left(\mathbf{g}^*\cdot\mathbf{g}+\mathbf{h}^*\cdot\mathbf{h}\right)\right],
\tag{5.35}
$$

which is an even Grassmann function. The proof of (5.35) forms an exercise. The orthogonality and normalisation results are of the same form for both boson and fermion coherent states. If one of the two coherent states, is the vacuum state, we have

$$
\begin{aligned}
\langle\mathbf{0}|\boldsymbol{\alpha},\boldsymbol{\alpha}^*\rangle &= \exp\left[-\frac{1}{2}(\boldsymbol{\alpha}^*\cdot\boldsymbol{\alpha})\right], \\
\langle\mathbf{0}|\mathbf{g},\mathbf{g}^*\rangle &= \exp\left[-\frac{1}{2}(\mathbf{g}^*\cdot\mathbf{g})\right],
\end{aligned}
\tag{5.36}
$$

useful results that we will require later.

The coherent states are complete in the sense that we can represent any operator in terms of them. This property is essential in the development of phase space methods, as we need to represent any possible quantum state in this way. The *completeness* condition is usually presented as a resolution of the identity operator. For the boson coherent states, this takes the form

$$\begin{aligned}\int \frac{d^2\boldsymbol{\alpha}}{\pi^n}|\boldsymbol{\alpha},\boldsymbol{\alpha}^*\rangle\langle\boldsymbol{\alpha},\boldsymbol{\alpha}^*| &= \prod_i \int \frac{d^2\alpha_i}{\pi}\hat{D}_{\mathrm{b}}(\alpha_i,\alpha_i^*)|0\rangle\langle 0|\hat{D}_{\mathrm{b}}(\alpha_i,\alpha_i^*)^\dagger \\ &= \hat{1}\,. \end{aligned} \tag{5.37}$$

Here the factor $d^2\alpha_i$ indicates integration over the complex α_i plane. The proof is left as an exercise. The corresponding completeness relation for the fermionic coherent states is expressed in the form of a Grassmann integral:

$$\begin{aligned}\int d^2\mathbf{g}\;|\mathbf{g},\mathbf{g}^*\rangle\langle\mathbf{g},\mathbf{g}^*| &= \prod_i \int d^2 g_i\;\hat{D}_{\mathrm{f}}(g_i,g_i^*)|0\rangle\langle 0|\hat{D}_{\mathrm{f}}(g_i,g_i^*)^\dagger \\ &= \hat{1}, \end{aligned} \tag{5.38}$$

where $d^2\mathbf{g} = \prod_i dg_i^*\;dg_i$. The proof of the fermion completeness result is set out in Appendix D.

5.4 Bargmann States

For bosons, a class of states related to the coherent states is the *Bargmann states*, which are defined as

$$\begin{aligned}|\boldsymbol{\alpha}\rangle_{\mathrm{B}} &= \hat{R}_{\mathrm{b}}(\boldsymbol{\alpha})|0\rangle \\ &= \exp\left(\frac{1}{2}\boldsymbol{\alpha}^*\cdot\boldsymbol{\alpha}\right)|\boldsymbol{\alpha},\boldsymbol{\alpha}^*\rangle, \end{aligned} \tag{5.39}$$

where

$$\hat{R}_{\mathrm{b}}(\boldsymbol{\alpha}) = \exp\left(\sum_{i=1}^{n}\hat{a}_i^\dagger\alpha_i\right) = \prod_i \exp\left(\hat{a}_i^\dagger\alpha_i\right). \tag{5.40}$$

Unlike the coherent states themselves, the bosonic Bargmann states depend only on the c-number variables $\boldsymbol{\alpha}$ and *not* on the complex conjugates $\boldsymbol{\alpha}^*$, hence the notation $|\boldsymbol{\alpha}\rangle_{\mathrm{B}}$ without the $\boldsymbol{\alpha}^*$. The non-appearance of the conjugate c-numbers $\alpha_1^*,\cdots,\alpha_i^*,\cdots,\alpha_n^*$ is relevant to analyticity considerations.

The Bargmann states are eigenstates of the annihilation operator:

$$\hat{a}_i|\boldsymbol{\alpha}\rangle_{\mathrm{B}} = \alpha_i|\boldsymbol{\alpha}\rangle_{\mathrm{B}}, \qquad {}_{\mathrm{B}}\langle\boldsymbol{\alpha}|\hat{a}_i^\dagger = {}_{\mathrm{B}}\langle\boldsymbol{\alpha}|\alpha_i^*\,. \tag{5.41}$$

Naturally, from (5.26), (5.39) and (5.27) they are neither normalised,

$${}_{\mathrm{B}}\langle\boldsymbol{\alpha}|\boldsymbol{\alpha}\rangle_{\mathrm{B}} = \exp(\boldsymbol{\alpha}^*\cdot\boldsymbol{\alpha}), \tag{5.42}$$

nor orthogonal,

$$_{\mathrm{B}}\langle \boldsymbol{\alpha}|\boldsymbol{\beta}\rangle_{\mathrm{B}} = \exp(\boldsymbol{\alpha}^* \cdot \boldsymbol{\beta}), \tag{5.43}$$

as the Bargmann states are unnormalised versions of the coherent states.

The multi-mode Bargmann states may also be written as the products of Bargmann states for the separate modes:

$$\begin{aligned} |\boldsymbol{\alpha}\rangle_{\mathrm{B}} &= \prod_i |\alpha_i\rangle_{\mathrm{B}}, \\ |\alpha_i\rangle_{\mathrm{B}} &= \exp\left(\frac{1}{2}\alpha_i^*\alpha_i\right)|\alpha_i, \alpha_i^*\rangle. \end{aligned} \tag{5.44}$$

We can determine the well-known expansion of the coherent states in terms of Fock states. For each mode, we have

$$\begin{aligned} |\alpha_i, \alpha_i^*\rangle &= \exp\left(-\frac{1}{2}\alpha_i^*\alpha_i\right)|\alpha_i\rangle_{\mathrm{B}} = \exp\left(-\frac{1}{2}\alpha_i^*\alpha_i\right)\exp\left(\hat{a}_i^\dagger\alpha_i\right)|0_i\rangle \\ &= \exp\left(-\frac{1}{2}\alpha_i^*\alpha_i\right)\sum_{\nu_i=0}^{\infty}\frac{(\alpha_i)^{\nu_i}}{\sqrt{\nu_i!}}|\nu_i\rangle. \end{aligned} \tag{5.45}$$

This c-number expansion represents a Poisson distribution of number states with a mean boson number $\langle \hat{n}_i\rangle = |\alpha_i|^2$ and a variance $\langle \Delta\hat{n}_i^2\rangle = \langle \hat{n}_i\rangle$. By using these results, we find the Fock state expansion for the multi-mode coherent state:

$$\begin{aligned} |\boldsymbol{\alpha}, \boldsymbol{\alpha}^*\rangle &= \exp\left(-\frac{1}{2}\boldsymbol{\alpha}^* \cdot \boldsymbol{\alpha}\right)|\boldsymbol{\alpha}\rangle_{\mathrm{B}}, \\ |\boldsymbol{\alpha}\rangle_{\mathrm{B}} &= \prod_i \sum_{\nu_i=0}^{\infty}\frac{(\alpha_i)^{\nu_i}}{\sqrt{\nu_i!}}|\nu_i\rangle. \end{aligned} \tag{5.46}$$

In this form, it is clear that the Bargmann state $|\boldsymbol{\alpha}\rangle_{\mathrm{B}}$ is an *analytic function* of the α_i. The bra vector $\langle\boldsymbol{\alpha}|_{\mathrm{B}}$ is an analytic function of the α_i^*. The operation of a creation operator on a Bargmann ket vector or an annihilation operator on a Bargmann bra vector can be written using derivatives of these vectors:

$$\hat{a}_i^\dagger|\boldsymbol{\alpha}\rangle_{\mathrm{B}} = \left(\frac{\partial}{\partial\alpha_i}\right)|\boldsymbol{\alpha}\rangle_{\mathrm{B}}, \qquad {}_{\mathrm{B}}\langle\boldsymbol{\alpha}|\hat{a}_i = \left(\frac{\partial}{\partial\alpha_i^*}\right){}_{\mathrm{B}}\langle\alpha|, \tag{5.47}$$

there being no distinction, of course, between left and right differentiation. Complex differentiation is well defined for analytic functions such as the Bargmann vectors.

For fermions, we define the Bargmann states via

$$|\mathbf{g}\rangle_{\mathrm{B}} = \hat{R}_{\mathrm{f}}(\mathbf{g})|0\rangle = \exp\left(\frac{1}{2}\mathbf{g}^* \cdot \mathbf{g}\right)|\mathbf{g}, \mathbf{g}^*\rangle, \tag{5.48}$$

where

$$\hat{R}_{\mathrm{f}}(\mathbf{g}) = \exp\left(\sum_{i=1}^{n} \hat{c}_i^\dagger g_i\right) = \prod_i \exp\left(\hat{c}_i^\dagger g_i\right). \tag{5.49}$$

The Bargmann states are clearly even Grassmann vectors. They are related to the fermion coherent states in a way similar to the boson case. Note that the fermion Bargmann states $|\mathbf{g}\rangle_{\mathrm{B}}$ now depend only on n Grassmann numbers $\mathbf{g}$, hence the notation $|\mathbf{g}\rangle_{\mathrm{B}}$ without the $\mathbf{g}^*$. The non-appearance of the conjugate g-numbers $\mathbf{g}^*$ is relevant to integration and differentiation considerations. It is not difficult to show that

$$\hat{R}_{\mathrm{f}}(\mathbf{g}) = \prod_i \left(1 + \hat{c}_i^\dagger g_i\right) = \prod_i \left(1 - g_i \hat{c}_i^\dagger\right). \tag{5.50}$$

The fermion Bargmann states are eigenstates of the annihilation operator:

$$\hat{c}_i |\mathbf{g}\rangle_{\mathrm{B}} = g_i |\mathbf{g}\rangle_{\mathrm{B}}, \qquad {}_{\mathrm{B}}\langle \mathbf{g}| \hat{c}_i^\dagger =_{\mathrm{D}} \langle \mathbf{g}| g_i^*; \tag{5.51}$$

but they are neither normalised nor orthogonal, from (5.34), (5.48) and (5.35):

$$\begin{aligned} {}_{\mathrm{B}}\langle \mathbf{g}|\mathbf{g}\rangle_{\mathrm{B}} &= \exp(\mathbf{g}^* \cdot \mathbf{g}), \\ {}_{\mathrm{B}}\langle \mathbf{g}|\mathbf{h}\rangle_{\mathrm{B}} &= \exp(\mathbf{g}^* \cdot \mathbf{h}). \end{aligned} \tag{5.52}$$

Note the similarity of these results to those for the boson states.

The Bargmann states may also be written as the products of single-mode states

$$\begin{aligned} |\mathbf{g}\rangle_{\mathrm{B}} &= \prod_i |g_i\rangle_{\mathrm{B}}, \\ |g_i\rangle_{\mathrm{B}} &= \exp\left(\frac{1}{2} g_i^* g_i\right) |g_i, g_i^*\rangle. \end{aligned} \tag{5.53}$$

Having related the fermion coherent states to the Bargmann states, we can use the above to determine the expansion of a coherent state in terms of Fock states. For each mode, we have

$$\begin{aligned} |g_i, g_i^*\rangle &= \exp\left(-\frac{1}{2} g_i^* g_i\right) |g_i\rangle_{\mathrm{B}} = \exp\left(-\frac{1}{2} g_i^* g_i\right) \exp\left(\hat{c}_i^\dagger g_i\right) |0_i\rangle \\ &= \exp\left(-\frac{1}{2} g_i^* g_i\right) (|0_i\rangle - g_i |1_i\rangle). \end{aligned} \tag{5.54}$$

This differs from the corresponding bosonic expansion because it involves only a superposition of zero- and one-fermion states, a natural consequence of the Pauli exclusion principle that any mode can only be occupied by at most one fermion. The last result allows us to write the Fock state expansion for the multi-mode coherent state as

$$\begin{aligned} |\mathbf{g}, \mathbf{g}^*\rangle &= \exp\left(-\frac{1}{2} \mathbf{g}^* \cdot \mathbf{g}\right) |\mathbf{g}\rangle_{\mathrm{B}}, \\ |\mathbf{g}\rangle_{\mathrm{B}} &= \prod_i (|0_i\rangle - g_i |1_i\rangle), \end{aligned} \tag{5.55}$$

which involves superpositions of states with differing total numbers of fermions.

We emphasise again that, unlike the coherent states themselves, the bosonic and fermionic Bargmann states depend only on the c-number and Grassmann variables $\boldsymbol{\alpha}$ and $\mathbf{g}$ and not on the complex conjugates $\boldsymbol{\alpha}^*$ and $\mathbf{g}^*$. The bosonic Bargmann states have analytic features with respect to the c-numbers $\boldsymbol{\alpha}$ and the fermionic ones are subject only to Grassmann differentiation with respect to the $\mathbf{g}$.

The operation of a creation operator on a Bargmann ket or an annihilation operator on a fermion Bargmann bra can also be written in terms of derivatives of these vectors:

$$\hat{c}_i^\dagger |\mathbf{g}\rangle_{\mathrm{B}} = \left(-\frac{\overrightarrow{\partial}}{\partial g_i}\right) |\mathbf{g}\rangle_{\mathrm{B}} = |\mathbf{g}\rangle_{\mathrm{B}} \left(+\frac{\overleftarrow{\partial}}{\partial g_i}\right),$$

$$_{\mathrm{B}}\langle\mathbf{g}|\hat{c}_i = {}_{\mathrm{B}}\langle\mathbf{g}| \left(-\frac{\overleftarrow{\partial}}{\partial g_i^*}\right) = \left(+\frac{\overrightarrow{\partial}}{\partial g_i^*}\right) {}_{\mathrm{B}}\langle\mathbf{g}|, \tag{5.56}$$

where both left- and right-derivative forms apply for each of $\hat{c}_i^\dagger|\mathbf{g}\rangle_{\mathrm{B}}$ and $\langle\mathbf{g}|_{\mathrm{B}}\hat{c}_i$. Note that in (5.51) the eigenvalues can be placed on either side of the bra or ket Bargmann vector. Grassmann differentiation of $|\mathbf{g}\rangle_{\mathrm{B}}$ with respect to the g_i is well defined, as is Grassmann differentiation of $\langle\mathbf{g}|_{\mathrm{B}}$ with respect to the g_i^*. Note the sign difference relative to the boson cases (5.47) for the first forms of the result.

It is useful to be able to express the completeness relationships for the coherent states, (5.37) and (5.38), in terms of the Bargmann states. We find

$$\int d^2\mathbf{g}\, e^{-\mathbf{g}^*\cdot\mathbf{g}}\, |\mathbf{g}\rangle_{\mathrm{BB}}\langle\mathbf{g}| = \int d^2\mathbf{g} \frac{|\mathbf{g}\rangle_{\mathrm{BB}}\langle\mathbf{g}|}{_{\mathrm{B}}\langle\mathbf{g}|\mathbf{g}\rangle_{\mathrm{B}}} = \hat{1},$$

$$\int d^2\boldsymbol{\alpha}\, e^{-\boldsymbol{\alpha}^*\cdot\boldsymbol{\alpha}} |\boldsymbol{\alpha}\rangle_{\mathrm{BB}}\langle\boldsymbol{\alpha}| = \int d^2\boldsymbol{\alpha} \frac{|\boldsymbol{\alpha}\rangle_{\mathrm{BB}}\langle\boldsymbol{\alpha}|}{_{\mathrm{B}}\langle\boldsymbol{\alpha}|\boldsymbol{\alpha}\rangle_{\mathrm{B}}} = \hat{1} \tag{5.57}$$

for fermions and for bosons, respectively.

5.5 Examples of Fermion States

At this stage we pause in our development of coherent states and provide some examples of fermionic states. Let us consider just three quantum modes with eight possible Fock states

$$|\nu_1, \nu_2, \nu_3\rangle = \left(\hat{c}_1^\dagger\right)^{\nu_1} \left(\hat{c}_2^\dagger\right)^{\nu_2} \left(\hat{c}_3^\dagger\right)^{\nu_3} |0\rangle \qquad (\nu_i = 0, 1), \tag{5.58}$$

so that, for example, the fully occupied state is

$$|1, 1, 1\rangle = \hat{c}_1^\dagger \hat{c}_2^\dagger \hat{c}_3^\dagger |0\rangle. \tag{5.59}$$

The most general two-fermion state is

$$|\Psi\rangle = A(1,2)\hat{c}_1^\dagger\hat{c}_2^\dagger|0\rangle + A(2,3)\hat{c}_2^\dagger\hat{c}_3^\dagger|0\rangle + A(3,1)\hat{c}_3^\dagger\hat{c}_1^\dagger|0\rangle, \tag{5.60}$$

where the $A(i,j)$ are c-number probability amplitudes. The fermion coherent state for our three modes is

$$
\begin{aligned}
|\mathbf{g},\mathbf{g}^*\rangle &= \left(1-\frac{1}{2}g_1^*g_1\right)\left(1-\frac{1}{2}g_2^*g_2\right)\left(1-\frac{1}{2}g_3^*g_3\right)|0\rangle \\
&+\hat{c}_1^\dagger g_1\left(1-\frac{1}{2}g_2^*g_2\right)\left(1-\frac{1}{2}g_3^*g_3\right)|0\rangle+\hat{c}_2^\dagger g_2\left(1-\frac{1}{2}g_1^*g_1\right)\left(1-\frac{1}{2}g_3^*g_3\right)|0\rangle \\
&+\hat{c}_3^\dagger g_3\left(1-\frac{1}{2}g_1^*g_1\right)\left(1-\frac{1}{2}g_2^*g_2\right)|0\rangle+\hat{c}_1^\dagger g_1\hat{c}_2^\dagger g_2\left(1-\frac{1}{2}g_3^*g_3\right)|0\rangle \\
&+\hat{c}_2^\dagger g_2\hat{c}_3^\dagger g_3\left(1-\frac{1}{2}g_1^*g_1\right)|0\rangle+\hat{c}_3^\dagger g_3\hat{c}_1^\dagger g_1\left(1-\frac{1}{2}g_2^*g_2\right)|0\rangle+\hat{c}_1^\dagger g_1\hat{c}_2^\dagger g_2\hat{c}_3^\dagger g_3|0\rangle.
\end{aligned}
\tag{5.61}
$$

This is a superposition of zero-, one-, two- and three-fermion states. It is also a Grassmann vector and, of course, an unphysical state. Note that the Pauli exclusion principle is satisfied – there is never more than one fermion per mode.

The fermion Bargmann state is

$$
\begin{aligned}
|\mathbf{g}\rangle_{\mathrm{B}} &= \left(1+\hat{c}_1^\dagger g_1\right)\left(1+\hat{c}_2^\dagger g_2\right)\left(1+\hat{c}_3^\dagger g_3\right)|0\rangle \\
&= |0\rangle+\left(\hat{c}_1^\dagger g_1+\hat{c}_2^\dagger g_2+\hat{c}_3^\dagger g_3\right)|0\rangle \\
&\quad+\left(\hat{c}_1^\dagger g_1\hat{c}_2^\dagger g_2+\hat{c}_2^\dagger g_2\hat{c}_3^\dagger g_3+\hat{c}_3^\dagger g_3\hat{c}_1^\dagger g_1\right)|0\rangle+\hat{c}_1^\dagger g_1\hat{c}_2^\dagger g_2\hat{c}_3^\dagger g_3|0\rangle.
\end{aligned}
\tag{5.62}
$$

This is also a superposition of zero-, one-, two- and three-fermion states but, unlike the coherent state, it is a Grassmann vector depending only on $\mathbf{g}$.

5.6 State and Operator Representations via Coherent States

The power of phase space methods derives in part from the fact that we can represent both states and operators of interest in terms of c-numbers or Grassmann numbers. This allows us to map quantum dynamics onto stochastic equations.

5.6.1 State Representation

An essential prerequisite for developing the representations we require is the ability to evaluate overlaps between state vectors and coherent states. Consider an N-boson pure state in which the N particles are shared between n modes:

$$
|\Psi\rangle_{N\mathrm{b}} = \sum_{\nu_1,\cdots,\nu_n} B_N(\nu)\frac{(\hat{a}_1^\dagger)^{\nu_1}}{\sqrt{\nu_1!}}\cdots\frac{(\hat{a}_n^\dagger)^{\nu_n}}{\sqrt{\nu_n!}}|0\rangle, \tag{5.63}
$$

where $\sum_i \nu_i = N$. The overlap with our coherent state $|\boldsymbol{\alpha},\boldsymbol{\alpha}^*\rangle$ is readily obtained using the left-eigenvalue property of the coherent states and $\langle\boldsymbol{\alpha},\boldsymbol{\alpha}^*|0\rangle = \exp(-\frac{1}{2}\boldsymbol{\alpha}^*\cdot\,\boldsymbol{\alpha})$.

We find

$$\langle \boldsymbol{\alpha}, \boldsymbol{\alpha}^* | \Psi \rangle_{N\mathrm{b}} = \sum_{\nu_1, \cdots, \nu_n} B_N(\nu) \frac{\alpha_1^*}{\sqrt{\nu_1!}} \cdots \frac{\alpha_n^*}{\sqrt{\nu_n!}} \exp\left(-\frac{1}{2}\boldsymbol{\alpha}^* \cdot \boldsymbol{\alpha}\right). \tag{5.64}$$

There is a one-to-one correspondence between the state $|\Psi\rangle_{N\mathrm{b}}$ and this c-number function; each of the creation operators $\hat{a}_j^\dagger$ has been mapped onto the corresponding c-number α_j^*. It follows that we can use the c-number function $\langle \boldsymbol{\alpha}, \boldsymbol{\alpha}^* | \Psi \rangle_{N\mathrm{b}}$ as a *representation* of the N-boson state.

It is similarly straightforward to use the eigenvalue property to evaluate the overlap for an N-fermion state. Consider the N-fermion pure state

$$|\Psi\rangle_{N\mathrm{f}} = \sum_{\nu_1, \cdots, \nu_n} B_N(\nu) \left(\hat{c}_1^\dagger\right)^{\nu_1} \cdots \left(\hat{c}_n^\dagger\right)^{\nu_n} |0\rangle, \tag{5.65}$$

where $\sum_i \nu_i = N$. The overlap with our fermion coherent state $|\mathbf{g}, \mathbf{g}^*\rangle$ also follows from the left-eigenvalue property and $\langle \mathbf{g}, \mathbf{g}^* | 0 \rangle = \exp(-\frac{1}{2}\mathbf{g}^* \cdot \mathbf{g})$:

$$\langle \mathbf{g}, \mathbf{g}^* | \Psi \rangle_{N\mathrm{f}} = \sum_{\nu_1, \cdots, \nu_n} B_N(\nu) (g_1^*)^{\nu_1} \cdots (g_n^*)^{\nu_n} \exp\left(-\frac{1}{2}\mathbf{g}^* \cdot \mathbf{g}\right). \tag{5.66}$$

As with bosons, there is a one-to-one correspondence between this Grassmann function and the original state, with $(g_j^*)^{\nu_j} = g_j^*$ for an occupied mode and $(g_j^*)^{\nu_j} = 1$ for an unoccupied one. This Grassmann function is, therefore, a representation of the N-fermion state.

For a physical N-fermion state based on $n \geq N$ modes, there are ${}^nC_N$ ways of choosing the N occupied modes, so that the number of distinct coefficients $B_N(\nu)$ required to represent the physical state is ${}^nC_N$. Hence the Grassmann function representation for a physical state is specified by the same number of c-numbers as in the original Fock state basis, as expected. In the case of a non-physical state, the restriction $\sum_i \nu_i = N$ need not apply. As there are ${}^nC_p$ ways of choosing p modes out of a total of n, we see that for an arbitrary (non-physical) state the number of such coefficients will be $\sum_{p=0}^{n} {}^nC_p = 2^n$, so that this large number of c-number coefficients would be required to specify the Grassmann function used to represent a general non-physical state. This number is no different, though, from that in the original Fock state representation of non-physical states. Note that if the state includes only Fock states with even or odd fermion numbers, then the associated Grassmann function is even or odd, respectively.

In deriving the above results, the following useful expressions were obtained for the scalar product or overlap of a fermion Fock state with a coherent state:

$$\begin{aligned} \langle \mathbf{g}, \mathbf{g}^* | \nu_1, \cdots, \nu_n \rangle &= (g_1^*)^{\nu_1} \cdots (g_n^*)^{\nu_n} e^{-(1/2)\mathbf{g}^* \cdot \mathbf{g}}, \\ \langle \xi_1, \cdots, \xi_n | \mathbf{h}, \mathbf{h}^* \rangle &= (h_n)^{\nu_n} \cdots (h_1)^{\nu_1} e^{-(1/2)\mathbf{h}^* \cdot \mathbf{h}}. \end{aligned} \tag{5.67}$$

These reflect the more general and standard rule for the complex conjugate of a scalar product, which states that

$$(\langle \mathbf{g}, \mathbf{g}^* | \Psi \rangle_{N\mathrm{f}})^* = {}_{N\mathrm{f}}\langle \Psi | \mathbf{g}, \mathbf{g}^* \rangle. \tag{5.68}$$

The fermion coherent states are even, but the Fock states will be even if there are an even number of fermions present and odd if there are an odd number. It follows that $\langle \mathbf{g}, \mathbf{g}^* | \Psi \rangle_{N\mathrm{f}}$ and its conjugate will be even Grassmann functions if N is even and odd Grassmann functions if N is odd.

For the Bargmann states, our scalar products have a similar form to those for the coherent states but with the normalisation removed:

$$\begin{aligned} {}_{\mathrm{B}}\langle \mathbf{g} | \nu_1, \cdots, \nu_n \rangle &= (g_1^*)^{\nu_1} \cdots (g_n^*)^{\nu_n}, \\ \langle \xi_1, \cdots, \xi_n | \mathbf{h} \rangle_{\mathrm{B}} &= (h_n)^{\nu_n} \cdots (h_1)^{\nu_1}, \end{aligned} \tag{5.69}$$

or, more generally,

$$ {}_{\mathrm{B}}\langle \mathbf{g} | \Psi \rangle_{N\mathrm{f}} = \sum_{\nu_1, \cdots, \nu_n} B_N(\nu) g_1^*)^{\nu_1} \cdots (g_n^*)^{\nu_n}. \tag{5.70}$$

We note, once again, that if Bargmann states are used then the overlap ${}_{\mathrm{B}}\langle \mathbf{g} | \Psi \rangle_{N\mathrm{f}}$, depends only on the g_i^* and not on the g_i. For bosons we have, similarly,

$$ {}_{\mathrm{B}}\langle \boldsymbol{\alpha} | \nu_1, \cdots, \nu_n \rangle = \frac{(\alpha_1^*)^{\nu_1}}{\sqrt{\nu_1 !}} \cdots \frac{(\alpha_n^*)^{\nu_n}}{\sqrt{\nu_n !}} = \langle \nu_1, \cdots, \nu_n | \boldsymbol{\alpha} \rangle_{\mathrm{B}}^*, \tag{5.71}$$

which depends only on the α_i^* and not on the α_i.

5.6.2 Coherent-State Projectors

From the boson and fermion coherent states, we can define *projector operators* via their action on quantum states:

$$(|\boldsymbol{\alpha}, \boldsymbol{\alpha}^* \rangle \langle \boldsymbol{\beta}, \boldsymbol{\beta}^* |) | \Psi \rangle = |\boldsymbol{\alpha}, \boldsymbol{\alpha}^* \rangle (\langle \boldsymbol{\beta}, \boldsymbol{\beta}^* |) | \Psi \rangle) \tag{5.72}$$

for bosons and

$$(|\mathbf{g}, \mathbf{g}^* \rangle \langle \mathbf{h}, \mathbf{h}^* |) | \Psi \rangle = |\mathbf{g}, \mathbf{g}^* \rangle (\langle \mathbf{h}, \mathbf{h}^* | \Psi \rangle) \tag{5.73}$$

for fermions. As we shall see, such projectors are important in representations of operators for both bosonic and fermionic systems. In the fermion case, the coherent-state projectors are Grassmann operators and are even operators, as they involve only displacement operators and the vacuum state projector $|0\rangle\langle 0|$. Note that the term 'projector' is sometimes applied in a more restrictive sense than is employed here, to refer only to projectors which are Hermitian with eigenvalues 0 or 1.

The Bargmann states are useful in deriving the trace of the fermionic projector $|\mathbf{g}, \mathbf{g}^* \rangle \langle \mathbf{h}, \mathbf{h}^* |$,

$$\mathrm{Tr}(|\mathbf{g}, \mathbf{g}^* \rangle \langle \mathbf{h}, \mathbf{h}^* |) = \exp \left[\mathbf{g} \cdot \mathbf{h}^* - \frac{1}{2} (\mathbf{g}^* \cdot \mathbf{g} + \mathbf{h}^* \cdot \mathbf{h}) \right]. \tag{5.74}$$

The proof is given in Appendix D. Note that this trace is *not* the same as the scalar product, which is

$$\langle \mathbf{h}, \mathbf{h}^* | \mathbf{g}, \mathbf{g}^* \rangle = \exp\left[\mathbf{h}^* \cdot \mathbf{g} - \frac{1}{2}(\mathbf{h}^* \cdot \mathbf{h} + \mathbf{g}^* \cdot \mathbf{g}) \right], \tag{5.75}$$

and for Grassmann variables $\mathbf{g} \cdot \mathbf{h}^* \neq \mathbf{h}^* \cdot \mathbf{g}$. Comparing these leads us to write the trace in the form

$$\mathrm{Tr}(|\mathbf{g}, \mathbf{g}^*\rangle\langle \mathbf{h}, \mathbf{h}^*|) = \langle \mathbf{h}, \mathbf{h}^* | -\mathbf{g}, -\mathbf{g}^* \rangle = \langle -\mathbf{h}, -\mathbf{h}^* | \mathbf{g}, \mathbf{g}^* \rangle. \tag{5.76}$$

This is a specific case of the general result for the trace of a fermion operator discussed later.

Corresponding results for the trace of the bosonic projector $|\boldsymbol{\alpha}, \boldsymbol{\alpha}^*\rangle\langle \boldsymbol{\beta}, \boldsymbol{\beta}^*|$ are

$$\begin{aligned} \mathrm{Tr}(|\boldsymbol{\alpha}, \boldsymbol{\alpha}^*\rangle\langle \boldsymbol{\beta}, \boldsymbol{\beta}^*|) &= \exp\left[\boldsymbol{\alpha} \cdot \boldsymbol{\beta}^* - \frac{1}{2}(\boldsymbol{\alpha}^* \cdot \boldsymbol{\alpha} + \boldsymbol{\beta}^* \cdot \boldsymbol{\beta}) \right] \\ &= \langle \boldsymbol{\beta}, \boldsymbol{\beta}^* | \boldsymbol{\alpha}, \boldsymbol{\alpha}^* \rangle. \end{aligned} \tag{5.77}$$

The proof is given in Appendix D. In this case the trace of the projector $|\boldsymbol{\alpha}, \boldsymbol{\alpha}^*\rangle\langle \boldsymbol{\beta}, \boldsymbol{\beta}^*|$ is the same as the scalar product $\langle \boldsymbol{\beta}, \boldsymbol{\beta}^* | \boldsymbol{\alpha}, \boldsymbol{\alpha}^* \rangle$, because $\boldsymbol{\alpha} \cdot \boldsymbol{\beta}^* = \boldsymbol{\beta}^* \cdot \boldsymbol{\alpha}$ for c-numbers.

In terms of Bargmann states, the traces of the projectors have simpler forms:

$$\mathrm{Tr_f}(|\mathbf{g}\rangle_{\mathrm{BB}}\langle \mathbf{h}|) = \exp(\mathbf{g} \cdot \mathbf{h}^*) = {}_{\mathrm{B}}\langle -\mathbf{h} | \mathbf{g} \rangle_{\mathrm{B}} = {}_{\mathrm{B}}\langle \mathbf{h} | -\mathbf{g} \rangle_{\mathrm{B}}, \tag{5.78}$$

$$\mathrm{Tr_b}(|\boldsymbol{\alpha}\rangle_{\mathrm{BB}}\langle \boldsymbol{\beta}|) = \exp(\boldsymbol{\alpha} \cdot \boldsymbol{\beta}^*) = {}_{\mathrm{B}}\langle \boldsymbol{\beta} | \boldsymbol{\alpha} \rangle_{\mathrm{B}}, \tag{5.79}$$

which involve the same exponential form for both bosons and fermions.

Normalised forms of the bosonic and fermionic projectors can be constructed as

$$\begin{aligned} \hat{\Lambda}_{\mathrm{b}}(\boldsymbol{\alpha}, \boldsymbol{\beta}^*) &= \frac{|\boldsymbol{\alpha}, \boldsymbol{\alpha}^*\rangle\langle \boldsymbol{\beta}, \boldsymbol{\beta}^*|}{\mathrm{Tr_b}(|\boldsymbol{\alpha}, \boldsymbol{\alpha}^*\rangle\langle \boldsymbol{\beta}, \boldsymbol{\beta}^*|)} = \frac{|\boldsymbol{\alpha}\rangle_{\mathrm{BB}}\langle \boldsymbol{\beta}|}{\mathrm{Tr_b}(|\boldsymbol{\alpha}\rangle_{\mathrm{BB}}\langle \boldsymbol{\beta}|)}, \\ \hat{\Lambda}_{\mathrm{f}}(\mathbf{g}, \mathbf{h}^*) &= \frac{|\mathbf{g}, \mathbf{g}^*\rangle\langle \mathbf{h}, \mathbf{h}^*|}{\mathrm{Tr_f}(|\mathbf{g}, \mathbf{g}^*\rangle\langle \mathbf{h}, \mathbf{h}^*|)} = \frac{|\mathbf{g}\rangle_{\mathrm{BB}}\langle \mathbf{h}|}{\mathrm{Tr_f}(|\mathbf{g}\rangle_{\mathrm{BB}}\langle \mathbf{h}|)}, \end{aligned} \tag{5.80}$$

where the projectors have also been expressed in terms of Bargmann states. These normalised projectors are constructed so that their trace is unity:

$$\mathrm{Tr}\ \hat{\Lambda}_{\mathrm{b}}(\boldsymbol{\alpha}, \boldsymbol{\beta}^*) = 1, \qquad \mathrm{Tr}\ \hat{\Lambda}_{\mathrm{f}}(\mathbf{g}, \mathbf{h}^*) = 1. \tag{5.81}$$

Also, as can be seen from the Bargmann state expressions, any dependence on α_i^*, β_i or g_i^*, h_i has been removed. Hence $\hat{\Lambda}_{\mathrm{b}}(\boldsymbol{\alpha}, \boldsymbol{\beta}^*)$ depends only on α_i and β_i^*, whilst $\hat{\Lambda}_{\mathrm{f}}(\mathbf{g}, \mathbf{h}^*)$ depends only on g_i and h_i^*. These features are useful in constructing canonical forms of the density operator in terms of phase space distribution functions.

5.6.3 Fock-State Projectors

The simplest of the coherent states is the vacuum state $|0\rangle$. From this we can build, by the application of creation operators, any of the Fock states. A Fock-state projector formed from the n-mode number states $|\nu_1, \cdots, \nu_n\rangle$ and $|\xi_1, \cdots, \xi_n\rangle$ has the form

$$|\nu_1, \cdots, \nu_n\rangle\langle\xi_1, \cdots, \xi_n| = \frac{\left(\hat{a}_1^\dagger\right)^{\nu_1}}{\sqrt{\nu_1!}} \cdots \frac{\left(\hat{a}_n^\dagger\right)^{\nu_n}}{\sqrt{\nu_n!}} |0\rangle\langle 0| \frac{(\hat{a}_n)^{\xi_n}}{\sqrt{\xi_n!}} \cdots \frac{(\hat{a}_1^\dagger)^{\xi_1}}{\sqrt{\xi_1!}} \tag{5.82}$$

for bosons and

$$|\nu_1, \cdots, \nu_n\rangle\langle\xi_1, \cdots, \xi_n| = \left(\hat{c}_1^\dagger\right)^{\nu_1} \cdots \left(\hat{c}_n^\dagger\right)^{\nu_n} |0\rangle\langle 0|(\hat{c}_n)^{\xi_n} \cdots (\hat{c}_1)^{\xi_1} \tag{5.83}$$

for fermions.

It is helpful to write these projectors purely in terms of the creation and annihilation operators. Then we can then use the properties of the coherent states to evaluate expectation values of the operators. The only part not as yet expressed in this way is the vacuum projector $|0\rangle\langle 0|$, but from (2.97) and (2.98) we can write this as

$$|0\rangle\langle 0| = \mathcal{N} \exp\left(-\sum_i \hat{a}_i^\dagger \hat{a}_i\right) = \mathcal{N}\left(e^{-\hat{\mathbf{a}}^\dagger \cdot \hat{\mathbf{a}}}\right), \tag{5.84}$$

$$|0\rangle\langle 0| = \mathcal{N} \exp\left(-\sum_i \hat{c}_i^\dagger \hat{c}_i\right) = \mathcal{N}\left(e^{-\hat{\mathbf{c}}^\dagger \cdot \hat{\mathbf{c}}}\right) \tag{5.85}$$

for bosons and for fermions, respectively. Here the symbol $\mathcal{N}$ (another notation involves pairs of dots) represents normal ordering, which is an instruction to order the operators so that all of the creation operators are to the left of all of the annihilation operators. For example, $\mathcal{N}[(\hat{a})^2(\hat{a}^\dagger)^4(\hat{a})] = (\hat{a}^\dagger)^4(\hat{a})^3$. This means, in particular, that the single-mode forms for our vacuum projectors are

$$\begin{aligned} |0\rangle\langle 0| &= \mathcal{N}(e^{-\hat{a}^\dagger \hat{a}}) = 1 - \hat{a}^\dagger \hat{a} + \frac{1}{2!}\left(\hat{a}^\dagger\right)^2 (\hat{a})^2 - \frac{1}{3!}\left(\hat{a}^\dagger\right)^3 (\hat{a})^3 + \cdots, \\ |0\rangle\langle 0| &= \mathcal{N}(e^{-\hat{c}^\dagger \hat{c}}) = 1 - \hat{c}^\dagger \hat{c}. \end{aligned} \tag{5.86}$$

It is straightforward to confirm that these operators are indeed the single-mode vacuum-state projectors. We can use our operator forms for the vacuum projectors to write the general Fock-state projectors as normally ordered operators in the form

$$|\nu_1, \cdots, \nu_n\rangle\langle\xi_1, \cdots, \xi_n| = \frac{(\hat{a}_1^\dagger)^{\nu_1}}{\sqrt{\nu_1!}} \cdots \frac{(\hat{a}_n^\dagger)^{\nu_n}}{\sqrt{\nu_n!}} \mathcal{N}(e^{-\hat{\mathbf{a}}^\dagger \cdot \hat{\mathbf{a}}}) \frac{(\hat{a}_n)^{\xi_n}}{\sqrt{\xi_n!}} \cdots \frac{(\hat{a}_1^\dagger)^{\xi_1}}{\sqrt{\xi_1!}}, \tag{5.87}$$

$$|\nu_1, \cdots, \nu_n\rangle\langle\xi_1, \cdots, \xi_n| = (\hat{c}_1^\dagger)^{\nu_1} \cdots (\hat{c}_n^\dagger)^{\nu_n} \mathcal{N}(e^{-\hat{\mathbf{c}}^\dagger \cdot \hat{\mathbf{c}}})(\hat{c}_n)^{\xi_n} \cdots (\hat{c}_1)^{\xi_1} \tag{5.88}$$

for bosons and fermions, respectively.

5.6.4 Representation of Operators

The coherent-state representation of state vectors described in Section 5.6.1 may be extended readily to operators in a form using the coherent-state matrix elements of the operator. The operator may be an observable, a field operator or a density operator representing the state of the bosons or fermions.

We start by noting that we can expand any operator $\hat{\Omega}$ in the basis of Fock states:

$$\hat{\Omega} = \sum_{\nu_1,\cdots,\nu_n} \sum_{\xi_1,\cdots,\xi_n} \Omega(\boldsymbol{\nu},\boldsymbol{\xi})|\nu_1,\cdots,\nu_n\rangle\langle\xi_1,\cdots,\xi_n|. \tag{5.89}$$

We can exploit the expressions for our Fock state operators in terms of normally ordered operators to write $\hat{\Omega}$ as a normally ordered function of the creation and annihilation operators:

$$\begin{aligned}\hat{\Omega}_{\mathrm{b}} &= \sum_{\nu_1,\cdots,\nu_n} \sum_{\xi_1,\cdots,\xi_n} \Omega(\boldsymbol{\nu},\boldsymbol{\xi})\frac{(\hat{a}_1^\dagger)^{\nu_1}}{\sqrt{\nu_1!}}\cdots\frac{(\hat{a}_n^\dagger)^{\nu_n}}{\sqrt{\nu_n!}}\mathcal{N}(e^{-\hat{\mathbf{a}}^\dagger\cdot\hat{\mathbf{a}}})\frac{(\hat{a}_n)^{\xi_n}}{\sqrt{\xi_n!}}\cdots\frac{(\hat{a}_1)^{\xi_1}}{\sqrt{\xi_1!}}\\ &= :\hat{\Omega}_{\mathrm{b}}\left(\hat{\mathbf{a}}^\dagger,\hat{\mathbf{a}}\right):\end{aligned} \tag{5.90}$$

for bosons and

$$\begin{aligned}\hat{\Omega}_{\mathrm{f}} &= \sum_{\nu_1,\cdots,\nu_n} \sum_{\xi_1,\cdots,\xi_n} \Omega(\boldsymbol{\nu},\boldsymbol{\xi})(\hat{c}_1^\dagger)^{\nu_1}\cdots(\hat{c}_n^\dagger)^{\nu_n}\mathcal{N}(e^{-\hat{\mathbf{c}}^\dagger\cdot\hat{\mathbf{c}}})(\hat{c}_n)^{\xi_n}\cdots(\hat{c}_1)^{\xi_1}\\ &= :\hat{\Omega}_{\mathrm{f}}\left(\hat{\mathbf{c}}^\dagger,\hat{\mathbf{c}}\right):\end{aligned} \tag{5.91}$$

for fermions.

The coherent-state matrix elements of the operators follow directly from the eigenvalue properties of the coherent states. For bosons and fermions we find, using results of the form $\langle\boldsymbol{\alpha},\boldsymbol{\alpha}^*|0\rangle = \exp(-\frac{1}{2}\boldsymbol{\alpha}^*\cdot\boldsymbol{\alpha})$ and $\langle\mathbf{g},\mathbf{g}^*|0\rangle = \exp(-\frac{1}{2}\mathbf{g}^*\cdot\mathbf{g})$,

$$\langle\boldsymbol{\alpha},\boldsymbol{\alpha}^*|\hat{\Omega}_{\mathrm{b}}|\boldsymbol{\beta},\boldsymbol{\beta}^*\rangle = e^{-(1/2)(\boldsymbol{\alpha}^*\cdot\boldsymbol{\alpha}+\boldsymbol{\beta}^*\cdot\boldsymbol{\beta})}\Omega_{\mathrm{b}}(\boldsymbol{\alpha}^*,\boldsymbol{\beta})) \tag{5.92}$$

and

$$\langle\mathbf{g},\mathbf{g}^*|\hat{\Omega}_{\mathrm{f}}|\mathbf{h},\mathbf{h}^*\rangle = e^{-(1/2)(\mathbf{g}^*\cdot\mathbf{g}+\mathbf{h}^*\cdot\mathbf{h})}\Omega_{\mathrm{f}}(\mathbf{g}^*,\mathbf{h}), \tag{5.93}$$

respectively. Here there is a simple relationship between the functions $\Omega_{\mathrm{b}}(\boldsymbol{\alpha}^*,\boldsymbol{\beta}))$, $\Omega_{\mathrm{f}}(\mathbf{g}^*,\mathbf{h})$ and the Fock state coefficients $\Omega(\boldsymbol{\nu},\boldsymbol{\xi})$. For bosons we have

$$\Omega_{\mathrm{b}}(\boldsymbol{\alpha}^*,\boldsymbol{\beta}) = \sum_{\nu_1,\cdots,\nu_n} \sum_{\xi_1,\cdots,\xi_n} \Omega(\boldsymbol{\nu},\boldsymbol{\xi})\frac{(\alpha_1^*)^{\nu_1}}{\sqrt{\nu_1!}}\cdots\frac{(\alpha_n^*)^{\nu_n}}{\sqrt{\nu_n!}}\frac{(\beta_n)^{\xi_n}}{\sqrt{\xi_n!}}\cdots\frac{(\beta_1)^{\xi_1}}{\sqrt{\nu_1!}}, \tag{5.94}$$

and for fermions

$$\Omega_{\mathrm{f}}(\mathbf{g}^*,\mathbf{h}) = \sum_{\nu_1,\cdots,\nu_n} \sum_{\xi_1,\cdots,\xi_n} \Omega(\boldsymbol{\nu},\boldsymbol{\xi})(g_1^*)^{\nu_1}\cdots(g_n^*)^{\nu_n}(h_n)^{\xi_n}\cdots(h_1)^{\xi_1}. \tag{5.95}$$

These functions are uniquely determined from the Fock state representations of the operators, and vice versa.

In terms of Bargmann states,

$$ {}_{\mathrm{B}}\langle \boldsymbol{\alpha}|\hat{\Omega}_{\mathrm{b}}|\boldsymbol{\beta}\rangle_{\mathrm{B}} = \Omega_{\mathrm{b}}(\boldsymbol{\alpha}^*, \boldsymbol{\beta}) \tag{5.96} $$

and

$$ {}_{\mathrm{B}}\langle \mathbf{g}|\hat{\Omega}_{\mathrm{f}}|\mathbf{h}\rangle_{\mathrm{B}} = \Omega_{\mathrm{f}}(\mathbf{g}^*, \mathbf{h}). \tag{5.97} $$

From the matrix elements (5.97), it is immediately clear that the representation of an even (or odd) operator is an even (or odd) Grassmann function. For example, a physical density operator will be a mixture of states with fixed particle number. It follows that this operator will be a sum of operators each of which has the same number of creation and annihilation operators. This then means that the coherent-state representation of the density operator for any physical fermion state will be an even Grassmann function.

5.6.5 Equivalence of Operators

It is useful to have a simple method for establishing the equivalence of two operators, and the Bargmann states provide this. As we have seen, (5.96) and (5.97) constitute a one-to-one map between the operators $\hat{\Omega}_{\mathrm{b}}$ and $\hat{\Omega}_{\mathrm{f}}$ and their representations $\Omega_{\mathrm{b}}(\boldsymbol{\alpha}^*, \boldsymbol{\beta})$ and $\Omega_{\mathrm{f}}(\mathbf{g}^*, \mathbf{h})$. There is, in fact, a simpler one-to-one map in the form of the *diagonal* Bargmann-state matrix elements

$$ \begin{aligned} {}_{\mathrm{B}}\langle \boldsymbol{\alpha}|\hat{\Omega}_{\mathrm{b}}|\boldsymbol{\alpha}\rangle_{\mathrm{B}} &= \Omega_{\mathrm{b}}(\boldsymbol{\alpha}^*, \boldsymbol{\alpha}) = \sum_{\nu_1,\cdots,\nu_n}\sum_{\xi_1,\cdots,\xi_n} \Omega_{\mathrm{b}}(\boldsymbol{\nu},\boldsymbol{\xi}) \frac{(\alpha_1^*)^{\nu_1}}{\sqrt{\nu_1!}} \cdots \frac{(\alpha_n^*)^{\nu_n}}{\sqrt{\nu_n!}} \frac{(\alpha_n)^{\xi_n}}{\sqrt{\xi_n!}} \cdots \frac{(\alpha_1)^{\xi_1}}{\sqrt{\nu_1!}}, \\ {}_{\mathrm{B}}\langle \mathbf{g}|\hat{\Omega}_{\mathrm{f}}|\mathbf{g}\rangle_{\mathrm{B}} &= \Omega_{\mathrm{f}}(\mathbf{g}^*, \mathbf{g}) = \sum_{\nu_1,\cdots,\nu_n}\sum_{\xi_1,\cdots,\xi_n} \Omega_{\mathrm{f}}(\boldsymbol{\nu},\boldsymbol{\xi}) (g_1^*)^{\nu_1} \cdots (g_n^*)^{\nu_n} (g_n)^{\xi_n} \cdots (g_1)^{\xi_1}. \end{aligned} \tag{5.98} $$

Now $\Omega_{\mathrm{b}}(\boldsymbol{\alpha}^*, \boldsymbol{\alpha})$ is a function of the c-number variables $\boldsymbol{\alpha}, \boldsymbol{\alpha}^*$, and $\Omega_{\mathrm{f}}(\mathbf{g}^*, \mathbf{g})$ is a Grassmann function of $\mathbf{g}, \mathbf{g}^*$. If these functions are known, then they uniquely determine the Fock space quantities $\Omega_{\mathrm{b}}(\boldsymbol{\nu}, \boldsymbol{\xi})$ and $\Omega_{\mathrm{f}}(\boldsymbol{\nu}, \boldsymbol{\xi})$ and hence they uniquely specify the operators $\hat{\Omega}_{\mathrm{b}}$ and $\hat{\Omega}_{\mathrm{f}}$. It follows that if two operators have the same diagonal Bargmann-state matrix elements for all Bargmann states then they are identical. For bosons this is because $|\boldsymbol{\alpha}\rangle_{\mathrm{B}}$ depends only on $\boldsymbol{\alpha}$ and ${}_{\mathrm{B}}\langle \boldsymbol{\alpha}|$ depends only on $\boldsymbol{\alpha}^*$ and these are independent variables, with similar considerations applying for fermions. Hence we have

$$ {}_{\mathrm{B}}\langle \boldsymbol{\alpha}|\hat{\Omega}_{\mathrm{b}}|\boldsymbol{\alpha}\rangle_{\mathrm{B}} = {}_{\mathrm{B}}\langle \boldsymbol{\alpha}|\hat{\Theta}_{\mathrm{b}}|\boldsymbol{\alpha}\rangle_{\mathrm{B}} \quad \forall \boldsymbol{\alpha} \qquad \Rightarrow \quad \hat{\Omega}_{\mathrm{b}} = \hat{\Theta}_{\mathrm{b}}, \tag{5.99} $$

$$ {}_{\mathrm{B}}\langle \mathbf{g}|\hat{\Omega}_{\mathrm{f}}|\mathbf{g}\rangle_{\mathrm{B}} = {}_{\mathrm{B}}\langle \mathbf{g}|\hat{\Theta}_{\mathrm{f}}|\mathbf{g}\rangle_{\mathrm{B}} \quad \forall \mathbf{g} \qquad \Rightarrow \quad \hat{\Omega}_{\mathrm{f}} = \hat{\Theta}_{\mathrm{f}}, \tag{5.100} $$

for bosons and fermions in turn.

5.7 Canonical Forms for States and Operators

The completeness of the coherent states enables us to write states and operators as superpositions of coherent states or projectors based on coherent states, the so-called *canonical forms.* For bosons these superpositions are in the form of integrals over the complex plane, and for fermions they are Grassmann integrals.

5.7.1 Fermions

We can use the resolution of the identity in terms of coherent states (5.38) to write a general N-fermion pure state as

$$\begin{aligned} |\Psi\rangle_{N\mathrm{f}} &= \int d^2\mathbf{g}\ |\mathbf{g},\mathbf{g}^*\rangle\langle\mathbf{g},\mathbf{g}^*|\Psi\rangle_{N\mathrm{f}} \\ &= \int d^2\mathbf{g}\ |\mathbf{g},\mathbf{g}^*\rangle \sum_{\nu_1,\cdots,\nu_n} B_N(\nu)(g_1^*)^{\nu_1}\cdots(g_n^*)^{\nu_n} e^{-(1/2)\ \mathbf{g}^*\cdot\mathbf{g}}. \end{aligned} \tag{5.101}$$

Alternatively, if desired, we can write the state as a superposition of Bargmann states

$$|\Psi\rangle_{N\mathrm{f}} = \int d^2\mathbf{g} \frac{|\mathbf{g}\rangle_{\mathrm{B}}}{{}_{\mathrm{B}}\langle\mathbf{g}|\mathbf{g}\rangle_{\mathrm{B}}} \sum_{\nu_1,\cdots,\nu_n} B_N(\nu)(g_1^*)^{\nu_1}\cdots(g_n^*)^{\nu_n}. \tag{5.102}$$

Similarly, we can write any fermionic operator as a linear combination of coherent-state projectors by pre- and post-multiplying by the identity:

$$\begin{aligned} \hat{\Omega}_{\mathrm{f}} &= \int d^2\mathbf{g}\ |\mathbf{g},\mathbf{g}^*\rangle\langle\mathbf{g},\mathbf{g}^*|\hat{\Omega}_{\mathrm{f}} \int d^2\mathbf{h}\ |\mathbf{h},\mathbf{h}^*\rangle\langle\mathbf{h},\mathbf{h}^*| \\ &= \int d^2\mathbf{g}\ d^2\mathbf{h}\ |\mathbf{g},\mathbf{g}^*\rangle\langle\mathbf{h},\mathbf{h}^*| e^{-(1/2)(\mathbf{g}^*\cdot\mathbf{g}+\mathbf{h}^*\cdot\mathbf{h})}\Omega_{\mathrm{f}}(\mathbf{g}^*,\mathbf{h}). \end{aligned} \tag{5.103}$$

Here we have used the facts that the differentials d^2h_i commute with $\hat{\Omega}_{\mathrm{f}}$ and with the even Grassmann quantities $\langle\mathbf{g},\mathbf{g}^*|$ and $|\mathbf{g},\mathbf{g}^*\rangle$ and also that the Grassmann function $\langle\mathbf{g},\mathbf{g}^*|\hat{\Omega}_{\mathrm{f}}|\mathbf{h},\ \mathbf{h}^*\rangle$ commutes with the even quantity $\langle\mathbf{h},\mathbf{h}^*|$. If Bargmann states are used, then the fermionic operator is

$$\begin{aligned} \hat{\Omega}_{\mathrm{f}} &= \int d^2\mathbf{g}\ e^{-\mathbf{g}^*\cdot\mathbf{g}}|\mathbf{g}\rangle_{\mathrm{BB}}\langle\mathbf{g}|\hat{\Omega}_{\mathrm{f}} \int d^2\mathbf{h}\ e^{-\mathbf{h}^*\cdot\mathbf{h}}|\mathbf{h}\rangle_{\mathrm{BB}}\langle\mathbf{h}| \\ &= \int d^2\mathbf{g}\ d^2\mathbf{h}\ |\mathbf{g}\rangle_{\mathrm{BB}}\langle\mathbf{h}|e^{-\mathbf{g}^*\cdot\mathbf{g}-\mathbf{h}^*\cdot\mathbf{h}}\Omega_{\mathrm{f}}(\mathbf{g}^*,\mathbf{h}). \end{aligned} \tag{5.104}$$

We arrive at the canonical form for a fermionic operator $\hat{\Omega}_{\mathrm{f}}$ by using the Bargmann states and making the change of Grassmann variables $\mathbf{h} \to \mathbf{g}^{+*}$ so that

$$\hat{\Omega}_{\mathrm{f}} = \int d^2\mathbf{g}\ d^2\mathbf{g}^{+*}\ |\mathbf{g}\rangle_{\mathrm{BB}}\langle\mathbf{g}^{+*}|e^{-\mathbf{g}^*\cdot\mathbf{g}-\mathbf{g}^{+}\cdot\mathbf{g}^{+*}}\Omega_{\mathrm{f}}(\mathbf{g}^*,\mathbf{g}^{+*}), \tag{5.105}$$

where $d^2\mathbf{g} = \prod_i dg_i^* \; dg_i$ and $d^2\mathbf{g}^{+*} = \prod_i dg_i^+ \; dg_i^{+*}$. We can rewrite this in terms of the normalised operator

$$\Lambda_{\mathrm{f}}(\mathbf{g}.\mathbf{g}^+) = \frac{|\mathbf{g}\rangle_{\mathrm{BB}}\langle\mathbf{g}^{+*}|}{\mathrm{Tr}(|\mathbf{g}\rangle_{\mathrm{BB}}\langle\mathbf{g}^{+*}|)} = e^{-\mathbf{g}\cdot\mathbf{g}^+}\,|\mathbf{g}\rangle_{\mathrm{BB}}\langle\mathbf{g}^{+*}| \tag{5.106}$$

as (the convention is $d\mathbf{g}^+ \equiv dg_1^+ \cdots dg_n^+ = \prod_i dg_i^+$ and $d\mathbf{g} \equiv dg_n \cdots dg_1 = \prod_i dg_i$)

$$\hat{\Omega}_{\mathrm{f}} = \int d\mathbf{g}^+ \; d\mathbf{g}\; \Omega_{\mathrm{f\,canon}}(\mathbf{g},\mathbf{g}^+)\hat{\Lambda}(\mathbf{g},\mathbf{g}^+) = \int d\mathbf{g}^+ \; d\mathbf{g}\; \hat{\Lambda}(\mathbf{g},\mathbf{g}^+)\Omega_{\mathrm{f\,canon}}(\mathbf{g},\mathbf{g}^+), \tag{5.107}$$

where (with $d\mathbf{g}^{+*} \equiv dg_1^{+*} \cdots dg_n^{+*} = \prod_i dg_i^{+*}$ and $d\mathbf{g}^* \equiv dg_n^* \cdots dg_1^* = \prod_i dg_i^*$)

$$\begin{aligned}\Omega_{\mathrm{f\,canon}}(\mathbf{g},\mathbf{g}^+) &= \int d\mathbf{g}^{+*} d\mathbf{g}^* \; e^{\mathbf{g}\cdot\mathbf{g}^* + \mathbf{g}^{+*}\cdot\mathbf{g}^+ + \mathbf{g}\cdot\mathbf{g}^+}\; {}_{\mathrm{B}}\langle\mathbf{g}|\hat{\Omega}_{\mathrm{f}}|\mathbf{g}^{+*}\rangle_{\mathrm{B}} \\ &= \int d\mathbf{g}^{+*} d\mathbf{g}^* \; e^{\mathbf{g}\cdot\mathbf{g}^* + \mathbf{g}^{+*}\cdot\mathbf{g}^+ + \mathbf{g}\cdot\mathbf{g}^+}\,\Omega_{\mathrm{f}}(\mathbf{g}^*,\mathbf{g}^{+*})\end{aligned} \tag{5.108}$$

is the canonical representation of the fermionic operator $\hat{\Omega}_{\mathrm{f}}$. Note that the above representation is valid for any fermionic operator, including those, such as $\hat{D}_{\mathrm{f}}(\mathbf{h},\mathbf{h}^*)$, that include Grassmann variables. It applies, moreover, for operators for which ${}_{\mathrm{B}}\langle\mathbf{g}|\hat{\Omega}_{\mathrm{f}}|\mathbf{g}^{+*}\rangle_{\mathrm{B}}$ is an even, odd or mixed Grassmann function.

5.7.2 Bosons

A somewhat similar canonical form exists for bosonic operators. We have

$$\begin{aligned}\hat{\Omega}_{\mathrm{b}} &= \int d^2\boldsymbol{\alpha}^+ d^2\boldsymbol{\alpha}\; \Omega_{\mathrm{b\,canon}}(\boldsymbol{\alpha},\boldsymbol{\alpha}^+,\boldsymbol{\alpha}^*,\boldsymbol{\alpha}^{+*})\hat{\Lambda}_{\mathrm{b}}(\boldsymbol{\alpha},\boldsymbol{\alpha}^+) \\ &= \int d^2\boldsymbol{\alpha}^+ d^2\boldsymbol{\alpha}\; \hat{\Lambda}_{\mathrm{b}}(\boldsymbol{\alpha},\boldsymbol{\alpha}^+)\Omega_{\mathrm{b\,canon}}(\boldsymbol{\alpha},\boldsymbol{\alpha}^+,\boldsymbol{\alpha}^*,\boldsymbol{\alpha}^{+*}),\end{aligned} \tag{5.109}$$

where

$$\hat{\Lambda}_{\mathrm{b}}(\boldsymbol{\alpha},\boldsymbol{\alpha}^+) = \frac{|\boldsymbol{\alpha}\rangle_{\mathrm{BB}}\langle\boldsymbol{\alpha}^{+*}|}{\mathrm{Tr}(|\boldsymbol{\alpha}\rangle_{\mathrm{BB}}\langle\boldsymbol{\alpha}^{+*}|} = |\boldsymbol{\alpha}\rangle_{\mathrm{BB}}\langle\boldsymbol{\alpha}^{+*}|\; e^{-\boldsymbol{\alpha}\cdot\boldsymbol{\alpha}^+} \tag{5.110}$$

is a normalised projector. Remarkably, there is no unique form for $\Omega_{\mathrm{b\,canon}}$ and we are free to choose the canonical form, which we set to be

$$\Omega_{\mathrm{b\,canon}}(\boldsymbol{\alpha},\boldsymbol{\alpha}^+,\boldsymbol{\alpha}^*,\boldsymbol{\alpha}^{+*}) = \left(\frac{1}{4\pi^2}\right)^n e^{-(1/2)(\boldsymbol{\alpha}\cdot\boldsymbol{\alpha}^* + \boldsymbol{\alpha}^{+*}\cdot\boldsymbol{\alpha}^+)}\,{}_{\mathrm{B}}\langle\bar{\boldsymbol{\alpha}}|\hat{\Omega}_{\mathrm{b}}|\bar{\boldsymbol{\alpha}}\rangle_{\mathrm{B}}, \tag{5.111}$$

where $\bar{\boldsymbol{\alpha}} = (\boldsymbol{\alpha} + \boldsymbol{\alpha}^{+*})/2$. It is straightforward to show that this canonical representation provides a true representation of the bosonic operator $\hat{\Omega}_{\mathrm{b}}$. As we have seen, any two bosonic operators will be equivalent if their diagonal matrix elements for all Bargmann states are equal. If we introduce the new variable $\boldsymbol{\alpha}_{\mathrm{d}} = (\boldsymbol{\alpha} - \boldsymbol{\alpha}^{+*})/2$ and use this together with $\bar{\boldsymbol{\alpha}}$, then we find

$$
\begin{aligned}
&{}_{\mathrm{B}}\langle\boldsymbol{\beta}|\left(\int d^2\boldsymbol{\alpha}^+ d^2\boldsymbol{\alpha}\ \hat{\Lambda}_{\mathrm{b}}\Omega_{\mathrm{b\,canon}}\right)|\boldsymbol{\beta}\rangle_{\mathrm{B}} \\
&= \int \frac{d^2\bar{\boldsymbol{\alpha}}\ d^2\boldsymbol{\alpha}_{\mathrm{d}}}{\pi^{2n}}\ e^{\boldsymbol{\alpha}_{\mathrm{d}}\cdot(\boldsymbol{\beta}^*-\bar{\boldsymbol{\alpha}}^*)-\boldsymbol{\alpha}_{\mathrm{d}}^*\cdot(\boldsymbol{\beta}-\bar{\boldsymbol{\alpha}})} \times e^{\boldsymbol{\beta}^*\cdot\bar{\boldsymbol{\alpha}}+\boldsymbol{\beta}\cdot\bar{\boldsymbol{\alpha}}^*-2\bar{\boldsymbol{\alpha}}^*\cdot\bar{\boldsymbol{\alpha}}}\,{}_{\mathrm{B}}\langle\bar{\boldsymbol{\alpha}}|\hat{\Omega}_{\mathrm{b}}|\bar{\boldsymbol{\alpha}}\rangle_{\mathrm{B}} \\
&= \int d^2\bar{\boldsymbol{\alpha}}\ \delta^{(2n)}(\bar{\boldsymbol{\alpha}}-\boldsymbol{\beta})e^{\boldsymbol{\beta}^*\cdot\bar{\boldsymbol{\alpha}}+\boldsymbol{\beta}\cdot\bar{\boldsymbol{\alpha}}^*-2\bar{\boldsymbol{\alpha}}^*\cdot\bar{\boldsymbol{\alpha}}}\,{}_{\mathrm{B}}\langle\bar{\boldsymbol{\alpha}}|\hat{\Omega}_{\mathrm{b}}|\bar{\boldsymbol{\alpha}}\rangle_{\mathrm{B}} \\
&= {}_{\mathrm{B}}\langle\boldsymbol{\beta}|\hat{\Omega}_{\mathrm{b}}|\boldsymbol{\beta}\rangle_{\mathrm{B}},
\end{aligned}
\tag{5.112}
$$

and the validity of our canonical representation is established. An alternative longer proof is given by Drummond and Gardiner [38].

Our canonical form has a number of advantageous features, which we come to in the following paragraph. The bosonic canonical form is different in form from that for fermions, principally because $\Omega_{\mathrm{b\,canon}}(\boldsymbol{\alpha}, \boldsymbol{\alpha}^+, \boldsymbol{\alpha}^*, \boldsymbol{\alpha}^{+*})$ depends on $\boldsymbol{\alpha}^*$ and $\boldsymbol{\alpha}^{+*}$ as well as $\boldsymbol{\alpha}$ and $\boldsymbol{\alpha}^+$. This means that the canonical representation $\Omega_{\mathrm{b\,canon}}(\boldsymbol{\alpha}, \boldsymbol{\alpha}^+, \boldsymbol{\alpha}^*, \boldsymbol{\alpha}^{+*})$ is *not* an analytic function of $\boldsymbol{\alpha}$ and $\boldsymbol{\alpha}^+$. This non-analyticity does not prove to be a problem, as the four sets of complex variables $\boldsymbol{\alpha}^*$, $\boldsymbol{\alpha}^{+*}$, $\boldsymbol{\alpha}$ and $\boldsymbol{\alpha}^+$ depend on the real variables α_{ix}, α_{iy}, α_{ix}^+, α_{iy}^+ and it is these that are integrated over for each mode. Moreover, derivatives of $\Omega_{\mathrm{b\,canon}}(\boldsymbol{\alpha}, \boldsymbol{\alpha}^+, \boldsymbol{\alpha}^*, \boldsymbol{\alpha}^{+*})$ with respect to α_{ix}, α_{iy}, α_{ix}^+ and α_{iy}^+ can still be defined.

The advantage of our choice of the canonical form for the bosonic operator $\hat{\Omega}_{\mathrm{b}}$ lies in the fact that its properties mirror some of those of the operator itself. This is because it involves diagonal matrix elements of the operator with the bosonic coherent states, multiplied by a positive exponential factor. The properties include the fact that if $\hat{\Omega}_{\mathrm{b}}$ is a Hermitian operator, then $\Omega_{\mathrm{b\,canon}}(\boldsymbol{\alpha}, \boldsymbol{\alpha}^+, \boldsymbol{\alpha}^*, \boldsymbol{\alpha}^{+*})$ is *real*. Further, if $\hat{\Omega}_{\mathrm{b}}$ is a positive Hermitian operator (such as the density operator), then $\Omega_{\mathrm{b\,canon}}(\alpha, \alpha^+, \alpha^*, \alpha^{+*})$ is *positive*. It is because of this feature that the canonical form for the density operator is central to defining the positive P distribution function in the bosonic case. Finally, it should be noted that as the coherent states form an overcomplete set of basis vectors, their representation of the bosonic operator is *not* unique.

For bosonic operators, the Bargmann states may be used to define another representation of an operator, called the *R representation*. This is an analytic function of $\boldsymbol{\alpha}, \boldsymbol{\alpha}^+$, as the Bargmann vectors ${}_{\mathrm{B}}\langle\boldsymbol{\alpha}^*|$ and $|\boldsymbol{\alpha}^+\rangle_{\mathrm{B}}$ are given by convergent power series in $\boldsymbol{\alpha}$ and $\boldsymbol{\alpha}^+$, respectively, with no dependence on $\boldsymbol{\alpha}^*$ or $\boldsymbol{\alpha}^{+*}$:

$$
R(\boldsymbol{\alpha}, \boldsymbol{\alpha}^+) = {}_{\mathrm{B}}\langle\boldsymbol{\alpha}^*|\hat{\Omega}_{\mathrm{b}}|\boldsymbol{\alpha}^+\rangle_{\mathrm{B}}. \tag{5.113}
$$

5.8 Evaluating the Trace of an Operator

It is useful to be able to evaluate the traces of bosonic and fermionic operators in terms of their coherent-state representations. The trace is the sum of the diagonal elements in any basis. In the Fock basis, for example, we have

$$\mathrm{Tr}(\hat{\Omega}) = \sum_{\nu_1,\cdots,\nu_n} \langle \nu_1,\cdots,\nu_n|\hat{\Omega}|\nu_1,\cdots,\nu_n\rangle. \tag{5.114}$$

Evaluating the trace for bosonic operators using coherent states turns out to be straightforward, but for fermions there are mathematical subtleties and potential traps.

5.8.1 Bosons

We can express the trace of a bosonic operator $\hat{\Omega}_\mathrm{b}$ in terms of the coherent states by using the completeness property (5.37) to write

$$\begin{aligned}\mathrm{Tr}(\hat{\Omega}_\mathrm{b}) &= \sum_{\nu_1,\cdots,\nu_n} \langle \nu_1,\cdots,\nu_n|\hat{\Omega}_\mathrm{b}\left(\int \frac{d^2\boldsymbol{\alpha}}{\pi^n}|\boldsymbol{\alpha}\rangle\langle\boldsymbol{\alpha}|\right)|\nu_1,\cdots,\nu_n\rangle \\ &= \int \frac{d^2\boldsymbol{\alpha}}{\pi^n}\langle\boldsymbol{\alpha}| \sum_{\nu_1,\cdots,\nu_n} |\nu_1,\cdots,\nu_n\rangle\langle\nu_1,\cdots,\nu_n|\hat{\Omega}_\mathrm{b}|\boldsymbol{\alpha}\rangle \\ &= \int \frac{d^2\boldsymbol{\alpha}}{\pi^n}\langle\boldsymbol{\alpha}|\hat{\Omega}_\mathrm{b}|\boldsymbol{\alpha}\rangle,\end{aligned} \tag{5.115}$$

where we have used the fact that $\langle\boldsymbol{\alpha}|\nu_1,\cdots,\nu_n\rangle$ and $\langle\nu_1,\cdots,\nu_n|\hat{\Omega}_\mathrm{b}|\boldsymbol{\alpha}\rangle$ are c-numbers. We see that the trace of a bosonic operator is given simply by a suitably weighted integral over the diagonal coherent-state matrix elements. As simple examples, we note the following traces involving coherent-state projectors:

$$\begin{aligned}\mathrm{Tr}(|\boldsymbol{\alpha},\boldsymbol{\alpha}^*\rangle\langle\boldsymbol{\beta},\boldsymbol{\beta}^*|) &= \langle\boldsymbol{\beta},\boldsymbol{\beta}^*|\boldsymbol{\alpha},\boldsymbol{\alpha}^*\rangle, \\ \mathrm{Tr}(|\boldsymbol{\alpha},\boldsymbol{\alpha}^*\rangle\langle\boldsymbol{\beta},\boldsymbol{\beta}^*|\hat{\Omega}_\mathrm{b}) &= \langle\boldsymbol{\beta},\boldsymbol{\beta}^*|\hat{\Omega}_\mathrm{b}|\boldsymbol{\alpha},\boldsymbol{\alpha}^*\rangle.\end{aligned} \tag{5.116}$$

If we have the trace of a product of bosonic operators, then the value of the trace will be unchanged if we perform a cyclic permutation of the operators, so that, for example,

$$\mathrm{Tr}(\hat{\Omega}_\mathrm{b}\hat{\Theta}_\mathrm{b}\hat{\Xi}_\mathrm{b}) = \mathrm{Tr}(\hat{\Theta}_\mathrm{b}\hat{\Xi}_\mathrm{b}\hat{\Omega}_\mathrm{b}) = \mathrm{Tr}(\hat{\Xi}_\mathrm{b}\hat{\Omega}_\mathrm{b}\hat{\Theta}_\mathrm{b}). \tag{5.117}$$

It is straightforward to verify this by inserting the resolution of the identity between each of the pairs of operators. This cyclic property holds generally for bosonic operators as long as the traces are finite. It is not the case, for example, that the trace of $\hat{a}^\dagger\hat{a}$ is equal to the trace of $\hat{a}\hat{a}^\dagger$; the traces of both operators are infinite, and the cyclic-permutation rule does not hold in this case.

5.8.2 Fermions

For fermions, we can proceed as above but with the additional complication that we cannot simply commute Grassmann functions. As a preliminary to deriving a coherent-state trace rule for fermions, let us consider again the overlap between a coherent state and a Fock state, $\langle \mathbf{g}, \mathbf{g}^* | \nu_1, \cdots, \nu_N \rangle$. We recall that this Grassmann function is an even function if the number of fermions N is even, and an odd function if N is odd (where $N = \sum_i \nu_i$). Further, if N is *even*, then changing the signs of the Grassmann variables does not change the Grassmann function. If N is odd, however, then changing the signs of the Grassmann variables changes the sign of the Grassmann function. We can combine both of these into the single expression

$$\langle \mathbf{g}, \mathbf{g}^* | \Psi \rangle_{N\mathrm{f}} = (-1)^N \langle -\mathbf{g}, -\mathbf{g}^* | \Psi \rangle_{N\mathrm{f}}. \tag{5.118}$$

To begin with, let us start by computing the trace of a fermionic operator $\hat{\Omega}_\mathrm{f}$ that does not contain any Grassmann variables. To find a coherent-state-based expression for the trace of a fermion operator, we begin with the completeness relation for the fermion coherent states (5.38) to write

$$\begin{aligned} \mathrm{Tr}(\hat{\Omega}_\mathrm{f}) &= \sum_{\nu_1, \cdots, \nu_n} \langle \nu_1, \cdots, \nu_n | \int d^2\mathbf{g} \; |\mathbf{g}, \mathbf{g}^* \rangle \langle \mathbf{g}, \mathbf{g}^* | \hat{\Omega}_\mathrm{f} | \nu_1, \cdots, \nu_n \rangle \\ &= \int d^2\mathbf{g} \sum_{\nu_1, \cdots, \nu_n} \langle \nu_1, \cdots, \nu_n | \mathbf{g}, \mathbf{g}^* \rangle \langle \mathbf{g}, \mathbf{g}^* | \hat{\Omega}_\mathrm{f} | \nu_1, \cdots, \nu_n \rangle, \end{aligned} \tag{5.119}$$

where we have used the fact that $d^2\mathbf{g}$ contains an even number of Grassmann variables to commute it to the left. Let us consider separately the even and odd parts of our operator, $\hat{\Omega}_\mathrm{f}^\mathrm{E}$ and $\hat{\Omega}_\mathrm{f}^\mathrm{O}$. An odd operator has terms with only odd numbers of creation and annihilation operators. It follows directly from the trace (5.114) that

$$\mathrm{Tr}(\hat{\Omega}_\mathrm{f}^\mathrm{O}) = 0, \tag{5.120}$$

and we need consider only the trace of the even part, $\hat{\Omega}_\mathrm{f}^\mathrm{E}$. For even-particle-number Fock states, $\langle \nu_1, \cdots, \nu_n | \mathbf{g}, \mathbf{g}^* \rangle$ and $\langle \mathbf{g}, \mathbf{g}^* | \hat{\Omega}_\mathrm{f}^\mathrm{E} | \nu_1, \cdots, \nu_n \rangle$ are both even Grassmann functions and so will commute. For odd-particle-number Fock states, $\langle \nu_1, \cdots, \nu_n | \mathbf{g}, \mathbf{g}^* \rangle$ and $\langle \nu_1, \cdots, \nu_n | \hat{\Omega}_\mathrm{f}^\mathrm{E} | \mathbf{g}, \mathbf{g}^* \rangle$ are both odd Grassmann functions and so will *anticommute*. Combining these, we have

$$\begin{aligned} \langle \nu_1, \cdots, \nu_n | \mathbf{g}, \mathbf{g}^* \rangle \langle \mathbf{g}, \mathbf{g}^* | \hat{\Omega}_\mathrm{f}^\mathrm{E} | \nu_1, \cdots, \nu_n \rangle &= (-1)^N \langle \mathbf{g}, \mathbf{g}^* | \hat{\Omega}_\mathrm{f}^\mathrm{E} | \nu_1, \cdots, \nu_n \rangle \langle \nu_1, \cdots, \nu_n | \mathbf{g}, \mathbf{g}^* \rangle \\ &= \langle \mathbf{g}, \mathbf{g}^* | \hat{\Omega}_\mathrm{f}^\mathrm{E} | \nu_1, \cdots, \nu_n \rangle \langle \nu_1, \cdots, \nu_n | -\mathbf{g}, -\mathbf{g}^* \rangle. \end{aligned} \tag{5.121}$$

It follows, therefore, that our trace formula (5.119) may be written as

$$\begin{aligned}\mathrm{Tr}(\hat{\Omega}_{\mathrm{f}}) &= \int d^2\mathbf{g}\ \langle \mathbf{g}, \mathbf{g}^* |\ \hat{\Omega}_{\mathrm{f}}^{\mathrm{E}} \sum_{\nu_1,\cdots,\nu_n} |\nu_1, \cdots, \nu_n\rangle\langle \nu_1, \cdots, \nu_n| - \mathbf{g}, -\mathbf{g}^*\rangle \\ &= \int d^2\mathbf{g}\ \langle \mathbf{g}, \mathbf{g}^* |\hat{\Omega}_{\mathrm{f}}| - \mathbf{g}, -\mathbf{g}^*\rangle \\ &= \int d^2\mathbf{g}\ \langle -\mathbf{g}, -\mathbf{g}^* |\hat{\Omega}_{\mathrm{f}}| \mathbf{g}, \mathbf{g}^*\rangle. \end{aligned} \tag{5.122}$$

The antisymmetry properties of the fermionic operators are reflected in the fact the coherent-state matrix elements for the states $|\mathbf{g}, \mathbf{g}^*\rangle$ and $| - \mathbf{g}, -\mathbf{g}^*\rangle$ occur in the trace, rather than the diagonal matrix elements as found in the formula for the bosonic trace.

Note that the above proof of these results was given for ordinary fermion operators. It turns out, however, that a modified version of these results can still be obtained when $\hat{\Omega}_{\mathrm{f}}^{\mathrm{E}}(\mathbf{k})$ is an even Grassmann operator, depending on a set of Grassmann variables $k_i, \cdots, k_m$. For such an even operator, we find

$$\begin{aligned}\mathrm{Tr}\left[\hat{\Omega}_{\mathrm{f}}^{\mathrm{E}}(\mathbf{k})\right] &= \int d^2\mathbf{g}\ \langle \mathbf{g}, \mathbf{g}^* |\hat{\Omega}_{\mathrm{f}}^{\mathrm{E}}(\mathbf{k})| - \mathbf{g}, -\mathbf{g}^*\rangle \\ &= \int d^2\mathbf{g}\ \langle -\mathbf{g}, -\mathbf{g}^* |\hat{\Omega}_{\mathrm{f}}^{\mathrm{E}}(-\mathbf{k})| \mathbf{g}, \mathbf{g}^*\rangle. \end{aligned} \tag{5.123}$$

For the case of odd Grassmann operators, we can show that

$$\mathrm{Tr}\left[\hat{\Omega}_{\mathrm{f}}^{\mathrm{O}}(\mathbf{k})\right] = \int d^2\mathbf{g}\ \langle \mathbf{g}, \mathbf{g}^* |\hat{\Omega}_{\mathrm{f}}^{\mathrm{O}}(\mathbf{k})| \mathbf{g}, \mathbf{g}^*\rangle, \tag{5.124}$$

where now there are no sign changes involved. Note that both sides are zero for odd non-Grassmann operators. As an example of our trace formulae, we can evaluate the trace of a coherent-state projector, which is, of course, an even Grassmann operator. We find

$$\begin{aligned}\mathrm{Tr}(|\mathbf{g}, \mathbf{g}^*\rangle\langle \mathbf{h}, \mathbf{h}^*|) &= \int d^2\mathbf{k}\ \langle \mathbf{k}, \mathbf{k}^* | \mathbf{g}, \mathbf{g}^*\rangle\langle \mathbf{h}, \mathbf{h}^* | - \mathbf{k}, -\mathbf{k}^*\rangle \\ &= \int d^2\mathbf{k}\ \langle -\mathbf{h}, -\mathbf{h}^* | \mathbf{k}, \mathbf{k}^*\rangle\langle \mathbf{k}, \mathbf{k}^* | \mathbf{g}, \mathbf{g}^*\rangle \\ &= \langle -\mathbf{h}, -\mathbf{h}^* | \mathbf{g}, \mathbf{g}^*\rangle = \langle \mathbf{h}, \mathbf{h}^* | - \mathbf{g}, -\mathbf{g}^*\rangle, \end{aligned} \tag{5.125}$$

which is the same as (5.76). The scalar products of two coherent states are even Grassmann functions that can be commuted. The second result in (5.123) leads to (5.76).

5.8.3 Cyclic Properties of the Fermion Trace

Ordinary (non-Grassmann) fermionic operators satisfy the standard quantum results for the cyclic properties of the trace:

$$\mathrm{Tr}(\hat{\Omega}_{\mathrm{f}}\hat{\Delta}_{\mathrm{f}}) = \mathrm{Tr}(\hat{\Delta}_{\mathrm{f}}\hat{\Omega}_{\mathrm{f}}). \tag{5.126}$$

This result is consistent with the expressions for the trace in terms of coherent states. If $\hat{\Omega}_\mathrm{f}$ and $\hat{\Delta}_\mathrm{f}$ are Grassmann operators, however, the *cyclic property* requires modification. For example, consider the two single-mode even Grassmann operators $\hat{\Omega}_\mathrm{f}^-(h) = \exp(ih\hat{c}^\dagger) = (1 + ih\hat{c}^\dagger)$ and $\hat{\Omega}_\mathrm{f}^+(h^+) = \exp(i\hat{c}h^+) = (1 + i\hat{c}h^+)$, which occur later in the definition of characteristic functions. We find that

$$\begin{aligned}
\mathrm{Tr}\left[\hat{\Omega}_\mathrm{f}^+(h^+)\hat{\Omega}_\mathrm{f}^-(h)\right] &= \langle 0|(1 + i\hat{c}h^+)(1 + ih\hat{c}^\dagger)|0\rangle + \langle 1|(1 + i\hat{c}h^+)(1 + ih\hat{c}^\dagger)|1\rangle \\
&= \left(\langle 0| + i\langle 1|h^+\right)\left(|0\rangle + ih|1\rangle\right) + \langle 1|1\rangle \\
&= 2 - h^+h, \\
\mathrm{Tr}\left[\hat{\Omega}_\mathrm{f}^-(h)\hat{\Omega}_\mathrm{f}^+(h^+)\right] &= 2 - hh^+ \neq \mathrm{Tr}\left[\hat{\Omega}_\mathrm{f}^+(h^+)\hat{\Omega}_\mathrm{f}^-(h)\right].
\end{aligned} \tag{5.127}$$

A modified cyclic property of the trace can be derived for the case of *even* Grassmann operators $\hat{\Omega}_\mathrm{f}^\mathrm{E}(\mathbf{h}), \hat{\Delta}_\mathrm{f}^\mathrm{E}(\mathbf{k})$ in the form

$$\mathrm{Tr}\left[\hat{\Omega}_\mathrm{f}^\mathrm{E}(\mathbf{h})\hat{\Delta}_\mathrm{f}^\mathrm{E}(\mathbf{k})\right] = \mathrm{Tr}\left[\hat{\Delta}_\mathrm{f}^\mathrm{E}(\mathbf{k})\hat{\Omega}_\mathrm{f}^\mathrm{E}(-\mathbf{h})\right] = \mathrm{Tr}\left[\hat{\Delta}_\mathrm{f}^\mathrm{E}(-\mathbf{k})\hat{\Omega}_\mathrm{f}^\mathrm{E}(\mathbf{h})\right]. \tag{5.128}$$

The proof is as follows:

$$\begin{aligned}
\mathrm{Tr}\left[\hat{\Omega}_\mathrm{f}^\mathrm{E}(\mathbf{h})\hat{\Delta}_\mathrm{f}^\mathrm{E}(\mathbf{k})\right] &= \int d^2\mathbf{g}\ \langle \mathbf{g}, \mathbf{g}^*|\hat{\Omega}_\mathrm{f}^\mathrm{E}(\mathbf{h})\hat{\Delta}_\mathrm{f}^\mathrm{E}(\mathbf{k})| - \mathbf{g}, -\mathbf{g}^*\rangle \\
&= \int d^2\mathbf{g}\ d^2\mathbf{f}\ \langle \mathbf{g}, \mathbf{g}^*|\hat{\Omega}_\mathrm{f}^\mathrm{E}(\mathbf{h})|\mathbf{f}, \mathbf{f}^*\rangle\langle \mathbf{f}, \mathbf{f}^*|\hat{\Delta}_\mathrm{f}^\mathrm{E}(\mathbf{k})| - \mathbf{g}, -\mathbf{g}^*\rangle \\
&= \int d^2\mathbf{g}\ d^2\mathbf{f}\ \langle \mathbf{f}, \mathbf{f}^*|\hat{\Delta}_\mathrm{f}^\mathrm{E}(\mathbf{k})| - \mathbf{g}, -\mathbf{g}^*\rangle\langle\ \mathbf{g}, \mathbf{g}^*|\hat{\Omega}_\mathrm{f}^\mathrm{E}(\mathbf{h})|\mathbf{f}, \mathbf{f}^*\rangle \\
&= \int d^2\mathbf{g}\ d^2\mathbf{f}\ \langle \mathbf{f}, \mathbf{f}^*|\hat{\Delta}_\mathrm{f}^\mathrm{E}(\mathbf{k})| - \mathbf{g}, -\mathbf{g}^*\rangle\langle -\mathbf{g}, -\mathbf{g}^*|\hat{\Omega}_\mathrm{f}^\mathrm{E}(-\mathbf{h})| - \mathbf{f}, -\mathbf{f}^*\rangle \\
&= \int d^2\mathbf{f}\ \langle \mathbf{f}, \mathbf{f}^*|\hat{\Delta}_\mathrm{f}^\mathrm{E}(\mathbf{k})\hat{\Omega}_\mathrm{f}^\mathrm{E}(-\mathbf{h})| - \mathbf{f}, -\mathbf{f}^*\rangle \\
&= \mathrm{Tr}\left[\hat{\Delta}_\mathrm{f}^\mathrm{E}(\mathbf{k})\hat{\Omega}_f^E(-\mathbf{h})\right].
\end{aligned} \tag{5.129}$$

The product of two odd Grassmann operators is an even operator, and this allows us to derive a cyclic trace rule for the product of two odd Grassmann operators as

$$\mathrm{Tr}\left[\hat{\Omega}_\mathrm{f}^\mathrm{O}(\mathbf{h})\hat{\Delta}_\mathrm{f}^\mathrm{O}(\mathbf{k})\right] = \mathrm{Tr}\left[\hat{\Delta}_\mathrm{f}^\mathrm{O}(\mathbf{k})\hat{\Omega}_\mathrm{f}^\mathrm{O}(-\mathbf{h})\right] = \mathrm{Tr}\left[\hat{\Delta}_\mathrm{f}^\mathrm{O}(-\mathbf{k})\hat{\Omega}_\mathrm{f}^\mathrm{O}(\mathbf{h})\right]. \tag{5.130}$$

The proof is left as an exercise. For the product of an even and an odd Grassmann operator or the product of two odd Grassmann operators, we can use (5.124) for odd Grassmann operators to show that

$$\mathrm{Tr}\left(\hat{\Omega}_\mathrm{f}^\mathrm{E}(\mathbf{h})\hat{\Delta}_\mathrm{f}^\mathrm{O}(\mathbf{k})\right) = \mathrm{Tr}\left(\hat{\Delta}_\mathrm{f}^\mathrm{O}(\mathbf{k})\hat{\Omega}_\mathrm{f}^\mathrm{E}(\mathbf{h})\right), \tag{5.131}$$

$$\mathrm{Tr}\left(\hat{\Omega}_\mathrm{f}^\mathrm{O}(\mathbf{h})\hat{\Delta}_\mathrm{f}^\mathrm{E}(\mathbf{k})\right) = \mathrm{Tr}\left(\hat{\Delta}_\mathrm{f}^\mathrm{E}(\mathbf{k})\hat{\Omega}_\mathrm{f}^\mathrm{O}(\mathbf{h})\right). \tag{5.132}$$

The trace of the product of a fermionic Grassmann operator $\hat{\Omega}_{\rm f}(\mathbf{h})$ (which may also be partly bosonic) and any bosonic operator $\hat{\Delta}_{\rm b}$ *does* satisfy the cyclic condition:

$$\mathrm{Tr}\left[\hat{\Omega}_{\rm f}(\mathbf{h})\hat{\Delta}_{\rm b}\right] = \mathrm{Tr}\left[\hat{\Delta}_{\rm b}\hat{\Omega}_{\rm f}(\mathbf{h})\right]. \tag{5.133}$$

5.8.4 Differentiating and Multiplying a Fermion Trace

Grassmann differentiation is another process involving the trace of Grassmann operators in which the standard quantum result does not apply. The *Grassmann derivative* of the trace of a Grassmann operator is *not* necessarily the same as the trace of the Grassmann derivative of the operator itself. For example, consider the single-mode even Grassmann operator $\hat{\Omega}_{\rm f}(h,h^+) = \exp(h\hat{c}^\dagger\hat{c}h^+) = (1 + h\hat{c}^\dagger\hat{c}h^+)$. The derivative of the trace of this operator is

$$\begin{aligned}\frac{\overrightarrow{\partial}}{\partial h}\mathrm{Tr}\left[\hat{\Omega}_{\rm f}(h,h^+)\right] &= \frac{\overrightarrow{\partial}}{\partial h}\left[(1 + h\langle 0|\left(\hat{c}^\dagger\hat{c}\right)|0\rangle h^+) + (1 + h\langle 1|\hat{c}^\dagger\hat{c})\,|1\rangle h^+\right] \\ &= \langle 0|\left(\hat{c}^\dagger\hat{c}\right)|0\rangle h^+ + \langle 1|\hat{c}^\dagger\hat{c})|1\rangle h^+ \\ &= h^+,\end{aligned} \tag{5.134}$$

but the trace of the derivative is

$$\begin{aligned}\mathrm{Tr}\left[\frac{\overrightarrow{\partial}}{\partial h}\hat{\Omega}_{\rm f}(h,h^+)\right] &= \mathrm{Tr}\left(\hat{c}^\dagger\hat{c}h^+\right) \\ &= \langle 0|\left(\hat{c}^\dagger\hat{c}\right)|0\rangle h^+ - \langle 1|\left(\hat{c}^\dagger\hat{c}\right)|1\rangle h^+ \\ &= -h^+ \neq \frac{\overrightarrow{\partial}}{\partial h}\mathrm{Tr}\left[\hat{\Omega}_{\rm f}(h,h^+)\right].\end{aligned} \tag{5.135}$$

These examples show that standard quantum results involving the trace of operators do not always apply to Grassmann operators.

Multiplication of an operator *by a Grassmann number* is another process in which the trace of the product with the operator is *not* the same as the product with the trace of the operator. Consider, for example, the single-mode unit operator $\hat{1}$ and the Grassmann number g. We find that

$$\begin{aligned}\mathrm{Tr}(g\hat{1}) &= \langle 0|g\hat{1}|0\rangle + \langle 1|g\hat{1}|1\rangle = g\langle 0|\hat{1}|0\rangle - g\langle 1|\hat{1}|1\rangle = g - g = 0, \\ g\;\mathrm{Tr}(\hat{1}) &= g(\langle 0|\hat{1}|0\rangle + \langle 1|\hat{1}|1\rangle) = g(1+1) = 2g \neq \mathrm{Tr}(g\hat{1}).\end{aligned} \tag{5.136}$$

On the other hand, when there are *two* of the above processes involved, such as two differentiation processes, or a product of two Grassmann numbers, or one differentiation process combined with multiplication by one Grassmann number, then the combined process applied to the trace of a Grassmann operator is the same as the trace of the combined process applied to the Grassmann operator. The same applies if there is any even number of processes including multiplication or differentiation from the left or right. For example,

$$\begin{aligned}
\frac{\overrightarrow{\partial}}{\partial g_i}\frac{\overrightarrow{\partial}}{\partial g_j}\left[\mathrm{Tr}\hat{\Omega}_{\mathrm{f}}(g))\right] &= \mathrm{Tr}\left[\frac{\overrightarrow{\partial}}{\partial g_i}\frac{\overrightarrow{\partial}}{\partial g_j}\hat{\Omega}_{\mathrm{f}}(g)\right], \\
\frac{\overrightarrow{\partial}}{\partial g_i}\left[g_j\left(\mathrm{Tr}\left(\hat{\Omega}_{\mathrm{f}}(g)\right)\right)\right] &= \mathrm{Tr}\left[\frac{\overrightarrow{\partial}}{\partial g_i}\left(g_j\hat{\Omega}_{\mathrm{f}}(g)\right)\right], \\
\left[\left(\mathrm{Tr}\left(\hat{\Omega}_{\mathrm{f}}(g)\right)\right)g_j\right]\frac{\overleftarrow{\partial}}{\partial g_i} &= \mathrm{Tr}\left(\left(\hat{\Omega}_{\mathrm{f}}(g)g_j\right)\frac{\overleftarrow{\partial}}{\partial g_i}\right), \\
g_i\left[\mathrm{Tr}\left(\hat{\Omega}_{\mathrm{f}}(g)\right)\right]g_j &= \mathrm{Tr}\left[g_i\hat{\Omega}_{\mathrm{f}}(g)g_j\right].
\end{aligned} \tag{5.137}$$

Note that all the left sides involve calculating the trace first and then applying the two processes, whereas the right sides involve applying the processes first and then calculating the trace. These results are based on writing the trace as a sum of diagonal elements for fermion Fock states $|\{\nu\}\rangle$. When a derivative or multiplier is moved to the left of the bra vector $\langle\{\nu\}|$ or to the right of the ket vector $|\{\nu\}\rangle$, a factor $(-1)^{\sum_i \nu_i}$ occurs. With two such processes, the two factors produce a net result of $+1$ for every term in the trace; in other cases, the same factor is produced when the process occurs inside the trace or after the trace, enabling the results to be proved. These results can be applied to developing the correspondence rules for the distribution function when the density operator is replaced by the density operator multiplied on either side by two fermion annihilation or creation operators.

5.9 Field Operators and Field Functions

The multi-mode boson and fermion coherent states provide a suitable starting point for introducing the idea of representing the field operators via classical (or non-operator) quantities called *field functions.* Like the field operators, these are functions of a c-number variable x, which usually refers to position. The field functions are eigenvalues of the field operators, with the coherent states as the eigenvectors. As might be expected, the nature of the field functions is different for bosons and fermions.

5.9.1 Boson Fields

A boson field refers to a c-number function $\psi_{\mathrm{b}}(x)$ that also depends on a c-number variable x and, linearly, on a set of c-number variables. Boson fields $\psi_{\mathrm{b}}(x)$ can be expanded in terms of a suitable orthonormal set of mode functions with c-number expansion coefficients α_k:

$$\psi_{\mathrm{b}}(x) = \sum_k \alpha_k \phi_k(x), \tag{5.138}$$

where the orthonormality conditions and the completeness relationship are

$$\int dx\ \phi_k^*(x)\phi_l(x) = \delta_{kl}, \tag{5.139}$$

$$\sum_k \phi_k(x)\phi_k^*(y) = \delta(x-y). \tag{5.140}$$

This gives the following result for the expansion coefficients:

$$\alpha_k = \int dx\ \phi_k^*(x)\psi_{\mathrm{b}}(x). \tag{5.141}$$

Boson fields have the feature that they commute under multiplication. Thus, with $\eta_{\mathrm{b}}(x) = \sum_l \beta_l\phi_l(x)$ a second boson field, we find

$$\psi_{\mathrm{b}}(x)\eta_{\mathrm{b}}(y) - \eta_{\mathrm{b}}(y)\psi_{\mathrm{b}}(x) = 0, \tag{5.142}$$

using the result $\alpha_k\beta_l - \beta_l\alpha_k = 0$. It is this commutation feature that makes them a natural form for representing boson field operators. It is sometimes convenient to expand a boson field in terms of the complex conjugate modes $\phi_k^*(x)$. Thus $\psi_{\mathrm{b}}^+(x)$, given by

$$\psi_{\mathrm{b}}^+(x) = \sum_k \phi_k^*(x)\alpha_k^+, \tag{5.143}$$

is also a boson field, and if $\alpha_k^+ = \alpha_k^*$ then $\psi_{\mathrm{b}}^+(x) = \psi_{\mathrm{b}}^*(x)$ is the complex conjugate boson field.

The boson coherent states $|\boldsymbol{\alpha},\boldsymbol{\alpha}^*\rangle$ are right eigenstates of the field annihilation operator $\hat{\psi}_{\mathrm{b}}(x)$, and the boson fields are the eigenvalues:

$$\hat{\psi}_{\mathrm{b}}(x)|\boldsymbol{\alpha},\boldsymbol{\alpha}^*\rangle = \psi_{\mathrm{b}}(x)|\boldsymbol{\alpha},\boldsymbol{\alpha}^*\rangle, \tag{5.144}$$

where we have used the result $\hat{a}_k|\boldsymbol{\alpha},\boldsymbol{\alpha}^*\rangle = \alpha_k|\boldsymbol{\alpha},\boldsymbol{\alpha}^*\rangle$. Although the boson coherent states are not physical states (except perhaps for light), this eigenvalue equation result is mathematically useful. An analogous result applies for the field creation operator $\hat{\psi}_{\mathrm{b}}(x)^\dagger$:

$$\langle\boldsymbol{\alpha}^{+*},\boldsymbol{\alpha}^+|\hat{\psi}_{\mathrm{b}}(x)^\dagger = \langle\boldsymbol{\alpha}^{+*},\boldsymbol{\alpha}^+|\psi_{\mathrm{b}}^+(x), \tag{5.145}$$

using the left-eigenvalue property $\langle\boldsymbol{\alpha}^{+*},\boldsymbol{\alpha}^+|\hat{a}_k^\dagger = \langle\boldsymbol{\alpha}^{+*},\boldsymbol{\alpha}^+|\alpha_k^*$.

5.9.2 Fermion Fields

A Grassmann field refers to a Grassmann function $\psi_{\mathrm{f}}(x)$ that also depends on a c-number variable x and, linearly, on a set of Grassmann variables. Grassmann fields $\psi_{\mathrm{f}}(x)$ can be expanded in terms of a suitable orthonormal set of mode functions with g-number expansion coefficients g_k:

$$\psi_{\mathrm{f}}(x) = \sum_k g_k\phi_k(x), \tag{5.146}$$

where the mode functions satisfy orthogonality and completeness relationships as in (5.139) and (5.140). This gives the following result for the expansion coefficients:

$$g_k = \int dx\ \phi_k^*(x)\psi_{\rm f}(x), \tag{5.147}$$

which has the same form as for c-number fields.

Grassmann fields have the feature that they anticommute under multiplication. Thus, with $\eta_{\rm f}(x) = \sum_l h_l \phi_l(x)$ a second Grassmann field, we find

$$\psi_{\rm f}(x)\eta_{\rm f}(y) + \eta_{\rm f}(y)\psi_{\rm f}(x) = 0, \tag{5.148}$$

using the result $g_k h_l + h_l g_k = 0$. It is this anticommutation feature that makes them a natural form for representing fermion field operators.

In the special case where $\eta_{\rm f}(x) = \psi_{\rm f}(x)$, we see from the last result that the square and hence any higher power of a Grassmann field is zero:

$$(\psi_{\rm f}(x))^2 = (\psi_{\rm f}(x))^3 = \cdots = 0. \tag{5.149}$$

This feature is in contrast to that for boson fields and, as in the case of the similar feature for the Grassmann variables g_k, this feature enables simple definitions of Grassmann functional differentiation and integration for functionals of Grassmann fields to be introduced. It is sometimes convenient to expand a Grassmann field in terms of the complex conjugate modes $\phi_k^*(x)$. Thus $\psi_{\rm f}^+(x)$, given by

$$\psi_{\rm f}^+(x) = \sum_k \phi_k^*(x) g_k^+, \tag{5.150}$$

is also a Grassmann field, and if $g_k^+ = g_k^*$ then $\psi_{\rm f}^+(x) = \psi_{\rm f}^*(x)$ is the complex conjugate Grassmann field.

The fermion coherent states $|\mathbf{g}, \mathbf{g}^*\rangle$ are right eigenstates of the field annihilation operator $\hat{\psi}_{\rm f}(x)$, and the Grassmann fields are the eigenvalues:

$$\hat{\psi}_{\rm f}(x)|\mathbf{g}, \mathbf{g}^*\rangle = \psi_{\rm f}(x)|\mathbf{g}, \mathbf{g}^*\rangle, \tag{5.151}$$

where we have used the result $\hat{c}_k|\mathbf{g}, \mathbf{g}^*\rangle = g_k|\mathbf{g}, \mathbf{g}^*\rangle$. Although the fermion coherent states are not physical states, this eigenvalue equation result is useful in mathematical developments. An analogous result applies for the field creation operator $\hat{\psi}_{\rm f}(x)^\dagger$:

$$\langle \mathbf{g}^{+*}, \mathbf{g}^+|\hat{\psi}(x)^\dagger = \langle \mathbf{g}^{+*}, \mathbf{g}^+|\psi_{\rm f}^+(x), \tag{5.152}$$

using the left-eigenvalue property $\langle \mathbf{g}^{+*}, \mathbf{g}^+|\hat{c}_k^\dagger = \langle \mathbf{g}^{+*}, \mathbf{g}^+|g_k^*$.

5.9.3 Quantum Correlation Functions

It is of some interest to evaluate the normally ordered quantum correlation functions when the system is in a bosonic or fermionic coherent state. For the bosonic case, the pth-order coherence function is

$$\begin{aligned} G^{(p,p)}(x_1,\cdots,x_p;y_p,\cdots,y_1) &= \mathrm{Tr}\left[\hat{\psi}_\mathrm{b}(y_p)\cdots\hat{\psi}_\mathrm{b}(y_1)|\boldsymbol{\alpha},\boldsymbol{\alpha}^*\rangle\langle\boldsymbol{\alpha},\boldsymbol{\alpha}^*|\hat{\psi}_\mathrm{b}^\dagger(x_1)\cdots\hat{\psi}_\mathrm{b}^\dagger(x_p)\right] \\ &= \psi_\mathrm{b}(y_p)\cdots\psi_\mathrm{b}(y_2)\psi_\mathrm{b}(y_1)\psi_\mathrm{b}^*(x_1)\psi_\mathrm{b}^*(x_2)\cdots\psi_\mathrm{b}^*(x_p) \\ &= G^{(1,1)}(x_1;y_1)\cdots G^{(1,1)}(x_p;y_p), \end{aligned} \tag{5.153}$$

which is a product of first-order correlation functions. Analogous results apply for the fermionic case:

$$\begin{aligned} G^{(p,p)}(x_1,\cdots,x_p;y_p,\cdots,y_1) &= \mathrm{Tr}\left[\hat{\psi}_\mathrm{f}(y_p)\cdots\hat{\psi}_\mathrm{f}(y_1)|\mathbf{g},\mathbf{g}^*\rangle\langle\mathbf{g},\mathbf{g}^*|\hat{\psi}_\mathrm{f}^\dagger(x_1)\cdots\hat{\psi}_\mathrm{f}^\dagger(x_p)\right] \\ &= \psi_\mathrm{f}(y_p)\cdots\psi_\mathrm{f}(y_2)\psi_\mathrm{f}(y_1)\psi_\mathrm{f}^*(x_1)\psi_\mathrm{f}^*(x_2)\cdots\psi_\mathrm{f}^*(x_p) \\ &= G^{(1,1)}(x_1;y_1)\cdots G^{(1,1)}(x_p;y_p), \end{aligned} \tag{5.154}$$

which is also a product of first-order correlation functions. Note that because of fermionic antisymmetry,

$$G^{(1,1)}(x;y) = \psi_\mathrm{f}(y)\psi_\mathrm{f}(x)^* = -\psi_\mathrm{f}(x)^*\psi_\mathrm{f}(y). \tag{5.155}$$

The factoring of higher-order correlation functions into a product of first-order ones indicates coherence to all orders. It is this property, of course, that is the origin of the term '*coherent* states'.

Exercises

(5.1) Derive the adjoint rules (5.5).

(5.2) Show that there are no right eigenstates of $\hat{a}^\dagger$ but that right eigenstates of $\hat{c}^\dagger$ do exist. (*Hint:* Consider the fully occupied fermion Fock state $|1,1,\cdots,1\rangle$.)

(5.3) The Baker–Hausdorff theorem is

$$\exp(A)\exp(B) = \exp(A+B)\exp\left(\frac{1}{2}[A,B]\right).$$

Use this theorem to justify the results in (5.8) and (5.12) for the unitary displacement operators.

(5.4) Prove the transformation properties (5.18) and (5.19) for the boson and fermion annihilation operators.

(5.5) Use the Baker–Hausdorff theorem to prove the ray representation results (5.10) and (5.15).

(5.6) Use the ray representation results (5.10) and (5.15) to prove the orthogonality results (5.27) and (5.35) for the coherent states.

(5.7) Prove, by complex integration, the boson completeness relation for the coherent states (5.37).

(5.8) Prove the differentiation results (5.47) for boson Bargmann states and (5.56) for fermion Bargmann states.

(5.9) Prove the correctness of the operator expressions for the vacuum states (5.84) and (5.85). (*Hint:* You might do this using either the Fock-state or coherent-state matrix elements of the operators.)

(5.10) Prove the relationship (5.130) for the trace of the product of two odd Grassmann operators.

6
Canonical Transformations

Among all the possible transformations we might envisage, the *canonical transformations* [9] are special in that they preserve the fundamental commutation relations or, for fermions, anticommutation relations among the annihilation and creation operators. It is very much the case that many of the quantum phenomena we are interested in can be traced back to these and it is essential, therefore, that they retain their correct forms. All the models we construct and the approximations we make should retain this feature.

In general, for the systems of interest in this text, such canonical transformations will take our annihilation operators, $\hat{a}_i$ for bosons and $\hat{c}_i$ for fermions, and transform these into new operators, $\hat{a}'_i$ and $\hat{c}'_i$, such that

$$\left[\hat{a}'_i, \hat{a}'^{\dagger}_j\right] = \delta_{ij}, \tag{6.1}$$

$$\left[\hat{a}'_i, \hat{a}'_j\right] = 0 = \left[\hat{a}'^{\dagger}_i, \hat{a}'^{\dagger}_j\right], \tag{6.2}$$

$$\left\{\hat{c}'_i, \hat{c}'^{\dagger}_j\right\} = \delta_{ij}, \tag{6.3}$$

$$\left\{\hat{c}'_i, \hat{c}'_j\right\} = 0 = \left\{\hat{c}'^{\dagger}_i, \hat{c}'^{\dagger}_j\right\}. \tag{6.4}$$

Fortunately, the condition for being a canonical transformation is simple; we require the original operators and the transformed (primed) operators to be related by a unitary transformation [39]. This means that for each possible canonical transformation, there is a unitary transformation $\hat{U}$ such that

$$\begin{aligned} \hat{a}'_i &= \hat{U}^{\dagger}\hat{a}_i\hat{U}, \\ \hat{a}'^{\dagger}_i &= \hat{U}^{\dagger}\hat{a}^{\dagger}_i\hat{U} \end{aligned} \tag{6.5}$$

for bosons and

$$\begin{aligned} \hat{c}'_i &= \hat{U}^{\dagger}\hat{c}_i\hat{U}, \\ \hat{c}'^{\dagger}_i &= \hat{U}^{\dagger}\hat{c}^{\dagger}_i\hat{U} \end{aligned} \tag{6.6}$$

for fermions. Recall that an operator $\hat{U}$ is unitary if $\hat{U}^\dagger\hat{U} = 1 = \hat{U}\hat{U}^\dagger$ and that any unitary operator can be written in terms of a *generator*, which is a Hermitian operator $\hat{G}$, in the form

$$\hat{U} = \exp\left(i\hat{G}\right),$$
$$\hat{U}^\dagger = \exp\left(-i\hat{G}\right). \tag{6.7}$$

Unitary transformations are common in quantum theory because they are intimately associated with dynamics. If the Hamiltonian is $\hat{H}$, then the time evolution will be a unitary transformation with $\hat{U} = \exp(-i\hat{H}t/\hbar)$.

6.1 Linear Canonical Transformations

The simplest and most frequently encountered canonical transformations are linear in the annihilation and creation operators [9]. This means that the transformed operators are linear combinations of the untransformed operators:

$$\hat{a}'_i = \sum_j u^{\mathrm{b}}_{ij}\hat{a}_j + \sum_j v^{\mathrm{b}}_{ij}\hat{a}^\dagger_j,$$
$$\hat{c}'_i = \sum_j u^{\mathrm{f}}_{ij}\hat{c}_j + \sum_j v^{\mathrm{f}}_{ij}\hat{c}^\dagger_j, \tag{6.8}$$

where the coefficients u_{ij} and v_{ij} are c-numbers. The unitary nature of the transformation ensures that it is invertible, so that

$$\hat{a}_i = \hat{U}\hat{a}'_i\hat{U}^\dagger,$$
$$\hat{c}_i = \hat{U}\hat{c}'_i\hat{U}^\dagger. \tag{6.9}$$

Naturally, this means that the untransformed operators are also linear combinations of the primed ones:

$$\hat{a}_i = \sum_j \tilde{u}^{\mathrm{b}}_{ij}\hat{a}'_j + \sum_j \tilde{v}^{\mathrm{b}}_{ij}\hat{a}'^\dagger_j,$$
$$\hat{c}_i = \sum_j \tilde{u}^{\mathrm{f}}_{ij}\hat{c}'_j + \sum_j \tilde{v}^{\mathrm{f}}_{ij}\hat{c}'^\dagger_j. \tag{6.10}$$

We can use the commutators or anticommutators for the two sets of operators to relate the coefficients in our transformation and in its inverse; for example,

$$\left[\hat{a}'_i, \hat{a}^\dagger_j\right] = u^{\mathrm{b}}_{ij},$$
$$\left[\hat{a}_i, \hat{a}'^\dagger_j\right] = \tilde{u}^{\mathrm{b}}_{ij}. \tag{6.11}$$

These are not independent, however, as

$$\left[\hat{a}_i, \hat{a}'^\dagger_j\right] = \left(\left[\hat{a}'_j, \hat{a}^\dagger_i\right]\right)^\dagger$$
$$\Rightarrow \tilde{u}^{\mathrm{b}}_{ij} = u^{\mathrm{b}\star}_{ji}. \tag{6.12}$$

The coefficients in our transformed operators (6.8) are, of course, not mutually independent, as we must respect the fundamental commutation or anticommutation relations so that for bosons we have

$$\begin{aligned} \left[\hat{a}'_i, \hat{a}'^{\dagger}_k\right] = \delta_{ik} &\Rightarrow \sum_j u^{\mathrm{b}}_{ij} u^{\mathrm{b}\star}_{jk} - \sum_j v^{\mathrm{b}}_{ij} v^{\mathrm{b}\star}_{jk} = \delta_{ik}, \\ [\hat{a}'_i, \hat{a}'_k] = 0 &\Rightarrow \sum_j u^{\mathrm{b}}_{ij} v^{\mathrm{b}}_{kj} - \sum_j v^{\mathrm{b}}_{ij} u^{\mathrm{b}}_{kj} = 0, \end{aligned} \tag{6.13}$$

and for fermions

$$\begin{aligned} \left\{\hat{c}'_i, \hat{c}'^{\dagger}_k\right\} = \delta_{ik} &\Rightarrow \sum_j u^{\mathrm{f}}_{ij} u^{\mathrm{f}\star}_{jk} + \sum_j v^{\mathrm{f}}_{ij} v^{\mathrm{f}\star}_{jk} = \delta_{ik}, \\ \{\hat{c}'_i, \hat{c}'_k\} = 0 &\Rightarrow \sum_j u^{f}_{ij} v^{\mathrm{f}}_{kj} + \sum_i v^{\mathrm{f}}_{ij} u^{\mathrm{f}}_{kj} = 0. \end{aligned} \tag{6.14}$$

These are the fundamental constraints on the allowed canonical transformations. Applying them, however, still leaves a very wide variety of possible transformations and these can be used to investigate a wide range of quantum phenomena.

6.2 One- and Two-Mode Transformations

At their simplest, the linear canonical transformations produce new annihilation operators that are either combinations of untransformed annihilation operators only or superpositions of both untransformed annihilation and untransformed creation operators. It suffices, in order to demonstrate the properties of these two types of transformations, to consider systems with just two modes. For ease of presentation we shall treat bosonic and fermionic systems separately.

6.2.1 Bosonic Modes

The linear forms of our transformations (6.8) suggest that we seek generators for our unitary transformations that are bilinear in the annihilation and creation operators. For one or two modes, all such generators can be written as linear combinations of the ten generators [40]

$$\begin{aligned} \hat{T}_1 &= \frac{1}{2}\left(\hat{a}_1\hat{a}_1^{\dagger} + \hat{a}_1^{\dagger}\hat{a}_1\right), \\ \hat{T}_2 &= \frac{1}{2}\left(\hat{a}_2\hat{a}_2^{\dagger} + \hat{a}_2^{\dagger}\hat{a}_2\right), \\ \hat{T}_3 &= \hat{a}_1^{\dagger}\hat{a}_2 + \hat{a}_2^{\dagger}\hat{a}_1, \\ \hat{T}_4 &= -i\left(\hat{a}_1^{\dagger}\hat{a}_2 - \hat{a}_2^{\dagger}\hat{a}_1\right), \\ \hat{T}_5 &= \hat{a}_1\hat{a}_2 + \hat{a}_2^{\dagger}\hat{a}_1^{\dagger}, \\ \hat{T}_6 &= -i\left(\hat{a}_1\hat{a}_2 - \hat{a}_2^{\dagger}\hat{a}_1^{\dagger}\right), \\ \hat{T}_7 &= \frac{1}{2}\left(\hat{a}_1^2 + \hat{a}_1^{\dagger 2}\right), \end{aligned}$$

$$\hat{T}_8 = -\frac{i}{2}\left(\hat{a}_1^2 - \hat{a}_1^{\dagger 2}\right),$$
$$\hat{T}_9 = \frac{1}{2}\left(\hat{a}_2^2 + \hat{a}_2^{\dagger 2}\right),$$
$$\hat{T}_{10} = -\frac{i}{2}\left(\hat{a}_2^2 - \hat{a}_2^{\dagger 2}\right). \tag{6.15}$$

These ten generators form a *Lie algebra* in that the commutator of any two of them is, apart from a multiplicative constant, one of the others. It is convenient to think of these ten generators as being fundamental and to consider the effects of these on our annihilation and creation operators.

We can obtain a unitary operator from any of our ten generators by exponentiation:

$$\hat{U}_\mu(\gamma) = \exp\left(i\gamma\hat{T}_\mu\right), \qquad \mu = 1, \cdots, 10, \tag{6.16}$$

where γ is real, and determine the form of the transformed annihilation and creation operators using the identity

$$\hat{U}_\mu^\dagger(\gamma)\hat{a}_i\hat{U}_\mu(\gamma) = \hat{a}_i - i\gamma\left[\hat{T}_\mu, \hat{a}_i\right] - \frac{\gamma^2}{2!}\left[\hat{T}_\mu, \left[\hat{T}_\mu, \hat{a}_i\right]\right] + i\frac{\gamma^3}{3!}\left[\hat{T}_\mu, \left[\hat{T}_\mu, \left[\hat{T}_\mu, \hat{a}_i\right]\right]\right] + \cdots. \tag{6.17}$$

We can construct, in this way, linear canonical transformations for any of the generators (6.15), but we list here only three examples:

$$\begin{aligned}
\hat{a}_1' = \hat{U}_3^\dagger(\gamma)\hat{a}_1\hat{U}_3(\gamma) &= \hat{a}_1 + i\gamma\hat{a}_2 - \frac{\gamma^2}{2!}\hat{a}_1 - i\frac{\gamma^3}{3!}\hat{a}_2 + \cdots \\
&= \cos\gamma\ \hat{a}_1 + i\sin\gamma\ \hat{a}_2, \\
\hat{a}_1' = \hat{U}_6^\dagger(\gamma)\hat{a}_1\hat{U}_6(\gamma) &= \hat{a}_1 - \gamma\hat{a}_2^\dagger + \frac{\gamma^2}{2!}\hat{a}_1 - \frac{\gamma^3}{3!}\hat{a}_2^\dagger + \cdots \\
&= \cosh\gamma\ \hat{a}_1 - \sinh\gamma\ \hat{a}_2^\dagger, \\
\hat{a}_1' = \hat{U}_9^\dagger(\gamma)\hat{a}_1\hat{U}_9(\gamma) &= \hat{a}_1 + i\gamma\hat{a}_1^\dagger + \frac{\gamma^2}{2!}\hat{a}_1 + i\frac{\gamma^3}{3!}\hat{a}_2^\dagger + \cdots \\
&= \cosh\gamma\ \hat{a}_1 + i\sinh\gamma\ \hat{a}_1^\dagger.
\end{aligned} \tag{6.18}$$

It is also possible, of course, to apply more than one of the unitary operators (6.16) and, indeed, any two-mode linear canonical transformation can be constructed in this way.

Transforming the operators is fully analogous to solving for quantum evolution in the Heisenberg picture, in which operators evolve and states are unchanged. It is also possible to apply the transformation to the state of the system and leave the operators unchanged. This is analogous, of course, to working in the Schrödinger picture. The transformed states produced by the unitary operators (6.16) are simply

$$|\psi'\rangle = \hat{U}_\mu(\gamma)|\psi\rangle. \tag{6.19}$$

The unitary nature of this transformation means that the overlap between any two possible states is preserved:

$$\langle\phi'|\psi'\rangle = \left(\langle\phi|\hat{U}^\dagger\right)\left(\hat{U}|\psi\rangle\right) = \langle\phi|\psi\rangle. \tag{6.20}$$

It is natural to study the effects of our unitary transformations on the Fock states of fixed particle number. In doing so it is helpful to be able to order, or disentangle, our unitary operators so as to separate the effects of operators which act to change the numbers of particles in the modes:

$$\begin{aligned}
\hat{U}_1(\gamma) &= \exp\left(i\gamma\frac{1}{2}\hat{a}_1\hat{a}_1^\dagger\right)\exp\left(i\gamma\frac{1}{2}\hat{a}_1^\dagger\hat{a}_1\right),\\
\hat{U}_2(\gamma) &= \exp\left(i\gamma\frac{1}{2}\hat{a}_2\hat{a}_2^\dagger\right)\exp\left(i\gamma\frac{1}{2}\hat{a}_2^\dagger\hat{a}_2\right),\\
\hat{U}_3(\gamma) &= \exp\left(i\tan\gamma\,\hat{a}_1^\dagger\hat{a}_2\right)\exp\left[-\ln(\cos^2\gamma)\frac{1}{2}\left(\hat{a}_1^\dagger\hat{a}_1-\hat{a}_2^\dagger\hat{a}_2\right)\right]\exp\left(\tan\gamma\,\hat{a}_2^\dagger\hat{a}_1\right),\\
\hat{U}_4(\gamma) &= \exp\left(i\tan\gamma\,\hat{a}_1^\dagger\hat{a}_2\right)\exp\left[-\ln(\cos^2\gamma)\frac{1}{2}\left(\hat{a}_1^\dagger\hat{a}_1-\hat{a}_2^\dagger\hat{a}_2\right)\right]\exp\left(-\tan\gamma\,\hat{a}_2^\dagger\hat{a}_1\right),\\
\hat{U}_5(\gamma) &= \exp\left(i\tanh\gamma\,\hat{a}_2^\dagger\hat{a}_1^\dagger\right)\exp\left[-\ln(\cosh\gamma)\left(\hat{a}_1^\dagger\hat{a}_1+\hat{a}_2\hat{a}_2^\dagger\right)\right]\exp\left(i\tanh\gamma\,\hat{a}_1\hat{a}_2\right),\\
\hat{U}_6(\gamma) &= \exp\left(-\tanh\gamma\,\hat{a}_2^\dagger\hat{a}_1^\dagger\right)\exp\left[-\ln(\cosh\gamma)\left(\hat{a}_1^\dagger\hat{a}_1+\hat{a}_2\hat{a}_2^\dagger\right)\right]\exp\left(\tanh\gamma\,\hat{a}_1\hat{a}_2\right),\\
\hat{U}_7(\gamma) &= \exp\left(i\tanh\gamma\,\frac{1}{2}\hat{a}_1^{\dagger 2}\right)\exp\left[-\ln(\cosh\gamma)\frac{1}{2}\left(\hat{a}_1^\dagger\hat{a}_1+\hat{a}_1\hat{a}_1^\dagger\right)\right]\exp\left(i\tanh\gamma\,\frac{1}{2}\hat{a}_1^2\right),\\
\hat{U}_8(\gamma) &= \exp\left(-\tanh\gamma\,\frac{1}{2}\hat{a}_1^{\dagger 2}\right)\exp\left[-\ln(\cosh\gamma)\frac{1}{2}\left(\hat{a}_1^\dagger\hat{a}_1+\hat{a}_1\hat{a}_1^\dagger\right)\right]\exp\left(\tanh\gamma\,\frac{1}{2}\hat{a}_1^2\right),\\
\hat{U}_9(\gamma) &= \exp\left(i\tanh\gamma\,\frac{1}{2}\hat{a}_2^{\dagger 2}\right)\exp\left[-\ln(\cosh\gamma)\frac{1}{2}\left(\hat{a}_2^\dagger\hat{a}_2+\hat{a}_2\hat{a}_2^\dagger\right)\right]\exp\left(i\tanh\gamma\,\frac{1}{2}\hat{a}_2^2\right),\\
\hat{U}_{10}(\gamma) &= \exp\left(-\tanh\gamma\,\frac{1}{2}\hat{a}_2^{\dagger 2}\right)\exp\left[-\ln(\cosh\gamma)\frac{1}{2}\left(\hat{a}_2^\dagger\hat{a}_2+\hat{a}_2\hat{a}_2^\dagger\right)\right]\exp\left(\tanh\gamma\,\frac{1}{2}\hat{a}_2^2\right).
\end{aligned} \tag{6.21}$$

The simplest Fock state is the zero-particle or vacuum state. The first four operators, $\hat{U}_1, \cdots, \hat{U}_4$, clearly leave the vacuum state unchanged apart from a phase change:

$$\begin{aligned}
\hat{U}_1(\gamma)|0\rangle &= e^{i\gamma/2}|0\rangle = \hat{U}_2(\gamma)|0\rangle,\\
\hat{U}_3(\gamma)|0\rangle &= |0\rangle = \hat{U}_4(\gamma)|0\rangle;
\end{aligned} \tag{6.22}$$

but the remaining six generators can create particles in pairs, so that the associated unitary transformations produce states with a superposition of even numbers of particles:

$$\begin{aligned}
\hat{U}_5(\gamma)|0\rangle &= \frac{1}{\cosh\gamma}\sum_{n=0}^{\infty}(i\tanh\gamma)^n|n,n\rangle, \\
\hat{U}_6(\gamma)|0\rangle &= \frac{1}{\cosh\gamma}\sum_{n=0}^{\infty}(-\tanh\gamma)^n|n,n\rangle, \\
\hat{U}_7(\gamma)|0\rangle &= \frac{1}{\sqrt{\cosh\gamma}}\sum_{n=0}^{\infty}\frac{\sqrt{(2n)!}}{n!}\left(i\frac{\tanh\gamma}{2}\right)^n|2n,0\rangle, \\
\hat{U}_8(\gamma)|0\rangle &= \frac{1}{\sqrt{\cosh\gamma}}\sum_{n=0}^{\infty}\frac{\sqrt{(2n)!}}{n!}\left(-\frac{\tanh\gamma}{2}\right)^n|2n,0\rangle, \\
\hat{U}_9(\gamma)|0\rangle &= \frac{1}{\sqrt{\cosh\gamma}}\sum_{n=0}^{\infty}\frac{\sqrt{(2n)!}}{n!}\left(i\frac{\tanh\gamma}{2}\right)^n|0,2n\rangle, \\
\hat{U}_{10}(\gamma)|0\rangle &= \frac{1}{\sqrt{\cosh\gamma}}\sum_{n=0}^{\infty}\frac{\sqrt{(2n)!}}{n!}\left(-\frac{\tanh\gamma}{2}\right)^n|0,2n\rangle,
\end{aligned} \tag{6.23}$$

where $|n,n'\rangle$ is the state with n bosons in mode 1 and n' in mode 2. More generally, the first four generators commute with the total number of particles, i.e.

$$\left[\hat{T}_\mu, \hat{a}_1^\dagger\hat{a}_1 + \hat{a}_2^\dagger\hat{a}_2\right] = 0, \qquad \mu = 1,\cdots,4, \tag{6.24}$$

and it follows that the associated unitary transformations can only redistribute any existing particles among the modes. The fifth and sixth generators create or destroy particles in pairs, one in each mode, and it follows that they commute with the difference in the number of particles:

$$\left[\hat{T}_\mu, \hat{a}_1^\dagger\hat{a}_1 - \hat{a}_2^\dagger\hat{a}_2\right] = 0, \qquad \mu = 5, 6. \tag{6.25}$$

It follows that the associated unitary transformations preserve this difference in particle number. The final four generators inject or remove particles in pairs in a single mode. This means that the parity of the particle number is conserved; if the initial state has an even (or odd) number of particles then so will the transformed state. This conservation law means that the generators commute with the particle-number parity operator:

$$\left[\hat{T}_\mu, \exp\left[i\pi\left(\hat{a}_1^\dagger\hat{a}_1 + \hat{a}_2^\dagger\hat{a}_2\right)\right]\right] = 0, \qquad \mu = 7,\cdots,10. \tag{6.26}$$

The transformations of the Fock states are readily evaluated by using the inverses of the transformations (6.17). Consider, for example, the transformation of a Fock state in which the two modes have m_1 and m_2 particles, respectively:

$$|m_1, m_2\rangle = \frac{\hat{a}_1^{\dagger m_1}}{\sqrt{(m_1)!}} \frac{\hat{a}_2^{\dagger m_2}}{\sqrt{(m_2)!}} |0\rangle. \tag{6.27}$$

The transformed state produced by the unitary operator $\hat{U}_\mu(\gamma)$ is then

$$\hat{U}_\mu(\gamma)|m_1, m_2\rangle = (m_1! m_2!)^{-1/2} \left[\hat{U}_\mu(\gamma)\hat{a}_1^\dagger \hat{U}_\mu^\dagger(\gamma)\right]^{m_1} \left[\hat{U}_\mu(\gamma)\hat{a}_2^\dagger \hat{U}_\mu^\dagger(\gamma)\right]^{m_2} \hat{U}_\mu(\gamma)|0\rangle. \tag{6.28}$$

We shall discuss some specific examples of such states in Section 6.3.

6.2.2 Fermionic Modes

The linear transformations (6.9) again suggest that we focus on generators that are bilinear in the annihilation and creation operators. Not all of the ten bosonic generators (6.15) can be converted into fermionic forms; the last four, in particular, would be zero if the bosonic operators were replaced by fermionic ones. The natural generalisations of the first six bosonic generators are

$$\begin{aligned}
\hat{S}_1 &= \frac{1}{2}\left(\hat{c}_1\hat{c}_1^\dagger - \hat{c}_1^\dagger\hat{c}_1\right), \\
\hat{S}_2 &= \frac{1}{2}\left(\hat{c}_2\hat{c}_2^\dagger - \hat{c}_2^\dagger\hat{c}_2\right), \\
\hat{S}_3 &= \hat{c}_1^\dagger\hat{c}_2 + \hat{c}_2^\dagger\hat{c}_1, \\
\hat{S}_4 &= -i\left(\hat{c}_1^\dagger\hat{c}_2 - \hat{c}_2^\dagger\hat{c}_1\right), \\
\hat{S}_5 &= \hat{c}_1\hat{c}_2 + \hat{c}_2^\dagger\hat{c}_1^\dagger, \\
\hat{S}_6 &= -i\left(\hat{c}_1\hat{c}_2 - \hat{c}_2^\dagger\hat{c}_1^\dagger\right).
\end{aligned} \tag{6.29}$$

As with their bosonic counterparts, these are Hermitian and form a Lie algebra.

We can construct a unitary operator from any of our six generators by exponentiation:

$$\hat{U}_\nu(\gamma) = \exp\left(i\gamma\hat{S}_\nu\right), \qquad \nu = 1, \cdots, 6. \tag{6.30}$$

Here γ is a real c-number, and not a Grassmann number as encountered in our discussion of fermionic coherent states in the preceding chapter. The transformed annihilation operators may be determined using the identity

$$\hat{U}_\nu^\dagger(\gamma)\hat{c}_i\hat{U}_\nu(\gamma) = \hat{c}_i - i\gamma\left[\hat{S}_\nu, \hat{c}_i\right] - \frac{\gamma^2}{2!}\left[\hat{S}_\nu, \left[\hat{S}_\nu, \hat{c}_i\right]\right] + i\frac{\gamma^3}{3!}\left[\hat{S}_\nu, \left[\hat{S}_\nu, \left[\hat{S}_\nu, \hat{c}_i\right]\right]\right] + \cdots. \tag{6.31}$$

It may seem strange that this transformation, which involves commutators, should lead to a linear canonical transformation when we recall that the canonical properties for fermions are expressed in terms of anticommutators. That this is so is a consequence of the fact that the generators of the transformations are bilinear in the annihilation and creation operators and that operators for different modes anticommute. Two simple examples serve to illustrate this point:

$$\begin{aligned}
\left[\hat{c}_i, \hat{c}_j^\dagger \hat{c}_k\right] &= \hat{c}_i \hat{c}_j^\dagger \hat{c}_k - \hat{c}_j^\dagger \hat{c}_k \hat{c}_i \\
&= \hat{c}_i \hat{c}_j^\dagger \hat{c}_k + \hat{c}_j^\dagger \hat{c}_i \hat{c}_k \\
&= \left\{\hat{c}_i, \hat{c}_j^\dagger\right\} \hat{c}_k \\
&= \delta_{ij} \hat{c}_k, \\
\left[\hat{c}_i, \hat{c}_j^\dagger \hat{c}_k^\dagger\right] &= \hat{c}_i \hat{c}_j^\dagger \hat{c}_k^\dagger - \hat{c}_j^\dagger \hat{c}_k^\dagger \hat{c}_i \\
&= \delta_{ij} \hat{c}_k^\dagger - \delta_{ik} \hat{c}_j^\dagger.
\end{aligned} \tag{6.32}$$

These are both linear combinations of the annihilation and creation operators, as required.

The annihilation operator $\hat{c}_1$ transforms under unitary transformations generated by $\hat{S}_1, \cdots, \hat{S}_6$ as

$$\begin{aligned}
\hat{U}_1^\dagger(\gamma)\hat{c}_1\hat{U}_1(\gamma) &= e^{i\gamma}\hat{c}_1, \\
\hat{U}_2^\dagger(\gamma)\hat{c}_1\hat{U}_2(\gamma) &= \hat{c}_1, \\
\hat{U}_3^\dagger(\gamma)\hat{c}_1\hat{U}_3(\gamma) &= \cos\gamma\hat{c}_1 + i\sin\gamma\hat{c}_2, \\
\hat{U}_4^\dagger(\gamma)\hat{c}_1\hat{U}_4(\gamma) &= \cos\gamma\hat{c}_1 + \sin\gamma\hat{c}_2, \\
\hat{U}_5^\dagger(\gamma)\hat{c}_1\hat{U}_5(\gamma) &= \cos\gamma\hat{c}_1 - i\sin\gamma\hat{c}_2^\dagger, \\
\hat{U}_6^\dagger(\gamma)\hat{c}_1\hat{U}_6(\gamma) &= \cos\gamma\hat{c}_1 + \sin\gamma\hat{c}_2^\dagger.
\end{aligned} \tag{6.33}$$

As with the bosonic transformations, we can apply more than one of the unitary operators (6.30), and any linear canonical transformation can be constructed in this manner.

Instead of transforming the operators, we can apply the unitary transformation to the state. Such a transformation will, as with bosonic states, preserve the overlap between states. In order to calculate the form of these states, it is helpful to disentangle our unitary operators so as to separate the effects of the operators that change the particle numbers:

$$\begin{aligned}
\hat{U}_1(\gamma) &= \exp\left(i\gamma\frac{1}{2}\hat{c}_1\hat{c}_1^\dagger\right)\exp\left(i\gamma\frac{1}{2}\hat{c}_1^\dagger\hat{c}_1\right), \\
\hat{U}_2(\gamma) &= \exp\left(i\gamma\frac{1}{2}\hat{c}_2\hat{c}_2^\dagger\right)\exp\left(i\gamma\frac{1}{2}\hat{c}_2^\dagger\hat{c}_2\right), \\
\hat{U}_3(\gamma) &= \exp\left(i\tan\gamma\hat{c}_1^\dagger\hat{c}_2\right)\exp\left[-\ln(\cos^2\gamma)\frac{1}{2}\left(\hat{c}_1^\dagger\hat{c}_1 - \hat{c}_2^\dagger\hat{c}_2\right)\right]\exp\left(i\tan\gamma\hat{c}_2^\dagger\hat{c}_1\right), \\
\hat{U}_4(\gamma) &= \exp\left(\tan\gamma\hat{c}_1^\dagger\hat{c}_2\right)\exp\left[-\ln(\cos^2\gamma)\frac{1}{2}\left(\hat{c}_1^\dagger\hat{c}_1 - \hat{c}_2^\dagger\hat{c}_2\right)\right]\exp\left(-\tan\gamma\hat{c}_2^\dagger\hat{c}_1\right),
\end{aligned}$$

$$\hat{U}_5(\gamma) = \exp\left(i \tan\gamma \hat{c}_2^\dagger \hat{c}_1^\dagger\right) \exp\left[-\ln(\cos^2\gamma)\frac{1}{2}\left(\hat{c}_2^\dagger \hat{c}_2 - \hat{c}_1 \hat{c}_1^\dagger\right)\right] \exp\left(i \tan\gamma \hat{c}_1 \hat{c}_2\right),$$

$$\hat{U}_6(\gamma) = \exp\left(-\tan\gamma \hat{c}_2^\dagger \hat{c}_1^\dagger\right) \exp\left[-\ln(\cos^2\gamma)\frac{1}{2}\left(\hat{c}_2^\dagger \hat{c}_2 - \hat{c}_1 \hat{c}_1^\dagger\right)\right] \exp\left(\tan\gamma \hat{c}_1 \hat{c}_2\right). \tag{6.34}$$

The actions of these unitary operators on the zero-fermion, or vacuum, state follow directly from these. The first four leave the vacuum state unchanged apart from a phase change:

$$\begin{aligned} \hat{U}_1(\gamma)|0\rangle &= e^{i\gamma/2}|0\rangle = \hat{U}_2(\gamma)|0\rangle, \\ \hat{U}_3(\gamma)|0\rangle &= |0\rangle = \hat{U}_4(\gamma)|0\rangle; \end{aligned} \tag{6.35}$$

but the other two induce the creation of a pair of fermions:

$$\begin{aligned} \hat{U}_5(\gamma)|0\rangle &= |\cos\gamma| \left(1 + i \tan\gamma \hat{c}_2^\dagger \hat{c}_1^\dagger\right)|0\rangle \\ &= |\cos\gamma| \left(|1,0:2,0\rangle - i\tan\gamma\, |1,1;2,1\rangle\right), \\ \hat{U}_6(\gamma)|0\rangle &= |\cos\gamma| \left(1 - \tan\gamma\ \hat{c}_2^\dagger \hat{c}_1^\dagger\right)|0\rangle \\ &= |\cos\gamma| \left(|1,0:2,0\rangle + i\tan\gamma\, |1,1;2,1\rangle\right), \end{aligned} \tag{6.36}$$

where $|1,0:2,0\rangle = |0\rangle$ is the two-mode vacuum Fock state and $|1,1;2,1\rangle = \hat{c}_1^\dagger \hat{c}_2^\dagger |1,0:2,0\rangle$ is a two-fermion Fock state.

Each of the six generators commutes either with the total number of particles in the two modes or with the difference. In particular,

$$\left[\hat{S}_\nu, \hat{c}_1^\dagger \hat{c}_1 + \hat{c}_2^\dagger \hat{c}_2\right] = 0, \qquad \nu = 1, \cdots, 4, \tag{6.37}$$

so that the states produced by the unitary transformations generated by $\hat{S}_1, \cdots, \hat{S}_4$ have the same total number of particles as the untransformed state. The remaining two generators commute with the difference in the particle number:

$$\left[\hat{S}_\nu, \hat{c}_1^\dagger \hat{c}_1 - \hat{c}_2^\dagger \hat{c}_2\right] = 0, \qquad \nu = 5, 6. \tag{6.38}$$

It follows that states generated by $\hat{S}_5$ and $\hat{S}_6$ have the same difference in particle number between the two modes as the untransformed state. The effect of our unitary transformations on the two-mode fermionic Fock states can be calculated using the inverses of the transformations (6.21):

$$\hat{U}_\nu|m_1, m_2\rangle = \left(\hat{U}_\nu(\gamma)\hat{c}_1^\dagger \hat{U}_\nu^\dagger(\gamma)\right)^{m_1} \left(\hat{U}_\nu(\gamma)\hat{c}_2^\dagger \hat{U}_\nu^\dagger(\gamma)\right)^{m_2} \hat{U}_\nu(\gamma)|m_1, m_2\rangle. \tag{6.39}$$

6.3 Two-Mode Interference

The wave-like nature of quantum particles is revealed most simply in interference, and any quantum theory must account for such effects. We find that single-particle interference occurs in essentially the same manner for bosons and for fermions. When more than one particle is involved, however, the observed phenomena depend on the quantum statistics of the particles [41].

In its simplest form, interference occurs when two modes overlap and become superposed. In optics, for example, this is readily achieved by the use of a partially reflecting mirror, or *beam splitter* [42]. For the associated annihilation operators, this produces the transformation

$$\begin{aligned} \hat{a}_1' &= t_1\hat{a}_1 + r_1\hat{a}_2, \\ \hat{a}_2' &= r_2\hat{a}_1 + t_2\hat{a}_2 \end{aligned} \tag{6.40}$$

for bosons and

$$\begin{aligned} \hat{c}_1' &= t_1\hat{c}_1 + r_1\hat{c}_2, \\ \hat{c}_2' &= r_2\hat{c}_1 + t_2\hat{c}_2 \end{aligned} \tag{6.41}$$

for fermions. The requirement that the resulting primed operators satisfy the required canonical commutation or anticommutation relations leads to constraints on the r and t coefficients:

$$\begin{aligned} |t_1|^2 + |r_1|^2 &= 1, \\ |t_2|^2 + |r_2|^2 &= 1, \\ r_2 t_1^* + t_2 r_1^* &= 0. \end{aligned} \tag{6.42}$$

These conditions also lead to the requirements that $|t_1|^2 = |t_2|^2$ and, of course, $|r_1|^2 = |r_2|^2$. The simplest choice is symmetrical in the two modes and has t real with r being purely imaginary, so that

$$\begin{aligned} \hat{a}_1' &= t\hat{a}_1 + r\hat{a}_2, \\ \hat{a}_2' &= r\hat{a}_1 + t\hat{a}_2, \end{aligned} \tag{6.43}$$

with a similar expression for the fermionic operators. We can readily invert these expressions using the properties of t and r to obtain

$$\begin{aligned} \hat{a}_1 &= t^*\hat{a}_1' + r^*\hat{a}_2' \\ &= t\hat{a}_1' - r\hat{a}_2', \\ \hat{a}_2 &= r^*\hat{a}_1' + t^*\hat{a}_2' \\ &= -r\hat{a}_1' + t\hat{a}_2', \end{aligned} \tag{6.44}$$

again with similar expressions for the fermionic operators. We can use these to determine the change in an initial state caused by the interference of the two modes.

The simplest non-trivial state to consider is that of a single particle, which we take to be prepared in mode 1. The interference of the modes produces the state

$$\begin{aligned}\hat{U}\hat{a}_1^\dagger|0\rangle &= t\hat{a}_1^\dagger|0\rangle + r\hat{a}_2^\dagger|0\rangle \\ &= t|1,0\rangle + r|0,1\rangle \end{aligned}\tag{6.45}$$

for bosons and

$$\begin{aligned}\hat{U}\hat{c}_1^\dagger|0\rangle &= t\hat{c}_1^\dagger|0\rangle + r\hat{c}_2^\dagger|0\rangle \\ &= t|1,0\rangle + r|0,1\rangle \end{aligned}\tag{6.46}$$

for fermions. We can interpret $|t|^2$ as the probability that the particle remains in the first mode and $|r|^2$ as the probability that it is transferred to the second. It is more interesting to note, however, that the state $t|1,0\rangle + r|0,1\rangle$ is a coherent *superposition* of two states, in which the particle is to be found in one of the two modes. From the perspective of the two modes, moreover, it is an entangled state; it cannot be separated into a tensor product of two one-mode states. It is important to appreciate that the phase embodied in this superposition is not lost, and it can be accessed by undoing the interference process. If the two modes are subject to distinct phase shifts ϕ_1 and ϕ_2, then the state evolves to

$$t|1,0\rangle + r|0,1\rangle \to te^{i\phi_1}|1,0\rangle + re^{i\phi_2}|0,1\rangle. \tag{6.47}$$

If we then cause the modes to interfere once more by a second transformation of the form (6.43), then we are left with the state

$$\hat{U}\left(te^{i\phi_1}|1,0\rangle + re^{i\phi_2}|0,1\rangle\right) = \left(t^2e^{i\phi_1} + r^2e^{i\phi_2}\right)|1,0\rangle + tr\left(e^{i\phi_1} + e^{i\phi_2}\right)|0,1\rangle. \tag{6.48}$$

It is clear that the amplitude for the particle to be left in mode 1 is a superposition of the amplitudes for having always been in mode 1 and that for changing to mode 2 and back again. The amplitude for being transferred to mode 2 is a superposition of amplitudes corresponding to transferring to mode 2 in the first interference process and to changing modes in the second. That the interference processes are coherent is clearly demonstrated by the dependence of these amplitudes on the two phase shifts ϕ_1 and ϕ_2. As we change these, the probability of finding the particle in mode 2 varies between 0 (complete destructive interference) and $4|t|^2|r|^2$, which takes a maximum value of 1 when $|t|^2 = |r|^2 = 1/2$.

The bosonic and fermionic behaviours differ most dramatically when there is precisely one particle in each of the modes. The interference of the two modes then produces the state

$$\begin{aligned}\hat{U}\hat{a}_1^\dagger\hat{a}_2^\dagger|0\rangle &= \left(t\hat{a}_1^\dagger + r\hat{a}_2^\dagger\right)\left(t\hat{a}_2^\dagger + r\hat{a}_1^\dagger\right)|0\rangle \\ &= \left[tr\left(\hat{a}_1^{\dagger 2} + \hat{a}_2^{\dagger 2}\right) + (t^2 + r^2)\hat{a}_1^\dagger\hat{a}_2^\dagger\right]|0\rangle \end{aligned}\tag{6.49}$$

for bosons and

$$\begin{aligned}\hat{U}\hat{c}_1^\dagger\hat{c}_2^\dagger|0\rangle &= \left(t\hat{c}_1^\dagger + r\hat{c}_2^\dagger\right)\left(t\hat{c}_2^\dagger + r\hat{c}_1^\dagger\right)|0\rangle \\ &= \left[tr\left(\hat{c}_1^{\dagger 2} + \hat{c}_2^{\dagger 2}\right) + (t^2 - r^2)\hat{c}_1^\dagger\hat{c}_2^\dagger\right]|0\rangle \end{aligned} \tag{6.50}$$

for fermions, where the minus sign associated with the r^2 term arises from the anti-commutation relation of the creation operators $\hat{c}_1^\dagger$ and $\hat{c}_2^\dagger$. These states appear, at first sight, to be rather similar, but they are in fact very different. The bosonic state, when written in the Fock basis, is

$$\hat{U}\hat{a}_1^\dagger\hat{a}_2^\dagger|0\rangle = \sqrt{2}tr\left(|2,0\rangle + |0,2\rangle\right) + (t^2 + r^2)|1,1\rangle. \tag{6.51}$$

The particles have not behaved independently. Had they done so, then we would expect to find two particles in the first mode and none in the second with probability $|t|^2|r|^2$, but here the probability takes the larger value of $2|t|^2|r|^2$. For two single-particle interference events, moreover, the probability of finding one particle in each output mode would be $|t|^4 + |r|^4$ but, instead, we find the value $|t^2 + r^2|^2$, which is *less* than $|t|^4 + |r|^4$ because r^2 is negative. At its most extreme, if $|t| = |r|$ then the probability that the interference leaves one particle in each mode becomes zero, so that the particles end up in the same mode. This two-particle interference effect has been demonstrated for photons by Hong, Ou and Mandel [43]. By way of contrast, for fermions we have $\hat{c}_1^{\dagger 2} = 0 = \hat{c}_2^{\dagger 2}$ and $t^2 - r^2 = |t|^2 + |r|^2 = 1$ so that

$$\hat{U}\hat{c}_1^\dagger\hat{c}_2^\dagger|0\rangle = \hat{c}_1^\dagger\hat{c}_2^\dagger|0\rangle = |1,1\rangle, \tag{6.52}$$

so that we are always left with precisely one particle in each mode. Pauli's exclusion principle tells us that we cannot have two fermions in the same mode, and hence one fermion in each mode is the only possible two-mode, two-fermion state.

6.4 Particle-Pair Creation

The unitary transformations associated with interference, described in the previous section, are strictly particle-number-conserving. We have seen, however, that some of our canonical transformations result in a change in the total number of particles. In each case, particles are added or removed in pairs. It might seem surprising that such particle-number-changing transformations would be important but, in fact, they appear in a variety of areas of quantum physics. Important examples include the pairing of electrons in superconductivity, the generation of squeezed states in quantum optics and the properties of condensed Bose gases. The main ideas are best illustrated by examples and to this end we discuss, briefly, three of these.

6.4.1 Squeezed States of Light

In the quantum theory of light, the electromagnetic field is quantised and may be written as a linear combination of the annihilation and creation operators for each of the modes. The electric and magnetic fields are observable quantities, and this means that we can measure linear combinations of $\hat{a}$ and $\hat{a}^\dagger$. The electric-field operator for a single field mode, of angular frequency ω, may be written in the simple form [44]

$$\hat{\vec{E}}(\mathbf{r},t) = \vec{\mathcal{E}}(\mathbf{r})\hat{a}e^{-i\omega t} + \vec{\mathcal{E}}^*(\mathbf{r})\hat{a}^\dagger e^{i\omega t}. \tag{6.53}$$

It is convenient to write this at a given point in space in a dimensionless form by introducing the *quadrature operators*

$$\hat{x}_\theta = \frac{1}{\sqrt{2}}\left(\hat{a}e^{-i\theta} + \hat{a}^\dagger e^{i\theta}\right), \tag{6.54}$$

so that the electric field has the form at the selected point in space

$$\hat{\vec{E}} \propto \hat{x}_0 \cos(\omega t) + \hat{x}_{\pi/2} \sin(\omega t). \tag{6.55}$$

The operators $\hat{x}_0$ and $\hat{x}_{\pi/2}$ do not commute but, rather, have a commutation relation which is reminiscent of that for the position and momentum operators:

$$\left[\hat{x}_0, \hat{x}_{\pi/2}\right] = i. \tag{6.56}$$

It follows that for every state of the field mode, the in-phase ($\hat{x}_0$) and in-quadrature ($\hat{x}_{\pi/2}$) components are incompatible observables and we have uncertainties bounded by the Heisenberg uncertainty principle:

$$\Delta\hat{x}_0 \; \Delta\hat{x}_{\pi/2} \geq \frac{1}{2}, \tag{6.57}$$

where $\Delta x_\theta^2 = \langle x_\theta^2 \rangle - \langle x_\theta \rangle^2$. The equality is satisfied with both uncertainties equal to $2^{-1/2}$ for the vacuum state of the field and also for the coherent states. This means that such states are characterized by the minimum possible level of *phase-insensitive* noise.

Non-linear optical media allow us to combine fields and to produce light with a different frequency. In second- and third-harmonic generation, for example, intense laser light of frequency ω is converted, in part, to light of frequency 2ω or 3ω. The fundamental process is the combination of two or three laser photons, catalysed by the non-linear medium, to produce a photon of frequency 2ω or 3ω, respectively. The inverse process of photon splitting is also possible and, in particular, in a parametric oscillator, laser light of frequency 2ω is converted into pairs of photons of frequency ω. The process is a coherent one and produces a superposition of even-photon-number states in the form of a *squeezed state* [4, 44, 45]

$$|\zeta\rangle = \hat{S}(\zeta)|0\rangle, \tag{6.58}$$

where $\hat{S}(\zeta)$ is the squeezing operator

$$\hat{S}(\zeta) = \exp\left(-\frac{\zeta}{2}\hat{a}^{\dagger 2} + \frac{\zeta^*}{2}\hat{a}^2\right) \tag{6.59}$$

and ζ is the squeezing parameter, given by $\zeta = re^{i\varphi}$, the value of which is determined by the intensity and phase of the 2ω laser field.

The defining property of a squeezed state is a reduction in the noise or uncertainty associated with one quadrature at the expense of the other. It is simplest to calculate the properties of the squeezed states by performing a canonical transformation on the creation and annihilation operators,

$$\begin{aligned}\hat{S}^\dagger(\zeta)\hat{a}\hat{S}(\zeta) &= \hat{a}\cosh r - \hat{a}^\dagger e^{i\varphi}\sinh r,\\ \hat{S}^\dagger(\zeta)\hat{a}^\dagger\hat{S}(\zeta) &= \hat{a}^\dagger\cosh r - \hat{a}e^{-i\varphi}\sinh r,\end{aligned} \tag{6.60}$$

so that the expectation value of a quadrature operator $\hat{x}_\theta$ is zero:

$$\begin{aligned}\langle\zeta|\hat{x}_\theta|\zeta\rangle &= \frac{1}{\sqrt{2}}\langle 0|\hat{S}^\dagger(\zeta)\left(\hat{a}e^{-i\theta} + \hat{a}^\dagger e^{i\theta}\right)\hat{S}(\zeta)|0\rangle\\ &= 0.\end{aligned} \tag{6.61}$$

The expectation value of the square of the quadrature, however, is

$$\begin{aligned}\langle\zeta|\hat{x}_\theta^2|\zeta\rangle &= \frac{1}{2}\langle 0|\hat{S}^\dagger(\zeta)\left(\hat{a}e^{-i\theta} + \hat{a}^\dagger e^{i\theta}\right)^2\hat{S}(\zeta)|0\rangle\\ &= \frac{1}{2}\left[e^{2r}\sin^2(\theta - \varphi/2) + e^{-2r}\cos^2(\theta - \varphi/2)\right].\end{aligned} \tag{6.62}$$

The uncertainty oscillates as a function of θ between a minimum value of

$$\Delta x_{\varphi/2} = \frac{1}{\sqrt{2}}e^{-r} \tag{6.63}$$

and a maximum value of

$$\Delta x_{(\varphi+\pi)/2} = \frac{1}{\sqrt{2}}e^{r}. \tag{6.64}$$

If we measure the component of the electric field associated with the quadrature $\hat{x}_{\varphi/2}$, then we find an uncertainty lower than that associated with the vacuum field. This reduction in the noise level has been shown to improve the sensitivity of a variety of optical measurements.

There are also *two-mode squeezed states*, generated in non-degenerate parametric oscillators, in which the photon pairs are created in two separate modes. The states have the form

$$|\zeta\rangle_{12} = \hat{S}_{12}(\zeta)|0\rangle, \tag{6.65}$$

where $\hat{S}_{12}$ is the two-mode squeezing operator

$$\hat{S}_{12} = \exp\left(-\zeta\hat{a}_2^\dagger\hat{a}_1^\dagger + \zeta^*\hat{a}_1\hat{a}_2\right). \tag{6.66}$$

Note that this is not simply the product of two single-mode squeezing operators. The two-mode squeezed states are *entangled states*, in which the quantum properties manifest themselves not in the single-mode properties, but rather in correlations between the two modes.

6.4.2 Thermofields

Systems in thermodynamic equilibrium are naturally described by mixed states corresponding to the maximum possible entropy subject to environmental constraints, usually corresponding to the mean energy. It is also possible to describe such systems by a *pure* state in an enlarged Hilbert space. These states and their manipulation form the topic of thermofield dynamics, and the method has been employed widely in finite-temperature quantum field theory.

Before introducing thermofields [46], it is worthwhile reviewing briefly the more familiar treatment of *thermal states*. It is simplest to allow both energy and particles to be exchanged with the environment and so work with the grand canonical ensemble. The key function for determining most of the thermodynamic properties of our system is the partition function, which we can write in the form

$$Z(\beta) = \sum_{N=1}^{\infty} \sum_{i} e^{\beta \mu N} e^{-\beta E_i}, \tag{6.67}$$

where $\beta = (k_B T)^{-1}$ is the inverse temperature and the E_i are the allowed energies. The chemical potential μ fixes the mean, or equilibrium, number of particles. For a Fermi gas μ is positive, and for a Bose gas it is typically negative. For photons, we have black-body radiation, for which $\mu = 0$. If $\hat{N}$ is the particle number operator and $\hat{H}$ is the particle Hamiltonian, then we can describe the equilibrium state of our particles by the density operator

$$\hat{\rho} = Z^{-1}(\beta) e^{\beta(\mu \hat{N} - \hat{H})}. \tag{6.68}$$

All the properties of our gas can be evaluated using this state.

It suffices, for the purposes of illustration, to restrict our attention to ideal (or very weakly interacting) gases and to consider a single mode. In this case the energy will be $\hbar\omega\hat{N}$, and the single-mode density operator becomes

$$\hat{\rho} = \frac{e^{(\mu - \hbar\omega)\beta \hat{N}}}{\mathrm{Tr}\left[e^{(\mu - \hbar\omega)\beta \hat{N}} \right]}. \tag{6.69}$$

For bosons this reduces to

$$\hat{\rho} = \left[1 - e^{(\mu - \hbar\omega)\beta} \right] e^{(\mu - \hbar\omega)\beta \hat{a}^\dagger \hat{a}}, \tag{6.70}$$

and for fermions it becomes

$$\hat{\rho} = \frac{e^{(\mu - \hbar\omega)\beta \hat{c}^\dagger \hat{c}}}{1 + e^{(\mu - \hbar\omega)\beta}}. \tag{6.71}$$

These operators are diagonal in the Fock-state basis and so can be written in the simple form

$$\hat{\rho} = \sum_n P(n)|n\rangle\langle n|. \tag{6.72}$$

For bosons the occupation probabilities are

$$P(n) = \left[1 - e^{(\mu-\hbar\omega)\beta}\right] e^{(\mu-\hbar\omega)\beta n}, \qquad n = 0, 1, \cdots, \infty, \tag{6.73}$$

which is the familiar Bose–Einstein distribution. For fermions, of course, we have the Fermi–Dirac distribution

$$P(n) = \frac{e^{(\mu-\hbar\omega)\beta n}}{1 + e^{(\mu-\hbar\omega)\beta}}, \qquad n = 0, 1. \tag{6.74}$$

In this case we can identify μ with the Fermi energy.

The key idea in *thermofields* is the observation that the mixed-state properties of our mode are mimicked perfectly by the two-mode pure state

$$|0(\beta)\rangle = \sum_n \sqrt{P(n)}|n, n\rangle. \tag{6.75}$$

To see this, we need only show that the expectation value of any single-mode operator $\hat{A}$ is the same for this state as it is for the thermal density operator, (6.71). It is straightforward to demonstrate this by direct calculation:

$$\begin{aligned}
\langle 0(\beta)|\hat{A} \otimes \hat{I}|0(\beta)\rangle &= \sum_n \sum_{n'} \sqrt{P(n)P(n')}\langle n', n'|\hat{A} \otimes \hat{I}|n, n\rangle \\
&= \sum_n P(n)\langle n|\hat{A}|n\rangle \\
&= \mathrm{Tr}\left[\hat{A}\sum_n P(n)|n\rangle\langle n|\right].
\end{aligned} \tag{6.76}$$

This method of replacing a mixed state by a pure state in a doubled state space is quite general and works for any mixed state of any quantum system. When rediscovered in the field of quantum information, this technique became known as *purification.*

The power of thermofields derives from the fact that, both for bosons and for fermions, the pure state $|0(\beta)\rangle$ is related to the two-mode vacuum state by a linear transformation. For this reason, the state $|0(\beta)\rangle$ is referred to as the thermal vacuum state. It is conventional to refer to the additional field as the 'fictitious' field and to denote operators and states associated with the modes of this field with a tilde. We find that the thermal vacuum state has the form

$$|0(\beta)\rangle = \exp\left[\theta(\beta)\left(\hat{\tilde{a}}^\dagger \hat{a}^\dagger - \hat{a}\hat{\tilde{a}}\right)\right]|0, \tilde{0}\rangle \tag{6.77}$$

for bosons and

$$|0(\beta)\rangle = \exp\left[\theta(\beta)\left(\hat{\tilde{c}}^\dagger \hat{c}^\dagger - \hat{c}\hat{\tilde{c}}\right)\right]|0,\tilde{0}\rangle \tag{6.78}$$

for fermions. It is straightforward to show that the parameter $\theta(\beta)$ is related to the inverse temperature by

$$\tanh\theta(\beta) = e^{(\mu-\hbar\omega)\beta} \tag{6.79}$$

for bosons and

$$\tan\theta(\beta) = e^{(\mu-\hbar\omega)\beta} \tag{6.80}$$

for fermions.

It is straightforward to transform any thermal expectation value into a vacuum expectation value by performing the unitary transformation implicit in the definition of $|0(\beta)\rangle$ on the operators. In particular, we find

$$\langle 0(\beta)|f(\hat{a},\hat{a}^\dagger)\,|0(\beta)\rangle = \langle 0,\tilde{(0)}|f\left(\cosh\theta\;\hat{a} + \sinh\theta\;\hat{\tilde{a}}^\dagger, \cosh\theta\;\hat{a}^\dagger + \sinh\theta\;\hat{\tilde{a}}\right)|0,\tilde{0}\rangle \tag{6.81}$$

for bosons and

$$\langle 0(\beta)|f(\hat{c},\hat{c}^\dagger)\,|0(\beta)\rangle = \langle 0,\tilde{(0)}|f\left(\cos\theta\;\hat{c} - \sin\theta\;\hat{\tilde{c}}^\dagger, \cos\theta\;\hat{c}^\dagger - \sin\theta\;\hat{\tilde{c}}\right)|0,\tilde{0}\rangle \tag{6.82}$$

for fermions. A simple example will help to clarify the procedure. The particle-number statistics are equivalent to the factorial moments, which for bosons take the form $n^{(m)} = \langle(\hat{a}^\dagger)^m(\hat{a})^m\rangle$. For thermal states, we find

$$\begin{aligned} n^{(m)} &= \langle 0(\beta)|(\hat{a}^\dagger)^m(\hat{a})^m|0(\beta)\rangle \\ &= \langle 0,\tilde{0}|\left(\cosh\theta\;\hat{a}^\dagger + \sinh\theta\;\hat{\tilde{a}}\right)^m\left(\cosh\theta\;\hat{a} + \sinh\theta\;\hat{\tilde{a}}^\dagger\right)^m|0,\tilde{0}\rangle \\ &= \sinh^{2m}\theta\;\langle 0,\tilde{0}|\left(\hat{\tilde{a}}\right)^m\left(\hat{\tilde{a}}^\dagger\right)^m|0,\tilde{0}\rangle \\ &= m!\bar{n}^m. \end{aligned} \tag{6.83}$$

6.4.3 Bogoliubov Excitations of a Zero-Temperature Bose Gas

As a final example of the application of linear canonical transformations, we present Bogoliubov's calculation of the excitation spectrum of a zero-temperature Bose gas [47]. We follow the analysis given by Pethick and Smith [48], where further details and elaborations may be found. The Hamiltonian for our system has the simple form

$$\hat{H} = \int_{\mathcal{V}} d^3r\left[-\hat{\psi}^\dagger\frac{\hbar^2}{2m}\nabla^2\hat{\psi} + V(\mathbf{r})\hat{\psi}^\dagger\hat{\psi} + \frac{U_0}{2}\hat{\psi}^\dagger\hat{\psi}^\dagger\hat{\psi}\hat{\psi}\right]. \tag{6.84}$$

Here $V(\mathbf{r})$ is the external potential trapping the gas, and we have assumed only a short-range interaction between the bosons, modelled by a contact or delta-function potential. For the purposes of illustration, we suppose that the gas is confined within a box of volume $\mathcal{V}$ with the potential $V(\mathbf{r}) = 0$ inside the box. With these simplifications, the Hamiltonian given in (6.84) becomes

$$\hat{H} = \sum_{\mathbf{p}} \frac{p^2}{2m} \hat{a}_{\mathbf{p}}^\dagger \hat{a}_{\mathbf{p}} + \frac{U_0}{2\mathcal{V}} \sum_{\mathbf{p},\mathbf{p}',\mathbf{q}} \hat{a}_{\mathbf{p}+\mathbf{q}}^\dagger \hat{a}_{\mathbf{p}'-\mathbf{q}}^\dagger \hat{a}_{\mathbf{p}'} \hat{a}_{\mathbf{p}}. \tag{6.85}$$

The single-mode annihilation operator $\hat{a}_{\mathbf{p}}$ removes a single boson with momentum $\mathbf{p}$ and satisfies the usual commutation relations

$$\begin{aligned} \left[\hat{a}_{\mathbf{p}}, \hat{a}_{\mathbf{p}'}^\dagger\right] &= \delta_{\mathbf{p}\mathbf{p}'}, \\ \left[\hat{a}_{\mathbf{p}}, \hat{a}_{\mathbf{p}'}\right] &= 0 = \left[\hat{a}_{\mathbf{p}}^\dagger, \hat{a}_{\mathbf{p}'}^\dagger\right]. \end{aligned} \tag{6.86}$$

At low temperatures, condensation occurs and there is a macroscopic occupation of the lowest-energy, zero-momentum state. We proceed by replacing the corresponding annihilation and creation operators, $\hat{a}_0$ and $\hat{a}_0^\dagger$, by $\sqrt{N_0}$, where N_0 is the number of condensate particles. This will become a good approximation in the limit of large particle number, in which $\sqrt{N_0 + 1} \approx \sqrt{N_0}$. We are, in effect, replacing the operator $\hat{\psi}$ by $\sqrt{N_0/\mathcal{V}} + \delta\hat{\psi}$ and treating $\delta\hat{\psi}$ as small. With this simplification, our Hamiltonian becomes

$$\begin{aligned} \hat{H} &= \sum_{\mathbf{p}} \frac{p^2}{2m} \hat{a}_{\mathbf{p}}^\dagger \hat{a}_{\mathbf{p}} + \frac{U_0}{2\mathcal{V}} \hat{a}_0^{\dagger 2} \hat{a}_0^2 + \frac{U_0}{2\mathcal{V}} \sum_{\mathbf{p}\neq 0} \left(4\hat{a}_0^\dagger \hat{a}_{\mathbf{p}}^\dagger \hat{a}_{\mathbf{p}} \hat{a}_0 + \hat{a}_0^{\dagger 2} \hat{a}_{\mathbf{p}} \hat{a}_{\mathbf{p}} + \hat{a}_{\mathbf{p}}^\dagger \hat{a}_{\mathbf{p}}^\dagger \hat{a}_0^2\right) \\ &\approx \frac{N_0^2 U_0}{2\mathcal{V}} + \sum_{\mathbf{p}\neq 0} \left(\frac{p^2}{2m} + \frac{2N_0 U_0}{\mathcal{V}}\right) \hat{a}_{\mathbf{p}}^\dagger \hat{a}_{\mathbf{p}} + \frac{N_0 U_0}{2\mathcal{V}} \sum_{\mathbf{p}\neq 0} \left(\hat{a}_{\mathbf{p}}^\dagger \hat{a}_{-\mathbf{p}}^\dagger + \hat{a}_{\mathbf{p}} \hat{a}_{-\mathbf{p}}\right). \end{aligned} \tag{6.87}$$

In this form, we have interactions only between modes corresponding to equal and opposite momenta. These serve to create or remove pairs of particles in a way that corresponds, of course, to scattering particles out of or into the macroscopically occupied state.

The original Hamiltonian (6.84) is strictly particle-number-conserving but, in treating the condensate as a classical c-number field, we have lost this feature. We can at least impose approximate conservation on the mean particle number by noting that the total particle-number operator is

$$\begin{aligned} \hat{N} &= \hat{a}_0^\dagger \hat{a}_0 + \sum_{\mathbf{p}\neq 0} \hat{a}_{\mathbf{p}}^\dagger \hat{a}_{\mathbf{p}} \\ &\approx N_0 + \sum_{\mathbf{p}\neq 0} \hat{a}_{\mathbf{p}}^\dagger \hat{a}_{\mathbf{p}}, \end{aligned} \tag{6.88}$$

and using this to write N_0 in terms of the total particle number N:

$$\begin{aligned}\hat{H} &= \frac{N^2U_0}{2\mathcal{V}} + \sum_{\mathbf{p}\neq 0}\left(\frac{p^2}{2m} + \frac{N_0U_0}{\mathcal{V}}\right)\hat{a}^\dagger{}_{\mathbf{p}}\hat{a}_{\mathbf{p}} + \sum_{\mathbf{p}\neq 0}\frac{N_0U_0}{2\mathcal{V}}\left(\hat{a}^\dagger_{\mathbf{p}}\hat{a}^\dagger_{-\mathbf{p}} + \hat{a}_{-\mathbf{p}}\hat{a}_{\mathbf{p}}\right)\\ &= \frac{N^2U_0}{2\mathcal{V}} + \frac{1}{2}\sum_{\mathbf{p}\neq 0}\left[\left(\frac{p^2}{2m} + \frac{N_0U_0}{\mathcal{V}}\right)\left(\hat{a}^\dagger_{\mathbf{p}}\hat{a}_{\mathbf{p}} + \hat{a}^\dagger_{-\mathbf{p}}\hat{a}_{-\mathbf{p}}\right) + \frac{N_0U_0}{\mathcal{V}}\left(\hat{a}^\dagger_{\mathbf{p}}\hat{a}^\dagger_{-\mathbf{p}} + \hat{a}_{-\mathbf{p}}\hat{a}_{\mathbf{p}}\right)\right].\end{aligned} \tag{6.89}$$

Our final task is to diagonalise the Hamiltonian and so arrive at the spectrum of excitations. It is simplest to consider just a pair of coupled momenta, $\mathbf{p}$ and $-\mathbf{p}$, for which the Hamiltonian is

$$\hat{H}_{\mathbf{p}} = \epsilon_p\left(\hat{a}^\dagger_{\mathbf{p}}\hat{a}_{\mathbf{p}} + \hat{a}^\dagger_{-\mathbf{p}}\hat{a}_{-\mathbf{p}}\right) + \kappa\left(\hat{a}^\dagger_{\mathbf{p}}\hat{a}^\dagger_{-\mathbf{p}} + \hat{a}_{-\mathbf{p}}\hat{a}_{\mathbf{p}}\right), \tag{6.90}$$

where

$$\begin{aligned}\epsilon_p &= \frac{p^2}{2m} + \frac{N_0U_0}{\mathcal{V}},\\ \kappa &= \frac{N_0U_0}{\mathcal{V}}.\end{aligned} \tag{6.91}$$

The Hamiltonian $\hat{H}_{\mathbf{p}}$ is diagonalised by means of the *Bogoliubov transformation*, which produces the new operators

$$\begin{aligned}\hat{a}'_{\mathbf{p}} &= u_p\hat{a}_{\mathbf{p}} + v_p\hat{a}^\dagger_{-\mathbf{p}}\\ \Rightarrow \hat{a}_{\mathbf{p}} &= u_p\hat{a}'_{\mathbf{p}} - v_p\hat{a}^{\dagger'}_{-\mathbf{p}}\end{aligned} \tag{6.92}$$

and their Hermitian conjugates. This is one of our linear canonical transformations, and imposing the bosonic commutation relations on the transformed operators leads to the requirement that

$$u_p^2 - v_p^2 = 1. \tag{6.93}$$

If we substitute our new operators into (6.90), then we find

$$\begin{aligned}\hat{H}_{\mathbf{p}} = 2v_p^2\epsilon_p - 2u_pv_p\kappa &+ \left[\epsilon_p\left(u_p^2 + v_p^2\right) - 2\kappa u_pv_p\right]\left(\hat{a}^{\dagger'}_{\mathbf{p}}\hat{a}'_{\mathbf{p}} + \hat{a}^{\dagger'}_{-\mathbf{p}}\hat{a}'_{-\mathbf{p}}\right)\\ &+ \left[\kappa\left(u_p^2 + v_p^2\right) - 2\epsilon_p\kappa u_pv_p\right]\left(\hat{a}^{\dagger'}_{\mathbf{p}}\hat{a}^{\dagger'}_{-\mathbf{p}} + \hat{a}'_{-\mathbf{p}}\hat{a}'_{\mathbf{p}}\right),\end{aligned} \tag{6.94}$$

which we can reduce to diagonal form by choosing u_p and v_p so that

$$\kappa\left(u_p^2 + v_p^2\right) - 2\epsilon_p\kappa u_pv_p = 0. \tag{6.95}$$

The solution of this is

$$u_p^2 = \frac{1}{2}\left(\frac{\epsilon_p}{\sqrt{\epsilon_p^2 - \kappa^2}} + 1\right) = v_p^2 + 1, \tag{6.96}$$

which leads to the form

$$\hat{H}_{\mathbf{p}} = \sqrt{\epsilon_p^2 - \kappa^2}\left(\hat{a}_{\mathbf{p}}^{\dagger'}\hat{a}_{\mathbf{p}}' + \hat{a}_{-\mathbf{p}}^{\dagger'}\hat{a}_{-\mathbf{p}}'\right) + \sqrt{\epsilon_p^2 - \kappa^2} - \epsilon_p. \tag{6.97}$$

Note that this diagonalisation implicitly assumes that $\epsilon_p^2 > \kappa^2$. If this is not the case then the diagonalisation fails, as the energy becomes complex. This reflects the fact that under these circumstances the condensed state is *unstable* to the scattering of pairs of particles into other modes.

The complete diagonalised Hamiltonian is

$$\begin{aligned}\hat{H} &= \frac{N^2U_0}{2\mathcal{V}} + \frac{1}{2}\sum_{\mathbf{p}\neq 0}\hat{H}_{\mathbf{p}} \\ &= \frac{N^2U_0}{2\mathcal{V}} + \sum_{\mathbf{p}\neq 0}\sqrt{\epsilon_p^2 - \kappa^2}\hat{a}_{\mathbf{p}}^{\dagger'}\hat{a}_{\mathbf{p}}' - \frac{1}{2}\sum_{\mathbf{p}\neq 0}\epsilon_p - \sqrt{\epsilon_p^2 - \kappa^2}.\end{aligned} \tag{6.98}$$

This tells us that the elementary excitations created by the operators $\hat{a}_{\mathbf{p}}^{\dagger'}$ have energy

$$\begin{aligned}\sqrt{\epsilon_p^2 - \kappa^2} &= \left[\left(\frac{p^2}{2m} + \frac{N_0U_0}{\mathcal{V}}\right)^2 - \left(\frac{N_0U_0}{\mathcal{V}}\right)^2\right]^{1/2} \\ &= \left[\frac{p^2}{2m}\left(\frac{p^2}{2m} + \frac{2N_0U_0}{\mathcal{V}}\right)\right]^{1/2}.\end{aligned} \tag{6.99}$$

This means that the excitations have energies which are phonon-like (proportional to k) for small momentum and free-particle-like (proportional to k^2) for large values of the momentum.

Exercises

(6.1) Complete the set of examples (6.18) by calculating the forms of $\hat{a}_1'$ and $\hat{a}_2'$ for transformations associated with each of the ten generators (6.15).

(6.2) Prove the conservation laws embodied in the commutators (6.24), (6.25) and (6.26).

(6.3) Complete the set of examples (6.33) by calculating the actions of the six unitary transformations on the operator $\hat{c}_2$.

(6.4) Calculate the Fock-state expansion of each of the states (6.39) for the six generators $\hat{S}_\nu$.

(6.5) Derive the conditions (6.42) from the bosonic commutation or fermionic anticommutation relations. Are there any other independent constraints?

7
Phase Space Distributions

In this chapter, we relate normally ordered quantum correlation functions of annihilation and creation operators to phase space integrals in which the density operator is represented by a distribution function and the operators for each mode are represented by pairs of c-numbers and Grassmann numbers for bosons and fermions, respectively. For these cases, the distribution function is in turn a c-number function or a Grassmann function. Combined boson–fermion cases are also covered. Unless stated otherwise, the treatment is confined to processes where conversions of pairs of fermions into a boson do not occur. The formalism is based on first introducing characteristic functions, which are uniquely determined from and equivalent to the density operator and effectively are an encoding of the quantum correlation functions. The existence of the distribution functions is demonstrated via the canonical representation of the density operator in terms of Bargmann state projectors. The bosonic distribution function is not unique, though the fermion distribution function is. In addition, unnormalised distribution functions are introduced, based on unnormalised projectors using Bargmann coherent states. These give simple phase space integral expressions for Fock state populations and coherences. In the next chapter, the replacement of the Liouville–von Neumann or master equation for the density operator by a Fokker–Planck equation for the distribution function is obtained via the use of so-called correspondence rules. Finally, in the following chapter, the phase space integrals that determine the quantum correlation functions via the distribution function are shown to be equivalent to stochastic averages of stochastic variables that represent the bosonic and fermionic operators and which satisfy Langevin equations of the Ito stochastic type, involving terms that are obtained from the Fokker–Planck equation. Applications of the theory for few-mode systems are presented in Chapter 10.

In this chapter, the focus on normally ordered quantum correlation functions leads to distribution functions of the positive P type. The use of a double phase space enables Fokker–Planck equations with positive definite diffusion matrices to be obtained (see the next chapter), showing that distribution functions that are real and positive can be found. For the positive P representation, the Fokker–Planck equation for systems involving Hamiltonians containing one- or two-body interactions and with weak, Markovian couplings to the environment will involve only first- and second-order differentiation. This is a key requirement for developing Ito stochastic equations and is an outcome not always possible with other distributions. If Ito stochastic equations are not required, then normally ordered correlation functions can be studied using the standard Glauber–Sudarshan P representation, which uses only a single phase space. In addition, several other varieties of characteristic and distribution functions

Phase Space Methods for Degenerate Quantum Gases. First Edition. Bryan J. Dalton, John Jeffers, and Stephen M. Barnett.
Published in 2015 by Oxford University Press.

are used, depending on whether the quantum correlation functions involve symmetrically or antinormally ordered products of annihilation and creation operators (assuming the density operator is placed to the left or right of these operators in the trace). For single phase spaces these are the Wigner W and Husimi Q distributions, respectively. However, double-phase-space versions of the Wigner distribution [49] and hybrids involving the positive P and Wigner distributions [50–53] are now also used. The Wigner distribution is extensively used for bosonic systems where Bose–Einstein condensation can occur, and which results in most of the bosons occupying one or two modes. In this case mean-field theories are a good approximation, and these are conveniently treated via the Wigner distribution since the quantum fluctuations that can be described via the positive P distribution are less significant. For completeness, the double-space Wigner distribution functions will be presented here for bosons. More recently, Deuar and Drummond [54, 55] have extended the positive P distribution for bosons to include stochastic gauges. Finally, Corney and Drummond have developed phase space distribution methods for both bosons [56] and fermions [57, 58], based on Gaussian projector operators. This approach is fully dealt with in the articles cited and is briefly described in Chapter 16.

In later chapters (12–15), we consider distribution functionals and their applications in situations where large numbers of modes are involved.

7.1 Quantum Correlation Functions

7.1.1 Normally Ordered Expectation Values

The aim is to determine quantum correlation functions which are expectation values of *normally ordered* products of creation and annihilation operators. These are defined in terms of the density operators $\hat{\rho}_{\mathrm{b}}$ and $\hat{\rho}_{\mathrm{f}}$ for the boson and fermion cases, respectively, by

$$\begin{aligned} G_{\mathrm{b}}(l_1,\cdots,l_p;m_q,\cdots,m_1) &= \left\langle \hat{a}_{l_1}^{\dagger}\cdots\hat{a}_{l_p}^{\dagger}\hat{a}_{m_q}\cdots\hat{a}_{m_1}\right\rangle \\ &= \mathrm{Tr}\left(\hat{a}_{m_q}\cdots\hat{a}_{m_1}\hat{\rho}_b\,\hat{a}_{l_1}^{\dagger}\cdots\hat{a}_{l_p}^{\dagger}\right) \end{aligned} \tag{7.1}$$

for bosons and

$$\begin{aligned} G_{\mathrm{f}}(l_1,\cdots,l_p;m_q,\cdots,m_1) &= \left\langle \hat{c}_{l_1}^{\dagger}\cdots\hat{c}_{l_P}^{\dagger}\hat{c}_{m_q}\cdots\hat{c}_{m_1}\right\rangle \\ &= \mathrm{Tr}\left(\hat{c}_{m_q}\cdots\hat{c}_{m_1}\hat{\rho}_f\,\hat{c}_{l_1}^{\dagger}\cdots\hat{c}_{l_p}^{\dagger}\right) \end{aligned} \tag{7.2}$$

for fermions. The quantities $l_1,l_2,\cdots,l_p;m_q,\cdots,m_2,m_1$ are quantum modes as usual, but now in any order. 'Normally ordered' means that creation operators appear to the left of annihilation operators *when* the density operator is placed to the left (or right) of *all* of these operators. As discussed in Chapter 1, the third forms for the quantum correlation functions – with the density operator placed in the middle – occur in theories of multi-particle detection and link the quantum correlation functions to experiment. For physical states of simple bosonic or fermionic systems where bosons or fermions are not created or destroyed, the quantum correlation functions are zero unless $p=q$.

7.1.2 Symmetrically Ordered Expectation Values

Here we determine quantum correlation functions which are expectation values of *symmetrically ordered* products of creation and annihilation operators. These are defined in terms of the density operators $\hat{\rho}_{\text{b}}$ for bosonic cases by

$$\begin{aligned} G_{\text{b}}^{\text{W}}(l_1,\cdots,l_p;m_q,\cdots,m_1) &= \left\langle \left\{ \hat{a}_{l_1}^{\dagger}\cdots\hat{a}_{l_p}^{\dagger}\hat{a}_{m_q}\cdots\hat{a}_{m_1} \right\} \right\rangle \\ &= \text{Tr}\left(\hat{\rho}_{\text{b}} \left\{ \hat{a}_{l_1}^{\dagger}\cdots\hat{a}_{l_p}^{\dagger}\hat{a}_{m_q}\cdots\hat{a}_{m_1} \right\} \right). \end{aligned} \tag{7.3}$$

'Symmetrically ordered' means the average of the product of the operators taken in any order. Thus

$$\{\hat{a}_{l_1}^{\dagger}\cdots\hat{a}_{l_p}^{\dagger}\hat{a}_{m_q}\cdots\hat{a}_{m_1}\} = \frac{1}{(p+q)!}\sum_{R}\Re(\hat{a}_{l_1}^{\dagger}\cdots\hat{a}_{l_p}^{\dagger}\hat{a}_{mq}\cdots\hat{a}_{m_1}). \tag{7.4}$$

In (7.4), the sum over R is over all $(p+q)!$ orderings $\Re$ of the factors $\hat{a}_{l_1}^{\dagger}\cdots\hat{a}_{l_p}^{\dagger}\hat{a}_{m_q}\cdots\hat{a}_{m_1}$.

7.2 Characteristic Functions

In order to determine the quantum correlation functions via the phase space approach involving distribution functions, it is convenient to first introduce so-called *characteristic functions.* For each mode i, the bosonic characteristic function $\chi_{\text{b}}(\xi,\xi^+)$ depends analytically on pairs of c-number variables ξ_i,ξ_i^+, whilst the fermionic characteristic function $\chi_{\text{f}}(h,h^+)$ involves two Grassmann variables h_i,h_i^+. These functions are essentially a way of encoding all the quantum correlation functions in one overall function. They are uniquely determined by and are equivalent to the density operator. As mentioned previously, there are also several other varieties of characteristic function, depending on whether the quantum correlation functions involve normally ordered, antinormally ordered or symmetrically ordered products of annihilation and creation operators and/or whether only a single c-number or Grassmann variable is associated with each mode, but here we will focus mainly on the normally ordered characteristic functions involving double phase spaces, though the symmetrically ordered case for bosons will also be included. We initially focus on simple bosonic or fermionic systems where the total number of particles is conserved.

7.2.1 Bosons

For bosonic systems, we define the *normally ordered* characteristic function $\chi_{\text{b}}(\boldsymbol{\xi},\boldsymbol{\xi}^+)$ via

$$\chi_{\text{b}}(\boldsymbol{\xi},\boldsymbol{\xi}^+) = \text{Tr}[\hat{\Omega}_{\text{b}}^+(\boldsymbol{\xi}^+)\,\hat{\rho}_{\text{b}}\,\hat{\Omega}_{\text{b}}^-(\boldsymbol{\xi})], \tag{7.5}$$

where we introduce the operator functions

$$\hat{\Omega}_{\text{b}}^+(\boldsymbol{\xi}^+) = \exp\left(i\hat{\mathbf{a}}\cdot\boldsymbol{\xi}^+\right), \qquad \hat{\Omega}_{\text{b}}^-(\boldsymbol{\xi}) = \exp\left(i\boldsymbol{\xi}\cdot\hat{\mathbf{a}}^{\dagger}\right). \tag{7.6}$$

Here we adopt the bold face notation for multi-mode systems that we introduced in Chapter 5. Note that the two numbers associated with each mode, ξ_i, ξ_i^+, are each complex and are not related to each other. Hence the pair ξ_i, ξ_i^+ represents points in two different complex planes. The characteristic function $\chi_{\mathrm{b}}(\boldsymbol{\xi}, \boldsymbol{\xi}^+)$ depends only on the ξ_i, ξ_i^+ and not on their complex conjugates ξ_i^*, ξ_i^{+*}. Indeed, the characteristic function is an analytic function of the complex variables $\boldsymbol{\xi}, \boldsymbol{\xi}^+$. The cyclic properties of the trace for bosonic operators mean that we can also write the characteristic function in the equivalent forms $\mathrm{Tr}[\hat{\rho}_{\mathrm{b}}\, \hat{\Omega}_{\mathrm{b}}^-(\boldsymbol{\xi})\hat{\Omega}_{\mathrm{b}}^+(\boldsymbol{\xi}^+)]$ and $\mathrm{Tr}[\hat{\Omega}_{\mathrm{b}}^-(\boldsymbol{\xi})\hat{\Omega}_{\mathrm{b}}^+(\boldsymbol{\xi}^+)\, \hat{\rho}_{\mathrm{b}}]$.

The symmetrically ordered characteristic function $\chi_{\mathrm{b}}^{\mathrm{W}}(\boldsymbol{\xi}, \boldsymbol{\xi}^+)$ is defined by

$$\chi_{\mathrm{b}}^{\mathrm{W}}(\boldsymbol{\xi}, \boldsymbol{\xi}^+) = \mathrm{Tr}\left[\hat{\Omega}_{\mathrm{b}}^{\mathrm{W}}(\boldsymbol{\xi}, \boldsymbol{\xi}^+)\, \hat{\rho}_{\mathrm{b}}\right], \tag{7.7}$$

where the operator $\hat{\Omega}_{\mathrm{b}}^{\mathrm{W}}$ is simply related to the Glauber displacement operator

$$\hat{\Omega}_{\mathrm{b}}^{\mathrm{W}}(\boldsymbol{\xi}, \boldsymbol{\xi}^+) = \exp\left[i\left(\hat{\mathbf{a}} \cdot \boldsymbol{\xi}^+ + \boldsymbol{\xi} \cdot \hat{\mathbf{a}}^\dagger\right)\right] = \hat{D}_{\mathrm{b}}(i\boldsymbol{\xi}, -i\boldsymbol{\xi}^+). \tag{7.8}$$

Note, however, that as $\boldsymbol{\xi}^+ \neq \boldsymbol{\xi}^*$, this operator is not unitary. By applying the Baker–Hausdorff theorem, the symmetrically and normally ordered characteristic functions can be related:

$$\hat{\Omega}_{\mathrm{b}}^{\mathrm{W}}(\boldsymbol{\xi}, \boldsymbol{\xi}^+) = \hat{\Omega}_{\mathrm{b}}^-(\boldsymbol{\xi})\, \hat{\Omega}_{\mathrm{b}}^+(\boldsymbol{\xi}^+) \exp\left(-\frac{1}{2}\boldsymbol{\xi} \cdot \boldsymbol{\xi}^+\right), \tag{7.9}$$

so that

$$\chi_{\mathrm{b}}^{\mathrm{W}}(\boldsymbol{\xi}, \boldsymbol{\xi}^+) = \exp\left(-\frac{1}{2}\boldsymbol{\xi} \cdot \boldsymbol{\xi}^+\right) \chi_{\mathrm{b}}(\boldsymbol{\xi}, \boldsymbol{\xi}^+), \tag{7.10}$$

which enables us to obtain results for (symmetrically ordered) Wigner distribution functions from those for positive P distribution functions.

7.2.2 Fermions

For fermionic systems, we define the characteristic function $\chi_{\mathrm{f}}(\mathbf{h}, \mathbf{h}^+)$ via

$$\chi_{\mathrm{f}}(\mathbf{h}, \mathbf{h}^+) = \mathrm{Tr}[\hat{\Omega}_{\mathrm{f}}^+(\mathbf{h}^+)\, \hat{\rho}_{\mathrm{f}}\, \hat{\Omega}_{\mathrm{f}}^-(\mathbf{h})], \tag{7.11}$$

where we have introduced the fermionic analogues of the corresponding bosonic operator functions

$$\begin{aligned} \hat{\Omega}_{\mathrm{f}}^+(\mathbf{h}^+) &= \exp\left(i\hat{\mathbf{c}} \cdot \mathbf{h}^+\right) = \prod_i \left(1 + i\hat{c}_i h_i^+\right), \\ \hat{\Omega}_{\mathrm{f}}^-(\mathbf{h}^-) &= \exp\left(i\mathbf{h} \cdot \hat{\mathbf{c}}^\dagger\right) = \prod_i (1 + i h_i^+ \hat{c}_i). \end{aligned} \tag{7.12}$$

Note that the pairs h_i, h_i^+ are both complex Grassmann variables and are independent of each other. In contrast with the boson case, h_i, h_i^+ cannot be thought of as points in two complex planes. The characteristic function $\chi_{\mathrm{f}}(\mathbf{h},\mathbf{h}^{+})$ depends only on the h_i, h_i^{+} and not on their complex conjugates h_i^{*}, h_i^{+*}. We can use the rules for the fermionic trace in Section 5.8 to write our characteristic function in the equivalent forms $\mathrm{Tr}[\hat{\rho}_{\mathrm{f}}\,\hat{\Omega}_{\mathrm{f}}^{-}(-\mathbf{h})\hat{\Omega}_{\mathrm{f}}^{+}(\mathbf{h}^{+})]$ and $\mathrm{Tr}[\hat{\Omega}_{\mathrm{f}}^{-}(-\mathbf{h})\hat{\Omega}_{\mathrm{f}}^{+}(\mathbf{h}^{+})\hat{\rho}_{\mathrm{f}}]$.

For the number-conserving fermionic physical states given by (2.11) the only non-zero expectation values must have the same number of $\hat{c}_i^{\dagger}$ as there are $\hat{c}_j$ – expectation values such as $\left\langle \hat{c}_{l_1}^{\dagger}\right\rangle$ and $\langle \hat{c}_{m_2}\hat{c}_{m_1}\rangle$, etc. are zero. For the present, we confine ourselves to this situation. We use this feature after expanding out all the terms in

$$\chi_{\mathrm{f}}(\mathbf{h},\mathbf{h}^{+}) = \mathrm{Tr}[(1+i\hat{c}_n h_n^{+})\cdots(1+i\hat{c}_1 h_1^{+})\,\hat{\rho}_{\mathrm{f}}\,(1+ih_1\hat{c}_1^{\dagger})\cdots(1+ih_n\hat{c}_n^{\dagger})], \tag{7.13}$$

where we have applied an ordering convention to list the commuting factors in the order $n, n-1, \cdots$ for those in $\hat{\Omega}_{\mathrm{f}}^{-}$ and in the order $1, 2, \cdots, n$ for those in $\hat{\Omega}_{\mathrm{f}}^{+}$. Noting that moving the Grassmann numbers to the left or right of the fermion annihilation or creation operators requires the same number of anticommutes in either direction, we get

$$\begin{aligned}\chi_{\mathrm{f}}(\mathbf{h},\mathbf{h}^{+}) &= \mathrm{Tr}\left[\hat{\rho}_{\mathrm{f}} + \sum_{j,i} i\hat{c}_j h_j^{+}\hat{\rho}_{\mathrm{f}} i h_i \hat{c}_i^{\dagger} + \sum_{j>k}\sum_{i<h} i\hat{c}_j h_j^{+} i\hat{c}_k h_k^{+}\hat{\rho}_{\mathrm{f}}\, i h_i\hat{c}_i^{\dagger}\, i h_h \hat{c}_h^{\dagger} + \cdots\right.\\ &\qquad \left. + \, i\hat{c}_n h_n^{+}\cdots i\hat{c}_1 h_1^{+}\hat{\rho}_{\mathrm{f}}\, i h_1 \hat{c}_1^{\dagger}\cdots i h_n \hat{c}_n^{\dagger}\right] \qquad (7.14)\\ &= \mathrm{Tr}\left[\hat{\rho}_{\mathrm{f}} + i^2\sum_{j,i} h_j^{+}(\hat{c}_j\hat{\rho}_{\mathrm{f}}\hat{c}_i^{\dagger})h_i + i^4\sum_{j>k}\sum_{i<h} h_j^{+}h_k^{+}(\hat{c}_j\hat{c}_k\hat{\rho}_{\mathrm{f}}\hat{c}_i^{\dagger}\hat{c}_h^{\dagger})h_i h_h + \cdots\right.\\ &\qquad \left. + i^{2n}h_n^{+}\cdots h_1^{+}(\hat{c}_n\cdots\hat{c}_1\hat{\rho}_{\mathrm{f}}\hat{c}_1^{\dagger}\cdots\hat{c}_n^{\dagger})h_1\cdots h_n\right]. \qquad (7.15)\end{aligned}$$

The trace requires summing over sets of occupation numbers $\{\nu\}$ of the matrix element $\langle \nu_1, \cdots, \nu_n|\{\cdots\}|\nu_1, \cdots, \nu_n\rangle$. Moving the Grassmann numbers to the left of $\langle \nu_1, \cdots, \nu_n|$ or to the right of $|\nu_1, \cdots, \nu_n\rangle$ produces two equal $(-1)^p$ factors. Hence

$$\begin{aligned}\chi_{\mathrm{f}}(\mathbf{h},\mathbf{h}^{+}) &= \mathrm{Tr}(\hat{\rho}_{\mathrm{f}}) + i^2\sum_{j,i} h_j^{+}\ \mathrm{Tr}(\hat{c}_j\,\hat{\rho}_{\mathrm{f}}\hat{c}_i^{\dagger})\,h_i + \cdots\\ &\qquad + i^{2n}h_n^{+}\cdots h_1^{+}\ \mathrm{Tr}(\hat{c}_n\cdots\hat{c}_2\hat{c}_1\,\hat{\rho}_{\mathrm{f}}\,\hat{c}_1^{\dagger}\cdots\hat{c}_n^{\dagger})h_1\cdots h_n\\ &= 1 + i^2\sum_{j,i}\left\langle \hat{c}_i^{\dagger}\hat{c}_j\right\rangle h_j^{+}\,h_i + \cdots + i^{2n}\left\langle \hat{c}_1^{\dagger}\cdots\hat{c}_n^{\dagger}\hat{c}_n\cdots\hat{c}_1\right\rangle h_n^{+}\cdots h_1^{+}h_1\cdots h_n,\end{aligned} \tag{7.16}$$

where we have used the result that the traces of products of fermion annihilation and creation operators are c-numbers and the cyclic properties of the trace enable a normally ordered form to be obtained. It is easy to see that the characteristic function for physical states of a simple fermionic system is an *even Grassmann function* of the variables $h_1, h_1^{+}, h_2, h_2^{+}, \cdots, h_n, h_n^{+}$ of order 2^n.

The fermion characteristic function can thus be written as a linear combination of even-order monomials. Particular h_i and h_j^{+} can of course occur only once in any monomial, and the ordering convention avoids repetition of any monomial. Thus the characteristic function is

$$\chi_{\mathrm{f}}(\mathbf{h},\mathbf{h}^{+}) \\ = \chi_0 + \sum_{j,i} \chi_2^{i;j}\, h_j^{+}\, h_i + \sum_{j>k}\sum_{i<h} \chi_4^{ih,jk}\, h_j^{+} h_k^{+}\, h_i h_h \cdots + \chi_{2n}^{1\cdots n,n\cdots 1}\, h_n^{+} \cdots h_1^{+}\, h_1 \cdots h_n, \tag{7.17}$$

where the χ_0, $\chi_2^{i;j}$, $\chi_4^{ih,jk}$, $\cdots$ are all c-numbers related to the normally ordered correlation functions. Thus

$$\begin{aligned} \chi_0 &= 1, \\ \chi_2^{i,j} &= i^2 \left\langle \hat{c}_i^{\dagger} \hat{c}_j \right\rangle, \\ \chi_4^{ih,jk} &= i^4 \left\langle \hat{c}_i^{\dagger} \hat{c}_h^{\dagger} \hat{c}_j \hat{c}_k \right\rangle, \\ &\vdots \\ \chi_{2n}^{1\cdots n,n\cdots 1} &= i^{2n} \left\langle \hat{c}_1^{\dagger} \cdots \hat{c}_n^{\dagger} \hat{c}_n \cdots \hat{c}_1 \right\rangle, \end{aligned} \tag{7.18}$$

where we note the relationship between the order of the operators $\hat{c}_i^{\dagger}\hat{c}_h^{\dagger} \cdots \hat{c}_j \hat{c}_k \cdots$ and the order of the symbols in the coefficients $\chi_{2p}^{ih\cdots,jk\cdots}$. Clearly, the characteristic function is equivalent to the set of quantum correlation functions, hence the name.

For the case of n modes, the total number of c-number coefficients required to specify the fermion characteristic function is $2n!/(n!n!) - ({}^nC_0\, {}^nC_0 +^n C_1\, {}^nC_1 +^n C_2\, {}^nC_2 + \cdots +^n C_n\, {}^nC_n)$ [59]. However, if there are $N \leq n$ fermions in the system, then the correlation functions of order $2p$ with $p > N$ are all zero, so the characteristic function is now specified by ${}^nC_0\, {}^nC_0 +^n C_1\, {}^nC_1 +^n C_2\, {}^nC_2 + \cdots +^n C_n\, {}^nC_n$ coefficients. This is more than the number of c-number coefficients $({}^nC_n)^2$ required to specify the density operator in either the Fock state or fermion coherent-state representation. The reason for this is that the correlation functions are not independent; for example, $\sum_{i=1}^{n} \left\langle \hat{c}_i^{\dagger} \hat{c}_i \right\rangle = N$. The case of the vacuum state gives a particularly simple form for the characteristic function. As all the correlation functions give zero, we see that

$$\chi_{\mathrm{f}}(\mathbf{h},\mathbf{h}^{+}) = 1. \tag{7.19}$$

For bosonic systems, no simple expression for the characteristic function analogous to (7.17) exists.

7.3 Distribution Functions

In this section we relate the characteristic function to a *distribution function* via a phase space integral involving Fourier-like factors. For bosonic systems we associate a pair of c-number variables α_i, α_i^{+} with each mode, the phase space integral then involving c-number integration over complex planes for all of these variables in the distribution function $P_{\mathrm{b}}(\boldsymbol{\alpha},\boldsymbol{\alpha}^+,\boldsymbol{\alpha}^*,\boldsymbol{\alpha}^{+*})$, whilst for fermions we associate a pair of Grassmann variables g_i, g_i^+ with each mode, the phase space integral then involving Grassmann integration over all of these variables in the distribution function $P_{\mathrm{f}}(\mathbf{g},\mathbf{g}^+)$. The expressions involving analytic and even exponential functions that relate the distribution function to the characteristic function via a phase space integral would actually be the same if the normally ordered positive P distribution were replaced by a double-phase-space version of the Wigner or Q distribution. For both the boson and the fermion cases, the distribution function is not required to be unique, though the fermionic distribution is – the key requirement is that at least one distribution function exists such that its integral with the Fourier-like factors gives the correct characteristic function. The bosonic distribution function is also not analytic. It depends on α_i, α_i^+ and the complex conjugates $\alpha_i^*, \alpha_i^{+*}$. The existence of the distribution function is discussed in Section 7.4. The distribution function is of the positive P type for the bosonic variables and similar to the complex P type for the fermionic variables. The corresponding Fokker–Planck equations contain left and right Grassmann derivatives with respect to g, g^+ and c-number derivatives with respect to $\alpha_x, \alpha_x^+, \alpha_y, \alpha_y^+$ or, equivalently, α, α^+.

7.3.1 Bosons

For the bosonic case, we introduce the positive P distribution function $P_{\mathrm{b}}(\boldsymbol{\alpha},\boldsymbol{\alpha}^+,\boldsymbol{\alpha}^*,\boldsymbol{\alpha}^{+*})$, which is required to determine the normally ordered characteristic function via the following c-number phase space integral:

$$\chi_{\mathrm{b}}(\boldsymbol{\xi},\boldsymbol{\xi}^+) = \int d^2\boldsymbol{\alpha}^+ d^2\boldsymbol{\alpha}\,\exp(i\boldsymbol{\alpha}\cdot\boldsymbol{\xi}^+)P_{\mathrm{b}}(\boldsymbol{\alpha},\boldsymbol{\alpha}^+,\boldsymbol{\alpha}^*,\boldsymbol{\alpha}^{+*})\exp(i\boldsymbol{\xi}\cdot\boldsymbol{\alpha}^+). \tag{7.20}$$

Our bosonic distribution function depends on $\alpha_i, \alpha_i^+, \alpha_i^*, \alpha_i^{+*}$, which are specified by points α_i, α_i^+ in two independent complex planes, sometimes referred to as a *double phase space*. We can express both α_i and α_i^+ in terms of their real and imaginary parts as $\alpha_i = \alpha_{ix} + i\alpha_{iy}$ and $\alpha_i^+ = \alpha_{ix}^+ + i\alpha_{iy}^+$. In that case the bosonic distribution function can be thought of as depending on the real variables $\alpha_{ix}, \alpha_{ix}^+, \alpha_{iy}, \alpha_{iy}^+$.

A key point is that in general the bosonic distribution function P_{b} is a function of the α_i, α_i^+ *and* their complex conjugates $\alpha_i^*, \alpha_i^{+*}$. The bosonic distribution function, moreover, is not unique, although it is still required to determine the correct, unique characteristic function. In contrast to the characteristic function, the distribution function P_{b} is *not* an analytic function of α, α^+. This does not matter; the characteristic function is still related to the distribution function by two c-number integrals, which are over the α_i and α_i^+ complex planes and involve integrations over α_{ix}, α_{iy} and $\alpha_{ix}^+, \alpha_{iy}^+$. Fokker–Planck equations involving differentiation of P_{b} with respect to α_{ix}, α_{iy} and $\alpha_{ix}^+, \alpha_{iy}^+$ can still be obtained, where the distribution function is considered as a function $F_{\mathrm{b}}(\boldsymbol{\alpha}_x, \boldsymbol{\alpha}_y, \boldsymbol{\alpha}_x^+, \boldsymbol{\alpha}_y^+)$. Alternatively, we can also regard $\alpha_i, \alpha_i^+, \alpha_i^*, \alpha_i^{+*}$ as four *independent* (complex) variables and treat differentiation of P_{b} with respect to α_i, α_i^+ as independent of differentiation with respect to $\alpha_i^*, \alpha_i^{+*}$.

In this work, we shall make use of both approaches. Only differentiation with respect to α_i, α_i^+, however, will occur in deriving the Fokker–Planck equation, except in the case of the canonical distribution function. The bosonic distribution function is essentially of the *positive* P type, positivity being the feature of interest rather than the analyticity feature that can apply to *complex* P-type distributions [31, 38]. The Fokker–Planck equation for the positive P distribution function P_{b} may involve c-number derivatives of P_{b} with respect to $\alpha_{ix}, \alpha_{ix}^+, \alpha_{iy}, \alpha_{iy}^+$, rather than derivatives with respect to α_i, α_i^* as in the case of the P, W and Q distributions.

As only c-numbers are involved, the relationship between the characteristic and distribution functions can also be written

$$\chi_{\mathrm{b}}\left(\boldsymbol{\xi}, \boldsymbol{\xi}^+\right) = \int d^2\boldsymbol{\alpha}^+ d^2\boldsymbol{\alpha}\, P_{\mathrm{b}}\left(\boldsymbol{\alpha}, \boldsymbol{\alpha}^+, \boldsymbol{\alpha}^*, \boldsymbol{\alpha}^{+*}\right) \exp\left[i\left(\boldsymbol{\xi}\cdot\boldsymbol{\alpha}^+ + \boldsymbol{\alpha}\cdot\boldsymbol{\xi}^+\right)\right], \tag{7.21}$$

which is the form for the relationship that is often displayed.

We can introduce the double-space *Wigner* distribution function $W_{\mathrm{b}}(\boldsymbol{\alpha}, \boldsymbol{\alpha}^+, \boldsymbol{\alpha}^*, \boldsymbol{\alpha}^{+*})$, with which we can determine the symmetrically ordered characteristic function via the following c-number phase space integral:

$$\chi_{\mathrm{b}}^{\mathrm{W}}(\boldsymbol{\xi}, \boldsymbol{\xi}^+) = \int d^2\boldsymbol{\alpha}^+ d^2\boldsymbol{\alpha}\, W_{\mathrm{b}}(\boldsymbol{\alpha}, \boldsymbol{\alpha}^+, \boldsymbol{\alpha}^*, \boldsymbol{\alpha}^{+*}) \exp\left[i\left(\boldsymbol{\xi}\cdot\boldsymbol{\alpha}^+ + \boldsymbol{\alpha}\cdot\boldsymbol{\xi}^+\right)\right]. \tag{7.22}$$

Note that the same relationship between the characteristic and distribution functions applies as in the positive-P case.

7.3.2 Fermions

For fermionic systems, we introduce a distribution function $P_{\mathrm{f}}(\mathbf{g}, \mathbf{g}^+)$, which is required to determine the characteristic function via the following Grassmann phase space integral:

$$\chi_{\mathrm{f}}(\mathbf{h}, \mathbf{h}^+) = \int d\mathbf{g}^+ d\mathbf{g} \exp(i\mathbf{g}\cdot\mathbf{h}^+) P_{\mathrm{f}}(\mathbf{g}, \mathbf{g}^+) \exp(i\mathbf{h}\cdot\mathbf{g}^+). \tag{7.23}$$

Unlike its bosonic counterpart, the fermionic distribution function $P_{\mathrm{f}}(\mathbf{g}, \mathbf{g}^+)$ is unique. Here the fermionic distribution function depends on two independent complex Grassmann variables g_i, g_i^+, though of course these cannot be represented as points in two complex planes. However, as in the bosonic case, we do have a double phase space associated with the pairs of Grassmann 'coordinates' g_i, g_i^+. Unlike the case for c-numbers, g_i, g_i^+ do not have real and imaginary parts (although quantities such as $g_{ix} = (g_i + g_i^*)/2$ and $g_{iy} = (g_i - g_i^*)/2i$ could be defined, which do not alter under complex conjugation and where $g_i = g_{ix} + ig_{iy}$ and $g_i^* = g_{ix} - ig_{iy}$). However, the Grassmann numbers g_i, g_i^+ are both considered to be complex, and the complex conjugation process is incorporated into the algebra. As before, $d\mathbf{g}^+ = dg_1^+ \cdots dg_n^+ = \prod_i dg_i^+$ and $d\mathbf{g} = dg_n \cdots dg_1 = \prod_i dg_i$. Note that the exponential factors and the distribution function *can* be placed in any order, since they are all even Grassmann functions.

A key point is that the fermionic distribution function $P_{\mathrm{f}}(\mathbf{g}, \mathbf{g}^+)$ is a function of the g_i, g_i^+ only and is *not* a function of their complex conjugates g_i^*, g_i^{+*}. The characteristic function is related here to the distribution function by *two* Grassmann integrals, which are over g_i and g_i^+. The fermionic distribution is similar to the *complex* P type for bosons. The Fokker–Planck equation for the distribution function will involve left and right Grassmann derivatives with respect to g_i, g_i^+.

Although Grassmann functions are involved, both the exponential functions and (as we will see) the distribution function itself are even functions and hence can be put in any order. Thus the characteristic–distribution function relationship can also be written

$$\chi_{\mathrm{f}}(\mathbf{h},\mathbf{h}^+) = \int d\mathbf{g}^+ d\mathbf{g}\; P_{\mathrm{f}}(\mathbf{g},\mathbf{g}^+)\, \exp\left[i\left(\mathbf{h}\cdot\mathbf{g}^+ + \mathbf{g}\cdot\mathbf{h}^+\right)\right], \tag{7.24}$$

which is the form for the relationship that is often displayed. Note that the same cyclic interchange of factors does *not* apply to the definition (7.11) of the characteristic function in terms of fermionic operators.

The fermion distribution function can always be written as a power series in the variables $g_1, g_1^+, g_2, g_2^+, \cdots, g_n, g_n^+$ with c-number expansion coefficients. Because the square and higher powers of any Grassmann variable are zero, each term in the power series can contain any one of the $g_1, g_1^+, g_2, g_2^+, \cdots, g_n, g_n^+$ at most once. The highest-power term would be of the $2n^{\text{th}}$ order, $g_1 \cdots g_n g_1^+ \cdots g_n^+$. There could be lower-order terms, such as first-order terms $g_1, g_1^+, g_2, g_2^+, \cdots, g_n, g_n^+$ and a zero-order term not dependent on these variables. The total number of possible expansion coefficients is 2^{2n} $(= {}^{2n}C_0 + {}^{2n}C_1 + \cdots + {}^{2n}C_{2n})$. The distribution function is specified by these 2^{2n} c-number coefficients. In fact, we can show that many of these can be chosen to be zero, as they could make no contribution to determining the characteristic function.

The phase space integral (7.23) involving the fermion distribution function gives the characteristic function, and the factors inside the Grassmann integral that depend on the $g_1, g_1^+,\ g_2, g_2^+, \cdots, g_n, g_n^+$ are

$$\begin{aligned}\exp\left(i\mathbf{g}\cdot\mathbf{h}^+\right) P_{\mathrm{f}}(\mathbf{g},\mathbf{g}^+) \exp\left(i\mathbf{h}\cdot\mathbf{g}^+\right) &= (1+ig_n h_n^+)\cdots(1+ig_1 h_1^+) P_{\mathrm{f}}(\mathbf{g},\mathbf{g}^+)\\ &\quad\times(1+ih_1 g_1^+)\cdots(1+ih_n g_n^+)\,.\end{aligned} \tag{7.25}$$

The distribution function will be a Grassmann function of the $g_1, g_2, \cdots, g_n$ and $g_1^+, g_2^+, \cdots, g_n^+$, the highest-order term being $g_1 g_2 \cdots g_n g_n^+ \cdots g_2^+ g_1^+$. The term to the left of $P_{\mathrm{f}}(g, g^+)$ will be a Grassmann function of the $g_1, g_2, \cdots, g_n$ and the term to the right a Grassmann function of the $g_1^+, g_2^+, \cdots, g_n^+$. For simple fermion-number-conserving systems, the characteristic function of the form given in (7.17) involves only terms which contain the same number of h_i^+ as h_j. To obtain such a term after carrying out the Grassmann integration, the left term must have contributed the same number of $ig_i h_i^+$ factors as the right term contributed factors $ih_j g_j^+$. However, the only non-zero multiple Grassmann integral involving the g_i and g_j^+ is of the form

$$\int dg_1^+ \cdots dg_n^+ dg_n \cdots dg_1\, (g_1 g_2 \cdots g_n\, g_n^+ \cdots g_2^+ g_1^+) = +1, \tag{7.26}$$

so that if the product of the $ig_ih_i^+$'s from the left term has certain g_k's missing, these must be made up from a term in $P_{\text{f}}(g, g^+)$. The same number of g_j^+'s will be missing from the right term (as the number of h_i^+ must equal the number of h_j), and hence these also must be made up from the *same* term in $P_{\text{f}}(\mathbf{g}, \mathbf{g}^+)$. This means that every term in $P(\mathbf{g}, \mathbf{g}^+)$ must contain the same number of g_k's as g_j^+'s, and thus the distribution function must be of the form

$$\begin{aligned} P_{\text{f}}(\mathbf{g}, \mathbf{g}^+) = P_0 &+ \sum_{i;j} P_2^{i,j} g_i g_j^+ + \sum_{i<j;k>l} P_4^{ij,kl} g_i g_j g_k^+ g_l^+ \\ &+ \sum_{i<j<k,l>m>p} P_6^{ijk,lmp} g_i g_j g_k g_l^+ g_m^+ g_p^+ + \cdots + P_{2n}^{12\cdots n,n\cdots 21} g_1 \cdots g_n g_n^+ \cdots g_1^+ \end{aligned} \tag{7.27}$$

for the case of simple number-conserving fermion systems. An ordering convention in which the g_i are arranged in ascending order and the g_j^+ in descending order has been used. This has a similar structure to the characteristic function. This involves $2n!/(n!n!)$ c-number coefficients to define the distribution function $P_{\text{f}}(\mathbf{g}, \mathbf{g}^+)$ as a Grassmann function, the same number of course as for the characteristic function that it determines. The distribution function for a general physical state of a simple fermion system is thus an *even Grassmann function* of order 2^n in the variables $g_1, g_1^+, g_2, g_2^+, \cdots, g_n, g_n^+$, a feature that is needed later.

7.4 Existence of Distribution Functions and Canonical Forms for Density Operators

It is important to show that the distribution function actually exists, and in this section we prove this for number-conserving systems. The proof here will be based on defining *canonical forms* for the distribution function and then showing that phase space integrals involving exponential functions of the form (7.20) or (7.23) determine the actual characteristic function. Another approach used is based on determining Fokker–Planck equations for the distribution function (see Chapter 8), showing that solutions exist for these equations and then demonstrating that the corresponding equation for the characteristic equation obtained via (7.20) or (7.23) is in accord with the equation governing the density operator.

7.4.1 Fermions

To demonstrate the existence of the distribution function for the fermionic case, a distribution function $P_{\text{c}}^{\text{f}}(\mathbf{g}, \mathbf{g}^+)$ is *defined* via a Grassmann integral involving the characteristic function $\chi_{\text{f}}(\mathbf{h}, \mathbf{h}^+)$:

$$P_{\text{c}}^{\text{f}}(\mathbf{g}, \mathbf{g}^+) = \int d\mathbf{h}^+ d\mathbf{h} \exp(-i\mathbf{g} \cdot \mathbf{h}^+)\, \chi_{\text{f}}(\mathbf{h}, \mathbf{h}^+) \exp(-i\mathbf{h} \cdot \mathbf{g}^+). \tag{7.28}$$

Given that the characteristic function exists, it then follows that $P_{\text{c}}^{\text{f}}(\mathbf{g}, \mathbf{g}^+)$ exists as a Grassmann function of the g_i, g_i^+.

We can immediately show that the distribution function $P_c^f(\mathbf{g},\mathbf{g}^+)$ satisfies the relationship in (7.23) that connects the distribution function to the characteristic function via the required Grassmann integral. For we have

$$
\begin{aligned}
&\int d\mathbf{g}^+ d\mathbf{g}\exp(i\mathbf{g}\cdot\mathbf{h}^+)\, P_c^f(\mathbf{g},\mathbf{g}^+)\exp(i\mathbf{h}\cdot\mathbf{g}^+) \\
&\quad = \int d\mathbf{g}^+ d\mathbf{g}\exp(i\mathbf{g}\cdot\mathbf{h}^+) \\
&\qquad \times \left(\int d\mathbf{k}^+ d\mathbf{k}\exp(-i\mathbf{g}\cdot\mathbf{k}^+)\,\chi_f(\mathbf{k},\mathbf{k}^+)\exp\left(-i\mathbf{k}\cdot\mathbf{g}^+\right)\right)\exp(i\mathbf{h}\cdot\mathbf{g}^+) \\
&\quad = \int d\mathbf{k}^+ d\mathbf{k}\chi_f(\mathbf{k},\mathbf{k}^+)\int d\mathbf{g}^+ d\mathbf{g}\exp(i\mathbf{g}\cdot[\mathbf{h}^+ - \mathbf{k}^+])\,\exp(i[\mathbf{h}\text{-}\mathbf{k}]\cdot\mathbf{g}^+) \\
&\quad = \chi_f(\mathbf{h},\mathbf{h}^+),
\end{aligned}
\tag{7.29}
$$

where we have used the Grassmann Fourier integral result in (4.72) and used the evenness of the exponentials, the products of Grassmann differentials and the characteristic function to commute various terms.

These considerations show that a Grassmann distribution function $P_f(\mathbf{g},\mathbf{g}^+)$ for fermions exists which is related to the characteristic function $\chi_f(\mathbf{h},\mathbf{h}^+)$ as in (7.23). A particular case of such a distribution function is the canonical distribution function defined in (7.28). However, this result implies that the fermion distribution function is in fact unique. Using (7.23) with a different distribution function $P^f(\mathbf{g},\mathbf{g}^+)$, it is easy to show that

$$0 = \int d\mathbf{g}^+ d\mathbf{g}\exp(i\mathbf{g}\cdot\mathbf{h}^+)\left[P^f(\mathbf{g},\mathbf{g}^+) - P_c^f(\mathbf{g},\mathbf{g}^+)\right]\exp(i\mathbf{h}\cdot\mathbf{g}^+) \tag{7.30}$$

for all $\mathbf{h},\mathbf{h}^+$. It then follows that

$$P^f(\mathbf{g},\mathbf{g}^+) = P_c^f(\mathbf{g},\mathbf{g}^+), \tag{7.31}$$

which establishes the uniqueness of the fermion distribution function.

Another approach for fermion systems is to use the *canonical form* of the density operator, which is a particular case of (5.107). We have

$$
\begin{aligned}
\hat{\rho}_f &= \int d\mathbf{g}^+ d\mathbf{g}\; P_{\text{canon}}^f(\mathbf{g},\mathbf{g}^+)\,\hat{\Lambda}_f(\mathbf{g},\mathbf{g}^+) \\
&= \int d\mathbf{g}^+ d\mathbf{g}\,\hat{\Lambda}_f(\mathbf{g},\mathbf{g}^+)\,P_{\text{canon}}^f(\mathbf{g},\mathbf{g}^+),
\end{aligned}
\tag{7.32}
$$

where

$$\hat{\Lambda}_f(\mathbf{g},\mathbf{g}^+) = \frac{{}_B|\mathbf{g}\rangle\,\langle\mathbf{g}^{+*}|_B}{\mathrm{Tr}(|\mathbf{g}\rangle_B\,{}_B\langle\mathbf{g}^{+*}|)} \tag{7.33}$$

is a normalised projector and where

$$P_{\text{canon}}^f(\mathbf{g},\mathbf{g}^+) = \int d\mathbf{k}^{+*} d\mathbf{k}^*\exp(\mathbf{g}\cdot\mathbf{k}^* + \mathbf{k}^{+*}\cdot\mathbf{g}^+ + \mathbf{g}\cdot\mathbf{g}^+)\,{}_B\langle\mathbf{k}|\,\hat{\rho}_f\,|\mathbf{k}^{+*}\rangle_B \tag{7.34}$$

is a *canonical distribution function* for the fermion density operator $\hat{\rho}_{\mathrm{f}}$ in terms of its matrix elements for Bargmann coherent states. Note that ${}_{\mathrm{B}}\langle\mathbf{k}|\,\hat{\rho}_{\mathrm{f}}\,|\mathbf{k}^{+*}\rangle_{\mathrm{B}}$ is a Grassmann function of $\mathbf{k}^*$ and $\mathbf{k}^{+*}$, not $\mathbf{k}$ and $\mathbf{k}^+$. The two forms for the density operator in (7.32) are the same because $\hat{\Lambda}_{\mathrm{f}}(\mathbf{g},\mathbf{g}^+)$ is an even Grassmann operator and therefore commutes with $P^{\mathrm{f}}_{\mathrm{canon}}(\mathbf{g},\mathbf{g}^+)$.

Using the canonical form of the density operator (7.32), we can then show that the characteristic function is related to the fermion distribution function as in (7.23), as must be the case since the fermion distribution function is unique. We have

$$\begin{aligned}
&\mathrm{Tr}\left[\hat{\Omega}^+_{\mathrm{f}}(\mathbf{h}^+)\int d\mathbf{g}^+ d\mathbf{g}\; P^{\mathrm{f}}_{\mathrm{canon}}(\mathbf{g},\mathbf{g}^+)\,\hat{\Lambda}_{\mathrm{f}}(\mathbf{g},\mathbf{g}^+)\,\hat{\Omega}^-_{\mathrm{f}}(\mathbf{h})\right] \\
&\quad= \int d\mathbf{g}^+ d\mathbf{g}\,\mathrm{Tr}\left[P^{\mathrm{f}}_{\mathrm{canon}}(\mathbf{g},\mathbf{g}^+)\,\hat{\Omega}^+_{\mathrm{f}}(\mathbf{h}^+)\,\hat{\Lambda}_{\mathrm{f}}(\mathbf{g},\mathbf{g}^+)\,\hat{\Omega}^-_{\mathrm{f}}(\mathbf{h})\right],
\end{aligned} \tag{7.35}$$

where even Grassmann functions, differentials and operators have been placed in different orders. It is straightforward to evaluate the operator products using the eigenvalue properties of the Bargmann states:

$$\begin{aligned}
\hat{\Omega}^+_{\mathrm{f}}(\mathbf{h}^+)\,|\mathbf{g}\rangle_{\mathrm{B}} &= \exp\big(i\hat{\mathbf{c}}\cdot\mathbf{h}^+\big)\,|\mathbf{g}\rangle_{\mathrm{B}} \\
&= \exp\big(-i\mathbf{h}^+\cdot\hat{\mathbf{c}}\big)\,|\mathbf{g}\rangle_{\mathrm{B}} \\
&= \exp\big(-i\mathbf{h}^+\cdot\mathbf{g}\big)\,|\mathbf{g}\rangle_{\mathrm{B}} \\
&= \exp\big(i\mathbf{g}\cdot\mathbf{h}^+\big)\,|\mathbf{g}\rangle_{\mathrm{B}}.
\end{aligned} \tag{7.36}$$

Similarly, we have the following for conjugate operation:

$${}_{\mathrm{B}}\langle\mathbf{g}^{+*}|\,\hat{\Omega}^-_{\mathrm{f}}(\mathbf{h}) = {}_{\mathrm{B}}\langle\mathbf{g}^{+*}|\exp\big(i\mathbf{h}\cdot\mathbf{g}^+\big). \tag{7.37}$$

It follows, therefore, that

$$\begin{aligned}
&\mathrm{Tr}\left[\hat{\Omega}^+_{\mathrm{f}}(\mathbf{h}^+)\int d\mathbf{g}^+ d\mathbf{g}\; P^{\mathrm{f}}_{\mathrm{canon}}(\mathbf{g},\mathbf{g}^+)\,\hat{\Lambda}_{\mathrm{f}}(\mathbf{g},\mathbf{g}^+)\,\hat{\Omega}^-_{\mathrm{f}}(\mathbf{h})\right] \\
&\quad= \int d\mathbf{g}^+ d\mathbf{g}\exp\big(i\mathbf{g}\cdot\mathbf{h}^+\big)\,P^{\mathrm{f}}_{\mathrm{canon}}(\mathbf{g},\mathbf{g}^+)\exp\big(i\mathbf{h}\cdot\mathbf{g}^+\big),
\end{aligned} \tag{7.38}$$

which is the required relationship (7.23) between the characteristic function and the distribution function, with the distribution function given via the canonical representation function for the fermion density operator $\hat{\rho}_{\mathrm{f}}$.

The two canonical forms of the distribution function in (7.28) and (7.34) are actually the same:

$$P^{\mathrm{f}}_{\mathrm{canon}}(\mathbf{g},\mathbf{g}^+) = P^{\mathrm{f}}_{\mathrm{c}}(\mathbf{g},\mathbf{g}^+), \tag{7.39}$$

since the fermion distribution function is unique. This equality is confirmed directly in Appendix E.

Both expressions for the canonical form of the distribution function are useful, depending on whether the density operator or the characteristic function is available.

The expression (7.32) for the canonical distribution function is useful for determining distribution functions from the density operator. As an example, the canonical distribution function for the *vacuum state* can easily be obtained by substituting the density operator $|0_1, 0_2, \cdots, 0_n\rangle\langle 0_1, 0_2, \cdots, 0_n|$ into (7.32). We find that

$$P^{\mathrm{f}}_{\mathrm{canon}}(\mathbf{g}, \mathbf{g}^+) = g_1 g_2 \cdots g_n g_n^+ \cdots g_2^+ g_1^+ . \tag{7.40}$$

Hence the only non-zero coefficient is $P_{2n}^{1\cdots n,n\cdots 1} = 1$.

The expression (7.34) for the canonical distribution function is useful for determining distribution functions from the characteristic function. As an example, the canonical distribution function for the vacuum state can easily be obtained by substituting the characteristic function (7.19) into (7.34). Thus

$$P^{\mathrm{f}}_{\mathrm{c}}(g_i, g_i^+) = g_1 g_2 \cdots g_n g_n^+ \cdots g_2^+ g_1^+ , \tag{7.41}$$

which is the same as the result obtained from the canonical representation (7.32) for the density operator. Hence the only non-zero coefficient is $P_{2n}^{12\cdots n,n\cdots 21} = 1$. For general states, the same procedures can be used to find the form of the canonical distribution function in terms of quantum correlation functions.

7.4.2 Bosons

A similar result applies in the bosonic case. For the boson case, the *canonical form* of the density operator is a particular case of (5.109). We have

$$\begin{aligned}\hat{\rho}_{\mathrm{b}} &= \int d^2\boldsymbol{\alpha}^+ d^2\boldsymbol{\alpha}\; P_{\mathrm{b\,canon}}(\boldsymbol{\alpha}, \boldsymbol{\alpha}^+, \boldsymbol{\alpha}^*, \boldsymbol{\alpha}^{+*})\hat{\Lambda}_{\mathrm{b}}(\boldsymbol{\alpha}, \boldsymbol{\alpha}^+) \\ &= \int d^2\boldsymbol{\alpha}^+ d^2\boldsymbol{\alpha}\; \hat{\Lambda}_{\mathrm{b}}(\boldsymbol{\alpha}, \boldsymbol{\alpha}^+) P_{\mathrm{b\,canon}}(\boldsymbol{\alpha}, \boldsymbol{\alpha}^+, \boldsymbol{\alpha}^*, \boldsymbol{\alpha}^{+*}),\end{aligned} \tag{7.42}$$

where

$$\hat{\Lambda}_{\mathrm{b}}(\boldsymbol{\alpha}, \boldsymbol{\alpha}^+) = \frac{|\boldsymbol{\alpha}\rangle_{\mathrm{B}}\,{}_{\mathrm{B}}\langle \boldsymbol{\alpha}^{+*}|}{\mathrm{Tr}({}_{\mathrm{B}}|\boldsymbol{\alpha}\rangle\langle \boldsymbol{\alpha}^{+*}|_{\mathrm{B}})} \tag{7.43}$$

is a normalised projector and

$$\begin{aligned}P_{\mathrm{b\,canon}}(\boldsymbol{\alpha}, \boldsymbol{\alpha}^+, \boldsymbol{\alpha}^*, \boldsymbol{\alpha}^{+*}) = \left(\frac{1}{4\pi^2}\right)^n \exp\!\left[-\frac{1}{2}\left(\boldsymbol{\alpha}\cdot\boldsymbol{\alpha}^* + \boldsymbol{\alpha}^{+*}\cdot\boldsymbol{\alpha}^+\right)\right] \\ \times{}_{\mathrm{B}}\big\langle(\boldsymbol{\alpha} + \boldsymbol{\alpha}^{+*})/2\big|\,\hat{\rho}_{\mathrm{b}}\,\big|(\boldsymbol{\alpha} + \boldsymbol{\alpha}^{+*})/2\big\rangle_{\mathrm{B}}\end{aligned} \tag{7.44}$$

is the canonical distribution function for the boson density operator $\hat{\rho}_{\mathrm{b}}$. The proof of the existence of such a canonical form was first given by Gardiner and Drummond (see [31, 38]). This result was established in Chapter 5 (see (5.111) for a general bosonic operator). The two forms for the density operator are equivalent, as $\hat{\Lambda}_{\mathrm{b}}(\boldsymbol{\alpha}, \boldsymbol{\alpha}^+)$ is a boson operator. As the density operator is Hermitian, the matrix element ${}_{\mathrm{B}}\langle(\boldsymbol{\alpha} + \boldsymbol{\alpha}^{+*}/2)|\,\hat{\rho}_{\mathrm{b}}\,|(\boldsymbol{\alpha} + \boldsymbol{\alpha}^{+*}/2)\rangle_B$ is real, and so the canonical representation $P_{\mathrm{b\,canon}}(\boldsymbol{\alpha}, \boldsymbol{\alpha}^+, \boldsymbol{\alpha}^*, \boldsymbol{\alpha}^{+*})$ is real. Furthermore, the density operator is also positive, so the matrix element is positive and hence the canonical representation is positive, and thus it is called the positive P representation. Using the canonical form of the density operator, we can determine the correct characteristic function via

$$\begin{aligned}\chi_{\mathrm{b}}(\boldsymbol{\xi},\boldsymbol{\xi}^{+}) &= \mathrm{Tr}\left[\hat{\Omega}_{\mathrm{b}}^{+}(\boldsymbol{\xi}^{+})\int d^2\boldsymbol{\alpha}^{+}d^2\boldsymbol{\alpha}\; P_{\mathrm{b\,canon}}(\boldsymbol{\alpha},\boldsymbol{\alpha}^{+},\boldsymbol{\alpha}^{*},\boldsymbol{\alpha}^{+*})\hat{\Lambda}_{\mathrm{b}}(\boldsymbol{\alpha},\boldsymbol{\alpha}^{+})\;\hat{\Omega}_{\mathrm{b}}^{-}(\boldsymbol{\xi})\right] \\ &= \int d^2\boldsymbol{\alpha}^{+}d^2\boldsymbol{\alpha}\;\exp\left(i\boldsymbol{\alpha}\cdot\boldsymbol{\xi}^{+}\right)P_{\mathrm{b\,canon}}(\boldsymbol{\alpha},\boldsymbol{\alpha}^{+},\boldsymbol{\alpha}^{*},\boldsymbol{\alpha}^{+*})\exp(i\boldsymbol{\xi}\cdot\boldsymbol{\alpha}^{+}),\end{aligned} \tag{7.45}$$

which is the required relationship between the characteristic function and the distribution function, but now with the distribution function given by the canonical representation function for the boson density operator $\hat{\rho}_{\mathrm{b}}$:

$$P_{\mathrm{b}}(\boldsymbol{\alpha},\boldsymbol{\alpha}^{+},\boldsymbol{\alpha}^{*},\boldsymbol{\alpha}^{+*}) = P_{\mathrm{b\,canon}}(\boldsymbol{\alpha},\boldsymbol{\alpha}^{+},\boldsymbol{\alpha}^{*},\boldsymbol{\alpha}^{+*}). \tag{7.46}$$

Thus the distribution has been shown to exist for the boson case. For bosons, only c-numbers are involved, so operators and functions can be commuted. In obtaining (7.45) we have used

$$\begin{aligned}\hat{\Omega}_{\mathrm{b}}^{+}(\boldsymbol{\xi}^{+})\,|\boldsymbol{\alpha}\rangle_{\mathrm{B}} &= \exp(i\boldsymbol{\alpha}\cdot\boldsymbol{\xi}^{+})\,|\boldsymbol{\alpha}\rangle_{\mathrm{B}}, \\ {}_{\mathrm{B}}\langle\boldsymbol{\alpha}^{+*}|\,\hat{\Omega}_{\mathrm{b}}^{-}(\boldsymbol{\xi}) &= {}_{\mathrm{B}}\langle\boldsymbol{\alpha}^{+*}|\exp\left(i\boldsymbol{\xi}\cdot\boldsymbol{\alpha}^{+}\right).\end{aligned} \tag{7.47}$$

For the existence of the double-space Wigner distribution function, we see from (7.10) and (7.45) and using integration by parts that

$$\begin{aligned}\chi_{\mathrm{b}}^{\mathrm{W}}(\boldsymbol{\xi},\boldsymbol{\xi}^{+}) &= \exp\left(\frac{1}{2}(i\boldsymbol{\xi}_i)\cdot(i\boldsymbol{\xi}^{+})\right)\chi_{\mathrm{b}}(\boldsymbol{\xi},\boldsymbol{\xi}^{+}) \\ &= \int d^2\boldsymbol{\alpha}^{+}d^2\boldsymbol{\alpha}\exp\left[\left(i\boldsymbol{\alpha}\cdot\boldsymbol{\xi}^{+}+\boldsymbol{\xi}\cdot\boldsymbol{\alpha}^{+}\right)\right]\prod_i\exp\left(\frac{1}{2}\frac{\partial}{\partial\alpha_i}\frac{\partial}{\partial\alpha_i^{+}}\right) \\ &\quad\times P_{\mathrm{b\,canon}}(\boldsymbol{\alpha},\boldsymbol{\alpha}^{+},\boldsymbol{\alpha}^{*},\boldsymbol{\alpha}^{+*}),\end{aligned} \tag{7.48}$$

giving a formal canonical expression for the double-space Wigner distribution function

$$W_{\mathrm{b}}^{\mathrm{canon}}(\boldsymbol{\alpha},\boldsymbol{\alpha}^{+},\boldsymbol{\alpha}^{*},\boldsymbol{\alpha}^{+*}) = \exp\left(\frac{1}{2}\sum_i\frac{\partial}{\partial\alpha_i}\frac{\partial}{\partial\alpha_i^{+}}\right)P_{\mathrm{b\,canon}}(\boldsymbol{\alpha},\boldsymbol{\alpha}^{+},\boldsymbol{\alpha}^{*},\boldsymbol{\alpha}^{+*}). \tag{7.49}$$

7.5 Combined Systems of Bosons and Fermions

In this section, we generalise from the cases of purely bosonic or fermionic systems to the treatment of combined systems of bosons and fermions. Here the treatment allows for the possibility that the numbers of bosons and fermions may change. However, as discussed previously, the processes envisaged involve the creation or destruction of fermions in pairs, with the corresponding destruction or creation of a boson. This restriction means that the expectation values of products of odd numbers of fermion annihilation or creation operators will remain zero. Thus expectation values such as $\langle \hat{c}^\dagger_{l_i} \rangle$, $\langle \hat{c}_{m_i} \rangle$, $\langle \hat{c}^\dagger_{l_i} \hat{c}^\dagger_{l_j} \hat{c}^\dagger_{l_k} \rangle$, $\langle \hat{c}^\dagger_{l_i} \hat{c}^\dagger_{l_j} \hat{c}_{m_k} \rangle$, $\langle \hat{c}^\dagger_{l_i} \hat{c}_{m_j} \hat{c}_{m_k} \rangle$ and $\langle \hat{c}_{m_i} \hat{c}_{m_j} \hat{c}_{m_k} \rangle$ are zero, but $\langle \hat{c}^\dagger_{l_i} \hat{c}^\dagger_{l_j} \rangle$ and $\langle \hat{c}_{m_i} \hat{c}_{m_j} \rangle$ may be non-zero. We may also find non-zero expectation values of the form $\langle \hat{c}^\dagger_{l_i} \hat{c}^\dagger_{l_j} \hat{a}_{m_k} \rangle$ and $\langle \hat{a}^\dagger_{l_k} \hat{c}_{m_i} \hat{c}_{m_j} \rangle$, corresponding to correlations between the destruction or creation of a boson and the creation or destruction of a fermion pair.

In the case of systems involving both bosons and fermions, we can introduce a *characteristic function* $\chi(\boldsymbol{\xi}, \boldsymbol{\xi}^+, \mathbf{h}, \mathbf{h}^+)$ for m bosonic modes and n fermionic modes:

$$\begin{aligned}\chi(\boldsymbol{\xi}, \boldsymbol{\xi}^+, \mathbf{h}, \mathbf{h}^+) &= \mathrm{Tr}\left[\hat{\Omega}^+_{\mathrm{b}}(\boldsymbol{\xi}^+)\hat{\Omega}^+_{\mathrm{f}}(\mathbf{h}^+)\hat{\rho}\,\hat{\Omega}^-_{\mathrm{f}}(\mathbf{h})\hat{\Omega}^-_{\mathrm{b}}(\boldsymbol{\xi})\right] \\ &= \mathrm{Tr}\left[\exp(i\hat{\mathbf{a}}\cdot\boldsymbol{\xi}^+)\exp(i\hat{\mathbf{c}}\cdot\mathbf{h}^+)\,\hat{\rho}\exp\left(i\mathbf{h}\cdot\hat{\mathbf{c}}^\dagger\right)\exp(i\boldsymbol{\xi}\cdot\hat{\mathbf{a}})\right]. \end{aligned} \tag{7.50}$$

This characteristic function may be related to a general phase space distribution function $P(\boldsymbol{\alpha}, \boldsymbol{\alpha}^+, \boldsymbol{\alpha}^*, \boldsymbol{\alpha}^{+*}, \mathbf{g}, \mathbf{g}^+)$ via a combination of c-number and Grassmann integrals:

$$\begin{aligned}\chi(\boldsymbol{\xi}, \boldsymbol{\xi}^+, \mathbf{h}, \mathbf{h}^+) = \int d\mathbf{g}^+ d\mathbf{g} \int d^2\boldsymbol{\alpha}^+ d^2\boldsymbol{\alpha} \exp\left(i\boldsymbol{\alpha}\cdot\boldsymbol{\xi}^+\right)\exp\left(i\mathbf{g}\cdot\mathbf{h}^+\right) \\ \times P(\boldsymbol{\alpha}, \boldsymbol{\alpha}^+, \boldsymbol{\alpha}^*, \boldsymbol{\alpha}^{+*}, \mathbf{g}, \mathbf{g}^+)\exp\left(i\mathbf{h}\cdot\mathbf{g}^+\right)\exp\left(i\boldsymbol{\xi}\cdot\boldsymbol{\alpha}^+\right). \end{aligned} \tag{7.51}$$

As previously, the characteristic function $\chi(\boldsymbol{\xi}, \boldsymbol{\xi}^+, \mathbf{h}, \mathbf{h}^+)$ is a function of the variables $\xi_i, \xi_i^+, h_i, h_i^+$ and $\alpha_i, \alpha_i^+, g_i, g_i^+$, and not of their complex conjugates, but in general the distribution function $P(\boldsymbol{\alpha}, \boldsymbol{\alpha}^+, \boldsymbol{\alpha}^*, \boldsymbol{\alpha}^{+*}, \mathbf{g}, \mathbf{g}^+)$ also depends on the $\alpha_i^*, \alpha_i^{+*}$. The distribution function is of the positive P type for the boson variables and is similar to the complex P type for fermion variables ('overall positive P' for short). As before, the distribution function is not required to be unique, nor is it an analytic function of $\boldsymbol{\alpha}, \boldsymbol{\alpha}^+$. The Grassmann variable dependence, is however, unique. The Fokker–Planck equation for the distribution function will involve left and right Grassmann derivatives with respect to g_i, g_i^+ and c-number derivatives with respect to $\alpha_{ix}, \alpha_{ix}^+, \alpha_{iy}, \alpha_{iy}^+$ or, equivalently, α_i, α_i^+.

The existence of the positive-P-type distribution function can be established using a canonical form for the density operator (again $d\mathbf{g}^+ \equiv dg_1^+ \cdots dg_n^+$ and $d\mathbf{g} \equiv dg_n \cdots dg_1$)

$$\begin{aligned}\hat{\rho} &= \int d\mathbf{g}^+ d\mathbf{g} \int d^2\boldsymbol{\alpha}^+ d^2\boldsymbol{\alpha}\; P_{\mathrm{canon}}(\boldsymbol{\alpha}, \boldsymbol{\alpha}^+, \boldsymbol{\alpha}^*, \boldsymbol{\alpha}^{+*}, \mathbf{g}, \mathbf{g}^+)\,\hat{\Lambda}(\mathbf{g}, \mathbf{g}^+, \boldsymbol{\alpha}, \boldsymbol{\alpha}^+) \\ &= \int d\mathbf{g}^+ d\mathbf{g} \int d^2\boldsymbol{\alpha}^+ d^2\boldsymbol{\alpha}\; \hat{\Lambda}(\mathbf{g}, \mathbf{g}^+, \boldsymbol{\alpha}, \boldsymbol{\alpha}^+) P_{\mathrm{canon}}(\boldsymbol{\alpha}, \boldsymbol{\alpha}^+, \boldsymbol{\alpha}^*, \boldsymbol{\alpha}^{+*}, \mathbf{g}, \mathbf{g}^+), \end{aligned} \tag{7.52}$$

where

$$\hat{\Lambda}(\mathbf{g}, \mathbf{g}^+, \boldsymbol{\alpha}, \boldsymbol{\alpha}^+) = \hat{\Lambda}_{\mathrm{b}}(\boldsymbol{\alpha}, \boldsymbol{\alpha}^+) \otimes \hat{\Lambda}_{\mathrm{f}}(\mathbf{g}, \mathbf{g}^+) \tag{7.53}$$

is a normalised projector and ($d\mathbf{g}^{+*} \equiv dg_1^{+*} \cdots dg_n^{+*}$ and $d\mathbf{g}^* \equiv dg_n^* \cdots dg_1^*$)

$$P_{\text{canon}}(\boldsymbol{\alpha}, \boldsymbol{\alpha}^+, \boldsymbol{\alpha}^*, \boldsymbol{\alpha}^{+*}, \mathbf{g}, \mathbf{g}^+) = \left(\frac{1}{4\pi^2}\right)^n \int d\mathbf{g}^{+*} d\mathbf{g}^* \exp(\mathbf{g} \cdot \mathbf{g}^* + \mathbf{g}^{+*} \cdot \mathbf{g}^+ + \mathbf{g} \cdot \mathbf{g}^+)$$
$$\times \exp\left[-\frac{1}{2}\left(\boldsymbol{\alpha} \cdot \boldsymbol{\alpha}^* + \boldsymbol{\alpha}^{+*} \cdot \boldsymbol{\alpha}^+\right)\right]$$
$$\times_{\text{B}} \langle \mathbf{g}| \left\langle (\boldsymbol{\alpha} + \boldsymbol{\alpha}^{+*})/2 \right|_{\text{B}} \hat{\rho} \left| (\boldsymbol{\alpha} + \boldsymbol{\alpha}^{+*})/2 \right\rangle_{\text{B}} |\mathbf{g}^{+*}\rangle_{\text{B}} \tag{7.54}$$

is the canonical distribution function for the density operator $\hat{\rho}$. The existence of this form for the density operator has been established in the previous section. Note $d\mathbf{g}^{+*} d\mathbf{g}^* \equiv \prod_i dg_i^{+*} \prod_i dg_i^*$ and $d\mathbf{g}^+ d\mathbf{g} \equiv \prod_i dg_i^+ \prod_i dg_i$. The two forms for the density operator are the same because $\hat{\Lambda}(\mathbf{g}, \mathbf{g}^+, \boldsymbol{\alpha}, \boldsymbol{\alpha}^+)$ is an even Grassmann operator and thus commutes with $P_{\text{canon}}(\boldsymbol{\alpha}, \boldsymbol{\alpha}^+, \boldsymbol{\alpha}^*, \boldsymbol{\alpha}^{+*}, \mathbf{g}, \mathbf{g}^+)$. By substituting the canonical form of the density operator (7.52) into the expression (7.50) for the characteristic function, we can easily show that the characteristic function and distribution function are related as in (7.51), with the distribution function given by the canonical form (7.54):

$$P(\boldsymbol{\alpha}, \boldsymbol{\alpha}^+, \boldsymbol{\alpha}^*, \boldsymbol{\alpha}^{+*}, \mathbf{g}, \mathbf{g}^+) = P_{\text{canon}}(\boldsymbol{\alpha}, \boldsymbol{\alpha}^+, \boldsymbol{\alpha}^*, \boldsymbol{\alpha}^{+*}, \mathbf{g}, \mathbf{g}^+). \tag{7.55}$$

Note also that, irrespective of whether or not the detailed form of P is given by the canonical form (7.54), the expression (7.52) is often used to define the positive-P-type distribution function [31], rather than the characteristic-function expression (7.51). If the density operator can be written as

$$\hat{\rho} = \int d\mathbf{g}^+ d\mathbf{g} \int d^2\boldsymbol{\alpha}^+ d^2\boldsymbol{\alpha}\, \hat{\Lambda}(\mathbf{g}, \mathbf{g}^+, \boldsymbol{\alpha}, \boldsymbol{\alpha}^+)\, P(\boldsymbol{\alpha}, \boldsymbol{\alpha}^+, \boldsymbol{\alpha}^*, \boldsymbol{\alpha}^{+*}, \mathbf{g}, \mathbf{g}^+), \tag{7.56}$$

then this defines *a* positive P distribution function, which is non-unique in terms of the bosonic variables. It is straightforward to confirm that this form of the density operator leads to the previous relationship (7.51) between the distribution and characteristic functions. The specific form (7.54) for the canonical distribution function shows that a positive P distribution function actually exists.

In the case where the total number of fermions does not change, these general characteristic and distribution functions will have the same features as those for pure fermion systems, namely requiring the same numbers of h_i as h_i^+ and the same numbers of g_i as g_i^+. Using the ordering convention applied to (7.17) and (7.27), we can write these functions as

$$\chi(\boldsymbol{\xi}, \boldsymbol{\xi}^+, \mathbf{h}, \mathbf{h}^+) = \chi_0(\boldsymbol{\xi}, \boldsymbol{\xi}^+) + \sum_{j,i} \chi_2^{i,j}(\boldsymbol{\xi}, \boldsymbol{\xi}^+)\, h_j^+ + \cdots$$
$$+ \chi_{2n}^{1\cdots n, n\cdots 1}(\boldsymbol{\xi}, \boldsymbol{\xi}^+)\, h_n^+ \cdots h_1^+\, h_1 \cdots h_n, \tag{7.57}$$

$$P(\boldsymbol{\alpha}, \boldsymbol{\alpha}^+, \boldsymbol{\alpha}^*, \boldsymbol{\alpha}^{+*}, \mathbf{g}, \mathbf{g}^+) = P_0(\boldsymbol{\alpha}, \boldsymbol{\alpha}^+, \boldsymbol{\alpha}^*, \boldsymbol{\alpha}^{+*}) + \sum_{i;j} P_2^{i,j}(\boldsymbol{\alpha}, \boldsymbol{\alpha}^+, \boldsymbol{\alpha}^*, \boldsymbol{\alpha}^{+*})\, g_i g_j^+ + \cdots$$
$$+ P_{2n}^{1\cdots n, n\cdots 1}(\boldsymbol{\alpha}, \boldsymbol{\alpha}^+, \boldsymbol{\alpha}^*, \boldsymbol{\alpha}^{+*})\, g_1 \cdots g_n g_n^+ \cdots g_1^+, \qquad (7.58)$$

where now the coefficients depend on the bosonic variables.

The situation changes somewhat in the case of combined boson and fermion systems where pairs of fermions can be converted to or from bosons. The discussion in Sections 7.2.2 and 7.3.2 must now allow for the possibility that the expectation value of any product of an *even* number of fermion annihilation and creation operators is non-zero. The expression (7.16) becomes

$$\chi(\boldsymbol{\xi}, \boldsymbol{\xi}^+, \mathbf{h}, \mathbf{h}^+) = \mathrm{Tr}_\mathrm{f}(\hat{\Theta}_\mathrm{f}) + i^2 \sum_{j>k} h_j^+ h_k^+ \mathrm{Tr}_\mathrm{f}(\hat{c}_j \hat{c}_k \hat{\Theta}_\mathrm{f}) + i^2 \sum_{j,i} h_j^+ \mathrm{Tr}_\mathrm{f}(\hat{c}_j \hat{\Theta}_\mathrm{f} \hat{c}_i^\dagger) h_i$$
$$+ i^2 \sum_{i<h} \mathrm{Tr}_\mathrm{f}\left(\hat{\Theta}_\mathrm{f} \hat{c}_i^\dagger \hat{c}_h^\dagger\right) h_i h_h + \cdots$$
$$+ i^{2n} h_n^+ \cdots h_1^+ \mathrm{Tr}_\mathrm{f}\left(\hat{c}_n \cdots \hat{c}_1 \hat{\Theta}_\mathrm{f} \hat{c}_1^\dagger \cdots \hat{c}_n^\dagger\right) h_1 \cdots h_n, \qquad (7.59)$$

where we write the fermionic operator given by the bosonic trace over $\hat{\Omega}_\mathrm{b}^+(\boldsymbol{\xi}^+)\, \hat{\rho}\, \hat{\Omega}_\mathrm{b}^-(\boldsymbol{\xi})$ as

$$\hat{\Theta}_\mathrm{f} = \mathrm{Tr}_\mathrm{b}\left[\hat{\Omega}_\mathrm{b}^+(\boldsymbol{\xi}^+)\, \hat{\rho}\, \hat{\Omega}_\mathrm{b}^-(\boldsymbol{\xi})\right], \qquad (7.60)$$

leaving the $\boldsymbol{\xi}, \boldsymbol{\xi}^+$ dependence implicit. Note that there are no terms involving an odd number of fermion creation or annihilation operators. If we define

$$\begin{aligned}
\chi_0(\boldsymbol{\xi}, \boldsymbol{\xi}^+) &= \mathrm{Tr}_\mathrm{f}(\hat{\Theta}_\mathrm{f}), \\
\chi_2^{jk\square}(\boldsymbol{\xi}, \boldsymbol{\xi}^+) &= \mathrm{Tr}_\mathrm{f}(\hat{c}_j\, \hat{c}_k \hat{\Theta}_\mathrm{f}), \\
\chi_2^{i\square j}(\boldsymbol{\xi}, \boldsymbol{\xi}^+) &= \mathrm{Tr}_\mathrm{f}(\hat{c}_j\, \hat{\Theta}_\mathrm{f}\, \hat{c}_i^\dagger), \\
\chi_2^{\square ih}(\boldsymbol{\xi}, \boldsymbol{\xi}^+) &= \mathrm{Tr}_\mathrm{f}(\hat{\Theta}_\mathrm{f}\, \hat{c}_i^\dagger\, \hat{c}_h^\dagger), \\
&\vdots \\
\chi_{2n}^{1\cdots n\square n\cdots 1}(\boldsymbol{\xi}, \boldsymbol{\xi}^+) &= \mathrm{Tr}_\mathrm{f}(\hat{c}_n \cdots \hat{c}_1\, \hat{\Theta}_\mathrm{f}\, \hat{c}_1^\dagger \cdots \hat{c}_n^\dagger\,),
\end{aligned} \qquad (7.61)$$

then the characteristic function becomes

$$\chi(\boldsymbol{\xi}, \boldsymbol{\xi}^+, \mathbf{h}, \mathbf{h}^+) = \chi_0(\boldsymbol{\xi}, \boldsymbol{\xi}^+) + i^2 \sum_{j>k} \chi_2^{jk\square}(\boldsymbol{\xi}, \boldsymbol{\xi}^+) h_j^+ h_k^+$$
$$+ i^2 \sum_{j,i} \chi_2^{i\square j}(\boldsymbol{\xi}, \boldsymbol{\xi}^+) h_j^+\, h_i + i^2 \sum_{i<h} \chi_2^{\square ih}(\boldsymbol{\xi}, \boldsymbol{\xi}^+)\, h_i h_h + \cdots$$
$$+ i^{2n} \chi_{2n}^{1\cdots n\square n\cdots 21}(\boldsymbol{\xi}, \boldsymbol{\xi}^+)\, h_n^+ \cdots h_1^+\, h_1 \cdots h_n, \qquad (7.62)$$

which is analogous to (7.57). The block notation $\square$ has been introduced to separate indices associated with creation operators from those associated with annihilation operators. We see that although the characteristic function is now more complicated, it still retains the feature of being an even Grassmann function. This means that the same argument as in Section 7.3.2 applies and the distribution function remains an even Grassmann function. This feature is important, since many of our later results will remain valid, as the only requirement is that the distribution function is even. The difference now will be that the distribution function now contains terms where the numbers of g_i and g_j^+ differ by an even number, rather than being equal as before. The distribution function will be of the form

$$\begin{aligned} P(\boldsymbol{\alpha},\boldsymbol{\alpha}^+,\boldsymbol{\alpha}^*,\boldsymbol{\alpha}^{+*},\mathbf{g},\mathbf{g}^+) &= P_0(\boldsymbol{\alpha},\boldsymbol{\alpha}^+,\boldsymbol{\alpha}^*,\boldsymbol{\alpha}^{+*}) + \sum_{i>j} P_2^{\square ij}(\boldsymbol{\alpha},\boldsymbol{\alpha}^+,\boldsymbol{\alpha}^*,\boldsymbol{\alpha}^{+*}) g_i^+ g_j^+ \\ &\quad + \sum_{i,j} P_2^{i\square j}(\boldsymbol{\alpha},\boldsymbol{\alpha}^+,\boldsymbol{\alpha}^*,\boldsymbol{\alpha}^{+*}) g_i g_j^+ \\ &\quad + \sum_{i<j} P_2^{ij\square}(\boldsymbol{\alpha},\boldsymbol{\alpha}^+,\boldsymbol{\alpha}^*,\boldsymbol{\alpha}^{+*})\, g_i g_j + \cdots \\ &\quad + P_{2n}^{1\cdots n\square n\cdots 1}(\boldsymbol{\alpha},\boldsymbol{\alpha}^+,\boldsymbol{\alpha}^*,\boldsymbol{\alpha}^{+*})\, g_1 \cdots g_n g_n^+ \cdots g_1^+ . \end{aligned} \tag{7.63}$$

The total number of c-number functions now required to specify the Bose–Fermi distribution function is now increased.

A case of particular interest is when the bosonic and fermionic systems are *uncorrelated*, in which case the density operator is a product of density operators $\hat{\rho}_{\mathrm{b}}, \hat{\rho}_{\mathrm{f}}$ for the bosonic and fermionic systems:

$$\hat{\rho} = \hat{\rho}_{\mathrm{b}}\,\hat{\rho}_{\mathrm{f}} = \hat{\rho}_{\mathrm{f}}\,\hat{\rho}_{\mathrm{b}} \,. \tag{7.64}$$

In this case the characteristic function also factorises:

$$\chi(\boldsymbol{\xi},\boldsymbol{\xi}^+,\mathbf{h},\mathbf{h}^+) = \chi_{\mathrm{b}}(\boldsymbol{\xi},\boldsymbol{\xi}^+)\,\chi_{\mathrm{f}}(\mathbf{h},\mathbf{h}^+) = \chi_{\mathrm{f}}(\mathbf{h},\mathbf{h}^+)\,\chi_{\mathrm{b}}(\boldsymbol{\xi},\boldsymbol{\xi}^+), \tag{7.65}$$

where the bosonic and fermionic characteristic functions $\chi_{\mathrm{b}}(\boldsymbol{\xi},\boldsymbol{\xi}^+)$ and $\chi_{\mathrm{f}}(\mathbf{h},\mathbf{h}^+)$ are defined as in (7.5) and (7.11). The proof is given in Appendix E. Naturally, the distribution function also factorises:

$$\begin{aligned} P(\boldsymbol{\alpha},\boldsymbol{\alpha}^+,\boldsymbol{\alpha}^*,\boldsymbol{\alpha}^{+*},\mathbf{g},\mathbf{g}^+) &= P_{\mathrm{b}}(\boldsymbol{\alpha},\boldsymbol{\alpha}^+,\boldsymbol{\alpha}^*,\boldsymbol{\alpha}^{+*})\,P_{\mathrm{f}}(\mathbf{g},\mathbf{g}^+) \\ &= P_{\mathrm{f}}(\mathbf{g},\mathbf{g}^+)\,P_{\mathrm{b}}(\boldsymbol{\alpha},\boldsymbol{\alpha}^+,\boldsymbol{\alpha}^*,\boldsymbol{\alpha}^{+*}), \end{aligned} \tag{7.66}$$

where the bosonic and fermionic distribution functions $P_{\mathrm{b}}(\boldsymbol{\alpha},\boldsymbol{\alpha}^+,\boldsymbol{\alpha}^*,\boldsymbol{\alpha}^{+*})$ and $P_{\mathrm{f}}(\mathbf{g},\mathbf{g}^+)$ are related to their corresponding characteristic functions as in (7.20) and (7.23). This result follows from substituting for $P(\boldsymbol{\alpha},\boldsymbol{\alpha}^+,\boldsymbol{\alpha}^*,\boldsymbol{\alpha}^{+*},\mathbf{g},\mathbf{g}^+)$ from (7.66) into the right side of (7.51) and then using (7.20) and (7.23) to give the characteristic function (7.65).

7.6 Hermiticity of the Density Operator

In this section, we obtain important constraints on the distribution function that follow from the density operator being Hermitian. These will be developed for the combined boson–fermion case using the canonical form (7.54) of the distribution function.

Firstly, with $\boldsymbol{\beta} = (\boldsymbol{\alpha} + \boldsymbol{\alpha}^{+*})/2$ for short, we have the following, using (5.39) and (5.48):

$$ {}_{\mathrm{B}}\langle \mathbf{k}| \langle \boldsymbol{\beta}|_{\mathrm{B}}\, \hat{\rho}\, |\boldsymbol{\beta}\rangle_{\mathrm{B}} \left|\mathbf{k}^{+*}\right\rangle_{\mathrm{B}} = \langle 0| \exp(\mathbf{k}^* \cdot \hat{\mathbf{c}}) \exp(\boldsymbol{\beta}^* \cdot \hat{\mathbf{a}}) \hat{\rho} \exp(\hat{\mathbf{a}}^\dagger \cdot \boldsymbol{\beta}) (\exp \hat{\mathbf{c}}^\dagger \cdot \mathbf{k}^{+*}) |0\rangle . \tag{7.67} $$

Using (5.5), we then have

$$ \begin{aligned} \left({}_{\mathrm{B}}\langle \mathbf{k}| \langle \boldsymbol{\beta}|_{\mathrm{B}}\, \hat{\rho}\, |\boldsymbol{\beta}\rangle_{\mathrm{B}} \left|\mathbf{k}^{+*}\right\rangle_{\mathrm{B}}\right)^* &= \langle 0| \exp(\mathbf{k}^+ \cdot \hat{\mathbf{c}}) \exp(\boldsymbol{\beta}^* \cdot \hat{\mathbf{a}}) (\hat{\rho})^\dagger \exp(\hat{\mathbf{a}}^\dagger \cdot \boldsymbol{\beta}) \exp(\hat{\mathbf{c}}^\dagger \cdot \mathbf{k}) |0\rangle \\ &= {}_{\mathrm{B}}\left\langle \mathbf{k}^{+*}\right| \langle \boldsymbol{\beta}|_{\mathrm{B}}\, \hat{\rho}\, |\boldsymbol{\beta}\rangle_{\mathrm{B}} |\mathbf{k}\rangle_{\mathrm{B}} \end{aligned} \tag{7.68} $$

because $\hat{\rho}$ is Hermitian. If we insert this into our expression for the canonical distribution function and use (3.43) and the result (4.66) for conjugating a Grassmann integral, we find

$$ \begin{aligned} (P_{\mathrm{canon}}(\boldsymbol{\alpha}, \boldsymbol{\alpha}^+, \boldsymbol{\alpha}^*, \boldsymbol{\alpha}^{+*}, \mathbf{g}, \mathbf{g}^+))^* &= \left(\frac{1}{4\pi^2}\right)^n \int dk_1 \cdots dk_n dk_n^+ \cdots dk_1^+ \\ &\times \exp(\mathbf{g}^{+*} \cdot \mathbf{k}^+ + \mathbf{k} \cdot \mathbf{g}^* + \mathbf{g}^{+*} \cdot \mathbf{g}^*) \exp\left[-\frac{1}{2}(\boldsymbol{\alpha} \cdot \boldsymbol{\alpha}^* + \boldsymbol{\alpha}^{+*} \cdot \boldsymbol{\alpha}^+)\right] \\ &\times \left(\langle \mathbf{k}|_{\mathrm{B}} \langle \boldsymbol{\beta}|_{\mathrm{B}}\, \hat{\rho}\, |\boldsymbol{\beta}\rangle_{\mathrm{B}} \left|\mathbf{k}^{+*}\right\rangle_{\mathrm{B}}\right)^*, \end{aligned} \tag{7.69} $$

where we have reordered the even Grassmann functions and substituted from (7.68) for $\left({}_{\mathrm{B}}\langle \mathbf{k}|_{\mathrm{B}} \langle \boldsymbol{\beta}|\, \hat{\rho}\, |\boldsymbol{\beta}\rangle_{\mathrm{B}} \left|\mathbf{k}^{+*}\right\rangle_{\mathrm{B}}\right)^*$. Also, the products of differentials $dk_1 \cdots dk_1^+$ commute with all of the Grassmann functions and can be placed on the left, hence restoring a left Grassmann integral. If we make the change of integration variables $k_i \to g_i^{+*}$, $k_i^+ \to g_i^*$, then reverse the order within the differentials, we find

$$ (P_{\mathrm{canon}}(\boldsymbol{\alpha}, \boldsymbol{\alpha}^+, \boldsymbol{\alpha}^*, \boldsymbol{\alpha}^{+*}, \mathbf{g}, \mathbf{g}^+))^* = P_{\mathrm{canon}}(\boldsymbol{\alpha}, \boldsymbol{\alpha}^+, \boldsymbol{\alpha}^*, \boldsymbol{\alpha}^{+*}, \mathbf{g}^{+*}, \mathbf{g}^*). $$

We can also arrive at this result using the characteristic function and so, more generally, we find that

$$ (P(\boldsymbol{\alpha}, \boldsymbol{\alpha}^+, \boldsymbol{\alpha}^*, \boldsymbol{\alpha}^{+*}, \mathbf{g}, \mathbf{g}^+))^* = P(\boldsymbol{\alpha}, \boldsymbol{\alpha}^+, \boldsymbol{\alpha}^*, \boldsymbol{\alpha}^{+*}, \mathbf{g}^{+*}, \mathbf{g}^*). \tag{7.70} $$

For the purely bosonic case, this shows the distribution function $P_{\mathrm{b}}(\boldsymbol{\alpha}, \boldsymbol{\alpha}^+, \boldsymbol{\alpha}^*, \boldsymbol{\alpha}^{+*})$ to be real, as noted previously. For the fermionic case, this establishes relationships between the coefficients that determine the Grassmann distribution function $P_{\mathrm{f}}(\mathbf{g}, \mathbf{g}^+)$. These are a special case of the general boson–fermion distribution function.

Applying the result (7.70) to the general form (7.58) for the distribution function in the case where the numbers of bosons and fermions are each conserved, we find that

$$ \begin{aligned} P_0(\boldsymbol{\alpha}, \boldsymbol{\alpha}^+, \boldsymbol{\alpha}^*, \boldsymbol{\alpha}^{+*})^* &= P_0(\boldsymbol{\alpha}, \boldsymbol{\alpha}^+, \boldsymbol{\alpha}^*, \boldsymbol{\alpha}^{+*}), \\ P_2^{i,j}(\boldsymbol{\alpha}, \boldsymbol{\alpha}^+, \boldsymbol{\alpha}^*, \boldsymbol{\alpha}^{+*})^* &= P_2^{j,i}(\boldsymbol{\alpha}, \boldsymbol{\alpha}^+, \boldsymbol{\alpha}^*, \boldsymbol{\alpha}^{+*}), \\ &\vdots \\ P_{2n}^{1\cdots n, n\cdots 1}(\boldsymbol{\alpha}, \boldsymbol{\alpha}^+, \boldsymbol{\alpha}^*, \boldsymbol{\alpha}^{+*})^* &= P_{2n}^{1\cdots n, n\cdots 1}(\boldsymbol{\alpha}, \boldsymbol{\alpha}^+, \boldsymbol{\alpha}^*, \boldsymbol{\alpha}^{+*}), \end{aligned} \tag{7.71} $$

which interrelates the c-number functions of bosonic variables that determine the boson–fermion distribution function. These mean, in particular, that certain of these coefficients, including P_0, $P_2^{i,i}$ and $P_4^{ij,ji}$, are real. A special case of these results applies to the pure fermion case, giving for the distribution function in (7.27)

$$\begin{aligned} P_0{}^* &= P_0, \\ P_2^{i,j*} &= P_2^{j,i}, \\ &\vdots \\ P_{2n}^{1\cdots n,n\cdots 1*} &= P_{2n}^{1\cdots n,n\cdots 1}. \end{aligned} \tag{7.72}$$

Results for the case where pairs of fermions are replaced by a boson can also be obtained.

7.7 Quantum Correlation Functions

In this section, we relate the quantum correlation functions to *derivatives* of the characteristic function and then show how the correlation functions are given as *phase space integrals* involving the distribution function. Quantum correlation functions involving operators that are not normally ordered are given by identical phase space integrals, but of course the distribution functions are different and satisfy different Fokker–Planck equations.

The quantum correlation functions can be obtained from differentiation of the characteristic function followed by making all of its variables zero. In the case of fermionic systems, Grassmann differentiation from the left and right is involved.

7.7.1 Bosons

For the boson case, the *normally ordered* quantum correlation function is given by

$$G_{\mathrm{b}}(l_1,\cdots,l_p;m_q,\cdots,m_1) = \left(\frac{\partial}{\partial(i\xi_{l_p})}\cdots\frac{\partial}{\partial(i\xi_{l_1})}\frac{\partial}{\partial(i\xi^+_{m_q})}\cdots\frac{\partial}{\partial(i\xi^+_{m_1})}\chi_{\mathrm{b}}(\boldsymbol{\xi},\boldsymbol{\xi}^+)\right)_{\boldsymbol{\xi},\boldsymbol{\xi}^+=0}. \tag{7.73}$$

Applying the differentiation to the expression involving the distribution function, we get

$$\begin{aligned} G_{\mathrm{b}}(l_1,\cdots,l_p;m_q,\cdots,m_1) &= \mathrm{Tr}(\hat{a}_{m_q}\cdots\hat{a}_{m_1}\hat{\rho}_b\,\hat{a}^\dagger_{l_1}\cdots\hat{a}^\dagger_{l_p}) \\ &= \int d^2\boldsymbol{\alpha}^+ d^2\boldsymbol{\alpha}(\alpha_{m_q}\cdots\alpha_{m_1})P_{\mathrm{b}}(\boldsymbol{\alpha},\boldsymbol{\alpha}^+,\boldsymbol{\alpha}^*,\boldsymbol{\alpha}^{+*})(\alpha^+_{l_1}\cdots\alpha^+_{l_p}), \end{aligned} \tag{7.74}$$

giving the phase space integral result in which the creation operators $\hat{a}_i^\dagger$ have been replaced by α_i^+, the annihilation operators $\hat{a}_i$ replaced by α_i and the density operator replaced by the distribution function $P_{\mathrm{b}}(\boldsymbol{\alpha},\boldsymbol{\alpha}^+,\boldsymbol{\alpha}^*,\boldsymbol{\alpha}^{+*})$. As all the quantities are c-numbers, the quantum correlation function is also given by

$$G_{\rm b}(l_1,\cdots,l_p;m_q,\cdots,m_1)=\int d^2\boldsymbol{\alpha}^+d^2\boldsymbol{\alpha}\,P_{\rm b}(\boldsymbol{\alpha},\boldsymbol{\alpha}^+,\boldsymbol{\alpha}^*,\boldsymbol{\alpha}^{+*})(\alpha_{l_1}^+\cdots\alpha_{l_p}^+)(\alpha_{m_q}\cdots\alpha_{m_1}),\tag{7.75}$$

in which form the c-numbers replacing the boson operators are placed to the right of the distribution function in the same order as the operators. A simple example is the normalisation integral for the Bose distribution function, which can be obtained from $\mathrm{Tr}\,\hat{\rho}_{\rm b}=\langle 1\rangle=1$ and is given by

$$\int d^2\boldsymbol{\alpha}^+d^2\boldsymbol{\alpha}\,P_{\rm b}(\boldsymbol{\alpha},\boldsymbol{\alpha}^+,\boldsymbol{\alpha}^*,\boldsymbol{\alpha}^{+*})=1\,.\tag{7.76}$$

Two key results used in the derivations are

$$\begin{aligned}\frac{\partial}{\partial(i\xi_j)}\chi_{\rm b}(\boldsymbol{\xi},\boldsymbol{\xi}^+)&=\mathrm{Tr}[\hat{\Omega}_{\rm b}^+(\boldsymbol{\xi}^+)\,\hat{\rho}\,\hat{a}_j^\dagger\,\hat{\Omega}_{\rm b}^-(\boldsymbol{\xi})]=\mathrm{Tr}[\hat{\Omega}_{\rm b}^+(\boldsymbol{\xi}^+)\,\hat{\rho}\,\hat{\Omega}_{\rm b}^-(\boldsymbol{\xi})\,\hat{a}_j^\dagger],\\ \frac{\partial}{\partial(i\xi_i^+)}\chi_{\rm b}(\boldsymbol{\xi},\boldsymbol{\xi}^+)&=\mathrm{Tr}[\hat{\Omega}_{\rm b}^+(\boldsymbol{\xi}^+)\,\hat{a}_i\,\hat{\rho}\,\hat{\Omega}_{\rm b}^-(\boldsymbol{\xi})]=\mathrm{Tr}[\hat{a}_i\,\hat{\Omega}_{\rm b}^+(\boldsymbol{\xi}^+)\,\hat{\rho}\,\hat{\Omega}_{\rm b}^-(\boldsymbol{\xi})].\end{aligned}\tag{7.77}$$

Combining these gives

$$\begin{aligned}\frac{\partial}{\partial(i\xi_j)}\frac{\partial}{\partial(i\xi_i^+)}\chi_{\rm b}(\boldsymbol{\xi},\boldsymbol{\xi}^+)&=\frac{\partial}{\partial(i\xi_i^+)}\frac{\partial}{\partial(i\xi_j)}\chi_{\rm b}(\boldsymbol{\xi},\boldsymbol{\xi}^+)\\ &=\mathrm{Tr}[\hat{\Omega}_{\rm b}^+(\boldsymbol{\xi}^+)\hat{a}_i\,\hat{\rho}\hat{a}_j^\dagger\hat{\Omega}_{\rm b}^-(\boldsymbol{\xi})]\\ &=\mathrm{Tr}[\hat{a}_i\,\hat{\Omega}_{\rm b}^+(\boldsymbol{\xi}^+)\hat{\rho}\,\hat{\Omega}_{\rm b}^-(\boldsymbol{\xi})\,\hat{a}_j^\dagger].\end{aligned}\tag{7.78}$$

We can also write the *symmetrically ordered* quantum correlation function in terms of the Wigner function

$$\begin{aligned}G_{\rm b}^{\rm W}(l_1,\cdots,l_p;m_q,\cdots,m_1)=\Bigg(&\frac{\partial}{\partial(i\xi_{l_p})}\cdots\frac{\partial}{\partial(i\xi_{l_1})}\frac{\partial}{\partial(i\xi_{m_q}^+)}\cdots\\ &\times\frac{\partial}{\partial(i\xi_{m_1}^+)}\chi_{\rm b}^{\rm W}(\boldsymbol{\xi},\boldsymbol{\xi}^+)\Bigg)_{\boldsymbol{\xi},\boldsymbol{\xi}^+=0}\end{aligned}\tag{7.79}$$

and, applying the differentiation to the expression involving the distribution function, we get

$$\begin{aligned}G_{\rm b}^{\rm W}(l_1,\cdots,l_p;m_q,\cdots,m_1)&=\mathrm{Tr}(\hat{\rho}_{\rm b}\{\hat{a}_{l_1}^\dagger\hat{a}_{l_2}^\dagger\cdots\hat{a}_{l_p}^\dagger\hat{a}_{m_q}\cdots\hat{a}_{m_2}\hat{a}_{m_1}\})\\ &=\int d^2\boldsymbol{\alpha}^+d^2\boldsymbol{\alpha}(\alpha_{m_q}\cdots\alpha_{m_1})W_{\rm b}(\boldsymbol{\alpha},\boldsymbol{\alpha}^+,\boldsymbol{\alpha}^*,\boldsymbol{\alpha}^{+*})(\alpha_{l_1}^+\cdots\alpha_{l_p}^+),\end{aligned}\tag{7.80}$$

giving the phase space integral result in which the creation operators $\hat{a}_i^\dagger$ have been replaced by α_i^+, the annihilation operators $\hat{a}_i$ replaced by α_i and the density operator replaced by the distribution function $W_{\rm b}(\boldsymbol{\alpha},\boldsymbol{\alpha}^+,\boldsymbol{\alpha}^*,\boldsymbol{\alpha}^{+*})$. The proof is given in Appendix E. The Wigner distribution function is also normalised to unity.

7.7.2 Fermions

For the fermion case, the correlation function can be written

$$G_{\mathrm{f}}(l_1,\cdots,l_p;m_q,\cdots,m_1) = \left(\frac{\overrightarrow{\partial}}{\partial(ih^+_{m_q})}\cdots\frac{\overrightarrow{\partial}}{\partial(ih^+_{m_1})}\chi_{\mathrm{f}}(\mathbf{h},\mathbf{h}^+)\frac{\overleftarrow{\partial}}{\partial(ih_{l_1})}\cdots\frac{\overleftarrow{\partial}}{\partial(ih_{l_p})}\right)_{\mathbf{h},\mathbf{h}^+=0}, \tag{7.81}$$

and, applying the differentiation to the expression involving the distribution function, we get

$$\begin{aligned} G(l_1,\cdots,l_p;m_q,\cdots,m_1) &= \mathrm{Tr}(\hat{c}_{m_q}\cdots\hat{c}_{m_1}\,\hat{\rho}_{\mathrm{f}}\,\hat{c}^\dagger_{l_1}\cdots\hat{c}^\dagger_{l_p}) \\ &= \int d\mathbf{g}^+ d\mathbf{g}\,(g_{m_q}\cdots g_{m_1})\,P_{\mathrm{f}}(\mathbf{g},\mathbf{g}^+)\,(g^+_{l_1}\cdots g^+_{l_p}), \\ & \quad p-q = 0,\pm 2,\pm 4,\cdots, \end{aligned} \tag{7.82}$$

giving the phase space integral result in which the creation operators $\hat{c}^\dagger_i$ have been replaced by g^+_i and the annihilation operators $\hat{c}_i$ replaced by g_i. The proof is given in Appendix E. Note that the result is restricted to the case where $p-q = 0,\pm 2,\pm 4,\cdots$, that is, equal to an even integer. As cases of the quantum correlation function with $p-q$ odd are equal to zero, this phase space result is all we need to consider. Clearly, if any two modes are the same then the result is zero. Note that in the fermion case it is important to keep the g^+_i and the g_i in the same order as the $\hat{c}^\dagger_i$ and the $\hat{c}_i$. See also the similarity to the result (7.74) for the boson case. A simple example is the fact that the normalisation integral for the Fermi distribution function can be obtained from $\mathrm{Tr}\;\hat{\rho}_{\mathrm{f}} = \langle 1\rangle = 1$ and is given by

$$\int d\mathbf{g}^+ d\mathbf{g}\,P_{\mathrm{f}}(\mathbf{g},\mathbf{g}^+) = 1. \tag{7.83}$$

If we substitute the specific ordered form (7.27) for the distribution function, we see that the normalisation integral gives

$$P^{1\cdots n,n\cdots 1}_{2n} = 1; \tag{7.84}$$

we leave the proof of this as an exercise.

Although all the quantities are Grassmann numbers, the result for the quantum correlation function can still be written in other forms. The distribution function is even and can be shifted to the left in the integrand. Placing each g_i to the right of all the g^+_j involves a factor of $(-1)^p$, so putting all the g_i to the right of all the g^+_j leads to a factor $(-1)^{q\times p}$. Thus the quantum correlation function is also given by

$$\begin{aligned} G(l_1,\cdots,l_P;m_q,\cdots,m_1) &= (-1)^{q\times p}\int d\mathbf{g}^+ d\mathbf{g}\,P_{\mathrm{f}}(\mathbf{g},\mathbf{g}^+)\,(g^+_{l_1}\cdots g^+_{l_p})(g_{m_q}\cdots g_{m_1}), \\ & \quad p-q = 0,\pm 2,\pm 4,\cdots, \end{aligned} \tag{7.85}$$

where the Grassmann numbers replacing the fermion operators are placed to the right of the distribution function in the same order as the operators; this is a form of the result analogous to that for the boson case.

Three key results used in the derivation are

$$\begin{aligned}
\frac{\overrightarrow{\partial}}{\partial(ih_i^+)}\chi_{\mathrm{f}}(\mathbf{h},\mathbf{h}^+)\frac{\overleftarrow{\partial}}{\partial(ih_j)} &= \mathrm{Tr}\left[\hat{c}_i\,\hat{\Omega}_{\mathrm{f}}^+(\mathbf{h}^+)\,\hat{\rho}_{\mathrm{f}}\,\hat{\Omega}_{\mathrm{f}}^-(\mathbf{h})\,\hat{c}_j^\dagger\right] \\
&= \mathrm{Tr}[\hat{\Omega}_{\mathrm{f}}^+(\mathbf{h}^+)\,\hat{c}_i\,\hat{\rho}_{\mathrm{f}}\,\hat{c}_j^\dagger\,\hat{\Omega}_{\mathrm{f}}^-(\mathbf{h})], \\
\chi_{\mathrm{f}}(\mathbf{h},\mathbf{h}^+)\frac{\overleftarrow{\partial}}{\partial(ih_j)}\frac{\overleftarrow{\partial}}{\partial(ih_i)} &= \mathrm{Tr}\left[\hat{\Omega}_{\mathrm{f}}^+(\mathbf{h}^+)\,\hat{\rho}_{\mathrm{f}}\,\hat{\Omega}_{\mathrm{f}}^-(\mathbf{h})\,\hat{c}_j^\dagger\hat{c}_i^\dagger\right] \\
&= \mathrm{Tr}\left[\hat{\Omega}_{\mathrm{f}}^+(\mathbf{h}^+)\,\hat{\rho}_{\mathrm{f}}\,\hat{c}_j^\dagger\hat{c}_i^\dagger\,\hat{\Omega}_{\mathrm{f}}^-(\mathbf{h})\right], \\
\frac{\overrightarrow{\partial}}{\partial(ih_i^+)}\frac{\overrightarrow{\partial}}{\partial(ih_j^+)}\chi_{\mathrm{f}}(\mathbf{h},\mathbf{h}^+) &= \mathrm{Tr}\left[\hat{c}_i\,\hat{c}_j\,\hat{\Omega}_{\mathrm{f}}^+(\mathbf{h}^+)\,\hat{\rho}_{\mathrm{f}}\,\hat{\Omega}_{\mathrm{f}}^-(\mathbf{h})\right] \\
&= \mathrm{Tr}\left[\hat{\Omega}_{\mathrm{f}}^+(\mathbf{h}^+)\,\hat{c}_i\,\hat{c}_j\,\hat{\rho}_{\mathrm{f}}\,\hat{\Omega}_{\mathrm{f}}^-(\mathbf{h})\right],
\end{aligned} \tag{7.86}$$

where there is differentiation from both the left and the right. Unlike the boson case, no simple result occurs for a single left or right differentiation alone. Note the similarity of (7.86) to the bosonic result (7.78).

It is not difficult to see that all of the quantum correlation functions are given by particular coefficients in the form (7.27) for the distribution function. From the result (7.82) that gives the correlation function $G(l_1,\cdots,l_p;m_q,\cdots,m_1)$ in terms of the Grassmann integral, the only contribution can come from the term in the expression for $P_{\mathrm{f}}(\mathbf{g},\mathbf{g}^+)$ in which the Grassmann variables in $(g_{m_q}\cdots g_{m_1})$ and $(g_{l_1}^+\cdots g_{l_p}^+)$ are absent, but in which all the remaining g_i and g_j^+ are present. This is because to get a non-zero result for the $2n$-fold multiple Grassmann integral, each g_i and g_j^+ must be present once and once only. However, for the particular correlation function $\langle\hat{c}_{l_1}^\dagger\cdots\hat{c}_{l_p}^\dagger\hat{c}_{m_q}\cdots\hat{c}_{m_1}\rangle$, there is only one term in the expression for $P_{\mathrm{f}}(\mathbf{g},\mathbf{g}^+)$ which has the required feature. For this term the Grassmann integral over all the g_i and g_j^+ will equal either $+1$ or -1, depending on the order of the factors, so the quantum correlation function will be equal to one of the expansion coefficients times ±1. Thus, for fermionic systems, we now see that the quantum correlation functions are essentially the same as the coefficients in the expansion (7.27) for the distribution function $P_{\mathrm{f}}(\mathbf{g},\mathbf{g}^+)$.

Two simple examples will serve to illustrate the point. Our first,

$$\begin{aligned}
\langle\hat{c}_i^\dagger\hat{c}_j\rangle &= \int d\mathbf{g}^+ d\mathbf{g}\,(g_j)\,P_{\mathrm{f}}(\mathbf{g},\mathbf{g}^+)\,(g_i^+) \\
&= (-1)^{j+i}\,P_{2n-2}^{1\cdots(j-1)(j+1)\cdots n,n\cdots(i+1)(i-1)\cdots 1},
\end{aligned} \tag{7.87}$$

shows that the first-order correlation function $\langle\hat{c}_i^\dagger\hat{c}_j\rangle$ is related to the coefficient $P_{2n-2}^{1\cdots(j-1)(j+1)\cdots n,n\cdots(i+1)(i-1)\cdots 1}$. Our second,

$$\begin{aligned}
\langle\hat{c}_k^\dagger\hat{c}_l^\dagger\hat{c}_i\hat{c}_j\rangle &= \int d\mathbf{g}^+ d\mathbf{g}\,(g_i\,g_j)\,P_{\mathrm{f}}(\mathbf{g},\mathbf{g}^+)\,(g_k^+\,g_l^+) \\
&= (-1)^{j+i+k+l}\,P_{2n-4}^{1\cdots(i-1)(i+1)\cdots(j-1)(j+1)\cdots n,n\cdots(k+1)(k-1)\cdots(l+1)(l-1)\cdots 1},
\end{aligned} \tag{7.88}$$

shows that the second-order correlation function $\langle \hat{c}_k^\dagger \hat{c}_l^\dagger \hat{c}_i \hat{c}_j \rangle$ $(i < j, k > l)$ is related to $P_{2n-4}^{1\cdots(i-1)(i+1)\cdots(j-1)(j+1)\cdots n,n\cdots(k+1)(k-1)\cdots(l+1)(l-1)\cdots 1}$.

7.7.3 Combined Case

The relationship between the quantum correlation function and the distribution function in the combined case is easily obtained by combining the boson and fermion results. The results apply to the general case where pairs of fermions can be created or destroyed from single bosons. We have

$$\begin{aligned} G(l_1, \cdots, l_p; m_q, \cdots, m_1;; p_1, \cdots, p_r; q_s, \cdots, q_1) &= \Bigg(\frac{\overrightarrow{\partial}}{\partial(ih^+_{m_q})} \cdots \frac{\overrightarrow{\partial}}{\partial(ih^+_{m_1})} \\ (\psi(p_1, \cdots, p_r; q_s, \cdots, q_1;; \boldsymbol{\xi}, \boldsymbol{\xi}^+, \mathbf{h}, \mathbf{h}^+) & \frac{\overleftarrow{\partial}}{\partial(ih_{l_1})} \cdots \frac{\overleftarrow{\partial}}{\partial(ih_{l_p})} \Bigg)_{\boldsymbol{\xi}, \boldsymbol{\xi}^+,\ \mathbf{h},\ \mathbf{h}^+=0} \\ p - q = 0, \pm 2, \pm 4, \cdots, & \end{aligned} \tag{7.89}$$

where

$$\begin{aligned} &\psi(p_1, \cdots, p_r; q_s, \cdots, q_1;; \boldsymbol{\xi}, \boldsymbol{\xi}^+, \mathbf{h}, \mathbf{h}^+) \\ &= \left(\frac{\partial}{\partial(i\xi_{p_r})} \cdots \frac{\partial}{\partial(i\xi_{p_1})} \frac{\partial}{\partial(i\xi^+_{q_s})} \cdots \frac{\partial}{\partial(i\xi^+_{q_1})} \chi(\boldsymbol{\xi}, \boldsymbol{\xi}^+, \mathbf{h}, \mathbf{h}^+) \right). \end{aligned} \tag{7.90}$$

The order of the c-number differentiation is irrelevant. Note that these results only apply when $p - q$ is an even integer.

For the combined boson and fermion case, the quantum correlation functions and their phase space integral result are given by

$$\begin{aligned} &G(l_1, \cdots, l_p; m_q, \cdots, m_1;; p_1, \cdots, p_r; q_s, \cdots, q_1) \\ &= \mathrm{Tr}(\hat{c}_{m_q} \cdots \hat{c}_{m_1} \hat{a}_{q_s} \cdots \hat{a}_{q_1} \hat{\rho} \hat{a}^\dagger_{p_1} \cdots \hat{a}^\dagger_{p_r} \hat{c}^\dagger_{l_1} \cdots \hat{c}^\dagger_{l_p}) \\ &= \int d^2\boldsymbol{\alpha}^+ d^2\boldsymbol{\alpha} \int d\mathbf{g}^+ d\mathbf{g}\, (g_{m_q} \cdots g_{m_1})\, (\alpha_{q_s} \cdots, \alpha_{q_1}) \\ &\quad \times P(\boldsymbol{\alpha}, \boldsymbol{\alpha}^+, \boldsymbol{\alpha}^*, \boldsymbol{\alpha}^{+*}, \mathbf{g}, \mathbf{g}^+)\, (\alpha^+_{p_1} \cdots \alpha^+_{p_r})\, (g^+_{l_1} \cdots g^+_{l_p}), \\ &\quad p - q = 0, \pm 2, \pm 4, \cdots, \end{aligned} \tag{7.91}$$

which involves both c-number and Grassmann number phase space integrals with the Bose–Fermi distribution function. This result is obtained by carrying out the differentiations on the formula (7.51) relating the characteristic and distribution functions. Again, alternative forms for the quantum correlation function with the distribution function as the left factor in the integrand and all the α_i and g_j placed to the right of the α_i^+ and g_j^+ can easily be found.

The normalisation integral for the combined Bose–Fermi distribution function can be obtained from $\mathrm{Tr}\, \hat{\rho} = \langle 1 \rangle = 1$ and is given by

$$\int d^2\boldsymbol{\alpha}^+ d^2\boldsymbol{\alpha} \int d\mathbf{g}^+ d\mathbf{g}\, P(\boldsymbol{\alpha}, \boldsymbol{\alpha}^+, \boldsymbol{\alpha}^*, \boldsymbol{\alpha}^{+*}, \mathbf{g}, \mathbf{g}^+) = 1\,. \tag{7.92}$$

If we substitute the specific ordered form (7.58) for the distribution function, we see that the normalisation integral gives

$$\int d^2\boldsymbol{\alpha}^+ d^2\boldsymbol{\alpha}\, P_{2n}^{1\cdots n,n\cdots 1}(\boldsymbol{\alpha},\boldsymbol{\alpha}^+,\boldsymbol{\alpha}^*,\boldsymbol{\alpha}^{+*}) = 1. \tag{7.93}$$

Another result of interest for combined Bose–Fermi systems involves quantum correlation functions with only fermion operators:

$$\begin{aligned}&\langle \hat{c}_{l_1}^\dagger \cdots \hat{c}_{l_p}^\dagger \hat{c}_{m_q} \cdots \hat{c}_{m_1}\rangle \\ &= \int d^2\boldsymbol{\alpha}^+ d^2\boldsymbol{\alpha}\, d\mathbf{g}^+ d\mathbf{g}\,(g_{m_q}\cdots g_{m_1})P(\boldsymbol{\alpha},\boldsymbol{\alpha}^+,\boldsymbol{\alpha}^*,\boldsymbol{\alpha}^{+*},\mathbf{g},\mathbf{g}^+)\,(g_{l_1}^+\cdots g_{l_p}^+), \\ &\quad p-q=0,\pm 2,\pm 4,\cdots. \end{aligned} \tag{7.94}$$

For example, after evaluating the Grassmann phase space integral and using (7.58), we obtain the result

$$\langle \hat{c}_1^\dagger \cdots \hat{c}_n^\dagger \hat{c}_n \cdots \hat{c}_1\rangle = \int d^2\boldsymbol{\alpha}^+ d^2\boldsymbol{\alpha}\, P_0(\boldsymbol{\alpha},\boldsymbol{\alpha}^+,\boldsymbol{\alpha}^*,\boldsymbol{\alpha}^{+*}). \tag{7.95}$$

This correlation function may be zero without $P_0(\boldsymbol{\alpha},\boldsymbol{\alpha}^+,\boldsymbol{\alpha}^*,\boldsymbol{\alpha}^{+*})$ being zero.

7.7.4 Uncorrelated Systems

A case of particular interest is when the bosonic and fermionic systems are uncorrelated, in which case the density operator is a product of density operators $\hat{\rho}_\mathrm{b}, \hat{\rho}_\mathrm{f}$ for the bosonic and fermionic systems:

$$\hat{\rho} = \hat{\rho}_\mathrm{b}\,\hat{\rho}_\mathrm{f} = \hat{\rho}_\mathrm{f}\,\hat{\rho}_\mathrm{b}. \tag{7.96}$$

As we have seen, the characteristic and distribution functions factorise (see (7.65) and (7.66)) and so does the quantum correlation function:

$$\begin{aligned}&\left\langle \hat{a}_{j_1}^\dagger \cdots \hat{a}_{j_r}^\dagger \hat{a}_{k_s}\cdots \hat{a}_{k_1}\hat{c}_{l_1}^\dagger \cdots \hat{c}_{l_p}^\dagger \hat{c}_{m_q}\cdots \hat{c}_{m_1}\right\rangle \\ &= \left\langle \hat{a}_{j_1}^\dagger \cdots \hat{a}_{j_r}^\dagger \hat{a}_{k_s}\cdots \hat{a}_{k_1}\right\rangle_\mathrm{b} \left\langle \hat{c}_{l_1}^\dagger \cdots \hat{c}_{l_p}^\dagger \hat{c}_{m_q}\cdots \hat{c}_{m_1}\right\rangle_\mathrm{f}, \end{aligned} \tag{7.97}$$

where the boson and fermion correlation functions are as in (7.1) and (7.2). This result may be shown either directly from the definition of the quantum correlation function or by using the general result (10.85) along with the factorisation of the distribution function (7.66).

7.8 Unnormalised Distribution Functions

The distribution functions $P_\mathrm{b}(\boldsymbol{\alpha},\boldsymbol{\alpha}^+,\boldsymbol{\alpha}^*,\boldsymbol{\alpha}^{+*})$ and $P_\mathrm{f}(\mathbf{g},\mathbf{g}^+)$ are both normalised to unity and, as we have seen, can be used to determine quantum correlation functions. However, it is sometimes convenient to consider unnormalised forms of the distribution functions, as these turn out to result in simpler Fokker–Planck and Langevin equations.

This is because the correspondence rules are simpler. This approach was introduced by Plimak et al.[14].

The unnormalised distribution functions can be introduced via canonical forms of the density operator involving unnormalised forms of the Bargmann state projectors. Thus we have

$$\begin{aligned}\hat{\rho}_{\mathrm{b}} &= \int d^2\boldsymbol{\alpha}^+ d^2\boldsymbol{\alpha}\; B_{\mathrm{b}}(\boldsymbol{\alpha},\boldsymbol{\alpha}^+,\boldsymbol{\alpha}^*,\boldsymbol{\alpha}^{+*})(|\boldsymbol{\alpha}\rangle_{\mathrm{B}}\ \langle\boldsymbol{\alpha}^{+*}|_{\mathrm{B}}) \\ &= \int d^2\boldsymbol{\alpha}^+ d^2\boldsymbol{\alpha}\, |\boldsymbol{\alpha}\rangle_{\mathrm{B}}\ \langle\boldsymbol{\alpha}^{+*}|_{\mathrm{B}}\, B_{\mathrm{b}}(\boldsymbol{\alpha},\boldsymbol{\alpha}^+,\boldsymbol{\alpha}^*,\boldsymbol{\alpha}^{+*})\end{aligned} \tag{7.98}$$

and

$$\begin{aligned}\hat{\rho}_{\mathrm{f}} &= \int d\mathbf{g}^+ d\mathbf{g}\; B_{\mathrm{f}}(\mathbf{g},\mathbf{g}^+)\,(|\mathbf{g}\rangle_{\mathrm{B}}\,\langle\mathbf{g}^{+*}|_{\mathrm{B}}) \\ &= \int d\mathbf{g}^+ d\mathbf{g}\,(|\mathbf{g}\rangle_{\mathrm{B}}\,\langle\mathbf{g}^{+*}|_{\mathrm{B}})\; B_{\mathrm{f}}(\mathbf{g},\mathbf{g}^+)\end{aligned} \tag{7.99}$$

for bosons and fermions, respectively. Clearly, the unnormalised distribution functions $B_{\mathrm{b}}(\boldsymbol{\alpha},\boldsymbol{\alpha}^+,\boldsymbol{\alpha}^*,\boldsymbol{\alpha}^{+*})$ and $B_{\mathrm{f}}(\mathbf{g},\mathbf{g}^+)$ are related to the normalised distribution functions via

$$\begin{aligned}B_{\mathrm{b}}(\boldsymbol{\alpha},\boldsymbol{\alpha}^+,\boldsymbol{\alpha}^*,\boldsymbol{\alpha}^{+*}) &= P_{\mathrm{b}}(\boldsymbol{\alpha},\boldsymbol{\alpha}^+,\boldsymbol{\alpha}^*,\boldsymbol{\alpha}^{+*}) \times \exp(-\boldsymbol{\alpha}\cdot\boldsymbol{\alpha}^+), \\ B_{\mathrm{f}}(\mathbf{g},\mathbf{g}^+) &= P_{\mathrm{f}}(\mathbf{g},\mathbf{g}^+) \times \exp(-\mathbf{g}\cdot\mathbf{g}^+).\end{aligned} \tag{7.100}$$

From (7.44) and (7.34), canonical forms for the unnormalised distribution function are given by

$$\begin{aligned}B_{\mathrm{b\ canon}}(\boldsymbol{\alpha},\boldsymbol{\alpha}^+,\boldsymbol{\alpha}^*,\boldsymbol{\alpha}^{+*}) &= \left(\frac{1}{4\pi^2}\right)^n \exp\left[-\frac{1}{2}\left(\boldsymbol{\alpha}\cdot\boldsymbol{\alpha}^* + \boldsymbol{\alpha}^{+*}\cdot\boldsymbol{\alpha}^+\right)\right] \\ &\quad\times \exp(-\boldsymbol{\alpha}\cdot\boldsymbol{\alpha}^+)\ {}_{\mathrm{B}}\langle(\boldsymbol{\alpha}+\boldsymbol{\alpha}^{+*})/2|\,\hat{\rho}_{\mathrm{b}}\,|(\boldsymbol{\alpha}+\boldsymbol{\alpha}^{+*})/2\rangle_{\mathrm{B}},\end{aligned} \tag{7.101}$$

$$B_{\mathrm{fcanon}}(\mathbf{g},\mathbf{g}^+) = \int d\mathbf{g}^{+*} d\mathbf{g}^* \exp\left(\mathbf{g}\cdot\mathbf{g}^* + \mathbf{g}^{+*}\cdot\mathbf{g}^+\right)\ {}_{\mathrm{B}}\langle\mathbf{g}|\,\hat{\rho}_{\mathrm{f}}\,|\mathbf{g}^{+*}\rangle_{\mathrm{B}}. \tag{7.102}$$

As usual, $d\mathbf{g}^{+*} = dg_1^{+*} dg_2^{+*}\cdots dg_n^{+*}$ and $d\mathbf{g}^* = dg_n^*\cdots dg_2^*\, dg_1^*$.

7.8.1 Quantum Correlation Functions

The expressions for the quantum correlation functions now become

$$\begin{aligned}G_{\mathrm{b}}(l_1,\cdots,l_p;m_q,\cdots,m_1) &= \mathrm{Tr}(\hat{a}_{m_q}\cdots\hat{a}_{m_1}\hat{\rho}_b\hat{a}_{l_1}^\dagger\cdots\hat{a}_{l_p}^\dagger) \\ &= \int d^2\boldsymbol{\alpha}^+ d^2\boldsymbol{\alpha}\,(\alpha_{m_q}\cdots\alpha_{m_1})\exp(\boldsymbol{\alpha}\cdot\boldsymbol{\alpha}^+)B_{\mathrm{b}}(\boldsymbol{\alpha},\boldsymbol{\alpha}^+,\boldsymbol{\alpha}^*,\boldsymbol{\alpha}^{+*}) \\ &\quad\times(\alpha_{l_1}^+\cdots\alpha_{l_p}^+)\end{aligned} \tag{7.103}$$

for bosons and

$$\begin{aligned} G(l_1, \cdots, l_p; m_q, \cdots, m_1) &= \mathrm{Tr}(\hat{c}_{m_q} \cdots \hat{c}_{m_1} \hat{\rho}_{\mathrm{f}} \hat{c}^\dagger_{l_1} \cdots \hat{c}^\dagger_{l_p}) \\ &= \int d\mathbf{g}^+ d\mathbf{g}\, (g_{m_q} \cdots g_{m_1}) \exp(\mathbf{g} \cdot \mathbf{g}^+) B_{\mathrm{f}}(\mathbf{g}, \mathbf{g}^+)\, (g^+_{l_1} \cdots g^+_{l_p}), \\ & \qquad p - q = 0, \pm 2, \pm 4, \cdots, \end{aligned} \tag{7.104}$$

for fermions. The consequence of using the unnormalised distribution functions is that the phase space averages now contain extra factors $\exp(\boldsymbol{\alpha} \cdot \boldsymbol{\alpha}^+)$ and $\exp(\mathbf{g} \cdot \mathbf{g}^+)$. In the fermion case,

$$\exp(\mathbf{g} \cdot \mathbf{g}^+) = \prod_i \exp(g_i\, g_i^+) = \prod_i (1 + g_i\, g_i^+) \tag{7.105}$$

and expanding the exponentials would result in a total of 2^n terms. However, the presence of any of the $g_{m_q} \cdots g_{m_1}$ or $g^+_{l_1} \cdots g^+_{l_p}$ would enable the corresponding $\exp(g_i\, g_i^+)$ to be replaced by unity, thereby reducing the overall number of terms that have to be considered. Nevertheless, we should point out here that the factors $(1 + g_i g_i^+)$ introduce an extra complication into determining quantum correlation functions, both in evaluating the phase space integral and in applying stochastic methods.

7.8.2 Populations and Coherences

Populations of Fock states and coherences between them can be expressed in terms of standard quantum correlation functions, but it turns out that simple expressions can be obtained in terms of phase space integrals involving the unnormalised distribution functions without the complications of the $\exp(\mathbf{g} \cdot \mathbf{g}^+)$ or $\exp(\boldsymbol{\alpha} \cdot \boldsymbol{\alpha}^+)$ factors. The main advantage of this occurs for fermions, as we will now see. Consider two p-fermion Fock states of the form

$$|\Phi\{l\}\rangle = \hat{c}^\dagger_{l_1} \cdots \hat{c}^\dagger_{l_p} |0\rangle\,, \qquad |\Phi\{m\}\rangle = \hat{c}^\dagger_{m_1} \cdots \hat{c}^\dagger_{m_p} |0\rangle\,. \tag{7.106}$$

The population of the state $|\Phi\{l\}\rangle$ and the coherence between the state $|\Phi\{l\}\rangle$ and the state $|\Phi\{m\}\rangle$ are given by

$$P(\Phi\{l\}) = \mathrm{Tr}(\hat{\Pi}(\{l\})\, \hat{\rho}), \tag{7.107}$$

$$C(\Phi\{m\}; \Phi\{l\}) = \mathrm{Tr}(\hat{\Xi}(\{m\}; \{l\})\, \hat{\rho}), \tag{7.108}$$

where the population and transition operators are

$$\hat{\Pi}(\{l\}) = \hat{c}^\dagger_{l_1} \cdots \hat{c}^\dagger_{l_p} |0\rangle \langle 0| \, \hat{c}_{l_p} \cdots \hat{c}_{l_1}, \tag{7.109}$$

$$\hat{\Xi}(\{m\}; \{l\}) = \hat{c}^\dagger_{m_1} \cdots \hat{c}^\dagger_{m_p} |0\rangle \langle 0| \, \hat{c}_{l_p} \cdots \hat{c}_{l_1}. \tag{7.110}$$

To determine the population, we substitute for $\hat{\rho}$ from (7.99) to give

$$\begin{aligned}
P(\Phi\{l\}) &= \mathrm{Tr}\left(\int d\mathbf{g}^{+}d\mathbf{g}\ B_{\mathrm{f}}(\mathbf{g},\mathbf{g}^{+})\hat{\Pi}(\{l\})(|\mathbf{g}\rangle_{\mathrm{B}}\left\langle \mathbf{g}^{+*}\right|_{\mathrm{B}})\right) \\
&= \mathrm{Tr}\left(\int d\mathbf{g}^{+}d\mathbf{g}\ B_{\mathrm{f}}(\mathbf{g},\mathbf{g}^{+})\hat{c}_{l_1}^{\dagger}\cdots\hat{c}_{l_p}^{\dagger}\left|0\right\rangle\left\langle 0\right| g_{l_p}\cdots g_{l_1}(|\mathbf{g}\rangle_{\mathrm{B}}\left\langle \mathbf{g}^{+*}\right|_{\mathrm{B}})\right) \\
&= \mathrm{Tr}\left(\int d\mathbf{g}^{+}d\mathbf{g}\ B_{\mathrm{f}}(\mathbf{g},\mathbf{g}^{+})(-1)^{p}g_{l_p}\cdots g_{l_1}\hat{c}_{l_1}^{\dagger}\cdots\hat{c}_{l_p}^{\dagger}\left|0\right\rangle\left\langle \mathbf{g}^{+*}\right|_{\mathrm{B}}\right), \qquad (7.111)
\end{aligned}$$

since as $\hat{\Pi}(\{l\})$ is an even operator it commutes with the product of the Grassmann differentials and with the even function $B_{\mathrm{f}}(\mathbf{g},\mathbf{g}^{+})$, enabling the fermion annihilation operators to act on $|\mathbf{g}\rangle_{\mathrm{B}}$ to give a product of eigenvalues $g_{l_p}\cdots g_{l_1}$. This product then commutes with the same number of fermion creation operators with a factor $(-1)^{p}$, and the factor $\langle 0|\mathbf{g}\rangle_{\mathrm{B}}$ is equal to unity. Then, evaluating the trace with Fock states $|\nu_1\nu_2\cdots\nu_n\rangle$ and noting that this Fock state is even or odd according to whether $(-1)^{\nu}=+1,-1$ ($\nu=\sum_i\nu_i$), we can commute the bra vector $\langle\nu_1\nu_2\cdots\nu_n|$ with the even product of Grassmann differentials, the even $B_{\mathrm{f}}(\mathbf{g},\mathbf{g}^{+})$ and the product $g_{l_p}\cdots g_{l_1}$, apart from a factor $(-1)^{\nu}$ if p is odd. Thus

$$\begin{aligned}
P(\Phi\{l\}) &= \int d\mathbf{g}^{+}d\mathbf{g}\ B_{\mathrm{f}}(\mathbf{g},\mathbf{g}^{+})(+1)\,g_{l_p}\cdots g_{l_1} \\
&\quad\times\sum_{\{\nu\}}\left\langle\nu_1\cdots\nu_n\right|\hat{c}_{l_1}^{\dagger}\cdots\hat{c}_{l_p}^{\dagger}\left|0\right\rangle_{\mathrm{B}}\left\langle\mathbf{g}^{+*}|\nu_1\cdots\nu_n\right\rangle \qquad (p\text{ even}) \\
&= \int d\mathbf{g}^{+}d\mathbf{g}\ B_{\mathrm{f}}(\mathbf{g},\mathbf{g}^{+})\,(-1)\,g_{l_p}\cdots g_{l_1} \\
&\quad\times\sum_{\{\nu\}}(-1)^{\nu}\left\langle\nu_1\cdots\nu_n\right|\hat{c}_{l_1}^{\dagger}\cdots\hat{c}_{l_p}^{\dagger}\left|0\right\rangle_{\mathrm{B}}\left\langle\mathbf{g}^{+*}|\nu_1\cdots\nu_n\right\rangle \qquad (p\text{ odd}), \qquad (7.112)
\end{aligned}$$

where we have inserted the result for $(-1)^{p}$ in the two cases. In the case where p is odd, we then use ${}_{\mathrm{B}}\left\langle\mathbf{g}^{+*}\,|\nu_1\cdots\nu_n\right\rangle=(-1)^{\nu}_{\mathrm{B}}\left\langle-\mathbf{g}^{+*}\,|\nu_1\cdots\nu_n\right\rangle$ from (5.118), which allows the two $(-1)^{\nu}$ to cancel out. We then reverse the order of the c-number $\left\langle\nu_1\cdots\nu_n\right|\hat{c}_{l_1}^{\dagger}\cdots\hat{c}_{l_p}^{\dagger}\left|0\right\rangle$ and the Grassmann function $\left\langle-\mathbf{g}^{+*}|\nu_1\cdots\nu_n\right\rangle$ and apply the result $\sum_{\{\nu\}}|\nu_1\cdots\nu_n\rangle\langle\nu_1\cdots\nu_n|=\hat{1}$. When p is even, no factor $(-1)^{\nu}$ occurs, so we just reverse the order of the c-number $\left\langle\nu_1\cdots\nu_n\right|\hat{c}_{l_1}^{\dagger}\cdots\hat{c}_{l_p}^{\dagger}\left|0\right\rangle$ and the Grassmann function ${}_{\mathrm{B}}\left\langle\mathbf{g}^{+*}|\nu_1\nu_2\cdots\nu_n\right\rangle$ and apply the completeness result. Hence we find that

$$\begin{aligned}
P(\Phi\{l\}) &= \int d\mathbf{g}^{+}d\mathbf{g}\ B_{\mathrm{f}}(\mathbf{g},\mathbf{g}^{+})\,(+1)\,g_{l_p}\cdots g_{l_1}\left\langle\mathbf{g}^{+*}\right|_{\mathrm{B}}\hat{c}_{l_1}^{\dagger}\cdots\hat{c}_{l_p}^{\dagger}\left|0\right\rangle \qquad (p\text{ even}) \\
&= \int d\mathbf{g}^{+}d\mathbf{g}\ B_{\mathrm{f}}(\mathbf{g},\mathbf{g}^{+})\,(-1)\,g_{l_p}\cdots g_{l_1}\left\langle-\mathbf{g}^{+*}\right|_{\mathrm{B}}\hat{c}_{l_1}^{\dagger}\cdots\hat{c}_{l_p}^{\dagger}\left|0\right\rangle \qquad (p\text{ odd}).
\end{aligned}$$

(7.113)

If p is odd, using the eigenvalue equation for $\langle -\mathbf{g}^{+*}|_{\mathrm{B}}$ leads to a further product of Grassmann variables $g^{+}_{l_1}\cdots g^{+}_{l_p}$ times (-1). If p is even, using the eigenvalue equation for $\langle \mathbf{g}^{+*}|_{\mathrm{B}}$ leads to the same product of Grassmann variables but with a $(+1)$ factor. As ${}_{\mathrm{B}}\langle \pm\mathbf{g}^{+*}|0\rangle = 1$ and noting that the two (-1) cancel in the odd-p case, we finally get the result

$$P(\Phi\{l\}) = \int d\mathbf{g}^{+}d\mathbf{g}\; B_{\mathrm{f}}(\mathbf{g},\mathbf{g}^{+})\, g_{l_p}\cdots g_{l_1} g^{+}_{l_1}\cdots g^{+}_{l_p} \tag{7.114}$$

for the population of the fermion Fock state $|\Phi\{l\}\rangle = \hat{c}^{\dagger}_{l_1}\cdots\hat{c}^{\dagger}_{l_p}|0\rangle$ for both of the cases of p even or odd. A similar treatment gives the result for the coherence,

$$C(\Phi\{m\};\Phi\{l\}) = \int d\mathbf{g}^{+}d\mathbf{g}\; B_{\mathrm{f}}(\mathbf{g},\mathbf{g}^{+}) g_{l_p}\cdots g_{l_1} g^{+}_{m_1}\cdots g^{+}_{m_p}. \tag{7.115}$$

In both cases, a phase space average with the unnormalised distribution function $B_{\mathrm{f}}(\mathbf{g},\mathbf{g}^{+})$ of a product of the Grassmann variables associated with the occupied modes is involved. The results are analogous to the quantum correlation function expressions in terms of the normalised distribution function $P_{\mathrm{f}}(\mathbf{g},\mathbf{g}^{+})$. Clearly, the unnormalised distribution function $B_{\mathrm{f}}(\mathbf{g},\mathbf{g}^{+})$ is more useful when calculating populations and coherences for Fock states than the normalised distribution $P_{\mathrm{f}}(\mathbf{g},\mathbf{g}^{+})$ is.

Exercises

(7.1) Prove the existence of the canonical form (7.54) by combining the approaches used for the existence of canonical forms of bosonic and fermionic operators.

(7.2) Confirm that the canonical distribution function (7.54) satisfies (7.58).

(7.3) Determine the number of c-number functions of the bosonic phase space variables that are needed to specify the distribution function (7.63) for a combined Bose–Fermi system.

(7.4) Using $P_{\mathrm{f}}(g,g^{+}) = g_1\cdots g_n g^{+}_n\cdots g^{+}_1$ for the true vacuum state of a fermion system, use the expressions in the last section to confirm that the population of the vacuum state is unity and that the population of any other Fock state is zero.

8
Fokker–Planck Equations

In this chapter, we develop the methods required for applying phase space distribution functions to physical problems. We obtain Fokker–Planck equations for the distribution function that are equivalent to the Liouville–von Neumann, master or Matsubara equation for the density operator for both bosons and fermions. This is accomplished via the application of a standard set of rules known as the correspondence rules, which relate to the effect of replacing the density operator by its product with annihilation or creation operators. For the distribution functions used to determine non-normally ordered quantum correlation functions, the correspondence rules and Fokker–Planck equations will be different. In addition, Fokker–Planck equations are also established for the unnormalised distribution functions. If the Fokker–Planck equation contains only first- and second-order derivatives of the phase variables, we show in Chapter 9 that the Fokker–Planck equation can be replaced by Ito stochastic equations for stochastic variables that replace the phase space c-number (for bosons) or Grassmann (for fermions) variables. Stochastic averages of products of the stochastic phase space variables then can be used to determine the quantum correlation functions instead of using phase space averages based on the distribution function. From the point of view of efficient numerical calculation of the quantum correlation functions of interest, the use of stochastic methods is preferable, as they in effect avoid having to sample the distribution function over the whole of phase space.

8.1 Correspondence Rules

We now wish to replace the Liouville–von Neumann equation or master equation for the density operator by a Fokker–Planck equation for the distribution function. To do this, we make use of so-called *correspondence rules*, which are presented here for the general case of a combined system of bosons and fermions. We also establish the correspondence rules for the unnormalised distribution functions. The correspondence rules state what happens to the distribution function $P(\alpha, \alpha^+, \alpha^*, \alpha^{+*}, g, g^+)$ when the density operator $\hat{\rho}$ is replaced by the product of the density operator with an annihilation or creation operator. In cases where the density operator is replaced by a product of the density operator with products of annihilation and creation operators, the correspondence rules are applied in succession. The proof of the correspondence rules is given in Appendix F. They are proved most simply by starting from the canonical form of the density operator in (7.56). For completeness, we also prove them by starting from the expression for the characteristic function. A key integration-by-parts

Phase Space Methods for Degenerate Quantum Gases. First Edition. Bryan J. Dalton, John Jeffers, and Stephen M. Barnett.
Published in 2015 by Oxford University Press.

step is involved, which for the bosonic case requires the distribution function to decrease to zero rapidly as the phase space boundary is approached. Here the proof using the canonical density operator relies on the Bargmann state projectors being analytic functions, whereas the proof using the characteristic function is based on analytic features of the characteristic function and the exponential functions involved. The step involving integration by parts occurs when differentiations with respect to α_i, α_i^+ are involved. For fermions, the proof using the canonical density operator also involves an integration-by-parts step. The characteristic-function proof requires considering the effects of fermion operators in pairs (12 cases) because differentiations of a fermionic trace only give simple results when carried out in pairs.

As the bosonic positive P distribution function is non-unique, it is not surprising that the correspondence rules are also non-unique. This implies that the form of the Fokker–Planck equations will also depend on which correspondence rules are applied, and this non-uniqueness will flow through to the Ito stochastic equations as well as to the distribution function. As we will see, arbitrary linear combinations of the derivatives $\partial/\partial\alpha_j^*$ and $\partial/\partial\alpha_j^{+*}$ may be added to each bosonic correspondence rule, leading to more general correspondence rules. Fundamentally, this arbitrariness results from both the bosonic Bargmann state projectors and the exponential functions involved being analytic functions of α_i, α_i^+, and therefore their derivatives with respect to α_i^*, α_i^{+*} are zero. This leads to two well-known options, namely $\partial/\partial\alpha_{ix}$ or $\partial/\partial(i\alpha_{iy})$ on the one hand and $\partial/\partial\alpha_{ix}^+$ or $\partial/\partial(i\alpha_{iy}^+)$ on the other, when $\partial/\partial\alpha_i$ or $\partial/\partial\alpha_i^+$ is applied to the non-analytic distribution function, results which are widely used to generate bosonic Fokker–Planck equations with positive definite diffusion matrices [31, 38]. Although the bosonic correspondence rules are in general arbitrary, this is not true for those associated with the canonical distribution function – which involves a diagonal matrix element of the density operator between Bargmann coherent states. The canonical rules are unique, and their derivation does not involve an integration-by-parts step nor any consideration of whether the distribution function approaches zero on the phase space boundary rapidly enough. These correspondence rules are a special case of the more general set of correspondence rules referred to above. For the fermionic correspondence rules, there are two options for each density operator replacement, one of which depends on the distribution function being an even Grassmann function. For successive applications of the correspondence rules, the other option is safest to use.

8.2 Bosonic Correspondence Rules

In this section, we set out the correspondence rules for bosons, emphasising the range of possibilities that is related to the non-uniqueness of the positive P distribution function.

8.2.1 Standard Correspondence Rules for Bosonic Annihilation and Creation Operators

For bosonic operators, the standard correspondence rules for the positive P distribution function are

$$\hat{\rho} \Rightarrow \hat{a}_i\hat{\rho}, \qquad P(\alpha, \alpha^+, \alpha^*, \alpha^{+*}, g, g^+) \Rightarrow \alpha_i P, \tag{8.1}$$

$$\hat{\rho} \Rightarrow \hat{\rho}\hat{a}_i, \qquad P(\alpha, \alpha^+, \alpha^*, \alpha^{+*}, g, g^+) \Rightarrow \left(-\frac{\partial}{\partial \alpha_i^+} + \alpha_i\right) P, \tag{8.2}$$

$$\hat{\rho} \Rightarrow \hat{a}_i^\dagger \hat{\rho}, \qquad P(\alpha, \alpha^+, \alpha^*, \alpha^{+*}, g, g^+) \Rightarrow \left(-\frac{\partial}{\partial \alpha_i} + \alpha_i^+\right) P, \tag{8.3}$$

$$\hat{\rho} \Rightarrow \hat{\rho}\hat{a}_i^\dagger, \qquad P(\alpha, \alpha^+, \alpha^*, \alpha^{+*}, g, g^+) \Rightarrow \alpha_i^+ P, \tag{8.4}$$

where the $\alpha, \alpha^+, \alpha^*, \alpha^{+*}$ are regarded as independent complex variables, an interpretation which is perfectly valid for the non-analytic function $P(\alpha, \alpha^+, \alpha^*, \alpha^{+*}, g, g^+)$. The correspondence rules are derived in Appendix F.

8.2.2 General Bosonic Correspondence Rules

More general correspondence rules can also be obtained. If $\Theta(\alpha, \alpha^+)$ is an analytic function of α_i and α_i^+, which may be a Bargmann projector $\hat{\Lambda}(g, g^+, \alpha, \alpha^+)$ or a product of exponential functions $\exp i \sum_i \{g_i h_i^+\} \exp i \sum_j \{\alpha_j \xi_j^+\}$ or $\exp i \sum_j \{\xi_j \alpha_j^+\} \exp i \sum_i \{h_i g_i^+\}$, then as

$$\frac{\partial}{\partial \alpha_i^*}\Theta(\alpha, \alpha^+) = \frac{\partial}{\partial \alpha_i^{+*}}\Theta(\alpha, \alpha^+) = 0, \tag{8.5}$$

it follows using integration by parts (and assuming that the distribution function goes to zero on the bosonic phase space boundary) that

$$\begin{aligned} 0 &= \int d^2\alpha^+ d^2\alpha \left(\frac{\partial}{\partial \alpha_i^*}\Theta(\alpha, \alpha^+)\right) P(\alpha, \alpha^+, \alpha^*, \alpha^{+*}, g, g^+) \\ &= -\int d^2\alpha^+ d^2\alpha\, \Theta(\alpha, \alpha^+) \left(\frac{\partial}{\partial \alpha_i^*} P(\alpha, \alpha^+, \alpha^*, \alpha^{+*}, g, g^+)\right), \end{aligned} \tag{8.6}$$

$$\begin{aligned} 0 &= \int d^2\alpha^+ d^2\alpha \left(\frac{\partial}{\partial \alpha_i^{+*}}\Theta(\alpha, \alpha^+)\right) P(\alpha, \alpha^+, \alpha^*, \alpha^{+*}, g, g^+) \\ &= -\int d^2\alpha^+ d^2\alpha\, \Theta(\alpha, \alpha^+) \left(\frac{\partial}{\partial \alpha_i^{+*}} P(\alpha, \alpha^+, \alpha^*, \alpha^{+*}, g, g^+)\right). \end{aligned} \tag{8.7}$$

Hence arbitrary linear combinations

$$\mu \frac{\partial}{\partial \alpha_i^*} + \mu^+ \frac{\partial}{\partial \alpha_i^{+*}} \tag{8.8}$$

of the derivatives with respect to $\alpha_i^*, \alpha_i^{+*}$ may be added to each of the standard bosonic correspondence rules in (8.1)–(8.4) without making any difference to either the new characteristic function or the new density operator associated with the replacement $\hat{\rho} \Rightarrow \hat{a}_i\hat{\rho}$, $\hat{\rho} \Rightarrow \hat{\rho}\hat{a}_i$, $\hat{\rho} \Rightarrow \hat{a}_i^\dagger\hat{\rho}$ or $\hat{\rho} \Rightarrow \hat{\rho}\hat{a}_i^\dagger$. Thus, for example,

$$\begin{aligned}\hat{\rho}\hat{a}_i &= \int dg^+dg \int d^2\alpha^+d^2\alpha\,\hat{\Lambda}(g,g^+,\alpha,\alpha^+)\left(-\frac{\partial}{\partial\alpha_i^+}+\alpha_i\right)P(\alpha,\alpha^+,\alpha^*,\alpha^{+*},g,g^+)\\ &= \int dg^+dg \int d^2\alpha^+d^2\alpha\;\hat{\Lambda}(g,g^+,\alpha,\alpha^+)\left(-\frac{\partial}{\partial\alpha_i^+}+\alpha_i+\mu\frac{\partial}{\partial\alpha_i^*}+\mu^+\frac{\partial}{\partial\alpha_i^{+*}}\right)P,\end{aligned} \tag{8.9}$$

using the proof based on the canonical form of the density operator.

In particular, the correspondence rule (8.2) can be replaced by

$$\hat{\rho} \Rightarrow \hat{\rho}\hat{a}_i,$$
$$P(\alpha,\alpha^+,\alpha^*,\alpha^{+*}g,g^+) \Rightarrow \left(\alpha_i - \frac{\partial}{\partial\alpha_i^+} - \frac{\partial}{\partial\alpha_i^{+*}}\right)P = \left(\alpha_i - \frac{\partial}{\partial\alpha_{ix}^+}\right)P, \tag{8.10}$$

$$\hat{\rho} \Rightarrow \hat{\rho}\hat{a}_i,$$
$$P(\alpha,\alpha^+,\alpha^*,\alpha^{+*},g,g^+) \Rightarrow \left(\alpha_i - \frac{\partial}{\partial i\alpha_i^+} + \frac{\partial}{\partial\alpha_i^{+*}}\right)P = \left(\alpha_i - \frac{\partial}{\partial(i\alpha_{iy}^+)}\right)P, \tag{8.11}$$

choosing $\mu = 0, \mu^+ = -1$ or $\mu = 0, \mu^+ = +1$, respectively. Although P is not an analytic function of α_i, α_i^+, it can be regarded as a function $F(\alpha_x, \alpha_x^+, \alpha_y, \alpha_y^+, g, g^+)$ involving the real and imaginary components $\alpha_{ix}, \alpha_{ix}^+, \alpha_{iy}, \alpha_{iy}^+$ of the bosonic phase space variables and the derivatives interpreted via

$$\begin{aligned}\frac{\partial}{\partial\alpha_i} &\Rightarrow \frac{\partial}{\partial\alpha_{ix}} \quad\text{or}\quad \frac{\partial}{\partial\alpha_i} \Rightarrow \frac{\partial}{\partial(i\alpha_{iy})},\\ \frac{\partial}{\partial\alpha_i^+} &\Rightarrow \frac{\partial}{\partial\alpha_{ix}^+} \quad\text{or}\quad \frac{\partial}{\partial\alpha_i^+} \Rightarrow \frac{\partial}{\partial(i\alpha_{iy}^+)},\end{aligned} \tag{8.12}$$

when acting on the distribution function (or its products with other functions). Thus the validity of the two interpretations in (8.12) follows from the analytic features of the Bargmann projectors or the characteristic function.

Furthermore, the flexibility in the correspondence rules is even greater than merely the ability to replace α_i, α_i^+, $\partial/\partial\alpha_i$ or $\partial/\partial\alpha_i^+$ by these quantities plus a particular linear combination of $\partial/\partial\alpha_i^*$ and $\partial/\partial\alpha_i^{+*}$, such as in (8.8) for every term when $\hat{\rho} \Rightarrow \hat{a}_i\hat{\rho}$, $\hat{\rho}\hat{a}_i$, $\hat{a}_i^\dagger\hat{\rho}$ or $\hat{\rho}\hat{a}_i^\dagger$. In fact, a different linear combination can be used for each α_i, α_i^+, $\partial/\partial\alpha_i$ or $\partial/\partial\alpha_i^+$ wherever it occurs! So, for example, if in one term where $\hat{\rho} \Rightarrow \hat{\rho}\hat{a}_i$ we replace $(\alpha_i - \partial/\partial\alpha_i^+)$ by $(\alpha_i - \partial/\partial\alpha_i^+ - \partial/\partial\alpha_i^{+*})$ to give $(\alpha_i - \partial/\partial\alpha_{ix}^+)$, in another term where $\hat{\rho} \Rightarrow \hat{\rho}\hat{a}_i$ we may replace $(\alpha_i - \partial/\partial\alpha_i^+)$ by $(\alpha_i - \partial/\partial\alpha_i^+ + \partial/\partial\alpha_i^{+*})$ to give $(\alpha_i - \partial/\partial(i\alpha_{iy}^+))$. This is possible because the only requirement is that the equation for the distribution function gives the correct equation for the characteristic function (or the density operator). Additional terms of the form (8.8) acting on the distribution function produce zero when, in the integration-by-parts step, they act back on either the exponential factors $\exp i\sum_j\{\alpha_j\,\xi_j^+\}$ and $\exp i\sum_j\{\xi_j\alpha_j^+\}$ or the Bargmann state projector $\hat{\Lambda}_{\mathrm{b}}(\alpha,\alpha^+)$, both of which are analytic functions of the α_i, α_i^+ and hence yield zero when $\partial/\partial\alpha_i^*$ or $\partial/\partial\alpha_i^{+*}$ is applied. This flexibility is important for being able to convert a Fokker–Planck equation based on the standard correspondence rules into a form with a *positive definite* diffusion matrix [31, 38], as will be seen below. It is not available for the correspondence rules associated with the canonical distribution function.

8.2.3 Canonical Bosonic Correspondence Rules

There are, however, further possibilities – a feature not widely commented upon in other work but which ultimately reflects the non-uniqueness of the positive P distribution for bosons. In particular, for the canonical distribution function defined in (7.54), the bosonic correspondence rules can be found directly from the properties of the Bargmann coherent states to give

$$\begin{aligned}
&\hat{\rho} \Rightarrow \hat{a}_i\hat{\rho},\\
P_{\text{canon}}(\alpha,\alpha^+,\alpha^*,\alpha^{+*},g,g^+) &\Rightarrow \left(\frac{\partial}{\partial\alpha_i^*}+\frac{\partial}{\partial\alpha_i^+}+\frac{1}{2}(\alpha_i+\alpha_i^{+*})\right)P_{\text{canon}},\\
&\hat{\rho} \Rightarrow \hat{\rho}\hat{a}_i,\\
P_{\text{canon}}(\alpha,\alpha^+,\alpha^*,\alpha^{+*},g,g^+) &\Rightarrow \left(\frac{1}{2}(\alpha_i+\alpha_i^{+*})\right)P_{\text{canon}},\\
&\hat{\rho} \Rightarrow \hat{a}_i^\dagger\hat{\rho},\\
P_{\text{canon}}(\alpha,\alpha^+,\alpha^*,\alpha^{+*},g,g^+) &\Rightarrow \left(\frac{1}{2}(\alpha_i^*+\alpha_i^+)\right)P_{\text{canon}},\\
&\hat{\rho} \Rightarrow \hat{\rho}\hat{a}_i^\dagger,\\
P_{\text{canon}}(\alpha,\alpha^+,\alpha^*,\alpha^{+*},g,g^+) &\Rightarrow \left(\frac{\partial}{\partial\alpha_i}+\frac{\partial}{\partial\alpha_i^{+*}}+\frac{1}{2}(\alpha_i^*+\alpha_i^+)\right)P_{\text{canon}}.
\end{aligned} \tag{8.13}$$

Note that these rules do not involve any integration-by-parts step, so the behaviour of the distribution function as the phase space boundary is approached is not involved. Here (as noted above), there is no flexibility in the correspondence rules. These rules were obtained by Schack and Schenzle [60] (see the Appendix in [60]).

In view of the form (7.54) for the canonical positive P distribution, it is convenient to introduce new complex variables γ_i, γ_i^*, δ_i, δ_i^* via

$$\begin{aligned}
\gamma_i &= \frac{1}{2}(\alpha_i+\alpha_i^{+*}), & \gamma_i^* &= \frac{1}{2}(\alpha_i^*+\alpha_i^+),\\
\delta_i &= \frac{1}{2}(\alpha_i-\alpha_i^{+*}), & \delta_i^* &= \frac{1}{2}(\alpha_i^*-\alpha_i^+),\\
\alpha_i &= \gamma_i+\delta_i, & \alpha_i^* &= \gamma_i^*+\delta_i^*,\\
\alpha_i^+ &= \gamma_i^*-\delta_i^*, & \alpha_i^{+*} &= \gamma_i-\delta_i.
\end{aligned} \tag{8.14}$$

The canonical distribution function can now be written in the form

$$P_{\rm canon}(\gamma,\gamma^*,\delta,\delta^*,g,g^+) = \left(\frac{1}{4\pi^2}\right)^n \exp\left(-\sum_i \delta_i\delta_i^*\right)\exp\left(-\sum_i \gamma_i\gamma_i^*\right)$$
$$\times \int\int dg^{+*}dg^* \exp\left(\sum_i (g_i g_i^* + g_i^{+*}g_i^+ + g_i g_i^+)\right)$$
$$\times \langle g|_{\rm B}\,\langle\gamma|_{\rm B}\,\hat{\rho}\,|\gamma\rangle_{\rm B}\,|g^{+*}\rangle_{\rm B}. \tag{8.15}$$

Note that the phase space integration is changed to

$$\int d^2\alpha^+\,d^2\alpha \Rightarrow 4\int d^2\delta\,d^2\gamma. \tag{8.16}$$

Using the results

$$\langle\gamma|_{\rm B}\,(\hat{a}_i\hat{\rho})\,|\gamma\rangle_{\rm B} = \frac{\partial}{\partial\gamma_i^*}\langle\gamma|_{\rm B}\,\hat{\rho}\,|\gamma\rangle_{\rm B}, \qquad \langle\gamma|_{\rm B}\,(\hat{\rho}\,\hat{a}_i)\,|\gamma\rangle_{\rm B} = \gamma_i\,\langle\gamma|_{\rm B}\,\hat{\rho}\,|\gamma\rangle_{\rm B},$$
$$\langle\gamma|_{\rm B}\,(\hat{a}_i^\dagger\hat{\rho})\,|\gamma\rangle_{\rm B} = \gamma_i^*\,\langle\gamma|_{\rm B}\,\hat{\rho}\,|\gamma\rangle_{\rm B}, \qquad \langle\gamma|_{\rm B}\,(\hat{\rho}\hat{a}_i^\dagger)\,|\gamma\rangle_{\rm B} = \frac{\partial}{\partial\gamma_i}\langle\gamma|_{\rm B}\,\hat{\rho}\,|\gamma\rangle_{\rm B}, \tag{8.17}$$

in (8.15) we find that

$$\hat{\rho} \Rightarrow \hat{a}_i\hat{\rho}, \qquad P_{\rm canon}(\gamma,\gamma^*,\delta,\delta^*,g,g^+) \Rightarrow \left(\frac{\partial}{\partial\gamma_i^*} + \gamma_i\right)P_{\rm canon},$$
$$\hat{\rho} \Rightarrow \hat{\rho}\hat{a}_i, \qquad P_{\rm canon}(\gamma,\gamma^*,\delta,\delta^*,g,g^+) \Rightarrow (\gamma_i)P_{\rm canon},$$
$$\hat{\rho} \Rightarrow \hat{a}_i^\dagger\hat{\rho}, \qquad P_{\rm canon}(\gamma,\gamma^*,\delta,\delta^*,g,g^+) \Rightarrow (\gamma_i^*)P_{\rm canon},$$
$$\hat{\rho} \Rightarrow \hat{\rho}\hat{a}_i^\dagger, \qquad P_{\rm canon}(\gamma,\gamma^* < \delta,\delta^*,g,g^+) \Rightarrow \left(\frac{\partial}{\partial\gamma_i} + \gamma_i^*\right)P_{\rm canon}. \tag{8.18}$$

As

$$\frac{\partial}{\partial\gamma_i^*} = \frac{\partial}{\partial\alpha_i^*} + \frac{\partial}{\partial\alpha_i^+}, \qquad \frac{\partial}{\partial\gamma_i} = \frac{\partial}{\partial\alpha_i} + \frac{\partial}{\partial\alpha_i^{+*}}, \tag{8.19}$$

it follows that (8.13) and (8.18) are the same correspondence rules. The same results can be obtained from the standard correspondence rules by adding particular choices of the additional terms (8.8). Some terms involving the δ_i, δ_i^* and their derivatives give zero when applied to the canonical distribution function and hence can be ignored. Details are given in Appendix F.

For bosonic operators, the correspondence rules for the Wigner function are

$$\hat{\rho} \Rightarrow \hat{a}_i\hat{\rho}, \tag{8.20}$$
$$W(\alpha,\alpha^+,\alpha^*,\alpha^{+*},g,g^+) \Rightarrow \left(\alpha_i + \frac{1}{2}\frac{\partial}{\partial\alpha_{ix}^+}\right)W \equiv \left(\alpha_i + \frac{1}{2}\frac{\partial}{\partial(i\alpha_{iy}^+)}\right)W,$$

$$\hat{\rho} \Rightarrow \hat{\rho}\hat{a}_i, \tag{8.21}$$
$$W(\alpha, \alpha^+, \alpha^*, \alpha^{+*}, g, g^+) \Rightarrow \left(\alpha_i - \frac{1}{2}\frac{\partial}{\partial \alpha_{ix}^+}\right) W \equiv \left(\alpha_i - \frac{1}{2}\frac{\partial}{\partial (i\alpha_{iy}^+)}\right) W,$$

$$\hat{\rho} \Rightarrow \hat{a}_i^\dagger \hat{\rho}, \tag{8.22}$$
$$W(\alpha, \alpha^+, \alpha^*, \alpha^{+*}, g, g^+) \Rightarrow \left(\alpha_i^+ - \frac{1}{2}\frac{\partial}{\partial \alpha_{ix}}\right) W \equiv \left(\alpha_i^+ - \frac{1}{2}\frac{\partial}{\partial (i\alpha_{iy})}\right) W,$$

$$\hat{\rho} \Rightarrow \hat{\rho}\hat{a}_i^\dagger, \tag{8.23}$$
$$W(\alpha, \alpha^+, \alpha^*, \alpha^{+*}, g, g^+) \Rightarrow \left(\alpha_i^+ + \frac{1}{2}\frac{\partial}{\partial \alpha_{ix}}\right) W \equiv \left(\alpha_i^+ + \frac{1}{2}\frac{\partial}{\partial (i\alpha_{iy})}\right) W,$$

where $\alpha, \alpha^+, \alpha^*, \alpha^{+*}$ are regarded as four independent complex variables. These are proved in Appendix F using the symmetrically ordered characteristic function, which is related to the normally ordered characteristic function in (7.10) and enables results for the latter to be used.

8.3 Fermionic Correspondence Rules

In this section, we set out the correspondence rules for fermions. In the fermionic case, there are two possibilities associated with left and right multiplication or differentiation of the fermionic positive P distribution function, which is an even Grassmann function. However, as we will see, great care is required when applying correspondence rules in succession because the previous application will change the evenness or oddness of the Grassmann function involved. Fortunately, one of the two choices of correspondence rule does not depend on the evenness or oddness of the Grassmann function obtained from the previous step, and this choice will be highlighted.

8.3.1 Fermionic Correspondence Rules for Annihilation and Creation Operators

For fermionic operators, the correspondence rules are

$$\hat{\rho} \Rightarrow \hat{c}_i\hat{\rho}, \tag{8.24}$$
$$P(\alpha, \alpha^+, \alpha^*, \alpha^{+*}, g, g^+) \Rightarrow g_i P = P g_i,$$

$$\hat{\rho} \Rightarrow \hat{\rho}\hat{c}_i, \tag{8.25}$$
$$P(\alpha, \alpha^+, \alpha^*, \alpha^{+*}, g, g^+) \Rightarrow P\left(+\frac{\overleftarrow{\partial}}{\partial g_i^+} - g_i\right) = \left(-\frac{\overrightarrow{\partial}}{\partial g_i^+} - g_i\right) P,$$

$$\hat{\rho} \Rightarrow \hat{c}_i^\dagger \hat{\rho}, \tag{8.26}$$
$$P(\alpha, \alpha^+, \alpha^*, \alpha^{+*}, g, g^+) \Rightarrow \left(+\frac{\overrightarrow{\partial}}{\partial g_i} - g_i^+\right) P = P\left(-\frac{\overleftarrow{\partial}}{\partial g_i} - g_i^+\right),$$

$$\hat{\rho} \Rightarrow \hat{\rho}\hat{c}_i^\dagger,$$
$$P(\alpha, \alpha^+, \alpha^*, \alpha^{+*}, g, g^+) \Rightarrow P g_i^+ = g_i^+ P. \tag{8.27}$$

Also,

$$\frac{\partial \hat{\rho}}{\partial t} \rightarrow \frac{\partial P(\alpha, \alpha^+, \alpha^*, \alpha^{+*}, g, g^+)}{\partial t}. \tag{8.28}$$

Note that the first forms of the fermion and boson results differ by an overall minus sign.

The proofs of the first expressions for the fermion results do not depend on the distribution function being an even Grassmann function. The proofs depend only on the existence of the canonical form (7.54) for the density operator and on the Bargmann projector $\hat{\Lambda}(g, g^+, \alpha, \alpha^+)$ (or, rather, its fermion factor $\hat{\Lambda}_{\mathrm{f}}(g, g^+)$) being an even Grassmann operator. However, the second form of the fermion results uses the feature that the distribution function is an even Grassmann function. If it were odd, then the second forms would have all their signs reversed. The first form of the rules for the fermion case obtained here is equivalent to those presented by Plimak et al. [14]. (See equation (57) in that paper, noting that these authors use a representation with $g^+ \Rightarrow \overline{g} \equiv g^*$ and, in terms of the present notation, $B_a(\overline{g}, g) \Rightarrow P(g, g^+)\exp(-\sum_i g_i g_i^+)$; see equation (53). Note also their notation for the ket and bra vectors; their $\langle \overline{g}| \Rightarrow \langle g|_{\mathrm{B}}$ and $|g\rangle \Rightarrow |g\rangle_{\mathrm{B}}$.)

8.4 Derivation of Bosonic and Fermionic Correspondence Rules

The standard bosonic results agree with those of Gardiner and Zoller [31]. The derivation of the bosonic correspondence rules starting from the canonical form of the density operator relies on the analyticity of the bosonic projector $\hat{\Lambda}_{\mathrm{b}}(\alpha, \alpha^+)$. The results are based on certain identities for derivatives of the projector operators $\hat{\Lambda}_{\mathrm{b}}(\alpha, \alpha^+)$. We have

$$\begin{aligned}
\hat{a}_i \hat{\Lambda}_{\mathrm{b}}(\alpha, \alpha^+) &= \alpha_i \hat{\Lambda}_{\mathrm{b}}(\alpha, \alpha^+), \\
\hat{\Lambda}_{\mathrm{b}}(\alpha, \alpha^+)\, \hat{a}_i^\dagger &= \alpha_i^+ \hat{\Lambda}_{\mathrm{b}}(\alpha, \alpha^+), \\
\hat{a}_i^\dagger \hat{\Lambda}_{\mathrm{b}}(\alpha, \alpha^+) &= \left(\frac{\partial}{\partial \alpha_i} + \alpha_i^+\right) \hat{\Lambda}_{\mathrm{b}}(\alpha, \alpha^+), \\
\hat{\Lambda}_{\mathrm{b}}(\alpha, \alpha^+)\, \hat{a}_i &= \left(\frac{\partial}{\partial \alpha_i^+} + \alpha_i\right) \hat{\Lambda}_{\mathrm{b}}(\alpha, \alpha^+).
\end{aligned} \tag{8.29}$$

The derivation also uses an integration-by-parts step, which relies on P going to zero fast enough on the boundaries of the α_i and α_i^+ planes to enable boundary terms to be ignored. The results do not require the distribution function to be analytic, but it is assumed that it is differentiable with respect to α_i, α_i^+ or α_{ix}, α_{iy}, α_{ix}^+ and α_{iy}^+. For the second and third bosonic cases, there are two different results depending on whether $(\partial/\partial\alpha_i)P$, is replaced by $(\partial/\partial\alpha_{ix})P$ or $(\partial/\partial(i\alpha_{iy}))P$, or $(\partial/\partial\alpha_i^+)P$ is replaced by $(\partial/\partial\alpha_{ix}^+)P$ or $(\partial/\partial(i\alpha_{iy}^+))P$. These results are equivalent rather than equal, in that either form leads to a Fokker–Planck equation for a distribution function that gives the correct characteristic function. However, the function $P(\alpha, \alpha^+, \alpha^*, \alpha^{+*}, g, g^+)$ is not an analytic function of α_i and α_i^+, though it is still differentiable with respect to α_i, α_i^+ or α_{ix}, α_{iy}, α_{ix}^+ and α_{iy}^+, and expressions like $(\partial/\partial\alpha_i)P$ and $(\partial/\partial\alpha_i^+)P$ can also be interpreted in terms of derivatives with respect to α_{ix}, α_{iy}, α_{ix}^+ and α_{iy}^+.

In obtaining these results from the characteristic function (see Appendix F), certain identities for derivatives of the operators $\hat{\Omega}_{\mathrm{b}}^+(\xi^+)$ and $\hat{\Omega}_{\mathrm{b}}^-(\xi)$ are used, which, along with the characteristic function itself, are analytic functions of ξ, ξ^+. We have

$$
\begin{aligned}
\hat{\Omega}_{\mathrm{b}}^+(\xi^+)\,\hat{a}_i &= \frac{\partial}{\partial(i\xi_i^+)}\hat{\Omega}_{\mathrm{b}}^+(\xi^+) = \hat{a}_i\,\hat{\Omega}_{\mathrm{b}}^+(\xi^+),\\
\hat{\Omega}_{\mathrm{b}}^-(\xi)\,\hat{a}_i^\dagger &= \frac{\partial}{\partial(i\xi_i)}\hat{\Omega}_{\mathrm{b}}^-(\xi) = \hat{a}_i^\dagger\,\hat{\Omega}_{\mathrm{b}}^-(\xi).
\end{aligned}
\tag{8.30}
$$

These lead to the following for the changes to the normally ordered characteristic function:

$$
\begin{aligned}
\hat{\rho} &\Rightarrow \hat{a}_i\hat{\rho}, & \chi(\xi,\xi^+,h,h^+) &\Rightarrow \frac{\partial}{\partial(i\xi_i^+)}\chi,\\
\hat{\rho} &\Rightarrow \hat{\rho}\hat{a}_i, & \chi(\xi,\xi^+,h,h^+) &\Rightarrow \left(\frac{\partial}{\partial(i\xi_i^+)} + i\xi_i\right)\chi,\\
\hat{\rho} &\Rightarrow \hat{a}_i^\dagger\hat{\rho}, & \chi(\xi,\xi^+,h,h^+) &\Rightarrow \left(\frac{\partial}{\partial(i\xi_i)} + i\xi_i^+\right)\chi,\\
\hat{\rho} &\Rightarrow \hat{\rho}\hat{a}_i^\dagger, & \chi(\xi,\xi^+,h,h^+) &\Rightarrow \frac{\partial}{\partial(i\xi_i)}\chi,
\end{aligned}
\tag{8.31}
$$

from which the corresponding changes to the distribution function can be deduced. This involves integration by parts. The analyticity of the functions $\exp i\sum_{i=1}^n\{\xi_i\alpha_i^+\}$ and $\exp i\sum_{i=1}^n\{\alpha_i\xi_i^+\}$ enables the two options for partial differentiation (α_{ix} or $i\alpha_{iy}$ and α_{ix}^+ or $i\alpha_{iy}^+$) to be obtained.

For the symmetrically ordered characteristic function, the rules are

$$
\begin{aligned}
\hat{\rho} &\Rightarrow \hat{a}_i\hat{\rho}, & \chi^{\mathrm{W}}(\xi,\xi^+,h,h^+) &\Rightarrow \left(\frac{\partial}{\partial(i\xi_i^+)} - \frac{1}{2}i\xi_i\right)\chi^{\mathrm{W}},\\
\hat{\rho} &\Rightarrow \hat{\rho}\hat{a}_i, & \chi^{\mathrm{W}}(\xi,\xi^+,h,h^+) &\Rightarrow \left(\frac{\partial}{\partial(i\xi_i^+)} + \frac{1}{2}i\xi_i\right)\chi^{\mathrm{W}},\\
\hat{\rho} &\Rightarrow \hat{a}_i^\dagger\hat{\rho}, & \chi^{\mathrm{W}}(\xi,\xi^+,h,h^+) &\Rightarrow \left(\frac{\partial}{\partial(i\xi_i)} + \frac{1}{2}i\xi_i^+\right)\chi^{\mathrm{W}},\\
\hat{\rho} &\Rightarrow \hat{\rho}\hat{a}_i^\dagger, & \chi^{\mathrm{W}}(\xi,\xi^+,h,h^+) &\Rightarrow \left(\frac{\partial}{\partial(i\xi_i)} - \frac{1}{2}i\xi_i^+\right)\chi^{\mathrm{W}}.
\end{aligned}
\tag{8.32}
$$

For fermions, the derivation of the correspondence rules from the canonical form of the density operator is based on certain identities for derivatives of the projector operators $\hat{\Lambda}_{\mathrm{f}}(g, g^{+})$. We have

$$\begin{aligned}
\hat{c}_i \hat{\Lambda}_{\mathrm{f}}(g, g^{+}) &= g_i \hat{\Lambda}_{\mathrm{f}}(g, g^{+}) = \hat{\Lambda}_{\mathrm{f}}(g, g^{+})\, g_i, \\
\hat{\Lambda}_{\mathrm{f}}(g, g^{+})\, \hat{c}_i^{\dagger} &= \hat{\Lambda}_{\mathrm{f}}(g, g^{+})\, g_i^{+} = g_i^{+} \hat{\Lambda}_{\mathrm{f}}(g, g^{+}), \\
\hat{c}_i^{\dagger} \hat{\Lambda}_{\mathrm{f}}(g, g^{+}) &= \left(-\frac{\overrightarrow{\partial}}{\partial g_i} - g_i^{+} \right) \hat{\Lambda}_{\mathrm{f}}(g, g^{+}) = \hat{\Lambda}_{\mathrm{f}}(g, g^{+}) \left(+\frac{\overleftarrow{\partial}}{\partial g_i} - g_i^{+} \right), \\
\hat{\Lambda}_{\mathrm{f}}(g, g^{+})\, \hat{c}_i &= \hat{\Lambda}_{\mathrm{f}}(g, g^{+}) \left(-\frac{\overleftarrow{\partial}}{\partial g_i^{+}} - g_i \right) = \left(+\frac{\overrightarrow{\partial}}{\partial g_i^{+}} - g_i \right) \hat{\Lambda}_{\mathrm{f}}(g, g^{+}),
\end{aligned} \tag{8.33}$$

the second form following from $\hat{\Lambda}_{\mathrm{f}}(g, g^{+})$ being an even Grassmann operator.

The correspondence rules for the distribution function in the fermion case lead to the following correspondences for the characteristic function, as determined from the distribution function via the Grassmann integral expression (7.51):

$$\begin{aligned}
&(\hat{\rho} \Rightarrow \hat{c}_i \hat{\rho}) \\
&P \Rightarrow g_i\, P, \qquad \chi(\xi, \xi^{+}, h, h^{+}) \Rightarrow \left(-\frac{\overrightarrow{\partial}}{\partial (i h_i^{+})} \right) \chi, \\
&(\hat{\rho} \Rightarrow \hat{\rho} \hat{c}_i) \\
&P \Rightarrow P \left(+\frac{\overleftarrow{\partial}}{\partial g_i^{+}} - g_i \right), \qquad \chi(\xi, \xi^{+}, h, h^{+}) \Rightarrow \chi \left(-\frac{\overleftarrow{\partial}}{\partial (i h_i^{+})} - i h_i \right), \\
&(\hat{\rho} \Rightarrow \hat{c}_i^{\dagger} \hat{\rho}) \\
&P \Rightarrow \left(+\frac{\overrightarrow{\partial}}{\partial g_i} - g_i^{+} \right) P, \qquad \chi(\xi, \xi^{+}, h, h^{+}) \Rightarrow \left(-\frac{\overrightarrow{\partial}}{\partial (i h_i)} - i h_i^{+} \right) \chi, \\
&(\hat{\rho} \Rightarrow \hat{\rho} \hat{c}_i^{\dagger}) \\
&P \Rightarrow P g_i^{+}, \qquad \chi(\xi, \xi^{+}, h, h^{+}) \Rightarrow \chi \left(-\frac{\overleftarrow{\partial}}{\partial (i h_i)} \right),
\end{aligned} \tag{8.34}$$

where none of the results depends on whether P is even or odd. The derivation is carried out in Appendix F. The expected changes to the density operator are in brackets. Just as for the distribution function, the fact that the characteristic function is even leads to other correspondences with the differentiation and multiplication applied from the opposite side, but these will not be exhibited here. The fermionic correspondence rules for the characteristic function are similar to those for bosons, but here there is a minus sign and a distinction between right and left derivatives. Provided that the correspondences in (8.24)–(8.27) for the distribution function lead to correspondences for the characteristic function that are the correct ones, then these distribution function correspondences are justified. The treatment in which the distribution function correspondences are obtained directly from the canonical form of the density operator in terms of Bargmann state projectors is, of course, more direct.

Thus, the correspondence rules for fermions can also be justified by assuming their effect on the distribution function P as in (8.24)–(8.27) and then deriving their effect on the characteristic function $\chi(\xi, \xi^+, h, h^+)$ as obtained in (8.34), given that the latter is required to be related to the distribution function via the phase space integral form (7.51). We can then return to the basic definition of the characteristic function (7.50) and confirm that the new characteristic function *does* correspond to the characteristic function that would apply when the density operator $\hat{\rho}$ is replaced by its product with a fermion annihilation or creation operator. However, processes of multiplying or differentiating a trace with Grassmann variables are involved, and to carry out these processes inside the trace we require there to be *two* Grassmann processes involved (see (5.137)). Hence the correspondence rules can only be confirmed in this way for cases where the density operator $\hat{\rho}$ is replaced by its product with two fermion operators. Unfortunately, though, there are a total of 12 different replacements that need to be considered, namely $\hat{\rho} \Rightarrow (\hat{c}_i\hat{c}_j)\hat{\rho}, (\hat{c}_i\hat{c}_j^\dagger)\hat{\rho}, (\hat{c}_i^\dagger\hat{c}_j)\hat{\rho}, (\hat{c}_i^\dagger\hat{c}_j^\dagger)\hat{\rho}, \hat{\rho}(\hat{c}_i\hat{c}_j), \cdots, \hat{\rho}(\hat{c}_i^\dagger\hat{c}_j^\dagger), \hat{c}_i\hat{\rho}\hat{c}_j, \cdots, \hat{c}_i^\dagger\hat{\rho}\hat{c}_j^\dagger$. In Appendix F, the correspondence rules are confirmed by this process in two typical cases. These are (1) $\hat{\rho} \Rightarrow \hat{c}_i\hat{c}_j\hat{\rho}$ and (11) $\hat{\rho} \Rightarrow \hat{c}_i^\dagger\hat{\rho}\hat{c}_j$:

$$
\begin{aligned}
&P(\alpha, \alpha^+, \alpha^*, \alpha^{+*}, g, g^+) \Rightarrow g_i g_j P, \qquad \hat{\rho} \Rightarrow \hat{c}_i\hat{c}_j\hat{\rho} \qquad \text{Case 1,} \\
&P(\alpha, \alpha^+, \alpha^*, \alpha^{+*}, g, g^+) \Rightarrow \left(+\frac{\overrightarrow{\partial}}{\partial g_i} - g_i^+\right) P \left(+\frac{\overleftarrow{\partial}}{\partial g_i^+} - g_i\right), \qquad \hat{\rho} \Rightarrow \hat{c}_i^\dagger\hat{\rho}\hat{c}_j \qquad \text{Case 11.}
\end{aligned}
\tag{8.35}
$$

8.5 Effect of Several Operators

In terms of applying the correspondence rules in physical situations involving fermions, kinetic-energy and one-body interaction terms in the Hamiltonian always involve pairs of fermion operators $\hat{c}_i^\dagger\hat{c}_j$, two-body interaction terms involve two pairs $\hat{c}_i^\dagger\hat{c}_j^\dagger\hat{c}_k\hat{c}_l$, terms in master equations involve forms such as $[\hat{c}_i^\dagger\hat{c}_j, \hat{\rho}]$, $[\hat{c}_i^\dagger\hat{c}_j^\dagger\hat{c}_k\hat{c}_l, \hat{\rho}]$, $[\hat{c}_i^\dagger, \hat{\rho}\hat{c}_j]$ and $[\hat{c}_i, \hat{\rho}\hat{c}_j^\dagger]$, and so on. All of these require the density operator $\hat{\rho}$ to be replaced in succession by a product of the density operator with two fermion annihilation or creation operators. Hence it is unnecessary to consider, in physical situations, the replacement of the density operator by its product with only one operator. However, in view of the direct derivation of the fermion correspondence rules for single-fermion operators via the canonical form of the density operator, we can still treat any replacement of the density operator by its product with these operators as a succession of single replacements.

The derivation of the correspondence rules is analogous to that in Appendix F for the basic rules given in (8.1)–(8.27). To derive the Fokker–Planck equation when several operators are involved, we simply apply the correspondence rules in succession, taking care with any sign changes that may occur. The same result should be obtained irrespective of whether left or right derivatives are used, provided care is taken with the evenness or oddness of the distribution functions. This creates problems, however.

The standard correspondence rules for single annihilation and creation operators have been set out in (8.1)–(8.4) for bosons and (8.24)–(8.27) for fermions. There are two forms for each of the fermionic cases, the second depending on the distribution function being an even Grassmann function, while the first form does not. It is this feature which can cause problems when more than one fermion annihilation or creation operator is applied to the density operator. After the correspondence rule is applied, the resultant expression involves changes from an even to an odd Grassmann function. Thus, for the replacement $\hat{\rho} \Rightarrow \hat{\rho}\hat{c}_i, P \Rightarrow P\left(+\overleftarrow{\partial}/\partial g_i^+ - g_i\right)$, the latter is an odd function (as is the alternative result $\left(-\overrightarrow{\partial}/\partial g_i^+ - g_i\right)P$). The question arises then of what the correspondence rule should be if, for example, $\hat{\rho} \Rightarrow \hat{\rho}\hat{c}_i\hat{c}_j$. Consideration of the two alternative one-fermion correspondence rules gives four possibilities:

$$P(\alpha,\alpha^+,\alpha^*,\alpha^{+*},g,g^+) \Rightarrow \left[P\left(+\frac{\overleftarrow{\partial}}{\partial g_i^+} - g_i\right)\right]\left(+\frac{\overleftarrow{\partial}}{\partial g_j^+} - g_j\right),$$

$$P(\alpha,\alpha^+,\alpha^*,\alpha^{+*},g,g^+) \Rightarrow \left(-\frac{\overrightarrow{\partial}}{\partial g_j^+} - g_j\right)\left[P\left(+\frac{\overleftarrow{\partial}}{\partial g_i^+} - g_i\right)\right],$$

$$P(\alpha,\alpha^+,\alpha^*,\alpha^{+*},g,g^+) \Rightarrow \left(-\frac{\overrightarrow{\partial}}{\partial g_j^+} - g_j\right)\left[\left(-\frac{\overrightarrow{\partial}}{\partial g_i^+} - g_i\right)P\right],$$

$$P(\alpha,\alpha^+,\alpha^*,\alpha^{+*},g,g^+) \Rightarrow \left[\left(-\frac{\overrightarrow{\partial}}{\partial g_i^+} - g_i\right)P\right]\left(+\frac{\overleftarrow{\partial}}{\partial g_j^+} - g_j\right), \qquad (8.36)$$

where the square brackets indicating the first correspondence process are included and the terms are evaluated using the Grassmann differentiation rules (4.23), (4.24) and (4.27). These four results are not the same. On simplification using the Grassmann differentiation rules (4.21) and (4.22), we find that the first expression is equal to

$$\left[P\left(\frac{\overleftarrow{\partial}}{\partial g_i^+}\right)\right]\left(\frac{\overleftarrow{\partial}}{\partial g_j^+}\right) - [P(g_i)]\left(\frac{\overleftarrow{\partial}}{\partial g_j^+}\right) - \left[P\left(\frac{\overleftarrow{\partial}}{\partial g_i^+}\right)\right](g_j) + [P(g_i)](g_j), \qquad (8.37)$$

whilst the second expression is minus this result. The first expression is actually the correct one. The second one is invalid because the result after the first replacement $\hat{\rho} \Rightarrow \hat{\rho}\hat{c}_i$ is $P\left(+\overleftarrow{\partial}/\partial g_i^+ - g_i\right)$, which is an odd function – and therefore, in the second replacement, $\hat{\rho}\hat{c}_i \Rightarrow \hat{\rho}\hat{c}_i\hat{c}_j$, it is not valid to use $\left(-\overrightarrow{\partial}/\partial g_j^+ - g_j\right)$. A similar problem arises for the third expression. The fourth expression is correct because although $\left(-\overrightarrow{\partial}/\partial g_i^+ - g_i\right)P$ is an odd function, the use of $\left(+\overleftarrow{\partial}/\partial g_j^+ - g_j\right)$ in the second replacement is valid irrespective of whether the first expression is even or odd.

Clearly, the consideration of which correspondence rule to use, depending on the evenness or oddness of the result at each stage of a successive replacement process, complicates the application of the rules to fermion operators. It is therefore prudent to apply only that form of the single-fermion operator correspondence rule that will apply irrespective of whether the previous function is even or odd. This means that the first form of the fermion correspondence rules in (8.24)–(8.27) should be used. Thus for a general operator $\hat{\sigma}$, where $\hat{\sigma} = \hat{\rho}, \hat{c}_i\hat{\rho}, \hat{\rho}\hat{c}_i, \hat{c}_i^\dagger\hat{\rho}, \hat{\rho}\hat{c}_i^\dagger, \hat{c}_i\hat{\rho}\hat{c}_j^\dagger, \hat{c}_j^\dagger\hat{c}_i\hat{\rho}, \hat{\rho}\hat{c}_j^\dagger\hat{c}_i, \cdots$, involving cases where $\hat{\rho}$ is replaced by successive products and $S(\alpha,\alpha^+,\alpha^*,\alpha^{+*},g,g^+)$ refers to the modified distribution function obtained after the previous replacement process, the safe forms of the fermion correspondence rules are

$$\hat{\sigma} \Rightarrow \hat{c}_i\hat{\sigma}, \qquad S(\alpha,\alpha^+,\alpha^*,\alpha^{+*},g,g^+) \Rightarrow g_i S, \tag{8.38}$$

$$\hat{\sigma} \Rightarrow \hat{\sigma}\hat{c}_i, \qquad S(\alpha,\alpha^+,\alpha^*,\alpha^{+*},g,g^+) \Rightarrow S\left(+\frac{\overleftarrow{\partial}}{\partial g_i^+} - g_i\right), \tag{8.39}$$

$$\hat{\sigma} \Rightarrow \hat{c}_i^\dagger\hat{\sigma}, \qquad S(\alpha,\alpha^+,\alpha^*,\alpha^{+*},g,g^+) \Rightarrow \left(+\frac{\overrightarrow{\partial}}{\partial g_i} - g_i^+\right)S, \tag{8.40}$$

$$\hat{\sigma} \Rightarrow \hat{\sigma}\hat{c}_i^\dagger, \qquad S(\alpha,\alpha^+,\alpha^*,\alpha^{+*},g,g^+) \Rightarrow S g_i^+. \tag{8.41}$$

For the bosonic case, this issue does not arise. The bosonic rules are

$$\hat{\sigma} \Rightarrow \hat{a}_i\hat{\sigma} \qquad S(\alpha,\alpha^+,\alpha^*,\alpha^{+*},g,g^+) \Rightarrow \alpha_i S, \tag{8.42}$$

$$\hat{\sigma} \Rightarrow \hat{\sigma}\hat{a}_i \qquad S(\alpha,\alpha^+,\alpha^*,\alpha^{+*},g,g^+) \Rightarrow \left(-\frac{\partial}{\partial \alpha_i^+} + \alpha_i\right)S, \tag{8.43}$$

$$\hat{\sigma} \Rightarrow \hat{a}_i^\dagger\hat{\sigma} \qquad S(\alpha,\alpha^+,\alpha^*,\alpha^{+*},g,g^+) \Rightarrow \left(-\frac{\partial}{\partial \alpha_i} + \alpha_i^+\right)S, \tag{8.44}$$

$$\hat{\sigma} \Rightarrow \hat{\sigma}\hat{a}_i^\dagger \qquad S(\alpha,\alpha^+,\alpha^*,\alpha^{+*},g,g^+) \Rightarrow \alpha_i^+ S. \tag{8.45}$$

A more formal justification of the last correspondence rules in given in Appendix F. Note the symmetry between the left- and right-derivative forms for fermions and the sign change between the boson and fermion results.

In cases where the density operator is replaced by products with annihilation and creation operators, we again apply these rules in succession. One or two examples are given to illustrate the approach:

$$\hat{\rho} \Rightarrow \hat{c}_j^\dagger\hat{c}_i\hat{\rho}, \qquad P(\alpha,\alpha^+,\alpha^*,\alpha^{+*},g,g^+) \Rightarrow \left(+\frac{\overrightarrow{\partial}}{\partial g_j} - g_j^+\right)(g_i P),$$

$$\begin{aligned}\hat{\rho} \Rightarrow \hat{c}_j^\dagger\hat{\rho}\hat{c}_i, \qquad P(\alpha,\alpha^+,\alpha^*,\alpha^{+*},g,g^+) &\Rightarrow \left(+\frac{\overrightarrow{\partial}}{\partial g_j} - g_j^+\right)\left(P\left(+\frac{\overleftarrow{\partial}}{\partial g_i^+} - g_i\right)\right) \\ &= \left(\left(+\frac{\overrightarrow{\partial}}{\partial g_j} - g_j^+\right)P\right)\left(+\frac{\overleftarrow{\partial}}{\partial g_i^+} - g_i\right) \\ &= \left(+\frac{\overrightarrow{\partial}}{\partial g_j} - g_j^+\right)P\left(+\frac{\overleftarrow{\partial}}{\partial g_i^+} - g_i\right),\end{aligned} \tag{8.46}$$

where the Grassmann differentiation rules show that the result in the second case is the same irrespective of which order the replacement for the distribution function is carried out in.

In the case of bosons, the differentiation or multiplication of the distribution function that occurs when a correspondence rule is applied can occur only in one way, there being no separate left and right differentiation or multiplication processes. Consequently, there is no ambiguity involved when the correspondence rules are applied in succession.

8.6 Correspondence Rules for Unnormalised Distribution Functions

The correspondence rules for the new distribution functions $B_{\mathrm{b}}(\alpha, \alpha^{+}, \alpha^{*}, \alpha^{+*})$ and $B_{\mathrm{f}}(g, g^{+})$ defined in (7.100) may be obtained either from those for the standard distribution functions or from the expressions in (7.98) and (7.99) for the density operator in terms of unnormalised Bargmann state projectors. For these projectors,

$$\begin{aligned}
\hat{a}_i(|\boldsymbol{\alpha}\rangle_{\mathrm{B}}\,\langle\boldsymbol{\alpha}^{+*}|_{\mathrm{B}}) &= \alpha_i(|\boldsymbol{\alpha}\rangle_{\mathrm{B}}\,\langle\boldsymbol{\alpha}^{+*}|_{\mathrm{B}}),\\
(|\boldsymbol{\alpha}\rangle_{\mathrm{B}}\,\langle\boldsymbol{\alpha}^{+*}|_{\mathrm{B}})\hat{a}_i^{\dagger} &= \alpha_i^{+}(|\boldsymbol{\alpha}\rangle_{\mathrm{B}}\,\langle\boldsymbol{\alpha}^{+*}|_{\mathrm{B}}),\\
\hat{a}_i^{\dagger}(|\boldsymbol{\alpha}\rangle_{\mathrm{B}}\,\langle\boldsymbol{\alpha}^{+*}|_{\mathrm{B}}) &= \left(\frac{\partial}{\partial\alpha_i}\right)(|\boldsymbol{\alpha}\rangle_{\mathrm{B}}\,\langle\boldsymbol{\alpha}^{+*}|_{\mathrm{B}}),\\
(|\boldsymbol{\alpha}\rangle_{\mathrm{B}}\,\langle\boldsymbol{\alpha}^{+*}|_{\mathrm{B}})\hat{a}_i &= \left(\frac{\partial}{\partial\alpha_i^{+}}\right)(|\boldsymbol{\alpha}\rangle_{\mathrm{B}}\,\langle\boldsymbol{\alpha}^{+*}|_{\mathrm{B}})
\end{aligned} \tag{8.47}$$

and

$$\begin{aligned}
\hat{c}_i(|\mathbf{g}\rangle_{\mathrm{B}}\,\langle\mathbf{g}^{+*}|_{\mathrm{B}}) &= g_i(|\mathbf{g}\rangle_{\mathrm{B}}\,\langle\mathbf{g}^{+*}|_{\mathrm{B}}),\\
(|\mathbf{g}\rangle_{\mathrm{B}}\,\langle\mathbf{g}^{+*}|_{\mathrm{B}})\hat{c}_i^{\dagger} &= (|\mathbf{g}\rangle_{\mathrm{B}}\,\langle\mathbf{g}^{+*}|_{\mathrm{B}})g_i^{+},\\
\hat{c}_i^{\dagger}(|\mathbf{g}\rangle_{\mathrm{B}}\,\langle\mathbf{g}^{+*}|_{\mathrm{B}}) &= \left(-\frac{\overrightarrow{\partial}}{\partial g_i}\right)(|\mathbf{g}\rangle_{\mathrm{B}}\,\langle\mathbf{g}^{+*}|_{\mathrm{B}}),\\
(|\mathbf{g}\rangle_{\mathrm{B}}\,\langle\mathbf{g}^{+*}|_{\mathrm{B}})\hat{c}_i &= (|\mathbf{g}\rangle_{\mathrm{B}}\,\langle\mathbf{g}^{+*}|_{\mathrm{B}})\left(-\frac{\overleftarrow{\partial}}{\partial g_i^{+}}\right),
\end{aligned} \tag{8.48}$$

and the correspondence rules can be obtained from the density operator using integration by parts. These are

$$\begin{aligned}
\hat{\rho} &\Rightarrow \hat{a}_i\hat{\rho}, & B_{\mathrm{b}}(\alpha,\alpha^{+},\alpha^{*},\alpha^{+*}) &\Rightarrow \alpha_i B_{\mathrm{b}},\\
\hat{\rho} &\Rightarrow \hat{\rho}\hat{a}_i, & B_{\mathrm{b}}(\alpha,\alpha^{+},\alpha^{*},\alpha^{+*}) &\Rightarrow \left(-\frac{\partial}{\partial\alpha_i^{+}}\right)B_{\mathrm{b}},\\
\hat{\rho} &\Rightarrow \hat{a}_i^{\dagger}\hat{\rho}, & B_{\mathrm{b}}(\alpha,\alpha^{+},\alpha^{*},\alpha^{+*}) &\Rightarrow \left(-\frac{\partial}{\partial\alpha_i}\right)B_{\mathrm{b}},\\
\hat{\rho} &\Rightarrow \hat{\rho}\hat{a}_i^{\dagger}, & B_{\mathrm{b}}(\alpha,\alpha^{+},\alpha^{*},\alpha^{+*}) &\Rightarrow \alpha_i^{+}B_{\mathrm{b}}
\end{aligned} \tag{8.49}$$

for bosons and

$$\begin{aligned}
\hat{\rho} &\Rightarrow \hat{c}_i\hat{\rho}, & B_{\mathrm{f}}(g,g^+) &\Rightarrow g_i B_{\mathrm{f}}, \\
\hat{\rho} &\Rightarrow \hat{\rho}\hat{c}_i, & B_{\mathrm{f}}(g,g^+) &\Rightarrow B_{\mathrm{f}}\left(+\frac{\overleftarrow{\partial}}{\partial g_i^+}\right), \\
\hat{\rho} &\Rightarrow \hat{c}_i^\dagger\hat{\rho}, & B_{\mathrm{f}}(g,g^+) &\Rightarrow \left(+\frac{\overrightarrow{\partial}}{\partial g_i}\right)B_{\mathrm{f}}, \\
\hat{\rho} &\Rightarrow \hat{\rho}\hat{c}_i^\dagger, & B_{\mathrm{f}}(g,g^+) &\Rightarrow B_{\mathrm{f}}g_i^+
\end{aligned} \tag{8.50}$$

for fermions [14]. In the case of fermions, we state here only the rules that can be applied in succession without having to allow for Grassmann functions being even or odd. The non-appearance of sums of derivatives and variables leads to much simpler Fokker–Planck equations, as pointed out in [14]. The correspondence rules for unnormalised distribution functions of joint boson and fermion systems are a straightforward combination of these rules.

8.7 Dynamical Processes and Fokker–Planck Equations

In this section, the general forms of the Fokker–Planck equations will be set out. The particular forms depend on the dynamical equations governing the density operator, and this will differ from system to system and depend on whether the density operator satisfies a Liouville–von Neumann equation for Hamiltonian dynamics, a master equation, as occurs when dissipation and noise effects due to an environment are involved, or a Matsubara equation when the effect of temperature changes are considered. In all cases there will be terms in the Fokker–Planck equation that involve differentiations and products (left and right for fermions) with respect to the four real variables $\alpha_{ix}, \alpha_{ix}^+, \alpha_{iy}, \alpha_{iy}^+$ or the two complex variables α_i, α_i^+ (boson case), or the two Grassmann variables g_i, g_i^+ (fermion case). Using the differentiation rules, the derivatives can be made to apply after the product operation. Ultimately, there will be a number of terms in which the distribution function is replaced by its product with functions of the variables, with the result being differentiated one or more times with respect to the variables. The Fokker–Planck equation also depends on the correspondence rules that are applied.

8.7.1 General Issues

For Hamiltonian evolution with a Hamiltonian given by (2.51) or (2.52), we find that in the case of bosonic positive P distribution functions, Fokker–Planck equations with only first- and second-order derivatives occur when the standard correspondence rules are applied. It might be thought that derivatives up to the fourth order could appear from the two-body interaction terms in these Hamiltonians; however, these all involve two creation operators and two annihilation operators, and then for $\hat{\rho}$ times the operator products $\hat{c}_i^\dagger\hat{c}_j^\dagger\hat{c}_l\hat{c}_k$ or $\hat{a}_i^\dagger\hat{a}_j^\dagger\hat{a}_l\hat{a}_k$ (on either side), the replacement involves only two factors containing derivatives. Furthermore, there are no terms involving the distribution function with no derivatives – these cancel out when the commutator is considered.

Terms in master equations such as $\hat{c}_j^\dagger \hat{c}_i \hat{\rho} \hat{c}_k^\dagger \hat{c}_l$ or $\hat{a}_j^\dagger \hat{a}_i \hat{\rho} \hat{a}_k^\dagger \hat{a}_l$ will lead to second-order derivatives, as may be seen from (8.46). The outcome is that, for Hamiltonians involving only one- and two-body interaction terms, for master equations with weak Markovian environmental coupling and for Matsubara equations, the Fokker–Planck equation contains only terms with first- and second-order derivatives. This is a key feature of the positive P function used here. For other distribution functions, generalised forms of Fokker–Planck equations containing higher-order derivatives can be involved, the Wigner distribution involving third-order derivatives being such a case that applies even for Hamiltonians involving only one- and two-body interaction terms.

Also, it should be noted that generalised Fokker–Planck equations can also occur in other situations. One case is that of a driven non-linear absorber, where the master equation for the mode with annihilation and creation operators $\hat{a}$, $\hat{a}^\dagger$ is

$$\frac{\partial}{\partial t}\hat{\rho} = \epsilon[\hat{a}^\dagger - \hat{a}, \hat{\rho}] + \frac{1}{2}K(2(\hat{a})^2\hat{\rho}(\hat{a}^\dagger)^2 - (\hat{a}^\dagger)^2(\hat{a})^2\hat{\rho} - \hat{\rho}(\hat{a}^\dagger)^2(\hat{a})^2), \tag{8.51}$$

where ϵ is the driving term and K determines the non-linear absorption. As pointed out by Schack and Schenzle [60], the generalised Fokker–Planck equation for the bosonic *canonical* positive P distribution function involves third- and fourth-order derivatives. This equation is given by

$$\begin{aligned}\frac{\partial}{\partial t}P_{\text{canon}}(\gamma,\gamma^*,\delta,\delta^*) &= \epsilon\left(-\frac{\partial}{\partial\gamma} - \frac{\partial}{\partial\gamma^*}\right)P_{\text{canon}}\\ &+ 2K\left(\frac{\partial}{\partial\gamma}\gamma\left(\frac{1}{2}\gamma\gamma^* - 1\right) + \frac{\partial}{\partial\gamma^*}\gamma^*\left(\frac{1}{2}\gamma\gamma^* - 1\right)\right)P_{\text{canon}}\\ &+ \frac{1}{2}K\left(\frac{\partial^2}{\partial\gamma^2}\gamma^2 + 8\frac{\partial}{\partial\gamma}\frac{\partial}{\partial\gamma^*}(\gamma^*\gamma - 1) + \frac{\partial^2}{\partial\gamma^{*2}}\gamma^{*2}\right)P_{\text{canon}}\\ &+ 2K\left(\frac{\partial^3}{\partial\gamma^2\partial\gamma^*}\gamma + \frac{\partial^3}{\partial\gamma^{*2}\partial\gamma}\gamma^*\right)P_{\text{canon}}\\ &+ K\left(\frac{\partial^4}{\partial\gamma^2\partial\gamma^{*2}}\right)P_{\text{canon}}\end{aligned} \tag{8.52}$$

in terms of the new phase space variables (8.14). The derivation is left as an exercise. The canonical positive P distribution involves an overall δ dependence in the form of a factor $(1/4\pi^2)\exp(-\delta\delta^*)$, and there is another factor $\exp(-\gamma\gamma^*)$ that can be removed. However, the presence of higher-order derivatives means that an equivalent Ito stochastic equation cannot be obtained, apart from parameter regimes where the important regime of phase space is such that γ, γ^* are $O(N)$, with the mean number of photons $N \sim \sqrt[3]{(\epsilon/K)}$ being much greater than 1. In that case truncation procedures enable the third- and fourth-order derivative terms to be discarded, as they are $O(1/N)$ and $O(1/N^2)$ times smaller than the second-derivative terms. On the other hand, the Fokker–Planck equation obtained using the standard bosonic correspondence rules is

$$\begin{aligned}\frac{\partial}{\partial t}P(\alpha,\alpha^+,\alpha^*,\alpha^{+*}) &= \varepsilon\left(-\frac{\partial}{\partial\alpha}-\frac{\partial}{\partial\alpha^+}\right)P \\ &\quad + 2K\left(\frac{\partial}{\partial\alpha}\alpha\left(\frac{1}{2}\alpha\alpha^+\right)+\frac{\partial}{\partial\alpha^+}\alpha^+\left(\frac{1}{2}\alpha\alpha^+\right)\right)P \\ &\quad - \frac{1}{2}K\left(\frac{\partial^2}{\partial\alpha^2}\alpha^2+\frac{\partial^2}{\partial\alpha^{+2}}\alpha^{+2}\right)P \end{aligned} \tag{8.53}$$

in terms of the original phase space variables. In this case a standard Fokker–Planck equation occurs, involving only derivatives up to second order. The two Fokker–Planck equations are fundamentally different, as are the two distribution functions. Both should, of course, still determine the same characteristic function. However, stochastic simulations based on the latter Fokker–Planck equation [61] indicate a failure of this equation to replicate the behaviour of the original master equation, presumably because the distribution function does not decrease fast enough as the phase space boundary is approached. Fortunately, as will be seen below, the situation where either the Fokker–Planck equation contains higher-order derivatives or it is suspect because of boundary condition problems is not common.

8.8 Boson Fokker–Planck Equations

In this section, we will obtain the bosonic Fokker–Planck equations for the positive P and Wigner distribution functions, showing how the former can be written in an explicitly positive definite form.

8.8.1 Bosonic Positive P Distribution

For the Liouville equation (2.82) associated with the Hamiltonian in (2.52), the terms in the Fokker–Planck equation for the positive P distribution function $P_{\mathrm{b}}(\alpha,\alpha^+,\alpha^*,\alpha^{+*})$ can be found from

$$\begin{aligned} &\hat{\rho}\Rightarrow\hat{\rho}\hat{a}_i^\dagger\hat{a}_j, && P_{\mathrm{b}}\Rightarrow\left(\alpha_j-\frac{\partial}{\partial\alpha_j^+}\right)\alpha_i^+P_{\mathrm{b}}, \\ &\hat{\rho}\Rightarrow\hat{a}_i^\dagger\hat{a}_j\hat{\rho}, && P_{\mathrm{b}}\Rightarrow\left(\alpha_i^+-\frac{\partial}{\partial\alpha_i}\right)\alpha_jP_{\mathrm{b}}, \\ &\hat{\rho}\Rightarrow\hat{\rho}\hat{a}_i^\dagger\hat{a}_j^\dagger\hat{a}_l\hat{a}_k, && P_{\mathrm{b}}\Rightarrow\left(\alpha_k-\frac{\partial}{\partial\alpha_k^+}\right)\left(\alpha_l-\frac{\partial}{\partial\alpha_l^+}\right)\alpha_j^+\alpha_i^+P_{\mathrm{b}}, \\ &\hat{\rho}\Rightarrow\hat{a}_i^\dagger\hat{a}_j^\dagger\hat{a}_l\hat{a}_k\hat{\rho}, && P_{\mathrm{b}}\Rightarrow\left(\alpha_i^+-\frac{\partial}{\partial\alpha_i}\right)\left(\alpha_j^+-\frac{\partial}{\partial\alpha_j}\right)\alpha_l\alpha_kP_{\mathrm{b}}, \end{aligned} \tag{8.54}$$

where at present we leave unspecified whether the derivatives will be with respect to α_{ix} or $i\alpha_{iy}$, or α_{ix}^+, or $i\alpha_{iy}^+$, or just left as derivatives with respect to α_i,α_i^+. It is clear that only derivatives up to second order will be involved. Furthermore, there will be no terms in the Fokker–Planck equation just involving P_{b} without derivatives, since when the commutators are considered, pairs of terms such as $(\alpha_j)(\alpha_i^+)P_{\mathrm{b}}$ and $(\alpha_i^+)\alpha_jP_{\mathrm{b}}$ or $(\alpha_k)(\alpha_l)\alpha_j^+\alpha_i^+P_{\mathrm{b}}$ and $(\alpha_i^+)(\alpha_j^+)\alpha_l\alpha_kP_{\mathrm{b}}$ cancel out. If the canonical P representation and its correspondence rules (8.18) are used, we find in this case that the Fokker–Planck equation contains only derivatives up to second order. In this case the derivatives will be with respect to the new variables γ_i and γ_i^*. The Fokker–Planck equation, however, may not be convertible to one with a positive definite diffusion matrix.

For the master equation (2.83) for the boson case, the terms in the Fokker–Planck equation arising from the relaxation term can be found from

$$
\begin{aligned}
\hat{\rho} &\Rightarrow (2\hat{a}_k^\dagger \hat{a}_l \hat{\rho} \hat{a}_j^\dagger \hat{a}_i - \hat{\rho} \hat{a}_j^\dagger \hat{a}_i \hat{a}_k^\dagger \hat{a}_l - \hat{a}_j^\dagger \hat{a}_i \hat{a}_k^\dagger \hat{a}_l \hat{\rho}), \\
P_{\mathrm{b}} &\Rightarrow 2\left(-\frac{\partial}{\partial \alpha_i^+} + \alpha_i\right)\left(-\frac{\partial}{\partial \alpha_k} + \alpha_k^+\right)(\alpha_j^+)(\alpha_l)\,P_{\mathrm{b}} \\
&\quad - \left(-\frac{\partial}{\partial \alpha_l^+} + \alpha_l\right)(\alpha_k^+)\left(-\frac{\partial}{\partial \alpha_i^+} + \alpha_i\right)(\alpha_j^+)\,P_{\mathrm{b}} \\
&\quad - \left(-\frac{\partial}{\partial \alpha_j} + \alpha_j^+\right)(\alpha_i)\left(-\frac{\partial}{\partial \alpha_k} + \alpha_k^+\right)(\alpha_l)\,P_{\mathrm{b}},
\end{aligned} \tag{8.55}
$$

so here only first- and second-order derivatives will be involved. There are initially no terms involving P_{b} without derivatives, since the terms in $2(\alpha_i)(\alpha_k^+)(\alpha_j^+)(\alpha_l)P_{\mathrm{b}} - (\alpha_l)(\alpha_k^+)(\alpha_i)(\alpha_j^+)P_{\mathrm{b}} - (\alpha_j^+)(\alpha_i)(\alpha_k^+)(\alpha_l)P_{\mathrm{b}}$ cancel out. If the canonical P representation and its correspondence rules (8.18) are utilised, we find that only derivatives up to second order occur in the Fokker–Planck equation. In this case the derivatives will be with respect to the new variables γ_i and γ_i^*. The derivation is left as an exercise.

A straightforward application of the results (8.54) and (8.55) together with the symmetry properties of the matrix elements (2.55) and applying the rules for c-number differentiation enables the Fokker–Planck equation for bosons to be derived from the master equation (2.83). The Fokker–Planck equation for the positive P distribution is

$$
\begin{aligned}
\frac{\partial}{\partial t} P_{\mathrm{b}}(\alpha, \alpha^+, \alpha^*, \alpha^{+*}) &= -\sum_i \frac{\partial}{\partial \alpha_i}(C_i^- P_{\mathrm{b}}) - \sum_i \frac{\partial}{\partial \alpha_i^+}(C_i^+ P_{\mathrm{b}}) \\
&\quad + \frac{1}{2}\sum_{ij} \frac{\partial}{\partial \alpha_i}\frac{\partial}{\partial \alpha_j}(F_{ij}^{--} P_{\mathrm{b}}) + \frac{1}{2}\sum_{ij} \frac{\partial}{\partial \alpha_i^+}\frac{\partial}{\partial \alpha_j^+}(F_{ij}^{++} P_{\mathrm{b}}) \\
&\quad + \frac{1}{2}\sum_{ij} \frac{\partial}{\partial \alpha_i}\frac{\partial}{\partial \alpha_j^+}(F_{ij}^{-+} P_{\mathrm{b}}) + \frac{1}{2}\sum_{ij} \frac{\partial}{\partial \alpha_i^+}\frac{\partial}{\partial \alpha_j}(F_{ij}^{+-} P_{\mathrm{b}})
\end{aligned} \tag{8.56}
$$

where the quantities C_i^-, C_i^+, F_{ij}^{--}, F_{ij}^{++}, F_{ij}^{-+} and F_{ij}^{+-} are c-number analytic functions of α_i, α_i^+, but not of $\alpha_i^*, \alpha_i^{+*}$. Because of the interpretation of the original Fokker–Planck equation describing Brownian motion, the C are referred to as *drift* terms, and the F are known as *diffusion* terms. For the moment we leave undecided whether the differentiations are with respect to α_{ix} or $i\alpha_{iy}$, or α_{ix}^+ or $i\alpha_{iy}^+$, or just left as derivatives with respect to α_i, or α_i^+. The distribution function is real and positive and, though non-analytic, can be considered as a function of α_{ix}, α_{iy}, α_{ix}^+ and α_{iy}^+ or of $\alpha_i, \alpha_i^+, \alpha_i^*, \alpha_i^{+*}$. We see that only derivatives up to second order are involved, a characteristic feature of positive P distribution functions. The quantities C_i^-, C_i^+, F_{ij}^{--}, F_{ij}^{++}, F_{ij}^{-+} and F_{ij}^{+-} depend on the parameters in the master equation and on $\hbar$ and are given by

$$\begin{aligned}
C_i^- &= \frac{1}{i\hbar}\sum_j k_{ij}\alpha_j + \frac{1}{i\hbar}\sum_{jkl}\xi_{ijkl}\alpha_j^+\alpha_l\alpha_k + \sum_{jkl}\left[\frac{(\Gamma_{jkil}-\Gamma_{likj})}{2}\alpha_l\alpha_k^+ - \frac{\Gamma_{likj}}{2}\delta_{lk}\right]\alpha_j, \\
C_i^+ &= \frac{-1}{i\hbar}\sum_j k_{ij}^*\alpha_j^+ - \frac{1}{i\hbar}\sum_{jkl}\xi_{ijkl}^*\alpha_k^+\alpha_l^+\alpha_j + \sum_{jkl}\alpha_j^+\left[\frac{(\Gamma_{jkil}^*-\Gamma_{likj}^*)}{2}\alpha_k\alpha_l^+ - \frac{\Gamma_{likj}^*}{2}\delta_{lk}\right], \\
F_{ij}^{--} &= \frac{1}{i\hbar}\sum_{kl}\xi_{ijkl}\alpha_l\alpha_k - \sum_{kl}\frac{\{\Gamma_{lijk}+\Gamma_{kjil}\}}{2}\alpha_l\alpha_k, \\
F_{ij}^{++} &= -\frac{1}{i\hbar}\sum_{kl}\xi_{ijkl}^*\alpha_k^+\alpha_l^+ - \sum_{kl}\frac{\{\Gamma_{lijk}^*+\Gamma_{kjil}^*\}}{2}\alpha_k^+\alpha_l^+, \\
F_{ij}^{-+} &= \sum_{kl}\{\Gamma_{iljk}\}\alpha_l^+\alpha_k, \qquad F_{ij}^{+-} = \sum_{kl}\{\Gamma_{iljk}^*\}\alpha_l\alpha_k^+.
\end{aligned} \tag{8.57}$$

To obtain the relaxation contributions to the diffusion terms, it is useful to consider terms with the same type of derivative ($\partial\alpha_i\ \partial\alpha_j$, $\partial\alpha_i^+\ \partial\alpha_j^+$, $\partial\alpha_i\ \partial\alpha_j^+$, $\partial\alpha_i^+\ \partial\alpha_j$) for each pair of modes i, j. It is seen that in the drift term, the relaxation rates Γ_{ijkl} appear as a difference. This feature is present in the case of a quantum harmonic oscillator coupled to a bath of two-level atoms (see [31, p. 188]).

From these forms, we can obtain the following symmetry relationships:

$$F_{ij}^{--} = F_{ji}^{--}, \qquad F_{ij}^{++} = F_{ji}^{++}, \qquad F_{ji}^{-+} = F_{ij}^{+-}, \tag{8.58}$$

so clearly the matrices F^{--} and F^{++} are symmetric, whilst the matrices F^{-+} and F^{+-} are related. Other interrelationships between the forms of the matrix elements are

$$\begin{aligned}
C_i^+ &= (C_i^-)^{(*)}, \\
F_{ij}^{++} &= (F_{ij}^{--})^{(*)}, \qquad F_{ij}^{-+} = (F_{ij}^{+-})^{(*)},
\end{aligned} \tag{8.59}$$

where $(*)$ denotes complex conjugation and replacement of α_i^* by α_i^+ and α_i^{+*} by α_i.

We now list the α_i, α_i^+ as $\{\alpha_p\} \equiv \{\alpha_1, \alpha_2, \cdots, \alpha_n, \alpha_1^+, \alpha_2^+, \cdots, \alpha_n^+\}$ and for the moment leave undecided whether the differentiations are with respect to α_{ix} or $i\alpha_{iy}$, or α_{ix}^+ or $i\alpha_{iy}^+$. The Fokker–Planck equation is still left in terms of the complex variables as

$$\frac{\partial}{\partial t}P_{\mathrm{b}}(\alpha, \alpha^+, \alpha^*, \alpha^{+*}) = -\sum_p \frac{\partial}{\partial\alpha_p}(A_p P_{\mathrm{b}}) + \frac{1}{2}\sum_{pq}\frac{\partial}{\partial\alpha_p}\frac{\partial}{\partial\alpha_q}(F_{pq}P_{\mathrm{b}}), \tag{8.60}$$

where the new drift and diffusion matrices A and F are made up from the previous quantities and become $2n \times 1$ and $2n \times 2n$ in size. They are given by

$$[A] = \begin{bmatrix} C^- \\ C^+ \end{bmatrix} \tag{8.61}$$

and

$$[F] = \begin{bmatrix} F^{--} & F^{-+} \\ F^{+-} & F^{++} \end{bmatrix}. \tag{8.62}$$

Hence, because of the construction, the new diffusion matrix F is symmetrical even though it is not real:

$$F_{pq} = F_{qp}. \tag{8.63}$$

Note that although for convenience we have used particular forms for C_i^-, C_i^+, F_{ij}^{--}, F_{ij}^{++}, F_{ij}^{-+} and F_{ij}^{+-} to demonstrate the last result, the latter can also be obtained from any form of the Fokker–Planck equation in which derivatives of only first and second order are involved.

8.8.2 Bosonic Wigner Distribution

For the case of the Wigner distribution function, the Fokker–Planck equation derived from the master equation (2.83) but with the relaxation terms ignored turns out to have no second-order derivatives, and only those of first and third order. It is given by

$$\begin{aligned}\frac{\partial}{\partial t}W_{\mathrm{b}}(\alpha,\alpha^+,\alpha^*,\alpha^{+*}) = &-\sum_i \frac{\partial}{\partial \alpha_i}(C_i^- W_{\mathrm{b}}) - \sum_i \frac{\partial}{\partial \alpha_i^+}(C_i^+ W_{\mathrm{b}}) \\ &+\sum_{ijl} \frac{\partial}{\partial \alpha_i}\frac{\partial}{\partial \alpha_j}\frac{\partial}{\partial \alpha_l^+}(G_{ijl}^{--+} W_{\mathrm{b}}) \\ &+\sum_{ijl} \frac{\partial}{\partial \alpha_i^+}\frac{\partial}{\partial \alpha_j^+}\frac{\partial}{\partial \alpha_l}(G_{ijl}^{++-} W_{\mathrm{b}}),\end{aligned} \tag{8.64}$$

where

$$\begin{aligned}C_i^- &= \frac{1}{i\hbar}\sum_j k_{ij}\alpha_j + \frac{1}{i\hbar}\sum_{jkl}\xi_{ijkl}((\alpha_j^+\alpha_l - \delta_{jl})\alpha_k + (\alpha_j^+\alpha_k - \delta_{jk})\alpha_l), \\ C_i^+ &= -\frac{1}{i\hbar}\sum_j k_{ij}^*\alpha_j^+ - \frac{1}{i\hbar}\sum_{jkl}\xi_{ijkl}^*((\alpha_j\alpha_l^+ - \delta_{jl})\alpha_k^+ + (\alpha_j\alpha_k^+ - \delta_{jk})\alpha_l^+), \\ G_{ijl}^{--+} &= \frac{1}{i\hbar}\sum_k \xi_{ijkl}\frac{1}{2}\alpha_k, \\ G_{ijl}^{++-} &= -\frac{1}{i\hbar}\sum_k \xi_{ijkl}^*\frac{1}{2}\alpha_k^+.\end{aligned} \tag{8.65}$$

The proof is left as an exercise. Wigner Fokker–Planck equations can only be converted into Ito stochastic equations if the third-order derivatives are ignored, such as when the mode occupancy is very large and the important regions of phase space are where α_i, α_i^+ are $O(N)$. In this situation there are no noise terms, as only drift terms appear in the Ito equations. The quantum noise is embodied in the initial distribution function.

The solution of these deterministic Ito equations is equivalent to solving the Fokker–Planck equation via the method of characteristics. Note the different drift term from that occurring for the positive P distribution.

8.8.3 Fokker–Planck Equation in Positive Definite Form

In order to establish the equivalence between the approaches based on the Fokker–Planck equation and the Ito stochastic differential equation for determining the behaviour of time-dependent quantum averages, it is often thought to be necessary to start from a Fokker–Planck equation free from ambiguity with regard to derivatives with respect to phase space variables. The Fokker–Planck equation is first changed to a form (8.78) involving the quantities α_{ix}, α_{iy}, α^+_{ix} and α^+_{iy}. However, as will be established below, the original Fokker–Planck equation (8.60) can be used directly to obtain a useful Ito stochastic differential equation from the original drift and diffusion matrices A and F without introducing the real and imaginary parts of α_i and α^+_i.

To write the Fokker–Planck equation in terms of the real variables α_{ix}, α_{iy}, α^+_{ix} and α^+_{iy}, we follow the approach of Drummond and Gardiner [31, 38]. The complex symmetric matrix F may be factorised in the form $BB^T = F$. This result is known as the *Takagi factorisation* [62]. The proof is given in [63, Section 4.4]. The construction involves the eigenvectors of the matrix $FF^* = FF^\dagger$, which is Hermitian because F is symmetric, and which also has non-negative real eigenvalues. This result is not well known and does not require F to be positive semi-definite, as is sometimes thought to be the case. We have

$$F = BB^T. \tag{8.66}$$

Note that in general B is a *complex* $2n \times 2n$ matrix. Note also that B is not unique, since with any orthogonal matrix R we also have $(BR)(BR)^T = F$.

We now divide the drift and diffusion terms into their real and imaginary parts, and do the same for the matrix B. Thus

$$\begin{aligned}
A_p &= A^p_x + iA^p_y, & p &= 1, \cdots, 2n, \\
B_{pq} &= B^{pq}_x + iB^{pq}_y, & p, q &= 1, \cdots, 2n, \\
F_{pq} &= F^{pq}_x + iF^{pq}_y, & p, q &= 1, \cdots, 2n,
\end{aligned} \tag{8.67}$$

where A^p_x, A^p_y, B^{pq}_x, B^{pq}_y, F^{pq}_x and F^{pq}_y are all real. It is easy to see that

$$\begin{aligned}
F^{pq}_x &= (B_x B^T_x)^{pq} - (B_y B^T_y)^{pq}, \\
F^{pq}_y &= (B_x B^T_y)^{pq} + (B_y B^T_x)^{pq}.
\end{aligned} \tag{8.68}$$

Note that the real and imaginary components F_x, F_y of the matrix F are themselves symmetrical as well as real.

The choice of which component is used for differentiation can now be made. We first make a new list containing the real and imaginary components of the α_i, α^+_i. Thus we write $\{\alpha_{p\mu}\} \equiv \{\alpha_{1x}, \cdots, \alpha_{nx}, \alpha^+_{1x}, \cdots, \alpha^+_{nx}, \alpha_{1y}, \cdots, \alpha_{ny}, \alpha^+_{1y}, \cdots, \alpha^+_{ny}\}$, so $\mu = x, y$ and $p = 1, \cdots, n, n+1, \cdots, 2n$ with $p = 1, \cdots, n$ listing the α and $p = n+1, \cdots, 2n$

listing the α^+. At this stage, we use the flexibility in the standard correspondence rules discussed in Section 8.2.2 on a term-by-term basis to convert the Fokker–Planck equation into a form where the diffusion matrix is positive definite. This step is not possible for the canonical distribution function.

For the drift terms,

$$\begin{aligned}\frac{\partial}{\partial\alpha_p}(A_p P_b) &= \frac{\partial}{\partial\alpha_p}(A_x^p P_b) + \frac{\partial}{\partial\alpha_p}(iA_y^p P_b)\\ &= \frac{\partial}{\partial\alpha_{px}}(A_x^p P_b) + \frac{\partial}{\partial\alpha_{py}}(A_y^p P_b)\\ &= \sum_\mu \frac{\partial}{\partial\alpha_{p\mu}}(A_\mu^p P_b),\end{aligned} \tag{8.69}$$

where we note that the quantities are all real. In the second line, the differentiation has been carried out with respect to α_{px} in the first term and with respect to $i\alpha_{py}$ in the second. The imaginary i thus cancels out.

For the diffusion terms, we have using (8.67) and (8.68)

$$\begin{aligned}\frac{\partial}{\partial\alpha_p}\frac{\partial}{\partial\alpha_q}(F_{pq}P_b) &= \frac{\partial}{\partial\alpha_p}\frac{\partial}{\partial\alpha_q}(F_x^{pq}P_b) + \frac{\partial}{\partial\alpha_p}\frac{\partial}{\partial\alpha_q}(iF_y^{pq}P_b)\\ &= \frac{\partial}{\partial\alpha_p}\frac{\partial}{\partial\alpha_q}\left((B_xB_x^T)^{pq} - (B_yB_y^T)^{pq}\right)P_b\\ &\quad + i\frac{\partial}{\partial\alpha_p}\frac{\partial}{\partial\alpha_q}\left((B_xB_y^T)^{pq} + (B_yB_x^T)^{pq}\right)P_b\\ &= \frac{\partial}{\partial\alpha_{px}}\frac{\partial}{\partial\alpha_{qx}}\left((B_xB_x^T)^{pq}\right)P_b - \frac{\partial}{\partial(i\alpha_{py})}\frac{\partial}{\partial(i\alpha_{qy})}\left((B_yB_y^T)^{pq}\right)P_b\\ &\quad + i\frac{\partial}{\partial\alpha_{px}}\frac{\partial}{\partial(i\alpha_{qy})}\left((B_xB_y^T)^{pq}\right)P_b + i\frac{\partial}{\partial(i\alpha_{py})}\frac{\partial}{\partial\alpha_{qx}}\left((B_yB_x^T)^{pq}\right)P_b\\ &= \frac{\partial}{\partial\alpha_{px}}\frac{\partial}{\partial\alpha_{qx}}\left((B_xB_x^T)^{pq}\right)P_b + \frac{\partial}{\partial\alpha_{py}}\frac{\partial}{\partial\alpha_{qy}}\left((B_yB_y^T)^{pq}\right)P_b\\ &\quad + \frac{\partial}{\partial\alpha_{px}}\frac{\partial}{\partial\alpha_{qy}}\left((B_xB_y^T)^{pq}\right)P_b + \frac{\partial}{\partial\alpha_{py}}\frac{\partial}{\partial\alpha_{qx}}\left((B_yB_x^T)^{pq}\right)P_b\\ &= \sum_{\mu\xi}\frac{\partial}{\partial\alpha_{p\mu}}\frac{\partial}{\partial\alpha_{q\xi}}\left((B_\mu B_\xi^T)^{pq}\right)P_b.\end{aligned} \tag{8.70}$$

In the third line, the differentiation has been carried out with respect to α_{px} and α_{qx} in the first term, with respect to $i\alpha_{py}$ and $i\alpha_{qy}$ in the second, with respect to α_{px} and $i\alpha_{qy}$ in the third, and with respect to $i\alpha_{py}$ and α_{qx} in the fourth.

If we now define the final diffusion matrix D so that

$$D_{\mu\xi}^{pq} = (B_\mu B_\xi^T)^{pq}, \qquad \mu,\xi = x,y, \tag{8.71}$$

then in the drift term

$$\frac{\partial}{\partial \alpha_p}(A_p P_{\rm b}) = \sum_{p=1}^{2n} \sum_{\mu=x,y} \frac{\partial}{\partial \alpha_{p\mu}}(A^p_\mu P_{\rm b}) \tag{8.72}$$

and in the diffusion term

$$\frac{1}{2}\sum_{pq} \frac{\partial}{\partial \alpha_p}\frac{\partial}{\partial \alpha_q}(F_{pq}P_{\rm b}) = \frac{1}{2}\sum_{p,q=1}^{2n} \sum_{\mu,\xi=x,y} \frac{\partial}{\partial \alpha_{p\mu}}\frac{\partial}{\partial \alpha_{q\xi}}(D^{pq}_{\mu\xi}P_{\rm b}). \tag{8.73}$$

We can write out the drift matrix A in terms of its two real $2n \times 1$ submatrices and the diffusion matrix D in terms of its four $2n \times 2n$ real submatrices as

$$[A] = \begin{bmatrix} A_x \\ A_y \end{bmatrix} \tag{8.74}$$

and

$$[D] = \begin{bmatrix} (B_x B_x^T) & (B_x B_y^T) \\ (B_y B_x^T) & (B_y B_y^T) \end{bmatrix} \tag{8.75}$$

We see from the symmetry of the matrices $(B_\mu B_\xi^T)$ that the matrix D is symmetric. It is also real. Thus

$$D^{pq}_{\mu\xi} = D^{qp}_{\xi\mu}, \tag{8.76}$$

$$D^{pq}_{\mu\xi} = (D^{pq}_{\mu\xi})^*, \tag{8.77}$$

and the Fokker–Planck equation is given by

$$\frac{\partial}{\partial t}P_{\rm b}(\alpha, \alpha^+, \alpha^*, \alpha^{+*}) = -\sum_{p=1}^{2n} \sum_{\mu=x,y} \frac{\partial}{\partial \alpha_{p\mu}}(A^p_\mu P_{\rm b}) + \frac{1}{2}\sum_{p,q=1}^{2n} \sum_{\mu,\xi=x,y} \frac{\partial}{\partial \alpha_{p\mu}}\frac{\partial}{\partial \alpha_{q\xi}}(D^{pq}_{\mu\xi}P_{\rm b}), \tag{8.78}$$

where the drift and diffusion matrices A and D are both real. The diffusion matrix is symmetric and is given by $D^{pq}_{\mu\xi} = (B_\mu B_\xi^T)^{pq}$, with $\mu, \xi = x, y$. In full, we have

$$D^{pq}_{\mu\xi} = \sum_{a=1}^{2n} B^{pa}_\mu B^{qa}_\xi. \tag{8.79}$$

Note that A is a $4n \times 1$ matrix (a vector) and D is a $4n \times 4n$ matrix. Furthermore, the matrix D is positive semi-definite, as for any vector with real components X_μ^p we have

$$\begin{aligned}
X^T D X &= \sum_{pq\mu\xi} X_\mu^p D_{\mu\xi}^{pq} X_\xi^q \\
&= \sum_{pq\mu\xi} \sum_a X_\mu^p B_\mu^{pa} B_\xi^{qa} X_\xi^q \\
&= \sum_a \left(\sum_{q\xi} B_\xi^{qa} X_\xi^q \right)^2 \geq 0.
\end{aligned} \tag{8.80}$$

In the Fokker–Planck equation (8.78), all quantities and derivatives are defined in terms of the real quantities α_{ix}, α_{iy}, α_{ix}^+ and α_{iy}^+. In this form, the Fokker–Planck equation enables Ito stochastic equations involving real quantities to be derived. The distribution function $P_\mathrm{b}(\alpha, \alpha^+, \alpha^*, \alpha^{+*}) = F_\mathrm{b}(\alpha_{ix}, \alpha_{iy}, \alpha_{ix}^+, \alpha_{iy}^+)$ is real. As will be seen in the next section, Ito stochastic differential equations for complex α_i, α_i^+ can also be obtained directly from the form (8.60) of the Fokker–Planck equation that involves the complex phase variables.

8.9 Fermion Fokker–Planck Equations

For the Liouville equation (2.82) associated with the Hamiltonian in (2.51), the terms in the Fokker–Planck equation for $P_\mathrm{f}(g, g^+)$ can be found from

$$\begin{aligned}
\hat\rho &\Rightarrow \hat\rho \hat c_i^\dagger \hat c_j, & P_\mathrm{f} &\Rightarrow P_\mathrm{f} g_i^+ \left(+ \frac{\overleftarrow{\partial}}{\partial g_j^+} - g_j \right), \\
\hat\rho &\Rightarrow \hat c_i^\dagger \hat c_j \hat\rho, & P_\mathrm{f} &\Rightarrow \left(+ \frac{\overrightarrow{\partial}}{\partial g_i} - g_i^+ \right) g_j P_\mathrm{f}, \\
\hat\rho &\Rightarrow \hat\rho \hat c_i^\dagger \hat c_j^\dagger \hat c_l \hat c_k, & P_\mathrm{f} &\Rightarrow P_\mathrm{f} g_i^+ g_j^+ \left(+ \frac{\overleftarrow{\partial}}{\partial g_l^+} - g_l \right) \left(+ \frac{\overleftarrow{\partial}}{\partial g_k^+} - g_k \right), \\
\hat\rho &\Rightarrow \hat c_i^\dagger \hat c_j^\dagger \hat c_l \hat c_k \hat\rho, & P_\mathrm{f} &\Rightarrow \left(+ \frac{\overrightarrow{\partial}}{\partial g_i} - g_i^+ \right) \left(+ \frac{\overrightarrow{\partial}}{\partial g_j} - g_j^+ \right) g_l g_k P_\mathrm{f},
\end{aligned} \tag{8.81}$$

so it is clear that only derivatives up to second order will be involved. Furthermore, there will be no terms in the Fokker–Planck equation just involving $P_\mathrm{f}(g, g^+)$ without derivatives, since when the commutators are considered pairs of terms such as $P_\mathrm{f} g_i^+ (-g_j)$ and $(-g_i^+) g_j P_\mathrm{f}$ or $P_\mathrm{f} g_i^+ g_j^+ (-g_l)(-g_k)$ and $(-g_i^+)(-g_j^+) g_l g_k P_\mathrm{f}$ cancel out.

For the master equation (2.83) for fermions, the terms in the Fokker–Planck equation arising from the relaxation term can be found from

$$
\begin{aligned}
\hat{\rho} &\Rightarrow (2\hat{c}_k^\dagger \hat{c}_l \hat{\rho} \hat{c}_j^\dagger \hat{c}_i - \hat{\rho} \hat{c}_j^\dagger \hat{c}_i \hat{c}_k^\dagger \hat{c}_l - \hat{c}_j^\dagger \hat{c}_i \hat{c}_k^\dagger \hat{c}_l \hat{\rho}),\\
P_{\mathrm{f}} &\Rightarrow 2\left(\left(\frac{\overrightarrow{\partial}}{\partial g_k} - g_k^+\right)(g_l)\, P_{\mathrm{f}}\,(g_j^+)\right)\left(\frac{\overleftarrow{\partial}}{\partial g_i^+} - g_i\right)\\
&\quad - \left(\left(P_{\mathrm{f}}\,(g_j^+)\left(\frac{\overleftarrow{\partial}}{\partial g_i^+} - g_i\right)\right)(g_k^+)\right)\left(\frac{\overleftarrow{\partial}}{\partial g_l^+} - g_l\right)\\
&\quad - \left(\frac{\overrightarrow{\partial}}{\partial g_j} - g_j^+\right)\left((g_i)\left(\left(\frac{\overrightarrow{\partial}}{\partial g_k} - g_k^+\right)(g_l)\, P_{\mathrm{f}}\right)\right),
\end{aligned}
\tag{8.82}
$$

so only first and second derivatives are involved. The brackets are needed to define the correct order for carrying out the processes in (8.82). As $P_{\mathrm{f}}(g, g^+)$ is an even Grassmann function, the terms without derivatives $2\,(-g_k^+)\,(g_l)\,P_{\mathrm{f}}\,(g_j^+)\,(-g_i) - P_{\mathrm{f}}\,(g_j^+)\,(-g_i)\,(g_k^+)\,(-g_l) - (-g_j^+)\,(g_i)\,(-g_k^+)\,(g_l)\,P_{\mathrm{f}}$ cancel out. From (4.26), the order in which the left and right differentiations are carried out is irrelevant.

A straightforward application of the results (8.81) and (8.82) together with the symmetry properties of the matrix elements (2.55) and (2.86) and applying the rules for Grassmann differentiation enables the Fokker–Planck equation for fermions to be derived from the master equation (2.83). The Fokker–Planck equation is of the form

$$
\begin{aligned}
\frac{\partial}{\partial t} P_{\mathrm{f}}(g, g^+) &= -\sum_i \frac{\overrightarrow{\partial}}{\partial g_i}(C_i^- P_{\mathrm{f}}) - \sum_i (P_{\mathrm{f}} C_i^+)\frac{\overleftarrow{\partial}}{\partial g_i^+}\\
&\quad + \frac{1}{2}\sum_{ij} \frac{\overrightarrow{\partial}}{\partial g_i}\frac{\overrightarrow{\partial}}{\partial g_j}(F_{ij}^{--} P_{\mathrm{f}}) + \frac{1}{2}\sum_{ij}(P_{\mathrm{f}} F_{ij}^{++})\frac{\overleftarrow{\partial}}{\partial g_j^+}\frac{\overleftarrow{\partial}}{\partial g_i^+}\\
&\quad + \frac{1}{2}\sum_{ij} \frac{\overrightarrow{\partial}}{\partial g_i}(F_{ij}^{-+} P_{\mathrm{f}})\frac{\overleftarrow{\partial}}{\partial g_j^+} + \frac{1}{2}\sum_{ij} \frac{\overrightarrow{\partial}}{\partial g_i^+}(F_{ij}^{+-} P_{\mathrm{f}})\frac{\overleftarrow{\partial}}{\partial g_j},
\end{aligned}
\tag{8.83}
$$

where the quantities C_i^-, C_i^+, F_{ij}^{--}, F_{ij}^{++}, F_{ij}^{-+} and F_{ij}^{+-} are Grassmann functions of g_i, g_i^+. They depend on the parameters in the master equation and on $\hbar$ and are

$$
\begin{aligned}
C_i^- &= -\frac{1}{i\hbar}\sum_j h_{ij} g_j + \frac{1}{i\hbar}\sum_{jkl} \nu_{ijkl} g_j^+ g_l g_k + \sum_{jkl}\left[\frac{(\Gamma_{jkil} - \Gamma_{likj})}{2} g_l g_k^+ + \frac{\Gamma_{likj}}{2}\delta_{lk}\right] g_j,\\
C_i^+ &= \frac{1}{i\hbar}\sum_j h_{ij}^* g_j^+ - \frac{1}{i\hbar}\sum_{jkl} \nu_{ijkl}^* g_k^+ g_l^+ g_j + \sum_{jkl} g_j^+\left[\frac{\left(\Gamma_{jkil}^* - \Gamma_{likj}^*\right)}{2} g_k g_l^+ + \frac{\Gamma_{likj}^*}{2}\delta_{lk}\right],\\
F_{ij}^{--} &= \frac{1}{i\hbar}\sum_{kl} \nu_{ijkl} g_l g_k + \sum_{kl}\frac{\{\Gamma_{lijk} + \Gamma_{kjil}\}}{2} g_l g_k,\\
F_{ij}^{++} &= -\frac{1}{i\hbar}\sum_{kl} \nu_{ijkl}^* g_k^+ g_l^+ + \sum_{kl}\frac{\{\Gamma_{lijk}^* + \Gamma_{kjil}^*\}}{2} g_k^+ g_l^+,\\
F_{ij}^{-+} &= \sum_{kl}\{\Gamma_{jkil}\} g_l g_k^+ \qquad F_{ij}^{+-} = \sum_{kl}\{\Gamma_{jkil}^*\} g_l^+ g_k.
\end{aligned}
\tag{8.84}
$$

To obtain the relaxation contributions to the diffusion terms, it is useful to decompose the terms arising from (8.82) for the derivatives $(\overrightarrow{\partial}g_i\overrightarrow{\partial}g_jF,\ F\overleftarrow{\partial}g_j^+\overleftarrow{\partial}g_i^+,\ \overrightarrow{\partial}g_iF\overleftarrow{\partial}g_j^+)$ into contributions with $i > j$, $i < j$ and $i = j$. The results (4.28), (4.25) and (4.27) for first and second derivatives are also required. The similarity of the results to those for the boson case in (8.57) is striking, though the forms are not identical. Note that for fermions only ν_{ijkl} with $i \neq j$ and $k \neq l$ are involved, as explained in Chapter 2.

From above, we see that C_i^- and C_i^+ are odd functions containing first- and third-order monomials, whereas F_{ij}^{--}, F_{ij}^{++}, F_{ij}^{-+} and F_{ij}^{+-} are even functions involving second-order monomials. Again, the C_i^- and C_i^+ are drift terms and the F_{ij}^{--}, F_{ij}^{++}, F_{ij}^{-+} and F_{ij}^{+-} are diffusion terms. The successive second-derivative terms could equally be written as $P_fF_{ij}^{--}$, $F_{ij}^{++}P_f$, $P_fF_{ij}^{-+}$ and $P_fF_{ij}^{+-}$. We see that only derivatives up to second order are involved, a characteristic feature of positive P distribution functions.

From these forms, we can obtain the following symmetry relationships,

$$F_{ij}^{--} = -F_{ji}^{--}, \qquad F_{ij}^{++} = -F_{ji}^{++}, \qquad F_{ij}^{+-} = -F_{ji}^{-+}, \tag{8.85}$$

so clearly the matrices F^{--} and F^{++} are antisymmetric, whilst the matrices F^{-+} and F^{+-} are related antisymmetrically. The matrices are also interrelated:

$$\begin{aligned} C_i^+ &= (C_i^-)^{(*)}, \\ F_{ij}^{++} &= (F_{ij}^{--})^{(*)}, \qquad F_{ij}^{+-} = -(F_{ij}^{-+})^{(*)}, \end{aligned} \tag{8.86}$$

where $(*)$ means taking the complex conjugate and replacing g_i^* by g_i^+ and g_i^{+*} by g_i.

The Fokker–Planck equation can be put into a more symmetrical form with all first-derivative terms involving both left and right derivatives and all second-derivative terms involving one left derivative and one right derivative. This form depends on $P_fF_{ij}^{--}$ and $F_{ij}^{++}P_f$ all being even Grassmann functions, with any first derivative being odd and $C_i^-P_f$ and $P_fC_i^+$ being odd functions. The derivation is left as an exercise.

It turns out that it is more useful to write the Fokker–Planck equation with right derivatives only. This form – given below as (8.91) – will be useful in considering Ito stochastic differential equations that are equivalent to the Fokker–Planck equation for determining quantum averages. The latter initially involve Grassmann phase space averages which are expressed as left Grassmann integrals, and it turns out that right Grassmann derivative forms of the Fokker–Planck equation are then more convenient. We use (4.21) and (4.22) to see that

$$\frac{\overrightarrow{\partial}}{\partial g_i}(C_i^-P_f) = (C_i^-P_f)\frac{\overleftarrow{\partial}}{\partial g_i}, \tag{8.87}$$

$$\frac{\overrightarrow{\partial}}{\partial g_i}\frac{\overrightarrow{\partial}}{\partial g_j}(F_{ij}^{--}P_f) = -\frac{\overrightarrow{\partial}}{\partial g_i}(F_{ij}^{--}P_f)\frac{\overleftarrow{\partial}}{\partial g_j} = -(F_{ij}^{--}P_f)\frac{\overleftarrow{\partial}}{\partial g_j}\frac{\overleftarrow{\partial}}{\partial g_i}, \tag{8.88}$$

$$\frac{\overrightarrow{\partial}}{\partial g_i}(F_{ij}^{-+}P_f)\frac{\overleftarrow{\partial}}{\partial g_j^+} = (F_{ij}^{-+}P_f)\frac{\overleftarrow{\partial}}{\partial g_j^+}\frac{\overleftarrow{\partial}}{\partial g_i}, \tag{8.89}$$

$$\frac{\overrightarrow{\partial}}{\partial g_i^+}(F_{ij}^{+-}P_f)\frac{\overleftarrow{\partial}}{\partial g_j} = (F_{ij}^{+-}P_f)\frac{\overleftarrow{\partial}}{\partial g_j}\frac{\overleftarrow{\partial}}{\partial g_i^+}, \tag{8.90}$$

and, using these results, we see that (8.83) becomes

$$\begin{aligned}\frac{\partial}{\partial t}P_{\rm f}(g,g^+) &= -\sum_i (C_i^- P_{\rm f})\frac{\overleftarrow{\partial}}{\partial g_i} - \sum_i (P_{\rm f}C_i^+)\frac{\overleftarrow{\partial}}{\partial g_i^+},\\ &+ \frac{1}{2}\sum_{ij}(-F_{ij}^{--}P_{\rm f})\frac{\overleftarrow{\partial}}{\partial g_j}\frac{\overleftarrow{\partial}}{\partial g_i} + \frac{1}{2}\sum_{ij}(P_{\rm f}F_{ij}^{++})\frac{\overleftarrow{\partial}}{\partial g_j^+}\frac{\overleftarrow{\partial}}{\partial g_i^+}\\ &+ \frac{1}{2}\sum_{ij}(F_{ij}^{-+}P_{\rm f})\frac{\overleftarrow{\partial}}{\partial g_j^+}\frac{\overleftarrow{\partial}}{\partial g_i} + \frac{1}{2}\sum_{ij}(F_{ij}^{+-}P_{\rm f})\frac{\overleftarrow{\partial}}{\partial g_j}\frac{\overleftarrow{\partial}}{\partial g_i^+}. \end{aligned} \tag{8.91}$$

If we now list the g_i, g_i^+ as $\{g_p\} \equiv \{g_1, g_2, \cdots, g_n, g_1^+, g_2^+, \cdots, g_n^+\}$ and we define the final drift and diffusion matrices A and D via

$$[A] = \begin{bmatrix} C^- \\ C^+ \end{bmatrix}, \tag{8.92}$$

$$[D] = \begin{bmatrix} -F^{--} & +F^{-+} \\ -(F^{-+})^T & +F^{++} \end{bmatrix}, \tag{8.93}$$

where T is the transpose, then the Fokker–Planck equation may be written in the right-derivative form

$$\frac{\partial}{\partial t}P_{\rm f}(g,g^+) = -\sum_{p=1}^{2n}(A_p P_{\rm f})\frac{\overleftarrow{\partial}}{\partial g_p} + \frac{1}{2}\sum_{p,q=1}^{2n}(D_{pq}P_{\rm f})\frac{\overleftarrow{\partial}}{\partial g_q}\frac{\overleftarrow{\partial}}{\partial g_p}. \tag{8.94}$$

We have also used the result $F^{+-} = -(F^{-+})^T$. Because the distribution function is even, $A_p P_{\rm f} = P_{\rm f} A_p$ and $D_{pq}P_{\rm f} = P_{\rm f} D_{pq}$. Note that because of the construction and (8.85), the diffusion matrix is antisymmetric:

$$D_{pq} = -D_{qp}, \tag{8.95}$$

a feature opposite to that for the bosonic case. Thus the fermion Fokker–Planck equation involves a drift $1 \times 2n$ matrix A whose elements are odd Grassmann functions of up to third order and an antisymmetric $2n \times 2n$ diffusion matrix D whose elements are even Grassmann functions of second order. Note that although for convenience we have used particular forms for C_i^-, C_i^+, F_{ij}^{--}, F_{ij}^{++}, F_{ij}^{-+} and F_{ij}^{+-} to demonstrate the last result, the latter can also be obtained from any form of the Fokker–Planck equation in which derivatives of only first and second order are involved.

In the fermion case, the right side of the Fokker–Planck equation involves products of the distribution function $P_{\rm f}(g,g^+)$ with functions of the Grassmann variables g_i, g_i^+ followed by Grassmann left or right differentiations. If the distribution function is then replaced by its expansion (7.27) and all the Grassmann products and differentiations are carried out, we must end up with a Grassmann function of the g_i, g_i^+ in which the various c-number coefficients ($P_0, P_2^{i;j}, \cdots, P_{2n}^{1\cdots n;n\cdots 1}$ etc.) in the expansion appear linearly. The left side of the Fokker–Planck equation will also be a Grassmann function, but one involving the time derivatives of the expansion coefficients. Equating the monomials on each side of the Fokker–Planck equation then leads to a coupled set of linear c-number differential equations for the coefficients, considered as functions of time. As we have seen, these coefficients are each equal to a quantum correlation function times ± 1, and thus we end up with a set of linear coupled differential equations for the time-dependent correlation functions. Solving these amounts to solving the fermion Fokker–Planck equation directly and gives the required correlation functions.

This hierarchy of equations could of course have been derived directly from the expression for the correlation functions by substituting for the time derivative of the density operator using either the Liouville–von Neumann equation or the master equation and simplifying the various products of fermion annihilation and creation operators using the anticommutation rules. The advantage of using a phase space distribution function involving Grassmann variables is that these steps are embodied in the correspondence rules and the application of Grassmann algebra and calculus.

However, if the Grassmann Fokker–Planck equation can be replaced by Ito stochastic equations and the quantum correlation functions obtained by stochastic averaging, then we can avoid having to solve the hierarchy of equations for the c-number coefficients $(P_0, P_2^{i;j}, \cdots$ etc.) in the expansion for the distribution function $P_{\mathrm{f}}(g, g^+)$.

8.10 Fokker–Planck Equations for Unnormalised Distribution Functions

In this section, the Fokker–Planck equations for unnormalised distribution functions are set out.

8.10.1 Boson Unnormalised Distribution Function

The Fokker–Planck equation for the unnormalised boson distribution function $B_{\mathrm{b}}(\alpha, \alpha^+, \alpha^*, \alpha^{+*})$ is obtained by applying the correspondence rules in (8.49):

$$\begin{aligned}\frac{\partial}{\partial t} B_{\mathrm{b}}(\alpha, \alpha^+, \alpha^*, \alpha^{+*}) = &-\sum_i \frac{\partial}{\partial \alpha_i}(\mathcal{C}_i^- B_{\mathrm{b}}) - \sum_i \frac{\partial}{\partial \alpha_i^+}(\mathcal{C}_i^+ B_{\mathrm{b}}) \\ &+ \frac{1}{2}\sum_{ij} \frac{\partial}{\partial \alpha_i}\frac{\partial}{\partial \alpha_j}(\mathcal{F}_{ij}^{--} B_{\mathrm{b}}) + \frac{1}{2}\sum_{ij} \frac{\partial}{\partial \alpha_i^+}\frac{\partial}{\partial \alpha_j^+}(\mathcal{F}_{ij}^{++} B_{\mathrm{b}}) \\ &+ \frac{1}{2}\sum_{ij} \frac{\partial}{\partial \alpha_i}\frac{\partial}{\partial \alpha_j^+}(\mathcal{F}_{ij}^{-+} B_{\mathrm{b}}) + \frac{1}{2}\sum_{ij} \frac{\partial}{\partial \alpha_i^+}\frac{\partial}{\partial \alpha_j}(\mathcal{F}_{ij}^{+-} B_{\mathrm{b}}),\end{aligned} \tag{8.96}$$

where the quantities $\mathcal{C}_i^-, \mathcal{C}_i^+, \mathcal{F}_{ij}^{--}, \mathcal{F}_{ij}^{++}, \mathcal{F}_{ij}^{-+}$ and $\mathcal{F}_{ij}^{+-}$ are c-number analytic functions of α_i, α_i^+ but not of α_i^*, α_i^{+*}, and are given by

$$
\begin{aligned}
\mathcal{C}_i^- &= \frac{1}{i\hbar}\sum_j k_{ij}\alpha_j - \sum_{jk}\left(\frac{\Gamma_{kikj}}{2}\right)\alpha_j,\\
\mathcal{C}_i^+ &= -\frac{1}{i\hbar}\sum_j k_{ij}^*\alpha_j^+ - \sum_{jk}\left(\frac{\Gamma_{kikj}^*}{2}\right)\alpha_j,\\
\mathcal{F}_{ij}^{--} &= \frac{1}{i\hbar}\sum_{kl}\xi_{ijkl}\alpha_l\alpha_k - \sum_{kl}\frac{\{\Gamma_{lijk}+\Gamma_{kjil}\}}{2}\alpha_l\alpha_k,\\
\mathcal{F}_{ij}^{++} &= -\frac{1}{i\hbar}\sum_{kl}\xi_{ijkl}^*\alpha_k^+\alpha_l^+ - \sum_{kl}\frac{\{\Gamma_{lijk}^*+\Gamma_{kjil}^*\}}{2}\alpha_k^+\alpha_l^+,\\
\mathcal{F}_{ij}^{-+} &= \sum_{kl}\{\Gamma_{iljk}\}\alpha_l^+\alpha_k, \qquad \mathcal{F}_{ij}^{+-} = \sum_{kl}\{\Gamma_{iljk}^*\}\alpha_l\alpha_k^+.
\end{aligned}
\tag{8.97}
$$

The coefficients in (8.97) are the same as those in (8.57) for the standard distribution function except for the absence of all the non-linear terms in the drift vector. The diffusion matrix is a quadratic function of the α_i, α_i^+. Symmetry results analogous to those in (8.58) still apply.

8.10.2 Fermion Unnormalised Distribution Function

The Fokker–Planck equation for the unnormalised fermion distribution function $B_{\mathrm{f}}(g, g^+)$ can also be obtained by applying the correspondence rules in (8.50). We find that

$$
\begin{aligned}
\frac{\partial}{\partial t}B_{\mathrm{f}}(g,g^+) &= -\sum_i \frac{\overrightarrow{\partial}}{\partial g_i}(\mathcal{C}_i^- B_{\mathrm{f}}) - \sum_i (B_{\mathrm{f}}\mathcal{C}_i^+)\frac{\overleftarrow{\partial}}{\partial g_i^+}\\
&\quad + \frac{1}{2}\sum_{ij}\frac{\overrightarrow{\partial}}{\partial g_i}\frac{\overrightarrow{\partial}}{\partial g_j}(\mathcal{F}_{ij}^{--}B_{\mathrm{f}}) + \frac{1}{2}\sum_{ij}(B_{\mathrm{f}}\mathcal{F}_{ij}^{++})\frac{\overleftarrow{\partial}}{\partial g_j^+}\frac{\overleftarrow{\partial}}{\partial g_i^+}\\
&\quad + \frac{1}{2}\sum_{ij}\frac{\overrightarrow{\partial}}{\partial g_i}(\mathcal{F}_{ij}^{-+}B_{\mathrm{f}})\frac{\overleftarrow{\partial}}{\partial g_j^+} + \frac{1}{2}\sum_{ij}\frac{\overrightarrow{\partial}}{\partial g_i^+}(\mathcal{F}_{ij}^{+-}B_{\mathrm{f}})\frac{\overleftarrow{\partial}}{\partial g_j},
\end{aligned}
\tag{8.98}
$$

where the quantities $\mathcal{C}_i^-, \mathcal{C}_i^+, \mathcal{F}_{ij}^{--}, \mathcal{F}_{ij}^{++}, \mathcal{F}_{ij}^{-+}$ and $\mathcal{F}_{ij}^{+-}$ are Grassmann functions of g_i, g_i^+ but not of g_i^*, g_i^{+*}, and are given by

$$
\begin{aligned}
\mathcal{C}_i^- &= -\frac{1}{i\hbar}\sum_j h_{ij}g_j + \sum_{jk}\left(\frac{\Gamma_{kikj}}{2}\right)g_j,\\
\mathcal{C}_i^+ &= +\frac{1}{i\hbar}\sum_j h_{ij}^*g_j^+ + \sum_{jk}g_j^+\left(\frac{\Gamma_{kikj}^*}{2}\right),\\
\mathcal{F}_{ij}^{--} &= \frac{1}{i\hbar}\sum_{kl}\nu_{ijkl}g_lg_k + \sum_{kl}\frac{\{\Gamma_{lijk}+\Gamma_{kjil}\}}{2}g_lg_k,\\
\mathcal{F}_{ij}^{++} &= -\frac{1}{i\hbar}\sum_{kl}\nu_{ijkl}^*g_k^+g_l^+ + \sum_{kl}\frac{\{\Gamma_{lijk}^*+\Gamma_{kjil}^*\}}{2}g_k^+g_l^+,\\
\mathcal{F}_{ij}^{-+} &= \sum_{kl}\{\Gamma_{jkil}\}g_lg_k^+, \qquad \mathcal{F}_{ij}^{+-} = \sum_{kl}\{\Gamma_{jkil}^*\}g_l^+g_k.
\end{aligned}
\tag{8.99}
$$

The coefficients in (8.99) are the same as those in (8.84) for the standard distribution function except for the absence of all the non-linear terms in the drift vector. The diffusion matrix is a quadratic function of the g_i, g_i^+. Antisymmetry results analogous to those in (8.85) still apply.

Exercises

(8.1) Justify the correspondence rules in (8.24)–(8.27) for the other ten cases referred to in Section 8.4.

(8.2) Derive the bosonic Fokker–Planck equation (8.56).

(8.3) Derive the fermion Fokker–Planck equation (8.83).

(8.4) Derive the generalised Fokker–Planck equation (8.52) using the correspondence rules (8.18).

(8.5) From the master equation (2.83), derive the bosonic Fokker–Planck equation (8.64) for the *Wigner* distribution function, ignoring the relaxation terms.

(8.6) From the master equation (8.51), derive the bosonic Fokker–Planck equation (8.52) for the *canonical positive P* distribution function using the correspondence rules (8.18).

(8.7) Determine whether the bosonic Fokker–Planck equation for the canonical positive P distribution function for the non-linear absorber has a positive definite diffusion matrix.

(8.8) Show using (4.21) and (4.22) that the fermionic Fokker–Planck equation (8.83) can be rewritten in the symmetrical form

$$\begin{aligned}\frac{\partial}{\partial t}P_{\mathrm{f}}(g,g^+) = &-\sum_i \frac{\overrightarrow{\partial}}{\partial g_i}\left(\frac{1}{2}C_i^- P_{\mathrm{f}}\right) - \sum_i \frac{\overrightarrow{\partial}}{\partial g_i^+}\left(P_{\mathrm{f}}\frac{1}{2}C_i^+\right)\\ &-\sum_i \left(\frac{1}{2}C_i^- P_{\mathrm{f}}\right)\frac{\overleftarrow{\partial}}{\partial g_i} - \sum_i \left(P_{\mathrm{f}}\frac{1}{2}C_i^+\right)\frac{\overleftarrow{\partial}}{\partial g_i^+}\\ &-\frac{1}{2}\sum_{ij}\frac{\overrightarrow{\partial}}{\partial g_i}(F_{ij}^{--}P_{\mathrm{f}})\frac{\overleftarrow{\partial}}{\partial g_j} + \frac{1}{2}\sum_{ij}\frac{\overrightarrow{\partial}}{\partial g_i^+}(P_{\mathrm{f}}F_{ij}^{++})\frac{\overleftarrow{\partial}}{\partial g_j^+}\\ &+\frac{1}{2}\sum_{ij}\frac{\overrightarrow{\partial}}{\partial g_i}(F_{ij}^{-+}P_{\mathrm{f}})\frac{\overleftarrow{\partial}}{\partial g_j^+} + \frac{1}{2}\sum_{ij}\frac{\overrightarrow{\partial}}{\partial g_i^+}(F_{ij}^{+-}P_{\mathrm{f}})\frac{\overleftarrow{\partial}}{\partial g_j}.\end{aligned} \tag{8.100}$$

9
Langevin Equations

In this chapter, we determine the Langevin equations that are equivalent to the Fokker–Planck equation for the distribution function. These will be in the form of Ito stochastic differential equations containing c-number Wiener increments. For boson systems, at most a total of $2n$ Wiener increments are required to determine the stochastic evolution of $2n$ c-number phase space variables. For fermions, at worst a total of $2n^2$ Wiener increments are needed to determine the stochastic evolution of $2n$ Grassmann phase space variables. Unfortunately, the fermion Ito equations are non-linear, making numerical calculations unfeasible for large mode numbers, though the Ito equations are still useful for analytical purposes such as deriving equations of motion for quantum correlation functions. The basic concepts related to stochastic averages are outlined in Appendix G. For the bosonic case, the derivation is based on that given by Gardiner and Zoller [31]. The final stochastic averages would then determine normally ordered quantum correlation functions using Ito equations based on the positive P Fokker–Planck equation. As mentioned previously, the Fokker–Planck equations for distribution functions associated with non-normally ordered quantum correlation functions are different from those for the positive P type considered here, the phase space averages involving products of phase space variables and the distribution function then determining the non-normally ordered quantum correlation functions. However, the relationship of the equivalent Langevin equation to the Fokker–Planck equation is the *same* as that presented here, depending only on the general form of the Fokker–Planck equation. The final stochastic averages would then determine non-normally ordered quantum correlation functions. Note also that the derivation of the Langevin equation does not depend on the distribution function having any particular properties, such as being real or positive. For the bosonic case, we must of course assume it to be non-analytic in general.

For Fock state populations and coherences, the unnormalised phase space distribution functions provide simpler expressions, so the Langevin equations for the stochastic phase space variables are also determined from the relevant Fokker–Planck equation. This case is of particular importance for fermion systems [14] as the Langevin equations are linear in the stochastic phase space variables, and this enables stochastic averages of phase space variables to be obtained from stochastic averages of *c-number* stochastic functions and stochastic averages of Grassmann phase space variables at an initial time. The latter quantities are determined from initial conditions for the Fock state populations and coherences. This could allow fermion cases to be treated numerically, since only c-number quantities need to be represented on the computer. Since

Phase Space Methods for Degenerate Quantum Gases. First Edition. Bryan J. Dalton, John Jeffers, and Stephen M. Barnett.
Published in 2015 by Oxford University Press.

at worst a total of $2n^2$ Wiener increments are involved in determining the stochastic evolution of $2n$ Grassmann phase space variables, numerical work would be feasible for moderate-size mode numbers. The fermion case is essentially of the same order of numerical complexity as the boson case.

9.1 Boson Ito Stochastic Equations

The phase space average of the function $F(\alpha,\alpha^+)$ is given by

$$\langle F(\alpha,\alpha^+)\rangle_t = \int\int d^2\alpha^+ d^2\alpha,\, F(\alpha,\alpha^+) P_{\mathrm{b}}(\alpha,\alpha^+,\alpha^*,\alpha^{+*},t). \tag{9.1}$$

In order to determine the normally ordered quantum correlation function $G_{\mathrm{b}}(l_1,\cdots,l_p;m_q,\cdots,m_1)$, the function $F(\alpha,\alpha^+)$ is given by

$$F(\alpha,\alpha^+) = (\alpha_{m_q}\cdots\alpha_{m_1})(\alpha^+_{l_1}\cdots\alpha^+_{l_p}), \tag{9.2}$$

where the distribution function has been placed on the right in (7.74) for convenience. Note that to be related to quantum averages of normally ordered operators the function $F(\alpha,\alpha^+)$ needs to be written in a form with the α_i to the left of the α^+_j, which we will refer to as antinormal order.

The change in the phase space average from time t to time $t+\delta t$ is given by

$$\langle F(\alpha,\alpha^+)\rangle_{t+\delta t} - \langle F(\alpha,\alpha^+)\rangle_t = \left(\int\int d^2\alpha^+ d^2\alpha\, F(\alpha,\alpha^+)\frac{\partial}{\partial t}P_{\mathrm{b}}\right)\delta t.$$

Substituting from the Fokker–Planck equation (8.78), we get the following, using integration by parts and assuming that the distribution function goes to zero fast enough on the phase space boundary:

$$\begin{aligned}
&\langle F(\alpha,\alpha^+)\rangle_{t+\delta t} - \langle F(\alpha,\alpha^+)\rangle_t \\
&\quad = \left(\int\int d^2\alpha^+ d^2\alpha\, F(\alpha,\alpha^+)\left\{-\sum_{p=1}^{2n}\sum_{\mu=x,y}\frac{\partial}{\partial\alpha_{p\mu}}(A^p_\mu P_{\mathrm{b}})\right.\right.\\
&\qquad \left.\left. + \frac{1}{2}\sum_{p,q=1}^{2n}\sum_{\mu,\xi=x,y}\frac{\partial}{\partial\alpha_{p\mu}}\frac{\partial}{\partial\alpha_{q\xi}}(D^{pq}_{\mu\xi}P_{\mathrm{b}})\right\}\right)\delta t\\
&\quad = \int\int d^2\alpha^+ d^2\alpha\left\{\sum_{p=1}^{2n}\sum_{\mu=x,y}\left[\frac{\partial}{\partial\alpha_{p\mu}}F(\alpha,\alpha^+)\right]A^p_\mu\right.\\
&\qquad \left. + \frac{1}{2}\sum_{p,q=1}^{2n}\sum_{\mu,\xi=x,y}\left[\frac{\partial}{\partial\alpha_{p\mu}}\frac{\partial}{\partial\alpha_{q\xi}}F(\alpha,\alpha^+)\right]D^{pq}_{\mu\xi}\right\}P_{\mathrm{b}}\,\delta t,
\end{aligned} \tag{9.3}$$

where the real and imaginary parts of the α_i, α_i^+ are listed as $\{\alpha_{p\mu}\} \equiv \{\alpha_{1x}, \cdots, \alpha_{nx}, \alpha_{1x}^+, \cdots, \alpha_{nx}^+, \alpha_{1y}, \cdots, \alpha_{ny}, \alpha_{1y}^+, \cdots, \alpha_{ny}^+\}$. The function $F(\alpha, \alpha^+) \equiv f(\alpha_{p\mu})$. Hence

$$\begin{aligned}
&\frac{d}{dt}\left\langle F(\alpha, \alpha^+)\right\rangle_t \\
&= \left\langle \left\{ \sum_{p=1}^{2n} \sum_{\mu=x,y} \left[\frac{\partial}{\partial \alpha_{p\mu}} F(\alpha, \alpha^+)\right] A_\mu^p + \frac{1}{2} \sum_{p,q=1}^{2n} \sum_{\mu,\xi=x,y} \left[\frac{\partial}{\partial \alpha_{p\mu}} \frac{\partial}{\partial \alpha_{q\xi}} F(\alpha, \alpha^+)\right] D_{\mu\xi}^{pq} \right\} \right\rangle,
\end{aligned} \tag{9.4}$$

giving the time derivative of the phase space average of $F(\alpha, \alpha^+)$ as the phase space average of the quantity in the brackets $\{\}$. Note that no specific properties of the distribution function were needed. This result will be used later to derive results for time derivatives of quantum correlation functions.

If we replace the variables $\alpha_{p\mu}$ by stochastic variables $\widetilde{\alpha}_{p\mu}$, then on the other hand the stochastic average at time t of $F(\alpha, \alpha^+)$ is given by

$$\overline{F(\widetilde{\alpha}(t), \widetilde{\alpha}^+(t))} = \frac{1}{m} \sum_{i=1}^{m} f(\widetilde{\alpha}_{p\mu i}(t)), \tag{9.5}$$

where $\widetilde{\alpha}_{p\mu i}(t)$ is the ith member of a stochastic ensemble of m samples. The key idea is that the phase space average at any time t of an arbitrary function $F(\alpha, \alpha^+)$ and the stochastic average of such a function are made to coincide when the stochastic equations for the $\widetilde{\alpha}_{p\mu}(t)$ are suitably related to the Fokker–Planck equation for the distribution function $P_b(\alpha, \alpha^+, \alpha^*, \alpha^{+*}, t)$. Thus

$$\begin{aligned}
\left\langle F(\alpha, \alpha^+)\right\rangle_t &= \int\!\!\int d^2\alpha^+ d^2\alpha \, F(\alpha, \alpha^+) P_b(\alpha, \alpha^+, \alpha^*, \alpha^{+*}, t) \\
&= \overline{F(\widetilde{\alpha}(t), \widetilde{\alpha}^+(t))} \\
&= \frac{1}{m} \sum_{i=1}^{m} f(\widetilde{\alpha}_{p\mu i}(t)).
\end{aligned} \tag{9.6}$$

In turn, the phase space averages are related to quantum averages. In particular, the normally ordered quantum correlation functions are given by the stochastic average of the product of stochastic c-number phase space variables:

$$\left\langle \hat{a}_{l_1}^\dagger \hat{a}_{l_2}^\dagger \cdots \hat{a}_{l_p}^\dagger \hat{a}_{m_q} \cdots \hat{a}_{m_2} \hat{a}_{m_1} \right\rangle_t = \overline{(\widetilde{\alpha}_{m_q}(t) \cdots \widetilde{\alpha}_{m_1}(t))(\widetilde{\alpha}_{l_1}^+(t) \cdots \widetilde{\alpha}_{l_p}^+(t))}. \tag{9.7}$$

For bosonic systems, such stochastic averages involving c-numbers can be carried out numerically, and this method is often more efficient than having to determine the full distribution function.

Now the difference in the stochastic average from time t to time $t+\delta t$ is given by

$$
\begin{aligned}
&\overline{F(\widetilde{\alpha}(t+\delta t),\widetilde{\alpha}^{+}(t+\delta t))} - \overline{F(\widetilde{\alpha}(t),\widetilde{\alpha}^{+}(t))} \\
&\quad = \overline{\{F(\widetilde{\alpha}(t+\delta t),\widetilde{\alpha}^{+}(t+\delta t)) - F(\widetilde{\alpha}(t),\widetilde{\alpha}^{+}(t))\}} \\
&\quad = \overline{\{f(\widetilde{\alpha}_{p\mu i}(t+\delta t)) - f(\widetilde{\alpha}_{p\mu i}(t))\}} \\
&\quad = \overline{\left\{\sum_{p\mu}\left[\frac{\partial}{\partial\alpha_{p\mu}}F(\alpha,\alpha^{+})\right]\delta\widetilde{\alpha}_{p\mu}(t) + \frac{1}{2}\sum_{p\mu}\sum_{q\xi}\left[\frac{\partial}{\partial\alpha_{p\mu}}\frac{\partial}{\partial\alpha_{q\xi}}F(\alpha,\alpha^{+})\right]\delta\widetilde{\alpha}_{p\mu}(t)\delta\widetilde{\alpha}_{q\xi}(t) + \cdots\right\}},
\end{aligned}
\tag{9.8}
$$

where we have used a Taylor expansion for $f(\widetilde{\alpha}_{p\mu}(t+\delta t))$ with the notation

$$
\delta\widetilde{\alpha}_{p\mu}(t) = \widetilde{\alpha}_{p\mu}(t+\delta t) - \widetilde{\alpha}_{p\mu}(t). \tag{9.9}
$$

In obtaining the last result, theorems in Appendix G relating the stochastic average of a sum to the sum of the stochastic averages have been used.

Now suppose $\widetilde{\alpha}_{p\mu}(t)$ satisfies an Ito stochastic equation of the form

$$
\widetilde{\alpha}_{p\mu}(t+\delta t) - \widetilde{\alpha}_{p\mu}(t) = \mathcal{A}^{p}_{\mu}(\widetilde{\alpha}_{q\xi}(t))\,\delta t + \sum_{a}\mathcal{B}^{p}_{\mu a}(\widetilde{\alpha}_{q\xi}(t))\int_{t}^{t+\delta t} dt_1\,\Gamma_a(t_1). \tag{9.10}
$$

The $\Gamma_a(t)$ are real Gaussian–Markov random noise terms ($a = 1, 2, \ldots$) whose stochastic averages are given by

$$
\begin{aligned}
\overline{\Gamma_a(t_1)} &= 0, \\
\overline{\Gamma_a(t_1)\Gamma_b(t_2)} &= \delta_{ab}\delta(t_1 - t_2), \\
\overline{\Gamma_a(t_1)\Gamma_b(t_2)\Gamma_c(t_3)} &= 0, \\
\overline{\Gamma_a(t_1)\Gamma_b(t_2)\Gamma_c(t_3)\Gamma_d(t_4)} &= \overline{\Gamma_a(t_1)\Gamma_b(t_2)}\,\overline{\Gamma_c(t_3)\Gamma_d(t_4)} + \overline{\Gamma_a(t_1)\Gamma_c(t_3)}\,\overline{\Gamma_b(t_2)\Gamma_d(t_4)} \\
&\quad + \overline{\Gamma_a(t_1)\Gamma_d(t_4)}\,\overline{\Gamma_b(t_2)\Gamma_c(t_3)}, \\
&\quad \cdots,
\end{aligned}
\tag{9.11}
$$

so that the stochastic averages of products of odd numbers of Γ vanishes and the stochastic averages of products of even numbers of Γ are the sums of products of stochastic averages of pairs of Γ. We choose the $\Gamma_a(t)$ to be *real*, as this will be convenient later for obtaining Ito stochastic equations for the complex $\widetilde{\alpha}_p$. For the moment we leave the number of Γ_a unspecified, but the number will turn out to be $2n$. It is also assumed that the $\widetilde{\alpha}_{p\mu}(t)$ and the $\Gamma_a(t)$ at later times are uncorrelated Thus

$$
\begin{aligned}
&\overline{G(\widetilde{\alpha}(t_1))\Gamma_a(t_2)\Gamma_b(t_3)\Gamma_c(t_4)\cdots\Gamma_k(t_l)} \\
&\quad = \overline{G(\widetilde{\alpha}(t_1))}\;\overline{\Gamma_a(t_2)\Gamma_b(t_3)\Gamma_c(t_4)\cdots\Gamma_k(t_l)}, \qquad t_1 < t_2, t_3, \cdots, t_l,
\end{aligned}
\tag{9.12}
$$

for any quantity G that depends only on the stochastic variables at time t_1.

With these results, we can now obtain expressions for the stochastic averages. For the first-order derivative terms, we have

$$
\begin{aligned}
&\overline{\left\{\sum_{p\mu}\left[\frac{\partial}{\partial\alpha_{p\mu}}F(\alpha,\alpha^{+})\right]\delta\widetilde{\alpha}_{p\mu}(t)\right\}} \\
&\quad=\overline{\left\{\sum_{p\mu}\left[\frac{\partial}{\partial\alpha_{p\mu}}F(\alpha,\alpha^{+})\right]\left[\mathcal{A}^{p}_{\mu}(\widetilde{\alpha}_{q\xi}(t))\,\delta t+\sum_{a}\mathcal{B}^{p}_{\mu a}(\widetilde{\alpha}_{q\xi}(t))\int_{t}^{t+\delta t}dt_{1}\,\Gamma_{a}(t_{1})\right]\right\}} \\
&\quad=\overline{\sum_{p\mu}\left[\frac{\partial}{\partial\alpha_{p\mu}}F(\alpha,\alpha^{+})\right]\mathcal{A}^{p}_{\mu}(\widetilde{\alpha}_{q\xi}(t))\,\delta t} \\
&\qquad+\overline{\sum_{p\mu}\left[\frac{\partial}{\partial\alpha_{p\mu}}F(\alpha,\alpha^{+})\right]\sum_{a}\mathcal{B}^{p}_{\mu a}(\widetilde{\alpha}_{q\xi}(t))}\;\overline{\int_{t}^{t+\delta t}dt_{1}\,\Gamma_{a}(t_{1})} \\
&\quad=\overline{\sum_{p\mu}\left[\frac{\partial}{\partial\alpha_{p\mu}}F(\alpha,\alpha^{+})\right]\mathcal{A}^{p}_{\mu}(\widetilde{\alpha}_{q\xi}(t))}\,\delta t,
\end{aligned}
\tag{9.13}
$$

where the stochastic-average rules for sums and products have been used, the non-correlation between the averages of functions of $\widetilde{\alpha}_{p\mu}(t)$ at time t and the Γ at later times between t and $t+\delta t$ has been applied and the term involving $\overline{\Gamma_{a}(t_{1})}$ is equal to zero from (9.11). Note that this term is proportional to δt.

For the second-order derivative terms,

$$
\begin{aligned}
&\overline{\left\{\frac{1}{2}\sum_{p\mu q\xi}\left[\frac{\partial}{\partial\alpha_{p\mu}}\frac{\partial}{\partial\alpha_{q\xi}}F(\alpha,\alpha^{+})\right]\delta\widetilde{\alpha}_{p\mu}(t)\,\delta\widetilde{\alpha}_{q\xi}(t)\right\}} \\
&\quad=\overline{\left\{\begin{array}{l}\frac{1}{2}\sum_{p\mu q\xi}\left[\frac{\partial}{\partial\alpha_{p\mu}}\frac{\partial}{\partial\alpha_{q\xi}}F(\alpha,\alpha^{+})\right]\left[\mathcal{A}^{p}_{\mu}(\widetilde{\alpha}_{r\lambda}(t))\,\delta t+\sum_{a}\mathcal{B}^{p}_{\mu a}(\widetilde{\alpha}_{r\lambda}(t))\int_{t}^{t+\delta t}dt_{1}\,\Gamma_{a}(t_{1})\right] \\ \qquad\times\left[\mathcal{A}^{q}_{\xi}(\widetilde{\alpha}_{r\lambda}(t))\,\delta t+\sum_{\mathrm{b}}\mathcal{B}^{q}_{\xi\mathrm{b}}(\widetilde{\alpha}_{r\lambda}(t))\int_{t}^{t+\delta t}dt_{2}\,\Gamma_{\mathrm{b}}(t_{2})\right]\end{array}\right\}} \\
&\quad=\overline{\frac{1}{2}\sum_{p\mu q\xi}\left[\frac{\partial}{\partial\alpha_{p\mu}}\frac{\partial}{\partial\alpha_{q\xi}}F(\alpha,\alpha^{+})\right]\left[\mathcal{A}^{p}_{\mu}(\widetilde{\alpha}_{r\lambda}(t))\delta t\mathcal{A}^{q}_{\xi}(\widetilde{\alpha}_{r\lambda}(t))\,\delta t\right]} \\
&\qquad+\overline{\frac{1}{2}\sum_{p\mu q\xi}\left[\frac{\partial}{\partial\alpha_{p\mu}}\frac{\partial}{\partial\alpha_{q\xi}}F(\alpha,\alpha^{+})\right]\left[\mathcal{A}^{p}_{\mu}(\widetilde{\alpha}_{r\lambda}(t))\,\delta t\sum_{\mathrm{b}}\mathcal{B}^{q}_{\xi\mathrm{b}}(\widetilde{\alpha}_{r\lambda}(t))\int_{t}^{t+\delta t}dt_{2}\,\Gamma_{\mathrm{b}}(t_{2})\right]} \\
&\qquad+\overline{\frac{1}{2}\sum_{p\mu q\xi}\left[\frac{\partial}{\partial\alpha_{p\mu}}\frac{\partial}{\partial\alpha_{q\xi}}F(\alpha,\alpha^{+})\right]\left[\sum_{a}\mathcal{B}^{p}_{\mu a}(\widetilde{\alpha}_{r\lambda}(t))\int_{t}^{t+\delta t}dt_{1}\,\Gamma_{a}(t_{1})\,\mathcal{A}^{q}_{\xi}(\widetilde{\alpha}_{r\lambda}(t))\,\delta t\right]} \\
&\qquad+\overline{\begin{array}{c}\frac{1}{2}\sum_{p\mu q\xi}\left[\frac{\partial}{\partial\alpha_{p\mu}}\frac{\partial}{\partial\alpha_{q\xi}}F(\alpha,\alpha^{+})\right] \\ \times\left[\sum_{a}\mathcal{B}^{p}_{\mu a}(\widetilde{\alpha}_{r\lambda}(t))\int_{t}^{t+\delta t}dt_{1}\,\Gamma_{a}(t_{1})\,\sum_{\mathrm{b}}\mathcal{B}^{q}_{\xi\mathrm{b}}(\widetilde{\alpha}_{r\lambda}(t))\int_{t}^{t+\delta t}dt_{2}\,\Gamma_{\mathrm{b}}(t_{2})\right]\end{array}}\,.
\end{aligned}
\tag{9.14}
$$

Using the result that the stochastic averages of functions of the $\alpha^s_{r\lambda}(t)$ and the Γ_a are uncorrelated, we find that

$$\begin{aligned}
&\overline{\left\{\frac{1}{2}\sum_{p\mu q\xi}\left[\frac{\partial}{\partial\alpha_{p\mu}}\frac{\partial}{\partial\alpha_{q\xi}}F(\alpha,\alpha^+)\right]\delta\widetilde{\alpha}_{p\mu}(t)\,\delta\widetilde{\alpha}_{q\xi}(t)\right\}}\\
&=\overline{\frac{1}{2}\sum_{p\mu q\xi}\left[\frac{\partial}{\partial\alpha_{p\mu}}\frac{\partial}{\partial\alpha_{q\xi}}F(\alpha,\alpha^+)\right]\left[\mathcal{A}^p_\mu(\widetilde{\alpha}_{r\lambda}(t))\mathcal{A}^q_\xi(\widetilde{\alpha}_{r\lambda}(t))\right]\delta t^2}\\
&+\overline{\frac{1}{2}\sum_{p\mu q\xi}\left[\frac{\partial}{\partial\alpha_{p\mu}}\frac{\partial}{\partial\alpha_{q\xi}}F(\alpha,\alpha^+)\right]\left[\mathcal{A}^p_\mu(\widetilde{\alpha}_{r\lambda}(t))\sum_{\mathrm{b}}\mathcal{B}^q_{\xi b}(\widetilde{\alpha}_{r\lambda}(t))\right]}\;\overline{\int_t^{t+\delta t}dt_2\,\Gamma_{\mathrm{b}}(t_2)\,\delta t}\\
&+\overline{\frac{1}{2}\sum_{p\mu q\xi}\left[\frac{\partial}{\partial\alpha_{p\mu}}\frac{\partial}{\partial\alpha_{q\xi}}F(\alpha,\alpha^+)\right]\left[\sum_a\mathcal{B}^p_{\mu a}(\widetilde{\alpha}_{r\lambda}(t))\,\mathcal{A}^q_\xi(\widetilde{\alpha}_{r\lambda}(t))\right]}\;\overline{\int_t^{t+\delta t}dt_1\,\Gamma_a(t_1)\,\delta t}\\
&+\overline{\frac{1}{2}\sum_{p\mu q\xi}\left[\frac{\partial}{\partial\alpha_{p\mu}}\frac{\partial}{\partial\alpha_{q\xi}}F(\alpha,\alpha^+)\right]\left[\sum_a\mathcal{B}^p_{\mu a}(\widetilde{\alpha}_{r\lambda}(t))\,\sum_{\mathrm{b}}\mathcal{B}^q_{\xi b}(\widetilde{\alpha}_{r\lambda}(t))\right]}\\
&\times\overline{\int_t^{t+\delta t}dt_1\,\Gamma_a(t_1)\int_t^{t+\delta t}dt_2\,\Gamma_{\mathrm{b}}(t_2)}.
\end{aligned}\tag{9.15}$$

Now the terms involving a single Γ have a zero stochastic average, whilst the terms with two Γ give a stochastic average proportional to δt:

$$\begin{aligned}
\overline{\int_t^{t+\delta t}dt_1\,\Gamma_a(t_1)\int_t^{t+\delta t}dt_2\,\Gamma_{\mathrm{b}}(t_2)}&=\int_t^{t+\delta t}dt_1\int_t^{t+\delta t}dt_2\,\overline{\Gamma_a(t_1)\Gamma_{\mathrm{b}}(t_2)}\\
&=\int_t^{t+\delta t}dt_1\int_t^{t+\delta t}dt_2\,\delta_{ab}\delta(t_1-t_2)\\
&=\delta_{ab}\,\delta t,
\end{aligned}\tag{9.16}$$

so that correct to order δt the second-order derivative term is

$$\begin{aligned}
&\overline{\left\{\frac{1}{2}\sum_{p\mu q\xi}\left[\frac{\partial}{\partial\alpha_{p\mu}}\frac{\partial}{\partial\alpha_{q\xi}}F(\alpha,\alpha^+)\right]\delta\widetilde{\alpha}_{p\mu}(t)\,\delta\widetilde{\alpha}_{q\xi}(t)\right\}}\\
&=\overline{\frac{1}{2}\sum_{p\mu q\xi}\left[\frac{\partial}{\partial\alpha_{p\mu}}\frac{\partial}{\partial\alpha_{q\xi}}F(\alpha,\alpha^+)\right]\left[\sum_a\mathcal{B}^p_{\mu a}(\widetilde{\alpha}_{r\lambda}(t))\mathcal{B}^q_{\xi a}(\widetilde{\alpha}_{r\lambda}(t))\right]}\,\delta t\\
&=\overline{\frac{1}{2}\sum_{p\mu q\xi}\left[\frac{\partial}{\partial\alpha_{p\mu}}\frac{\partial}{\partial\alpha_{q\xi}}F(\alpha,\alpha^+)\right]\left[[\mathcal{B}(\widetilde{\alpha}_{r\lambda}(t))\mathcal{B}^T(\widetilde{\alpha}_{r\lambda}(t))]^{p,q}_{\mu,\xi}\right]}\,\delta t.
\end{aligned}\tag{9.17}$$

The remaining terms give stochastic averages correct to order δt^2 or higher, so that we have the following, correct to first order in δt:

$$\begin{aligned}
&\overline{F(\widetilde{\alpha}(t+\delta t),\widetilde{\alpha}^+(t+\delta t))} - \overline{F(\widetilde{\alpha}(t),\widetilde{\alpha}^+(t))} \\
&\quad = \overline{\sum_{p\mu}\left[\frac{\partial}{\partial\alpha_{p\mu}}F(\alpha,\alpha^+)\right]\mathcal{A}^p_\mu(\widetilde{\alpha}_{q\xi}(t))\,\delta t} \\
&\quad + \overline{\frac{1}{2}\sum_{p\mu q\xi}\left[\frac{\partial}{\partial\alpha_{p\mu}}\frac{\partial}{\partial\alpha_{q\xi}}F(\alpha,\alpha^+)\right]\left[[\mathcal{B}(\widetilde{\alpha}_{r\lambda}(t))\,\mathcal{B}^T(\widetilde{\alpha}_{r\lambda}(t))]^{p,q}_{\mu,\xi}\right]\delta t}
\end{aligned} \tag{9.18}$$

or

$$\begin{aligned}
&\frac{d}{dt}\overline{F(\widetilde{\alpha}(t),\widetilde{\alpha}^+(t))} \\
&\quad = \overline{\sum_{p\mu}\left[\frac{\partial}{\partial\alpha_{p\mu}}F(\alpha,\alpha^+)\right]\mathcal{A}^p_\mu(\widetilde{\alpha}_{q\xi}(t))}. \\
&\quad + \overline{\frac{1}{2}\sum_{p\mu q\xi}\left[\frac{\partial}{\partial\alpha_{p\mu}}\frac{\partial}{\partial\alpha_{q\xi}}F(\alpha,\alpha^+)\right]\left[[\mathcal{B}(\widetilde{\alpha}_{r\lambda}(t))\mathcal{B}^T(\widetilde{\alpha}_{r\lambda}(t))]^{p,q}_{\mu,\xi}\right]}.
\end{aligned} \tag{9.19}$$

9.1.1 Relationship between Fokker–Planck and Ito Equations

The above result will be the same as that based on the phase space average if we have the following relationships between the matrices A and D in the Fokker–Planck equation and the matrices $\mathcal{A}$ and $\mathcal{B}$ occurring in the Ito stochastic differential equation:

$$\begin{aligned}
\mathcal{A}^p_\mu(\alpha_{q\xi}) &= A^p_\mu(\alpha_{q\xi}), \\
[\mathcal{B}(\alpha_{r\lambda})\mathcal{B}^T(\alpha_{r\lambda})]^{p,q}_{\mu,\xi} &= D^{pq}_{\mu\xi}(\alpha_{r\lambda}).
\end{aligned} \tag{9.20}$$

But, from (8.71), the form of D is such that $D^{pq}_{\mu\xi} = (B_\mu B^T_\xi)^{pq}$. Hence we can clearly satisfy the required relationship between D in the Fokker–Planck equation and $\mathcal{B}$ in the Ito stochastic differential equation by the choice

$$\mathcal{B}^p_{\mu a}(\alpha_{r\lambda}) = B^{pa}_\mu(\alpha_{r\lambda}), \qquad a = 1,\cdots,2n. \tag{9.21}$$

Note that this relationship requires that the number of real Gaussian–Markov noise terms $\Gamma_a(t_+)$ is $2n$, the number of α_i and α_i^+.

The Ito stochastic differential equation can then be written in the form

$$\frac{d}{dt}\widetilde{\alpha}_{p\mu}(t) = A^p_\mu(\widetilde{\alpha}_{r\lambda}(t)) + \sum_{a=1}^{2n} B^{pa}_\mu(\widetilde{\alpha}_{r\lambda}(t))\,\Gamma_a(t_+) \tag{9.22}$$

or in matrix form

$$\frac{d}{dt}\begin{bmatrix}\widetilde{\alpha}_x(t)\\ \widetilde{\alpha}_y(t)\end{bmatrix} = \begin{bmatrix}A_x(\widetilde{\alpha}_{r\lambda}(t))\\ A_y(\widetilde{\alpha}_{r\lambda}(t))\end{bmatrix} + \begin{bmatrix}B_x(\widetilde{\alpha}_{r\lambda}(t)) & 0\\ B_y(\widetilde{\alpha}_{r\lambda}(t)) & 0\end{bmatrix}\begin{bmatrix}\Gamma(t_+)\\ \Gamma(t_+)\end{bmatrix}, \tag{9.23}$$

where A_μ, B_μ and Γ are $2n \times 1$, $2n \times 2n$ and $2n \times 1$ real matrices whose elements are A^p_μ, B^{pa}_μ and Γ_q. $BB^T = D$. These quantities appeared in the Fokker–Planck equation. Note that the number of real Gaussian–Markov stochastic variables $\Gamma^a_\xi(t_+)$ is still $2n$, which is the number of complex phase space variables $\{\alpha_1, \alpha_2, \cdots, \alpha_n, \alpha_1^+, \alpha_2^+, \cdots, \alpha_n^+\}$.

9.1.2 Boson Stochastic Differential Equation in Complex Form

Having established the Ito stochastic differential equation for the $\widetilde{\alpha}_{p\mu}(t)$, we can now form the Ito stochastic differential equation for the original complex forms $\widetilde{\alpha}_p(t) = \widetilde{\alpha}_{px}(t) + i\widetilde{\alpha}_{py}(t)$, where $\{\alpha_p\} \equiv \{\alpha_1, \cdots, \alpha_n, \alpha_1^+, \cdots \alpha_n^+\}$ as before. In terms of the matrix elements given in (8.67),

$$\frac{d}{dt}\widetilde{\alpha}_p(t) = A_p(\widetilde{\alpha}_q(t)) + \sum_a B_{pa}(\widetilde{\alpha}_q(t))\, \Gamma_q(t_+), \tag{9.24}$$

where

$$\begin{aligned}
\widetilde{\alpha}_p(t) &= \widetilde{\alpha}_{px}(t) + i\widetilde{\alpha}_{py}(t), \\
A_p(\widetilde{\alpha}_q(t)) &= A^p_x(\widetilde{\alpha}_{q\xi}(t)) + iA^p_y(\widetilde{\alpha}_{q\xi}(t)), \\
B_{pa}(\widetilde{\alpha}_q(t)) &= B^{pa}_x(\widetilde{\alpha}_{q\xi}(t)) + iB^{pa}_y(\widetilde{\alpha}_{q\xi}(t)),
\end{aligned} \tag{9.25}$$

as before in (8.67). Note that A, B are $2n \times 1$ and $2n \times 2n$ matrices, respectively. The matrix A is the drift vector and $F = BB^T$ is the diffusion matrix in the *original* Fokker–Planck equation (8.60) written in terms of the complex α_p, where $\{\alpha_p\} \equiv \{\alpha_1, \cdots, \alpha_n, \alpha_1^+, \cdots, \alpha_n^+\}$. The importance of these last results is that we have now established how to write down the Ito stochastic equation for complex stochastic α_p directly from the original complex Fokker–Planck equation (8.60) by obtaining the drift and diffusion quantities A, B. The complex Ito stochastic equation (9.24) will be used as the basis for later developments in relation to Ito stochastic field equations.

9.1.3 Summary of Boson Stochastic Equations

For convenience, the related Ito and Fokker–Planck equations are repeated here. We have reverted to the notation in (9.10):

$$\begin{aligned}
\frac{\partial}{\partial t}P_\mathrm{b}(\alpha, \alpha^+, \alpha^*, \alpha^{+*}) &= -\sum_{p=1}^{2n} \frac{\partial}{\partial \alpha_p}(A_p P_\mathrm{b}) + \frac{1}{2}\sum_{p,q=1}^{2n} \frac{\partial}{\partial \alpha_p}\frac{\partial}{\partial \alpha_q}(F_{pq}P_\mathrm{b}), \\
\frac{d}{dt}\widetilde{\alpha}_p(t) &= \mathcal{A}^p(\widetilde{\alpha}_q(t)) + \sum_a \mathcal{B}^p_a(\widetilde{\alpha}_q(t))\, \Gamma_a(t_+), \\
\mathcal{B}\mathcal{B}^T &= F, \qquad \mathcal{A}^p = A_p,
\end{aligned} \tag{9.26}$$

where the distribution function P_b depends on complex phase variables α_p. There are $a = 1, \cdots, 2n$ Gaussian–Markov noise terms. In terms of real and imaginary components, the related Ito and Fokker–Planck equations are

$$\frac{\partial}{\partial t}P_{\mathrm{b}}(\alpha,\alpha^{+},\alpha^{*},\alpha^{+*}) = -\sum_{p=1}^{2n}\sum_{\mu=x,y}\frac{\partial}{\partial\alpha_{p\mu}}(A_{\mu}^{p}P_{\mathrm{b}}) + \frac{1}{2}\sum_{p,q=1}^{2n}\sum_{\mu,\xi=x,y}\frac{\partial}{\partial\alpha_{p\mu}}\frac{\partial}{\partial\alpha_{q\xi}}(D_{\mu\xi}^{pq}P_{\mathrm{b}}),$$

$$\frac{d}{dt}\widetilde{\alpha}_{p\mu}(t) = A_{\mu}^{p}(\widetilde{\alpha}_{q\xi}(t)) + \sum_{q=1}^{2n}B_{\mu}^{pq}(\widetilde{\alpha}_{q\xi}(t))\,\Gamma_{q}(t_{+}),$$

$$D_{\mu\xi}^{pq} = \sum_{a=1}^{2n}B_{\mu}^{pa}B_{\xi}^{qa}, \tag{9.27}$$

where the distribution function is written in the form $F_{\mathrm{b}}(\alpha_{ix},\alpha_{iy},\alpha_{ix}^{+},\alpha_{iy}^{+})$ and is real. Note, however, that the $B_{\mu}^{pq}(\alpha_{q\xi}(t))$ are real and that D is positive semi-definite.

These bosonic Ito stochastic equations are in general non-linear but as only c-numbers are involved, the solutions could be found numerically via standard computational methods. Unlike the fermionic case, there would seem to be no particular advantage in removing the non-linearity by applying Fokker–Planck and Langevin equations based on unnormalised distribution functions.

9.2 Wiener Stochastic Functions

Finally, we note that the Γ functions are related to *Wiener stochastic functions* defined by (see Appendix G)

$$\widetilde{w}_{a}(t) = \int_{t_0}^{t}dt_1\,\Gamma_{a}(t_1). \tag{9.28}$$

Hence the Wiener stochastic functions satisfy the Ito stochastic differential equation

$$\delta\widetilde{w}_{a}(t) = \widetilde{w}_{a}(t+\delta t) - \widetilde{w}_{a}(t) = \int_{t}^{t+\delta t}dt_1\,\Gamma_{a}(t_1). \tag{9.29}$$

This is a particular case of the integral form of the stochastic differential equation in (9.26), but with $\mathcal{A}^{p}=0$ and $\mathcal{B}_{a}^{p}=\delta_{ap}$. The quantity $\delta\widetilde{w}_{a}(t)$ is known as the *Wiener increment.*

If we now consider the w_a as phase space variables with a quasi-distribution function $P(w,t)$ with $w\equiv\{w_1,w_2,\cdots,w_a,\cdots w_{2n}\}$, it then follows from the Fokker–Planck equation in (9.26) with $A_a=0$, $F_{ab}=\sum_p\mathcal{B}_{a}^{p}\mathcal{B}_{a}^{p}=\delta_{ab}$ that the Fokker–Planck equation must be

$$\frac{\partial}{\partial t}P(w,t) = \frac{1}{2}\sum_{a}\frac{\partial^2}{\partial\alpha_a^2}(P(w,t)). \tag{9.30}$$

This is a Fokker–Planck equation with no drift terms and equal diffusion terms for all components. In its simplest form, it was used by Einstein in his first description of Brownian motion.

9.3 Fermion Ito Stochastic Equations

The phase space average of the function $F(g, g^+)$ is given by

$$\langle F(g, g^+) \rangle_t = \int\int dg^+ dg\, F(g, g^+) P_{\mathrm{f}}(g, g^+, t). \qquad (9.31)$$

In order to determine the normally ordered quantum correlation function $G_{\mathrm{f}}(l_1, \cdots l_p; m_q, \cdots, m_1)$ (for $p - q$ equal to an even integer), the function is given by

$$F(g, g^+) = (g_{m_q} \cdots g_{m_1})(g^+_{l_1} \cdots g^+_{l_p}), \qquad (9.32)$$

where the even distribution function has been placed on the right in (7.82) for convenience. As the Grassmann integral would vanish anyway if $F(g, g^+)$ was an odd function, we will assume the $F(g, g^+)$ is even (as in determining non-zero quantum correlation functions). Note that to be related to quantum averages of normally ordered operators the function $F(g, g^+)$ needs to be in a form where the g_i are to the left of the g^+_j, which we will refer to as antinormal order.

The change in the phase space average from time t to time $t + \delta t$ is given by

$$\langle F(g, g^+) \rangle_{t+\delta t} - \langle F(g, g^+) \rangle_t = \left(\int\int dg^+ dg\, F(g, g^+) \frac{\partial}{\partial t} P_{\mathrm{f}}(g, g^+, t) \right) \delta t. \qquad (9.33)$$

Substituting from the Fokker–Planck equation (8.94), we get the following, using integration by parts (F.9), noting the oddness of $A_p P_{\mathrm{f}}$ and $(D_{pq} P_{\mathrm{f}}) \overleftarrow{\partial} / \partial g_q$ and the evenness of $D_{pq} P_{\mathrm{f}}$:

$$\begin{aligned}
&\langle F(g, g^+) \rangle_{t+\delta t} - \langle F(g, g^+) \rangle_t \\
&= \left(\int\int dg^+ dg\, F(g, g^+) \left\{ -\sum_{p=1}^{2n} (A_p P_{\mathrm{f}}) \frac{\overleftarrow{\partial}}{\partial g_p} + \frac{1}{2} \sum_{p,q=1}^{2n} (D_{pq} P_{\mathrm{f}}) \frac{\overleftarrow{\partial}}{\partial g_q} \frac{\overleftarrow{\partial}}{\partial g_p} \right\} \right) \delta t \\
&= \left(\int\int dg^+ dg \left\{ -\sum_{p=1}^{2n} \left[F(g, g^+) \frac{\overleftarrow{\partial}}{\partial g_p} \right] (A_p) \right.\right. \\
&\qquad \left.\left. - \frac{1}{2} \sum_{p,q=1}^{2n} \left[F(g, g^+) \frac{\overleftarrow{\partial}}{\partial g_p} \frac{\overleftarrow{\partial}}{\partial g_q} \right] (D_{pq}) \right\} P_{\mathrm{f}}(g, g^+) \right) \delta t, \qquad (9.34)
\end{aligned}$$

where $\{g_p\} \equiv \{g_1, \cdots, g_n, g^+_1, \cdots, g^+_n\}$. Hence, using $D_{pq} = -D_{qp}$,

$$\frac{d}{dt} \langle F(g, g^+) \rangle_t = \left\langle \left\{ -\sum_{p=1}^{2n} \left[F \frac{\overleftarrow{\partial}}{\partial g_p} \right] (A_p) + \frac{1}{2} \sum_{p,q=1}^{2n} \left[F \frac{\overleftarrow{\partial}}{\partial g_p} \frac{\overleftarrow{\partial}}{\partial g_q} \right] (D_{qp}) \right\} \right\rangle, \qquad (9.35)$$

giving the time derivative of the Grassmann phase space average of $F(g, g^+)$ as the phase space average of the quantity in the brackets $\{\}$. Note that no specific properties of the distribution function were needed. This result will be used later to derive results for time derivatives of quantum correlation functions.

If we replace the variables g_p by stochastic variables $\widetilde{g}_p$, then on the other hand the stochastic average at time t of $F(g, g^+)$ is given by

$$\overline{F(\widetilde{g}(t), \widetilde{g}^+(t))} = \frac{1}{m}\sum_{i=1}^{m} f(\widetilde{g}_{pi}(t)), \tag{9.36}$$

where $\widetilde{g}_{pi}(t)$ is the ith member of a stochastic ensemble and $F(\widetilde{g}_i(t), \widetilde{g}_i^+(t)) = f(\widetilde{g}_{pi}(t))$ for short. As in the bosonic case, the key idea is that the phase space average at any time t of an arbitrary function $F(g, g^+)$ and the stochastic average of such a function are made to coincide when the stochastic equations for the $\widetilde{g}_p(t)$ are suitably related to the Fokker–Planck equation for the distribution function $P_{\mathrm{f}}(g, g^+, t)$. Thus

$$\begin{aligned}\langle F(g, g^+)\rangle_t &= \int\int dg^+ dg\, F(g, g^+) P_{\mathrm{f}}(g, g^+, t) \\ &= \overline{F(\widetilde{g}(t), \widetilde{g}^+(t))} \\ &= \frac{1}{m}\sum_{i=1}^{m} f(\widetilde{g}_{pi}(t)).\end{aligned} \tag{9.37}$$

In turn, the phase space averages are related to quantum averages. In particular, the normally ordered quantum correlation functions are given by the stochastic average of the product of stochastic Grassmann phase space variables:

$$\left\langle \hat{c}_{l_1}^\dagger \cdots \hat{c}_{l_p}^\dagger \hat{c}_{m_q} \cdots \hat{c}_{m_1} \right\rangle_t = \overline{(\widetilde{g}_{m_q}(t) \cdots \widetilde{g}_{m_1}(t))(\widetilde{g}_{l_1}^+(t) \cdots \widetilde{g}_{l_p}^+(t))}. \tag{9.38}$$

A key issue is whether such stochastic averages involving Grassmann numbers can be carried out numerically.

Now the difference in the stochastic average from time t to time $t + \delta t$ is given by

$$\begin{aligned}&\overline{F(\widetilde{g}(t+\delta t), \widetilde{g}^+(t+\delta t))} - \overline{F(\widetilde{g}(t), \widetilde{g}^+(t))} \\ &\quad = \overline{\{F(\widetilde{g}(t+\delta t), \widetilde{g}^+(t+\delta t)) - F(\widetilde{g}(t), \widetilde{g}^+(t))\}} \\ &\quad = \overline{\{f(\widetilde{g}_{pi}(t+\delta t)) - f(\widetilde{g}_{pi}(t))\}} \\ &\quad = \overline{\left\{\sum_p \left[F(g, g^+)\frac{\overleftarrow{\partial}}{\partial g_p}\right] \delta\widetilde{g}_p(t) + \frac{1}{2}\sum_p\sum_q \left[F(g, g^+)\frac{\overleftarrow{\partial}}{\partial g_p}\frac{\overleftarrow{\partial}}{\partial g_q}\right] \delta\widetilde{g}_q(t)\,\delta\widetilde{g}_p(t) + \cdots\right\}},\end{aligned} \tag{9.39}$$

where we have used the Taylor expansion for $f(\widetilde{g}_p(t + \delta t))$ given in (4.38) with the notation

$$\delta\widetilde{g}_p(t) = \widetilde{g}_p(t + \delta t) - \widetilde{g}_p(t). \tag{9.40}$$

Note here that the $\delta\widetilde{g}_p(t)$ can be any odd Grassmann functions and the Taylor expansion will still apply. If the $\delta\widetilde{g}_p(t)$ are odd Grassmann functions then $\widetilde{g}_p(t + \delta t)$ will behave as a Grassmann variable, as would be required. In obtaining the last result, theorems in Appendix G relating the stochastic average of a sum to the sum of the stochastic averages have also been used. The quantities in the brackets $[\cdots]$ are functions of the $\widetilde{g}_p(t)$ and do not depend on the $\delta\widetilde{g}_p(t)$.

Now suppose $\widetilde{g}_p(t)$ satisfies an Ito stochastic equation of the form

$$\widetilde{g}_p(t+\delta t) - \widetilde{g}_p(t) = \mathcal{A}^p(\widetilde{g}_q(t))\,\delta t + \sum_a \mathcal{B}^p_a(\widetilde{g}_q(t)) \int_t^{t+\delta t} dt_1\,\Gamma_a(t_1). \tag{9.41}$$

The $\Gamma_a(t)$ are c-number Gaussian–Markov random noise terms $(a = 1, 2, \ldots, i(n))$ whose stochastic averages are given by

$$\begin{aligned}
\overline{\Gamma_a(t_1)} &= 0, \\
\overline{\Gamma_a(t_1)\Gamma_b(t_2)} &= \delta_{ab}\delta(t_1 - t_2), \\
\overline{\Gamma_a(t_1)\Gamma_b(t_2)\Gamma_c(t_3)} &= 0, \\
\overline{\Gamma_a(t_1)\Gamma_b(t_2)\Gamma_c(t_3)\Gamma_d(t_4)} &= \overline{\Gamma_a(t_1)\Gamma_b(t_2)}\,\overline{\Gamma_c(t_3)\Gamma_d(t_4)} + \overline{\Gamma_a(t_1)\Gamma_c(t_3)}\,\overline{\Gamma_b(t_2)\Gamma_d(t_4)} \\
&\quad + \overline{\Gamma_a(t_1)\Gamma_d(t_4)}\,\overline{\Gamma_b(t_2)\Gamma_c(t_3)}, \\
&\quad \ldots,
\end{aligned} \tag{9.42}$$

so that the stochastic averages of products of odd numbers of Γ are zero and the stochastic averages of products of even numbers of Γ are the sums of products of stochastic averages of pairs of Γ. For the moment we leave unspecified how many Γ_a there are, but it will turn out that there are $2n^2$. It is also assumed that the $\widetilde{g}_p(t)$ and the $\Gamma_a(t)$ at later times are uncorrelated Thus

$$\begin{aligned}
&\overline{G(\widetilde{g}(t_1))\Gamma_a(t_2)\Gamma_b(t_3)\Gamma_c(t_4)\cdots\Gamma_k(t_l)} \\
&\quad = \overline{G(\widetilde{g}(t_1))}\;\overline{\Gamma_a(t_2)\Gamma_b(t_3)\Gamma_c(t_4)\cdots\Gamma_k(t_l)}, \qquad t_1 < t_2, t_3, \cdots, t_l,
\end{aligned} \tag{9.43}$$

for any Grassmann function $G(\widetilde{g}(t_1))$ that depends only on the stochastic variables at time t_1. These features are the same as for bosons. However, in the fermion case $\mathcal{A}^p(\widetilde{g}_q(t))$ and $\mathcal{B}^p_a(\widetilde{g}_q(t))$ are not c-numbers, but instead are odd Grassmann functions. Overall, then, the $\delta\widetilde{g}_p(t)$ will be odd Grassmann functions, as required. At this stage the required number of Gaussian–Markov random noise terms $\Gamma_a(t)$ is unspecified, but this will be some integer function $i(n)$ of the number of modes.

With these results, we can now obtain expressions for the stochastic averages. For the first-order derivative terms, we have

$$\begin{aligned}
&\overline{\left\{\sum_p \left[F(g, g^+)\frac{\overleftarrow{\partial}}{\partial g_p}\right] \delta\widetilde{g}_p(t)\right\}} \\
&= \overline{\left\{\sum_p \left[F\frac{\overleftarrow{\partial}}{\partial g_p}\right]\left[\mathcal{A}^p(\widetilde{g}_q(t))\,\delta t + \sum_a \mathcal{B}^p_a(\widetilde{g}_q(t)) \int_t^{t+\delta t} dt_1\,\Gamma_a(t_1)\right]\right\}} \\
&= \overline{\sum_p \left[F\frac{\overleftarrow{\partial}}{\partial g_p}\right] \mathcal{A}^p(\widetilde{g}_q(t))\,\delta t} + \overline{\sum_P \left[F\frac{\overleftarrow{\partial}}{\partial g_p}\right] \sum_a \mathcal{B}^p_a(\widetilde{g}_q(t))}\;\overline{\int_t^{t+\delta t} dt_1\,\Gamma_a(t_1)} \\
&= \overline{\sum_p \left[F\frac{\overleftarrow{\partial}}{\partial g_p}\right] \mathcal{A}^p(\widetilde{g}_q(t))\,\delta t},
\end{aligned} \tag{9.44}$$

where the stochastic-average rules for sums and products have been used, the non-correlation between the averages of functions of $\widetilde{g}_p(t)$ at time t and the Γ at later times between t and $t+\delta t$ has been applied and the term involving $\overline{\Gamma_a(t_1)}$ is equal to zero from (9.11). Note that this term is proportional to δt.

For the second-order derivative terms,

$$
\begin{aligned}
&\overline{\left\{\frac{1}{2}\sum_{p,q}\left[F(g,g^+)\frac{\overleftarrow{\partial}}{\partial g_p}\frac{\overleftarrow{\partial}}{\partial g_q}\right]\delta\widetilde{g}_q(t)\,\delta\widetilde{g}_p(t)\right\}} \\
&= \overline{\left\{\begin{array}{l} \frac{1}{2}\sum_{p,q}\left[F\frac{\overleftarrow{\partial}}{\partial g_p}\frac{\overleftarrow{\partial}}{\partial g_q}\right]\left[\mathcal{A}^q(\widetilde{g}_r(t))\,\delta t+\sum_a \mathcal{B}_a^q(\widetilde{g}_r(t))\int_t^{t+\delta t}dt_1\,\Gamma_a(t_1)\right] \\ \qquad\times\left[\mathcal{A}^p(\widetilde{g}_r(t))\,\delta t+\sum_{\mathrm{b}}\mathcal{B}_{\mathrm{b}}^p(\widetilde{g}_r(t))\int_t^{t+\delta t}dt_2\,\Gamma_{\mathrm{b}}(t_2)\right]\end{array}\right\}} \\
&= \overline{\frac{1}{2}\sum_{p,q}\left[F\frac{\overleftarrow{\partial}}{\partial g_p}\frac{\overleftarrow{\partial}}{\partial g_q}\right][\mathcal{A}^q(\widetilde{g}_r(t))\,\delta t\,\mathcal{A}^p(\widetilde{g}_r(t))\,\delta t]} \\
&\quad+\overline{\frac{1}{2}\sum_{p,q}\left[F\frac{\overleftarrow{\partial}}{\partial g_p}\frac{\overleftarrow{\partial}}{\partial g_q}\right]\left[\mathcal{A}^q(\widetilde{g}_r(t))\,\delta t\sum_{\mathrm{b}}\mathcal{B}_{\mathrm{b}}^p(\widetilde{g}_r(t))\int_t^{t+\delta t}dt_2\,\Gamma_{\mathrm{b}}(t_2)\right]} \\
&\quad+\overline{\frac{1}{2}\sum_{p,q}\left[F(g,g^+)\frac{\overleftarrow{\partial}}{\partial g_p}\frac{\overleftarrow{\partial}}{\partial g_q}\right]\left[\sum_a\mathcal{B}_a^q(\widetilde{g}_r(t))\int_t^{t+\delta t}dt_1\,\Gamma_a(t_1)\,\mathcal{A}^p(\widetilde{g}_r(t))\,\delta t\right]} \\
&\quad+\overline{\frac{1}{2}\sum_{p,q}\left[F\frac{\overleftarrow{\partial}}{\partial g_p}\frac{\overleftarrow{\partial}}{\partial g_q}\right]\left[\sum_a\mathcal{B}_a^q(\widetilde{g}_r(t))\int_t^{t+\delta t}dt_1\,\Gamma_a(t_1)\sum_{\mathrm{b}}\mathcal{B}_{\mathrm{b}}^p(\widetilde{g}_r(t))\int_t^{t+\delta t}dt_2\,\Gamma_{\mathrm{b}}(t_2)\right]}.
\end{aligned}
\tag{9.45}
$$

Using the result that the stochastic averages of the $\widetilde{g}_p$ and the Γ_a are uncorrelated, we find that

$$
\begin{aligned}
&\overline{\left\{\frac{1}{2}\sum_{p,q}\left[F(g,g^+)\frac{\overleftarrow{\partial}}{\partial g_p}\frac{\overleftarrow{\partial}}{\partial g_q}\right]\delta\widetilde{g}_q(t)\,\delta\widetilde{g}_p(t)\right\}} \\
&= \overline{\frac{1}{2}\sum_{p,q}\left[F\frac{\overleftarrow{\partial}}{\partial g_p}\frac{\overleftarrow{\partial}}{\partial g_q}\right][\mathcal{A}^q(\widetilde{g}_r(t))\mathcal{A}^p(\widetilde{g}_r(t))]}\,\delta t^2 \\
&\quad+\overline{\frac{1}{2}\sum_{p,q}\left[F\frac{\overleftarrow{\partial}}{\partial g_p}\frac{\overleftarrow{\partial}}{\partial g_q}\right]\left[\mathcal{A}^q(\widetilde{g}_r(t))\,\delta t\sum_{\mathrm{b}}\mathcal{B}_{\mathrm{b}}^p(\widetilde{g}_r(t))\right]}\;\overline{\int_t^{t+\delta t}dt_2\,\Gamma_{\mathrm{b}}(t_2)} \\
&\quad+\overline{\frac{1}{2}\sum_{p,q}\left[F\frac{\overleftarrow{\partial}}{\partial g_p}\frac{\overleftarrow{\partial}}{\partial g_q}\right]\left[\sum_a\mathcal{B}_a^q(\widetilde{g}_r(t))\mathcal{A}^p(\widetilde{g}_r(t))\,\delta t\right]}\;\overline{\int_t^{t+\delta t}dt_1\,\Gamma_a(t_1)} \\
&\quad+\overline{\frac{1}{2}\sum_{p,q}\left[F\frac{\overleftarrow{\partial}}{\partial g_p}\frac{\overleftarrow{\partial}}{\partial g_q}\right]\left[\sum_a\mathcal{B}_a^q(\widetilde{g}_r(t))\sum_{\mathrm{b}}\mathcal{B}_{\mathrm{b}}^p(\widetilde{g}_r(t))\right]}\;\overline{\int_t^{t+\delta t}dt_1\,\Gamma_a(t_1)\int_t^{t+\delta t}dt_2\,\Gamma_{\mathrm{b}}(t_2)}.
\end{aligned}
\tag{9.46}
$$

Now the terms involving a single Γ have a zero stochastic average, whilst the terms with two Γ give a stochastic average proportional to δt:

$$\begin{aligned}\overline{\int_t^{t+\delta t} dt_1\,\Gamma_a(t_1)\int_t^{t+\delta t} dt_2\,\Gamma_b(t_2)} &= \int_t^{t+\delta t} dt_1 \int_t^{t+\delta t} dt_2\,\overline{\Gamma_a(t_1)\Gamma_b(t_2)} \\ &= \int_t^{t+\delta t} dt_1 \int_t^{t+\delta t} dt_2\,\delta_{ab}\delta(t_1-t_2) \\ &= \delta_{ab}\,\delta t, \end{aligned} \tag{9.47}$$

so that correct to order δt the second-order derivative term is

$$\begin{aligned}&\overline{\left\{\frac{1}{2}\sum_{p,q}\left[F(g,g^+)\frac{\overleftarrow{\partial}}{\partial g_p}\frac{\overleftarrow{\partial}}{\partial g_q}\right]\delta\widetilde{g}_q(t)\,\delta\widetilde{g}_p(t)\right\}} \\ &= \frac{1}{2}\sum_{p,q}\overline{\left[F(g,g^+)\frac{\overleftarrow{\partial}}{\partial g_p}\frac{\overleftarrow{\partial}}{\partial g_q}\right]}\,\overline{\left[\sum_a \mathcal{B}_a^q(g_r^s(t))\mathcal{B}_a^p(g_r^s(t))\right]}\delta t \\ &= \frac{1}{2}\sum_{p,q}\overline{\left[F(g,g^+)\frac{\overleftarrow{\partial}}{\partial g_p}\frac{\overleftarrow{\partial}}{\partial g_q}\right][[\mathcal{B}(g_r^s(t))\mathcal{B}^T(g_r^s(t))]_{qp}]}\,\delta t. \end{aligned} \tag{9.48}$$

The remaining terms give stochastic averages correct to order δt^2 or higher, so that we have the following, correct to first order in δt:

$$\begin{aligned}&\overline{F(\widetilde{g}(t+\delta t),\widetilde{g}^+(t+\delta t))} - \overline{F(\widetilde{g}(t),\widetilde{g}^+(t))} \\ &= \overline{\sum_p\left[F(g,g^+)\frac{\overleftarrow{\partial}}{\partial g_p}\right]\mathcal{A}^p(\widetilde{g}_q(t))}\,\delta t \\ &\quad + \overline{\frac{1}{2}\sum_p\sum_q\left[F(g,g^+)\frac{\overleftarrow{\partial}}{\partial g_p}\frac{\overleftarrow{\partial}}{\partial g_q}\right][\mathcal{B}(\widetilde{g}_r(t))\mathcal{B}^T(\widetilde{g}_r(t))]_{qp}}\,\delta t \end{aligned} \tag{9.49}$$

or

$$\begin{aligned}&\frac{d}{dt}\overline{F(\widetilde{g}(t),\widetilde{g}^+(t))} \\ &= \overline{\sum_p\left[F(g,g^+)\frac{\overleftarrow{\partial}}{\partial g_p}\right]\mathcal{A}^p(\widetilde{g}_q(t))} \\ &\quad + \overline{\frac{1}{2}\sum_p\sum_q\left[F(g,g^+)\frac{\overleftarrow{\partial}}{\partial g_p}\frac{\overleftarrow{\partial}}{\partial g_q}\right][\mathcal{B}(\widetilde{g}_r(t))\mathcal{B}^T(\widetilde{g}_r(t))]_{qp}}. \end{aligned} \tag{9.50}$$

9.3.1 Relationship between Fokker–Planck and Ito Equations

The above result will be the same as that based on the phase space average if we have the following relationships between the matrices A and D in the Fokker–Planck equation and the matrices $\mathcal{A}$ and $\mathcal{B}$ in the Ito stochastic differential equation:

$$\mathcal{A}^p(g_q) = -A^p(g_q),$$
$$[\mathcal{B}(g_r)\mathcal{B}^T(g_r)]_{qp} = D_{qp}(g_r). \tag{9.51}$$

Note the sign difference for the drift term compared with the boson case. This sign difference ultimately arises in the first term of (9.34) from the differentiation rule (4.28) for the product of an odd function with an arbitrary function, which unlike the c-number case has a minus sign in one of the terms. The question then arises – can a matrix of odd Grassmann functions $\mathcal{B}$ be found such that $\mathcal{B}\mathcal{B}^T = D$, where D involves even Grassmann functions and is an antisymmetric matrix? The answer is yes, we can construct such a matrix [64].

9.3.2 Existence of Coupling Matrix for Fermions

Firstly, we note that since the $\mathcal{B}_a^p$ are odd Grassmann functions,

$$[\mathcal{B}\mathcal{B}^T]_{qp} = \sum_a \mathcal{B}_a^q \mathcal{B}_a^p = -\sum_a \mathcal{B}_a^p \mathcal{B}_a^q = -[\mathcal{B}\mathcal{B}^T]_{pq}, \tag{9.52}$$

showing that $\mathcal{B}\mathcal{B}^T$ is antisymmetric, as required.

Secondly, we now show a construction by which a suitable matrix $\mathcal{B}$ can be found. We first note from (8.93), (8.84) and (2.86) that the submatrices in the diffusion matrix D are as follows:

$$\begin{aligned}
-F_{ij}^{--} &= \frac{1}{i\hbar}\sum_{kl} \nu_{ijkl} g_k g_l + \sum_{kl} \frac{\{\Gamma_{lijk} + \Gamma_{kjil}\}}{2} g_k g_l = \sum_{kl} M_{ikjl}^{--} g_k g_l, \\
F_{ij}^{++} &= -\frac{1}{i\hbar}\sum_{kl} \nu_{ijkl}^* g_k^+ g_l^+ + \sum_{kl} \frac{\{\Gamma_{lijk}^* + \Gamma_{kjil}^*\}}{2} g_k^+ g_l^+ = \sum_{kl} M_{ikjl}^{++} g_k^+ g_l^+, \\
F_{ij}^{-+} &= \sum_{kl} \{\Gamma_{ikjl}^*\} g_k g_l^+ = \sum_{kl} M_{ikjl}^{-+} g_k g_l^+, \\
F_{ij}^{+-} &= \sum_{kl} \{\Gamma_{ikjl}\} g_k^+ g_l = \sum_{kl} M_{ikjl}^{+-} g_k^+ g_l,
\end{aligned} \tag{9.53}$$

where

$$\begin{aligned}
M_{ikjl}^{--} &= \frac{1}{i\hbar}\nu_{ijkl} + \frac{\{\Gamma_{lijk} + \Gamma_{kjil}\}}{2}, \\
M_{ikjl}^{++} &= -\frac{1}{i\hbar}\nu_{ijkl}^* + \frac{\{\Gamma_{lijk}^* + \Gamma_{kjil}^*\}}{2} = (M_{ikjl}^{--})^*, \\
M_{ikjl}^{-+} &= \Gamma_{ikjl}^*, \\
M_{ikjl}^{+-} &= \Gamma_{ikjl} = (M_{ikjl}^{-+})^*
\end{aligned} \tag{9.54}$$

define the four $n^2 \times n^2$ c-number matrices M^{--}, M^{++}, M^{-+} and M^{+-}. In this construction we include all ν_{ijkl}, with those for which $i = j$ or $k = l$ being set to zero. From (2.55) and (2.86), the complex matrices M^{--} and M^{++} are symmetric and the matrices M^{-+} and M^{+-} are Hermitian:

$$M^{--}_{jlik} = \frac{1}{i\hbar}\nu_{jilk} + \frac{\{\Gamma_{kjil} + \Gamma_{lijk}\}}{2} = M^{--}_{ikjl},$$
$$(M^{-+}_{jlik})^* = (\Gamma^*_{jlik})^* = \Gamma_{jlik} = \Gamma^*_{ikjl} = M^{-+}_{ikjl}. \tag{9.55}$$

It is now convenient to use the notation where the $2n$ Grassmann variables are listed as $g_p \equiv \{g_1, \cdots, g_n, g_1^+, \cdots, g_n^+\}$ and the diffusion matrix elements are D_{pq}, which are related to the F_{ij}^{AB} $(A, B = -, +)$ as in (8.92). We can write

$$D_{pq} = \sum_{r,s} Q^{pq}_{rs}\, g_r g_s, \tag{9.56}$$

where the Q^{pq}_{rs} are all c-numbers and are related to the M^{AB}_{ikjl} in (9.54). This relationship is conveniently set out in a table:

$\mathbf{g}_p$	$\mathbf{g}_q$	$\mathbf{g}_r$	$\mathbf{g}_s$	$\mathbf{Q}^{pq}_{rs}$
g_i	g_j	g_k	g_l	M^{--}_{ikjl}
g_i	g_j^+	g_k	g_l^+	M^{-+}_{ikjl}
g_i^+	g_j	g_k^+	g_l	M^{+-}_{ikjl}
g_i^+	g_j^+	g_k^+	g_l^+	M^{++}_{ikjl}

, (9.57)

where $i, j, k, l = 1, \cdots, n$ and $p, q, r, s = 1, \cdots, 2n$. For each row with a given form for g_p and g_q, the matrix element $\mathbf{Q}^{pq}_{rs}$ is given by the stated M^{AB}_{ikjl} when g_r and g_s are given in the same row, and zero otherwise. For example, with p and q both in the range $1, \cdots, n$ the matrix element M^{pq}_{rs} vanishes if either r or s is in the range $n+1, n+2, \cdots, 2n$, since $-F^{--}_{ij}$ does not involve any g_k^+ or g_l^+. Although the matrix Q could in principle have $(2n)^2$ rows and $(2n)^2$ columns, with row and column indices as the joint quantities p, r or q, s, most of the elements would be zero. Taking this into account, the matrix Q is only required to have $2n^2$ rows and $2n^2$ columns and can be formatted as

$$[Q] = \begin{bmatrix} M^{--} & M^{-+} \\ M^{+-} & M^{++}, \end{bmatrix} \tag{9.58}$$

where the rows are listed in each $n^2 \times n^2$ submatrix as ik and the columns are listed as jl. For each submatrix M^{AB}, specifying an element by ik (row) and jl (column) uniquely specifies the element for Q via the pr (row) and qs (column) indices using (9.57). An example for $n = 2$ illustrates the procedure:

$p, r \downarrow, q, s \rightarrow$	**11**	**12**	**21**	**22**	**33**	**34**	**43**	**44**
11								
12								
21								
22								
33								
34								
43								
44								

. (9.59)

A genuine 8×8 matrix with consistently indexed entries results.

It is easy to see that Q is symmetric, which follows from the antisymmetry of the matrix D, i.e. $D_{pq} = -D_{qp}$:

$$\begin{aligned}
-D_{qp} &= -\sum_{r,s} Q^{qp}_{sr} g_s g_r \\
&= \sum_{r,s} Q^{qp}_{sr} g_r g_s \\
&= D_{pq}, \\
Q^{pq}_{rs} &= Q^{qp}_{sr}.
\end{aligned} \tag{9.60}$$

Hence, using Takagi factorisation as in (8.66) (see Horn and Johnson [63]), we can find a $2n^2 \times 2n^2$ c-number matrix K with $2n^2$ rows indexed by subsets of p, r (for each p, only half of the possible r are needed) and $2n^2$ columns indexed by a such that

$$\begin{aligned}
Q &= KK^T, \\
Q^{pq}_{rs} &= \sum_a K^p_{ra} K^q_{sa}.
\end{aligned} \tag{9.61}$$

We then have

$$\begin{aligned}
D_{pq} &= \sum_{r,s} \sum_a K^p_{ra} K^q_{sa} \, g_r g_s \\
&= \sum_a \mathcal{B}^p_a \mathcal{B}^q_a, \\
D &= \mathcal{B}(g_r) \mathcal{B}^T(g_r),
\end{aligned} \tag{9.62}$$

where

$$\mathcal{B}^p_a = \sum_r K^p_{ra} \, g_r \tag{9.63}$$

is an odd Grassmann function, which is linear in the Grassmann variables. $\mathcal{B}$ is a matrix with $2n$ rows indexed by p and $2n^2$ columns indexed by a. In general, the $\mathcal{B}^p_a$ are linear combinations of all $g_1, \cdots, g_n, g_1^+, \cdots, g_n^+$.

The Ito stochastic equation can now be written in the useful form

$$\widetilde{g}_p(t+\delta t) - \widetilde{g}_p(t) = -A^p(\widetilde{g}_q(t))\, \delta t + \sum_r \left(\sum_a K^p_{ra} \, \delta \widetilde{w}_a(t) \right) \widetilde{g}_r(t). \tag{9.64}$$

In this form, the Wiener increments are separated from the stochastic Grassmann variables. The number of Wiener increments (or Gaussian–Markov random noise terms) is $i(n) = 2n^2$. Note that the actual number of Wiener increments needed may be smaller if enough elements Q^{pq}_{rs} are zero, and this leads to a smaller non-zero submatrix for Q to be factorised. We note that $i(n)$ increases as the square of the number of modes. A similar feature applies in the Gaussian phase-space treatment of fermion systems developed by Corney and Drummond [57, 58], which involves pairs of fermion annihilation and creation operators. The total number of such pairs – $\hat{c}_i\hat{c}_j, \hat{c}_i^\dagger\hat{c}_j, \hat{c}_i\hat{c}_j^\dagger$ and $\hat{c}_i^\dagger\hat{c}_j^\dagger$, discarding those that are zero such as $\hat{c}_i^2$, or related such as $\hat{c}_j\hat{c}_i = -\hat{c}_i\hat{c}_j$ or $\hat{c}_i\hat{c}_j^\dagger = \delta_{ij} - \hat{c}_j^\dagger\hat{c}_i$ – is given by $2n^2 - n$, which is one less than the number of parameters used to specify their density operator.

9.3.3 Summary of Fermion Stochastic Equations

The Ito stochastic differential equation for fermionic systems can also be written as

$$\frac{d}{dt}\widetilde{g}_p(t) = \mathcal{A}^p(\widetilde{g}_q(t)) + \sum_a \mathcal{B}_a^p(\widetilde{g}_q(t))\,\Gamma_a(t_+), \tag{9.65}$$

where $\mathcal{B}\mathcal{B}^T = D$ and $\mathcal{A} = -A$. The difference from the bosonic case is that the $\mathcal{A}^p, \mathcal{B}_a^p$ are odd Grassmann functions rather than c-number functions and the D_{pq} are even Grassmann functions rather than c-number functions, with the matrix D being anti-symmetric rather than symmetric. Also there is a sign reversal in the drift term. For convenience, the related Ito and Fokker–Planck equations are repeated here:

$$\begin{aligned}
\frac{\partial}{\partial t}P_{\mathrm{f}}(g, g^+) &= -\sum_{p=1}^{2n}(A_p P_{\mathrm{f}})\frac{\overleftarrow{\partial}}{\partial g_p} + \frac{1}{2}\sum_{p,q=1}^{2n}(D_{pq}P_{\mathrm{f}})\frac{\overleftarrow{\partial}}{\partial g_q}\frac{\overleftarrow{\partial}}{\partial g_p},\\
\frac{d}{dt}\widetilde{g}_p(t) &= \mathcal{A}^p(\widetilde{g}_q(t)) + \sum_a \mathcal{B}_a^p(\widetilde{g}_q(t))\,\Gamma_a(t_+), \qquad \mathcal{B}\mathcal{B}^T = D, \qquad \mathcal{A}^p = -A_p,\\
D_{pq} &= \sum_{r,s} Q_{rs}^{pq}\, g_r g_s, \qquad Q = KK^T,\\
\mathcal{B}_a^p &= \sum_r K_{ra}^p\, g_r.
\end{aligned} \tag{9.66}$$

These may be compared to (9.26). Q is a c-number symmetric matrix. There are $a = 1, \cdots, 2n^2$ Gaussian–Markov noise terms.

As in the bosonic case, these fermionic Ito stochastic equations are in general non-linear. However, Grassmann variables are now involved, so the solutions cannot be found numerically via standard computational methods, because representing Grassmann variables on a computer does not seem feasible. As proposed by Plimak et al. [14], this problem can be resolved by using Fokker–Planck and Langevin equations based on unnormalised Grassmann distribution functions, as we will see below.

9.4 Ito Stochastic Equations for Fermions – Unnormalised Distribution Functions

In the previous section, Ito stochastic equations for fermions were obtained which were based on standard normalised distribution functions. In this case the Fokker–Planck equation contains non-linear drift terms, which result in non-linear Ito equations for the stochastic Grassmann variables. In this section, we follow the general approach of Plimak et al. [14] and derive new Ito stochastic equations based on the unnormalised fermion distribution functions $B_{\rm f}(g, g^+)$ and the Fokker–Planck equations that were set out in Sections 7.8 and 8.10. We present the detailed forms of these Ito equations and describe how the Ito equations for stochastic Grassmann variables can be solved numerically in terms of stochastic c-numbers.

The Ito stochastic equations are of the same general form as in (9.41),

$$\widetilde{g}_p(t+\delta t) - \widetilde{g}_p(t) = \mathcal{M}^p(\widetilde{g}_q(t))\,\delta t + \sum_a \mathcal{N}_a^p(\widetilde{g}_q(t))\,\delta\widetilde{w}_a(t), \tag{9.67}$$

involving stochastic Wiener increments $\delta\widetilde{w}_a(t)$ with different quantities $\mathcal{M}^p(\widetilde{g}_q(t))$ and $\mathcal{N}_a^p(\widetilde{g}_q(t))$ involved. The connection between the Ito stochastic equations and the Fokker–Planck equations is established in exactly the same way as in Section 9.3, but now of course the Fokker–Planck equation analogous to (8.83) is given by (8.98) and the drift vector and diffusion matrix quantities are given by (8.99) rather than (8.84). We then have that, instead of (9.51), the requirements for the results for phase space and stochastic averages to be the same are given by

$$\begin{aligned} \mathcal{M}^p(g_q) &= -A^p(g_q), \\ [\mathcal{N}(g_r)\mathcal{N}^T(g_r)]_{pq} &= D_{pq}(g_r), \end{aligned} \tag{9.68}$$

where the Fokker–Planck equation (8.98) has first been changed to the form as given in (8.91), in which only right Grassmann differentiation is involved.

In this case of an unnormalised fermion distribution, the vector A and matrix D are now given by

$$[A] = \left[\begin{array}{|c|} \hline \mathcal{C}^- \\ \hline \mathcal{C}^+ \\ \hline \end{array}\right] \tag{9.69}$$

and

$$[D] = \left[\begin{array}{|c|c|} \hline -\mathcal{F}^{--} & +\mathcal{F}^{-+} \\ \hline -(\mathcal{F}^{-+})^T & +\mathcal{F}^{++} \\ \hline \end{array}\right], \tag{9.70}$$

where the $\mathcal{C}^A$ and $\mathcal{F}^{AB}$ $(A, B = +, -)$ are given in (8.99).

There is another important difference from the normalised-distribution-function case, that the quantity $F(g, g^+)$ is different when the quantum correlation functions are to be determined. From (7.104), the function for determining the normally ordered quantum correlation function $G_{\rm f}(l_1, \cdots, l_p; m_q, \cdots, m_1)$ (for $p - q$ equal to an even integer) is now given by

$$F(g, g^+) = (g_{m_q} \cdots g_{m_1}) \exp(\mathbf{g} \cdot \mathbf{g}^+)(g_{l_1}^+ \cdots l_p{}^+), \tag{9.71}$$

where the even distribution function $B_{\rm f}(g, g^+)$ has been placed on the right in (7.104) for convenience. The product of exponential factors $\exp(\mathbf{g} \cdot \mathbf{g}^+) = \prod_i (1 + g_i g_i^+)$ has an important consequence for the stochastic averages that are needed. Clearly, if i is the same as any of the $l_1, l_2, \ldots, l_p, m_1, m_2, \ldots, m_q$, then the exponential factor can be replaced by unity and the expansion of the remaining exponential factors will result in fewer than 2^n terms.

However, the diffusion matrix elements $\mathcal{F}^{AB}$ are given by exactly the same expressions as the diffusion matrix elements F^{AB} given in (8.84) for the normalised distribution function P_{f}. Consequently the expressions for the $\mathcal{N}(g_r)$ are the *same* as for the previous $\mathcal{B}(g_r)$. Thus, from (9.56),

$$D_{pq} = \sum_{r,s} Q^{pq}_{rs} g_r g_s, \tag{9.72}$$

where from (9.57) the non-zero elements of Q are

$\mathbf{g}_p$	$\mathbf{g}_q$	$\mathbf{g}_r$	$\mathbf{g}_s$	$\mathbf{Q}^{pq}_{rs}$
g_i	g_j	g_k	g_l	M^{--}_{ikjl}
g_i	g_j^+	g_k	g_l^+	M^{-+}_{ikjl}
g_i^+	g_j	g_k^+	g_l	M^{+-}_{ikjl}
g_i^+	g_j^+	g_k^+	g_l^+	M^{++}_{ikjl}

, (9.73)

with the M^{AB}_{ikjl} given by (9.54), so Q is of the same form as in (9.58):

$$[Q] = \left[\begin{array}{|c|c|} \hline M^{--} & M^{-+} \\ \hline M^{+-} & M^{++} \\ \hline \end{array} \right]. \tag{9.74}$$

We now have from (9.61) the Takagi factorisation of Q,

$$\begin{aligned} Q &= KK^T, \\ Q^{pq}_{rs} &= \sum_a K^p_{ra} K^q_{sa}, \end{aligned} \tag{9.75}$$

and from (9.62) the diffusion matrix can then be written as

$$\begin{aligned} D_{pq} &= \sum_{r,s} \sum_a K^p_{ra} K^q_{sa} g_r g_s \\ &= \sum_a \mathcal{B}^p_a \mathcal{B}^q_a, \\ D &= \mathcal{B}(g_r) \mathcal{B}^T(g_r), \end{aligned} \tag{9.76}$$

where from (9.63)

$$\mathcal{B}^p_a = \sum_r K^p_{ra} g_r, \tag{9.77}$$

$$[\mathcal{N}] = [\mathcal{B}]. \tag{9.78}$$

Consequently, the diffusion terms $\mathcal{B}^p_a$ in the Ito stochastic equations, are still linear Grassmann functions of the g_r. As before, there will be a total of $i(n) = 2n^2$ c-number Gaussian–Markov random noise terms.

For the drift terms in the Ito stochastic equations, we have from (9.68) and (9.69)

$$[\mathcal{M}] = -\left[\begin{array}{|c|}\hline \mathcal{C}^{-} \\ \hline \mathcal{C}^{+} \\ \hline \end{array}\right], \tag{9.79}$$

where from (8.99)

$$\begin{aligned} \mathcal{C}_i^{-} &= -\frac{1}{i\hbar}\sum_j h_{ij} g_j + \sum_{jk}\left(\frac{\Gamma_{kikj}}{2}\right) g_j = -\sum_j L_{ij}^{-} g_j, \\ \mathcal{C}_i^{+} &= +\frac{1}{i\hbar}\sum_j h_{ij}^{*} g_j^{+} + \sum_{jk} g_j^{+}\left(\frac{\Gamma_{kikj}^{*}}{2}\right) = -\sum_j L_{ij}^{+} g_j^{+}. \end{aligned} \tag{9.80}$$

Note that $L_{ij}^{+} = (L_{ij}^{-})^{*}$. The drift terms are also all linear functions of the g_r with c-number coefficients, and these may be written as

$$\mathcal{M}^p = \sum_r L_r^p g_r, \tag{9.81}$$

where the L_r^p can be obtained from (9.80). The relationship can be tabulated:

$$\begin{array}{|c|c|c|}\hline \mathbf{g}_p & \mathbf{g}_r & \mathbf{L}_r^p \\ \hline g_i & g_j & L_{ij}^{-} \\ \hline g_i^{+} & g_j^{+} & L_{ij}^{+} \\ \hline \end{array}, \tag{9.82}$$

where $i, j = 1, \cdots, n$ and $p, r = 1, \cdots, 2n$. For each row with a given form for g_p, the matrix element $\mathbf{L}_r^p$ is given by the stated L_{ij}^A when g_r is given in the same row, and zero otherwise. For example, with p in the range $1, \cdots, n$ the matrix element $\mathbf{L}_r^p$ is zero if r is in the range $n+1, n+2, \cdots, 2n$, since $-\mathcal{C}_i^{-}$ does not involve any g_j^{+}. The matrix L has $2n$ rows and $2n$ columns, as the row and column indices are the joint quantities p and r. Thus the matrix L may be formatted as

$$[L] = \left[\begin{array}{|c|c|}\hline L^{-} & 0 \\ \hline 0 & L^{+} \\ \hline \end{array}\right], \tag{9.83}$$

where the rows are listed in each $n \times 1$ submatrix as i and the columns are listed as j. Again, for each submatrix an element specified by i (row) and j (column) can also be listed for L via p (row) and r (column), where for each submatrix ij uniquely specifies pr. An example for $n = 2$ illustrates the procedure:

$$[L] = \left[\begin{array}{|c|c|c|c|c|}\hline p\downarrow, r\rightarrow & \mathbf{1} & \mathbf{2} & \mathbf{3} & \mathbf{4} \\ \hline \mathbf{1} & & & 0 & 0 \\ \hline \mathbf{2} & & & 0 & 0 \\ \hline \mathbf{3} & 0 & 0 & & \\ \hline \mathbf{4} & 0 & 0 & & \\ \hline \end{array}\right]. \tag{9.84}$$

From the expressions for M and N, it is clear that each row of these matrices is a linear function of the Grassmann variables. Transferring the $\widetilde{g}_p(t)$ over to the other side of the Ito equation, we now have

$$\widetilde{g}_p(t+\delta t) = \sum_r G_{pr}(t)\widetilde{g}_r(t), \tag{9.85}$$

where

$$G_{pr}(t) = \delta_{pr} + L_r^p + \sum_{a=1}^{2n^2} K_{ra}^p \, \delta\widetilde{w}_a(t). \tag{9.86}$$

The $2n^2$ Wiener increments $\delta\widetilde{w}_a(t)$ are all independent and are taken to be real. Thus the set of Grassmann stochastic variables at time $t+\delta t$ are related to those at time t via linear transformation matrices $G_{pr}(t)$ which, although stochastic, involve only c-number variables. This is the key result needed to obtain expressions for stochastic averages of products of stochastic $\widetilde{g}_i, \widetilde{g}_i^+$.

Dividing the time interval between the initial time t_0 and the measurement time t_f into small intervals $\delta t_a = t_a - t_{a-1}$, we see that

$$\begin{aligned} \widetilde{g}_p(t_f) &= \sum_{rs\cdots yz} G_{pr}(t_{f-1})G_{rs}(t_{f-2})\cdots G_{yz}(t_0)\,\widetilde{g}_z(t_0) \\ &= \sum_z \mathcal{G}_{pz}(t_f;t_0)\,\widetilde{g}_z(t_0), \end{aligned} \tag{9.87}$$

so there is a linear relation between $\widetilde{g}_p(t_f)$ and $\widetilde{g}_z(t_0)$ that involves only the products of c-number (though stochastic) matrices.

In an obvious notation, the stochastic average of a product of stochastic $\widetilde{g}_i, \widetilde{g}_i^+$ at time t_f will be a stochastic average of products of c-number matrix elements $G_{iz}(t_f;t_0)$ times products of stochastic $\widetilde{g}_j, \widetilde{g}_j^+$ at time t_0. Thus, with $a, b, \cdots, u \neq l_1, \cdots l_p, m_1, \cdots m_q$ and where $a, b, \cdots, u$ list terms still remaining after the exponentials in $\exp(\mathbf{g}\cdot\mathbf{g}^+)$ are expanded, the general result for a quantum correlation function will involve terms such as

$$\begin{aligned} &\left(\overline{(g_{m_q}\cdots g_{m_1})(g_a g_a^+)\cdots(g_u g_u^+)(g_{l_1}^+\cdots g_{l_p}^+)}\right)_{t_f} \\ &= \prod \overline{(\mathcal{G}_{AB}(t_f;t_0)\mathcal{G}_{CD}(t_f;t_0)\cdots\mathcal{G}_{YZ}(t_f;t_0))(g_a(t_0)^\gamma g_b(t_0)^\delta\cdots g_z(t_0)^\eta)} \\ &= \prod \overline{(\mathcal{G}_{AB}(t_f;t_0)\mathcal{G}_{CD}(t_f;t_0)\cdots\mathcal{G}_{YZ}(t_f;t_0))}\;\overline{(g_a(t_0)^\gamma g_b(t_0)^\delta\cdots g_z(t_0)^\eta)}, \end{aligned} \tag{9.88}$$

where $\gamma, \delta, \cdots = -, +$ and we have used the feature that the $G_{AB}(t_f;t_0)$ etc. all involve Wiener increments (and hence Gaussian–Markov random variables) at times later than t_0. For ease of notation, we have left the tildes denoting stochastic quantities understood. Hence the stochastic averages factorise into c-number stochastic averages involving the $G_{AB}(t_f;t_0)$ and stochastic averages of the g_i, g_i^+ at time t_0.

Expressions for the latter stochastic averages are not calculated, but inserted from initial values of the quantum correlation functions. This final expression shows how quantum correlation functions for fermions can be calculated in terms of stochastic quantities solely involving c-numbers, even though the original Ito equations contain Grassmann variables. However, the determination of the initial stochastic averages $\overline{(g_a(t_0)^\gamma g_b(t_0)^\delta \cdots g_z(t_0)^\eta)}$ from the initial conditions is a non-trivial problem, owing largely to the presence of the $(1 + g_i g_i^+)$ factors in expressions for the correlation functions. But for Fock state populations and coherences, these complications due to the exponential factors are absent, facilitating stochastic numerical calculations of these Fock state quantities.

9.5 Fluctuations and Time Dependence of Quantum Correlation Functions

Having established that Fokker–Planck equations for the distribution functions of the phase space variables can be linked to Ito stochastic equations for stochastic replacements of these variables, we can use these results to establish the time dependence of the quantum correlation functions. The time dependence can be expressed in terms of stochastic averages involving the drift and diffusion terms arising from the Fokker–Planck equations. As we will see, these results do not require knowledge of the B or $\mathcal{B}$ matrices themselves, only the products BB^T or $\mathcal{B}\mathcal{B}^T$, which give the diffusion matrix D. For both the bosonic and the fermionic cases, we first consider the stochastic averages of single-fluctuation terms and products of two fluctuation terms. For the single-fluctuation terms, the result is given in terms of stochastic averages of the drift term, whilst the results for products of stochastic fluctuations relate to stochastic averages of the diffusion term. The former average can be interpreted in terms of correlation functions. For bosonic systems with c-number variables, the results for the stochastic fluctuations and the drift effects are analogous to those applying in Brownian motion. There, the Brownian particle drifts slowly along, driven by the large-scale external forces that are acting, whilst at the same time exhibiting random fluctuation effects due to its coupling with the local environment of solute molecules. As in the earlier analyses of Brownian motion by Einstein and Langevin, the drift effects are related to the drift term in the Fokker–Planck equations, whilst the fluctuation effects are related to the diffusion term D. We next consider the time dependences of quantum correlation functions of single annihilation and creation operators or normally ordered pairs of these operators. These results can be developed to show how the quantum correlations of higher order are related to the time derivatives of the lower-order correlation functions.

9.5.1 Boson Fluctuations

For the fluctuation terms, we see from (9.26) that

$$\begin{aligned}\delta\widetilde{\alpha}_p(t) &= \widetilde{\alpha}_p(t+\delta t) - \widetilde{\alpha}_p(t) \\ &= \mathcal{A}^p(\widetilde{\alpha}_q(t))\,\delta t + \sum_a \mathcal{B}^p_a(\widetilde{\alpha}_q(t)) \int_t^{t+\delta t} dt_1\, \Gamma_a(t_1).\end{aligned} \tag{9.89}$$

Hence, on taking the stochastic average using the results in Appendix G and (9.12), we find that the single-fluctuation term is

$$\overline{\delta\widetilde{\alpha}_p(t)} = \overline{\mathcal{A}^p(\widetilde{\alpha}_q(t))}\,\delta t, \tag{9.90}$$

where the stochastic averages (9.11) have been applied. Thus the average of $\delta\widetilde{\alpha}_p(t)$ changes linearly with time for short times δt and the rate of change is given by the stochastic average of the drift term in the Fokker–Planck equation.

For the product of two fluctuations, we see that

$$\begin{aligned}\delta\widetilde{\alpha}_p(t)\,\delta\widetilde{\alpha}_q(t) &= \mathcal{A}^p(\widetilde{\alpha}_q(t))\,\mathcal{A}^q(\widetilde{\alpha}_q(t))\,\delta t^2 + \mathcal{A}^p(\widetilde{\alpha}_q(t))\sum_a \mathcal{B}^q_a(\widetilde{\alpha}_q(t))\int_t^{t+\delta t} dt_1\,\Gamma_a(t_1)\,\delta t\\ &\quad+\sum_a \mathcal{B}^p_a(\widetilde{\alpha}_q(t))\int_t^{t+\delta t} dt_1\,\Gamma_a(t_1)\,\mathcal{A}^q(\widetilde{\alpha}_q(t))\,\delta t\\ &\quad+\sum_a \mathcal{B}^p_a(\widetilde{\alpha}_q(t))\int_t^{t+\delta t} dt_1\,\Gamma_a(t_1)\sum_b \mathcal{B}^q_b(\widetilde{\alpha}_q(t))\int_t^{t+\delta t} dt_2\,\Gamma_b(t_2),\end{aligned} \tag{9.91}$$

so, taking the stochastic average and using the results in Appendix G, (9.12) and (9.11), we have

$$\begin{aligned}\overline{\delta\widetilde{\alpha}_p(t)\,\delta\widetilde{\alpha}_q(t)} &= \overline{\mathcal{A}^p(\widetilde{\alpha}_q(t))\,\mathcal{A}^q(\widetilde{\alpha}_q(t))}\,\delta t^2 + \sum_a \overline{\mathcal{A}^p(\widetilde{\alpha}_q(t))\,\mathcal{B}^q_a(\widetilde{\alpha}_q(t))}\int_t^{t+\delta t} dt_1\,\overline{\Gamma_a(t_1)}\,\delta t\\ &\quad+\sum_a \overline{\mathcal{B}^p_a(\widetilde{\alpha}_q(t))\mathcal{A}^q(\widetilde{\alpha}_q(t))}\int_t^{t+\delta t} dt_1\,\overline{\Gamma_a(t_1)}\,\delta t\\ &\quad+\sum_{ab} \overline{\mathcal{B}^p_a(\widetilde{\alpha}_q(t))\mathcal{B}^q_b(\widetilde{\alpha}_q(t))}\int_t^{t+\delta t} dt_1\int_t^{t+\delta t} dt_2\,\overline{\Gamma_a(t_1)\Gamma_b(t_2)}\\ &= \overline{F_{pq}(\widetilde{\alpha}_q(t))}\,\delta t,\end{aligned} \tag{9.92}$$

where the result $\mathcal{B}\mathcal{B}^T = F$ has been used and the terms of $O(\delta t^2)$ have been discarded. These results for the product of the stochastic fluctuations $\delta\widetilde{\alpha}_p(t)\,\delta\widetilde{\alpha}_q(t)$ show that the stochastic average changes linearly with time for short times δt and the rate of change is given by the stochastic average of the diffusion term in the Fokker–Planck equation. The results (9.90) and (9.92) are interesting, but only the former is directly related to quantum correlation functions. However, they do confirm the well-known result from the theory of Brownian motion that the drift term is related to the average motion of the Brownian particle and the diffusion term is related to the random fluctuations about its trajectory.

9.5.2 Boson Correlation Functions

We may conveniently list the bosonic annihilation and creation operators as $\{\hat{a}_p\} \equiv \{\hat{a}_1, \cdots, \hat{a}_n, \hat{a}_1^\dagger, \cdots, \hat{a}_n^\dagger\}$, $p = 1, \cdots, 2n$, with the corresponding phase space variables $\{\alpha_p\} \equiv \{\alpha_1, \cdots, \alpha_n, \alpha_1^+, \cdots, \alpha_n^+\}$, as before. To avoid any issues regarding ambiguities in differentiation with respect to complex α_p, the following derivations treat the real and imaginary components α_x^p, α_y^p using the Fokker–Planck equation (9.27) and recover the complex forms at the end.

First, we consider $F(\alpha, \alpha^+) = \alpha_p = \alpha_x^p + i\alpha_y^p$. From (9.4) and with

$$\frac{\partial}{\partial \alpha_{rx}}(\alpha_x^p + i\alpha_y^p) = \delta_{rp}, \qquad \frac{\partial}{\partial \alpha_{ry}}(\alpha_x^p + i\alpha_y^p) = i\delta_{rp}, \tag{9.93}$$

$$\frac{\partial}{\partial \alpha_{r\mu}}\frac{\partial}{\partial \alpha_{s\xi}}(\alpha_x^p + i\alpha_y^p) = 0, \tag{9.94}$$

we find for the time derivative of the phase space average

$$\frac{d}{dt}\langle \alpha_p \rangle_t = \langle (A_x^p + iA_y^p) \rangle = \langle A_p \rangle, \tag{9.95}$$

where (8.67) has been used. Then, using the equivalence (9.6) to the stochastic average, the time derivative of the phase space average is

$$\frac{d}{dt}\langle \alpha_p \rangle_t = \overline{\mathcal{A}_p(\alpha_q^s(t))}, \tag{9.96}$$

relating the time derivative to the stochastic average of the drift term. For the quantum averages of the annihilation and creation operators, we have

$$\langle \hat{a}_p \rangle_t = \int d^2\alpha^+ d^2\alpha\, \alpha_p P_{\mathrm{b}}(\alpha, \alpha^+, \alpha^*, \alpha^{+*}, t) = \langle \alpha_p \rangle_t\,, \tag{9.97}$$

so that for the time derivative of the quantum correlation function,

$$\frac{d}{dt}\langle \hat{a}_p \rangle_t = \overline{\mathcal{A}_p(\widetilde{\alpha}_q(t))}. \tag{9.98}$$

This result could be used numerically in conjunction with the Ito stochastic equations for the $\widetilde{\alpha}_q(t)$ to give the time derivative as a stochastic average if this was needed.

To obtain equations of motion for the correlation function, we use the particular expressions (8.57) and (8.61) for A_p and then reinterpret the stochastic average as a phase space average, placing the phase space variables in antinormal order. For c-number phase variables, no sign changes are required. We can then interpret the terms on the right side as quantum correlation functions using (7.74). Thus, for the case of $\langle \hat{a}_i \rangle_t$, we require C_i^- on reverting to the original α_i, α_i^+ notation and identifying quantities via (8.57):

$$
\begin{aligned}
\frac{d}{dt}\langle \hat{a}_i\rangle_t &= \frac{1}{i\hbar}\sum_j k_{ij}\langle \alpha_j\rangle + \frac{1}{i\hbar}\sum_{jkl}\xi_{ij\,kl}\,\langle \alpha_j^+\alpha_l\alpha_k\rangle \\
&\quad + \sum_{jkl}\left(\{\Gamma_{jk\,il} - \Gamma_{li\,kj}\}\langle \alpha_l\alpha_k^+\alpha_j\rangle - \Gamma_{li\,kj}\delta_{lk}\langle \alpha_j\rangle\right) \\
&= \frac{1}{i\hbar}\sum_j k_{ij}\langle \alpha_j\rangle + \frac{1}{i\hbar}\sum_{jkl}\xi_{ij\,kl}\,\langle \alpha_l\alpha_k\alpha_j^+\rangle \\
&\quad + \sum_{jkl}\left(\{\Gamma_{jk\,il} - \Gamma_{li\,kj}\}\langle \alpha_l\alpha_j\alpha_k^+\rangle - \Gamma_{li\,kj}\delta_{lk}\langle \alpha_j\rangle\right) \\
&= \frac{1}{i\hbar}\sum_j k_{ij}\langle \hat{a}_j\rangle + \frac{1}{i\hbar}\sum_{jkl}\xi_{ij\,kl}\,\langle \hat{a}_j^\dagger\hat{a}_l\hat{a}_k\rangle \\
&\quad + \sum_{jkl}\left(\{\Gamma_{jk\,il} - \Gamma_{li\,kj}\}\langle \hat{a}_k^\dagger\hat{a}_l\hat{a}_j\rangle - \Gamma_{li\,kj}\delta_{lk}\langle \hat{a}_j\rangle\right),
\end{aligned}
\tag{9.99}
$$

with a similar equation for $d/dt\langle \hat{a}_i^\dagger\rangle_t$. This shows that the first-order quantum correlation function is coupled to third-order quantum correlation functions.

Second, we consider $F(\alpha,\alpha^+) = \alpha_p\alpha_q = (\alpha_x^p + i\alpha_y^p)(\alpha_x^q + i\alpha_y^q)$. We find for the time derivative of the phase space average

$$
\frac{d}{dt}\langle \alpha_p\alpha_q\rangle_t = \langle(\alpha_p A_q + A_p\alpha_q)\rangle + \langle F_{pq}\rangle. \tag{9.100}
$$

Details are given in Appendix G. Then, using the equivalence (9.6) to the stochastic average and $\mathcal{A} = A$, the time derivative of the phase space average is

$$
\frac{d}{dt}\left\langle \alpha_p\alpha_q\right\rangle_t = \overline{\widetilde{\alpha}_p(t)C_q(\widetilde{\alpha}_r(t)) + C_q(\widetilde{\alpha}_r(t))\widetilde{\alpha}_q(t)} + \overline{F_{pq}(\widetilde{\alpha}_r(t))}, \tag{9.101}
$$

relating the time derivative to a sum of stochastic averages of products of the stochastic variable with the drift terms together with the stochastic average of the diffusion term. For the quantum averages of the three cases of normally ordered products of the annihilation and creation operators, we have

$$
\begin{aligned}
\langle \hat{a}_i\hat{a}_j\rangle_t &= \int d^2\alpha^+ d^2\alpha\,\alpha_i\alpha_j P_{\mathrm{b}}(\alpha,\alpha^+,\alpha^*,\alpha^{+*},t) = \langle \alpha_i\alpha_j\rangle_t, \\
\langle \hat{a}_i^\dagger\hat{a}_j^\dagger\rangle_t &= \int d^2\alpha^+ d^2\alpha\,\alpha_i^+\alpha_j^+ P_{\mathrm{b}}(\alpha,\alpha^+,\alpha^*,\alpha^{+*},t) = \langle \alpha_i^+\alpha_j^+\rangle_t, \\
\langle \hat{a}_j^\dagger\hat{a}_i\rangle_t &= \int d^2\alpha^+ d^2\alpha\,\alpha_i\alpha_j^+ P_{\mathrm{b}}(\alpha,\alpha^+,\alpha^*,\alpha^{+*},t) = \langle \alpha_i\alpha_j^+\rangle_t,
\end{aligned}
\tag{9.102}
$$

so, for example, the time derivative of the quantum correlation function $\langle \hat{a}_j^\dagger\hat{a}_i\rangle_t$ is

$$
\frac{d}{dt}\langle \hat{a}_j^\dagger\hat{a}_i\rangle_t = \overline{\widetilde{\alpha}_i(t)C_j^+(\widetilde{\alpha}_r(t)) + C_i^-(\widetilde{\alpha}_r(t))\widetilde{\alpha}_j^+(t)} + \overline{F_{ij}^{-+}(\widetilde{\alpha}_r(t))}, \tag{9.103}
$$

where we have reverted to the original α_i, α_i^+ notation and identified quantities via (8.56). This result could be used numerically in conjunction with the Ito stochastic equations for the $\widetilde{\alpha}_q(t)$ to give the time derivative as a stochastic average if required.

To obtain equations of motion for the quantum correlation function, we use the particular expressions (8.57) for C_i^-, C_j^+ and F_{ij}^{-+} and then reinterpret the stochastic average as a phase space average, placing the phase space variables in antinormal order. For c-number phase variables, no sign changes are required. We can then interpret the terms on the right side as quantum correlation functions using (7.74). Using a similar approach as for the previous example, we finally have

$$
\begin{aligned}
&\frac{d}{dt}\langle \hat{a}_j^\dagger \hat{a}_i \rangle_t \\
&= -\frac{1}{i\hbar}\sum_k k_{jk}^* \langle \hat{a}_k^\dagger \hat{a}_i \rangle_t - \frac{1}{i\hbar}\sum_{mkl} \xi_{jm\,kl}^* \langle \hat{a}_k^\dagger \hat{a}_l^\dagger \hat{a}_i \hat{a}_m \rangle_t \\
&\quad + \sum_{mkl} (\{\Gamma_{mk\,jl}^* - \Gamma_{lj\,km}^*\} \langle \hat{a}_l^\dagger \hat{a}_m^\dagger \hat{a}_i \hat{a}_k \rangle_t - \Gamma_{lj\,km}^* \delta_{lk} \langle \hat{a}_m^\dagger \hat{a}_i \rangle_t) \\
&\quad + \frac{1}{i\hbar}\sum_k k_{ik} \langle \hat{a}_j^\dagger \hat{a}_k \rangle_t + \frac{1}{i\hbar}\sum_{mkl} \xi_{im\,kl} \langle \hat{a}_m^\dagger \hat{a}_j^\dagger \hat{a}_l \hat{a}_k \rangle_t \\
&\quad + \sum_{mkl} (\{\Gamma_{mk\,il} - \Gamma_{li\,km}\} \langle \hat{a}_j^\dagger \hat{a}_k^\dagger \hat{a}_l \hat{a}_m \rangle_t - \Gamma_{li\,km} \delta_{lk} \langle \hat{a}_j^\dagger \hat{a}_m \rangle_t) \\
&\quad + 2\sum_{kl} \{\Gamma_{il\,jk}\} \langle \hat{a}_l^\dagger \hat{a}_k \rangle_t.
\end{aligned}
\tag{9.104}
$$

Details are given in Appendix G. This shows that the second-order and fourth-order quantum correlation functions are coupled. This feature of coupling in higher-order correlation functions is quite general, and a hierarchy of equations for the functions results. This may be solved via truncation methods.

It should be pointed out that the evolution equations for bosonic correlation functions can be derived without using the intermediate step of introducing the stochastic forms. Thus one can pass directly from (9.100) to (9.104) without using (9.103). However, the application of the stochastic equations does perhaps represent the simpler approach. The stochastic expressions are particularly useful in numerical calculations.

9.5.3 Fermion Fluctuations

For the fluctuation terms, we see from (9.66) that

$$
\begin{aligned}
\delta\widetilde{g}_p(t) &= \widetilde{g}_p(t+\delta t) - \widetilde{g}_p(t) \\
&= \mathcal{A}^p(\widetilde{\alpha}_q(t))\,\delta t + \sum_a \mathcal{B}_a^p(\widetilde{\alpha}_q(t)) \int_t^{t+\delta t} dt_1\, \Gamma_a(t_1).
\end{aligned}
\tag{9.105}
$$

Hence, on taking the stochastic average using the results in Appendix G and (9.43), we find that the stochastic average of the single-fluctuation term is

$$
\overline{\delta\widetilde{g}_p(t)} = \overline{\mathcal{A}^p(\widetilde{g}_q(t))}\,\delta t, \tag{9.106}
$$

where the stochastic averages (9.11) have been applied. Thus the stochastic average of $\delta\widetilde{g}_p(t)$ changes linearly with time for short times δt and the rate of change is given by the stochastic average of the drift term in the Fokker–Planck equation.

For the product of two fluctuations, we see that

$$\begin{aligned}\delta\widetilde{g}_p(t)\,\delta\widetilde{g}_q(t) &= \mathcal{A}^p(\widetilde{g}_q(t))\,\mathcal{A}^q(\widetilde{g}_q(t))\,\delta t^2 + \mathcal{A}^p(\widetilde{g}_q(t))\sum_a \mathcal{B}^q_a(\widetilde{\alpha}_q(t))\int_t^{t+\delta t} dt_1\,\Gamma_a(t_1)\,\delta t \\ &\quad + \sum_a \mathcal{B}^p_a(\widetilde{g}_q(t))\int_t^{t+\delta t} dt_1\,\Gamma_a(t_1)\,\mathcal{A}^q(\widetilde{g}_q(t))\,\delta t \\ &\quad + \sum_a \mathcal{B}^p_a(\widetilde{g}_q(t))\int_t^{t+\delta t} dt_1\,\Gamma_a(t_1)\sum_b \mathcal{B}^q_b(\widetilde{g}_q(t))\int_t^{t+\delta t} dt_2\,\Gamma_b(t_2),\end{aligned} \tag{9.107}$$

so, taking the stochastic average and using the results in Appendix G, (9.43) and (9.11), we have

$$\begin{aligned}\overline{\delta\widetilde{g}_p(t)\,\delta\widetilde{g}_q(t)} &= \overline{\mathcal{A}^p(\widetilde{g}_q(t))\,\mathcal{A}^q(\widetilde{g}_q(t))}\,\delta t^2 + \sum_a \overline{\mathcal{A}^p(\widetilde{g}_q(t))\,\mathcal{B}^q_a(\widetilde{g}_q(t))}\int_t^{t+\delta t} dt_1\,\overline{\Gamma_a(t_1)}\,\delta t \\ &\quad + \sum_a \overline{\mathcal{B}^p_a(\widetilde{g}_q(t))\mathcal{A}^q(\widetilde{g}_q(t))}\int_t^{t+\delta t} dt_1\,\overline{\Gamma_a(t_1)}\,\delta t \\ &\quad + \sum_{ab} \overline{\mathcal{B}^p_a(\widetilde{g}_q(t))\mathcal{B}^q_b(\widetilde{g}_q(t))}\int_t^{t+\delta t} dt_1\int_t^{t+\delta t} dt_2\,\overline{\Gamma_a(t_1)\Gamma_b(t_2)} \\ &= \overline{D_{pq}(\widetilde{g}_q(t))}\,\delta t,\end{aligned} \tag{9.108}$$

where the result $\mathcal{B}\mathcal{B}^T = D$ has been used and the terms of $O(\delta t^2)$ have been discarded. These results for the product of the stochastic fluctuations $\delta\widetilde{g}_p(t)\ \delta\widetilde{g}_q(t)$ show that the stochastic average changes linearly with time for short times δt and the rate of change is given by the stochastic average of the diffusion term in the Fokker–Planck equation. The results (9.106) and (9.108) are interesting, but only the former is directly related to quantum correlation functions, and that correlation function will be zero anyway.

9.5.4 Fermion Correlation Functions

We may conveniently list the fermionic annihilation and creation operators as $\{\hat{c}_p\} \equiv \{\hat{c}_1, \cdots, \hat{c}_n, \hat{c}_1^\dagger, \cdots, \hat{c}_n^\dagger\}$, $p = 1, \cdots, 2n$, with the corresponding phase space variables $\{g_p\} \equiv \{g_1, \cdots, g_n, g_1^+, \cdots, g_n^+\}$, as before. The Fokker–Planck equation is taken from (9.66).

First, we consider $F(g, g^+) = g_p$. From (9.35) and with

$$(g_p)\frac{\overleftarrow{\partial}}{\partial g_r} = \delta_{rp}, \tag{9.109}$$

$$(g_p)\frac{\overleftarrow{\partial}}{\partial g_s}\frac{\overleftarrow{\partial}}{\partial g_r} = 0, \tag{9.110}$$

we find for the time derivative of the phase space average

$$\frac{d}{dt}\langle g_p \rangle_t = -\langle A_p \rangle. \tag{9.111}$$

Then, using the equivalence (9.37) to the stochastic average, the time derivative of the phase space average is

$$\frac{d}{dt}\langle g_p \rangle_t = \overline{\mathcal{A}^p(\widetilde{g}_q(t))}, \tag{9.112}$$

relating the time derivative to the stochastic average of the drift term. For the quantum averages of the annihilation and creation operators, we have

$$\langle \hat{c}_p \rangle_t = \int dg^+ dg\, g_p P_{\mathrm{f}}(g, g^+, t) = \langle g_p \rangle_t \,, \tag{9.113}$$

so that for the time derivative of the quantum correlation function,

$$\frac{d}{dt}\langle \hat{c}_p \rangle_t = \overline{\mathcal{A}^p(\widetilde{g}_q(t))}. \tag{9.114}$$

To obtain equations of motion for the correlation function, we use the particular expressions (8.84) for C^p and then reinterpret the stochastic average as a phase space average, placing the phase space variables in antinormal order. For Grassmann number phase variables, sign changes are required. We can then interpret the terms on the right side as quantum correlation functions using (7.82). Thus, for the case of $\langle \hat{c}_i \rangle_t$, we require C_i^- on reverting to the original g_i, g_i^+ notation and identifying quantities via (8.84):

$$\begin{aligned}
\frac{d}{dt}\langle \hat{c}_i \rangle_t &= -\frac{1}{i\hbar}\sum_j h_{ij}\langle g_j \rangle + \frac{1}{i\hbar}\sum_{jkl} \nu_{ij\,kl}\,\langle g_j^+ g_l\, g_k \rangle \\
&\quad + \sum_{jkl} \left(\{\Gamma_{jk\,il} - \Gamma_{li\,kj}\}\,\langle g_l g_k^+ g_j \rangle + \Gamma_{li\,kj}\delta_{lk}\,\langle g_j \rangle \right) \\
&= -\frac{1}{i\hbar}\sum_j h_{ij}\langle g_j \rangle + \frac{1}{i\hbar}\sum_{jkl} \nu_{ij\,kl}\,\langle g_l\, g_k g_j^+ \rangle \\
&\quad + \sum_{jkl} \left(\{\Gamma_{jk\,il} - \Gamma_{li\,kj}\}\,\langle g_j g_l g_k^+ \rangle + \Gamma_{li\,kj}\delta_{lk}\,\langle g_j \rangle \right) \\
&= -\frac{1}{i\hbar}\sum_j h_{ij}\langle \hat{c}_j \rangle + \frac{1}{i\hbar}\sum_{jkl} \nu_{ij\,kl}\,\langle \hat{c}_j^\dagger \hat{c}_l \hat{c}_k \rangle \\
&\quad + \sum_{jkl} \Big(\{\Gamma_{jk\,il} - \Gamma_{li\,kj}\}\,\langle \hat{c}_k^\dagger \hat{c}_j \hat{c}_l \rangle + \Gamma_{li\,kj}\delta_{lk}\,\langle \hat{c}_j \rangle \Big),
\end{aligned} \tag{9.115}$$

with a similar equation for $d/dt\langle \hat{c}_i^\dagger \rangle_t$. This shows that the first-order quantum correlation function is coupled to third order quantum correlation functions. However, in the fermionic case all odd-order quantum correlation functions are zero – the fermion distribution function is an even Grassmann function, and therefore the Grassmann phase space integral is of an odd function overall. Hence both sides of (9.115) are zero.

Second, we consider $F(g, g^+) = g_p g_q$. From (9.35) and with

$$\begin{aligned}
(g_p g_q)\frac{\overleftarrow{\partial}}{\partial g_r} &= -\delta_{rp}(g_q) + (g_p)\delta_{rq}, \\
(g_p g_q)\frac{\overleftarrow{\partial}}{\partial g_s}\frac{\overleftarrow{\partial}}{\partial g_r} &= (-\delta_{sp}(g_q) + (g_p)\delta_{sq})\frac{\overleftarrow{\partial}}{\partial g_r} \\
&= -\delta_{sp}\delta_{rq} + \delta_{sq}\delta_{rp},
\end{aligned} \tag{9.116}$$

we find for the time derivative of the phase space average

$$\begin{aligned}
\frac{d}{dt}\langle g_p g_q\rangle_t &= \langle g_q A_p - g_p A_q\rangle \\
&\quad + \frac{1}{2}\langle D_{pq}\rangle - \frac{1}{2}\langle D_{qp}\rangle \\
&= -\langle (g_p A_q + A_p \alpha_q)\rangle + \langle D_{pq}\rangle,
\end{aligned} \tag{9.117}$$

where (8.95) and the anticommuting of the Grassmann variables have all been used. Then, using the equivalence (9.37) to the stochastic average and $\mathcal{A} = -A$, the time derivative of the phase space average is

$$\frac{d}{dt}\langle g_p g_q\rangle_t = \overline{\widetilde{g}_p(t)\mathcal{A}^q(\widetilde{g}_r(t)) + \mathcal{A}^p(\widetilde{g}_r(t))\widetilde{g}_q(t)} + \overline{D_{pq}(\widetilde{g}_r(t))}, \tag{9.118}$$

relating the time derivative to a sum of stochastic averages of stochastic-variable products with the drift terms together with the stochastic average of the diffusion term. The result has the same form as for bosons. The quantum averages of the three cases of normally ordered products of the annihilation and creation operators are

$$\begin{aligned}
\langle \hat{c}_i \hat{c}_j\rangle_t &= \int dg^+ dg\, g_i g_j P_{\mathrm{f}}(g, g^+, t) = \langle g_i g_j\rangle_t, \\
\langle \hat{c}_i^\dagger \hat{c}_j^\dagger\rangle_t &= \int dg^+ dg\, g_i^+ g_j^+ P_{\mathrm{f}}(g, g^+, t) = \langle g_i^+ g_j^+\rangle_t, \\
\langle \hat{c}_j^\dagger \hat{c}_i\rangle_t &= \int dg^+ dg\, g_i g_j^+ P_{\mathrm{f}}(g, g^+, t) = \langle g_i g_j^+\rangle_t,
\end{aligned} \tag{9.119}$$

so, for example, the time derivative of the quantum correlation function $\langle \hat{c}_j^\dagger \hat{c}_i\rangle_t$ is

$$\frac{d}{dt}\langle \hat{c}_j^\dagger \hat{c}_i\rangle_t = \overline{\widetilde{g}_i(t)C_j^+(\widetilde{g}_r(t)) + C_i^-(\widetilde{g}_r(t))\widetilde{g}_j^+(t)} + \overline{F_{ij}^{-+}(\widetilde{g}_r(t))}, \tag{9.120}$$

where we have reverted to the original g_i, g_i^+ notation and identified quantities via (8.83).

To obtain equations of motion for the quantum correlation function, we use the particular expressions (8.84) for C_i^-, C_j^+ and F_{ij}^{-+} and then reinterpret the stochastic average as a phase space average, placing the phase space variables in antinormal order. For Grassmann phase variables, sign changes are required. We can then interpret the terms on the right side as quantum correlation functions using (7.74). Using a similar approach as for the previous example, we finally have

$$\begin{aligned}
\frac{d}{dt}\langle \hat{c}_j^\dagger \hat{c}_i \rangle_t = &+ \frac{1}{i\hbar}\sum_k h_{jk}^* \langle \hat{c}_k^\dagger \hat{c}_i \rangle_t - \frac{1}{i\hbar}\sum_{mkl} \nu_{jm\,kl}^* \langle \hat{c}_k^\dagger \hat{c}_l^\dagger \hat{c}_i \hat{c}_m \rangle_t \\
&+ \sum_{mkl} (\{\Gamma_{mk\,jl}^* - \Gamma_{lj\,km}^*\} \langle \hat{c}_m^\dagger \hat{c}_l^\dagger \hat{c}_k \hat{c}_i \rangle_t + \Gamma_{lj\,km}^* \delta_{lk} \langle \hat{c}_m^\dagger \hat{c}_i \rangle_t) \\
&- \frac{1}{i\hbar}\sum_k h_{ik} \langle \hat{c}_j^\dagger \hat{c}_k \rangle_t + \frac{1}{i\hbar}\sum_{mkl} \nu_{im\,kl} \langle \hat{c}_m^\dagger \hat{c}_j^\dagger \hat{c}_l \hat{c}_k \rangle_t \\
&+ \sum_{mkl} \left(\{\Gamma_{mk\,il} - \Gamma_{li\,km}\} \langle \hat{c}_k^\dagger \hat{c}_j^\dagger \hat{c}_m \hat{c}_l \rangle_t + \Gamma_{li\,km} \delta_{lk} \langle \hat{c}_j^\dagger \hat{c}_m \rangle_t \right) \\
&+ 2\sum_{kl} \{\Gamma_{jk\,il}\} \langle \hat{c}_k^\dagger \hat{c}_l \rangle_t .
\end{aligned} \tag{9.121}$$

Details are given in Appendix G. This shows that the second-order quantum correlation functions are coupled to fourth-order quantum correlation functions. The result is similar to the bosonic case. This feature of coupling in higher-order quantum correlation functions is quite general, and a hierarchy of equations for the quantum correlation functions results. This may be solved via truncation methods.

It should be pointed out that the equations for the time derivatives of fermionic quantum correlation functions can be derived without using the intermediate step of introducing the stochastic forms. Thus one can pass directly from (9.117) to (9.121) without using (9.120). However, the application of the stochastic equations does perhaps represent the simpler approach.

Exercises

(9.1) Derive the equation of motion for the bosonic third-order quantum correlation functions $(d/dt)\langle \hat{a}_j^\dagger \hat{a}_l \hat{a}_k \rangle$ and $(d/dt)\langle \hat{a}_k^\dagger \hat{a}_l \hat{a}_j \rangle$.

(9.2) Derive the equation of motion for the fermionic third-order quantum correlation functions $(d/dt)\ \langle \hat{c}_j^\dagger \hat{c}_l \hat{c}_k \rangle$ and $(d/dt)\langle \hat{c}_k^\dagger \hat{c}_j \hat{c}_l \rangle$.

10
Application to Few-Mode Systems

As a preamble to the later multi-mode applications, which we cover in Chapters 12–15, it is instructive to look at the treatment of some problems with a limited number of modes using the formalism covered in the previous three chapters. Many systems have been previously studied by these methods in standard quantum optics textbooks, so we concentrate on three that are less well studied, but which are simple enough to elucidate the general method. These are two-mode BEC interferometry, Cooper pairing in a two-fermion system and the Jaynes–Cummings model. The first two involve bosons and fermions in turn; the last involves both.

10.1 Boson Case – Two-Mode BEC Interferometry

10.1.1 Introduction

Bose–Einstein condensates (BECs) are quantum systems on a macroscopic scale. When bosonic atoms are trapped in a single-well potential and cooled enough so that their de Broglie waves overlap significantly, all the atoms eventually occupy the lowest-energy single-particle state or mode, and hence macroscopic numbers of atoms are all described by a single atomic wave [27, 47]. The coherence properties of such a BEC have led to applications of BECs in high-precision interferometry [65–67]. One form of BEC interferometry requires a two-mode system, as in a near-symmetric double-well potential – which may be produced either by magnetic fields or by standing-wave light fields. These modes may each be localised in different wells or may be symmetric and antisymmetric delocalised modes spread over both wells. In both cases the quantum state can involve bosonic atoms split between the two modes, leading to interferometry using a BEC based on quantum interference in a two-mode system. This may depend on asymmetries in the double well, such as from gravitational potentials, and the interference patterns are affected by collisions between the atoms – the latter result in dephasing effects. The interferometric process itself depends on the initial state chosen, and can result from changes to the double well potential – for example, changes in the barrier height will alter the coupling associated with quantum tunnelling between the two localised modes. This leads to effects similar to those found for an optical beam splitter. We consider here a simple theory of double-well BEC interferometry based on the Josephson Hamiltonian (see [29] for a review), in which the mode functions are assumed to be localised in the two wells and where changes in the mode functions during each stage of the interferometric process are ignored. More elaborate theoretical treatments of two-mode BEC interferometry are available. These include theories that

Phase Space Methods for Degenerate Quantum Gases. First Edition. Bryan J. Dalton, John Jeffers, and Stephen M. Barnett.
Published in 2015 by Oxford University Press.

only allow for condensate modes and involve coupled Gross–Pitaevskii equations for the two mode functions and matrix mechanics equations for the various Fock state amplitudes [29, 68], as well as full phase space theories involving both condensate and non-condensate modes [52, 53] or just treating all modes together [69, 70]. An example of the former with just a single condensate mode is covered in Chapter 15.

10.1.2 Modes and Hamiltonian

If the mode functions in the left and right wells, ϕ_{L} and ϕ_{R}, are spatially separate and almost completely non-overlapping, we can treat them as two separate weakly interacting modes. Thus two localised modes are involved.

In the Josephson model (see [29] for details), we can write the Hamiltonian in simplified form in terms of creation and annihilation operators for the left and right modes,

$$\hat{H} = \hat{H}_0 + \hat{H}_1, \tag{10.1}$$

as the sum of one-boson and two-boson terms where

$$\begin{aligned} \hat{H}_0 &= \hbar\left(\omega_{\mathrm{L}}\hat{a}_{\mathrm{L}}^\dagger\hat{a}_{\mathrm{L}} + \omega_{\mathrm{R}}\hat{a}_{\mathrm{R}}^\dagger\hat{a}_{\mathrm{R}} + \Omega_{\mathrm{LR}}\hat{a}_{\mathrm{L}}^\dagger\hat{a}_{\mathrm{R}} + \Omega_{\mathrm{RL}}\hat{a}_{\mathrm{R}}^\dagger\hat{a}_{\mathrm{L}}\right), \\ \hat{H}_1 &= \hbar\left(\chi_{\mathrm{L}}\hat{a}_{\mathrm{L}}^\dagger\hat{a}_{\mathrm{L}}^\dagger\hat{a}_{\mathrm{L}}\hat{a}_{\mathrm{L}} + \chi_{\mathrm{R}}\hat{a}_{\mathrm{R}}^\dagger\hat{a}_{\mathrm{R}}^\dagger\hat{a}_{\mathrm{R}}\hat{a}_{\mathrm{R}}\right). \end{aligned} \tag{10.2}$$

The Hamiltonian contains parameters related to the mode functions in the left and right wells by

$$\begin{aligned} &\int dx\phi_{\mathrm{L}}^*\left(-\frac{\hbar^2}{2m}\nabla^2 + V\right)\phi_{\mathrm{L}} = \hbar\omega_{\mathrm{L}}, \qquad \int dx\phi_{\mathrm{R}}^*\left(-\frac{\hbar^2}{2m}\nabla^2 + V\right)\phi_{\mathrm{R}} = \hbar\omega_{\mathrm{R}}, \\ &\int dx\phi_{\mathrm{L}}^*\left(-\frac{\hbar^2}{2m}\nabla^2 + V\right)\phi_{\mathrm{R}} = \hbar\Omega_{\mathrm{LR}}, \qquad \int dx\phi_{\mathrm{R}}^*\left(-\frac{\hbar^2}{2m}\nabla^2 + V\right)\phi_{\mathrm{L}} = \hbar\Omega_{\mathrm{RL}}, \\ &\frac{g}{2}\int dx\,|\phi_{\mathrm{L}}|^4 = \hbar\chi_{\mathrm{L}}, \qquad \frac{g}{2}\int dx\,|\phi_{\mathrm{R}}|^4 = \hbar\chi_{\mathrm{R}}, \end{aligned} \tag{10.3}$$

where $\hbar\omega_{\mathrm{L}}, \hbar\omega_{\mathrm{R}}$ are single-atom mode energies, $\hbar\Omega_{\mathrm{LR}}, \hbar\Omega_{\mathrm{RL}}$ are inter-mode coupling energies due to quantum tunnelling and $\hbar\chi_{\mathrm{L}}, \hbar\chi_{\mathrm{R}}$ are boson–boson collision energies for bosons in the same mode. The trap potential is V, and g is the boson collision coupling constant for the zero-range approximation. Mean-field shifts of the single-atom energies and other terms involving smaller overlap integrals such as $\int dx\,|\phi_{\mathrm{L}}|^2\,|\phi_{\mathrm{R}}|^2$ are omitted.

10.1.3 Fokker–Planck and Ito Equations – $P+$

By using the formulae derived in Chapter 8, we can write down the various terms of the Fokker–Planck equation for the *positive* P representation. The system is closed, and so there is no external reservoir to take into account and the density operator satisfies the Liouville–von Neumann equation (2.82). We use the equations (8.57) (with relaxation terms Γ_{ijkl} set to zero) to write the c-number quantities in the Fokker–Planck equation (8.56) as

$$
\begin{aligned}
C_{\mathrm{L}}^{-} &= -i\left(\omega_{\mathrm{L}}+\chi_{\mathrm{L}}\alpha_{\mathrm{L}}^{+}\alpha_{\mathrm{L}}\right)\alpha_{\mathrm{L}}-i\Omega_{\mathrm{LR}}\alpha_{\mathrm{R}},\\
C_{\mathrm{R}}^{-} &= -i\left(\omega_{\mathrm{R}}+\chi_{\mathrm{R}}\alpha_{\mathrm{R}}^{+}\alpha_{\mathrm{R}}\right)\alpha_{\mathrm{R}}-i\Omega_{\mathrm{RL}}\alpha_{\mathrm{L}},\\
C_{\mathrm{L}}^{+} &= i\left(\omega_{\mathrm{L}}+\chi_{\mathrm{L}}\alpha_{\mathrm{L}}^{+}\alpha_{\mathrm{L}}\right)\alpha_{\mathrm{L}}^{+}+i\Omega_{\mathrm{RL}}\alpha_{\mathrm{R}}^{+},\\
C_{\mathrm{R}}^{+} &= i\left(\omega_{\mathrm{R}}+\chi_{\mathrm{R}}\alpha_{\mathrm{R}}^{+}\alpha_{\mathrm{R}}\right)\alpha_{\mathrm{R}}^{+}+i\Omega_{\mathrm{LR}}\alpha_{\mathrm{L}}^{+}
\end{aligned}
\tag{10.4}
$$

for the drift terms and

$$
\begin{aligned}
F_{\mathrm{LL}}^{--} &= -i\chi_{\mathrm{L}}\alpha_{\mathrm{L}}^{2},\; F_{\mathrm{RR}}^{--} = -i\chi_{\mathrm{R}}\alpha_{\mathrm{R}}^{2},\\
F_{\mathrm{LL}}^{++} &= i\chi_{\mathrm{L}}(\alpha_{\mathrm{L}}^{+})^{2},\; F_{\mathrm{RR}}^{++} = i\chi_{\mathrm{R}}(\alpha_{\mathrm{R}}^{+})^{2}
\end{aligned}
\tag{10.5}
$$

for the diffusion terms. We can combine these into a diffusion matrix

$$
[F]=\begin{pmatrix}
-i\chi_{\mathrm{L}}\alpha_{\mathrm{L}}^{2} & 0 & 0 & 0\\
0 & -i\chi_{\mathrm{R}}\alpha_{\mathrm{R}}^{2} & 0 & 0\\
0 & 0 & i\chi_{\mathrm{L}}(\alpha_{\mathrm{L}}^{+})^{2} & 0\\
0 & 0 & 0 & i\chi_{\mathrm{R}}(\alpha_{\mathrm{R}}^{+})^{2}
\end{pmatrix},
\tag{10.6}
$$

where the rows and columns correspond to $\alpha_{\mathrm{L}}, \alpha_{\mathrm{R}}, \alpha_{\mathrm{L}}^{+}$ and α_{R}^{+}. The matrix is symmetric and diagonal, and so the Takagi factorisation $[F]=[B][B]^{T}$ is performed trivially to give

$$
[B]=\begin{pmatrix}
\frac{1}{\sqrt{2}}(1-i)\sqrt{\chi_{\mathrm{L}}}\alpha_{\mathrm{L}} & 0 & 0 & 0\\
0 & \frac{1}{\sqrt{2}}(1-i)\sqrt{\chi_{\mathrm{R}}}\alpha_{\mathrm{R}} & 0 & 0\\
0 & 0 & \frac{1}{\sqrt{2}}(1+i)\sqrt{\chi_{\mathrm{L}}}\alpha_{\mathrm{L}}^{+} & 0\\
0 & 0 & 0 & \frac{1}{\sqrt{2}}(1+i)\sqrt{\chi_{\mathrm{R}}}\alpha_{\mathrm{R}}^{+}
\end{pmatrix}.
\tag{10.7}
$$

This matrix gives the noise terms in the Ito stochastic differential equations, which we now write down from (9.24) as

$$
\begin{aligned}
\frac{\partial\widetilde{\alpha}_{\mathrm{L}}}{\partial t} &= -i\left(\omega_{\mathrm{L}}+\chi_{\mathrm{L}}\widetilde{\alpha}_{\mathrm{L}}^{+}\widetilde{\alpha}_{\mathrm{L}}\right)\widetilde{\alpha}_{\mathrm{L}}-i\Omega_{\mathrm{LR}}\widetilde{\alpha}_{\mathrm{R}}+\frac{1}{\sqrt{2}}(1-i)\sqrt{\chi_{\mathrm{L}}}\widetilde{\alpha}_{\mathrm{L}}\Gamma_{\mathrm{L}},\\
\frac{\partial\widetilde{\alpha}_{\mathrm{R}}}{\partial t} &= -i\left(\omega_{\mathrm{R}}+\chi_{\mathrm{R}}\widetilde{\alpha}_{\mathrm{R}}^{+}\widetilde{\alpha}_{\mathrm{R}}\right)\widetilde{\alpha}_{\mathrm{R}}-i\Omega_{\mathrm{RL}}\widetilde{\alpha}_{\mathrm{L}}+\frac{1}{\sqrt{2}}(1-i)\sqrt{\chi_{\mathrm{R}}}\widetilde{\alpha}_{\mathrm{R}}\Gamma_{\mathrm{R}},\\
\frac{\partial\widetilde{\alpha}_{\mathrm{L}}^{+}}{\partial t} &= i\left(\omega_{\mathrm{L}}+\chi_{\mathrm{L}}\widetilde{\alpha}_{\mathrm{L}}^{+}\widetilde{\alpha}_{\mathrm{L}}\right)\widetilde{\alpha}_{\mathrm{L}}^{+}+i\Omega_{\mathrm{RL}}\widetilde{\alpha}_{\mathrm{R}}^{+}+\frac{1}{\sqrt{2}}(1+i)\sqrt{\chi_{\mathrm{L}}}\widetilde{\alpha}_{\mathrm{L}}^{+}\Gamma_{\mathrm{L}}^{+},\\
\frac{\partial\widetilde{\alpha}_{\mathrm{R}}^{+}}{\partial t} &= i\left(\omega_{\mathrm{R}}+\chi_{\mathrm{R}}\widetilde{\alpha}_{\mathrm{R}}^{+}\widetilde{\alpha}_{\mathrm{R}}\right)\widetilde{\alpha}_{\mathrm{R}}^{+}+i\Omega_{\mathrm{LR}}\widetilde{\alpha}_{\mathrm{L}}^{+}+\frac{1}{\sqrt{2}}(1+i)\sqrt{\chi_{\mathrm{R}}}\widetilde{\alpha}_{\mathrm{R}}^{+}\Gamma_{\mathrm{R}}^{+}
\end{aligned}
\tag{10.8}
$$

where the Γ's are Gaussian–Markov noise sources required for stochastic simulation of these equations. Their properties are given in (9.11). The Ito stochastic equations have several interesting features. Each phase space stochastic variable contains terms representing a free oscillation of the phase space variable at a mode frequency such as ω_{L}, the latter being shifted by a mean-field correction such as $\chi_{\mathrm{L}}\widetilde{\alpha}_{\mathrm{L}}^{+}\widetilde{\alpha}_{\mathrm{L}}$ associated with bosons in the same mode. In addition, there is a deterministic coupling to a related phase variable via tunnelling terms such as $\Omega_{\mathrm{LR}}\widetilde{\alpha}_{\mathrm{R}}$. Finally, each phase variable is affected by stochastic noise, such as $(1/\sqrt{2})(1-i)\sqrt{\chi_{\mathrm{L}}}\widetilde{\alpha}_{\mathrm{L}}\Gamma_{\mathrm{L}}$, which is phase-dependent through factors such as $\widetilde{\alpha}_{\mathrm{L}}$ and whose importance increases with collision energy factors such as χ_{L}. In this case, dephasing effects originate from boson–boson collisions.

10.1.4 Fokker–Planck and Ito Equations – Wigner

It is instructive to perform the same derivation for the Wigner function as a comparison. The new Fokker–Planck equation is of the form (8.64), where from (8.65) the analogues of the equations (10.4) are

$$\begin{aligned}
C_{\mathrm{L}}^{-} &= -i\left(\omega_{\mathrm{L}} + 2\chi_{\mathrm{L}}(\alpha_{\mathrm{L}}^{+}\alpha_{\mathrm{L}} - 1)\right)\alpha_{\mathrm{L}} - i\Omega_{\mathrm{LR}}\alpha_{\mathrm{R}},\\
C_{\mathrm{R}}^{-} &= -i\left(\omega_{\mathrm{R}} + 2\chi_{\mathrm{R}}(\alpha_{\mathrm{R}}^{+}\alpha_{\mathrm{R}} - 1)\right)\alpha_{\mathrm{R}} - i\Omega_{\mathrm{RL}}\alpha_{\mathrm{L}},\\
C_{\mathrm{L}}^{+} &= i\left(\omega_{\mathrm{L}} + 2\chi_{\mathrm{L}}(\alpha_{\mathrm{L}}^{+}\alpha_{\mathrm{L}} - 1)\right)\alpha_{\mathrm{L}}^{+} + i\Omega_{\mathrm{RL}}\alpha_{\mathrm{R}}^{+},\\
C_{\mathrm{R}}^{+} &= i\left(\omega_{\mathrm{R}} + 2\chi_{\mathrm{R}}(\alpha_{\mathrm{R}}^{+}\alpha_{\mathrm{R}} - 1)\right)\alpha_{\mathrm{R}}^{+} + i\Omega_{\mathrm{LR}}\alpha_{\mathrm{L}}^{+}.
\end{aligned}\tag{10.9}$$

Note that the drift terms are different from the $P+$ distribution case. The diffusion terms corresponding to (10.5) all vanish, and the Fokker–Planck equation contains third-order terms. Ito equations can only be obtained if the third-order terms are neglected, which is standard in Wigner-based stochastic methods. This amounts to ignoring some parts of the evolution which produce the most quantum behaviour, but is valid in the limit where the mean boson numbers in each well are significantly greater than 1. In the case of BEC interferometry based on highly occupied modes, this ought to be a good approximation. Ignoring the third-order terms leads to Ito stochastic equations

$$\begin{aligned}
\frac{\partial\widetilde{\alpha}_{\mathrm{L}}}{\partial t} &= -i\left(\omega_{\mathrm{L}} + 2\chi_{\mathrm{L}}(\widetilde{\alpha}_{\mathrm{L}}^{+}\widetilde{\alpha}_{\mathrm{L}} - 1)\right)\widetilde{\alpha}_{\mathrm{L}} - i\Omega_{\mathrm{LR}}\widetilde{\alpha}_{\mathrm{R}},\\
\frac{\partial\widetilde{\alpha}_{\mathrm{R}}}{\partial t} &= -i\left(\omega_{\mathrm{R}} + 2\chi_{\mathrm{R}}(\widetilde{\alpha}_{\mathrm{R}}^{+}\widetilde{\alpha}_{\mathrm{R}} - 1)\right)\widetilde{\alpha}_{\mathrm{R}} - i\Omega_{\mathrm{RL}}\widetilde{\alpha}_{\mathrm{L}},\\
\frac{\partial\widetilde{\alpha}_{\mathrm{L}}^{+}}{\partial t} &= i\left(\omega_{\mathrm{L}} + 2\chi_{\mathrm{L}}(\widetilde{\alpha}_{\mathrm{L}}^{+}\widetilde{\alpha}_{\mathrm{L}} - 1)\right)\widetilde{\alpha}_{\mathrm{L}}^{+} + i\Omega_{\mathrm{RL}}\widetilde{\alpha}_{\mathrm{R}}^{+},\\
\frac{\partial\widetilde{\alpha}_{\mathrm{R}}^{+}}{\partial t} &= i\left(\omega_{\mathrm{R}} + 2\chi_{\mathrm{R}}(\widetilde{\alpha}_{\mathrm{R}}^{+}\widetilde{\alpha}_{\mathrm{R}} - 1)\right)\widetilde{\alpha}_{\mathrm{R}}^{+} + i\Omega_{\mathrm{LR}}\widetilde{\alpha}_{\mathrm{L}}^{+}.
\end{aligned}\tag{10.10}$$

There are of course no noise terms for the Wigner-based Ito equations. This is offset by the fact that the stochastic averages determined from these equations will provide symmetrically ordered correlation functions, and the quantum noise is naturally already contained in these. The Ito stochastic equations have several interesting features, apart from the lack of noise terms. Each phase space stochastic variable contains terms representing a free oscillation at a mode frequency such as ω_{L}, the latter being shifted by a mean-field correction such as $2\chi_{\mathrm{L}}(\widetilde{\alpha}_{\mathrm{L}}^{+}\widetilde{\alpha}_{\mathrm{L}} - 1)$ associated with bosons in the same mode. This mean-field correction is about twice that for the $P+$ case and involves expressions such as $g\int dx\,|\phi_{\mathrm{L}}|^{4}/\hbar = 2\chi_{\mathrm{L}}$. Since the other factor, such as $(\widetilde{\alpha}_{\mathrm{L}}^{+}\widetilde{\alpha}_{\mathrm{L}} - 1)$

$\sim (n_{\mathrm{L}} - 1)$, gives the number of bosons in the mode minus one, this expression for the mean-field correction reflects the physical situation more accurately than in the $P+$ case. In addition, there is also a deterministic coupling to a related phase variable via tunnelling terms such as $\Omega_{\mathrm{LR}}\widetilde{\alpha}_{\mathrm{R}}$, as in the $P+$ case. However, the lack of any noise terms raises the question of how to account for dephasing effects.

10.1.5 Conclusion

Whether the equations based on the Wigner or the positive P distribution are used to model this system is to some extent a matter of personal choice, though the Wigner approach would be questionable if the number of bosonic atoms was small. If the normally ordered qualities of the positive P distribution are used, the correlation functions correspond to proper boson number and field quantities, but stochastic noise must modelled as an integral part of the equations describing the stochastic quantities. Alternatively, if the symmetrically ordered Wigner distribution is used, the stochasticity vanishes in the modelled equations, but the noise expectation values will contribute to the correlations. In either case the intrinsic stochasticity will limit the visibility of any interference effects that arise from the interaction of BECs in the two wells.

10.2 Fermion Case – Cooper Pairing in a Two-Fermion System

10.2.1 Introduction

Superconductors are a further example of a quantum system on a macroscopic scale. In this case the particles involved are fermions – electrons – and the physics of superconductivity is based on the presence of an interaction between pairs of electrons, with opposite spins and momenta. This interaction leads to the formation of so-called *Cooper pairs* of electrons, which extend over macroscopic distances and result in the long-range coherence that is characteristic of superconductors. In normal superconductors, this interaction results from couplings between the electrons and the phonons associated with lattice vibrations in the metal. Here, however, we will see that similar Cooper pairing can occur for any fermionic particles which interact via short-range spin-conserving potentials. The spin-conserving interactions lead to non-zero two-fermion-number correlations of the form $\langle \hat{n}_{\mathbf{K}(+)}\hat{n}_{\mathbf{L}(-)} \rangle - \langle \hat{n}_{\mathbf{K}(+)} \rangle \langle \hat{n}_{\mathbf{L}(+)} \rangle$ for the four cases $(\mathbf{K}, \mathrm{L}) = (\mathbf{k}, \mathbf{k}), (\mathbf{k}, -\mathbf{k}), (-\mathbf{k}, -\mathbf{k})$ and $(-\mathbf{k}, \mathbf{k})$. Non-zero correlations for the $(\mathbf{k}, \mathbf{k})$ and $(-\mathbf{k}, -\mathbf{k})$ cases are unexpected. The treatment follows the approach of Plimak et al. [14], who established this result using unnormalised Grassmann distributions and Ito stochastic equations for Grassmann phase space variables. Their approach was based on the Fokker–Planck equation obtained from the Matsubara equation (2.89), since they investigated the effect of temperature change on the equilibrium state. For our purposes, we will consider the dynamical evolution of coherences between two distinct Cooper pair states in the situation where the system is initially in one of these states and where relaxation processes and external potentials are ignored. The essential features of how the two-fermion-number correlations develop owing to coupling between the two distinct Cooper pair states via the short-range spin-conserving potentials can be demonstrated in a system with two spin-1/2 fermions involving four free-space modes with two opposite momenta $\mathbf{k}$, $-\mathbf{k}$ and each with spins $+$ and $-$.

10.2.2 Modes and Hamiltonian

Designating the modes as $|\phi_1\rangle = |\phi_{\mathbf{k}\,(+)}\rangle, |\phi_2\rangle = |\phi_{\mathbf{k}\,(-)}\rangle, |\phi_3\rangle = |\phi_{-\mathbf{k}(+)}\rangle, |\phi_4\rangle = |\phi_{-\mathbf{k}\,(-)}\rangle$ and the corresponding fermion annihilation operators as $\hat{c}_1 \equiv \hat{c}_{\mathbf{k}\,(+)}, \hat{c}_2 \equiv \hat{c}_{\mathbf{k}\,(-)}, \hat{c}_3 \equiv \hat{c}_{-\mathbf{k}\,(+)}, \hat{c}_4 \equiv \hat{c}_{-\,\mathbf{k}\,(-)}$, the spatial mode functions are

$$\phi_1(\mathbf{r}) = (\langle +| \langle \mathbf{r}|)\,|\phi_1\rangle = \frac{1}{\sqrt{V}} \exp(i\mathbf{k}\cdot\mathbf{r}), \quad \phi_2(\mathbf{r}) = (\langle -| \langle \mathbf{r}|)\,|\phi_2\rangle = \frac{1}{\sqrt{V}} \exp(i\mathbf{k}\cdot\mathbf{r}),$$
$$\phi_3(\mathbf{r}) = (\langle +| \langle \mathbf{r}|)\,|\phi_3\rangle = \frac{1}{\sqrt{V}} \exp(-i\mathbf{k}\cdot\mathbf{r}), \quad \phi_4(\mathbf{r}) = (\langle -| \langle \mathbf{r}|)\,|\phi_4\rangle = \frac{1}{\sqrt{V}} \exp(-i\mathbf{k}\cdot\mathbf{r}), \tag{10.11}$$

where the mode functions are box-normalised in a volume V. From (2.73), the field operators for spin-$+$ and spin-$-$ fermions are then

$$\hat{\Psi}_+(\mathbf{r}) = \hat{c}_1\phi_1(\mathbf{r}) + \hat{c}_3\phi_3(\mathbf{r}), \quad \hat{\Psi}_-(\mathbf{r}) = \hat{c}_2\phi_2(\mathbf{r}) + \hat{c}_4\phi_4(\mathbf{r}). \tag{10.12}$$

The Hamiltonian is as in (2.79). Substituting for the field operators, we find that the Hamiltonian can be written as

$$\hat{H} = \hat{H}_0 + \hat{H}_1, \tag{10.13}$$

as the sum of one-fermion and two-fermion terms where

$$\hat{H}_0 = \hbar\omega(\hat{c}_1^\dagger\hat{c}_1 + \hat{c}_2^\dagger\hat{c}_2 + \hat{c}_3^\dagger\hat{c}_3 + \hat{c}_4^\dagger\hat{c}_4), \tag{10.14}$$

$$\begin{aligned}\hat{H}_1 = \frac{g}{2V}(&\hat{c}_1^\dagger\hat{c}_2^\dagger\hat{c}_2\hat{c}_1 + \hat{c}_2^\dagger\hat{c}_1^\dagger\hat{c}_1\hat{c}_2 + \hat{c}_1^\dagger\hat{c}_4^\dagger\hat{c}_4\hat{c}_1 + \hat{c}_4^\dagger\hat{c}_1^\dagger\hat{c}_1\hat{c}_4 \\ &+\hat{c}_3^\dagger\hat{c}_2^\dagger\hat{c}_2\hat{c}_3 + \hat{c}_2^\dagger\hat{c}_3^\dagger\hat{c}_3\hat{c}_2 + \hat{c}_3^\dagger\hat{c}_4^\dagger\hat{c}_4\hat{c}_3 + \hat{c}_4^\dagger\hat{c}_3^\dagger\hat{c}_3\hat{c}_4 \\ &+\hat{c}_1^\dagger\hat{c}_4^\dagger\hat{c}_2\hat{c}_3 + \hat{c}_4^\dagger\hat{c}_1^\dagger\hat{c}_3\hat{c}_2 + \hat{c}_3^\dagger\hat{c}_2^\dagger\hat{c}_4\hat{c}_1 + \hat{c}_2^\dagger\hat{c}_3^\dagger\hat{c}_1\hat{c}_4)\end{aligned} \tag{10.15}$$

and $\hbar\omega = \hbar^2\mathbf{k}^2/2M$. Expressions for h_{ij} and ν_{ijkl} are obtained using (2.53) and (2.54). The one-body terms are diagonal and all equal. Several of the two-body integrals are zero in box normalisation and the remainder are equal. In terms of the general notation in (2.51), the non-zero h_{ij} and ν_{ijkl} are

$$\begin{aligned}&h_{11} = h_{22} = h_{33} = h_{44} = \hbar\omega, \\ &\nu_{12\,12} = \nu_{21\,21} = \nu_{14\,14} = \nu_{41\,41} = \nu_{32\,32} = \nu_{23\,23} = \nu_{34\,34} = \nu_{43\,43} = \kappa, \\ &\nu_{14\,32} = \nu_{41\,23} = \nu_{32\,14} = \nu_{23\,41} = \kappa,\end{aligned} \tag{10.16}$$

with $\kappa = g/V$.

10.2.3 Initial Conditions

For the case where there are $N = 2$ fermions, there are six different Fock states, designated $|\Phi_a\rangle$:

$$\begin{aligned}
&|\Phi_1\rangle = \hat{c}_1^\dagger \hat{c}_2^\dagger \, |0\rangle \,, \quad |\Phi_2\rangle = \hat{c}_3^\dagger \hat{c}_4^\dagger \, |0\rangle \,, \quad |\Phi_3\rangle = \hat{c}_1^\dagger \hat{c}_4^\dagger \, |0\rangle \,, \quad |\Phi_4\rangle = \hat{c}_2^\dagger \hat{c}_3^\dagger \, |0\rangle \,, \\
&|\Phi_5\rangle = \hat{c}_1^\dagger \hat{c}_3^\dagger \, |0\rangle \,, \quad |\Phi_6\rangle = \hat{c}_2^\dagger \hat{c}_4^\dagger \, |0\rangle \,.
\end{aligned} \tag{10.17}$$

The states $|\Phi_1\rangle$ to $|\Phi_4\rangle$ are non-magnetic, having one fermion in a spin-$+$ mode and one fermion in a spin-$-$ mode. The states $|\Phi_5\rangle$ and $|\Phi_6\rangle$ are magnetic, having both fermions in a spin-$+$ mode or in a spin-$-$ mode. The states $|\Phi_3\rangle$ and $|\Phi_4\rangle$ both describe Cooper pairs. We will consider the case where the initial state is the pure Cooper pair state $|\Phi_3\rangle$,

$$\hat{\rho}(0) = |\Phi_3\rangle \, \langle \Phi_{,3} | , \tag{10.18}$$

so there is one fermion in the mode $\left|\phi_{\mathbf{k}\,(+)}\right\rangle$ and the other in the mode $\left|\phi_{-\mathbf{k}\,(-)}\right\rangle$. The interaction term $(g/2V)\,\hat{c}_3^\dagger \hat{c}_2^\dagger \hat{c}_4 \hat{c}_1$ couples the Cooper pair state $|\Phi_3\rangle$ to the other Cooper pair state $|\Phi_4\rangle$, which has one fermion in the mode $\left|\phi_{\mathbf{k}\,(-)}\right\rangle$ and the other in the mode $\left|\phi_{-\mathbf{k}\,(+)}\right\rangle$, so a non-zero coherence between the state $|\Phi_3\rangle$ and the state $|\Phi_4\rangle$ should develop. It is this coherence that we will determine, since its presence indicates that coupling between the two Cooper pair states, each with opposite momenta and spins, is taking place. This coupling leads to the two-fermion-number correlation results in [14], and is due to the true energy eigenstates being of the form $(|\Phi_3\rangle \pm |\Phi_4\rangle)/\sqrt{2}$ rather than just $|\Phi_3\rangle$ or $|\Phi_4\rangle$. These eigenstates are entangled states.

10.2.4 Fokker–Planck Equations – Unnormalised B

From (8.99), the quantities in the Fokker–Planck equation are

$$\begin{aligned}
&C_1^- = i\omega g_1, \quad C_2^- = i\omega g_2, \quad C_3^- = i\omega g_3, \quad C_4^- = i\omega g_4, \\
&C_1^+, = -i\omega g_1^+, \quad C_2^+ = -i\omega g_2^+, \quad C_3^+ = -i\omega g_3^+ \quad C_4^+ = -i\omega g_4^+
\end{aligned} \tag{10.19}$$

for the drift terms and

$$\begin{aligned}
&\mathcal{F}_{11}^{--} = 0, \quad \mathcal{F}_{12}^{--} = \frac{\kappa}{i\hbar} g_2 g_1, \quad \mathcal{F}_{13}^{--} = 0, \quad \mathcal{F}_{14}^{--} = \frac{\kappa}{i\hbar}(g_4 g_1 + g_2 g_3), \\
&\mathcal{F}_{21}^{--} = \frac{\kappa}{i\hbar} g_1 g_2, \quad \mathcal{F}_{22}^{--} = 0, \quad \mathcal{F}_{23}^{--} = \frac{\kappa}{i\hbar}(g_3 g_2 + g_1 g_4), \quad \mathcal{F}_{24}^{--} = 0, \\
&\mathcal{F}_{31}^{--} = 0, \quad \mathcal{F}_{32}^{--} = \frac{\kappa}{i\hbar}(g_2 g_3 + g_4 g_1), \quad \mathcal{F}_{33}^{--} = 0, \quad \mathcal{F}_{34}^{--} = \frac{\kappa}{i\hbar} g_4 g_3, \\
&\mathcal{F}_{41}^{--} = \frac{\kappa}{i\hbar}(g_1 g_4 + g_3 g_2), \quad \mathcal{F}_{42}^{--} = 0, \quad \mathcal{F}_{43}^{--} = \frac{\kappa}{i\hbar} g_3 g_4, \quad \mathcal{F}_{44}^{--} = 0,
\end{aligned} \tag{10.20}$$

together with

$$\begin{aligned}
&\mathcal{F}_{11}^{++} = 0, \quad \mathcal{F}_{12}^{++} = -\frac{\kappa}{i\hbar} g_1^+ g_2^+, \quad \mathcal{F}_{13}^{++} = 0, \quad \mathcal{F}_{14}^{++} = -\frac{\kappa}{i\hbar}(g_1^+ g_4^+ + g_3^+ g_2^+), \\
&\mathcal{F}_{21}^{++} = -\frac{\kappa}{i\hbar} g_2^+ g_1^+, \quad \mathcal{F}_{22}^{++} = 0, \quad \mathcal{F}_{23}^{++} = -\frac{\kappa}{i\hbar}(g_2^+ g_3^+ + g_4^+ g_1^+), \quad \mathcal{F}_{24}^{++} = 0, \\
&\mathcal{F}_{31}^{++} = 0, \quad \mathcal{F}_{32}^{++} = -\frac{\kappa}{i\hbar}(g_3^+ g_2^+ + g_1^+ g_4^+), \quad \mathcal{F}_{33}^{++} = 0, \quad \mathcal{F}_{34}^{++} = -\frac{\kappa}{i\hbar} g_3^+ g_4^+, \\
&\mathcal{F}_{41}^{++} = -\frac{\kappa}{i\hbar}(g_4^+ g_1^+ + g_2^+ g_3^+), \quad \mathcal{F}_{42}^{++} = 0, \quad \mathcal{F}_{43}^{++} = -\frac{\kappa}{i\hbar} g_4^+ g_3^+, \quad \mathcal{F}_{44}^{++} = 0
\end{aligned} \tag{10.21}$$

and

$$\mathcal{F}_{ij}^{-+} = 0, \qquad \mathcal{F}_{ij}^{+-} = 0, \qquad i, j = 1, 2, 3, 4, \tag{10.22}$$

for the diffusion terms. The diffusion matrices are antisymmetric Grassmann functions, as shown in the general theory.

10.2.5 Ito Equations – Unnormalised *B*

The form of the Ito stochastic equations is determined via the matrices M^{AB} in (9.54). The matrix M^{--} has elements $M_{ikjl}^{--} = \nu_{ijkl}/i\hbar$, so, using (10.16), we have

$$\begin{aligned}
&M_{11\,22}^{--} = M_{22\,11}^{--} = M_{11\,44}^{--} = M_{44\,11}^{--} = M_{33\,22}^{--} = M_{22\,33}^{--} = M_{33\,44}^{--} = M_{44\,33}^{--} = \lambda, \\
&M_{13\,42}^{--} = M_{42\,13}^{--} = M_{31\,24}^{--} = M_{24\,31}^{--} = \lambda,
\end{aligned} \tag{10.23}$$

where $\lambda = \kappa/i\hbar$. Note the different order of the indices. The elements of the matrix M^{++} are given by $M_{ikjl}^{++} = (M_{ik\,jl}^{--})^*$ and those of the matrices M^{-+} and M^{+-} are zero, i.e. $M_{ikjl}^{-+} = M_{ikjl}^{+-} = 0$.

In this case the diffusion matrix (9.70) has the simple form

$$[D] = \left[\begin{array}{|c|c|} \hline -\mathcal{F}^{--} & 0 \\ \hline 0 & +\mathcal{F}^{++} \\ \hline \end{array} \right], \tag{10.24}$$

where from (9.53)

$$\begin{aligned}
[-\mathcal{F}^{--}]_{ij} &= \sum_{kl} M_{ikjl}^{--}\, g_k\, g_l, \\
[+\mathcal{F}^{++}]_{ij} &= \sum_{kl} M_{ikjl}^{++}\, g_k^+\, g_l^+
\end{aligned} \tag{10.25}$$

in terms of the symmetric c-number matrices M^{--} and M^{++}, with $M^{--} = (K^{--})(K^{--})^T$ and $M^{++} = (K^{++})(K^{++})^T$, where $K^{++} = (K^{--})^*$.

From (9.56), the diffusion matrix is of the form

$$D_{pq} = \sum_{r,s} Q_{rs}^{pq}\, g_r g_s, \tag{10.26}$$

and from (9.58) the matrix Q is of the form

$$[Q] = \left[\begin{array}{|c|c|} \hline M^{--} & 0 \\ \hline 0 & M^{++} \\ \hline \end{array} \right], \tag{10.27}$$

where we list the rows and columns of M^{--} and M^{++} as $11, 12, 13, 14, 21, \cdots, 44$. In terms of the p, r (rows) and q, s (columns) listing for Q, the 16 rows are $11, 12, 13, 14, 21, 22, \cdots, 43, 44; 55, 56, 57, 58, 65, 66, \cdots, 87, 88$; as are the 16 columns. We can also just list the rows and columns as $1, 2, 3, \cdots, 16$, and that listing is used below.

To obtain the form of the Ito equations we must write $Q = KK^T$ via Takagi factorisation, as in (9.61). This can be accomplished via separate factorisations of M^{--} and M^{++}. Thus $M^{--} = (K^{--})(K^{--})^T$. The matrix M^{--} is just λ times a real symmetric matrix $\tilde{M}^{--}$, and if the real eigenvalues μ_a for $\tilde{M}^{--}$ are all non-negative then K^{--} can be obtained from the real, orthonormal column eigenvectors X_a of $\tilde{M}^{--}$ as

$$K^{--} = \sqrt{\lambda} \sum_{\mu_a > 0} \sqrt{\mu_a}\, X_a\, X_a^T, \tag{10.28}$$

where only the non-zero eigenvalues need be included. A straightforward application of matrix theory establishes that $\tilde{M}^{--}$ has eight zero eigenvalues $\mu_2, \mu_4, \mu_5, \mu_7, \mu_{10}, \mu_{12}, \mu_{13}, \mu_{15}$ with eigenvectors with a 1 in the ath row and 0 elsewhere, and another four zero eigenvalues $\mu_9, \mu_{11}, \mu_{14}, \mu_{16}$ with eigenvectors

$$\begin{aligned}
X_9^T &= (1/\sqrt{2}) \left[\begin{array}{|c|c|c|c|c|c|c|c|c|c|c|c|c|c|c|c|} \hline 0&0&0&0&0&0&0&1&-1&0&0&0&0&0&0&0 \\ \hline \end{array} \right], \\
X_{11}^T &= (1/\sqrt{2}) \left[\begin{array}{|c|c|c|c|c|c|c|c|c|c|c|c|c|c|c|c|} \hline 1&0&0&0&0&0&0&0&0&0&-1&0&0&0&0&0 \\ \hline \end{array} \right], \\
X_{14}^T &= (1/\sqrt{2}) \left[\begin{array}{|c|c|c|c|c|c|c|c|c|c|c|c|c|c|c|c|} \hline 0&0&1&0&0&0&0&0&0&0&0&0&0&-1&0&0 \\ \hline \end{array} \right], \\
X_{16}^T &= (1/\sqrt{2}) \left[\begin{array}{|c|c|c|c|c|c|c|c|c|c|c|c|c|c|c|c|} \hline 0&0&0&0&0&1&0&0&0&0&0&0&0&0&0&-1 \\ \hline \end{array} \right].
\end{aligned} \tag{10.29}$$

Finally, there are four eigenvectors with eigenvalues $\mu_1, \mu_3, \mu_6, \mu_8$ all equal to two.

$$\begin{aligned}
X_8^T &= (1/\sqrt{2}) \left[\begin{array}{|c|c|c|c|c|c|c|c|c|c|c|c|c|c|c|c|} \hline 0&0&0&0&0&0&0&1&1&0&0&0&0&0&0&0 \\ \hline \end{array} \right], \\
X_1^T &= (1/\sqrt{2}) \left[\begin{array}{|c|c|c|c|c|c|c|c|c|c|c|c|c|c|c|c|} \hline 1&0&0&0&0&0&0&0&0&0&1&0&0&0&0&0 \\ \hline \end{array} \right], \\
X_3^T &= (1/\sqrt{2}) \left[\begin{array}{|c|c|c|c|c|c|c|c|c|c|c|c|c|c|c|c|} \hline 0&0&1&0&0&0&0&0&0&0&0&0&0&1&0&0 \\ \hline \end{array} \right], \\
X_6^T &= (1/\sqrt{2}) \left[\begin{array}{|c|c|c|c|c|c|c|c|c|c|c|c|c|c|c|c|} \hline 0&0&0&0&0&1&0&0&0&0&0&0&0&0&0&1 \\ \hline \end{array} \right].
\end{aligned} \tag{10.30}$$

This results in the following matrix for K^{--}:

$$
\begin{aligned}
&\left[K^{--}\right] \\
&= \frac{\lambda}{\sqrt{2}} \times \begin{bmatrix}
1&0&0&0&0&0&0&0&0&0&1&0&0&0&0&0 \\
0&0&0&0&0&0&0&0&0&0&0&0&0&0&0&0 \\
0&0&1&0&0&0&0&0&0&0&0&0&0&1&0&0 \\
0&0&0&0&0&0&0&0&0&0&0&0&0&0&0&0 \\
0&0&0&0&0&0&0&0&0&0&0&0&0&0&0&0 \\
0&0&0&0&0&1&0&0&0&0&0&0&0&0&0&1 \\
0&0&0&0&0&0&0&0&0&0&0&0&0&0&0&0 \\
0&0&0&0&0&0&0&1&1&0&0&0&0&0&0&0 \\
0&0&0&0&0&0&0&1&1&0&0&0&0&0&0&0 \\
0&0&0&0&0&0&0&0&0&0&0&0&0&0&0&0 \\
1&0&0&0&0&0&0&0&0&0&1&0&0&0&0&0 \\
0&0&0&0&0&0&0&0&0&0&0&0&0&0&0&0 \\
0&0&0&0&0&0&0&0&0&0&0&0&0&0&0&0 \\
0&0&1&0&0&0&0&0&0&0&0&0&0&1&0&0 \\
0&0&0&0&0&0&0&0&0&0&0&0&0&0&0&0 \\
0&0&0&0&0&1&0&0&0&0&0&0&0&0&0&1.
\end{bmatrix}.
\end{aligned}
\tag{10.31}
$$

The matrix K^{++} is given by $K^{++} = (K^{--})^*$ and $K^{+-} = K^{-+} = 0$.

From (9.62), the diffusion matrix is then obtained in the form

$$
D_{pq} = \sum_a \mathcal{B}^p_a \, \mathcal{B}^q_a, \tag{10.32}
$$

where from (9.63)

$$
\mathcal{B}^p_a = \sum_r K^p_{ra} \, g_r. \tag{10.33}
$$

From (9.85), and (9.86), the Ito stochastic equations are

$$
\begin{aligned}
\widetilde{g}_1(t+\delta t) &= \left\{(1 - i\omega\,\delta t) + \frac{\lambda}{\sqrt{2}}(\delta\widetilde{w}_1 + \delta\widetilde{w}_{11})\right\}\widetilde{g}_1(t) + \left\{\frac{\lambda}{\sqrt{2}}(\delta\widetilde{w}_3 + \delta\widetilde{w}_{14})\right\}\widetilde{g}_3(t), \\
\widetilde{g}_2(t+\delta t) &= \left\{(1 - i\omega\,\delta t) + \frac{\lambda}{\sqrt{2}}(\delta\widetilde{w}_6 + \delta\widetilde{w}_{16})\right\}\widetilde{g}_2(t) + \left\{\frac{\lambda}{\sqrt{2}}(\delta\widetilde{w}_8 + \delta\widetilde{w}_9)\right\}\widetilde{g}_4(t), \\
\widetilde{g}_3(t+\delta t) &= \left\{(1 - i\omega\,\delta t) + \frac{\lambda}{\sqrt{2}}(\delta\widetilde{w}_1 + \delta\widetilde{w}_{11})\right\}\widetilde{g}_3(t) + \left\{\frac{\lambda}{\sqrt{2}}(\delta\widetilde{w}_8 + \delta\widetilde{w}_9)\right\}\widetilde{g}_1(t), \\
\widetilde{g}_4(t+\delta t) &= \left\{(1 - i\omega\,\delta t) + \frac{\lambda}{\sqrt{2}}(\delta\widetilde{w}_6 + \delta\widetilde{w}_{16})\right\}\widetilde{g}_4(t) + \left\{\frac{\lambda}{\sqrt{2}}(\delta\widetilde{w}_3 + \delta\widetilde{w}_{14})\right\}\widetilde{g}_2(t)
\end{aligned}
\tag{10.34}
$$

and

$$\begin{aligned}
\widetilde{g}_1^+(t+\delta t) &= \left\{(1+i\omega\,\delta t) + \frac{\lambda^*}{\sqrt{2}}(\delta\widetilde{w}_1^+ + \delta\widetilde{w}_{11}^+)\right\}\widetilde{g}_1^+(t) + \left\{\frac{\lambda^*}{\sqrt{2}}(\delta\widetilde{w}_3^+ + \delta\widetilde{w}_{14}^+)\right\}\widetilde{g}_3^+(t),\\
\widetilde{g}_2^+(t+\delta t) &= \left\{(1+i\omega\,\delta t) + \frac{\lambda^*}{\sqrt{2}}(\delta\widetilde{w}_6^+ + \delta\widetilde{w}_{16}^+)\right\}\widetilde{g}_2^+(t) + \left\{\frac{\lambda^*}{\sqrt{2}}(\delta\widetilde{w}_8^+ + \delta\widetilde{w}_9^+)\right\}\widetilde{g}_4^+(t),\\
\widetilde{g}_3^+(t+\delta t) &= \left\{(1+i\omega\,\delta t) + \frac{\lambda^*}{\sqrt{2}}(\delta\widetilde{w}_1^+ + \delta\widetilde{w}_{11}^+)\right\}\widetilde{g}_3^+(t) + \left\{\frac{\lambda^*}{\sqrt{2}}(\delta\widetilde{w}_8^+ + \delta\widetilde{w}_9^+)\right\}\widetilde{g}_1^+(t),\\
\widetilde{g}_4^+(t+\delta t) &= \left\{(1+i\omega\,\delta t) + \frac{\lambda^*}{\sqrt{2}}(\delta\widetilde{w}_6^+ + \delta\widetilde{w}_{16}^+)\right\}\widetilde{g}_4^+(t) + \left\{\frac{\lambda^*}{\sqrt{2}}(\delta\widetilde{w}_3^+ + \delta\widetilde{w}_{14}^+)\right\}\widetilde{g}_2^+(t),
\end{aligned} \tag{10.35}$$

where the Grassmann variables are stochastic. The matrix G in (9.86) can easily be read off. We see that the $\widetilde{g}_i(t+\delta t)$, are linear combinations of only the $\widetilde{g}_i(t)$, with c-number coefficients involving Wiener increments. Also, the $\widetilde{g}_i^+(t+\delta t)$ are linear combinations of only the $\widetilde{g}_i^+(t)$ with c-number coefficients – in this physical system there is no cross-coupling between $\widetilde{g}_i$ and $\widetilde{g}_i^+$. Overall, there are a total of 16 different Wiener increments involved here. This is less than the maximum number of $2n^2 = 32$.

10.2.6 Populations and Coherences

Using (7.114) and (7.115), expressions for the populations and coherences can be obtained. The population for the Fock state $|\Phi_3\rangle$ is

$$P(\Phi_3) = \int d\mathbf{g}^+ d\mathbf{g}\; B^{\mathrm{f}}(\mathbf{g},\mathbf{g}^+)\, g_4 g_1\, g_1^+ g_4^+ \tag{10.36}$$

$$= \overline{\widetilde{g}_4\widetilde{g}_1\,\widetilde{g}_1^+\widetilde{g}_4^+}, \tag{10.37}$$

and the coherence between the Fock state $|\Phi_3\rangle$ and the Fock state $|\Phi_4\rangle$ is

$$C(\Phi_4;\Phi_3) = \int d\mathbf{g}^+ d\mathbf{g}\; B^{\mathrm{f}}(\mathbf{g},\mathbf{g}^+)\, g_4 g_1\, g_2^+ g_3^+ \tag{10.38}$$

$$= \overline{\widetilde{g}_4\widetilde{g}_1\,\widetilde{g}_2^+\widetilde{g}_3^+}, \tag{10.39}$$

where both the phase space and the equivalent stochastic expressions are given. The initial condition shows that

$$\left(\overline{\widetilde{g}_4\widetilde{g}_1\,\widetilde{g}_1^+\widetilde{g}_4^+}\right)_{t=0} = 1 \tag{10.40}$$

and all other initial stochastic averages are zero.

The coherence $C(\Phi_4;\Phi_3)$ at time δt can be obtained from (10.34) and (10.35) via performing the stochastic averages. We have

$$
\begin{aligned}
&C(\Phi_4;\Phi_3)_{t=\delta t} \\
&= \Bigg[\Bigg[\left\{(1-i\omega\,\delta t)+\frac{\lambda}{\sqrt{2}}(\delta\widetilde{w}_6+\delta\widetilde{w}_{16})\right\}\widetilde{g}_4(0)+\left\{\frac{\lambda}{\sqrt{2}}(\delta\widetilde{w}_3+\delta\widetilde{w}_{14})\right\}\widetilde{g}_2(0)\Bigg] \\
&\quad\times\Bigg[\left\{(1-i\omega\,\delta t)+\frac{\lambda}{\sqrt{2}}(\delta\widetilde{w}_1+\delta\widetilde{w}_{11})\right\}\widetilde{g}_1(0)+\left\{\frac{\lambda}{\sqrt{2}}(\delta\widetilde{w}_3+\delta\widetilde{w}_{14})\right\}\widetilde{g}_3(0)\Bigg] \\
&\quad\times\Bigg[\left\{(1+i\omega\,\delta t)+\frac{\lambda^*}{\sqrt{2}}(\delta\widetilde{w}_6^++\delta\widetilde{w}_{16}^+)\right\}\widetilde{g}_2^+(0)+\left\{\frac{\lambda^*}{\sqrt{2}}(\delta\widetilde{w}_8^++\delta\widetilde{w}_9^+)\right\}\widetilde{g}_4^+(0)\Bigg] \\
&\quad\times\Bigg[\left\{(1+i\omega\,\delta t)+\frac{\lambda^*}{\sqrt{2}}(\delta\widetilde{w}_1^++\delta\widetilde{w}_{11}^+)\right\}\widetilde{g}_3^+(0)+\left\{\frac{\lambda^*}{\sqrt{2}}(\delta\widetilde{w}_8^++\delta\widetilde{w}_9^+)\right\}\widetilde{g}_1^+(0)\Bigg]\Bigg]_{\text{StochAv}}.
\end{aligned}
\tag{10.41}
$$

Multiplying out this stochastic average results in 16 terms. The Grassmann and c-number variables can be factored out and the stochastic average of any term is just the product of the stochastic average of the c-number factor and the Grassmann variable factor, as the Wiener increments involve a Gaussian–Markov Γ at times greater than zero. However, the only non-zero Grassmann stochastic average is $\left(\overline{\widetilde{g}_4\widetilde{g}_1\,\widetilde{g}_1^+\widetilde{g}_4^+}\right)_{t=0}=1$. This only appears in one of the 16 terms, so we have

$$
\begin{aligned}
C(\Phi_4;\Phi_3)_{t=\delta t} &= \Bigg[\left\{(1-i\omega\,\delta t)+\frac{\lambda}{\sqrt{2}}(\delta\widetilde{w}_6+\delta\widetilde{w}_{16})\right\}\left\{(1-i\omega\,\delta t)+\frac{\lambda}{\sqrt{2}}(\delta\widetilde{w}_1+\delta\widetilde{w}_{11})\right\} \\
&\quad\times\left\{\frac{\lambda^*}{\sqrt{2}}(\delta\widetilde{w}_8^++\delta\widetilde{w}_9^+)\right\}\left\{\frac{\lambda^*}{\sqrt{2}}(\delta\widetilde{w}_8^++\delta\widetilde{w}_9^+)\right\}\Bigg]_{\text{StochAv}}\left(\overline{\widetilde{g}_4\widetilde{g}_1\,\widetilde{g}_4^+\widetilde{g}_1^+}\right)_{t=0}.
\end{aligned}
\tag{10.42}
$$

There are terms involving the stochastic averages of products of two, three and four Wiener increments. Those involving three are always zero. Those involving four are sums of products of stochastic averages of two Wiener increments. These are zero because those products involving $\delta\widetilde{w}_6,\cdots,\delta\widetilde{w}_{11}$ such as $\overline{\delta\widetilde{w}_6\,\delta\widetilde{w}_1}=\overline{\delta\widetilde{w}_6\,\delta\widetilde{w}_{11}}=\cdots=\overline{\delta\widetilde{w}_{11}\,\delta\widetilde{w}_9^+}$ are all zero. From (9.16), the only non-zero contributions are

$$\overline{\delta\widetilde{w}_8^+\delta\widetilde{w}_8^+}=\overline{\delta\widetilde{w}_9^+\delta\widetilde{w}_9^+}=\delta t, \tag{10.43}$$

so, noting that $\left(\overline{\widetilde{g}_4\widetilde{g}_1\,\widetilde{g}_4^+\widetilde{g}_1^+}\right)_{t=0}=-\left(\overline{\widetilde{g}_4\widetilde{g}_1\,\widetilde{g}_1^+\widetilde{g}_4^+}\right)_{t=0}=-1$, we finally obtain

$$C(\Phi_4;\Phi_3)_{t=\delta t}=(g/i\hbar V)\,\delta t \tag{10.44}$$

correct to order δt.

10.2.7 Conclusion

The same result can of course be obtained via standard first-order perturbation theory, but here we have obtained it from Ito equations for stochastic Grassmann variables. Numerical methods could be used to determine other coherences and populations at any finite time t, since we have shown that the stochastic evolution only involves taking stochastic averages of quantities that, though involving Wiener increments, are just c-numbers combined of course with stochastic averages of the Grassmann variables at the initial time. The latter are just obtained from the initial conditions.

10.3 Combined Case – Jaynes–Cummings Model

10.3.1 Introduction

The simplest theoretical model in quantum optics is a two-state atomic system coupled to a single-mode quantum electromagnetic field, in which the system transition frequency is in near resonance with the mode frequency. This system was first treated using the rotating-wave approximation and is referred to as the Jaynes–Cummings model [71]. The model can be used not only in cavity quantum electrodynamics (QED) to describe a two-level atom (TLA) in a high-Q cavity – which could be two Rydberg atom states in a microwave cavity, or two lower-lying atomic states in an optical cavity – but also in laser spectroscopy for treating a two-level atom in free space coupled to a single-mode laser field. The Jaynes–Cummings model is of fundamental importance both in quantum optics and in quantum physics generally, involving the interaction of two simple quantum systems – one being fermionic (the TLA) and the other bosonic (the cavity mode). In laser spectroscopy, the prediction [72], interpretation [73] and observation [74–76] of a three-peak resonance fluorescence spectrum for a TLA strongly driven by a narrow-bandwidth laser field stimulated much interest in quantum optics. In the cavity QED case, a variety of interesting effects occur depending on the initial conditions, ranging from ongoing oscillations of the atomic population difference at the Rabi frequency when the atom is excited and the cavity is in an n-photon Fock state, to collapses and revivals of these oscillations starting with the atom unexcited and the cavity mode in a coherent state [4, 77]. The observation of revivals [78, 79] for Rydberg atoms in a high-Q microwave cavity is key experimental evidence for quantisation of the electromagnetic field. Although the Jaynes–Cummings model and the prediction of collapses and revivals can be treated by standard matrix mechanics methods [4, 77] based on photon number states for the field, or by the approach of Stenholm [80], where the field is described by Bargmann states, the system provides a test case for applying phase space methods to a combined boson–fermion system. Phase space treatments of the Jaynes–Cummings model are themselves unusual; however, Eiselt and Risken [81] used such an approach, though without using Grassmann phase space variables to describe the atomic system.

The present treatment [82] is based on using the canonical form of the positive-P-type Grassmann phase space distribution function, and the distribution function is obtained analytically from the Fokker–Planck equation via an adaption of the Stenholm treatment [80]. We show that this solution gives results equivalent to those obtained by standard quantum optics methods, thereby demonstrating the applicability of the Grassmann phase space approach to this important coupled boson–fermion system. It turns out, however, that applying the standard correspondence rules results in a Fokker–Planck equation whose solution via similar methods gives an invalid distribution function that diverges on the bosonic phase space boundary, as well as increasing exponentially at long time. This different distribution function reflects the non-uniqueness of the bosonic positive P distribution function and illustrates how invalid Fokker–Planck equations may occur when distribution functions do not vanish quickly enough on the bosonic phase space boundary.

10.3.2 Physics of One-Atom Cavity Mode System

The basic atomic system treated in the Jaynes–Cummings model is a *single-atom* system with just two internal states, denoted $|1\rangle$, $|2\rangle$, with energies E_1, E_2. Centre-of-mass degrees of freedom are ignored. There are therefore just four atomic operators to consider, denoted $\hat{\sigma}_{ij} = |i\rangle\langle j|$. Two of these are *population* operators, $\hat{P}_1 = |1\rangle\langle 1|$ and $\hat{P}_2 = |2\rangle\langle 2|$, and two are atomic *transition* operators $\hat{\sigma}_+ = |2\rangle\langle 1|$ and $\hat{\sigma}_- = |1\rangle\langle 2|$. In the simple Jaynes–Cummings model, the two-level atom interacts with a cavity mode. Let $\hat{a}, \hat{a}^\dagger$ be the boson annihilation and creation operators for the field mode. The states for the cavity mode are the usual n-photon states, denoted $|n\rangle$, with $n = 0, 1, 2, \cdots$. If the atomic transition frequency is $\omega_0 = (E_2 - E_1)/\hbar$, the frequency of the field mode is ω and the atom–field mode-coupling constant (one-photon Rabi frequency) is Ω, then the Hamiltonian for the one-atom Jaynes–Cummings model is given by

$$\hat{H}_{\mathrm{JC}} = E_{\mathrm{a}}(\hat{P}_2 + \hat{P}_1) + \frac{1}{2}\hbar\omega_0(\hat{P}_2 - \hat{P}_1) + \hbar\omega(\hat{a}^\dagger\hat{a}) + \frac{1}{2}\hbar\Omega(\hat{\sigma}_+\hat{a} + \hat{a}^\dagger\hat{\sigma}_-), \tag{10.45}$$

where the average atomic energy is $E_{\mathrm{a}} = (E_2 + E_1)/2$. The first term is usually ignored as it merely introduces a phase factor $\exp(-E_{\mathrm{a}}t/\hbar)$ into the evolution of all states.

The dynamics of the Jaynes–Cummings model is treated in many quantum optics textbooks and papers (see [83]). The state vector is expanded in terms of n-photon states for the field and the two internal atomic states as

$$\begin{aligned}|\Psi\rangle = \exp(-E_{\mathrm{a}}t/\hbar)\sum_N \Bigg(& A_{\overline{n-1}2}(t)\exp\left(-i\left(\overline{n-1}\omega + \frac{1}{2}\omega_0\right)t\right)|2\rangle|n-1\rangle \\ & + A_{n1}(t)\exp\left(-i\left(n\omega - \frac{1}{2}\omega_0\right)t\right)|1\rangle|n\rangle\Bigg),\end{aligned} \tag{10.46}$$

involving interaction picture amplitudes $A_{n1}, A_{\overline{n-1}2}$. The coupled equations for the amplitudes,

$$\begin{aligned} i\frac{\partial}{\partial t}A_{\overline{n-1}2} &= \frac{1}{2}\Omega\sqrt{n}\exp(+i\Delta t)\; A_{n1}, \\ i\frac{\partial}{\partial t}A_{n1} &= \frac{1}{2}\Omega\sqrt{n}\exp(-i\Delta t)\; A_{\overline{n-1}2}, \end{aligned} \tag{10.47}$$

also contain detuning terms to represent the relative dephasing of the atom and field,

$$\Delta = \omega_0 - \omega, \tag{10.48}$$

and may be solved using Laplace transforms. The solution is

$$\begin{aligned}A_{n1}(t) &= \exp(-i\Delta t/2)\left(\cos(\omega_n t/2)\left\{A_{n1}(0)\right\}\right.\\&\quad\left.+ i\sin(\omega_n t/2)\left\{\Delta A_{n1}(0) - \Omega\sqrt{n}\,A_{\overline{n-1}2}(0)\omega_n\right\}/\omega_n\right),\\A_{\overline{n-1}2}(t) &= \exp(i\Delta t/2)\left(\cos(\omega_n t/2)\left\{A_{\overline{n-1}2}(0)\right\}\right.\\&\quad\left.- i\sin(\omega_n t/2)\left\{\Omega\sqrt{n}\,A_{n1}(0) + \Delta A_{\overline{n-1}2}(0)\right\}/\omega_n\right),\end{aligned} \tag{10.49}$$

where

$$\omega_n = \sqrt{\Delta^2 + n\Omega^2} \tag{10.50}$$

is the Rabi frequency. Population and coherence oscillations at various frequencies ω_n are therefore predicted, and which frequencies are observed depend on the initial conditions. For example, if the atom is not excited and the field is in an m-photon state, then $A_{n1}(0) = \delta_{nm}$, $A_{\overline{n-1}2}(0) = 0$ and hence oscillations at a single frequency ω_m will occur. On the other hand, if the field is in a coherent state with amplitude η, then oscillation frequencies clustered around the mean frequency $\omega_{\overline{n}}$ associated with the mean photon number $\overline{n} = |\eta|^2$ occur, the range of frequencies being associated with the standard deviation $\Delta n = \sqrt{\overline{n}}$ of the photon number. This results in the collapse and revival effects discussed previously. In an alternative approach, Stenholm [80] expanded the state vector in products of Bargmann states for the field and spinor states for the TLA and also demonstrated collapse and revival effects. As we see, his solution can be adapted to determining the Grassmann distribution function in the present treatment.

10.3.3 Fermionic and Bosonic Modes

Instead of treating the Jaynes–Cummings model via the usual elementary quantum optics approach – for example via matrix mechanics with basis states $|1\rangle\,|n\rangle$, $|2\rangle\,|n-1\rangle$ – we can replace the original system of a two-level atom plus a cavity mode with a somewhat *enlarged system* consisting of two fermionic modes 1, 2 interacting with one bosonic mode. This enlarged system includes states that are in one-to-one correspondence with the states of the original system, and the general dynamics of the larger Fermi–Bose system incorporates all possible behaviour of the original Jaynes–Cummings model. Note that relating the two-state atomic system to two fermion modes has nothing to do with whether the atom itself is a fermion or a boson. This process requires the introduction of standard fermion annihilation and creation operators $\hat{c}_i, \hat{c}_i^\dagger$ $(i = 1, 2)$ for the two fermion modes. Note that the fermion and boson operators commute.

The full basis states for the Jaynes–Cummings model are Fock states of the form

$$|m_1; m_2; n\rangle = (\hat{c}_1^\dagger)^{m_1}(\hat{c}_2^\dagger)^{m_2}\frac{(\hat{a}^\dagger)^n}{\sqrt{n!}}\,|0\rangle\,, \tag{10.51}$$

where $m_i = 0, 1$ gives the numbers of atoms in states $i = 1, 2$ and $n = 0, 1, 2, \cdots$ gives the number of photons in the field mode. $|0\rangle$ is the vacuum state. With n photons in the field, the state $|0; 0; n\rangle$ has no fermions in either mode; $|1; 0; n\rangle$ and $|0; 1; n\rangle$ are states with one fermion in modes 1 and 2, respectively; and $|1;1;n\rangle$ is a two-fermion state with one fermion in each mode. Obviously, since $(\hat{c}_i^\dagger)^2 = 0$, there cannot be more than one atom in state i, as the Pauli exclusion principle requires.

10.3.4 Quantum States

The most general mixed state for which there may be either zero, one or two atoms has a density operator

$$\begin{aligned}
\hat{\rho} = &\sum_{n,m}(\rho_{00n;00m}\,|0;0;n\rangle\,\langle 0;0,m| \\
&+\sum_{n,m}(\rho_{10n;10m}\,|1;0;n\rangle\,\langle 1;0,m|\,+\rho_{10n;01m}\,|1;0;n\rangle\,\langle 0;1,m| \\
&+\,\rho_{01n;10m}\,|0;1;n\rangle\,\langle 1;0,m|\,+\rho_{01n;01m}\,|0;1;n\rangle\,\langle 0;1,m|) \\
&+\,\sum_{n,m}(\rho_{11n;11m}\,|1;1;n\rangle\,\langle 1;1,m|\,,
\end{aligned} \tag{10.52}$$

where the $\rho_{ijn;klm}$ $(i,j,k,l = 0,1)$ are density matrix elements. This corresponds to a statistical mixture of states in which there are no atoms, one atom or two atoms. Pure states which are quantum superpositions of states with differing numbers of atoms are forbidden by the super-selection rule based on conservation of atom number. Statistical mixtures of such forbidden states would, in general, lead to the presence of coherences between states with differing numbers of atoms. Thus there are no such coherences in (10.52).

Note that in addition to atomic states corresponding to the one-atom Jaynes–Cummings model, we now have two extra fermionic states in our enlarged system, and we need to consider whether these have any physical significance. The vacuum state $|0;0\rangle$ corresponds to an experiment with no atoms inside the cavity – not a situation of great interest. The two-fermion state $|1;1\rangle$ corresponds to a two-atom system inside the cavity, with one atom in state $|1\rangle$ and the other in state $|2\rangle$. All four fermion states $|m_1;m_2\rangle$ also have a mathematical interpretation in terms of spin states in our enlarged system.

10.3.5 Population and Transition Operators

The vacuum state $|0\rangle$, which contains no bosons or fermions in any mode, is associated with the vacuum state projector $|0\rangle\,\langle 0|$. As seen in Section 2.8, the vacuum projector can be written in terms of normally ordered forms [34] of products of exponential operators based on mode number operators.

If we identify the two one-fermion Fock states with the two internal atomic states in the one-atom Jaynes–Cummings model as $\hat{c}_1^\dagger|0\rangle \Longleftrightarrow |1\rangle$, $\hat{c}_2^\dagger|0\rangle \Longleftrightarrow |2\rangle$, then the four one-atom atomic population or *projector operators* $\hat{P}_{1,2}$ and *transition operators* $\hat{\sigma}_\pm$ are

$$\begin{aligned}
&\hat{c}_1^\dagger|0\rangle\,\langle 0|\hat{c}_1 \Longleftrightarrow \hat{P}_1, \qquad &&\hat{c}_2^\dagger|0\rangle\,\langle 0|\hat{c}_2 \Longleftrightarrow \hat{P}_2, \\
&\hat{c}_1^\dagger|0\rangle\,\langle 0|\hat{c}_2 \Longleftrightarrow \hat{\sigma}_-, \qquad &&\hat{c}_2^\dagger|0\rangle\,\langle 0|\hat{c}_1 \Longleftrightarrow \hat{\sigma}_+.
\end{aligned} \tag{10.53}$$

For the expanded Jaynes–Cummings model, the two-atom state is given by $\hat{c}_1^\dagger \hat{c}_2^\dagger |0\rangle$, so the two-atom state population or projector operator $\hat{P}_{12}$ with one atom in each of the atomic states is

$$\hat{c}_1^\dagger \hat{c}_2^\dagger |0\rangle \langle 0| \hat{c}_2 \hat{c}_1 \Longleftrightarrow \hat{P}_{12}, \tag{10.54}$$

and as the zero-atom state is $|0\rangle$, the zero-atom population or projector operator $\hat{P}_0$ with no atoms in either of the internal atomic states is

$$|0\rangle \langle 0| \Longleftrightarrow \hat{P}_0. \tag{10.55}$$

This of course is the same as the vacuum state projector.

Using (2.98) and the result (2.94), we can obtain expressions for all of the atom state operators just in terms of fermion annihilation and creation operators. We find for the one-atom population and transition operators

$$\hat{P}_1 = \hat{c}_1^\dagger \hat{c}_1 - \hat{c}_1^\dagger \hat{c}_2^\dagger \hat{c}_2 \hat{c}_1, \qquad \hat{P}_2 = \hat{c}_2^\dagger \hat{c}_2 - \hat{c}_1^\dagger \hat{c}_2^\dagger \hat{c}_2 \hat{c}_1, \tag{10.56}$$

$$\hat{\sigma}_- = \hat{c}_1^\dagger \hat{c}_2, \qquad \hat{\sigma}_+ = \hat{c}_2^\dagger \hat{c}_1, \tag{10.57}$$

and for the two-atom and zero-atom population operators

$$\hat{P}_{12} = \hat{c}_1^\dagger \hat{c}_2^\dagger \hat{c}_2 \hat{c}_1, \tag{10.58}$$

$$\hat{P}_0 = 1 - \hat{c}_1^\dagger \hat{c}_1 - \hat{c}_2^\dagger \hat{c}_2 + \hat{c}_1^\dagger \hat{c}_2^\dagger \hat{c}_2 \hat{c}_1. \tag{10.59}$$

The population operators sum to unity, showing that the total probability for any general mixed state (10.52) that the system is found in a zero-, one- or two- atom state is unity.

10.3.6 Hamiltonian and Number Operators

The Hamiltonian needs to be modified for the enlarged system to allow for states with two or zero atoms. The one-atom Jaynes–Cummings model Hamiltonian $\hat{H}_{JC}$ is replaced by an enlarged Hamiltonian $\hat{H}$ by adding a two-atom energy term $(E_2 + E_1)\hat{P}_{12}$ to the Hamiltonian in (10.45). This Hamiltonian has zero matrix elements between the two-atom states $|1;1\rangle$, one-atom states $|0;1\rangle$, $|1;0\rangle$ and zero-atom states $|0;0\rangle$. The two- and zero-atom states are eigenstates of the Hamiltonian with atomic energies $E_2 + E_1$ and 0, as expected. The atom–field coupling term also cannot cause a transition between states with differing total atom number, consistent with not violating the super-selection rule.

The enlarged Hamiltonian for the Jaynes–Cummings model may be written as

$$\hat{H} = E_a(\hat{c}_2^\dagger \hat{c}_2 + \hat{c}_1^\dagger \hat{c}_1) + \frac{1}{2}\hbar\omega_0(\hat{c}_2^\dagger \hat{c}_2 - \hat{c}_1^\dagger \hat{c}_1) + \hbar\omega(\hat{a}^\dagger \hat{a}) + \frac{1}{2}\hbar\Omega(\hat{c}_2^\dagger \hat{c}_1 \hat{a} + \hat{a}^\dagger \hat{c}_1^\dagger \hat{c}_2), \tag{10.60}$$

where the results (10.56), (10.57) and (10.58) for the one- and two-atom operators have been substituted into the enlarged Hamiltonian. Terms involving $\hat{c}_1^\dagger \hat{c}_2^\dagger \hat{c}_2 \hat{c}_1$ cancel out.

The number of photons present is determined from the operator

$$\hat{n} = \hat{a}^{\dagger}\hat{a}, \tag{10.61}$$

whose eigenvalues are $n = 0, 1, 2, \cdots$. We can also introduce number operators for the two atomic states via

$$\hat{n}_i = \hat{c}_i^{\dagger}\hat{c}_i \qquad (i = 1, 2). \tag{10.62}$$

The total number of atoms present is given by the number operator $\hat{N} = (\hat{c}_2^{\dagger}\hat{c}_2 + \hat{c}_1^{\dagger}\hat{c}_1)$, and

$$\hat{N} = 0 \times \hat{P}_0 + 1 \times (\hat{P}_1 + \hat{P}_2) + 2 \times \hat{P}_{12}, \tag{10.63}$$

showing that the number operator is related to the projectors $\hat{P}_0$, $\hat{P}_1$, $\hat{P}_2$ and $\hat{P}_{12}$ for zero-, one-, one- and two-atom states, respectively. The number operator can have eigenvalues $0, 1, 2$, and clearly

$$\hat{N}\, |m_1; m_2; n\rangle = (m_1 + m_2)\, |m_1; m_2; n\rangle\,. \tag{10.64}$$

The number operator commutes with the Hamiltonian, so if initially we have a physical state with a specific number of atoms, then state evolution does not change this. Thus one-atom states do not become two-atom states, and if the initial density operator has the form in (10.52) with $\rho_{00n;00m} = \rho_{11n;11m} = 0$ for one-atom states it will remain in this form. There is, of course, no conservation law for the photon number.

10.3.7 Probabilities and Coherences

We can obtain expressions for the physical quantities in terms of the annihilation and creation operators. The mean number of photons is

$$\overline{n} = \langle\hat{a}^{\dagger}\hat{a}\rangle, \tag{10.65}$$

where $\langle\hat{\Xi}\rangle = \mathrm{Tr}(\hat{\rho}\hat{\Xi})$.

For the one-atom probability of finding one atom in state $|1\rangle$ and none in state $|2\rangle$ and vice versa, we have

$$P_1 = \mathrm{Tr}(\hat{P}_1\hat{\rho}) = \langle\hat{c}_1^{\dagger}\hat{c}_1\rangle - \langle\hat{c}_1^{\dagger}\hat{c}_2^{\dagger}\hat{c}_2\hat{c}_1\rangle = \mathrm{Tr}(\hat{c}_1\hat{\rho}\hat{c}_1^{\dagger}) - \mathrm{Tr}(\hat{c}_2\hat{c}_1\hat{\rho}\hat{c}_1^{\dagger}\hat{c}_2^{\dagger}), \tag{10.66}$$

$$P_2 = \mathrm{Tr}(\hat{P}_2\hat{\rho}) = \langle\hat{c}_2^{\dagger}\hat{c}_2\rangle - \langle\hat{c}_1^{\dagger}\hat{c}_2^{\dagger}\hat{c}_2\hat{c}_1\rangle = \mathrm{Tr}(\hat{c}_2\hat{\rho}\hat{c}_2^{\dagger}) - \mathrm{Tr}(\hat{c}_2\hat{c}_1\hat{\rho}\hat{c}_1^{\dagger}\hat{c}_2^{\dagger}). \tag{10.67}$$

Note that these probabilities are not the same as $\langle\hat{c}_i^{\dagger}\hat{c}_i\rangle$, as might be expected. The one-atom coherences between state $|1\rangle$ and state $|2\rangle$ are given by

$$\begin{aligned} \rho_{12} &= \mathrm{Tr}(\hat{\sigma}_{-}\hat{\rho}) = \langle\hat{c}_1^{\dagger}\hat{c}_2\rangle = \mathrm{Tr}(\hat{c}_2\hat{\rho}\,\hat{c}_1^{\dagger}), \\ \rho_{21} &= \mathrm{Tr}(\hat{\sigma}_{+}\hat{\rho}) = \langle\hat{c}_2^{\dagger}\hat{c}_1\rangle = \mathrm{Tr}(\hat{c}_1\hat{\rho}\,\hat{c}_2^{\dagger}). \end{aligned} \tag{10.68}$$

Also, the two-atom probability P_{12} for finding one atom in state $|1\rangle$ and one in state $|2\rangle$, and the zero-atom probability P_0 for finding no atom either in state $|1\rangle$ or in state $|2\rangle$ are given by

$$P_{12} = \mathrm{Tr}(\hat{P}_{12}\hat{\rho}) = \langle \hat{c}_1^\dagger \hat{c}_2^\dagger \hat{c}_2 \hat{c}_1 \rangle = \mathrm{Tr}(\hat{c}_2 \hat{c}_1 \hat{\rho} \hat{c}_1^\dagger \hat{c}_2^\dagger), \tag{10.69}$$

$$P_0 = \mathrm{Tr}(\hat{P}_0\hat{\rho}) = 1 - P_1 - P_2 - P_{12}, \tag{10.70}$$

the last result just expressing the fact that all the probabilities must add up to one. Expressions involving the density matrix elements for general mixed states can easily be obtained for all of these results.

For general mixed states as in (10.52), we see that the expectation values of odd numbers of fermionic creation and destruction operators are zero. Thus

$$\langle \hat{c}_i \rangle = \langle \hat{c}_i^\dagger \rangle = \langle \hat{c}_1^\dagger \hat{c}_2^\dagger \hat{c}_i \rangle = \langle \hat{c}_i^\dagger \hat{c}_2 \hat{c}_1 \rangle = 0. \tag{10.71}$$

Clearly, for one-atom states where $\rho_{00n,00m} - \rho_{11n;11m} = 0$, we have, using $Tr(\hat{\rho}) = 1$,

$$P_0 = P_{12} = 0, \qquad P_1 + P_2 = 1, \tag{10.72}$$

$$\overline{n} = \sum_N n(\rho_{10n;10n} + \rho_{01n;01n}). \tag{10.73}$$

For one-atom states, the only non-zero expectation values are those that involve one fermion annihilation operator and one creation operator, of the form $\langle \hat{c}_i^\dagger \hat{c}_j \rangle$. Thus, in addition to the last results, the expectation value with four fermion operators vanishes:

$$\langle \hat{c}_1^\dagger \hat{c}_2^\dagger \hat{c}_2 \hat{c}_1 \rangle = P_{12} = 0, \tag{10.74}$$

since $\rho_{11n;11n} = 0$ for one-atom states. This is related to the probability of finding two atoms being zero in the one-atom Jaynes–Cummings model. For general mixed states the quantity $\langle \hat{c}_1^\dagger \hat{c}_2^\dagger \hat{c}_2 \hat{c}_1 \rangle$ is conserved, since $\hat{c}_1^\dagger \hat{c}_2^\dagger \hat{c}_2 \hat{c}_1$ commutes with the Hamiltonian

Although we are mainly focused on states corresponding to one-atom states, it is convenient to introduce non-physical states involving coherent superpositions of different total numbers of atoms, and even states where the expansion coefficients are Grassmann numbers. For non-physical states involving superpositions of different atom numbers, there may be off-diagonal density matrix elements between states of different atom numbers, and the results in (10.71) may not apply.

10.3.8 Characteristic Function

We define the characteristic function $\chi(\xi, \xi^+, h_i, h_i^+)$ via

$$\chi(\xi, \xi^+, h, h^+) = \mathrm{Tr}(\hat{\Omega}_b^+(\xi^+)\,\hat{\Omega}_f^+(h^+)\,\hat{\rho}\,\hat{\Omega}_f^-(h)\,\hat{\Omega}_b^-(\xi)), \tag{10.75}$$

$$\hat{\Omega}_b^+(\xi^+) = \exp i\hat{a}\xi^+, \qquad \hat{\Omega}_b^-(\xi) = \exp i\xi\hat{a}^\dagger, \tag{10.76}$$

$$\hat{\Omega}_f^+(h^+) = \exp i\sum_{i=1,2} \hat{c}_i h_i^+, \qquad \hat{\Omega}_f^-(h) = \exp i\sum_{i=1,2} h_i \hat{c}_i^\dagger. \tag{10.77}$$

For the bosonic and fermionic modes, respectively, we associate a pair of c-numbers ξ, ξ^+, and for $i = 1, 2$ we associate a pair of Grassmann numbers h_i, h_i^+ and use $h = \{h_1, h_2\}$, $h^+ = \{h_1^+, h_2^+\}$. The characteristic function will be a c-number function of ξ, ξ^+ and a Grassmann function of h_1, h_1^+, h_2, h_2^+. The factors for the different modes in the last two expressions for $\hat{\Omega}_{\mathrm{f}}^+(h^+)$ and $\hat{\Omega}_{\mathrm{f}}^-(h)$ may be put in any order, since they commute, but by convention the order will be $2, 1$ for $\hat{\Omega}_{\mathrm{f}}^+(h^+)$ and $1, 2$ for $\hat{\Omega}_{\mathrm{f}}^-(h)$. This leads to a quasi-distribution function of the positive P type.

For the physical states where the state vector involves a single number of atoms ($N = 0, 1, 2$), expectation values vanish if the numbers of fermion annihilation and creation operators differ. Using $\hat{\Omega}_{\mathrm{f}}^+(h^+) = (1 + i\hat{c}_2 h_2^+)(1 + i\hat{c}_1 h_1^+)$ and $\hat{\Omega}_{\mathrm{f}}^-(h) = (1 + ih_1\hat{c}_1^\dagger)(1 + ih_2\hat{c}_2^\dagger)$, the characteristic function takes the form

$$\chi(\xi, \xi^+, h, h^+) = \chi_0(\xi, \xi^+) + \sum_{i,j=1,2} \chi_2^{i;j}(\xi, \xi^+)\, h_j^+ h_i + \chi_4^{12;21}(\xi, \xi^+)\, h_2^+ h_1^+ h_1 h_2, \tag{10.78}$$

and the relationship with the coefficients for $\xi, \xi^+ = 0, 0$ is

$$\begin{aligned} \chi_0(0,0) &= 1, \\ \chi_2^{i,j}(0,0) &= i^2 \langle \hat{c}_i^\dagger \hat{c}_j \rangle, \\ \chi_4^{12,21}(0,0) &= i^4 \langle \hat{c}_1^\dagger \hat{c}_2^\dagger \hat{c}_2 \hat{c}_1 \rangle. \end{aligned} \tag{10.79}$$

Thus, for one-atom physical states, at most six $(2.2!/(2!2!) = ({}^2C_0)^2 + ({}^2C_1)^2 + ({}^2C_2)^2$ [59]) c-number coefficients are required to define the characteristic function $\chi(\xi, \xi^+, h, h^+)$ as a Grassmann function.

The various normally ordered quantum correlation functions can be expressed as c-number and Grassmann derivatives of the characteristic function:

$$\begin{aligned} &G(m_1, m_2, n;\, p, l_2, l_1) \\ &= \langle (\hat{c}_1^\dagger)^{m_1} (\hat{c}_2^\dagger)^{m_2} (\hat{a}^\dagger)^n (\hat{a})^p (\hat{c}_2)^{l_2} (\hat{c}_1)^{l_1} \rangle \\ &= \mathrm{Tr}((\hat{c}_2)^{l_2} (\hat{c}_1)^{l_1} (\hat{a})^p \hat{\rho}\, (\hat{a}^\dagger)^n\, (\hat{c}_1^\dagger)^{m_1} (\hat{c}_2^\dagger)^{m_2}) \end{aligned} \tag{10.80}$$

$$= \left[\frac{(\overrightarrow{\partial})^{l_2}}{\partial(ih_2^+)^{l_2}} \frac{(\overrightarrow{\partial})^{l_1}}{\partial(ih_1^+)^{l_1}} \left[\frac{\partial^n}{\partial(i\xi)^n} \frac{\partial^p}{\partial(i\xi^+)^p} \chi(\xi, \xi^+, h, h^+) \right] \frac{(\overleftarrow{\partial})^{m_1}}{\partial(ih_1)^{m_1}} \frac{(\overleftarrow{\partial})^{m_2}}{\partial(ih_2)^{m_2}} \right]_{\xi, \xi^+, h, h^+ = 0}, \tag{10.81}$$

where $m_i, l_i = 0, 1$ only, and if l_i or m_i is zero, then no differentiation takes place.

10.3.9 Distribution Function

The characteristic function $\chi(\xi, \xi^+, h, h^+)$ is related to the distribution function $P(\alpha, \alpha^+, \alpha^*, \alpha^{+*}, g, g^+)$ via phase space integrals, which are c-number integrals for the field mode and Grassmann integrals for the two atomic modes. The formula is

$$\chi(\xi, \xi^+, h, h^+) = \int \prod_{i=1,2} dg_i^+ \prod_{i=1,2} dg_i \int d^2\alpha^+ d^2\alpha \exp i \sum_{i=1}^{2} \{g_i h_i^+\} \exp i\{\alpha\,\xi^+\}$$
$$P(\alpha, \alpha^+, \alpha^*, \alpha^{+*}, g, g^+) \cdot \exp i\{\xi\alpha^+\} \exp i \sum_{i=1}^{2} \{h_i g_i^+\}, \tag{10.82}$$

with the usual associations and functional dependencies. As previously, the characteristic function $\chi(\xi, \xi^+, h, h^+)$ is a function of the variables ξ, ξ^+, h_i, h_i^+ and g_i, g_i^+ but *not* of their complex conjugates, whereas the distribution function P also depends on the complex conjugates α^*, α^{+*}. The c-number integrations $d^2\alpha^+$ and $d^2\alpha$ are over the two complex planes α, α^+ and thus $d^2\alpha \equiv d\alpha_x\, d\alpha_y$, etc. The Grassmann integrals dg_i^+ and dg_i are over the single variables g_i^+ and g_i only, and not over the conjugates. The distribution function is of the positive P type for the bosonic variables α, α^+ and similar to the complex P type for the Grassmann variables g_i, g_i^+. Although the differentials can be placed in reverse order, the convention used here is to write $\prod_i dg_i^+ \prod_i dg_i = dg_1^+ dg_2^+\, dg_2 dg_1$.

The distribution function is of the form

$$P = P_0(\tilde{\boldsymbol{\alpha}}) + \sum_{i;j} P_2^{i,j}(\tilde{\boldsymbol{\alpha}}) g_i g_j^+ + P_4^{12,21}(\tilde{\boldsymbol{\alpha}}) g_1 g_2 g_2^+ g_1^+, \tag{10.83}$$

where for short we write $\tilde{\boldsymbol{\alpha}} \equiv \{\alpha, \alpha^+, \alpha^*, \alpha^{+*}\}$. Other possible forms do not lead to the characteristic function given by (10.78) when the Grassmann integrations in (7.51) are carried out. An ordering convention in which the g_i are arranged in ascending order and the g_j^+ in descending order has been used. The coefficients are functions of the bosonic variables α, α^+. There are six $((2.2)!/(2!2!) = (^2C_0)^2 + (^2C_1)^2 + (^2C_2)^2$ [59]) c-number coefficients to define the distribution function $P(\alpha, \alpha^+, g, g^+)$ as a Grassmann function, the same number of course as the characteristic function that it determines. The distribution function is thus an even Grassmann function of order $2^2 = 4$ in the variables g_1, g_1^+, g_2, g_2^+, a feature that is needed later.

The Hermiticity of the density operator leads to relationships between the coefficients

$$\begin{aligned} &P_0(\tilde{\boldsymbol{\alpha}})^* = P_0(\tilde{\boldsymbol{\alpha}}), && P_2^{1,1}(\tilde{\boldsymbol{\alpha}})^* = P_2^{1,1}(\tilde{\boldsymbol{\alpha}}), && P_2^{1,2}(\tilde{\boldsymbol{\alpha}})^* = P_2^{2,1}(\tilde{\boldsymbol{\alpha}}), \\ &P_2^{2,1}(\tilde{\boldsymbol{\alpha}})^* = P_2^{1;2}(\tilde{\boldsymbol{\alpha}}), && P_2^{2,2}(\tilde{\boldsymbol{\alpha}})^* = P_2^{2,2}(\tilde{\boldsymbol{\alpha}}), && P_4^{12,21}(\tilde{\boldsymbol{\alpha}})^* = P_4^{12,21}(\tilde{\boldsymbol{\alpha}}). \end{aligned} \tag{10.84}$$

This shows that four of the bosonic coefficients are real and the other two, $P_2^{1,2}(\tilde{\boldsymbol{\alpha}})$ and $P_2^{2,1}(\tilde{\boldsymbol{\alpha}})$, are complex conjugates of one another.

For the Jaynes–Cummings model case, the quantum correlation functions are obtained from the characteristic function using (10.81). Applying this result by carrying out the differentiations on the formula (7.51) relating the characteristic and distribution functions gives the quantum correlation functions in terms of *phase space integrals:*

$$G(m_1, m_2, n; p, l_2, l_1) = \int d^2\alpha^+ d^2\alpha \int \prod_{i=1,2} dg_i^+ \prod_{i=1,2} dg_i$$
$$\times (g_2)^{l_2} (g_1)^{l_1} (\alpha)^p \, P(\alpha, \alpha^+, \alpha^*, \alpha^{+*}, g, g^+) \, (\alpha^+)^n \, (g_1^+)^{m_1} (g_2^+)^{m_2}, \qquad (10.85)$$

which involves both c-number and Grassmann number phase space integrals with the Bose–Fermi distribution function. Note that the numbers of fermion annihilation and creation operators are the same. Alternative forms for the quantum correlation function with the distribution function as the left factor in the integrand and all the α_i and g_j placed to the right of the α_i^+ and g_j^+ can easily be found.

10.3.10 Probabilities and Coherences as Phase Space Integrals

The Grassmann phase space integrals can be evaluated using the non-zero result $\int \prod_i dg_i^+ dg_i \, g_1 g_2 g_2^+ g_1^+ = 1$, giving the probabilities and coherences as bosonic phase space integrals involving the six coefficients that determine the distribution function (10.83).

For the two-atom probability which is given by the fourth-order quantum correlation function, we obtain the result

$$P_{12} = \int d^2\alpha^+ d^2\alpha \, P_0(\tilde{\boldsymbol{\alpha}}). \qquad (10.86)$$

This correlation function may be zero without $P_0(\tilde{\boldsymbol{\alpha}})$ being zero.

The one-atom probabilities for states 1, 2 are given by

$$P_1 = \int d^2\alpha^+ d^2\alpha \left(P_2^{2;2}(\tilde{\boldsymbol{\alpha}}) - P_0(\tilde{\boldsymbol{\alpha}}) \right), \qquad (10.87)$$

$$P_2 = \int d^2\alpha^+ d^2\alpha \left(P_2^{1;1}(\tilde{\boldsymbol{\alpha}}) - P_0(\tilde{\boldsymbol{\alpha}}) \right). \qquad (10.88)$$

These quantities are of course real, consistent with (10.84).

For the atomic coherences, we have the results

$$\rho_{12} = -\int d^2\alpha^+ d^2\alpha \, P_2^{1;2}(\tilde{\boldsymbol{\alpha}}), \qquad (10.89)$$

$$\rho_{21} = -\int d^2\alpha^+ d^2\alpha \, P_2^{2;1}(\tilde{\boldsymbol{\alpha}}). \qquad (10.90)$$

These quantities are complex conjugates, consistent with (10.84). The mean number of photons is

$$\overline{n} = \int d^2\alpha^+ d^2\alpha \, (\alpha) \, P_4^{12;21}(\alpha, \alpha^+) \, (\alpha^+). \qquad (10.91)$$

This quantity is of course real, consistent with (10.84).

Note that the mean-photon-number result involves the fourth-order expansion coefficient $P_4^{12;21}(\tilde{\boldsymbol{\alpha}})$, and the one-atom probabilities for states 1, 2 involve the opposite second-order expansion coefficients $P_2^{2;2}(\tilde{\boldsymbol{\alpha}})$ and $P_2^{1;1}(\tilde{\boldsymbol{\alpha}})$, respectively, minus the quantity $P_0(\tilde{\boldsymbol{\alpha}})$ that determines the two-atom probability. The coherences contain the second-order expansion coefficients $P_2^{1;2}(\tilde{\boldsymbol{\alpha}})$ and $P_2^{2;1}(\tilde{\boldsymbol{\alpha}})$. Thus, in their final form, only c-number integrations are needed to determine the quantum correlation functions.

The normalisation integral for the combined Bose–Fermi distribution function is given by

$$\mathrm{Tr}\hat{\rho} = \int\int d^2\alpha^+ d^2\alpha \prod_{i=1,2} dg_i^+ \prod_{i=1,2} dg_i\, P(\alpha,\alpha^+,\alpha^*,\alpha^{+*},g,g^+) = 1. \tag{10.92}$$

If we substitute the specific ordered form (10.83) for the distribution function, we see that the normalisation integral gives

$$\int d^2\alpha^+ d^2\alpha\, P_4^{12;21}(\alpha,\alpha^+,\alpha^*,\alpha^{+*}) = 1. \tag{10.93}$$

10.3.11 Fokker–Planck Equation

The derivation of the Fokker–Planck equation is based on using the correspondence rules. As has been noted previously, the distribution function is not unique and is a non-analytic function of the bosonic phase space variables. The correspondence rules are also non-unique, but the application of the standard correspondence rules (8.1)–(8.4) and (8.24)–(8.27) is expected to lead to a distribution function which correctly determines the quantum correlation functions. It turns out, however, that the Fokker–Planck equation based on the standard rules leads to a distribution function that is not satisfactory. However, the canonical form of the distribution function (8.15) always exists and can be applied to the initial conditions. Furthermore, it turns out to lead to a Fokker–Planck equation that can be solved analytically, so the Fokker–Planck equation for the canonical distribution function will now be obtained. The relevant correspondence rules are those given in (8.18) and the distribution function is written in terms of the new variables $\gamma,\gamma^*,\delta,\delta^*$ as defined in (8.14).

The Fokker–Planck equation for the canonical distribution function P_{canon} is

$$\begin{aligned}
&\frac{\partial P_{\mathrm{canon}}(\gamma,\gamma^*,\delta,\delta^*,g,g^+)}{\partial t} = \\
&-i\frac{E_a}{\hbar}\left(\frac{\overrightarrow{\partial}}{\partial g_2}(g_2)P_{\mathrm{canon}} + \frac{\overrightarrow{\partial}}{\partial g_1}(g_1)P_{\mathrm{canon}}\right) + i\frac{E_a}{\hbar}\left(P_{\mathrm{canon}}\frac{\overleftarrow{\partial}}{\partial g_2^+} + P_{\mathrm{canon}}\frac{\overleftarrow{\partial}}{\partial g_1^+}\right) \\
&-\frac{1}{2}i\omega_0\left(\frac{\overrightarrow{\partial}}{\partial g_2}(g_2)P_{\mathrm{canon}} - \frac{\overrightarrow{\partial}}{\partial g_1}(g_1)P_{\mathrm{canon}}\right) \\
&+\frac{1}{2}i\omega_0\left(P_{\mathrm{canon}}(g_2^+)\frac{\overleftarrow{\partial}}{\partial g_2^+} - P_{\mathrm{canon}}(g_1^+)\frac{\overleftarrow{\partial}}{\partial g_1^+}\right) - i\omega\left(\frac{\partial}{\partial\gamma^*}(\gamma^*)\,P_{\mathrm{canon}} - \frac{\partial}{\partial\gamma}(\gamma)\,P_{\mathrm{canon}}\right) \\
&-\frac{1}{2}i\Omega\left(\frac{\overrightarrow{\partial}}{\partial g_1}(g_2)(\gamma^*)P_{\mathrm{canon}}\right) + \frac{1}{2}i\Omega\left((\gamma)\,P_{\mathrm{canon}}(g_2^+)\frac{\overleftarrow{\partial}}{\partial g_1^+}\right) \\
&-\frac{1}{2}i\Omega\frac{\overrightarrow{\partial}}{\partial g_2}(g_1)\left(\left(\gamma+\frac{\partial}{\partial\gamma^*}\right)P_{\mathrm{canon}}\right) + \frac{1}{2}i\Omega\left(\left(\left(\gamma^*+\frac{\partial}{\partial\gamma}\right)P_{\mathrm{canon}}\right)(g_1^+)\frac{\overleftarrow{\partial}}{\partial g_2^+}\right) \\
&+\frac{1}{2}i\Omega\,(g_2^+g_1)\left(\left(\frac{\partial}{\partial\gamma^*}\right)P_{\mathrm{canon}}\right) - \frac{1}{2}i\Omega\left(\left(\left(\frac{\partial}{\partial\gamma}\right)P_{\mathrm{canon}}\right)(g_1^+g_2)\right),
\end{aligned} \tag{10.94}$$

where the variables $(\gamma, \gamma^*, \delta, \delta^*, g, g^+)$ are left understood. The derivation is left as an exercise.

10.3.12 Coupled Distribution Function Coefficients

Writing γ, δ for $\gamma, \gamma^*, \delta, \delta^*$, the expression (10.83) for the distribution function can now be substituted into the Fokker–Planck equation (10.94) for the *canonical* $P+$ distribution to obtain coupled equations for the six c-number coefficients $P_0(\gamma,\delta)$, $P_2^{i;;j}(\gamma,\delta)$ and $P_4^{12;21}(\gamma,\delta)$ that specify the distribution function. For convenience, we use the same terminology for these coefficients as in the general case and leave the 'canonical' label understood. In the derivation, Grassmann differentiations of various products of Grassmann variables are first carried out, and we then equate terms involving the zeroth-, second- and fourth-order monomials in the Grassmann variables g_1, g_2, g_2^+ and g_1^+ to arrive at six separate coupled equations for the c-number coefficients $P_0(\gamma,\delta)$, $P_2^{i;;j}(\gamma,\delta)$ and $P_4^{12;21}(\gamma,\delta)$ that specify the distribution function (the derivation is left as an exercise).

For the zeroth-order terms, we have

$$\frac{\partial}{\partial t}P_0(\gamma,\delta) = -i\omega\left(\frac{\partial}{\partial\gamma^*}\gamma^* - \frac{\partial}{\partial\gamma}\gamma\right)P_0(\gamma,\delta). \tag{10.95}$$

For the second-order terms, we have four equations:

$$\begin{aligned}\frac{\partial}{\partial t}P_2^{1;1}(\gamma,\delta) &= -i\omega\left(\frac{\partial}{\partial\gamma^*}\gamma^* - \frac{\partial}{\partial\gamma}\gamma\right)P_2^{1;1}(\gamma,\delta)\\ &+\frac{1}{2}i\Omega\left(\gamma+\frac{\partial}{\partial\gamma^*}\right)P_2^{2;1}(\gamma,\delta) - \frac{1}{2}i\Omega\left(\gamma^*+\frac{\partial}{\partial\gamma}\right)P_2^{1;2}(\gamma,\delta),\end{aligned} \tag{10.96}$$

$$\begin{aligned}&\frac{\partial}{\partial t}P_2^{1;2}(\gamma,\delta) = -i\omega\left(\frac{\partial}{\partial\gamma^*}\gamma^* - \frac{\partial}{\partial\gamma}\gamma\right)P_2^{1;2}(\gamma,\delta) - i\omega_0 P_2^{1;2}(\gamma,\delta)\\ &-\frac{1}{2}i\Omega\gamma\left(P_2^{1;1}(\gamma,\delta) - P_2^{2;2}(\gamma,\delta)\right) + \frac{1}{2}i\Omega\left(\frac{\partial}{\partial\gamma^*}\right)\left(P_2^{2;2}(\gamma,\delta) - P_0(\gamma,\delta)\right),\end{aligned} \tag{10.97}$$

$$\begin{aligned}&\frac{\partial}{\partial t}P_2^{2;1}(\gamma,\delta) = -i\omega\left(\frac{\partial}{\partial\gamma^*}\gamma^* - \frac{\partial}{\partial\gamma}\gamma\right)P_2^{2;1}(\gamma,\delta) + i\omega_0 P_2^{2;1}(\gamma,\delta)\\ &+\frac{1}{2}i\Omega\gamma^*\left(P_2^{1;1}(\gamma,\delta) - P_2^{2;2}(\gamma,\delta)\right) - \frac{1}{2}i\Omega\left(\frac{\partial}{\partial\gamma}\right)\left(P_2^{2;2}(\gamma,\delta) - P_0(\gamma,\delta)\right),\end{aligned} \tag{10.98}$$

$$\begin{aligned}&\frac{\partial}{\partial t}P_2^{2;2}(\gamma,\delta) = -i\omega\left(\frac{\partial}{\partial\gamma^*}\gamma^* - \frac{\partial}{\partial\gamma}\gamma\right)P_2^{2;2}(\gamma,\delta)\\ &+\frac{1}{2}i\Omega\left(\gamma^*P_2^{1;2}(\gamma,\delta)\right) - \frac{1}{2}i\Omega\left(\gamma P_2^{2;1}(\gamma,\delta)\right).\end{aligned} \tag{10.99}$$

For the fourth-order term,

$$\frac{\partial}{\partial t}P_4^{12;21}(\gamma,\delta) = -i\omega\left(\frac{\partial}{\partial\gamma^*}\gamma^* - \frac{\partial}{\partial\gamma}\gamma\right)P_4^{12;21}(\gamma,\delta)$$
$$+\frac{1}{2}i\Omega\left(\frac{\partial}{\partial\gamma^*}\right)P_2^{2;1}(\gamma,\delta) - \frac{1}{2}i\Omega\left(\frac{\partial}{\partial\gamma}\right)P_2^{1;2}(\gamma,\delta). \tag{10.100}$$

10.3.13 Initial Conditions for Uncorrelated Case

For the case of uncorrelated atom and cavity initial states, the density operator is a product $\hat{\rho} = \hat{\rho}_{\mathrm{f}}\hat{\rho}_{\mathrm{b}}$ and the overall *initial distribution function* is just the product of the atomic term $P_{\mathrm{f}}(g, g^+)$ with the cavity term $P_{\mathrm{b}}(\tilde{\boldsymbol{\alpha}})$, so that at the initial time the distribution function is

$$P(\alpha,\alpha^+,\alpha^*,\alpha^{+*},g,g^+)_0 = P_{\mathrm{f}}(g,g^+)P_{\mathrm{b}}(\tilde{\boldsymbol{\alpha}}), \tag{10.101}$$

and therefore the coefficients are given by

$$\begin{aligned}
P_0 &= \langle\hat{c}_1^\dagger\hat{c}_2^\dagger\hat{c}_2\hat{c}_1\rangle P_{\mathrm{b}}(\tilde{\boldsymbol{\alpha}}) = 0,\\
P_2^{1;1} &= \langle\hat{c}_2^\dagger\hat{c}_2\rangle P_{\mathrm{b}}(\tilde{\boldsymbol{\alpha}}), \qquad P_2^{1;2} = -\langle\hat{c}_1^\dagger\hat{c}_2\rangle P_{\mathrm{b}}(\tilde{\boldsymbol{\alpha}}),\\
P_2^{2;1} &= -\langle\hat{c}_2^\dagger\hat{c}_1\rangle P_{\mathrm{b}}(\tilde{\boldsymbol{\alpha}}), \qquad P_2^{2;2} = \langle\hat{c}_1^\dagger\hat{c}_1\rangle P_{\mathrm{b}}(\tilde{\boldsymbol{\alpha}}),\\
P_4^{12;21} &= (1 - \langle\hat{c}_1^\dagger\hat{c}_2^\dagger\hat{c}_2\hat{c}_1\rangle)P_b(\tilde{\alpha}) = P_b(\tilde{\alpha}),
\end{aligned} \tag{10.102}$$

where the initial distribution function for the cavity mode is

$$P_{\mathrm{b}}(\tilde{\boldsymbol{\alpha}}) = \left(\frac{1}{4\pi^2}\right)\exp\left(-\frac{1}{4}|\alpha-\alpha^{+*}|^2\right)\left\langle\frac{\alpha+\alpha^{+*}}{2},\frac{\alpha^*+\alpha^+}{2}\right|\hat{\rho}_{\mathrm{b}}\left|\frac{\alpha+\alpha^{+*}}{2},\frac{\alpha^*+\alpha^+}{2}\right\rangle. \tag{10.103}$$

10.3.14 Rotating Phase Variables and Coefficients

The final equations to be used are obtained from (10.95)–(10.100) via a series of transformations, both of the phase variables and of the canonical distribution coefficients. Firstly, rotating phase variables are defined, leading to new variables β, β^*, β^+, β^{+*} instead of the original $\alpha,\alpha^*,\alpha^+,\alpha^{+*}$, and this leads to $\tilde{\gamma},\tilde{\gamma}^*,\tilde{\delta},\tilde{\delta}^*$ (or $\tilde{\gamma},\tilde{\delta}$ for short) replacing the $\gamma,\gamma^*,\delta,\delta^*$ in the coefficients for the canonical $P+$ distribution function, now denoted $\tilde{P}(\tilde{\gamma},\tilde{\delta})$. This transformation enables the elimination of the cavity frequency terms from the equations for the coefficients. Note that in this section (Section 10.3), the tilde ˜ does *not* refer to stochastic variables.

Secondly, equations with no explicit time dependence can be obtained via a change of coefficients from $\tilde{P}$ to $\tilde{S}$ by incorporating $\exp(\pm i\omega t)$ into the coefficients $\tilde{P}_2^{1;2}(\tilde{\gamma},\tilde{\delta})$, $\tilde{P}_2^{2;1}(\tilde{\gamma},\tilde{\delta})$, and we can also factor out the explicit dependence on $\tilde{\delta}$ via the overall factor $\exp(-\tilde{\delta}\tilde{\delta}^*) = \exp(-\delta\delta^*)$.

The rotating phase variables are defined via the transformation

$$\begin{aligned}
\alpha &= \beta\exp(-i\omega t), \qquad \alpha^* = \beta^*\exp(+i\omega t),\\
\alpha^+ &= \beta^+\exp(+i\omega t), \qquad \alpha^{+*} = \beta^{+*}\exp(-i\omega t).
\end{aligned} \tag{10.104}$$

This then gives

$$\begin{aligned}
\gamma &= \frac{1}{2}(\beta + \beta^{+*})\exp(-i\omega t) = \tilde{\gamma}\exp(-i\omega t), \\
\gamma^* &= \frac{1}{2}(\beta^* + \beta^{+})\exp(+i\omega t) = \tilde{\gamma}^*\exp(+i\omega t), \\
\delta &= \frac{1}{2}(\beta - \beta^{+*})\exp(-i\omega t) = \tilde{\delta}\exp(-i\omega t), \\
\delta^* &= \frac{1}{2}(\beta^* - \beta^{+})\exp(+i\omega t) = \tilde{\delta}^*\exp(+i\omega t).
\end{aligned} \tag{10.105}$$

The new coefficients are

$$\begin{aligned}
\tilde{P}_2^{1;1}(\tilde{\gamma},\tilde{\delta}) &= \tilde{S}_2^{1;1}(\tilde{\gamma})\,\exp(-\tilde{\gamma}\tilde{\gamma}^*)\,\exp(-\tilde{\delta}\,\tilde{\delta}^*), \\
\tilde{P}_2^{2;2}(\tilde{\gamma},\tilde{\delta}) &= \tilde{S}_2^{2;2}(\tilde{\gamma})\,\exp(-\tilde{\gamma}\tilde{\gamma}^*)\,\exp(-\tilde{\delta}\,\tilde{\delta}^*), \\
\tilde{P}_2^{1;2}(\tilde{\gamma},\tilde{\delta}) &= \tilde{S}_2^{1;2}(\tilde{\gamma})\,\exp(-\tilde{\gamma}\tilde{\gamma}^*)\,\exp(-\tilde{\delta}\,\tilde{\delta}^*)\exp(-i\omega t), \\
\tilde{P}_2^{2;1}(\tilde{\gamma},\tilde{\delta}) &= \tilde{S}_2^{2;1}(\tilde{\gamma})\,\exp(-\tilde{\gamma}\tilde{\gamma}^*)\,\exp(-\tilde{\delta}\,\tilde{\delta}^*)\exp(+i\omega t), \\
\tilde{P}_0(\tilde{\gamma},\tilde{\delta}) &= \tilde{S}_0(\tilde{\gamma})\,\exp(-\tilde{\gamma}\tilde{\gamma}^*)\,\exp(-\tilde{\delta}\,\tilde{\delta}^*), \\
\tilde{P}_4^{12;21}(\tilde{\gamma},\tilde{\delta}) &= \tilde{S}_4^{12;21}(\tilde{\gamma})\,\exp(-\tilde{\gamma}\tilde{\gamma}^*)\,\exp(-\tilde{\delta}\,\tilde{\delta}^*).
\end{aligned} \tag{10.106}$$

The variable $\tilde{\delta}$ plays no further role in the dynamics and the $\tilde{S}$ depend only on $\tilde{\gamma},\tilde{\gamma}^*$. The separate cavity and atomic transition frequencies are then incorporated into the detuning

$$\Delta = \omega_0 - \omega. \tag{10.107}$$

With these substitutions, the equations for the coefficients are as follows. For the zeroth-order term,

$$\frac{\partial}{\partial t}\tilde{S}_0(\tilde{\gamma}) = 0. \tag{10.108}$$

The four second-order equations are

$$\frac{\partial}{\partial t}\left(\tilde{S}_2^{1;1}(\tilde{\gamma}) - \tilde{S}_0(\tilde{\gamma})\right) = +\frac{1}{2}i\Omega\left(\frac{\partial}{\partial\tilde{\gamma}^*}\right)\tilde{S}_2^{2;1}(\tilde{\gamma}) - \frac{1}{2}i\Omega\left(\frac{\partial}{\partial\tilde{\gamma}}\right)\tilde{S}_2^{1;2}(\tilde{\gamma}), \tag{10.109}$$

$$\frac{\partial}{\partial t}\tilde{S}_2^{1;2}(\tilde{\gamma}) = -i\Delta\tilde{S}_2^{1;2}(\tilde{\gamma}) - \frac{1}{2}i\Omega\tilde{\gamma}\left(\tilde{S}_2^{1;1}(\tilde{\gamma}) - \tilde{S}_0(\tilde{\gamma})\right) + \frac{1}{2}i\Omega\left(\frac{\partial}{\partial\tilde{\gamma}^*}\right)\left(\tilde{S}_2^{2;2}(\tilde{\gamma}) - \tilde{S}_0(\tilde{\gamma})\right), \tag{10.110}$$

$$\frac{\partial}{\partial t}\tilde{S}_2^{2;1}(\tilde{\gamma}) = +i\Delta\tilde{S}_2^{2;1}(\tilde{\gamma}) + \frac{1}{2}i\Omega\tilde{\gamma}^*\left(\tilde{S}_2^{1;1}(\tilde{\gamma}) - \tilde{S}_0(\tilde{\gamma})\right) - \frac{1}{2}i\Omega\left(\frac{\partial}{\partial\tilde{\gamma}}\right)\left(\tilde{S}_2^{2;2}(\tilde{\gamma}) - \tilde{S}_0(\tilde{\gamma})\right), \tag{10.111}$$

$$\frac{\partial}{\partial t}\left(\tilde{S}_2^{2;2}(\tilde{\gamma}) - \tilde{S}_0(\tilde{\gamma})\right) = +\frac{1}{2}i\Omega\left(\tilde{\gamma}^*\tilde{S}_2^{1;2}(\tilde{\gamma})\right) - \frac{1}{2}i\Omega\left(\tilde{\gamma}\tilde{S}_2^{2;1}(\tilde{\gamma})\right), \tag{10.112}$$

where we have subtracted the zero quantity $(\partial/\partial t)\tilde{S}_0(\tilde{\gamma})$ from each side of the first and the fourth equation. From Section 10.3.10, we see that the quantities $\tilde{S}_2^{i;i}(\tilde{\gamma}) - \tilde{S}_0(\tilde{\gamma})$ determine the one-atom probabilities. However, from (10.108) and the initial conditions (10.102) we see that $\tilde{S}_0(\tilde{\gamma})$ will be zero for the one-atom Jaynes–Cummings model, so we can ignore $\tilde{S}_0(\tilde{\gamma})$ henceforth. The fourth-order equation is

$$\frac{\partial}{\partial t}\tilde{S}_4^{12;21}(\tilde{\gamma}) = \frac{1}{2}i\Omega\left(-\tilde{\gamma} + \frac{\partial}{\partial\tilde{\gamma}^*}\right)\tilde{S}_2^{2;1}(\tilde{\gamma}) - \frac{1}{2}i\Omega\left(-\tilde{\gamma}^* + \frac{\partial}{\partial\tilde{\gamma}}\right)\tilde{S}_2^{1;2}(\tilde{\gamma}).$$

10.3.15 Solution to Fokker–Planck Equation

The approach used by Stenholm [80] can be adapted to provide an analytical solution for the canonical distribution function for the one-atom Jaynes–Cummings model with any initial conditions. The equations (10.109)–(10.112) for the $\tilde{S}_2^{i;j}(\tilde{\gamma})$ can be solved via the substitution

$$\tilde{S}_2^{i;j}(\tilde{\gamma}) = \Psi_i^*(\tilde{\gamma}^*)\,\Psi_j(\tilde{\gamma}), \tag{10.113}$$

where the $\Psi_i^*(\tilde{\gamma}^*)$ are functions of the $\tilde{\gamma}^*$ and the $\Psi_i(\tilde{\gamma})$ are functions of the $\tilde{\gamma}$, and where the $\Psi_i(\tilde{\gamma})$ satisfy the coupled equations

$$\begin{aligned}
\frac{\partial}{\partial t}\Psi_1(\tilde{\gamma}) &= \frac{1}{2}i\Delta\Psi_1(\tilde{\gamma}) - \frac{1}{2}i\Omega\left(\frac{\partial}{\partial\tilde{\gamma}}\right)\Psi_2(\tilde{\gamma}),\\
\frac{\partial}{\partial t}\Psi_2(\tilde{\gamma}) &= -\frac{1}{2}i\Delta\Psi_2(\tilde{\gamma}) - \frac{1}{2}i\Omega\,\tilde{\gamma}\,\Psi_1(\tilde{\gamma}).
\end{aligned} \tag{10.114}$$

This ansatz is consistent with the original equations (10.109)–(10.112) for the $\tilde{S}_2^{i;j}(\tilde{\gamma})$. As indicated previously, we have set $\tilde{S}_0(\tilde{\gamma}) = 0$ for the one-atom case.

Differentiating the second equation and substituting from the first gives

$$\begin{aligned}
\frac{\partial^2}{\partial t^2}\Psi_2(\tilde{\gamma}) + \frac{1}{4}\Delta^2\,\Psi_2(\tilde{\gamma}) &= -\frac{1}{4}\Omega^2\,\tilde{\gamma}\left(\frac{\partial}{\partial\tilde{\gamma}}\right)\Psi_2(\tilde{\gamma})\\
&= -\frac{1}{4}\Omega^2\left(\frac{\partial}{\partial s}\right)\Psi_2(\tilde{\gamma}),
\end{aligned} \tag{10.115}$$

where we have made the substitution

$$s = \lg\tilde{\gamma}, \qquad \tilde{\gamma} = \exp s. \tag{10.116}$$

A solution of the equation (10.115) can be obtained using separation of the variables,

$$\Psi_2(\tilde{\gamma}) = T(t)\,K(s), \tag{10.117}$$

whence we find that

$$\frac{1}{T(t)}\frac{d^2}{dt^2}T(t) + \frac{1}{4}\Delta^2 = -\frac{1}{4}\Omega^2\,\frac{1}{K(s)}\left(\frac{\partial}{\partial s}\right)K(s) = -\lambda, \tag{10.118}$$

where, since the left side is a function of t and the right side is a function of s, the quantity λ must be a constant. The solution of these two equations is straightforward. We have

$$
\begin{aligned}
K(s) &= C \exp\left(\frac{4\lambda}{\Omega^2} s\right) \\
&= C\,(\tilde{\gamma})^{(4\lambda/\Omega^2)}\,,
\end{aligned}
\tag{10.119}
$$

where C is a constant, and

$$
T(t) = A\cos\left(\sqrt{\lambda + \Delta^2/4}\,t\right) + B\sin\left(\sqrt{\lambda + \Delta^2/4}\,t\right), \tag{10.120}
$$

with A and B also constant.

Since we require the overall distribution function to be a non-singular single-valued function of the phase space variables, we see from (10.119) that there is a restriction on λ such that

$$
\frac{4\lambda}{\Omega^2} = n \qquad (n = 0, 1, 2, \cdots), \tag{10.121}
$$

where n is an integer.

Combining the variables to eliminate λ and absorbing C into the other constants, we see that a solution for $\Psi_2(\tilde{\gamma})$ for a particular integer n is

$$
\Psi_2(\tilde{\gamma}) = \tilde{\gamma}^n \left(A_n \cos\frac{1}{2}\omega_n t + B_n \sin\frac{1}{2}\omega_n t\right), \tag{10.122}
$$

where

$$
\omega_n = \sqrt{n\Omega^2 + \Delta^2} \tag{10.123}
$$

is the frequency associated with population and coherence oscillations in the one-atom Jaynes–Cummings model. The corresponding solution for $\Psi_1(\tilde{\gamma})$ is then obtained from (10.114), and thus

$$
\Psi_1(\tilde{\gamma}) = i\tilde{\gamma}^{(n-1)}\left(\cos\frac{1}{2}\omega_n t\left\{\frac{\omega_n B_n + i\Delta A_n}{\Omega}\right\} + \sin\frac{1}{2}\omega_n t\left\{\frac{-\omega_n A_n + i\Delta B_n}{\Omega}\right\}\right). \tag{10.124}
$$

However, a solution with $n = 0$ leads to a singular $\tilde{\gamma}^{-1}$ behaviour, so it follows that n is restricted to the positive integers. Also, as the ansatz equations are linear, the general solution is a sum of terms with differing n, so that we finally have the solution in the form

$$
\begin{aligned}
\Psi_1(\tilde{\gamma}) &= i\sum_{n=1}^{\infty}\tilde{\gamma}^{n-1}\left[\left\{\frac{\omega_n B_n + i\Delta A_n}{\Omega}\right\}\cos\frac{1}{2}\omega_n t + \left\{\frac{-\omega_n A_n + i\Delta B_n}{\Omega}\right\}\sin\frac{1}{2}\omega_n t\right], \\
\Psi_2(\tilde{\gamma}) &= \sum_{n=1}^{\infty}\tilde{\gamma}^n\left[A_n\cos\frac{1}{2}\omega_n t + B_n\sin\frac{1}{2}\omega_n t\right].
\end{aligned}
\tag{10.125}
$$

The constants A_n, B_n are chosen to fit the initial conditions.

We can now express the original distribution function coefficients in terms of these function using (10.113) and (10.106). Expressions for $\tilde{S}_4^{12;21}(\tilde{\gamma})$ could also be obtained, but these are of little interest. The results can be written in terms of the original phase variables by substituting

$$\tilde{\gamma} = \frac{1}{2}(\alpha + \alpha^{+*})\exp(+i\omega t), \qquad \tilde{\delta} = \frac{1}{2}(\alpha - \alpha^{+*})\exp(+i\omega t) \tag{10.126}$$

from (10.104) and (10.105) into the above results.

It turns out that the Fokker–Planck equation based on the standard correspondence rules leads to equations for the coefficients that can also be solved by a similar ansatz. However, the solutions lead to a distribution function that diverges on the phase space boundary and in general diverges at large t, with dependences on hyperbolic functions of $\frac{1}{2}\sqrt{n}\Omega t$ in the case of zero detuning. This then throws the original derivation of the standard Fokker–Planck equation into doubt because the integration-by-parts step fails. Other cases where this occurs have been studied by Gilchrist et al. [61].

Finally, as we will see in Section 10.3.16, the solutions based on the canonical distribution function (10.106) agree with the standard quantum optics result, and in particular the quantities A_n, B_n can be chosen so that the initial conditions are the same as for the canonical distribution function determined from the standard quantum optics solution.

10.3.16 Comparison with Standard Quantum Optics Result

As a comparison, we now calculate the canonical distribution function (7.54) as determined from the state vector given in (10.46) and (10.49), obtained from standard quantum optics methods.

The density operator is

$$\begin{aligned}
\hat{\rho} = & \sum_{n,m} B_{n1}(t)B^*_{m1}(t)\exp(-i\overline{n-m}\omega t)\hat{c}_1^\dagger\left|0\right\rangle\left\langle 0\right|\hat{c}_1\ \left|n\right\rangle\left\langle m\right| \\
& + \sum_{n,m} B_{\overline{n-1}2}(t)B^*_{\overline{m-1}2}(t)\exp(-i\overline{n-m}\omega t)\hat{c}_2^\dagger\left|0\right\rangle\left\langle 0\right|\hat{c}_2\ \left|\overline{n-1}\right\rangle\left\langle\overline{m-1}\right| \\
& + \sum_{n,m} B_{n1}(t)B^*_{\overline{m-1}2}(t)\exp(-i\overline{n-m}\omega t)\hat{c}_1^\dagger\left|0\right\rangle\left\langle 0\right|\hat{c}_2\ \left|n\right\rangle\left\langle\overline{m-1}\right| \\
& + \sum_{n,m} B_{\overline{n-1}2}(t)B^*_{m1}(t)\exp(-i\overline{n-m}\omega t)\hat{c}_2^\dagger\left|0\right\rangle\left\langle 0\right|\hat{c}_1\ \left|\overline{n-1}\right\rangle\left\langle m\right|,
\end{aligned} \tag{10.127}$$

where the new amplitudes are given by

$$\begin{aligned}
B_{n1}(t) &= \left(\cos\left(\frac{1}{2}\omega_n t\right)\{A_{n1}(0)\} + i\sin\left(\frac{1}{2}\omega_n t\right)\left\{\frac{\Delta A_{n1}(0) - \Omega\sqrt{n}A_{\overline{n-1}2}(0)}{\omega_n}\right\}\right), \\
B_{\overline{n-1}2}(t) &= \left(\cos\left(\frac{1}{2}\omega_n t\right)\{A_{\overline{n-1}2}(0)\} - i\sin\left(\frac{1}{2}\omega_n t\right)\left\{\frac{\Omega\sqrt{n}A_{n1}(0) + \Delta A_{\overline{n-1}2}(0)}{\omega_n}\right\}\right),
\end{aligned} \tag{10.128}$$

where

$$\exp(-\frac{1}{2}i\Delta t)\exp(-i(n\omega - \frac{1}{2}\omega_0)t) = \exp(-i(n - \frac{1}{2})\omega t)$$

and

$$\exp(+\frac{1}{2}i\Delta t)\exp(-i(\overline{n-1}\omega + \frac{1}{2}\omega_0)t) = \exp(-i(n - \frac{1}{2})\omega t)$$

have been used.

The derivation of the canonical distribution function from the density operator is obtained via the application of (7.54). We find

$$\begin{aligned}
&P_{\text{canon}}(\alpha, \alpha^+, \alpha^*, \alpha^{+*}, g, g^+) \\
&= \left(\frac{1}{4\pi^2}\right)\exp(-\tilde{\gamma}\,\tilde{\gamma}^*)\exp(-\tilde{\delta}\,\tilde{\delta}^*) \\
&\times\left[\sum_{n,m} B_{n1}(t)B^*_{m1}(t)\left\{g_2g_2^+ + g_1g_2g_2^+g_1^+\right\}\frac{(\tilde{\gamma}^*)^n}{\sqrt{n!}}\frac{(\tilde{\gamma})^m}{\sqrt{m!}}\right. \\
&+\sum_{n,m} B_{\overline{n-1}2}(t)B^*_{\overline{m-1}2}(t)\left\{g_1g_1^+ + g_1g_2g_2^+g_1^+\right\}\frac{(\tilde{\gamma}^*)^{(n-1)}}{\sqrt{(n-1)!}}\frac{(\tilde{\gamma})^{(m-1)}}{\sqrt{(m-1)!}} \\
&+\sum_{n,m} B_{n1}(t)B^*_{\overline{m-1}2}(t)\left\{-g_2g_1^+\right\}\frac{(\tilde{\gamma}^*)^n}{\sqrt{n!}}\frac{(\tilde{\gamma})^{(m=1)}}{\sqrt{(m-1)!}}\exp(+i\omega t) \\
&\left.+\sum_{n,m} B_{\overline{n-1}2}(t)B^*_{m1}(t)\left\{-g_1g_2^+\right\}\frac{(\tilde{\gamma}^*)^{(n-1)}}{\sqrt{(n-1)!}}\frac{(\tilde{\gamma})^m}{\sqrt{m!}}\exp(-i\omega t)\right].
\end{aligned} \tag{10.129}$$

From this result, we can identify the coefficients

$$\begin{aligned}
\tilde{P}_2^{1;1}(\tilde{\gamma},\tilde{\delta}) &= \left[\frac{1}{4\pi^2}\right]\exp(-\tilde{\gamma}\tilde{\gamma}^*)\exp(-\tilde{\delta}\tilde{\delta}^*)\sum_{n,m} B_{\overline{n-1}2}(t)B^*_{\overline{m-1}2}(t)\frac{(\tilde{\gamma}^*)^{(n-1)}}{\sqrt{(n-1)!}}\frac{(\tilde{\gamma})^{(m-1)}}{\sqrt{(m-1)!}}, \\
\tilde{P}_2^{1;2} &= -\exp(-i\omega t)\left[\frac{1}{4\pi^2}\right]\exp(-\tilde{\gamma}\tilde{\gamma}^* - \tilde{\delta}\tilde{\delta}^*)\sum_{n,m} B_{\overline{n-1}2}(t)B^*_{m1}(t)\frac{(\tilde{\gamma}^*)^{(n-1)}}{\sqrt{(n-1)!}}\frac{(\tilde{\gamma})^m}{\sqrt{m!}}, \\
\tilde{P}_2^{2;1} &= -\exp(i\omega t)\left[\frac{1}{4\pi^2}\right]\exp(-\tilde{\gamma}\tilde{\gamma}^* - \tilde{\delta}\tilde{\delta}^*)\sum_{n,m} B_{n1}(t)B^*_{\overline{m-1}2}(t)\frac{(\tilde{\gamma}^*)^n}{\sqrt{n!}}\frac{(\tilde{\gamma})^{(m=1)}}{\sqrt{(m-1)!}}, \\
\tilde{P}_2^{2;2} &= \left[\frac{1}{4\pi^2}\right]\exp(-\tilde{\gamma}\tilde{\gamma}^* - \tilde{\delta}\tilde{\delta}^*)\sum_{n,m} B_{n1}(t)B^*_{m1}(t)\frac{(\tilde{\gamma}^*)^n}{\sqrt{n!}}\frac{(\tilde{\gamma})^m}{\sqrt{m!}}
\end{aligned} \tag{10.130}$$

and

$$\tilde{P}_0(\tilde{\gamma},\tilde{\delta}) = 0,$$
$$\tilde{P}_4^{12;21} = \left[\frac{1}{4\pi^2}\right] \exp(-\tilde{\gamma}\tilde{\gamma}^*)\exp(-\tilde{\delta}\tilde{\delta}^*) \sum_{n,m} B_{n1}(t)B_{m1}^*(t)\frac{(\tilde{\gamma}^*)^n}{\sqrt{n!}}\frac{(\tilde{\gamma})^m}{\sqrt{m!}}$$
$$+\sum_{n,m} B_{\overline{n-1}2}(t)B_{\overline{m-1}2}^*(t)\frac{(\tilde{\gamma}^*)^{(n-1)}}{\sqrt{(n-1)!}}\frac{(\tilde{\gamma})^{(m-1)}}{\sqrt{(m-1)!}}. \qquad (10.131)$$

If we write

$$\Phi_1(\tilde{\gamma}) = \left(\frac{1}{2\pi}\right)\sum_m B_{\overline{m-1}2}^*(t)\ \frac{(\tilde{\gamma})^{(m-1)}}{\sqrt{(m-1)!}},$$
$$\Phi_2(\tilde{\gamma}) = -\left(\frac{1}{2\pi}\right)\sum_m B_{m1}^*(t)\ \frac{(\tilde{\gamma})^m}{\sqrt{m!}}, \qquad (10.132)$$

then

$$\tilde{P}_2^{1;1}(\tilde{\gamma},\tilde{\delta}) = \exp(-\tilde{\gamma}\,\tilde{\gamma}^*)\ \exp(-\tilde{\delta}\,\tilde{\delta}^*)\ \Phi_1^*(\tilde{\gamma}^*)\,\Phi_1(\tilde{\gamma}),$$
$$\tilde{P}_2^{1;2}(\tilde{\gamma},\tilde{\delta}) = \exp(-\tilde{\gamma}\,\tilde{\gamma}^*)\ \exp(-\tilde{\delta}\,\tilde{\delta}^*)\ \Phi_1^*(\tilde{\gamma}^*)\,\Phi_2(\tilde{\gamma})\ \exp(-i\omega t),$$
$$\tilde{P}_2^{2;1}(\tilde{\gamma},\tilde{\delta}) = \exp(-\tilde{\gamma}\,\tilde{\gamma}^*)\ \exp(-\tilde{\delta}\,\tilde{\delta}^*)\ \Phi_2^*(\tilde{\gamma}^*)\,\Phi_1(\tilde{\gamma})\ \exp(+i\omega t),$$
$$\tilde{P}_2^{2;2}(\tilde{\gamma},\tilde{\delta}) = \exp(-\tilde{\gamma}\,\tilde{\gamma}^*)\ \exp(-\tilde{\delta}\,\tilde{\delta}^*)\ \Phi_2^*(\tilde{\gamma}^*)\,\Phi_2(\tilde{\gamma}). \qquad (10.133)$$

To have agreement with the previous results in (10.106) so that

$$\Phi_1(\tilde{\gamma}) = \Psi_1(\tilde{\gamma}),$$
$$\Phi_2(\tilde{\gamma}) = \Psi_2(\tilde{\gamma}), \qquad (10.134)$$

where $\Psi_1(\tilde{\gamma})$ and $\Psi_2(\tilde{\gamma})$ are as in (10.125), we require

$$i\left\{\frac{\omega_n B_n + i\Delta A_n}{\Omega}\right\} = \left(\frac{1}{2\pi}\right)\{A_{\overline{n-1}2}(0)\}^*\ \frac{1}{\sqrt{(n-1)!}},$$
$$i\left\{\frac{-\omega_n A_n + i\Delta B_n}{\Omega}\right\} = \left(\frac{1}{2\pi}\right)(-i)^*\left\{\frac{\Omega\sqrt{n}A_{n1}(0)+\Delta A_{\overline{n-1}2}(0)}{\omega_n}\right\}^*\ \frac{1}{\sqrt{(n-1)!}},$$
$$A_n = \left(\frac{1}{2\pi}\right)(-)\{A_{n1}(0)\}^*\ \frac{1}{\sqrt{(n)!}},$$
$$B_n = \left(\frac{1}{2\pi}\right)(-)(+i)^*\left\{\frac{\Delta A_{n1}(0)-\Omega\sqrt{n}A_{\overline{n-1}2}(0)}{\omega_n}\right\}^*\ \frac{1}{\sqrt{(n)!}}. \qquad (10.135)$$

The last two equations give explicit expressions for A_n and B_n. Substituting these expressions into the left side of the first two equations gives the right-hand sides, showing that the four equations are consistent. Hence we see that for *any* initial conditions for the one-atom Jaynes–Cummings model, the solution given by the Grassmann phase space approach is the same as that from the standard quantum optics treatment.

10.3.17 Application of Results

As an illustration of how to apply the above results for the canonical distribution function, we consider the case where the atom is initially in the *lower* state and the field is in a *coherent* state of amplitude η. In this case, we have from Section 10.3.13

$$
\begin{aligned}
P_2^{1;1} &= 0, \qquad P_2^{1;2} = 0, \\
P_2^{2;1} &= 0, \qquad P_2^{2;2} = P_b(\alpha, \alpha^+),
\end{aligned}
\tag{10.136}
$$

with

$$
P_b(\alpha, \alpha^+) = \frac{1}{4\pi^2} \exp\left(-\frac{|\alpha - \alpha^{+*}|^2}{4}\right) \exp\left(-\left|\frac{1}{2}(\alpha + \alpha^{+*}) - \eta\right|^2\right).
$$

Hence

$$
\begin{aligned}
\tilde{P}_2^{2;2}(\tilde{\gamma}, \tilde{\delta}) &= \frac{1}{4\pi^2} \exp(-\tilde{\delta}\,\tilde{\delta}^*) \exp\left(-|\tilde{\gamma} - \eta|^2\right) = \exp(-\tilde{\delta}\,\tilde{\delta}^*) \exp(-\tilde{\gamma}\tilde{\gamma}^*)\, \tilde{S}_2^{2;2}(\tilde{\gamma}), \\
\tilde{S}_2^{2;2}(\tilde{\gamma}) &= \frac{1}{4\pi^2} \exp\left(-(\tilde{\gamma} - \eta)(\tilde{\gamma}^* - \eta^*)\right) \exp(+\tilde{\gamma}\tilde{\gamma}^*) \\
&= \left(\frac{1}{2\pi} \exp\left(-\frac{1}{2}\eta\eta^*\right) \exp(-\tilde{\gamma}^*\eta)\right) \left(\frac{1}{2\pi} \exp\left(-\frac{1}{2}\eta\eta^*\right) \exp(-\tilde{\gamma}\eta^*)\right).
\end{aligned}
\tag{10.137}
$$

At $t = 0$ we have

$$
\begin{aligned}
\Psi_1(\tilde{\gamma}) &= i \sum_{n=1}^{\infty} \tilde{\gamma}^{(n-1)} \left(\left\{\frac{\omega_n B_n + i\Delta A_n}{\Omega}\right\}\right) = 0, \\
\Psi_2(\tilde{\gamma}) &= \sum_{n=1}^{\infty} \tilde{\gamma}^n (A_n) = \frac{1}{2\pi} \exp\left(-\frac{1}{2}\eta\eta^*\right) \exp(-\tilde{\gamma}\eta^*),
\end{aligned}
\tag{10.138}
$$

so that if we choose

$$
\begin{aligned}
A_n &= \frac{1}{2\pi} \exp\left(-\frac{1}{2}\eta\eta^*\right) \frac{(-\eta^*)^n}{n!}, \\
B_n &= \frac{-i\Delta}{\omega_n} A_n,
\end{aligned}
\tag{10.139}
$$

the solutions for $\Psi_1(\tilde{\gamma})$ and $\Psi_2(\tilde{\gamma})$ determine the time-dependent distribution function.

10.3.18 Conclusion

We have shown that a phase space approach using Grassmann variables to describe the atomic system and c-number variables to describe the cavity mode can be used to treat the Jaynes–Cummings model, and to obtain the same results when treating phenomena such as pure Rabi oscillations and collapse and revival effects as those obtained from standard quantum optics methods. The Liouville–von Neumann equation for the density operator was converted into a Fokker–Planck equation for the canonical positive P distribution function using the correspondence rules associated with this choice of distribution function. The distribution function is a Grassmann function involving Grassmann phase space variables g_1, g_1^+ and g_2, g_2^+ for the two fermionic modes associated with the two atomic states, with six c-number functions of the bosonic phase space variables α, α^+ associated with the cavity mode being involved as coefficients in specifying the distribution function. In the context of a general mixed state where there may be zero, one or two atoms present, expressions for the probabilities of finding one atom in one of the two atomic states, one atom in both atomic states and no atom in either atomic state were obtained as bosonic phase space integrals involving the six bosonic coefficients. Coupled equations for the six bosonic coefficients of the canonical distribution function were obtained from the Fokker–Planck equation These equations were solved for the one-atom Jaynes–Cummings model using an ansatz similar to that applied by Stenholm [80] in an earlier Bargmann state treatment of the Jaynes–Cummings model, and the results were shown to be equivalent to the standard quantum optics treatment based on state vectors and coupled amplitude equations.

Positive P distribution functions have the feature of being non-unique, with different correspondence rules applying in the derivation of the specific Fokker–Planck equation. In this application, we also found that applying the standard correspondence rules (rather than those for the canonical positive P case) leads to a Fokker–Planck equation where the solution of the coupled equations for the bosonic coefficients via a similar ansatz was quite unsuitable. Not only did the solutions diverge for large time t, but the distribution function diverged for large phase space variables α, α^+, thereby throwing into question the derivation of the Fokker–Planck equation. The standard treatment requires the distribution function to vanish on the phase space boundary. As the correspondence rules for the canonical positive P distribution do not require this feature, it is suggested that Fokker–Planck equations based on the canonical positive P distribution may be more reliable. Furthermore, in terms of matching a general solution of the Fokker–Planck equation to the initial conditions, the use of the canonical form of the distribution function is easiest, since initial conditions are usually specified via the initial density operator, from which the canonical distribution function is directly determined. However, more general Fokker–Planck equations involving derivatives higher than second order may occur using the canonical distribution function [60], so that no replacement by Langevin stochastic equations is then possible. In fact, even if the Fokker–Planck equation is only second order, the diffusion matrix may not be positive definite.

Exercises

(10.1) Derive the Fokker–Planck equation in (10.94) for the Jaynes–Cummings model.

(10.2) Derive the coupled equations in (10.95)–(10.100) for the distribution function coefficients in the Jaynes–Cummings model.

11

Functional Calculus for C-Number and Grassmann Fields

11.1 Features

Functional calculus has been applied in physics in many fields, ranging from classical mechanics to quantum mechanics, and the basic laws of physics can be expressed in terms of functionals [6, 84, 85]. For example, in classical mechanics the action is a functional of the position considered as a function of time, and Lagrange's equations of motion are based on the principle of least action, which leads to the functional derivative of the action with respect to position being equated to zero.

In quantum mechanics, we can define an action functional which involves the time-dependent state vector, and the time-dependent Schrödinger equation can be derived from a principle of least action. The path integral formulation of quantum mechanics, which is widely used in quantum field theory, involves writing the transition amplitude for a quantum process as a functional.

In the field of degenerate quantum gases (Bose–Einstein condensates and Fermi gases), the physics of large numbers of quantum modes – which can be occupied by many bosons or fermions – may be involved. For such multi-mode systems, it is often convenient to avoid considering individual modes explicitly, and to represent the quantum field operators via field functions with the quantum density operator then mapped onto a distribution functional. Functional methods arise naturally in this field.

The basic ideas of functional calculus are first outlined here for the case of c-number quantities. Then we consider the case of Grassmann number quantities. The two main processes of interest are *functional differentiation* and *functional integration*, but we begin by explaining what is meant by a functional.

It is also necessary to also consider functionals involving c-number and Grassmann field functions $\psi^K(x)$ which are still based on an expansion in terms of orthonormal mode functions $\phi_k(x)$, but where there is some restriction on the modes that are included. Such functions will be referred to as *restricted functions*. Examples include the fields $\psi_{\mathrm{C}}(\mathbf{r}), \psi_{\mathrm{C}}^{+}(\mathbf{r}), \psi_{\mathrm{NC}}(\mathbf{r}), \psi_{\mathrm{NC}}^{+}(\mathbf{r})$ used for condensate and non-condensate modes in the theory of Bose condensates, where even the combined condensate and non-condensate modes are subject to a restriction, in that modes associated with a momentum greater than a cut-off value are excluded. In this situation, restricted delta functions, such as $\delta_{\mathrm{C}}(r, r')$ and $\delta_{\mathrm{NC}}(r, r')$ (see (H.4)), will be involved. Again the main processes of interest are functional differentiation and integration. As much of the

Phase Space Methods for Degenerate Quantum Gases. First Edition. Bryan J. Dalton, John Jeffers, and Stephen M. Barnett.
Published in 2015 by Oxford University Press.

theory for restricted functions is similar to that presented in the present chapter, it is included as Appendix H.

11.2 Functionals of Bosonic C-Number Fields

11.2.1 Basic Idea

A *functional* $F[\psi(x)]$ maps a c-number function $\psi(x)$ onto a c-number that depends on *all* the values of $\psi(x)$ over its entire range. The independent variable x could in some cases refer to a position coordinate or to time. If x does refer to position, then $\psi(x)$ is referred to as a *field function*, as for example in the case of a *bosonic field.* Note that the functional is written with square brackets to distinguish it from a function, written with round brackets. As previously, we assume that c-number functions $\psi(x)$ can be expanded in terms of a suitable orthonormal set of mode functions with c-number expansion coefficients α_k,

$$\psi(x) = \sum_k \alpha_k \, \phi_k(x), \tag{11.1}$$

where the orthonormality conditions are

$$\int dx \, \phi_k^*(x)\phi_l(x) = \delta_{kl} \tag{11.2}$$

and the completeness relationship is

$$\sum_k \phi_k(x)\phi_k^*(y) = \delta(y - x). \tag{11.3}$$

This gives the well-known result for the expansion coefficients

$$\alpha_k = \int dx \, \phi_k^*(x)\psi(x). \tag{11.4}$$

As the value of the function at any point in the range for x is determined uniquely by the expansion coefficients $\alpha \equiv \{\alpha_k\}$, the functional $F[\psi(x)]$ must therefore also just depend on the c-number expansion coefficients, and hence may also be viewed as a function $f(\alpha_1, \cdots, \alpha_k, \cdots, \alpha_n)$ of the expansion coefficients, a useful equivalence when functional differentiation and integration are considered:

$$F[\psi(x)] \equiv f(\alpha_1, \cdots, \alpha_k, \cdots, \alpha_n). \tag{11.5}$$

Bosonic field functions have the feature that they commute with regard to multiplication. Thus, with $\eta(x) = \sum_l \beta_l \, \phi_l(x)$ a second field function, we have

$$\psi(x)\eta(y) - \eta(y)\psi(x) = 0 \tag{11.6}$$

using the result $\alpha_k\beta_l - \beta_l\alpha_k = 0$. It is this commutation feature that makes them a natural form for representing bosonic field operators.

It is sometimes convenient to expand a bosonic field function in terms of the complex conjugate modes $\phi_k^*(x)$. Thus $\psi^+(x)$ given by

$$\psi^+(x) = \sum_k \phi_k^*(x)\alpha_k^+ \tag{11.7}$$

is also a field function, and if $\alpha_k^+ = \alpha_k^*$ then $\psi^+(x) = \psi^*(x)$, the complex conjugate field. The expansion coefficient is given by

$$\alpha_k^+ = \int dx\, \phi_k(x)\psi^+(x). \tag{11.8}$$

Functionals of the form $F[\psi^+(x)]$ now occur, and these are also equivalent to a function of the expansion coefficients $\alpha^+ \equiv \{\alpha_k^+\}$:

$$F[\psi^+(x)] \equiv f^+(\alpha_1^+, \cdots, \alpha_k^+, \cdots \alpha_n^+). \tag{11.9}$$

Further generalisations can occur with functionals of the form $F[\psi(x), \psi^+(x)]$ or even $F[\psi(x), \psi^+(x), \psi^*(x), \psi^{+*}(x)]$. These functionals may also be viewed as functions of the expansion coefficients:

$$F[\psi(x), \psi^+(x)] \equiv f(\alpha, \alpha^+), \tag{11.10}$$

$$F[\psi(x), \psi^+(x), \psi^*(x), \psi^{+*}(x)] \equiv f(\alpha, \alpha^+, \alpha^*, \alpha^{+*}). \tag{11.11}$$

In the case where both $\psi(x)$ and $\psi^*(x)$ are present, we can also consider the real and imaginary components of the expansion coefficients as variables. Thus

$$F[\psi(x), \psi^+(x), \psi^*(x), \psi^{+*}(x)] \equiv g(\alpha_x, \alpha_x^+, \alpha_y, \alpha_y^+), \tag{11.12}$$

where $\alpha_x \equiv \{\alpha_{kx}\}, \alpha_y \equiv \{\alpha_{ky}\}$ etc.

The idea of a functional can be extended to cases of the form $F[\psi(x_1, \cdots, x_n)]$, where $\psi(x_1, \cdots, x_n)$ is a function of several variables. For the situation of 3D fields, the identification $x_1 = x, x_2 = y, x_3 = z$ is such an application. In addition, cases $F[\hat{\psi}(x)]$ where $\hat{\psi}(x)$ is an operator function rather than a c-number function occur. For example, $\hat{\psi}(x)$ may be a bosonic field operator. In this case $F[\hat{\psi}(x)]$ maps the operator function onto an operator. Also, functionals $F[\psi_1(x), \cdots, \psi_i(x), \cdots, \psi_n(x)]$ involving several functions $\psi_1(x), \cdots, \psi_i(x), \cdots, \psi_n(x)$ occur. For example, a bosonic field operator $\hat{\psi}(x)$ may be associated with a field function $\psi_1(x) = \psi(x)$, and the field operator $\hat{\psi}(x)^\dagger$ may be associated with a different field function $\psi_2(x) = \psi^+(x)$. Functional derivatives and functional integrals can be defined for all of these cases. The next section will now deal just with c-number functions and functionals, with the corresponding Grassmann entities being covered in a later section.

11.3 Examples of C-Number Functionals

A typical example of a functional involves an integration process:

$$F[\psi(x)] = \int_a^b dx\, \phi(\psi(x)), \tag{11.13}$$

where $\phi(\psi(x))$ is some function of $\psi(x)$.

The scalar product of $\psi(x)$ with a fixed function $\chi(x)$ is a typical example of a functional (written $\chi[\psi(x)]$), since

$$\chi[\psi(x)] = \int dx\, \chi^*(x)\, \psi(x). \tag{11.14}$$

Another example in physics is the action functional $S[q(t)]$, which contains a time integral of the Lagrangian function $L(q(t))$ involving the time-dependent position $q(t)$:

$$S[q(t)] = \int_{t_0}^{t_1} dt\, L(q(t)), \tag{11.15}$$

$$L(q(t)) = \frac{1}{2}m\left(\frac{dq}{dt}\right)^2 - V(q). \tag{11.16}$$

Here we consider a particle of mass m moving in a potential $V(q)$.

A functional $F[\psi(x)]$ may take the form of an integral of a function $\mathcal{F}(\psi(x), \partial\psi(x))$ involving the spatial derivative $\partial_x\psi(x)$ as well as $\psi(x)$:

$$F[\psi(x)] = \int dx\, \mathcal{F}(\psi(x), \partial_x\psi(x)). \tag{11.17}$$

A somewhat trivial application of the functional concept is to express a function $\psi(y)$ as a functional $F_y[\psi(x)]$ of $\psi(x)$:

$$\begin{aligned} F_y[\psi(x)] &\equiv \psi(y) \\ &= \int dx\, \delta(y-x)\, \psi(x). \end{aligned} \tag{11.18}$$

Here this specific functional involves the Dirac delta function as a kernel.

Another example involves the spatial derivative $\nabla_y\psi(y)$, which may also be expressed as a functional $F_{\nabla y}[\psi(x)]$:

$$\begin{aligned} F_{\nabla y}[\psi(x)] &\equiv \nabla_y\psi(y) \\ &= \int dx\, \delta(y-x)\, \nabla_x\psi(x) \\ &= -\int dx\, \{\nabla_x\delta(y-x)\}\, \psi(x) \\ &= \int dx\, \{\nabla_y\delta(y-x)\}\, \psi(x). \end{aligned} \tag{11.19}$$

Here the functional involves $\nabla_y \delta(y-x)$ as a kernel.

A functional is said to be linear if, for constant c_1, c_2,

$$F[c_1\psi_1(x) + c_2\psi_2(x)] = c_1 F[\psi_1(x)] + c_2 F[\psi_2(x)]. \tag{11.20}$$

The scalar product is a linear functional, but the action is not. Note that sums and products of functionals are also functionals. Overall, the c-number function $\psi(x)$ is still mapped onto a c-number that depends on all the values of $\psi(x)$ over its entire range. For sums or products, the process just involves combining the results of two mapping processes.

11.4 Functional Differentiation for C-Number Fields

11.4.1 Definition of Functional Derivative

The *functional derivative* $\delta F[\psi(x)]/\delta\psi(x)$ is defined by [84]

$$F[\psi(x) + \delta\psi(x)] \approx F[\psi(x)] + \int dx\, \delta\psi(x) \left(\frac{\delta F[\psi(x)]}{\delta\psi(x)}\right)_x, \tag{11.21}$$

where $\delta\psi(x)$ is small. The left side is a functional of $\psi(x) + \delta\psi(x)$, and the first term on the right side is a functional of $\psi(x)$. The second term on the right is a functional of $\delta\psi(x)$ and thus the functional derivative is a function of x, hence the subscript x. In most situations this subscript will be left understood. If we choose $\delta\psi(x) = \epsilon\delta(y-x)$ for small ϵ, then an equivalent result for the functional derivative at $x = y$ is [7]

$$\left(\frac{\delta F[\psi(x)]}{\delta\psi(x)}\right)_{x=y} = \lim_{\epsilon\to 0}\left(\frac{F[\psi(x) + \epsilon\delta(y-x)] - F[\psi(x)]}{\epsilon}\right). \tag{11.22}$$

We emphasise again: the functional derivative of a c-number functional is a function, not a functional. The specific examples below illustrate this feature.

For functionals of the form $F[\psi^+(x)]$, we have similar expressions:

$$F[\psi^+(x) + \delta\psi^+(x)] \approx F[\psi^+(x)] + \int dx\, \delta\psi^+(x) \left(\frac{\delta F[\psi^+(x)]}{\delta\psi^+(x)}\right)_x, \tag{11.23}$$

$$\left(\frac{\delta F[\psi^+(x)]}{\delta\psi^+(x)}\right)_{x=y} = \lim_{\epsilon\to 0}\left(\frac{F[\psi^+(x) + \epsilon\delta(y-x)] - F[\psi^+(x)]}{\epsilon}\right). \tag{11.24}$$

Further generalisations of functional derivatives can occur with functionals of the form $F[\psi(x), \psi^+(x)]$ or even $F[\psi(x), \psi^+(x), \psi^*(x), \psi^{+*}(x)]$. In the latter case $\psi^*(x), \psi^{+*}(x)$ are regarded as independent of $\psi(x), \psi^+(x)$, so functional derivatives with respect to the latter are also involved. Thus, using slightly abbreviated notation,

$$\begin{aligned}
&F[\psi(x) + \delta\psi(x), \psi^+(x) + \delta\psi^+(x), \psi^*(x) + \delta\psi^*(x), \psi^{+*}(x) + \delta\psi^{+*}(x)] - F[\,] \\
&\approx \int dx \left\{\delta\psi(x)\left(\frac{\delta F[\,]}{\delta\psi(x)}\right) + \delta\psi^+(x)\left(\frac{\delta F[\,]}{\delta\psi^+(x)}\right)\right\} \\
&\quad + \int dx \left\{\delta\psi^*(x)\left(\frac{\delta F[\,]}{\delta\psi^*(x)}\right) + \delta\psi^{+*}(x)\left(\frac{\delta F[\,]}{\delta\psi^{+*}(x)}\right)\right\}.
\end{aligned} \tag{11.25}$$

This definition of a functional derivative can be extended to cases where $\psi(x_1, \cdots, x_n)$ is a function of several variables or where $\hat{\psi}(x)$ is an operator function rather than a c-number function. Also, functionals $F[\psi_1(x), \cdots, \psi_i(x), \cdots, \psi_n(x)]$ involving several functions $\psi_1(x), \cdots, \psi_i(x), \cdots, \psi_n(x)$ occur, and functional derivatives with respect to any of these functions can be defined. Finally, higher-order functional derivatives can be defined by applying the basic definitions to lower-order functional derivatives.

11.4.2 Examples of Functional Derivatives

For the case of a functional based on the integral of a function involving the spatial derivative (as in $F[\psi(x)] = \int dx\, \mathcal{F}(\psi(x), \partial_x \psi(x))$), the functional derivative takes on a special form:

$$\begin{aligned}
F[\psi(x) + \delta\psi(x)] &= \int dx\, \mathcal{F}(\psi(x) + \delta\psi(x), \partial_x(\psi(x) + \delta\psi(x))) \\
&\approx \int dx \left\{ \mathcal{F}(\psi(x), \partial_x\psi(x)) + \left(\frac{\partial \mathcal{F}}{\partial \psi}\right) \delta\psi(x) + \left(\frac{\partial \mathcal{F}}{\partial(\partial_x \psi)}\right) \delta(\partial_x \psi(x)) \right\} \\
&\approx F[\psi(x)] + \int dx\, \delta\psi(x) \left\{ \left(\frac{\partial \mathcal{F}}{\partial \psi}\right) - \partial_x \left(\frac{\partial \mathcal{F}}{\partial(\partial_x \psi)}\right) \right\},
\end{aligned}$$

using integration by parts and assuming the function goes to zero on the boundary. Hence we obtain for the functional derivative

$$\left(\frac{\delta F[\psi(x)]}{\delta\psi(x)}\right)_{x=y} = \left(\frac{\partial \mathcal{F}}{\partial \psi}\right)_{x=y} - \left(\partial_x \left(\frac{\partial \mathcal{F}}{\partial(\partial_x \psi)}\right)\right)_{x=y}, \tag{11.26}$$

which in this case is obviously just a function.

For the case of the functional $F_y[\psi(x)]$ in (11.18) that gives the function $\psi(y)$,

$$\begin{aligned}
\left(\frac{\delta F_y[\psi(x)]}{\delta\psi(x)}\right)_x &= \left(\frac{\delta\psi(y)}{\delta\psi(x)}\right)_x \\
&= \lim_{\epsilon \to 0} \left(\frac{\int dz\, \delta(y-z)\, \{\psi(z) + \epsilon\delta(x-z)\} - \int dz\, \delta(y-z)\, \psi(z)}{\epsilon} \right) \\
&= \lim_{\epsilon \to 0} \left(\int dz\, \delta(y-z)\, \delta(x-z) \right) \\
&= \delta(y-x),
\end{aligned} \tag{11.27}$$

so here the functional derivative is a delta function.

A similar situation applies to the case of the functional $F_{\nabla y}[\psi(x)]$ in (11.19) that gives the spatial-derivative function $\nabla_y \psi(y)$. Using integration by parts,

$$\begin{aligned}
F_{\nabla y}[\psi(x)+\delta\psi(x)] &= \int dx\, \delta(y-x)\, \nabla_x(\psi(x)+\delta\psi(x)) \\
&= F_{\nabla y}[\psi(x)] + \int dx\, \delta(y-x)\, \nabla_x \delta\psi(x) \\
&= F_{\nabla y}[\psi(x)] - \int dx\, \nabla_x \delta(y-x)\, \delta\psi(x) \\
&= F_{\nabla y}[\psi(x)] + \int dx\, \nabla_y \delta(y-x)\, \delta\psi(x).
\end{aligned}$$

Hence

$$\begin{aligned}
\left(\frac{\delta F_{\nabla y}[\psi(x)]}{\delta\psi(x)}\right)_x &= \left(\frac{\delta \nabla_y \psi(y)}{\delta\psi(x)}\right)_x \\
&= \nabla_y \delta(y-x) = -\nabla_x \delta(y-x),
\end{aligned} \tag{11.28}$$

so here the functional derivative is the derivative of a delta function.

Similarly, with $F_y[\psi^+(x)] \equiv \psi^+(y)$ and $F_{\nabla y}[\psi^+(x)] \equiv \nabla_y \psi^+(y)$,

$$\left(\frac{\delta\psi^+(y)}{\delta\psi^+(x)}\right)_x = \delta(y-x), \tag{11.29}$$

$$\left(\frac{\delta\, \nabla_y \psi^+(y)}{\delta\psi^+(x)}\right)_x = \nabla_y \delta(y-x). \tag{11.30}$$

11.4.3 Functional Derivative and Mode Functions

If a mode expansion for $\psi(x)$ as in (11.1) is used, then we can obtain an expression for the functional derivative in terms of mode functions. With

$$\delta\psi(x) = \sum_k \delta\alpha_k\, \phi_k(x), \tag{11.31}$$

we see that

$$\begin{aligned}
F[\psi(x)+\delta\psi(x)] - F[\psi(x)] &\approx \int dx\, \delta\psi(x) \left(\frac{\delta F[\psi(x)]}{\delta\psi(x)}\right)_x \\
&\approx \sum_k \delta\alpha_k \int dx\, \phi_k(x) \left(\frac{\delta F[\psi(x)]}{\delta\psi(x)}\right)_x.
\end{aligned}$$

But the left side is the same as

$$f(\alpha_1+\delta\alpha_1,\cdots,\alpha_k+\delta\alpha_k,\cdots) - f(\alpha_1,\cdots,\alpha_k,\cdots) \approx \sum_k \delta\alpha_k \frac{\partial f(\alpha_1,\cdots,\alpha_k,\cdots)}{\partial\alpha_k}. \tag{11.32}$$

Equating the coefficients of the independent $\delta\alpha_k$ gives

$$\frac{\partial f(\alpha_1,\cdots,\alpha_k,\cdots)}{\partial\alpha_k} = \int dx\, \phi_k(x) \left(\frac{\delta F[\psi(x)]}{\delta\psi(x)}\right)_x, \tag{11.33}$$

giving the phase space derivative in terms of the functional derivative. Using the completeness relationship in (11.3) gives the key result

$$\left(\frac{\delta F[\psi(x)]}{\delta\psi(x)}\right)_x = \sum_k \phi_k^*(x)\,\frac{\partial f(\alpha_1,\cdots,\alpha_k,\cdots)}{\partial\alpha_k}. \tag{11.34}$$

This relates the functional derivative to the mode functions and to the ordinary partial derivatives of the function $f(\alpha_1,\cdots,\alpha_k,\cdots\alpha_n)$ that was equivalent to the original functional $F[\psi(x)]$. Again, we see that the result is a function of x. Note that the functional derivative involves an expansion in terms of the conjugate mode functions $\phi_k^*(x)$ rather than the original modes $\phi_k(x)$.

The last result may be put in the form of a useful *operational identity*

$$\left(\frac{\delta}{\delta\psi(x)}\right)_x = \sum_k \phi_k^*(x)\,\frac{\partial}{\partial\alpha_k}, \tag{11.35}$$

where the left side is understood to operate on an arbitrary functional $F[\psi(x)]$ and the right side is understood to operate on the equivalent function $f(\alpha_1,\cdots,\alpha_k,\cdots)$.

The equivalent results for functionals of the form $F[\psi^+(x)]$ are

$$\frac{\partial f^+(\alpha_1^+,\alpha_2^+,\cdots,\alpha_k^+,\cdots,\alpha_n^+)}{\partial\alpha_k^+} = \int dx\,\phi_k^*(x)\left(\frac{\delta F[\psi^+(x)]}{\delta\psi^+(x)}\right)_x, \tag{11.36}$$

$$\left(\frac{\delta F[\psi^+(x)]}{\delta\psi^+(x)}\right)_x = \sum_k \phi_k(x)\,\frac{\partial f^+(\alpha_1^+,\alpha_2^+,\cdots,\alpha_k^+,\cdots,\alpha_n^+)}{\partial\alpha_k^+}, \tag{11.37}$$

$$\left(\frac{\delta}{\delta\psi^+(x)}\right)_x = \sum_k \phi_k(x)\,\frac{\partial}{\partial\alpha_k^+}. \tag{11.38}$$

For functionals of the form $F[\psi(x),\psi^+(x),\psi^*(x),\psi^{+*}(x)] \equiv f(\alpha,\alpha^+,\alpha^*,\alpha^{+*})$, functional derivatives with respect to $\psi(x),\psi^+(x),\psi^*(x),\psi^{+*}(x)$ are all involved, since the functions $\psi^*(x),\psi^{+*}(x)$ are independent. Hence

$$\begin{aligned}\delta\psi(x) &= \sum_k \delta\alpha_k\,\phi_k(x), & \delta\psi^*(x) &= \sum_k \delta\alpha_k^*\,\phi_k^*(x),\\ \delta\psi^+(x) &= \sum_k \delta\alpha_k^+\,\phi_k^*(x), & \delta\psi^{+*}(x) &= \sum_k \delta\alpha_k^{+*}\,\phi_k(x),\end{aligned} \tag{11.39}$$

$$\begin{aligned}&F[\psi(x)+\delta\psi(x),\psi^+(x)+\delta\psi^+(x),\psi^*(x)+\delta\psi^*(x),\psi^{+*}(x)+\delta\psi^{+*}(x)]\\ &-F[\psi(x),\psi^+(x),\psi^*(x),\psi^{+*}(x)]\\ &\quad\approx \int dx\left\{\delta\psi(x)\left(\frac{\delta F[\,]}{\delta\psi(x)}\right)+\delta\psi^+(x)\left(\frac{\delta F[\,]}{\delta\psi^+(x)}\right)\right\}\\ &\qquad+\int dx\left\{\delta\psi^*(x)\left(\frac{\delta F[\,]}{\delta\psi^*(x)}\right)+\delta\psi^{+*}(x)\left(\frac{\delta F[\,]}{\delta\psi^{+*}(x)}\right)\right\}\\ &\quad\approx \sum_k \delta\alpha_k\int dx\,\phi_k(x)\left(\frac{\delta F[\,]}{\delta\psi(x)}\right)+\sum_k \delta\alpha_k^+\int dx\,\phi_k^*(x)\left(\frac{\delta F[\,]}{\delta\psi^+(x)}\right)\\ &\qquad+\sum_k \delta\alpha_k^*\int dx\,\phi_k^*(x)\left(\frac{\delta F[\,]}{\delta\psi^*(x)}\right)+i\sum_k \delta\alpha_k^{+*}\int dx\,\phi_k(x)\left(\frac{\delta F[\,]}{\delta\psi^{+*}(x)}\right).\end{aligned} \tag{11.40}$$

But the left side is the same as

$$
\begin{aligned}
& f(\alpha+\delta\alpha,\alpha^{+}+\delta\alpha^{+},\alpha^{*}+\delta\alpha^{*},\alpha^{+*}+\delta\alpha_{y}^{+*})-f(\alpha,\alpha^{+},\alpha^{*},\alpha^{+*}) \\
& \approx \sum_{k}\left\{\delta\alpha_{k}\,\frac{\partial f(\alpha,\alpha^{+},\alpha^{*},\alpha^{+*})}{\partial\alpha_{k}}+\delta\alpha_{k}^{+}\,\frac{\partial f(\alpha,\alpha^{+},\alpha^{*},\alpha^{+*})}{\partial\alpha_{k}^{+}}\right\} \\
& +\sum_{k}\left\{\delta\alpha_{k}^{*}\,\frac{\partial f(\alpha,\alpha^{+},\alpha^{*},\alpha^{+*})}{\partial\alpha_{k}^{*}}+\delta\alpha_{k}^{+*}\,\frac{\partial f(\alpha,\alpha^{+},\alpha^{*},\alpha^{+*})}{\partial\alpha_{k}^{+*}}\right\}.
\end{aligned} \tag{11.41}
$$

Equating the coefficients of the independent $\delta\alpha_k$, $\delta\alpha_k^*$, $\delta\alpha_k^+$, $\delta\alpha_k^{+*}$, we have

$$\frac{\partial f(\alpha,\alpha^{+},\alpha^{*},\alpha^{+*})}{\partial\alpha_{k}}=\int dx\,\phi_{k}(x)\left(\frac{\delta F[\psi(x),\psi^{+}(x),\psi^{*}(x),\psi^{+*}(x)]}{\delta\psi(x)}\right), \tag{11.42}$$

$$\frac{\partial f(\alpha,\alpha^{+},\alpha^{*},\alpha^{+*})}{\partial\alpha_{k}^{+}}=\int dx\,\phi_{k}^{*}(x)\left(\frac{\delta F[\psi(x),\psi^{+}(x),\psi^{*}(x),\psi^{+*}(x)]}{\delta\psi^{+}(x)}\right), \tag{11.43}$$

$$\frac{\partial f(\alpha,\alpha^{+},\alpha^{*},\alpha^{+*})}{\partial\alpha_{k}^{*}}=\int dx\,\phi_{k}^{*}(x)\left(\frac{\delta F[\psi(x),\psi^{+}(x),\psi^{*}(x),\psi^{+*}(x)]}{\delta\psi^{*}(x)}\right), \tag{11.44}$$

$$\frac{\partial f(\alpha,\alpha^{+},\alpha^{*},\alpha^{+*})}{\partial\alpha_{k}^{+*}}=\int dx\,\phi_{k}(x)\left(\frac{\delta F[\psi(x),\psi^{+}(x),\psi^{*}(x),\psi^{+*}(x)]}{\delta\psi^{+*}(x)}\right). \tag{11.45}$$

These equations give the phase space derivatives in terms of the functional derivatives. Then, using the completeness relationship in (11.3) gives the key results

$$\left(\frac{\delta F[\psi(x),\psi^{+}(x),\psi^{*}(x),\psi^{+*}(x)]}{\delta\psi(x)}\right)_{x}=\sum_{k}\phi_{k}^{*}(x)\frac{\partial f(\alpha,\alpha^{+},\alpha^{*},\alpha^{+*})}{\partial\alpha_{k}}, \tag{11.46}$$

$$\left(\frac{\delta F[\psi(x),\psi^{+}(x),\psi^{*}(x),\psi^{+*}(x)]}{\delta\psi^{+}(x)}\right)_{x}=\sum_{k}\phi_{k}(x)\frac{\partial f(\alpha,\alpha^{+},\alpha^{*},\alpha^{+*})}{\partial\alpha_{k}^{+}}, \tag{11.47}$$

$$\left(\frac{\delta F[\psi(x),\psi^{+}(x),\psi^{*}(x),\psi^{+*}(x)]}{\delta\psi^{*}(x)}\right)_{x}=\sum_{k}\phi_{k}(x)\frac{\partial f(\alpha,\alpha^{+},\alpha^{*},\alpha^{+*})}{\partial\alpha_{k}^{*}}, \tag{11.48}$$

$$\left(\frac{\delta F[\psi(x),\psi^{+}(x),\psi^{*}(x),\psi^{+*}(x)]}{\delta\psi^{+*}(x)}\right)_{x}=\sum_{k}\phi_{k}^{*}(x)\frac{\partial f(\alpha,\alpha^{+},\alpha^{*},\alpha^{+*})}{\partial\alpha_{k}^{+*}}. \tag{11.49}$$

These relate the functional derivatives to the phase space derivatives.

We can invert these relationships using the completeness results to give the phase space derivatives in terms of the functional derivatives:

$$\frac{\partial f(\alpha,\alpha^+,\alpha^*,\alpha^{+*})}{\partial \alpha_k} = \int dx\, \phi_k(x) \left(\frac{\delta F[\psi(x),\psi^+(x),\psi^*(x),\psi^{+*}(x)]}{\delta \psi(x)} \right)_x, \tag{11.50}$$

$$\frac{\partial f(\alpha,\alpha^+,\alpha^*,\alpha^{+*})}{\partial \alpha_k^+} = \int dx\, \phi_k^*(x) \left(\frac{\delta F[\psi(x),\psi^+(x),\psi^*(x),\psi^{+*}(x)]}{\delta \psi^+(x)} \right)_x, \tag{11.51}$$

$$\frac{\partial f(\alpha,\alpha^+,\alpha^*,\alpha^{+*})}{\partial \alpha_k^*} = \int dx\, \phi_k^*(x) \left(\frac{\delta F[\psi(x),\psi^+(x),\psi^*(x),\psi^{+*}(x)]}{\delta \psi^*(x)} \right)_x, \tag{11.52}$$

$$\frac{\partial f(\alpha,\alpha^+,\alpha^*,\alpha^{+*})}{\partial \alpha_k^{+*}} = \int dx\, \phi_k(x) \left(\frac{\delta F[\psi(x),\psi^+(x),\psi^*(x),\psi^{+*}(x)]}{\delta \psi^{+*}(x)} \right)_x. \tag{11.53}$$

Hence the operational identities for functionals of the form above are

$$\left(\frac{\delta}{\delta \psi(x)} \right)_x = \sum_k \phi_k^*(x) \frac{\partial}{\partial \alpha_k}, \tag{11.54}$$

$$\left(\frac{\delta}{\delta \psi^+(x)} \right)_x = \sum_k \phi_k(x) \frac{\partial}{\partial \alpha_k^+}, \tag{11.55}$$

$$\left(\frac{\delta}{\delta \psi^*(x)} \right)_x = \sum_k \phi_k(x) \frac{\partial}{\partial \alpha_k^*}, \tag{11.56}$$

$$\left(\frac{\delta}{\delta \psi^{+*}(x)} \right)_x = \sum_k \phi_k^*(x) \frac{\partial}{\partial \alpha_k^{+*}}, \tag{11.57}$$

and those for functions of the form $f(\alpha,\alpha^+,\alpha^*,\alpha^{+*})$ are

$$\frac{\partial}{\partial \alpha_k} = \int dx\, \phi_k(x) \left(\frac{\delta}{\delta \psi(x)} \right)_x, \tag{11.58}$$

$$\frac{\partial}{\partial \alpha_k^+} = \int dx\, \phi_k^*(x) \left(\frac{\delta}{\delta \psi^+(x)} \right)_x, \tag{11.59}$$

$$\frac{\partial}{\partial \alpha_k^*} = \int dx\, \phi_k^*(x) \left(\frac{\delta}{\delta \psi^*(x)} \right)_x, \tag{11.60}$$

$$\frac{\partial}{\partial \alpha_k^{+*}} = \int dx\, \phi_k(x) \left(\frac{\delta}{\delta \psi^{+*}(x)} \right)_x. \tag{11.61}$$

An example of the use of these results is to rederive the functional derivatives of the field functions:

$$
\begin{aligned}
\left(\frac{\delta}{\delta\psi(x)}\right)_x \psi(y) &= \sum_k \phi_k^*(x)\,\frac{\partial}{\partial\alpha_k}\sum_l \alpha_l\phi_l(y) \\
&= \sum_k \phi_k^*(x)\,\phi_k(y) \\
&= \delta(y-x), \\
\left(\frac{\delta}{\delta\psi^+(x)}\right)_x \psi^+(y) &= \delta(x-y), \\
\left(\frac{\delta}{\delta\psi(x)}\right)_x \psi^+(y) &= \sum_k \phi_k^*(x)\,\frac{\partial}{\partial\alpha_k}\sum_l \alpha_l^+\phi_l^*(y) \\
&= 0, \\
\left(\frac{\delta}{\delta\psi^+(x)}\right)_x \psi(y) &= 0,
\end{aligned} \tag{11.62}
$$

which are the same as before.

We may also introduce real and imaginary components into the fields via

$$
\begin{aligned}
\psi(x) &= \psi_x(x) + i\psi_y(x), \qquad & \psi^+(x) &= \psi_x^+(x) + i\psi_y^+(x), \\
\psi^*(x) &= \psi_x(x) - i\psi_y(x), \qquad & \psi^{+*}(x) &= \psi_x^+(x) - i\psi_y^+(x),
\end{aligned} \tag{11.63}
$$

and write these in terms of the $\alpha_{kx}, \alpha_{kx}^+, \alpha_{ky}, \alpha_{ky}^+$ and the real and imaginary components of the mode functions. Thus

$$
\begin{aligned}
\psi_x(x) &= \sum_k(\alpha_{kx}\phi_{kx}(x) - \alpha_{ky}\phi_{ky}(x)), \qquad & \psi_y(x) &= \sum_k(\alpha_{kx}\phi_{ky}(x) + \alpha_{ky}\phi_{kx}(x)), \\
\psi_x^+(x) &= \sum_k(\alpha_{kx}^+\phi_{kx}(x) + \alpha_{ky}^+\phi_{ky}(x)), \qquad & \psi_y^+(x) &= \sum_k(-\alpha_{kx}^+\phi_{ky}(x) + \alpha_{ky}^+\phi_{kx}(x)).
\end{aligned} \tag{11.64}
$$

The functional $G[\psi_x(x), \psi_y(x), \psi_x^+(x), \psi_y^+(x)] \equiv g(\alpha_x, \alpha_x^+, \alpha_y, \alpha_y^+)$ and functional derivatives with respect to $\psi_x(x), \psi_y(x), \psi_x^+(x), \psi_y^+(x)$ can be defined in the usual way. These also can be written in terms of derivatives with respect to the $\alpha_{kx}, \alpha_{kx}^+, \alpha_{ky}, \alpha_{ky}^+$. The results are

$$
\begin{aligned}
\left(\frac{\delta}{\delta\psi_x(x)}\right)_x &= \sum_k\left(\phi_{kx}(x)\,\frac{\partial}{\partial\alpha_{kx}} - \phi_{ky}(x)\,\frac{\partial}{\partial\alpha_{ky}}\right), \\
\left(\frac{\delta}{\delta\psi_y(x)}\right)_x &= \sum_k\left(\phi_{ky}(x)\,\frac{\partial}{\partial\alpha_{kx}} + \phi_{kx}(x)\,\frac{\partial}{\partial\alpha_{ky}}\right), \\
\left(\frac{\delta}{\delta\psi_x^+(x)}\right)_x &= \sum_k\left(\phi_{kx}(x)\,\frac{\partial}{\partial\alpha_{kx}^+} + \phi_{ky}(x)\,\frac{\partial}{\partial\alpha_{ky}^+}\right), \\
\left(\frac{\delta}{\delta\psi_y^+(x)}\right)_x &= \sum_k\left(-\phi_{ky}(x)\,\frac{\partial}{\partial\alpha_{kx}^+} + \phi_{kx}(x)\,\frac{\partial}{\partial\alpha_{ky}^+}\right).
\end{aligned} \tag{11.65}
$$

The proof of the last result is left as an exercise. Again, these results can be inverted.

11.4.4 Taylor Expansion for Functionals

The definition of the functional derivative in (11.21) is actually the leading term in a Taylor expansion for the functional $F[\psi(x)+\delta\psi(x)]$ which is equivalent to the function $f(\alpha_1+\delta\alpha_1,\cdots,\alpha_k+\delta\alpha_k,\cdots)$. Using the Taylor expansion

$$\begin{aligned}&f(\alpha_1+\delta\alpha_1,\cdots,\alpha_k+\delta\alpha_k,\cdots)\\&=f(\alpha_1,\cdots,\alpha_k,\cdots)+\sum_k \delta\alpha_k\,\frac{\partial f(\,)}{\partial\alpha_k}+\frac{1}{2}\sum_{kl}\delta\alpha_k\,\delta\alpha_l\frac{\partial^2 f(\,)}{\partial\alpha_k\,\partial\alpha_l}+\cdots\end{aligned}$$

and substituting from (11.33), we have

$$\begin{aligned}F[\psi(x)+\delta\psi(x)] &= F[\psi(x)]+\sum_k \delta\alpha_k\int dx\,\phi_k(x)\left(\frac{\delta F[\psi(x)]}{\delta\psi(x)}\right)\\&+\frac{1}{2}\sum_{kl}\delta\alpha_k\,\delta\alpha_l\int dx\,\phi_k(x)\left(\frac{\delta}{\delta\psi(x)}\right)\int dy\,\phi_l(y)\left(\frac{\delta}{\delta\psi(y)}\right)F[\psi(x)]+\cdots \qquad (11.66)\\&= F[\psi(x)]+\int dx\,\delta\psi(x)\left(\frac{\delta F[\psi(x)]}{\delta\psi(x)}\right)+\frac{1}{2}\int dx\,dy\,\delta\psi(x)\,\delta\psi(y)\left(\frac{\delta^2 F[\psi(x)]}{\delta\psi(x)\,\delta\psi(y)}\right)+\cdots, \qquad (11.67)\end{aligned}$$

which is a Taylor expansion for the functional $F[\psi(x)]$. Analogous Taylor expansions can be obtained for $F[\psi(x),\psi^+(x)]$ and $F[\psi(x),\psi^+(x),\psi^*(x),\psi^{+*}(x)]$.

11.4.5 Basic Rules for Functional Derivatives

Rules can be established for the functional derivative of the sum of two functionals. It is easily shown that

$$\left(\frac{\delta\{F[\psi(x)]+G[\psi(x)]\}}{\delta\psi(x)}\right)_{x=y}=\left(\frac{\delta\{F[\psi(x)]\}}{\delta\psi(x)}\right)_{x=y}+\left(\frac{\delta\{G[\psi(x)]\}}{\delta\psi(x)}\right)_{x=y}. \qquad (11.68)$$

Also, rules can be established for functional derivatives of the product of two functionals. We will keep these in order to cover the case where the functionals are operators:

$$\begin{aligned}\left(\frac{\delta\{F[\,]G[\,]\}}{\delta\psi(x)}\right)_{x=y} &= \lim_{\epsilon\to 0}\left(\frac{F[\psi(x)+\epsilon\delta(y-x)]G[\psi(x)+\epsilon\delta(y-x)]-F[\,]G[\,]}{\epsilon}\right)\\&=\lim_{\epsilon\to 0}\left(\frac{\begin{array}{c}F[\psi(x)+\epsilon\delta(y-x)]G[\psi(x)+\epsilon\delta(y-x)]\\-F[\psi(x)]G[\psi(x)+\epsilon\delta(y-x)]\end{array}}{\epsilon}\right)\\&+\lim_{\epsilon\to 0}\left(\frac{F[\psi(x)]G[\psi(x)+\epsilon\delta(y-x)]-F[\psi(x)]G[\psi(x)]}{\epsilon}\right)\\&=\left(\frac{\delta F[\psi(x)]}{\delta\psi(x)}\right)_{x=y}G[\psi(x)]+F[\psi(x)]\left(\frac{\delta G[\psi(x)]}{\delta\psi(x)}\right)_{x=y}. \qquad (11.69)\end{aligned}$$

This result can then be extended to products of more than two functionals, so that

$$\begin{aligned}\left(\frac{\delta\{F[\,]G[\,]H[\,]\}}{\delta\psi(x)}\right)_{x=y} &= \left(\frac{\delta F[\psi(x)]}{\delta\psi(x)}\right)_{x=y} G[\psi(x)]H[\psi(x)] \\ &\quad + F[\psi(x)]\left(\frac{\delta G[\psi(x)]}{\delta\psi(x)}\right)_{x=y} H[\psi(x)] \\ &\quad + F[\psi(x)]G[\psi(x)]\left(\frac{\delta H[\psi(x)]}{\delta\psi(x)}\right)_{x=y},\end{aligned} \tag{11.70}$$

and so on. These results are essentially the same as for ordinary differentiation.

A chain rule for functional differentiation can also be derived for the case where a functional $G[\psi_y(x)]$ involves not just one function $\psi(x)$, but a set of functions each labelled by a variable y. Since $G[\psi_y(x)]$ maps $\psi_y(x)$ onto a c-number which depends on y, we can regard the functional $G[\psi_y(x)]$ also as a function $G(y)$ of the variable y. Now consider a second functional $F[G(y)]$ of this function $G(y)$, and we could determine the functional derivative $(\delta F[G(y)]/\delta G(y))_y$. But $F[G(y)]$ is also a functional of the $\psi_y(x)$ via

$$F[G[\psi_y(x)]] \equiv F[G(y)], \tag{11.71}$$

so we want to find the total functional derivative $(\delta F[G[\psi_y(x)]]/\delta\psi_y(x))_x$. But from the definitions, and with $\delta G(y) = G[\psi_y(x) + \delta\psi_y(x)] - G[\psi_y(x)]$, we have

$$\begin{aligned}\delta F &= \int dy\,\delta G(y)\left(\frac{\delta F[G(y)]}{\delta G(y)}\right)_y \\ &= \int dy \int dx\,\delta\psi_y(x)\left(\frac{\delta G[\psi_y(x)]}{\delta\psi_y(x)}\right)_x \left(\frac{\delta F[G(y)]}{\delta G(y)}\right)_y \\ &= \int dx\,\delta\psi_y(x)\left(\frac{\delta F[G[\psi_y(x)]]}{\delta\psi_y(x)}\right)_x.\end{aligned}$$

Hence we obtain the chain rule

$$\left(\frac{\delta F[G[\psi_y(x)]]}{\delta\psi_y(x)}\right)_x = \int dy \left(\frac{\delta G[\psi_y(x)]}{\delta\psi_y(x)}\right)_x \left(\frac{\delta F[G(y)]}{\delta G(y)}\right)_y, \tag{11.72}$$

where we have left the order of the factors as they appeared in order to allow for operator cases.

We may also define the spatial derivative of the functional derivative, since the latter is just a function. Thus

$$
\begin{aligned}
&\partial_y \left(\frac{\delta F[\psi(x)]}{\delta \psi(x)} \right)_{x=y} \\
&= \lim_{\Delta y \to 0} \frac{1}{\Delta y} \left(\left(\frac{\delta F[\psi(x)]}{\delta \psi(x)} \right)_{x=y+\Delta y} - \left(\frac{\delta F[\psi(x)]}{\delta \psi(x)} \right)_{x=y} \right) \\
&= \lim_{\Delta y \to 0} \lim_{\epsilon \to 0} \frac{1}{\epsilon\, \Delta y} \begin{pmatrix} \{F[\psi(x) + \epsilon\, \delta(y + \Delta y - x)] - F[\psi(x)] \\ -F[\psi(x) + \epsilon\, \delta(y - x)] + F[\psi(x)]\} \end{pmatrix} \\
&\approx \lim_{\Delta y \to 0} \lim_{\epsilon \to 0} \frac{1}{\epsilon\, \Delta y} \begin{pmatrix} \{F[\psi(x) + \epsilon(\delta(y - x) + (\partial/\partial y \delta(y - x))\, \Delta y)] \\ -F[\psi(x)]\} \\ -\{F[\psi(x) + \epsilon\, \delta(y - x)] - F[\psi(x)]\} \end{pmatrix} \\
&\approx \lim_{\Delta y \to 0} \lim_{\epsilon \to 0} \frac{1}{\epsilon\, \Delta y} \begin{pmatrix} \{\int dx\, \epsilon \delta(y - x)\, (\delta F[\psi(x)]/\delta \psi(x))_x \\ + \int dx\, (\partial/\partial y \delta(y - x))\, \epsilon\, \Delta y\, (\delta F[\psi(x)]/\delta \psi(x))_x \\ - \int dx\, \epsilon\, \delta(y - x)\, (\delta F[\psi(x)]/\delta \psi(x))_x\} \end{pmatrix} \\
&= \int dx \left(\frac{\partial}{\partial y} \delta(y - x) \right) \left(\frac{\delta F[\psi(x)]}{\delta \psi(x)} \right)_x .
\end{aligned}
\tag{11.73}
$$

This expresses the spatial derivative as an integral involving the functional derivative and the spatial derivative of the delta function. The result will be a function of y. The same result could have been obtained by writing $(\delta F[\psi(x)]/\delta \psi(x))_{x=y}$ in the form

$$
\left(\frac{\delta F[\psi(x)]}{\delta \psi(x)} \right)_{x=y} = \int dx\, \delta(y - x) \left(\frac{\delta F[\psi(x)]}{\delta \psi(x)} \right)_x \tag{11.74}
$$

and then differentiating both sides with respect to y.

11.4.6 Other Rules for Functional Derivatives

A number of other rules may also be established:

(1) Power rule:

$$
\begin{aligned}
F[\psi(x)] &= \int dx\, \psi(x)^n, \\
\frac{\delta F[\psi(x)]}{\delta \psi(x)} &= n \psi(x)^{n-1}.
\end{aligned}
\tag{11.75}
$$

(2) Function rule:

$$
\begin{aligned}
F[\psi(x)] &= \int dx\, \phi(\psi(x)), \\
\frac{\delta F[\psi(x)]}{\delta \psi(x)} &= \phi'(\psi(x)).
\end{aligned}
\tag{11.76}
$$

(3) Power derivative rule:

$$\begin{aligned} F[\psi(x)] &= \int dx \left(\frac{d\psi(x)}{dx}\right)^n, \\ \frac{\delta F[\psi(x)]}{\delta\psi(x)} &= -n\frac{d}{dx}\left(\left(\frac{d\psi(x)}{dx}\right)^{n-1}\right). \end{aligned} \tag{11.77}$$

(4) Function derivative rule:

$$\begin{aligned} F[\psi(x)] &= \int dx\, \phi\left(\frac{d\psi(x)}{dx}\right), \\ \frac{\delta F[\psi(x)]}{\delta\psi(x)} &= -\frac{d}{dx}\left(\frac{d\phi}{d(d\psi/dx)}\right). \end{aligned} \tag{11.78}$$

(5) Convolution rule:

$$\begin{aligned} F_y[\psi(x)] &= \int dx\, K(y,x)\, \psi(x), \\ \left(\frac{\delta F_y[\psi(x)]}{\delta\psi(x)}\right)_x &= K(y,x). \end{aligned} \tag{11.79}$$

(6) Trivial rule:

$$\begin{aligned} F_y[\psi(x)] &= \psi(y), \\ \left(\frac{\delta F_y[\psi(x)]}{\delta\psi(x)}\right)_x &= \left(\frac{\delta\psi(y)}{\delta\psi(x)}\right)_x \\ &= \delta(y-x). \end{aligned} \tag{11.80}$$

This was proved above.

(7) Gradient rule:

$$\begin{aligned} F_{\nabla y}[\psi(x)] &= \nabla_y\psi(y), \\ \left(\frac{\delta F_{\nabla y}[\psi(x)]}{\delta\psi(x)}\right)_x &= \nabla_y\delta(y-x) = -\nabla_x\delta(y-x). \end{aligned} \tag{11.81}$$

This was proved above.

(8) Exponential rule:

$$\begin{aligned} F[\psi(x)] &= \exp G[\psi(x)], \\ \frac{\delta F[\psi(x)]}{\delta\psi(x)} &= \exp G[\psi(x)]\, \frac{\delta G[\psi(x)]}{\delta\psi(x)}. \end{aligned} \tag{11.82}$$

The exponential rule only applies in this form if $F[\psi(x)]$ and $G[\psi(x)]$ are c-numbers.

11.5 Functional Integration for C-Number Fields

11.5.1 Definition of Functional Integral

If the range for the function $\psi(x)$ is divided up into n small intervals $\Delta x_i = x_{i+1} - x_i$ (the ith interval), then we may specify the value ψ_i of the function $\psi(x)$ in the ith interval via the average

$$\psi_i = \frac{1}{\Delta x_i} \int\limits_{\Delta x_i} dx\, \psi(x), \tag{11.83}$$

and then the functional $F[\psi(x)]$ may be regarded as a function $F(\psi_1, \cdots, \psi_i, \cdots, \psi_n)$ of all the ψ_i.

Introducing a suitable weight function $w(\psi_1, \cdots, \psi_i, \cdots, \psi_n)$, we may then define the functional integral for the case of real functions as [84]

$$\int D\psi\, F[\psi(x)] = \lim_{n\to\infty} \lim_{\epsilon\to 0} \int d\psi_1 \cdots d\psi_n\, w(\psi_1, \cdots, \psi_n) F(\psi_1, \cdots, \psi_n), \tag{11.84}$$

where $\epsilon > \Delta x_i$. Thus the symbol $D\psi$ stands for $d\psi_1 \cdots d\psi_n\, w(\psi_1, \cdots, \psi_n)$.

If the functions are complex, then the functional integral is

$$\int D^2\psi\, F[\psi(x)] = \lim_{n\to\infty} \lim_{\epsilon\to 0} \int d^2\psi_1 \cdots d^2\psi_n\, w(\psi_1, \cdots, \psi_n) F(\psi_1, \cdots, \psi_n). \tag{11.85}$$

The symbol $D^2\psi$ stands for $d^2\psi_1 \cdots d^2\psi_n w(\psi_1, \cdots, \psi_n)$, where, with $\psi_i = \psi_{ix} + i\psi_{iy}$, the quantity $d^2\psi_i$ means $d\psi_{ix}\, d\psi_{iy}$, involving integration over the real and imaginary parts of the function. If the functional $F[\psi(x), \psi^*(x)]$ involves pairs of complex conjugate c-number fields, then the functional integral will be of the form $\int D^2\psi\, F[\psi(x), \psi^*(x)]$, where $D^2\psi = d^2\psi_n \cdots d^2\psi_1\, w(\psi_1, \cdots, \psi_n, \psi_1^*, \cdots, \psi_n^*)$, with $d^2\psi_i = d\psi_{ix}\, d\psi_{iy}$. If the functional $F[\psi(x), \psi^+(x)]$ involves pairs of c-number fields not related by complex conjugation, then the functional integral will be of the form $\int D^2\psi\, D^2\psi^+\, F[\psi(x), \psi^+(x)]$, where $D^2\psi\, D^2\psi^+ = d^2\psi_1 \cdots d^2\psi_n\, d^2\psi_1^+ \cdots d^2\psi_n^+\, w(\psi_1, \cdots, \psi_n, \psi_1^+, \cdots, \psi_n^+)$, and with $\psi_i^+ = \psi_{ix}^+ + i\psi_{iy}^+$, the quantity $d^2\psi_i^+$ means $d\psi_{ix}^+\, d\psi_{iy}^+$

For cases involving several complex functions such as $F[\psi(x), \psi^+(x), \psi^*(x), \psi^{+*}(x)]$ the functional integrals are of the form

$$\begin{aligned}
&\int\int D^2\psi\, D^2\psi^+\, F[\psi(x), \psi^+(x), \psi^*(x), \psi^{+*}(x)] \\
&= \lim_{n\to\infty} \lim_{\epsilon\to 0} \int d^2\psi_1 \cdots d^2\psi_n \int d^2\psi_1^+ \cdots d^2\psi_n^+ \\
&\quad \times w(\psi_1, \cdots, \psi_n, \psi_1^+, \cdots, \psi_n^+, \psi_1^*, \cdots, \psi_n^*, \psi_1^{+*}, \cdots, \psi_n^{+*}) \\
&\quad \times F(\psi_1, \cdots, \psi_n, \psi_1^+, \cdots, \psi_n^+, \psi_1^*, \cdots, \psi_n^*, \psi_1^{+*}, \cdots, \psi_n^{+*}),
\end{aligned} \tag{11.86}$$

where the weight function is now $w(\psi_1, \cdots, \psi_n, \psi_1^+, \cdots, \psi_n^+, \psi_1^*, \cdots, \psi_n^*, \psi_1^{+*}, \cdots, \psi_n^{+*})$. A functional integral of a functional of a c-number function gives a c-number. Unlike ordinary calculus, functional integration and differentiation are not inverse processes.

11.5.2 Functional Integrals and Phase Space Integrals

We first consider the case of a functional $F[\psi(x)]$ of a real function $\psi(x)$, which we expand in terms of real, orthogonal mode functions. The expansion coefficients in this case will be real also. If a mode expansion such as in (11.1) is used, then the value ϕ_{ki} of the mode function in the ith interval is also defined via the average

$$\phi_{ki} = \frac{1}{\Delta x_i} \int\limits_{\Delta x_i} dx\, \phi_k(x), \tag{11.87}$$

and hence

$$\psi_i = \sum_k \alpha_k \phi_{ki}. \tag{11.88}$$

This shows that the values in the ith interval of the function ψ_i and the mode function ϕ_{ki} are related via the expansion coefficients α_k. For simplicity, we will choose the same number n of intervals as mode functions. Using the expression (11.4) for the expansion coefficients, we then obtain the inverse formula to (11.88),

$$\alpha_k = \sum_i \Delta x_i\, \phi_{ki} \psi_i. \tag{11.89}$$

Note that this involves a sum over intervals i, and the interval size Δx_i is also involved.

The relationship (11.88) shows that the functions $F(\psi_1, \cdots, \psi_n)$ and $w(\psi_1, \cdots, \psi_n)$ of all the interval values ψ_i can also be regarded as functions of the expansion coefficients α_k, which we may write as

$$f(\alpha_1, \cdots, \alpha_n) \equiv F(\psi_1(\alpha_1, \cdots, \alpha_n), \cdots, \psi_i(\alpha_1, \cdots, \alpha_n), \cdots, \psi_n), \tag{11.90}$$

$$v(\alpha_1, \cdots, \alpha_n) \equiv w(\psi_1(\alpha_1, \cdots, \alpha_n), \cdots, \psi_i(\alpha_1, \cdots, \alpha_n), \cdots, \psi_n). \tag{11.91}$$

Thus the various values $\psi_1, \cdots, \psi_n$ that the function $\psi(x)$ takes on in the n intervals – and which are integrated over in the functional integration process – are *all* determined by the choice of the expansion coefficients $\alpha_1, \cdots, \alpha_n$. Hence integration over all the ψ_i will be equivalent to integration over all the α_k.

This enables us to express the functional integral in (11.84) as a phase space integral over the expansion coefficients $\alpha_1, \cdots, \alpha_n$. We have

$$\int D\psi\, F[\psi(x)] = \lim_{n\to\infty,\epsilon\to 0} \int d\alpha_1 \cdots d\alpha_n\, \|J(\alpha_1, \cdots, \alpha_n)\|\, v(\alpha_1, \cdots, \alpha_n) f(\alpha_1, \cdots, \alpha_n) \tag{11.92}$$

where the Jacobian is given by

$$||J(\alpha_1, \alpha_2, \cdots, \alpha_k, \cdots, \alpha_n)|| = \left\| \begin{array}{cccc} \dfrac{\partial\psi_1}{\partial\alpha_1} & \dfrac{\partial\psi_1}{\partial\alpha_2} & \cdots & \dfrac{\partial\psi_1}{\partial\alpha_n} \\ \dfrac{\partial\psi_2}{\partial\alpha_1} & \dfrac{\partial\psi_2}{\partial\alpha_2} & \cdots & \dfrac{\partial\psi_2}{\partial\alpha_n} \\ \cdots & \cdots & \cdots & \cdots \\ \dfrac{\partial\psi_n}{\partial\alpha_1} & \dfrac{\partial\psi_n}{\partial\alpha_2} & & \dfrac{\partial\psi_n}{\partial\alpha_n} \end{array} \right\| . \tag{11.93}$$

Now, using (11.88),

$$\frac{\partial\psi_i}{\partial\alpha_k} = \phi_{ki}, \tag{11.94}$$

and evaluating the Jacobian after showing that $(JJ^T)_{ik} = \delta_{ik}/\Delta x_i$ using the completeness relationship in (2.62), we find that

$$||J(\alpha_1, \alpha_2, \cdots, \alpha_k, \cdots, \alpha_n)|| = \prod_i (\Delta x_i)^{-1/2} \tag{11.95}$$

and thus

$$\int D\psi\, F[\psi(x)] = \lim_{n\to\infty} \lim_{\epsilon\to 0} \int d\alpha_1 \cdots d\alpha_n \prod_i \frac{1}{(\Delta x_i)^{1/2}} v(\alpha_1, \cdots, \alpha_n)\, f(\alpha_1, \cdots, \alpha_n). \tag{11.96}$$

This key result expresses the original functional integral as a phase space integral over the expansion coefficients α_k of the function $\psi(x)$ in terms of the mode functions $\phi_k(x)$ for the case where all quantities are real.

The general result can be simplified with a special choice of the weight function

$$w(\psi_1, \cdots, \psi_n) = \prod_i (\Delta x_i)^{1/2}, \tag{11.97}$$

and we then get a simple expression for the functional integral,

$$\int D\psi\, F[\psi(x)] = \lim_{n\to\infty} \lim_{\epsilon\to 0} \int d\alpha_1 \cdots d\alpha_n\; f(\alpha_1, \cdots, \alpha_n). \tag{11.98}$$

In this form of the functional integral, the original functional $F[\psi(x)]$ has been replaced by the equivalent function $f(\alpha_1, \cdots, \alpha_n)$ of the expansion coefficients α_k, and the functional integration is now replaced by a phase space integration over the coefficients.

The relationship between the functional integral and the phase space integral can be generalised to cases involving several complex functions. For the case of the functional $F[\psi(x), \psi^+(x), \psi^*(x), \psi^{+*}(x)]$, where $\psi(x), \psi^+(x)$ are expanded in terms of complex mode functions as in (11.1) and (11.7) and ψ_i, ψ_i^+ are defined in analogy to (11.83), we have

$$\psi_i = \sum_k \alpha_k \phi_{ki}, \qquad \alpha_k = \sum_i \Delta x_i \, \phi_{ki}^* \psi_i,$$
$$\psi_i^+ = \sum_k \alpha_k^+ \phi_{ki}^*, \qquad \alpha_k^+ = \sum_i \Delta x_i \, \phi_{ki} \psi_i^+.$$

For variety, we will turn the phase space integral into a functional integral. We first have the transformation involving real quantities:

$$\alpha_{kx} = \sum_i \Delta x_i \, (\phi_{kix}\psi_{ix} + \phi_{kiy}\psi_{iy}), \qquad \alpha_{ky} = \sum_i \Delta x_i \, (\phi_{kix}\psi_{iy} - \phi_{kiy}\psi_{ix}),$$
$$\alpha_{kx}^+ = \sum_i \Delta x_i \, (\phi_{kix}\psi_{ix}^+ - \phi_{kiy}\psi_{iy}^+), \qquad \alpha_{ky}^+ = \sum_i \Delta x_i \, (\phi_{kix}\psi_{iy}^+ + \phi_{kiy}\psi_{ix}^+).$$
$$(11.99)$$

In the standard notation with $\alpha_k = \alpha_{kx} + i\alpha_{ky}$, $\alpha_k^+ = \alpha_{kx}^+ + i\alpha_{ky}^+$ and $d^2\alpha_k = d\alpha_{kx}\, d\alpha_{ky}$, $d^2\alpha_k^+ = d\alpha_{kx}^+\, d\alpha_{ky}^+$, the phase space integral is of the form

$$\int d^2\alpha\, d^2\alpha^+ f(\alpha, \alpha^+, \alpha^*, \alpha^{+*}) = \int d^2\alpha_1 \cdots d^2\alpha_n \int d^2\alpha_1^+ \cdots d^2\alpha_n^+ \; f,$$

and after transforming to the new variables $\psi_{ix}, \psi_{iy}, \psi_{ix}^+, \psi_{iy}^+$ we get

$$\int d^2\alpha_1 \cdots d^2\alpha_n \int d^2\alpha_1^+ \cdots d^2\alpha_n^+ \; f = \int d^2\psi_1 \cdots d^2\psi_n \int d^2\psi_1^+ \cdots d^2\psi_n^+ \; ||J||$$
$$\times F(\psi_1, \cdots, \psi_n, \psi_1^+, \cdots, \psi_n^+, \psi_1^*, \cdots, \psi_n^*, \psi_1^{+*}, \cdots, \psi_n^{+*}), \qquad (11.100)$$

where the Jacobian can be written in terms of the notation $\alpha_{kx} \to \alpha_{k1}, \alpha_{ky} \to \alpha_{k2}, \alpha_{kx}^+ \to \alpha_{k3}, \alpha_{ky}^+ \to \alpha_{k4}$ and $\psi_{ix} \to \psi_{i1}, \psi_{iy} \to \psi_{i2}, \psi_{ix}^+ \to \psi_{i3}, \psi_{iy}^+ \to \psi_{i4}$. The Jacobian is the determinant of the matrix J, where

$$J_{k\mu\; i\nu} = \frac{\partial \alpha_{k\mu}}{\partial \psi_{i\nu}} \quad (k = 1, \cdots, n;\; i = 1, \cdots, n;\; \mu = 1, \cdots, 4;\; \nu = 1, \cdots, 4),$$
$$||J(\alpha_k, \alpha_k^+, \alpha_k^*, \alpha_k^{+*})|| = \|J_{k\mu\; i\nu}\| \,. \qquad (11.101)$$

The elements in the 4×4 submatrix J_{ki} are obtained from (11.99) and are

$$[J_{ki}] = \begin{bmatrix} \Delta x_i \, \phi_{kix} & \Delta x_i \, \phi_{kiy} & 0 & 0 \\ -\Delta x_i \, \phi_{kiy} & \Delta x_i \, \phi_{kix} & 0 & 0 \\ 0 & 0 & \Delta x_i \, \phi_{kix} & -\Delta x_i \, \phi_{kiy} \\ 0 & 0 & \Delta x_i \, \phi_{kiy} & \Delta x_i \, \phi_{kix} \end{bmatrix}. \qquad (11.102)$$

The completeness relationship (2.62) can then be used to show that

$$\Delta x_i \Delta x_j \sum_k (\phi_{kix}\phi_{kjx} + \phi_{kiy}\phi_{kjy}) = \Delta x_i \, \delta_{ij},$$
$$\Delta x_i \Delta x_j \sum_k (-\phi_{kix}\phi_{kjy} + \phi_{kiy}\phi_{kjx}) = 0, \tag{11.103}$$

which is the same as

$$\sum_{k\mu} J_{k\mu\; i\nu} J_{k\mu\; j\xi} = \Delta x_i \, \delta_{ij}\delta_{\nu\xi},$$
$$\left[J^T J\right]_{i\nu j\xi} = \Delta x_i \delta_{ij}\delta_{\nu\xi}. \tag{11.104}$$

Hence the Jacobian is

$$\|J_{k\mu\; i\nu}\| = \prod_{i=1}^{n} (\Delta x_i)^2, \tag{11.105}$$

so that we finally have, after letting $n \to \infty$ and $\Delta x_i \to 0$ and with $d^2\alpha = \prod_k d^2\alpha_k$ and $d^2\alpha^+ = \prod_k d^2\alpha_k^+$,

$$\begin{aligned}
&\int d^2\alpha \, d^2\alpha^+ \; f(\alpha, \alpha^+, \alpha^*, \alpha^{+*}) \\
&= \lim_{n\to\infty} \lim_{\epsilon\to 0} \int d^2\alpha_1 \cdots d^2\alpha_n \int d^2\alpha_1^+ \cdots d^2\alpha_n^+ \; f \\
&= \lim_{n\to\infty} \lim_{\epsilon\to 0} \int d^2\psi_1 \cdots d^2\psi_n \int d^2\psi_1^+ \cdots d^2\psi_n^+ \\
&\quad \times w(\psi_1, \cdots, \psi_n, \psi_1^+, \cdots, \psi_n^+, \psi_1^*, \cdots, \psi_n^*, \psi_1^{+*}, \cdots, \psi_n^{+*}) \\
&\quad \times F(\psi_1, \cdots, \psi_n, \psi_1^+, \cdots, \psi_n^+, \psi_1^*, \cdots, \psi_n^*, \psi_1^{+*}, \cdots, \psi_n^{+*}) \\
&= \int D^2\psi \; D^2\psi^+ \; F[\psi(x), \psi^+(x), \psi^*(x), \psi^{+*}(x)],
\end{aligned} \tag{11.106}$$

where $D^2\psi \, D^2\psi^+ = \prod_i d^2\psi_i \prod_i d^2\psi_i \; w(\psi_1, \cdots, \psi_n, \psi_1^+, \cdots, \psi_n^+, \psi_1^*, \cdots, \psi_n^*, \psi_1^{+*}, \cdots, \psi_n^{+*})$ and the weight function is chosen as

$$w(\psi_1, \cdots, \psi_n, \psi_1^+, \cdots, \psi_n^+, \psi_1^*, \cdots, \psi_n^*, \psi_1^{+*}, \cdots, \psi_n^{+*}) = \prod_{i=1}^{n} (\Delta x_i)^2, \tag{11.107}$$

which is independent of the functions. The power law $(\Delta x_i)^2$ is consistent with there being four real functions involved instead of the single function as previously.

11.5.3 Functional Integration by Parts

A useful integration-by-parts rule can often be established from (11.69). Consider the functional $H[\psi(x)] = F[\psi(x)]G[\psi(x)]$. Then

$$F[\psi(x)] \left(\frac{\delta G[\psi(x)]}{\delta \psi(x)} \right) = \left(\frac{\delta \{F[\psi(x)]G[\psi(x)]\}}{\delta \psi(x)} \right) - \left(\frac{\delta F[\psi(x)]}{\delta \psi(x)} \right) G[\psi(x)]. \tag{11.108}$$

Then

$$\int D\psi\, F[\psi(x)] \left(\frac{\delta G[\psi(x)]}{\delta\psi(x)}\right) = \int D\psi \left(\frac{\delta H[\psi(x)]}{\delta\psi(x)}\right) - \int D\psi \left(\frac{\delta F[\psi(x)]}{\delta\psi(x)}\right) G[\psi(x)]. \tag{11.109}$$

If we now introduce mode expansions and use (11.34) for the functional derivative of $H[\psi(x)]$ and (11.98) for the first of the two functional integrals on the right-hand side of the last equation, then

$$\begin{aligned}\int D\psi \left(\frac{\delta H[\psi(x)]}{\delta\psi(x)}\right) &= \lim_{n\to\infty}\lim_{\epsilon\to 0}\int d\alpha_1\cdots d\alpha_k\cdots d\alpha_n \sum_k \phi_k^*(x)\,\frac{\partial h(\alpha_1,\cdots,\alpha_k,\cdots)}{\partial\alpha_k} \\ &= \lim_{n\to\infty}\lim_{\epsilon\to 0}\sum_k \phi_k^*(x)\int\cdots\int d\alpha_1\, d\alpha_2\cdots \\ &\quad\times\{h(\alpha_1,\cdots,\alpha_k,\cdots)_{\alpha_k\to+\infty} - h(\alpha_1,\cdots,\alpha_k,\cdots)_{\alpha_k\to-\infty}\}\cdots d\alpha_n,\end{aligned}$$

so that the functional integral of this term reduces to contributions on the boundaries of phase space. Hence if $h(\alpha_1,\cdots,\alpha_k,\cdots)\to 0$ as all $\alpha_k\to\pm\infty$, then the functional integral involving the functional derivative of $H[\psi(x)]$ vanishes and we have the integration-by-parts result

$$\int D\psi\, F[\psi(x)] \left(\frac{\delta G[\psi(x)]}{\delta\psi(x)}\right) = -\int D\psi \left(\frac{\delta F[\psi(x)]}{\delta\psi(x)}\right) G[\psi(x)]. \tag{11.110}$$

All these rules have obvious generalisations for functionals such as $F[\psi(x),\psi^+(x),\psi^*(x),\psi^{+*}(x)]$ involving several fields.

11.5.4 Differentiating a Functional Integral

Functionals can be defined via functional integration processes, and it is of interest to find rules for their functional derivatives. This leads to a rule for differentiating a functional integral. Suppose we have a functional $G[\chi(x)]$ determined from another functional $F[\psi(x)]$ via a functional integral that involves a transfer functional $A_{GF}[\chi(x),\psi(x)]$ in the form

$$G[\chi(x)] = \int D\psi\, A_{GF}[\chi(x),\psi(x)]\, F[\psi(x)]. \tag{11.111}$$

Applying the definition of the functional derivatives of $G[\chi(x)]$ and $A_{GF}[\chi(x),\psi(x)]$ with respect to $\chi(x)$, we have

$$
\begin{aligned}
G[\chi(x) + \delta\chi(x)]&\\
&= \int D\psi\, A_{GF}[\chi(x) + \delta\chi(x), \psi(x)]\, F[\psi(x)]\\
&= \int D\psi \left\{ A_{GF}[\chi(x), \psi(x)] + \int dx\, \delta\chi(x) \left(\frac{\delta A_{GF}[\chi(x), \psi(x)]}{\delta\chi(x)} \right) \right\} F[\psi(x)]\\
&= \int D\psi \left\{ A_{GF}[\chi(x), \psi(x)] \right\} F[\psi(x)]\\
&\quad + \int D\psi \left\{ \int dx\, \delta\chi(x) \left(\frac{\delta A_{GF}[\chi(x), \psi(x)]}{\delta\chi(x)} \right) \right\} F[\psi(x)]\\
&= G[\chi(x)] + \int dx\, \delta\chi(x) \int D\psi \left\{ \left(\frac{\delta A_{GF}[\chi(x), \psi(x)]}{\delta\chi(x)} \right) \right\} F[\psi(x)],
\end{aligned}
$$

since (for reasonably well-behaved quantities) the functional integration over $D\psi$ and the ordinary integration over dx can be carried out in either order, given that both just involve processes that are limits of summations. Hence, from the definition of the functional derivative, we have

$$
\left(\frac{\delta G[\psi(x)]}{\delta\psi(x)} \right) = \int D\psi \left(\frac{\delta A_{GF}[\chi(x), \psi(x)]}{\delta\chi(x)} \right) F[\psi(x)], \tag{11.112}
$$

which is the required rule for differentiating a functional defined via a functional of another function. Clearly, the rule is to just differentiate the transfer functional under the functional integration sign, a rule similar to that applying in ordinary calculus.

As a particular case, consider the Fourier-like transfer functional

$$
\begin{aligned}
G[\chi(x)] &= \int D\psi\, A_{GF}[\chi(x), \psi(x)]\, F[\psi(x)],\\
A_{GF}[\chi(x), \psi(x)] &= \exp\left\{ i \int dx\, \chi(x)\psi(x) \right\}.
\end{aligned} \tag{11.113}
$$

In this case

$$
\begin{aligned}
A_{GF}[\chi(x) + \delta\chi(x), \psi(x)]&\\
&= \exp\left\{ i \int dx\, (\chi(x) + \delta\chi(x))\, \psi(x) \right\}\\
&= \exp\left\{ i \int dx\, \chi(x)\psi(x) \right\} \exp\left\{ i \int dx\, \delta\chi(x)\psi(x) \right\}\\
&\approx \exp\left\{ i \int dx \chi(x)\, \psi(x) \right\} \left(1 + i \int dx\, \delta\chi(x)\psi(x) \right)\\
&= A_{GF}[\chi(x), \psi(x)] + A_{GF}[\chi(x), \psi(x)]\, i \int dx\, \delta\chi(x)\psi(x)).
\end{aligned}
$$

Hence

$$
\left(\frac{\delta A_{GF}[\chi(x), \psi(x)]}{\delta\chi(x)} \right) = A_{GF}[\chi(x), \psi(x)] \times i\psi(x) \tag{11.114}
$$

and

$$\left(\frac{\delta G[\psi(x)]}{\delta\psi(x)}\right) = \int D\psi\,\{A_{GF}[\chi(x),\psi(x)]\,(i\psi(x))\,\}\,F[\psi(x)]. \tag{11.115}$$

All these rules have obvious generalisations for functionals such as $F[\psi(x),\psi^{+}(x),\psi^{*}(x),\psi^{+*}(x)]$ involving several fields.

11.5.5 Examples of Functional Integrals

An important case is that of the functional integral

$$\int D\psi\,\psi(y)\,P[\psi(x)], \tag{11.116}$$

where for simplicity we consider a single real field $\psi(x)$ involving only one variable. This involves the functional integral of a functional which is the *product* of the two functionals $F_y[\psi(x)] \equiv \psi(y)$ (see (11.18)) and $P[\psi(x)]$. This overall functional integral could be evaluated as a phase space integral via mode expansions (see (11.1) and (11.98)) to give

$$\int D\psi\,\psi(y)\,P[\psi(x)] = \lim_{n\to\infty}\lim_{\epsilon\to 0}\sum_k \phi_k(y)\int d\alpha_1\cdots d\alpha_k\cdots d\alpha_n\;\alpha_k p(\alpha_1,\cdots,\alpha_k,\cdots,\alpha_n), \tag{11.117}$$

where $p(\alpha_1,\cdots,\alpha_k,\cdots,\alpha_n)$ is the phase space function that is equivalent to $P[\psi(x)]$ (see (11.5)).

However, we can also express the result in terms of functional integrals. As in (11.84), with $P(\psi_1,\cdots,\psi_i,\cdots,\psi_n)$ the function of $\psi_1,\cdots,\psi_i,\cdots,\psi_n$ that is equivalent to $P[\psi(x)]$ (see (11.83)) and $w(\psi_1,\cdots,\psi_i,\cdots,\psi_n)$ the weight function, and noting that the function $F_y(\psi_1,\cdots,\psi_i,\cdots,\psi_n)$ that is equivalent to the functional $F_y[\psi(x)] \equiv \psi(y)$ is given by

$$\begin{aligned} F_y(\psi_1,\cdots,\psi_i,\cdots,\psi_n) &= \psi(y) = \psi_i \qquad \text{if } y \text{ is inside } \Delta x_i \\ &= 0 \qquad \text{if } y \text{ is not inside } \Delta x_i, \end{aligned} \tag{11.118}$$

we find that

$$\begin{aligned} \int D\psi\,\psi(y)\,P[\psi(x)] &= \lim_{n\to\infty}\lim_{\epsilon\to 0}\int d\psi_1\cdots d\psi_i\cdots d\psi_n\,w(\psi_1,\cdots,\psi_i,\cdots,\psi_n) \\ &\quad\times\psi_i P(\psi_1,\cdots,\psi_i,\cdots,\psi_n) \qquad \text{where } y \text{ is inside } \Delta x_i \\ &= \lim_{\epsilon\to 0}\int d\psi_i\,\psi_i\,P_y(\psi_i), \end{aligned} \tag{11.119}$$

where $\Delta x_i = x_{i+1} - x_i$ and $\varepsilon > \Delta x_i$, and where

$$\begin{aligned} P_y(\psi_i) = &\lim_{(n-1)\to\infty}\lim_{\epsilon\to 0}\int d\psi_1\cdots d\psi_{i-1}\,d\psi_{i+1}\cdots d\psi_n\,w(\psi_1,\cdots,\psi_i,\cdots,\psi_n) \\ &\times P(\psi_1,\cdots,\psi_i,\cdots,\psi_n) \qquad \text{where } y \text{ is inside } \Delta x_i \end{aligned} \tag{11.120}$$

is a reduced form of $P(\psi_1, \psi_2, \cdots, \psi_i, \cdots, \psi_n)$, which is only a function of ψ_i and where the interval Δx_i is the one which includes y.

Other cases involving single complex fields such as $\int D^2\psi\, \psi(y)\, P[\psi(x)]$, or pairs of complex fields, such as $\int\int D^2\psi\, D^2\psi^+ \psi(y)\, P[\psi(x), \psi^+(x)\,]$, etc. can be similarly treated.

11.6 Functionals of Fermionic Grassmann Fields

11.6.1 Basic Idea

The idea of a functional can be extended to cases where $\psi(x)$ is a Grassmann function rather than a c-number function, where the variable x may refer to position as previously, so in this case $\psi(x)$ is a Grassmann field. Grassmann fields $\psi(x)$ can be expanded in terms of a suitable orthonormal set of mode functions with g-number expansion coefficients g_k, where the mode functions satisfy the same equations as in (5.139) and (5.140). Thus we have for Grassmann functions

$$\psi(x) = \sum_i g_i \phi_i(x), \tag{11.121}$$

$$\psi^+(x) = \sum_i \phi_i^*(x) g_i^+. \tag{11.122}$$

Again, if $g_i^+ = g_i^*$ then $\psi^+(x) = \psi^*(x)$, the complex conjugate Grassmann field.

The Grassmann fields are odd Grassmann functions of the first order. The following results for the expansion coefficients can easily be obtained:

$$g_k = \int dx\, \phi_k^*(x)\psi(x), \tag{11.123}$$

$$g_k^+ = \int dx\, \phi_k(x)\psi^+(x), \tag{11.124}$$

which has the same form as for c-number fields.

As pointed out previously, Grassmann fields differ from bosonic fields in that they anticommute and their square and higher powers are zero. Thus, with $\eta_{\mathrm{f}}(x) = \sum_l h_l\phi_l(x)$ a second Grassmann field,

$$\psi_{\mathrm{f}}(x)\eta_{\mathrm{f}}(y) + \eta_{\mathrm{f}}(y)\psi_{\mathrm{f}}(x) = 0, \tag{11.125}$$

$$(\psi_{\mathrm{f}}(x))^2 = (\psi_{\mathrm{f}}(x))^3 = \cdots = 0. \tag{11.126}$$

The idea of a Grassmann functional is analogous to that for c-number functionals. A Grassmann functional $F[\psi(x)]$ maps a g-number function $\psi(x)$ onto a Grassmann function that depends on *all* the values of $\psi(x)$ over its entire range. As the value of the function at any point in the range for x is determined uniquely by the expansion coefficients $\{g_k\}$, the functional $F[\psi(x)]$ must therefore also depend on just the g-number expansion coefficients, and hence may also be viewed as a Grassmann function

$f(g_1, \cdots, g_k, \cdots, g_n)$ of the expansion coefficients, a useful equivalence when functional differentiation and integration are considered:

$$F[\psi(x)] \equiv f(g_1, \cdots, g_k, \cdots, g_n). \tag{11.127}$$

Functionals of the form $F[\psi^+(x)]$ also occur, and these are also equivalent to a function of the expansion coefficients:

$$F[\psi^+(x)] \equiv f^+(g_1^+, \cdots, g_k^+, \cdots, g_n^+). \tag{11.128}$$

The idea of a Grassmann functional can also be extended to cases of the form $F[\psi(x_1, \cdots, x_n)]$, where $\psi(x_1, \cdots, x_n)$ is a Grassmann function of several variables $x_1, \cdots, x_n$, or cases $F[\hat{\psi}(x)]$, where $\hat{\psi}(x)$ is an operator function involving g-numbers rather than a c-numbers, in which case $F[\hat{\psi}(x)]$ maps the operator Grassmann function onto a Grassmann operator. Also, Grassmann functionals $F[\psi_1(x), \cdots, \psi_i(x), \cdots, \psi_n(x)]$ involving several Grassmann fields $\psi_1(x), \cdots, \psi_i(x), \cdots, \psi_n(x)$ occur. For example, a fermionic field annihilation operator $\hat{\psi}(x)$ may be associated with a Grassmann field $\psi_1(x) = \psi(x)$, and the field creation operator $\hat{\psi}(x)^\dagger$ may be associated with a different Grassmann field $\psi_2(x) = \psi^+(x)$, and thus functionals of the form $F[\psi(x), \psi^+(x)]$ will be involved. In some cases we may choose $\psi^+(x)$ to be the conjugate Grassmann field $\psi^*(x)$ to give functionals of the form $F[\psi(x), \psi^*(x)]$. Grassmann functional derivatives and Grassmann functional integrals can be defined for all of these cases.

11.6.2 Examples of Grassmann-Number Functionals

A somewhat trivial application of the Grassmann functional concept is to express a Grassmann field $\psi(y)$ as a functional $F_y[\psi(x)]$ of $\psi(x)$:

$$F_y[\psi(x)] \equiv \psi(y) \tag{11.129}$$

$$= \int dx\, \delta(y - x)\, \psi(x). \tag{11.130}$$

Here this specific functional involves the Dirac delta function as a kernel.

Another example involves the spatial derivative $\nabla_y \psi(y)$, which may also be expressed as a functional $F_{\nabla y}[\psi(x)]$:

$$F_{\nabla y}[\psi(x)] \equiv \nabla_y \psi(y) \tag{11.131}$$

$$= \int dx\, \delta(y - x)\, \nabla_x \psi(x)$$

$$= -\int dx\, \{\nabla_x \delta(y - x)\}\, \psi(x)$$

$$= \int dx\, \{\nabla_y \delta(y - x)\}\, \psi(x). \tag{11.132}$$

Here the functional involves $\nabla_y \delta(y - x)$ as a kernel.

A functional is said to be linear if

$$F[c_1\psi_1(x) + c_2\psi_2(x)] = c_1F[\psi_1(x)] + c_2F[\psi_2(x)], \tag{11.133}$$

where c_1, c_2 are constants. Both the Grassmann function $\psi(y)$ and its spatial derivative are linear functionals.

An example of a non-linear Grassmann functional is

$$\begin{aligned} F_{\chi(2)} &= \left(\int dx\, \chi^*(x)\psi(x) \right)^2 \\ &= \int\int dx\, dy\, \chi^*(x)\chi^*(y)\psi(x)\psi(y). \end{aligned} \tag{11.134}$$

11.7 Functional Differentiation for Grassmann Fields

11.7.1 Definition of Functional Derivative

The left functional derivative $(\overrightarrow{\delta}/\delta\psi(x))F[\psi(x)]$ is defined by

$$F[\psi(x) + \delta\psi(x)] \approx F[\psi(x)] + \int dx\, \delta\psi(x) \left(\frac{\overrightarrow{\delta}}{\delta\psi(x)} F[\psi(x)] \right)_x, \tag{11.135}$$

where $F[\psi(x) + \delta\psi(x)] - F[\psi(x)]$ is evaluated correct to the first order in a change $\delta\psi(x)$ in the Grassmann field. Note that the idea of smallness does not apply to the Grassmann field change $\delta\psi(x)$. In (11.135), the left side is a functional of $\psi(x) + \delta\psi(x)$ and the first term on the right side is a functional of $\psi(x)$. Both these quantities and the second term on the right side are Grassmann functions. The latter term is a functional of the Grassmann field $\delta\psi(x)$ and thus the functional derivative must be a Grassmann function of x, hence the subscript x. In most situations this subscript will be left understood. Note that because Grassmann variables cannot be divided, there is no equivalent Grassmann functional derivative result equivalent to (11.22).

Note that $\Delta F = F[\psi(x) + \delta\psi(x)] - F[\psi(x)]$ will in general involve terms that are non-linear in $\delta\psi(x)$. For the example in (11.134), $\Delta F = \int\int dx\, dy\, \chi^*(x)\chi^*(y)\, \delta\psi(x)\, \delta\psi(y)$ and here the functional derivative is zero.

In addition, we note that the Grassmann function $\delta\psi(x)$ does not necessarily commute with the g-function $\left(\overrightarrow{\delta}/\delta\psi(x)F[\psi(x)]\right)_x$. This therefore means that there is also a right functional derivative $\left(F[\psi(x)]\overleftarrow{\delta}/\delta\psi(x)\right)_x$, defined by

$$F[\psi(x) + \delta\psi(x)] \approx F[\psi(x)] + \int dx\, \left(F[\psi(x)] \frac{\overleftarrow{\delta}}{\delta\psi(x)} \right)_x \delta\psi(x). \tag{11.136}$$

We emphasise again: the functional derivative of a g-number functional is a Grassmann function, not a functional. The specific examples below illustrate this feature.

For functionals of the form $F[\psi^+(x)]$, we have similar expressions for the left and right functional derivatives:

$$
\begin{aligned}
F[\psi^+(x) + \delta\psi^+(x)] & \qquad (11.137)\\
\approx F[\psi^+(x)] + \int dx\, \delta\psi^+(x) \left[\frac{\overrightarrow{\delta}}{\delta\psi^+(x)} F[\psi^+(x)] \right]_x & \\
\approx F[\psi^+(x)] + \int dx \left[F[\psi^+(x)] \frac{\overleftarrow{\delta}}{\delta\psi^+(x)} \right]_x \delta\psi^+(x). & \qquad (11.138)
\end{aligned}
$$

This definition of a functional derivative can be extended to cases where $\psi(x_1, \cdots, x_n)$ is a function of several variables or where $\hat{\psi}(x)$ is an operator function rather than a c-number function. Also, functionals $F[\psi_1(x), \cdots, \psi_i(x), \cdots, \psi_n(x)]$ involving several functions $\psi_1(x), \cdots, \psi_i(x), \cdots, \psi_n(x)$ occur, and functional derivatives with respect to any of these functions can be defined.

Finally, higher-order functional derivatives can be defined by applying the basic definitions to lower-order functional derivatives.

11.7.2 Examples of Grassmann Functional Derivatives

For the case of the functional $F_y[\psi(x)]$ in (11.130) that gives the function $\psi(y)$, as

$$
\begin{aligned}
F_y[\psi(x) + \delta\psi(x)] - F_y[\psi(x)] &= \psi(y) + \delta\psi(y) - \psi(y) \\
&= \int dx\, \delta\psi(x)\, \delta(y - x) \\
&= \int dx\, \delta\psi(x) \left(\frac{\overrightarrow{\delta}}{\delta\psi(x)} F_y[\psi(x)] \right)_x \\
&= \int dx\, \delta(y - x)\, \delta\psi(x) \\
&= \int dx \left(F_y[\psi(x)] \frac{\overleftarrow{\delta}}{\delta\psi(x)} \right)_x \delta\psi(x),
\end{aligned}
$$

we have the same result for the left and right Grassmann functional derivatives:

$$
\begin{aligned}
\left(\frac{\overrightarrow{\delta}}{\delta\psi(x)} F_y[\psi(x)] \right)_x &= \left(\frac{\overrightarrow{\delta}\, \psi(y)}{\delta\psi(x)} \right)_x & (11.139)\\
&= \delta(y - x), & (11.140)\\
\left(F_y[\psi(x)] \frac{\overleftarrow{\delta}}{\delta\psi(x)} \right)_x &= \left(\frac{\psi(y)\, \overleftarrow{\delta}}{\delta\psi(x)} \right)_x & (11.141)\\
&= \delta(y - x), & (11.142)
\end{aligned}
$$

so here the left and right functional derivatives are a delta function, just as for c-numbers.

A similar situation applies to the functional $F_{\nabla y}[\psi(x)]$ in (11.132) that gives the spatial-derivative function $\nabla_y\psi(y)$;

$$\begin{aligned}
F_{\nabla y}[\psi(x)+\delta\psi(x)] - F_{\nabla y}[\psi(x)] &= \nabla_y\psi(y) + \nabla_y\delta\psi(y) - \nabla_y\psi(y) \\
&= \int dx\, \delta\psi(x)\, \nabla_y\delta(y-x) \\
&= \int dx\, \delta\psi(x) \left(\frac{\overrightarrow{\delta}}{\delta\psi(x)} F_{\nabla y}[\psi(x)] \right)_x \\
&= \int dx \nabla_y\delta(y-x)\, \delta\psi(x) \\
&- \int dx \left(F_{\nabla y}[\psi(x)] \frac{\overleftarrow{\delta}}{\delta\psi(x)} \right)_x \delta\psi(x).
\end{aligned}$$

Also, we have the same result for the left and right Grassmann functional derivatives:

$$\begin{aligned}
\left(\frac{\overrightarrow{\delta}}{\delta\psi(x)} F_{\nabla y}[\psi(x)] \right)_x &= \left(\frac{\overrightarrow{\delta}\ \nabla_y\psi(y)}{\delta\psi(x)} \right)_x \\
&= \nabla_y\delta(y-x),
\end{aligned} \tag{11.143}$$

$$\begin{aligned}
\left(F_{\nabla y}[\psi(x)] \frac{\overleftarrow{\delta}}{\delta\psi(x)} \right)_x &= \left(\frac{\nabla_y\psi(y)\,\overleftarrow{\delta}}{\delta\psi(x)} \right)_x \\
&= \nabla_y\delta(y-x),
\end{aligned} \tag{11.144}$$

so again the functional derivatives are the spatial derivative of a delta function. Similarly, with the Grassmann functionals $F_y[\psi^+(x)] \equiv \psi^+(y)$ and $F_{\nabla y}[\psi^+(x)] \equiv \nabla_y\psi^+(y)$,

$$\left(\frac{\overrightarrow{\delta}\ \psi^+(y)}{\delta\psi^+(x)} \right)_x = \delta(y-x) = \left(\frac{\psi^+(y)\,\overleftarrow{\delta}}{\delta\psi^+(x)} \right)_x, \tag{11.145}$$

$$\left(\frac{\overrightarrow{\delta}\ \nabla_y\psi^+(y)}{\delta\psi^+(x)} \right)_x = \nabla_y\delta(y-x) = \left(\frac{\nabla_y\psi^+(y)\,\overleftarrow{\delta}}{\delta\psi^+(x)} \right)_x. \tag{11.146}$$

11.7.3 Grassmann Functional Derivative and Mode Functions

If a mode expansion of the Grassmann field $\psi(x)$ as in (11.121) is performed, then we can obtain an expression for the functional derivative in terms of mode functions. With

$$\delta\psi(x) = \sum_k \delta g_k\, \phi_k(x), \tag{11.147}$$

where $\delta g_1, \delta g_2, \cdots, \delta g_n$ are Grassmann variables that determine the change $\delta\psi(x)$ in the Grassmann field, we see that

$$\begin{aligned} F[\psi(x)+\delta\psi(x)] - F[\psi(x)] &\approx \int dx\, \delta\psi(x) \left(\frac{\overrightarrow{\delta}}{\delta\psi(x)} F[\psi(x)] \right)_x \\ &\approx \sum_k \delta g_k \int dx\, \phi_k(x) \left(\frac{\overrightarrow{\delta}}{\delta\psi(x)} F[\psi(x)] \right)_x . \end{aligned}$$

Suppose we write $F[\psi(x)]$ as a Grassmann function, making explicit the g_k dependence:

$$\begin{aligned} F[\psi(x)] &= f(g_1, \cdots, g_k, \cdots) \\ &= f_0 + \sum_{k_1} g_{k_1} f_1(k_1) + \sum_{k_1,k_2} g_{k_1} g_{k_2} f_2(k_1,k_2) + \sum_{k_1,k_2,k_3} g_{k_1} g_{k_2} g_{k_3} f_3(k_1,k_2,k_3) + \cdots , \end{aligned}$$

where the coefficients are c-numbers, and where there is a convention that these coefficients are zero unless $k_1 < k_2$ in $f_2(k_1,k_2)$, $k_1 < k_2 < k_3$ in $f_3(k_1,k_2,k_3)$ and so on. The summations are not restricted. We have shown previously (see (4.35)) that

$$f(g_1+\delta g_1, \cdots, g_k+\delta g_k, \cdots) - f(g_1, \cdots, g_k, \cdots) = \sum_k \delta g_k \left\{ \frac{\overrightarrow{\partial}}{\partial g_k} f(g_1, \cdots, g_k, \cdots) \right\}$$

correct to first order in the δg_k, so that we can write

$$F[\psi(x)+\delta\psi(x)] - F[\psi(x)] = \sum_k \delta g_k \frac{\overrightarrow{\partial}}{\partial g_k} f(g_1, \cdots, g_k, \cdots).$$

We have therefore found that

$$\sum_k \delta g_k \int dx\, \phi_k(x) \left(\frac{\overrightarrow{\delta}}{\delta\psi(x)} F[\psi(x)] \right)_x = \sum_k \delta g_k \frac{\overrightarrow{\partial}}{\partial g_k} f(g_1, \cdots, g_k, \cdots).$$

Equating the coefficients of the δg_k and then using the completeness relationship in (11.3) gives the key result

$$\left(\frac{\overrightarrow{\delta}}{\delta\psi(x)} F[\psi(x)] \right)_x = \sum_k \phi_k^*(x) \frac{\overrightarrow{\partial}}{\partial g_k} f(g_1, \cdots, g_k, \cdots). \tag{11.148}$$

This relates the left functional derivative to the mode functions and to the ordinary left Grassmann derivatives of the function $f(g_1, \cdots, g_k, \cdots; g_n)$ that was equivalent to the original Grassmann functional $F[\psi(x)]$. Again, we see that the result is a function of x. Note that the left functional derivative involves an expansion in terms of the conjugate mode functions $\phi_k^*(x)$ rather than the original modes $\phi_k(x)$. The last result may be put in the form of a useful operational identity

$$\left(\frac{\overrightarrow{\delta}}{\delta\psi(x)} \right)_x = \sum_k \phi_k^*(x) \frac{\overrightarrow{\partial}}{\partial g_k}, \tag{11.149}$$

where the left side is understood to operate on an arbitrary functional $F[\psi(x)]$ and the right side is understood to operate on the equivalent function $f(g_1, \cdots, g_k, \cdots)$.

Similar results can be obtained for the right functional derivative:

$$\left(F[\psi(x)]\frac{\overleftarrow{\delta}}{\delta\psi(x)}\right)_x = \sum_k \phi_k^*(x) f(g_1, \cdots, g_k, \cdots)\frac{\overleftarrow{\partial}}{\partial g_k} \tag{11.150}$$

and

$$\left(\frac{\overleftarrow{\delta}}{\delta\psi(x)}\right)_x = \sum_k \phi_k^*(x)\,\frac{\overleftarrow{\partial}}{\partial g_k}. \tag{11.151}$$

As an example of applying these rules, consider the Grassmann functional $F_y[\psi(x)] = \sum_k g_k\phi_k(y) = \psi(y)$. Since

$$\frac{\overrightarrow{\partial}}{\partial g_k}\psi(y) = \phi_k(y) = \psi(y)\frac{\overleftarrow{\partial}}{\partial g_k}, \tag{11.152}$$

then

$$\begin{aligned}\left(\frac{\overrightarrow{\delta}}{\delta\psi(x)}\psi(y)\right)_x &= \sum_k \phi_k^*(x)\,\phi_k(y) = \delta(y-x),\\ \psi(y)\left(\frac{\overleftarrow{\delta}}{\delta\psi(x)}\right)_x &= \sum_k \phi_k^*(x)\,\phi_k(y) = \delta(y-x),\end{aligned}$$

which are the same results as before. The equivalent results for functionals of the form $F[\psi^+(x)]$ are

$$\left(\frac{\overrightarrow{\delta}}{\delta\psi^+(x)}F[\psi^+(x)]\right)_x = \sum_k \phi_k(x)\,\frac{\overrightarrow{\partial} f^+(g_1^+, \cdots, g_k^+, \cdots g_n^+)}{\partial g_k^+}, \tag{11.153}$$

$$\left(\frac{\overrightarrow{\delta}}{\delta\psi^+(x)}\right)_x = \sum_k \phi_k(x)\,\frac{\overrightarrow{\partial}}{\partial g_k^+}, \tag{11.154}$$

$$\left(F[\psi^+(x)]\frac{\overleftarrow{\delta}}{\delta\psi^+(x)}\right)_x = \sum_k \phi_k(x) f^+(g_1^+, \cdots, g_k^+, \cdots)\frac{\overleftarrow{\partial}}{\partial g_k^+}, \tag{11.155}$$

$$\left(\frac{\overleftarrow{\delta}}{\delta\psi^+(x)}\right)_x = \sum_k \phi_k(x)\,\frac{\overleftarrow{\partial}}{\partial g_k^+}. \tag{11.156}$$

The results for left and right functional derivatives can be inverted to give

$$\frac{\overrightarrow{\partial}}{\partial g_k} = \int dx\,\phi_k(x)\left(\frac{\overrightarrow{\delta}}{\delta\psi(x)}\right)_x, \tag{11.157}$$

$$\frac{\overrightarrow{\partial}}{\partial g_k^+} = \int dx\,\phi_k^*(x)\left(\frac{\overrightarrow{\delta}}{\delta\psi^+(x)}\right)_x, \tag{11.158}$$

$$\frac{\overleftarrow{\partial}}{\partial g_k} = \int dx\, \phi_k(x) \left(\frac{\overleftarrow{\delta}}{\delta\psi(x)} \right)_x, \tag{11.159}$$

$$\frac{\overleftarrow{\partial}}{\partial g_k^+} = \int dx\, \phi_k^*(x) \left(\frac{\overleftarrow{\delta}}{\delta\psi^+(x)} \right)_x. \tag{11.160}$$

11.7.4 Basic Rules for Grassmann Functional Derivatives

It is possible to establish useful rules for the functional derivative of the sum of two Grassmann functionals. It is easily shown that

$$\left(\frac{\overrightarrow{\delta}}{\delta\psi(x)} \right)_x \{F[\psi(x)] + G[\psi(x)]\} = \left(\frac{\overrightarrow{\delta}}{\delta\psi(x)} \right)_x F[\,] + \left(\frac{\overrightarrow{\delta}}{\delta\psi(x)} \right)_x G[\,], \tag{11.161}$$

$$\{F[\,] + G[\,]\} \left(\frac{\overleftarrow{\delta}}{\delta\psi(x)} \right)_x = F[\,] \left(\frac{\overleftarrow{\delta}}{\delta\psi(x)} \right)_x + G[\,] \left(\frac{\overleftarrow{\delta}}{\delta\psi(x)} \right)_x, \tag{11.162}$$

with similar results for functionals of $\psi^+(x)$.

Rules can be established for the functional derivative of the product of two Grassmann functionals that depend on the parity of the functions equivalent to the functionals, and the proofs are quite different from the c-number case. For functionals that are neither even nor odd, results can be obtained by expressing the relevant functional as a sum of even and odd contributions. We will keep the functionals in order to cover the case where the functionals are operators.

Correct to first order in $\delta\psi(x)$, we have from the definitions

$$\begin{aligned}
&\int dx\, \delta\psi(x) \left(\frac{\overrightarrow{\delta}}{\delta\psi(x)} \right)_x \{F[\psi(x)]G[\psi(x)]\} \\
&\approx F[\psi(x) + \delta\psi(x)]G[\psi(x) + \delta\psi(x)] - F[\psi(x)]G[\psi(x)] \\
&\approx F[\psi(x) + \delta\psi(x)]G[\psi(x) + \delta\psi(x)] - F[\psi(x)]G[\psi(x) + \delta\psi(x)] \\
&\quad + F[\psi(x)]G[\psi(x) + \delta\psi(x)] - F[\psi(x)]G[\psi(x)] \\
&\approx \left\{ \int dx\, \delta\psi(x) \left(\frac{\overrightarrow{\delta}}{\delta\psi(x)} \right)_x F[\psi(x)] \right\} G[\psi(x) + \delta\psi(x)] \\
&\quad + F[\psi(x)] \left\{ \int dx\, \delta\psi(x) \left(\frac{\overrightarrow{\delta}}{\delta\psi(x)} \right)_x G[\psi(x)] \right\} \\
&\approx \int dx\, \delta\psi(x) \left\{ \left(\frac{\overrightarrow{\delta}}{\delta\psi(x)} \right)_x F[\psi(x)] \right\} G[\psi(x)] \\
&\quad + \int dx\, \delta\psi(x)\, \sigma(F) F[\psi(x)] \left\{ \left(\frac{\overrightarrow{\delta}}{\delta\psi(x)} \right)_x G[\psi(x)] \right\},
\end{aligned}$$

where $\sigma(F,G) = +1, -1$ depending on the parity of the function f or g that is equivalent to the functional F or G. In the first term, the factor $G[\psi(x) + \delta\psi(x)]$ is replaced by $G[\psi(x)]$ to discard second-order contributions. Hence

$$\begin{aligned}
&\left(\frac{\overrightarrow{\delta}}{\delta\psi(x)}\right)_x \{F[\psi(x)]G[\psi(x)]\} \\
&= \left\{\left(\frac{\overrightarrow{\delta}}{\delta\psi(x)}\right)_x F[\psi(x)]\right\} G[\psi(x)] + \sigma(F)F[\psi(x)]\left\{\left(\frac{\overrightarrow{\delta}}{\delta\psi(x)}\right)_x G[\psi(x)]\right\}. \qquad (11.163)
\end{aligned}$$

A similar derivation covers the right functional derivative. The result is

$$\begin{aligned}
&\{F[\psi(x)]G[\psi(x)]\}\left(\frac{\overleftarrow{\delta}}{\delta\psi(x)}\right)_x \\
&= F[\psi(x)]\left\{G[\psi(x)]\left(\frac{\overleftarrow{\delta}}{\delta\psi(x)}\right)_x\right\} + \sigma(G)\left\{F[\psi(x)]\left(\frac{\overleftarrow{\delta}}{\delta\psi(x)}\right)_x\right\} G[\psi(x)]. \qquad (11.164)
\end{aligned}$$

The proof is left as an exercise. These two rules are the functional derivative extensions of the previous left- and right-product rules (4.28) for Grassmann derivatives.

This result can then be extended to products of more than two functionals (each of which is either even or odd), so that

$$\begin{aligned}
&\left(\frac{\overrightarrow{\delta}}{\delta\psi(x)}\right)_x \{F[\psi(x)]G[\psi(x)]H[\psi(x)]\} \\
&= \left\{\left(\frac{\overrightarrow{\delta}}{\delta\psi(x)}\right)_x F[\,]\right\} G[\,]H[\,] + \sigma(F)F[\,]\left\{\left(\frac{\overrightarrow{\delta}}{\delta\psi(x)}\right)_x G[\,]\right\} H[\,] \\
&\quad + \sigma(F)\sigma(G)F[\,]G[\,]\left\{\left(\frac{\overrightarrow{\delta}}{\delta\psi(x)}\right)_x H[\,]\right\} \qquad (11.165)
\end{aligned}$$

and so on. Also,

$$\begin{aligned}
&\{F[\psi(x)]G[\psi(x)]H[\psi(x)]\}\left(\frac{\overleftarrow{\delta}}{\delta\psi(x)}\right)_x \qquad (11.166) \\
&= F[\,]G[\,]\left\{H[\,]\left(\frac{\overleftarrow{\delta}}{\delta\psi(x)}\right)_x\right\} + \sigma(H)F[\,]\left\{G[\,]\left(\frac{\overleftarrow{\delta}}{\delta\psi(x)}\right)_x\right\} H[\,] \\
&\quad + \sigma(H)\sigma(G)\left\{F[\,]\left(\frac{\overleftarrow{\delta}}{\delta\psi(x)}\right)_x\right\} G[\,]H[\,].
\end{aligned}$$

These results are essentially the same as for Grassmann ordinary differentiation.

The product rules can be used to establish results for Grassmann functional derivatives of products of field functions. Consider $F_{y_1,\cdots,y_n}[\psi(x)] = \psi(y_1)\cdots\psi(y_n)$:

$$\begin{aligned}
&\left(\frac{\overrightarrow{\delta}}{\delta\psi(x)}\right)_x \{\psi(y_1)\cdots\psi(y_n)\} \\
&= \delta(y_1 - x)\psi(y_2)\cdots\psi(y_n) + (-1)\delta(y_2 - x)\psi(y_1)\psi(y_3)\cdots\psi(y_n) + \cdots \\
&\quad +(-1)^{n-1}\delta(y_n - x)\psi(y_1)\psi(y_2)\cdots\psi(y_{n-1})
\end{aligned} \tag{11.167}$$

and

$$\begin{aligned}
&\{\psi(y_1)\cdots\psi(y_n)\}\left(\frac{\overleftarrow{\delta}}{\delta\psi(x)}\right)_x = \delta(y_n - x)\psi(y_1)\cdots\psi(y_{n-1}) + \cdots \\
&+(-1)^{n-2}\delta(y_2 - x)\psi(y_1)\psi(y_3)\cdots\psi(y_n) + (-1)^{n-1}\delta(y_1 - x)\psi(y_2)\cdots\psi(y_n).
\end{aligned} \tag{11.168}$$

These results are often taken as the basis for defining rules for Grassmann functional differentiation.

11.7.5 Other Rules for Grassmann Functional Derivatives

There are several rules that are needed because of the distinction between left and right functional differentiation. These are analogous to the rules applying to left and right differentiation of Grassmann functions and may be established using the mode-based expressions for Grassmann functional derivatives. With $\psi(x)$ a general Grassmann field, these include:

(1) Right- and left-functional-derivative relations for even and odd functionals:

$$\begin{aligned}
\frac{\overrightarrow{\delta}}{\delta\psi(x)} F_{\mathrm{E}}[\psi(x)] &= -F_{\mathrm{E}}[\psi(x)]\frac{\overleftarrow{\delta}}{\delta\psi(x)}, \\
\frac{\overrightarrow{\delta}}{\delta\psi(x)} F_{\mathrm{O}}[\psi(x)] &= +F_{\mathrm{O}}[\psi(x)]\frac{\overleftarrow{\delta}}{\delta\psi(x)}.
\end{aligned} \tag{11.169}$$

(2) Altering order of functional derivatives:

$$\frac{\overrightarrow{\delta}}{\delta\psi(x)}\frac{\overrightarrow{\delta}}{\delta\psi(y)} F[\psi(x)] = -\frac{\overrightarrow{\delta}}{\delta\psi(y)}\frac{\overrightarrow{\delta}}{\delta\psi(x)} F[\psi(x)]. \tag{11.170}$$

An analogous result applies for right functional derivatives.

(3) Mixed functional derivatives:

$$\begin{aligned}
\left(\frac{\overrightarrow{\delta}}{\delta\psi(x)} F[\psi(x)]\right)\frac{\overleftarrow{\delta}}{\delta\psi(y)} &= \frac{\overrightarrow{\delta}}{\delta\psi(x)}\left(F[\psi(x)]\frac{\overleftarrow{\delta}}{\delta\psi(y)}\right) \\
&= \frac{\overrightarrow{\delta}}{\delta\psi(x)} F[\psi(x)]\frac{\overleftarrow{\delta}}{\delta\psi(y)}.
\end{aligned} \tag{11.171}$$

The proof of these results is left as an exercise.

11.8 Functional Integration for Grassmann Fields

11.8.1 Definition of Functional Integral

If the range over the variable x for the Grassmann field $\psi(x)$ is divided up into n small intervals $\Delta x_i = x_{i+1} - x_i$ (the ith interval), then we may specify the value ψ_i of the function $\psi(x)$ in the ith interval via the average

$$\psi_i = \frac{1}{\Delta x_i} \int\limits_{\Delta x_i} dx\, \psi(x). \tag{11.172}$$

Averaging a Grassmann field over a position interval still results in a linear form involving the Grassmann variables $g_1, \cdots, g_k, \cdots, g_n$. As previously, for simplicity we will choose the same number n of intervals as mode functions.

The functional $F[\psi(x)]$ may be regarded as a function $F(\psi_1, \cdots, \psi_i, \cdots, \psi_n)$ of all of the n different ψ_i, which in the present case are a set of Grassmann variables. As we will see, these Grassmann variables $\psi_1, \cdots, \psi_i, \cdots, \psi_n$ just involve a linear transformation from the $g_1, \cdots, g_k, \cdots g_n$. Introducing a suitable weight function $w(\psi_1, \cdots, \psi_i, \cdots, \psi_n)$, we may then define the functional integral via the multiple Grassmann integral

$$\begin{aligned}\int D\psi\, F[\psi(x)] = \lim_{n\to\infty} \lim_{\epsilon\to 0} \int \cdots \int d\psi_n \cdots d\psi_i \cdots d\psi_1\, w(\psi_1, \cdots, \psi_i, \cdots, \psi_n) \\ \times F(\psi_1, \cdots, \psi_i, \cdots, \psi_n),\end{aligned} \tag{11.173}$$

where $\epsilon > \Delta x_i$. As previously, we use left integration and follow the convention in which the symbol $D\psi$ stands for $d\psi_n \cdots d\psi_i \cdots d\psi_1\, w(\psi_1, \cdots, \psi_i, \cdots, \psi_n)$. A total functional integral of a functional of a Grassmann field gives a c-number.

If the functional $F[\psi(x), \psi^*(x)]$ involves pairs of Grassmann fields related by complex conjugation, then the functional integral will be of the form $\int D^2\psi\, F[\psi(x), \psi^*(x)]$, where $D^2\psi = d^2\psi_n \cdots d^2\psi_i \cdots d^2\psi_1\, w(\psi_1, \cdots, \psi_i, \cdots, \psi_n)$, with $d^2\psi_i = d\psi_i^* d\,\psi_i$. If the functional $F[\psi(x), \psi^+(x)]$ involves pairs of Grassmann fields not related by complex conjugation, then the functional integral will take the form $\int D^2\psi\, F[\psi(x), \psi^+(x)]$, where $D^2\psi = d^2\psi_n \cdots d^2\psi_i \cdots d^2\psi_1\, w(\psi_1, \cdots, \psi_i, \cdots, \psi_n)$, but now with $d^2\psi_i = d\psi_i^+ d\psi_i$. Similarly to differentiation and integration in ordinary Grassmann calculus, functional integration and differentiation are not inverse processes.

11.8.2 Functional Integrals and Phase Space Integrals

For a mode expansion such as in (11.1) the value ϕ_{ki} of the mode function in the ith interval is also defined via the average

$$\phi_{ki} = \frac{1}{\Delta x_i} \int\limits_{\Delta x_i} dx\, \phi_k(x). \tag{11.174}$$

Unlike the Grassmann field, this is just a c-number. It is then easy to see that the Grassmann variables $\psi_1, \cdots, \psi_i, \cdots, \psi_n$ are related to the g-number expansion coefficients $g_1, \cdots, g_k, \cdots, g_n$, via the following linear transformation with c-number coefficients ϕ_{ki}:

$$\psi_i = \sum_k \phi_{ki} g_k \tag{11.175}$$

$$= \sum_k g_k \phi_{ki}. \tag{11.176}$$

This shows that the average values in the ith interval of the function ψ_i and the mode function ϕ_{ki} are related via the expansion coefficients g_k.

Using the expression (11.123) for the expansion coefficients, we then obtain the inverse formula to (11.175),

$$g_k = \sum_i \Delta x_i\, \phi_{ki}^* \psi_i. \tag{11.177}$$

The relationship in (11.175) shows that the functions $F(\psi_1, \cdots, \psi_i, \cdots, \psi_n)$ and $w(\psi_1, \cdots, \psi_i, \cdots, \psi_n)$ of all the interval values ψ_i can also be regarded as functions of the expansion coefficients g_k, which we may write as

$$f(g_1, \cdots, g_k, \cdots, g_n) \equiv F(\psi_1(g_1, \cdots, g_k, \cdots, g_n), \cdots, \psi_i(g_1, \cdots, g_k, \cdots, g_n), \cdots, \psi_n),$$
$$v(g_1, \cdots, g_k, \cdots, g_n) \equiv w(\psi_1(g_1, \cdots, g_k, \cdots, g_n), \cdots, \psi_i(g_1, \cdots, g_k, \cdots, g_n), \cdots, \psi_n). \tag{11.178}$$

Thus the various values $\psi_1, \cdots, \psi_1, \cdots, \psi_i, \cdots, \psi_n, \cdots, \psi_n$ that the function $\psi(x)$ takes on in the n intervals – and which are integrated over in the functional integration process – are all determined by the choice of the expansion coefficients $g_1, \cdots, g_k, \cdots, g_n$. Hence Grassmann integration over all the ψ_i is equivalent to Grassmann integration over all the g_k. This enables us to express the functional integral in (11.173) as a Grassmann phase space integral over the expansion coefficients $g_1, \cdots, g_k, \cdots, g_n$. However, the derivation of the result differs from the c-number case because the transformation of the product of Grassmann differentials $d\psi_n \cdots d\psi_i \cdots d\psi_1$ into the new product of Grassmann differentials $dg_n \cdots dg_i \cdots dg_2\, dg_1$ requires a treatment similar to that explained in Section 4.3, where the transformation between Grassmann integration variables is linear. We cannot just write $d\psi_i = \sum_k \phi_{ki}\, dg_k$, because the differentials are also Grassmann variables. Hence the usual c-number transformation involving the Jacobian does not apply. The required result can be obtained from Section 4.3 (see (4.48), (4.49) and (4.53)) by making the identifications $g_i \rightarrow \psi_i, h_k \rightarrow g_k, A_{ik} \rightarrow \phi_{ki}$, so

$$d\psi_n \cdots d\psi_i \cdots d\psi_1 = (\mathrm{Det}\, A)^{-1}\, dg_n \cdots dg_i \cdots dg_1, \tag{11.179}$$

using (4.57). Now, with $A_{ik} = \phi_{ki}$, we have the following, using the completeness relationship in (2.62):

$$
\begin{aligned}
(AA^{\dagger})_{ij} &= \sum_k \phi_{ki}\phi_{kj}^* \\
&= \sum_k \frac{1}{\Delta x_i}\int_{\Delta x_i} dx\, \phi_k(x)\frac{1}{\Delta x_j}\int_{\Delta x_j} dy\, \phi_k^*(y) \\
&= \frac{1}{\Delta x_i}\int_{\Delta x_i} dx\, \frac{1}{\Delta x_j}\int_{\Delta x_j} dy\, \delta(x-y) \\
&= \delta_{ij}\frac{1}{(\Delta x_i)^2}\int_{\Delta x_i} dx \\
&= \delta_{ij}\,(\Delta x_i)^{-1}.
\end{aligned}
$$

This result is the same as that obtained previously for the Jacobian. Thus

$$
\begin{aligned}
(\mathrm{Det}\, A)^2 &= \prod_i (\Delta x_i)^{-1}, \\
(\mathrm{Det}\, A)^{-1} &= \prod_i (\Delta x_i)^{+1/2}.
\end{aligned}
$$

Hence we have

$$
\begin{aligned}
\int D\psi\, F[\psi(x)] &= \lim_{n\to\infty}\lim_{\epsilon\to 0}\int\cdots\int dg_n\cdots dg_k\cdots dg_1\,(\mathrm{Det}\, A)^{-1} \\
&\quad \times v(g_1,\cdots,g_k,\cdots,g_n)\, f(g_1,\cdots,g_k,\cdots,g_n)
\end{aligned}
\tag{11.180}
$$

and thus

$$
\begin{aligned}
\int D\psi\, F[\psi(x)] &= \lim_{n\to\infty}\lim_{\epsilon\to 0}\int\cdots\int dg_n\cdots dg_k\cdots dg_1\,\prod_i (\Delta x_i)^{1/2} \\
&\quad \times v(g_1,\cdots,g_k,\cdots,g_n)\, f(g_1,\cdots,g_k,\cdots,g_n).
\end{aligned}
\tag{11.181}
$$

This key result expresses the original Grassmann functional integral as a Grassmann phase space integral over the g-number expansion coefficients g_k for the Grassmann field $\psi(x)$ in terms of the mode functions $\phi_k(x)$. Note that this result is different from the previous c-number case, where the factor is $\prod_i (\Delta x_i)^{-1/2}$ instead of $\prod_i (\Delta x_i)^{1/2}$. This is because the Grassmann differentials transform via $(\mathrm{Det}\, A)^{-1}$ instead of $(\mathrm{Det}\, A)^{+1}$.

The general result can be simplified with a special choice of the weight function

$$
w(\psi_1,\cdots,\psi_i,\cdots,\psi_n) = \prod_i (\Delta x_i)^{-1/2}, \tag{11.182}
$$

and we then get a simple expression for the Grassmann functional integral

$$
\int D\psi\, F[\psi(x)] = \lim_{n\to\infty}\lim_{\epsilon\to 0}\int\cdots\int dg_n\cdots dg_k\cdots dg_1\; f(g_1,\cdots,g_k,\cdots g_n). \tag{11.183}
$$

In this form of the Grassmann functional integral, the original Grassmann functional $F[\psi(x)]$ has been replaced by the equivalent function $f(g_1,\cdots,g_k,\cdots,g_n)$ of the g-number expansion coefficients g_k, and the functional integration is now replaced by a Grassmann phase space integration over the expansion coefficients. With the appropriate choice of weight function, the similarity between the cases of bosonic and fermionic fields has been restored.

For two Grassmann fields $\psi(x), \psi^+(x)$, a straightforward extension of the last result gives

$$\int D\psi^+ \, D\psi \, F[\psi(x), \psi^+(x)] = \lim_{n\to\infty} \lim_{\epsilon\to 0} \int \cdots \int dg_1^+ \cdots dg_k^+ \cdots dg_n^+ \, dg_n \cdots dg_k \cdots dg_1$$
$$\times f(g_1, \cdots, g_k, \cdots g_n, g_1^+, \cdots, g_k^+, \cdots g_n^+), \qquad (11.184)$$

where the weight function is now

$$w(\psi_1, \cdots, \psi_i, \cdots, \psi_n, \psi_1^+, \cdots, \psi_i^+, \cdots, \psi_n^+) = \prod_i (\Delta x_i)^{-1} \qquad (11.185)$$

and $\psi^+(x)$ is given via (5.150).

11.8.3 Functional Integration by Parts

A useful integration-by-parts rule can often be established from (11.163). Consider the Grassmann functional $H[\psi(x)] = F[\psi(x)]G[\psi(x)]$. Then

$$F[\psi(x)] \left\{ \left(\frac{\overrightarrow{\delta}}{\delta\psi(x)} \right)_x G[\psi(x)] \right\}$$
$$= \sigma(F) \left(\frac{\overrightarrow{\delta}}{\delta\psi(x)} \right)_x \{F[\psi(x)]G[\psi(x)]\} - \sigma(F) \left\{ \left(\frac{\overrightarrow{\delta}}{\delta\psi(x)} \right)_x F[\psi(x)] \right\} G[\psi(x)].$$

Then

$$\int D\psi \, F[\,] \left\{ \left(\frac{\overrightarrow{\delta}}{\delta\psi(x)} \right)_x G[\,] \right\} = \sigma(F) \int D\psi \left(\frac{\overrightarrow{\delta}}{\delta\psi(x)} \right)_x H[\,]$$
$$-\sigma(F) \int D\psi \left\{ \left(\frac{\overrightarrow{\delta}}{\delta\psi(x)} \right)_x F[\,] \right\} G[\,]. \qquad (11.186)$$

If we now introduce mode expansions and use (11.148) for the functional derivative of $H[\psi(x)]$ and (11.183) for the first of the two functional integrals on the right-hand side of the last equation, then

$$\int D\psi \left(\frac{\overrightarrow{\delta}}{\delta\psi(x)} \right)_x H[\psi(x)]$$
$$= \lim_{n\to\infty} \lim_{\epsilon\to 0} \int dg_n \cdots dg_k \cdots dg_1 \sum_k \phi_k^*(x) \frac{\overrightarrow{\partial}}{\partial g_k} h(g_1, \cdots, g_k, \cdots)$$
$$= \lim_{n\to\infty} \lim_{\epsilon\to 0} \sum_k \phi_k^*(x) \int \cdots \int dg_n \cdots dg_{k+1} \, dg_{k-1} \cdots dg_1$$
$$\times (-1)^{k-1} \left\{ \int dg_k \, \frac{\overrightarrow{\partial}}{\partial g_k} h(g_1, \cdots, g_k, \cdots) \right\},$$

so that the functional integral of this term reduces to the Grassmann integral of a Grassmann derivative. This is zero, since differentiation removes the g_k dependence. Hence the Grassmann functional integral involving the functional derivative of $H[\psi(x)]$ vanishes, and we have the integration-by-parts result

$$\int D\psi\, F[\psi(x)] \left\{ \left(\frac{\overrightarrow{\delta}}{\delta\psi(x)} \right)_x G[\psi(x)] \right\} = -\sigma(F) \int D\psi \left\{ \left(\frac{\overrightarrow{\delta}}{\delta\psi(x)} \right)_x F[\psi(x)] \right\} G[\psi(x)]. \tag{11.187}$$

A similar result involving right functional differentiation can be established.

11.8.4 Differentiating a Functional Integral

Functionals can be defined via functional integration processes, and it is useful to find rules for their functional derivatives. This leads to a rule for differentiating a functional integral.

Suppose we have a functional $G[\chi(x)]$ determined from another functional $F[\psi(x)]$ via a functional integral that involves a left transfer functional $A_{GF}[\chi(x), \psi(x)]$:

$$G[\chi(x)] = \int D\psi\, A_{GF}[\chi(x), \psi(x)]\, F[\psi(x)]. \tag{11.188}$$

Applying the definition of the left Grassmann functional derivatives of $G[\chi(x)]$ and $A_{GF}[\chi(x) + \delta\chi(x), \psi(x)]$, with respect to $\chi(x)$, we have

$$\begin{aligned}
& G[\chi(x) + \delta\chi(x)] \\
&= \int D\psi\, A_{GF}[\chi(x) + \delta\chi(x), \psi(x)]\, F[\psi(x)] \\
&= \int D\psi \left\{ A_{GF}[\chi(x), \psi(x)] + \int dx\, \delta\chi(x) \left\{ \left(\frac{\overrightarrow{\delta}}{\delta\psi(x)} \right)_x A_{GF}[\chi(x), \psi(x)] \right\} \right\} F[\psi(x)] \\
&= \int D\psi\, \{A_{GF}[\chi(x), \psi(x)]\}\, F[\psi(x)] \\
&\quad + \int D\psi \left\{ \int dx\, \delta\chi(x) \left\{ \left(\frac{\overrightarrow{\delta}}{\delta\psi(x)} \right)_x A_{GF}[\chi(x), \psi(x)] \right\} \right\} F[\psi(x)] \\
&= G[\chi(x)] + \int dx\, \delta\chi(x) \int D\psi \left\{ \left(\frac{\overrightarrow{\delta}}{\delta\psi(x)} \right)_x A_{GF}[\chi(x), \psi(x)] \right\} F[\psi(x)],
\end{aligned}$$

since (for reasonably well-behaved quantities) the functional integration over $D\psi$ and the ordinary integration over dx can be carried out in either order, given that both just involve processes that are limits of summations. Hence, from the definition of the functional derivative, we have

$$\left(\frac{\overrightarrow{\delta}}{\delta\chi(x)} \right)_x G[\chi(x)] = \int D\psi \left\{ \left(\frac{\overrightarrow{\delta}}{\delta\psi(x)} \right)_x A_{GF}[\chi(x), \psi(x)] \right\} F[\psi(x)], \tag{11.189}$$

which is the required rule for left differentiating a functional defined via a functional of another function. Clearly, the rule is to just differentiate the transfer functional under the functional integration sign, a rule similar to that applying in ordinary calculus.

A similar rule can be obtained for a functional $H[\chi(x)]$ determined from another functional $F[\psi(x)]$ via a functional integral that involves a right transfer functional $A_{GF}[\chi(x),\psi(x)]$ in the form

$$H[\chi(x)] = \int D\psi\, F[\psi(x)]\, A_{GF}[\chi(x),\psi(x)]. \tag{11.190}$$

We find that

$$H[\chi(x)]\left(\frac{\overleftarrow{\delta}}{\delta\chi(x)}\right)_x = \int D\psi\, F[\psi(x)]\left\{A_{GF}[\chi(x),\psi(x)]\left(\frac{\overleftarrow{\delta}}{\delta\psi(x)}\right)_x\right\}, \tag{11.191}$$

which is the required rule for right differentiating a functional defined via a functional of another function. The proof is left as an exercise. Clearly, the rule is to just differentiate the transfer functional under the functional integration sign, a rule similar to that applying in ordinary calculus (except that right and left differentiation are different).

As a particular case, consider the Fourier-like Grassmann transfer functional

$$A_{GF}[\chi(x),\psi(x)] = \exp\left\{i\int dx\,\chi(x)\,\psi(x)\right\}. \tag{11.192}$$

This Grassmann transfer functional is equivalent to a even Grassmann function of the expansion coefficients g_k for $\psi(x)$ and h_k for $\chi(x)$. In this case

$$\begin{aligned}
A_{GF}[\chi(x)+\delta\chi(x),\psi(x)] &= \exp\left\{i\int dx\,(\chi(x)+\delta\chi(x))\,\psi(x)\right\}\\
&= \exp\left\{i\int dx\,\chi(x)\,\psi(x)\right\}\exp\left\{i\int dx\,\delta\chi(x)\,\psi(x)\right\}\\
&\approx \exp\left\{i\int dx\,\chi(x)\,\psi(x)\right\}\left(1+i\int dx\,\delta\chi(x)\,\psi(x)\right)\\
&= A_{GF}[\chi(x),\psi(x)] + A_{GF}[\chi(x),\psi(x)]\,i\int dx\,\delta\chi(x)\,\psi(x)\\
&= A_{GF}[\chi(x),\psi(x)] + \int dx\,\delta\chi(x)\,iA_{GF}[\chi(x),\psi(x)]\,\psi(x),
\end{aligned}$$

where we have used the Baker–Hausdorff theorem (3.38) with $A = \int dx\,\chi(x)\psi(x)$ and $B = \int dx\,\delta\chi(x)\psi(x)$ together with the commutator result based on Grassmann fields anticommuting,

$$
\begin{aligned}
[A, B] &= \int dx\, \chi(x)\psi(x) \int dy\, \delta\chi(y)\, \psi(y) - \int dy\, \delta\chi(y)\, \psi(y) \int dx\, \chi(x)\psi(x) \\
&= \int dx \int dy\, \chi(x)\psi(x)\, \delta\chi(y)\, \psi(y) - (-1)^{2+2} \int dx \int dy\, \chi(x)\psi(x)\, \delta\chi(y)\, \psi(y) \\
&= 0,
\end{aligned}
$$

to establish the third line of the derivation. The last line follows from $A_{GF}[\chi(x), \psi(x)]$ being equivalent to an even Grassmann function and therefore commuting with the Grassmann field $\delta\chi(x)$. Hence, for the left functional derivative,

$$\left(\frac{\overrightarrow{\delta}}{\delta\psi(x)} \right)_x A_{GF}[\chi(x), \psi(x)] = A_{GF}[\chi(x), \psi(x)] \times i\psi(x) \tag{11.193}$$

and

$$G[\chi(x)] = \int D\psi\, A_{GF}[\chi(x), \psi(x)]\, F[\psi(x)], \tag{11.194}$$

$$\left(\frac{\overrightarrow{\delta}}{\delta\chi(x)} \right)_x G[\chi(x)] = \int D\psi\, \{A_{GF}[\chi(x), \psi(x)] \times (i\psi(x))\}\, F[\psi(x)]. \tag{11.195}$$

A similar result follows for the right-transfer-functional case. We have

$$H[\chi(x)] = \int D\psi\, F[\psi(x)]\, A_{GF}[\chi(x), \psi(x)], \tag{11.196}$$

$$H[\chi(x)] \left(\frac{\overleftarrow{\delta}}{\delta\chi(x)} \right)_x = \int D\psi\, F[\psi(x)]\, \{A_{GF}[\chi(x), \psi(x)] \times (-i\psi(x))\}\,. \tag{11.197}$$

The proof is left as an exercise.

Exercises

(11.1) Derive the Grassmann right-product rule (11.164).

(11.2) Derive the rule (11.191) for the right differentiation of a Grassmann functional integral.

(11.3) Derive the rule (11.197) for the right differentiation of a Grassmann Fourier functional integral.

(11.4) Derive the rules (11.169), (11.170) and (11.171) for Grassmann functional derivatives.

12

Distribution Functionals in Quantum Atom Optics

In this chapter, we relate normally ordered quantum correlation functions of field operators to phase space functional integrals in which the density operator is represented by a distribution functional and the field operators are represented by pairs of c-number fields (for bosons) and pairs of Grassmann fields (for fermions). For these cases the distribution functional is, respectively, a c-number functional or a Grassmann functional. Because we are dealing with normally ordered correlation functions, the distribution functionals are of the $P+$ type. Combined boson–fermion cases are also treated. For simplicity, the treatment is confined to the situation where the numbers of each type of particle remain fixed, and we will deal mainly with the case where only a single pair of field creation and annihilation operators is involved. The treatment is based on first introducing related characteristic functionals which are uniquely determined from and equivalent to the density operator, and effectively are an encoding of the correlation functions. The existence of the distribution functionals (which are not required to be unique) is based on the results in Chapter 7, in which the existence of the equivalent distribution functions has already been demonstrated. Similarly, the results for correlation functions in terms of phase space functional integrals are established from the phase space integral forms found in Chapter 7. As we have seen, unnormalised distribution functions are useful not only because such functions determine Fock state populations and coherences via phase space integrals but also because their Fokker–Planck equations are based on simpler correspondence rules, leading to Ito stochastic equations that involve only linear terms. This is particularly important in the fermion case, as it leads to Ito equations for the stochastic Grassmann variables that are computable. For these reasons, the corresponding unnormalised distribution functionals will also be treated in this chapter.

In the following chapter (Chapter 13), the replacement of the Liouville–von Neumann or master equation for the density operator by a functional Fokker–Planck equation for the distribution functional is found via correspondence rules. These are derived simply from the correspondence rules already established in Chapter 8. Finally, the phase space functional integrals that determine the quantum correlation functions via the distribution functional are shown in Chapter 14 to be equivalent to stochastic averages of stochastic fields that represent the bosonic and fermionic field operators, and which satisfy Langevin field equations of the Ito stochastic type involving terms that are obtained from the functional Fokker–Planck equation. The derivation of the

Phase Space Methods for Degenerate Quantum Gases. First Edition. Bryan J. Dalton, John Jeffers, and Stephen M. Barnett.
Published in 2015 by Oxford University Press.

Langevin stochastic field equations is developed from the Ito equations already obtained in Chapter 9. Note that although it is convenient to derive the general results starting from the previous results for mode-based characteristic and distribution functions, Fokker–Planck and Langevin equations, etc., the general results can also be obtained using purely functional calculus methods, basically by repeating the same derivations used for the mode-based results. Furthermore, the determination of the functional Fokker–Planck equations in the specific cases treated in Chapter 15 will be carried out directly from the correspondence rules given in terms of functional derivatives. Once these are found, the Ito stochastic field equations can be directly obtained.

12.1 Quantum Correlation Functions

The aim is to determine normally ordered quantum correlation functions, which now involve the field creation and annihilation operators $\hat{\Psi}^{\dagger}(\mathbf{r}), \hat{\Psi}(\mathbf{r})$ and are of the form (1.7)

$$\begin{aligned} &G^{(p,q)}(\mathbf{r}_1,\mathbf{r}_2,\cdots,\mathbf{r}_p;\mathbf{s}_q,\cdots,\mathbf{s}_2,\mathbf{s}_1) \\ &= \langle \hat{\Psi}(\mathbf{r}_1)^{\dagger}\cdots\hat{\Psi}(\mathbf{r}_p)^{\dagger}\hat{\Psi}(\mathbf{s}_q)\cdots\hat{\Psi}(\mathbf{s}_1)\rangle \\ &= \mathrm{Tr}(\hat{\rho}(t)\hat{\Psi}(\mathbf{r}_1)^{\dagger}\cdots\hat{\Psi}(\mathbf{r}_p)^{\dagger}\hat{\Psi}(\mathbf{s}_q)\cdots\hat{\Psi}(\mathbf{s}_1)) \qquad (12.1) \\ &= \mathrm{Tr}(\hat{\Psi}(\mathbf{s}_q)\cdots\hat{\Psi}(\mathbf{s}_1)\hat{\rho}(t)\hat{\Psi}(\mathbf{r}_1)^{\dagger}\cdots\hat{\Psi}(\mathbf{r}_p)^{\dagger}), \qquad (12.2) \end{aligned}$$

where for an N-particle system we require $p, q \leq N$ to give a non-zero result. For the fermion case, we also require $p - q = 0, \pm 2, \pm 4, \cdots$. As we will see, this result can be written in terms of phase space functional integrals for both bosons and fermions.

12.2 Characteristic Functionals

In order to determine the quantum correlation functions using the phase space approach involving distribution functionals, it is convenient to first introduce characteristic functionals. The bosonic characteristic functional $\chi_{\mathrm{b}}[\xi(x), \xi^{+}(x)]$ depends analytically on pairs of c-number fields $\xi(x), \xi^{+}(x)$, whilst the fermionic characteristic functional $\chi_{\mathrm{f}}[h(x), h^{+}(x)]$ involves two Grassmann fields $h(x), h^{+}(x)$. For convenience, the position variable is written as x. These functionals are essentially a way of encoding all the quantum correlation functions in one overall functional. They are uniquely determined by and are equivalent to the density operator. For the normally ordered case, the characteristic functional will be of the $P+$ type. As previously, there are several varieties of characteristic functional depending on whether the correlation functions involve normally ordered, antinormally ordered or symmetrically ordered products of field operators (assuming the density operator is placed on the left or right of these operators in the trace) and/or whether more than one pair of bosonic or fermionic field creation and annihilation operators is involved – as is the case when differing internal spin states are present. Here we will focus mainly on normally ordered characteristic functionals involving double phase spaces, but results for the double-space Wigner case will also be treated. We initially focus on simple bosonic or fermionic systems where the total number of bosons or fermions is conserved. The characteristic and distribution functionals are also equivalent to characteristic and distribution functions involving the individual modes, an equivalence we will exploit to avoid re-justifying existence theorems from first principles.

12.2.1 Boson Case

For the bosonic case, we define the $P+$ characteristic functional $\chi_{\rm b}[\xi(x), \xi^+(x)]$ via

$$\chi_{\rm b}[\xi(x), \xi^+(x)] = \mathrm{Tr}(\hat{\Omega}_{\rm b}^+[\xi^+(x)]\,\hat{\rho}_{\rm b}\,\hat{\Omega}_{\rm b}^-[\xi(x)]), \tag{12.3}$$

$$\hat{\Omega}_{\rm b}^+[\xi^+(x)] = \exp i\int dx\,\hat{\Psi}(x)\xi^+(x),$$

$$\hat{\Omega}_{\rm b}^-[\xi(x)] = \exp i\int dx\,\xi(x)\hat{\Psi}^\dagger(x), \tag{12.4}$$

$$\hat{\Omega}_{\rm b}^-[\xi(x)] = (\hat{\Omega}_b^+[-\xi^*(x)])^\dagger, \tag{12.5}$$

where for the field operators $\hat{\Psi}(x)$ and $\hat{\Psi}^\dagger(x)$ we associate a pair of c-number fields $\xi^+(x)$ and $\xi(x)$. Note that ξ and ξ^+ are both complex, and are not related to each other. Note also that $\chi_{\rm b}[\xi(x), \xi^+(x)]$ depends only on $\xi(x), \xi^+(x)$ and not on their complex conjugates $\xi^*(x), \xi^{+*}(x)$. The characteristic functional is an analytic functional of ξ, ξ^+. Owing to the cyclic properties of the trace for bosonic operators, the characteristic functional is also equal to $\mathrm{Tr}(\hat{\rho}_{\rm b}\,\hat{\Omega}_{\rm b}^-[\xi(x)]\hat{\Omega}_{\rm b}^+[\xi^+(x)])$ and $\mathrm{Tr}(\hat{\Omega}_{\rm b}^-[\xi(x)]\hat{\Omega}_{\rm b}^+[\xi^+(x)]\,\hat{\rho}_{\rm b})$.

Using mode expansions

$$\hat{\Psi}(x) = \sum_i \hat{a}_i\phi_i(x), \qquad \hat{\Psi}^\dagger(x) = \sum_i \hat{a}_i^\dagger\phi_i^*(x), \tag{12.6}$$

$$\xi(x) = \sum_i \xi_i\phi_i(x), \qquad \xi^+(x) = \sum_i \xi_i^+\phi_i^*(x), \tag{12.7}$$

it is easy to see using the orthogonality of the modes that

$$\int dx\,\hat{\Psi}(x)\xi^+(x) = \sum_i \hat{a}_i\xi_i^+, \qquad \int dx\,\xi(x)\hat{\Psi}^\dagger(x) = \sum_i \xi_i\hat{a}_i^\dagger, \tag{12.8}$$

so that from (7.5),

$$\chi_{\rm b}[\xi(x), \xi^+(x)] \equiv \chi_{\rm b}(\xi, \xi^+). \tag{12.9}$$

Thus the characteristic functional of the fields $\xi(x), \xi^+(x)$ is entirely equivalent to the original bosonic characteristic function of the expansion coefficients ξ_i, ξ_i^+.

The Wigner characteristic functional is defined by

$$\chi_{\rm b}^{\rm W}[\xi(x), \xi^+(x)] = \mathrm{Tr}(\hat{\rho}_{\rm b}\,\hat{\Omega}_{\rm b}^{\rm W}[\xi(x), \xi^+(x)]), \tag{12.10}$$

$$\hat{\Omega}_{\rm b}^{\rm W}[\xi(x)] = \exp i\int dx\,(\xi(x)\hat{\Psi}^\dagger(x) + \hat{\Psi}(x)\xi^+(x)) \tag{12.11}$$

and is the same as

$$\chi_{\rm b}^{\rm W}[\xi(x),\xi^+(x)] = {\rm Tr}\Big(\hat{\rho}_{\rm b}\,\exp i\sum_i(\xi_i\hat{a}_i^\dagger + \hat{a}_i\xi_i^+)\Big)$$
$$\equiv \chi_{\rm b}^{\rm W}(\xi,\xi^+), \tag{12.12}$$

which is the original bosonic Wigner characteristic function of the expansion coefficients ξ_i, ξ_i^+.

12.2.2 Fermion Case

For fermionic systems, we define the characteristic functional $\chi_{\rm f}[h(x),h^+(x)]$ via

$$\chi_{\rm f}[h(x),h^+(x)] = {\rm Tr}(\hat{\Omega}_{\rm f}^+[h^+(x)]\,\hat{\rho}_{\rm f}\hat{\Omega}_{\rm f}^-[h(x)]), \tag{12.13}$$

$$\hat{\Omega}_{\rm f}^+[h^+(x)] = \exp i\int dx\,\hat{\Psi}(x)h^+(x),$$
$$\hat{\Omega}_{\rm f}^-[h(x)] = \exp i\int dx\,h(x)\hat{\Psi}^\dagger(x), \tag{12.14}$$

$$\hat{\Omega}_{\rm f}^+[h^+(x)] = 1 + i\int dx\,\hat{\Psi}(x)h^+(x),$$
$$\hat{\Omega}_{\rm f}^-[h(x)] = 1 + i\int dx\,h(x)\hat{\Psi}^\dagger(x), \tag{12.15}$$

$$\hat{\Omega}_{\rm f}^-[h(x)] = \left(\hat{\Omega}_{\rm f}^+[-h^*(x)]\right)^\dagger, \tag{12.16}$$

where for the field operators $\hat{\Psi}(x)$ and $\hat{\Psi}^\dagger(x)$ we associate a pair of Grassmann fields $h^+(x)$ and $h(x)$. Note that h and h^+ are both complex, and are not related to each other. Note also that $\chi_{\rm f}[h(x),h^+(x)]$ depends only on $h(x), h^+(x)$ and not on their complex conjugates $h^*(x), h^{+*}(x)$. The simplification of the exponentials follows from the second and higher powers of the field operators being zero. Owing to the lack of cyclic properties of the trace for fermionic Grassmann operators, the characteristic functional is not equal to ${\rm Tr}(\hat{\rho}_{\rm f}\,\hat{\Omega}_{\rm f}^-[h(x)]\hat{\Omega}_{\rm f}^+[h^+(x)])$ or ${\rm Tr}(\hat{\Omega}_{\rm f}^-[h(x)]\hat{\Omega}_{\rm f}^+[h^+(x)]\hat{\rho}_{\rm f})$.

Using mode expansions

$$\hat{\Psi}(x) = \sum_i \hat{c}_i\phi_i(x), \qquad \hat{\Psi}^\dagger(x) = \sum_i \hat{c}_i^\dagger\phi_i^*(x), \tag{12.17}$$

$$h(x) = \sum_i h_i\phi_i(x), \qquad h^+(x) = \sum_i h_i^+\phi_i^*(x), \tag{12.18}$$

it is easy to see using the orthogonality of the modes that

$$\int dx\,\hat{\Psi}(x)h^+(x) = \sum_i \hat{c}_i h_i^+, \qquad \int dx\,h(x)\hat{\Psi}^\dagger(x) = \sum_i h_i\hat{c}_i^\dagger, \tag{12.19}$$

so that from (7.11),

$$\chi_{\rm f}[h(x),h^+(x)] \equiv \chi_{\rm f}(h,h^+). \tag{12.20}$$

Thus the characteristic functional of the fields $h(x), h^+(x)$ is entirely equivalent to the original fermionic characteristic function of the g-number expansion coefficients h_i, h_i^+.

12.3 Distribution Functionals

In this section, we relate the characteristic functional to a distribution functional via a phase space functional integral involving Fourier-like factors. For bosons, the phase space integral involves c-number functional integration over the complex fields $\psi(x), \psi^+(x)$ in the bosonic distribution functional $P_{\mathrm{b}}[\psi(x), \psi^+(x), \psi^*(x), \psi^{+*}(x)]$, whilst in the fermionic case the phase space integral involves Grassmann integration over all the Grassmann fields $\psi(x), \psi^+(x)$ in the fermionic functional $P_{\mathrm{f}}[\psi(x), \psi^+(x)]$. The expressions would be the same if the normally ordered positive-P-type distribution functional were replaced by a double-phase-space version of the Wigner or Q distribution functional. In both the boson and the fermion cases, the distribution functional is not required to be unique – the key requirement is that at least one distribution functional exists such that its functional integral with the Fourier-like factors gives the correct characteristic functional. In the bosonic case, the functional is also not analytic; it depends on $\psi(x), \psi^+(x)$ and the complex conjugate fields $\psi^*(x), \psi^{+*}(x)$. The existence of the distribution functional follows from the previously established existence of the equivalent distribution functions discussed in Chapter 7. In the present work the distribution functional is of the positive P type, and the corresponding functional Fokker–Planck equation involves left and right Grassmann functional derivatives with respect to $\psi(x), \psi^+(x)$ for the fermion case and c-number functional derivatives with respect to real and imaginary components $\psi_x(x), \psi_x^+(x), \psi_y(x), \psi_y^+(x)$ (or, equivalently, the complex pair $\psi(x), \psi^+(x)$) for the boson case.

12.3.1 Boson Case

For bosons, we know from (7.20) that there is a distribution function $P_{\mathrm{b}}(\alpha, \alpha^+, \alpha^*, \alpha^{+*})$ which determines the characteristic function via the c-number phase space integral

$$\chi_{\mathrm{b}}(\xi, \xi^+) = \int \prod_i d^2\alpha_i^+ d^2\alpha_i \exp i \sum_i \{\alpha_i\, \xi_i^+\}\, P_{\mathrm{b}}(\alpha, \alpha^+, \alpha^*, \alpha^{+*})\, \exp i \sum_i \{\xi_i\, \alpha_i^+\}, \tag{12.21}$$

where with each mode i we associate another pair of c-numbers α_i, α_i^+. However, using the mode expansions

$$\psi(x) = \sum_i \alpha_i \phi_i(x), \qquad \psi^+(x) = \sum_i \alpha_i^+ \phi_i^*(x) \tag{12.22}$$

and the results

$$\int dx\, \psi(x)\xi^+(x) = \sum_i \alpha_i \xi_i^+, \qquad \int dx\, \xi(x)\psi^+(x) = \sum_i \xi_i \alpha_i^+, \tag{12.23}$$

we can use the procedure in Section 11.5 (see (11.106)) to convert the right side into a phase space functional integral in which the distribution function $P_{\rm b}(\alpha, \alpha^+, \alpha^*, \alpha^{+*})$ is replaced by the distribution functional $P_{\rm b}[\psi(x), \psi^+(x), \psi^*(x), \psi^{+*}(x)]$:

$$P_{\rm b}(\alpha, \alpha^+, \alpha^*, \alpha^{+*}) \equiv P_{\rm b}[\psi(x), \psi^+(x), \psi^*(x), \psi^{+*}(x)]. \tag{12.24}$$

From (12.9), the left side is equal to the characteristic functional, so we now have

$$\chi_{\rm b}[\xi(x), \xi^+(x)] = \int D^2\psi\; D^2\psi^+ \;\exp i \int dx\, \psi(x)\xi^+(x)\, P_{\rm b}[\psi(x), \psi^+(x), \psi^*(x), \psi^{+*}(x)] \times \exp i \int dx\, \xi(x)\psi^+(x), \tag{12.25}$$

where

$$D^2\,\psi\; D^2\psi^+ = \prod_i d^2\psi_i \prod_i d^2\psi_i^+\; w(\psi_1, \cdots, \psi_n, \psi_1^+, \cdots, \psi_n^+, \psi_1^*, \cdots, \psi_n^*, \psi_1^{+*}, \cdots, \psi_n^{+*}),$$

and, with $\psi_i = \psi_{ix} + i\psi_{iy}$ and $\psi_i^+ = \psi_{ix}^+ + i\psi_{iy}^+$, the quantities $d^2\psi_i$ mean $d\psi_{ix}\, d\psi_{iy}$ and $d\psi_{ix}^+\, d\psi_{iy}^+$. The weight function is chosen as

$$w(\psi_1, \cdots, \psi_n, \psi_1^+, \cdots, \psi_n^+, \psi_1^*, \cdots, \psi_n^*, \psi_1^{+*}, \cdots, \psi_n^{+*}) = \prod_{i=1}^{n} (\Delta x_i)^2, \tag{12.26}$$

which is independent of the functions. The power law $(\Delta x_i)^2$ is consistent with there being four real functions involved, instead of the single function as previously. This then establishes the existence of the bosonic distribution functional.

A similar procedure relates the Wigner distribution functional to its characteristic functional:

$$\chi_{\rm b}^{\rm W}[\xi(x), \xi^+(x)] = \int D^2\psi\, D^2\psi^+ \exp i \int dx\, \psi(x)\xi^+(x) W_{\rm b}[\psi(x), \psi^+(x), \psi^*(x), \psi^{+*}(x)] \times \exp i \int dx\, \xi(x)\psi^+(x). \tag{12.27}$$

12.3.2 Fermion Case

For fermions, we know from (7.23) that there is a Grassmann distribution function $P_{\rm f}(g, g^+)$ which determines the characteristic function via the following Grassmann phase space integral:

$$\chi_{\rm f}(h, h^+) = \int \prod_i dg_i^+ \prod_i dg_i \exp i \sum_i \{g_i\, h_i^+\}\, P_{\rm f}(g, g^+)\, \exp i \sum_i \{h_i\, g_i^+\}, \tag{12.28}$$

where with each mode i we associate another pair of g-numbers g_i, g_i^+. However, using the mode expansions

$$\psi(x) = \sum_i g_i \phi_i(x), \qquad \psi^+(x) = \sum_i g_i^+ \phi_i^*(x) \tag{12.29}$$

and the results

$$\int dx\, \psi(x)h^{+}(x) = \sum_{i} g_i h_i^{+}, \qquad \int dx\, h(x)\psi^{+}(x) = \sum_{i} h_i g_i^{+}, \tag{12.30}$$

we can use the procedure set out in Section 11.8 (see (11.184)) to convert the right side into a phase space functional integral in which the distribution function $P_{\mathrm{g}}(g, g^{+})$ is replaced by the equivalent distribution functional $P_{\mathrm{f}}[\psi(x), \psi^{+}(x)]$:

$$P_{\mathrm{f}}(g, g^{+}) \equiv P_{\mathrm{f}}[\psi(x), \psi^{+}(x)]. \tag{12.31}$$

From (12.20), the left side is equal to the characteristic functional, so we now have

$$\begin{aligned}&\chi_{\mathrm{f}}[h(x), h^{+}(x)]\\ &= \int\int D\psi^{+}\, D\psi \exp i \int dx\, \psi(x)h^{+}(x)\, P_{\mathrm{f}}[\psi(x), \psi^{+}(x)] \exp i \int dx\, h(x)\psi^{+}(x),\end{aligned} \tag{12.32}$$

where $D\psi^{+}\, D\psi = \prod_i d\psi_i^{+} \prod_i d\psi_i\, w(\psi_1, \cdots, \psi_n, \psi_1^{+}, \cdots, \psi_n^{+})$. The weight function is chosen as

$$w(\psi_1, \cdots, \psi_i, \cdots, \psi_n, \psi_1^{+}, \cdots, \psi_i^{+}, \cdots, \psi_n^{+}) = \prod_i (\Delta x_i)^{-1}. \tag{12.33}$$

The power law $(\Delta x_i)^{-1}$ for two g-fields is different from the bosonic case, where there were four real fields. This result (12.32) then establishes the existence of the fermionic distribution functional.

12.3.3 Quantum Correlation Functions

For the bosonic case, if we substitute the mode expansions (12.6) for the field operators then the quantum correlation functions (12.2) are

$$\begin{aligned}&G^{(p,q)}(\mathbf{r}_1, \mathbf{r}_2, \cdots, \mathbf{r}_p; \mathbf{s}_q, \cdots, \mathbf{s}_2, \mathbf{s}_1)\\
&= \sum_{m_q \cdots m_1} \sum_{l_1 \cdots l_p} \phi_{m_q}(\mathbf{s}_q) \cdots \phi_{m_1}(\mathbf{s}_1)\phi_{l_1}^{*}(\mathbf{r}_1) \cdots \phi_{l_p}^{*}(\mathbf{r}_p)\, \mathrm{Tr}(\hat{a}_{m_q} \cdots \hat{a}_{m_1}\hat{\rho}(t)\hat{a}_{l_1}^{\dagger} \cdots \hat{a}_{l_p}^{\dagger})\\
&= \sum_{m_q \cdots m_1} \sum_{l_1 \cdots l_p} \phi_{m_q}(\mathbf{s}_q) \cdots \phi_{m_1}(\mathbf{s}_1)\phi_{l_1}^{*}(\mathbf{r}_1) \cdots \phi_{l_p}^{*}(\mathbf{r}_p)\\
&\quad \times \int \prod_i d^2\alpha_i^{+} d^2\alpha_i\, (\alpha_{m_q} \cdots \alpha_{m_2}\alpha_{m_1})\, P_{\mathrm{b}}(\alpha, \alpha^{+}, \alpha^{*}, \alpha^{+*})\, (\alpha_{l_1}^{+}\, \alpha_{l_2}^{+} \cdots \alpha_{l_p}^{+})\\
&= \int \prod_i d^2\alpha_i^{+} d^2\alpha_i\, (\psi(\mathbf{s}_q) \cdots \psi(\mathbf{s}_1))\, P_{\mathrm{b}}(\alpha, \alpha^{+}, \alpha^{*}, \alpha^{+*})\, (\psi^{+}(\mathbf{r}_1) \cdots \psi^{+}(\mathbf{r}_p))\\
&= \int\int D^2\psi\, D^2\psi^{+}\, \psi(\mathbf{s}_q) \cdots \psi(\mathbf{s}_1)\, P_{\mathrm{b}}[\psi(x), \psi^{+}(x), \psi^{*}(x), \psi^{+*}(x)]\, \psi^{+}(\mathbf{r}_1) \cdots \psi^{+}(\mathbf{r}_p)),\end{aligned} \tag{12.34}$$

using (7.74) and converting the phase space integral to a functional integral via (11.106) after using (12.22) to introduce the bosonic field functions.

For the fermionic case, if we substitute the mode expansions (12.17) for the field operators then the quantum correlation functions (12.2) are

$$
\begin{aligned}
&G^{(p,q)}(\mathbf{r}_1, \mathbf{r}_2, \cdots, \mathbf{r}_p; \mathbf{s}_q, \cdots, \mathbf{s}_2, \mathbf{s}_1) \\
&\quad = \sum_{m_q \cdots m_1} \sum_{l_1 \cdots l_p} \phi_{m_q}(\mathbf{s}_q) \cdots \phi_{m_1}(\mathbf{s}_1) \phi^*_{l_1}(\mathbf{r}_1) \cdots \phi^*_{l_p}(\mathbf{r}_p) \operatorname{Tr}(\hat{c}_{m_q} \cdots \hat{c}_{m_1} \hat{\rho}(t) \hat{c}^\dagger_{l_1} \cdots \hat{c}^\dagger_{l_p}) \\
&\quad = \sum_{m_q \cdots m_1} \sum_{l_1 \cdots l_p} \phi_{m_q}(\mathbf{s}_q) \cdots \phi_{m_1}(\mathbf{s}_1) \phi^*_{l_1}(\mathbf{r}_1) \cdots \phi^*_{l_p}(\mathbf{r}_p) \\
&\qquad \times \int \prod_i dg_i^+ \prod_i dg_i \, (g_{m_q} \cdots g_{m_2} g_{m_1}) P_f(g, g^+) (g^+_{l_1} g^+_{l_2} \cdots g^+_{l_p}) \\
&\quad = \int \prod_i dg_i^+ \prod_i dg_i \, (\psi(\mathbf{s}_q) \cdots \psi(\mathbf{s}_1)) \, P_f(g, g^+) (\psi^+(\mathbf{r}_1) \cdots \psi^+(\mathbf{r}_p)) \\
&\quad = \int \int D\psi^+ D\psi \, \psi(\mathbf{s}_q) \cdots \psi(\mathbf{s}_1) P_f[\psi(x), \psi^+(x)] \psi^+(\mathbf{r}_1) \cdots \psi^+(\mathbf{r}_p))
\end{aligned}
\tag{12.35}
$$

using (7.82) and converting the phase space integral to a functional integral via (11.106) after introducing fermion field functions via (12.29). Note that the results are zero unless $p - q = 0, \pm 2, \pm 4 \cdots$. Thus we have expressed a quantum correlation function involving bosonic or fermionic field creation and annihilation operators as a phase space functional integral in which the density operator is replaced by the distribution functional and the field operators are replaced by c-number or g-number fields.

12.4 Unnormalised Distribution Functionals – Fermions

As the importance of the unnormalised distribution functionals is much greater for fermions than for bosons, we will focus on fermions. The treatment for bosons is similar.

12.4.1 Distribution Functional

The unnormalised distribution function $B_{\mathrm{f}}(g, g^+)$ is related to the normalised function $P_{\mathrm{f}}(g, g^+)$ via (7.100). These distribution functions are equivalent to functionals $B_{\mathrm{f}}[\psi(x), \psi^+(x)]$ and $P_{\mathrm{f}}[\psi(x), \psi^+(x)]$ of the Grassmann fields defined in (12.29). Noting that $\exp(-\int dx\, \psi(x)\psi^+(x)) = \exp(-\sum_i g_i g_i^+)$, it follows that the relationship between the $B_{\mathrm{f}}[\psi(x), \psi^+(x)]$ and $P_{\mathrm{f}}[\psi(x), \psi^+(x)]$ is given by

$$
B_{\mathrm{f}}[\psi(x), \psi^+(x)] = P_{\mathrm{f}}[\psi(x), \psi^+(x)] \exp\left(-\int dx\, \psi(x)\psi^+(x)\right),
\tag{12.36}
$$

where the two even Grassmann functionals can be interchanged.

12.4.2 Populations and Coherences

Rather than focus on quantum correlation functions, we will concentrate on populations and coherences associated with fermion position eigenstates. We consider fermion position eigenstates as in (2.67):

$$|\Phi\{\mathbf{r}\}\rangle = |\mathbf{r}_1 \cdots \mathbf{r}_p\rangle = \hat{\Psi}_{\mathrm{f}}^{\dagger}(\mathbf{r}_1) \cdots \hat{\Psi}_{\mathrm{f}}^{\dagger}(\mathbf{r}_p)|0\rangle,$$
$$|\Phi\{\mathbf{s}\}\rangle = |\mathbf{s}_1 \cdots \mathbf{s}_p\rangle = \hat{\Psi}_{\mathrm{f}}^{\dagger}(\mathbf{s}_1) \cdots \hat{\Psi}_{\mathrm{f}}^{\dagger}(\mathbf{s}_p)|0\rangle. \tag{12.37}$$

The population for the state $|\Phi\{\mathbf{r}\}\rangle$ and the coherence between the state $|\Phi\{\mathbf{r}\}\rangle$ and the state $|\Phi\{\mathbf{s}\}\rangle$ are given by

$$P(\Phi\{\mathbf{r}\}) = \mathrm{Tr}(\hat{\Pi}(\{\mathbf{r}\})\hat{\rho}), \tag{12.38}$$
$$C(\Phi\{\mathbf{s}\};\Phi\{\mathbf{r}\}) = \mathrm{Tr}(\hat{\Xi}(\{\mathbf{s}\};\{\mathbf{r}\})\hat{\rho}), \tag{12.39}$$

where the population and transition operators are

$$\hat{\Pi}(\{\mathbf{r}\}) = \hat{\Psi}_{\mathrm{f}}^{\dagger}(\mathbf{r}_1) \cdots \hat{\Psi}_{\mathrm{f}}^{\dagger}(\mathbf{r}_p)\,|0\rangle\,\langle 0|\,\hat{\Psi}_{\mathrm{f}}(\mathbf{r}_p) \cdots \hat{\Psi}_{\mathrm{f}}(\mathbf{r}_1), \tag{12.40}$$
$$\hat{\Xi}(\{\mathbf{s}\};\{\mathbf{r}\}) = \hat{\Psi}_{\mathrm{f}}^{\dagger}(\mathbf{s}_1) \cdots \hat{\Psi}_{\mathrm{f}}^{\dagger}(\mathbf{s}_p)\,|0\rangle\,\langle 0|\,\hat{\Psi}_{\mathrm{f}}(\mathbf{r}_p) \cdots \hat{\Psi}_{\mathrm{f}}(\mathbf{r}_1). \tag{12.41}$$

Substituting for the field operators from (12.17) and using the results (7.114) and (7.115), we see that for the fermion position probability,

$$\begin{aligned} P(\Phi\{\mathbf{r}\}) &= \sum_{l_1,\cdots l_p} \sum_{m_1,\cdots m_p} \phi_{l_1}^*(\mathbf{r}_1) \cdots \phi_{l_p}^*(\mathbf{r}_p)\, \phi_{m_p}(\mathbf{r}_p) \cdots \phi_{m_1}(\mathbf{r}_1) \\ &\quad \times \int d\mathbf{g}^+ d\mathbf{g}\; g_{m_p} \cdots g_{m_1}\, B^{\mathrm{f}}(\mathbf{g},\mathbf{g}^+)\, g_{l_1}^+ \cdots g_{l_p}^+ \\ &= \int D\psi^+ D\psi\, \psi(\mathbf{r}_p) \cdots \psi(\mathbf{r}_1)\, B_{\mathrm{f}}[\psi(x),\psi^+(x)]\, \psi^+(\mathbf{r}_1) \cdots \psi^+(\mathbf{r}_p), \end{aligned} \tag{12.42}$$

where the phase space integral has been converted into a functional integral involving the fermion unnormalised distribution functional. This expression is useful for discussing simultaneous position measurements – note the probability density factors such as $\psi^+(\mathbf{r}_1)\psi(\mathbf{r}_1)$.

A similar treatment shows that the fermion position coherence is given by

$$C(\Phi\{\mathbf{s}\};\Phi\{\mathbf{r}\}) = \int D\psi^+ D\psi\, \psi(\mathbf{r}_p) \cdots \psi(\mathbf{r}_1) B_{\mathrm{f}}[\psi(x),\psi^+(x)]\, \psi^+(\mathbf{s}_1) \cdots \psi^+(\mathbf{s}_p), \tag{12.43}$$

again involving the unnormalised distribution functional. The position coherence is useful for discussing spatial-coherence effects in systems such as Fermi gases. In both cases the result is given as a functional average of products of field functions rather similar to equivalent classical formulae.

13
Functional Fokker–Planck Equations

In this chapter, we develop the methods required for applying phase space distribution functionals to physical problems for systems with large mode numbers. We obtain the functional Fokker–Planck equations equivalent to the Liouville–von Neumann, master or Matsubara equations for the density operator for both the boson and the fermion cases. For fermions, the unnormalised distribution functional will also be covered. This is accomplished via the aforementioned correspondence rules. We then show in Chapter 14 that the functional Fokker–Planck equation can be replaced by Ito stochastic field equations for stochastic fields that replace the phase space c-number (bosons) and Grassmann (fermions) fields. The Ito stochastic field equations associated with the functional Fokker–Planck equation for the unnormalised distribution functional are different from those for the standard situation. Stochastic averages of products of the stochastic phase space fields then can be used to determine the correlation functions. From the point of view of efficient numerical calculation of the quantum correlation functions of interest, the use of stochastic methods is preferable, as they in effect avoid having to sample the distribution function over the whole of phase space.

The results for the functional Fokker–Planck equation can be obtained directly from the dynamical equation for the density operator, with the Hamiltonian expressed in terms of the field operators by applying the correspondence rules for the effect of the field operators on the density operator. However, here we derive the functional Fokker–Planck equation from the ordinary Fokker–Planck equation for the distribution function because this is the simplest way to establish its form. In the applications to specific physical problems in Chapter 15, the functional Fokker–Planck equations will be obtained by applying the correspondence rules directly to the equation of motion for the density operator. Similarly, we will derive the Ito stochastic field equations from the ordinary Ito stochastic equations for the stochastic phase variables, though we will also present the derivation from the functional Fokker–Planck equations. Again, the former is the simplest way to establish the form of the Langevin field equations. In the applications to specific physical problems in Chapter 15, the Ito stochastic field equations will be obtained from the functional Fokker–Planck equations by using the relationships established in this chapter.

13.1 Correspondence Rules for Boson and Fermion Functional Fokker–Planck Equations

13.1.1 Boson Case

For bosonic distribution functions $P_{\mathrm{b}}(\alpha,\alpha^{+},\alpha^{*},\alpha^{+*})$, the correspondence rules are, from Chapter 8,

$$\hat{\rho} \Rightarrow \hat{a}_i\hat{\rho}, \qquad P_{\mathrm{b}}(\alpha,\alpha^{+},\alpha^{*},\alpha^{+*}) \Rightarrow \alpha_i P_{\mathrm{b}}, \tag{13.1}$$

$$\hat{\rho} \Rightarrow \hat{\rho}\hat{a}_i, \qquad P_{\mathrm{b}}(\alpha,\alpha^{+},\alpha^{*},\alpha^{+*}) \Rightarrow \left(-\frac{\partial}{\partial\alpha_i^{+}} + \alpha_i\right) P_{\mathrm{b}}, \tag{13.2}$$

$$\hat{\rho} \Rightarrow \hat{a}_i^{\dagger}\hat{\rho}, \qquad P_{\mathrm{b}}(\alpha,\alpha^{+},\alpha^{*},\alpha^{+*}) \Rightarrow \left(-\frac{\partial}{\partial\alpha_i} + \alpha_i^{+}\right) P_{\mathrm{b}}, \tag{13.3}$$

$$\hat{\rho} \Rightarrow \hat{\rho}\hat{a}_i^{\dagger}, \qquad P(\alpha,\alpha^{+},\alpha^{*},\alpha^{+*}) \Rightarrow \alpha_i^{+} P_{\mathrm{b}}. \tag{13.4}$$

Hence we see that

$$\hat{\rho} \Rightarrow \sum_i \phi_i(x)\hat{a}_i\hat{\rho}, \qquad P_{\mathrm{b}}(\alpha,\alpha^{+},\alpha^{*},\alpha^{+*}) \Rightarrow \sum_i \phi_i(x)\alpha_i P_{\mathrm{b}}, \tag{13.5}$$

$$\hat{\rho} \Rightarrow \hat{\rho}\sum_i \phi_i(x)\hat{a}_i, \qquad P_{\mathrm{b}}(\alpha,\alpha^{+},\alpha^{*},\alpha^{+*}) \Rightarrow \sum_i \phi_i(x)\left(-\frac{\partial}{\partial\alpha_i^{+}} + \alpha_i\right) P_{\mathrm{b}}, \tag{13.6}$$

$$\hat{\rho} \Rightarrow \sum_i \phi_i^{*}(x)\hat{a}_i^{\dagger}\hat{\rho}, \qquad P_{\mathrm{b}}(\alpha,\alpha^{+},\alpha^{*},\alpha^{+*}) \Rightarrow \sum_i \phi_i^{*}(x)\left(-\frac{\partial}{\partial\alpha_i} + \alpha_i^{+}\right) P_{\mathrm{b}}, \tag{13.7}$$

$$\hat{\rho} \Rightarrow \hat{\rho}\sum_i \phi_i^{*}(x)\hat{a}_i^{\dagger}, \qquad P_{\mathrm{b}}(\alpha,\alpha^{+},\alpha^{*},\alpha^{+*}) \Rightarrow \sum_i \phi_i^{*}(x)\alpha_i^{+} P_{\mathrm{b}}. \tag{13.8}$$

Using the mode expressions (2.60), (11.1), (11.7), (11.54) and (11.55) for the field operators, field functions and functional derivatives, we see that the correspondence rules for a bosonic distribution functional $P_{\mathrm{b}}[\psi(x),\psi^{+}(x),\psi^{*}(x),\psi^{+*}(x)]$ of the $P+$ type will be given by

$$\hat{\rho} \Rightarrow \hat{\Psi}(x)\hat{\rho}, \qquad P_{\mathrm{b}}[\psi(x),\psi^{+}(x),\psi^{*}(x),\psi^{+*}(x)] \Rightarrow \psi(x)P_{\mathrm{b}}, \tag{13.9}$$

$$\hat{\rho} \Rightarrow \hat{\rho}\hat{\Psi}(x), \qquad P_{\mathrm{b}}[\psi(x),\psi^{+}(x),\psi^{*}(x),\psi^{+*}(x)] \Rightarrow \left(-\frac{\delta}{\delta\psi^{+}(x)} + \psi(x)\right) P_{\mathrm{b}}, \tag{13.10}$$

$$\hat{\rho} \Rightarrow \hat{\Psi}^{\dagger}(x)\hat{\rho}, \qquad P_{\mathrm{b}}[\psi(x),\psi^{+}(x),\psi^{*}(x),\psi^{+*}(x)] \Rightarrow \left(-\frac{\delta}{\delta\psi(x)} + \psi^{+}(x)\right) P_{\mathrm{b}}, \tag{13.11}$$

$$\hat{\rho} \Rightarrow \hat{\rho}\hat{\Psi}^{\dagger}(x), \qquad P_{\mathrm{b}}[\psi(x),\psi^{+}(x),\psi^{*}(x),\psi^{+*}(x)] \Rightarrow \psi^{+}(x)\, P_{\mathrm{b}}. \tag{13.12}$$

We also need correspondence rules involving the spatial derivatives of the field operators $\partial_\mu\hat{\Psi}(x)$ and $\partial_\mu\hat{\Psi}^{\dagger}(x)$. By applying ∂_μ to each side of (13.5)–(13.8), we see using the mode expressions that

$$\hat{\rho} \Rightarrow \sum_i \partial_\mu \phi_i(x)\, \hat{a}_i \hat{\rho}, \quad P_{\rm b}(\alpha, \alpha^+, \alpha^*, \alpha^{+*}) \Rightarrow \sum_i \partial_\mu \phi_i(x)\, \alpha_i P_{\rm b}, \tag{13.13}$$

$$\hat{\rho} \Rightarrow \hat{\rho} \sum_i \partial_\mu \phi_i(x)\, \hat{a}_i, \quad P_{\rm b}(\alpha, \alpha^+, \alpha^*, \alpha^{+*}) \Rightarrow \sum_i \partial_\mu \phi_i(x) \left(-\frac{\partial}{\partial \alpha_i^+} + \alpha_i \right) P_{\rm b}, \tag{13.14}$$

$$\hat{\rho} \Rightarrow \sum_i \partial_\mu \phi_i^*(x) \hat{a}_i^\dagger\, \hat{\rho} \quad P_{\rm b}(\alpha, \alpha^+, \alpha^*, \alpha^{+*}) \Rightarrow \sum_i \partial_\mu \phi_i^*(x) \left(-\frac{\partial}{\partial \alpha_i} + \alpha_i^+ \right) P_{\rm b}, \tag{13.15}$$

$$\hat{\rho} \Rightarrow \hat{\rho} \sum_i \partial_\mu \phi_i^*(x)\, \hat{a}_i^\dagger, \quad P_{\rm b}(\alpha, \alpha^+, \alpha^*, \alpha^{+*}) \Rightarrow \sum_i \partial_\mu \phi_i^*(x)\, \alpha_i^+ P_{\rm b}, \tag{13.16}$$

and hence the correspondence rules are

$$\hat{\rho} \Rightarrow \partial_\mu \hat{\Psi}(x) \hat{\rho}, \qquad P_{\rm b}[\psi(x), \psi^+(x), \psi^*(x), \psi^{+*}(x)] \Rightarrow \partial_\mu \psi(x)\, P_{\rm b}, \tag{13.17}$$

$$\hat{\rho} \Rightarrow \hat{\rho} \partial_\mu \hat{\Psi}(x), \qquad P_{\rm b}[\,] \Rightarrow \left(-\partial_\mu \frac{\delta}{\delta \psi^+(x)} + \partial_\mu \psi(x) \right) P_{\rm b}, \tag{13.18}$$

$$\hat{\rho} \Rightarrow \partial_\mu \hat{\Psi}^\dagger(x) \hat{\rho}, \qquad P_{\rm b}[\,] \Rightarrow \left(-\partial_\mu \frac{\delta}{\delta \psi(x)} + \partial_\mu \psi^+(x) \right) P_{\rm b}, \tag{13.19}$$

$$\hat{\rho} \Rightarrow \hat{\rho} \partial_\mu \hat{\Psi}^\dagger(x), \qquad P_{\rm b}[\,] \Rightarrow \partial_\mu \psi^+(x)\, P_{\rm b}, \tag{13.20}$$

where, from mode expansions as in (11.54) and (11.55), the following apply:

$$\partial_\mu \frac{\delta}{\delta \psi(x)} \equiv \sum_k \partial_\mu \phi_k^*(x) \frac{\partial}{\partial \alpha_k}, \tag{13.21}$$

$$\partial_\mu \frac{\delta}{\delta \psi^+(x)} \equiv \sum_k \partial_\mu \phi_k(x) \frac{\partial}{\partial \alpha_k^+}. \tag{13.22}$$

In applications of these rules, the field operators occur in expressions for Hamiltonians involving spatial integrals (2.78) being a typical example. Expressions involving spatial derivatives of functional derivatives, such as those arising from (13.18), are then further developed by using spatial integration by parts. This is based on the assumption that the relevant mode functions become zero on the spatial boundary. An illustrative application is presented in Appendix I.

Similar considerations apply to other types of bosonic distribution functionals, such as the Wigner type. In that case the correspondence rules are

$$\hat{\rho} \Rightarrow \hat{\Psi}(x) \hat{\rho}, \;\; W_{\rm b}[\psi(x), \psi^+(x), \psi^*(x), \psi^{+*}(x)] \Rightarrow \left(+\frac{1}{2} \frac{\delta}{\delta \psi^+(x)} + \psi(x) \right) W_{\rm b}, \tag{13.23}$$

$$\hat{\rho} \Rightarrow \hat{\rho} \hat{\Psi}(x), \;\; W_{\rm b}[\psi(x), \psi^+(x), \psi^*(x), \psi^{+*}(x)] \Rightarrow \left(-\frac{1}{2} \frac{\delta}{\delta \psi^+(x)} + \psi(x) \right) W_{\rm b}, \tag{13.24}$$

$$\hat{\rho} \Rightarrow \hat{\Psi}^\dagger(x) \hat{\rho}, \;\; W_{\rm b}[\psi(x), \psi^+(x), \psi^*(x), \psi^{+*}(x)] \Rightarrow \left(-\frac{1}{2} \frac{\delta}{\delta \psi(x)} + \psi^+(x) \right) W_{\rm b}, \tag{13.25}$$

$$\hat{\rho} \Rightarrow \hat{\rho} \hat{\Psi}^\dagger(x), \;\; W_{\rm b}[\psi(x), \psi^+(x), \psi^*(x), \psi^{+*}(x)] \Rightarrow \left(+\frac{1}{2} \frac{\delta}{\delta \psi(x)} + \psi^+(x) \right) W_{\rm b}. \tag{13.26}$$

The derivation of the latter result is left as an exercise. Those involving the spatial derivatives of the field operators are analogous to (13.17)–(13.20).

13.1.2 Fermion Case

For fermionic distribution functions $P_{\mathrm{f}}(g, g^{+})$, the correspondence rules, are from Chapter 8,

$$\hat{\rho} \Rightarrow \hat{c}_i \hat{\rho}, \qquad P_{\mathrm{f}}(g, g^{+}) \Rightarrow g_i \, P_{\mathrm{f}}, \tag{13.27}$$

$$\hat{\rho} \Rightarrow \hat{\rho}\hat{c}_i, \qquad P_{\mathrm{f}}(g, g^{+}) \Rightarrow P_{\mathrm{f}} \left(+ \frac{\overleftarrow{\partial}}{\partial g_i^{+}} - g_i \right), \tag{13.28}$$

$$\hat{\rho} \Rightarrow \hat{c}_i^{\dagger} \hat{\rho}, \qquad P_{\mathrm{f}}(g, g^{+}) \Rightarrow \left(+ \frac{\overrightarrow{\partial}}{\partial g_i} - g_i^{+} \right) P_{\mathrm{f}}, \tag{13.29}$$

$$\hat{\rho} \Rightarrow \hat{\rho}\hat{c}_i^{\dagger}, \qquad P_{\mathrm{f}}(g, g^{+}) \Rightarrow P_{\mathrm{f}} g_i^{+}. \tag{13.30}$$

Hence we see that

$$\hat{\rho} \Rightarrow \sum_i \phi_i(x) \hat{c}_i \hat{\rho}, \qquad P_{\mathrm{f}}(g, g^{+}) \Rightarrow \sum_i \phi_i(x) g_i \, P_{\mathrm{f}} \tag{13.31}$$

$$\hat{\rho} \Rightarrow \hat{\rho} \sum_i \phi_i(x) \hat{c}_i, \qquad P_{\mathrm{f}}(g, g^{+}) \Rightarrow P_{\mathrm{f}} \sum_i \phi_i(x) \left(+ \frac{\overleftarrow{\partial}}{\partial g_i^{+}} - g_i \right), \tag{13.32}$$

$$\hat{\rho} \Rightarrow \sum_i \phi_i^{*}(x) \hat{c}_i^{\dagger} \hat{\rho}, \qquad P_{\mathrm{f}}(g, g^{+}) \Rightarrow \sum_i \phi_i^{*}(x) \left(+ \frac{\overrightarrow{\partial}}{\partial g_i} - g_i^{+} \right) P_{\mathrm{f}}, \tag{13.33}$$

$$\hat{\rho} \Rightarrow \hat{\rho} \sum_i \phi_i^{*}(x) \hat{c}_i^{\dagger}, \qquad P_{\mathrm{f}}(g, g^{+}) \Rightarrow P_{\mathrm{f}} \sum_i \phi_i^{*}(x) g_i^{+}. \tag{13.34}$$

Using the mode expressions (2.59), (5.146), (5.150), (11.149), (11.151), (11.154) and (11.156) for the field operators, field functions and functional derivatives, we see that the correspondence rules for a fermionic distribution functional $P_{\mathrm{f}}[\psi(x), \psi^{+}(x)]$ of the $P+$ type will be given by

$$\hat{\rho} \Rightarrow \hat{\Psi}(x)\hat{\rho}, \qquad P_{\mathrm{f}}[\psi(x), \psi^{+}(x)] \Rightarrow \psi(x) P_{\mathrm{f}}, \tag{13.35}$$

$$\hat{\rho} \Rightarrow \hat{\rho}\hat{\Psi}(x), \qquad P_{\mathrm{f}}[\psi(x), \psi^{+}(x)] \Rightarrow P_{\mathrm{f}} \left(+ \frac{\overleftarrow{\delta}}{\delta \psi^{+}(x)} - \psi(x) \right), \tag{13.36}$$

$$\hat{\rho} \Rightarrow \hat{\Psi}^{\dagger}(x)\hat{\rho}, \qquad P_{\mathrm{f}}[\psi(x), \psi^{+}(x)] \Rightarrow \left(+ \frac{\overrightarrow{\delta}}{\delta \psi(x)} - \psi^{+}(x) \right) P_{\mathrm{f}}, \tag{13.37}$$

$$\hat{\rho} \Rightarrow \hat{\rho}\hat{\Psi}^{\dagger}(x), \qquad P_{\mathrm{f}}[\psi(x), \psi^{+}(x)] \Rightarrow P_{\mathrm{f}} \psi^{+}(x). \tag{13.38}$$

It is noteworthy that the first forms of the fermion and boson results differ by an overall minus sign. Here we only state the correspondence rules that can be applied in succession. The correspondence rules involving spatial derivatives of the field operators can be established by applying ∂_μ to each side of (13.31)–(13.34). We see that

$$\hat{\rho} \Rightarrow \sum_i \partial_\mu \phi_i(x) \hat{c}_i \hat{\rho}, \qquad P_{\mathrm{f}}(g, g^+) \Rightarrow \sum_i \partial_\mu \phi_i(x) g_i P_{\mathrm{f}}, \tag{13.39}$$

$$\hat{\rho} \Rightarrow \hat{\rho} \sum_i \partial_\mu \phi_i(x)\, \hat{c}_i, \qquad P_{\mathrm{f}}(g, g^+) \Rightarrow P_{\mathrm{f}} \sum_i \partial_\mu \phi_i(x) \left(+\frac{\overleftarrow{\partial}}{\partial g_i^+} - g_i \right), \tag{13.40}$$

$$\hat{\rho} \Rightarrow \sum_i \partial_\mu \phi_i^*(x)\, \hat{c}_i^\dagger\, \hat{\rho}, \qquad P_{\mathrm{f}}(g, g^+) \Rightarrow \sum_i \partial_\mu \phi_i^*(x) \left(+\frac{\overrightarrow{\partial}}{\partial g_i} - g_i^+ \right) P_{\mathrm{f}}, \tag{13.41}$$

$$\hat{\rho} \Rightarrow \hat{\rho} \sum_i \partial_\mu \phi_i^*(x)\, \hat{c}_i^\dagger, \qquad P_{\mathrm{f}}(g, g^+) \Rightarrow P_{\mathrm{f}} \sum_i \partial_\mu \phi_i^*(x)\, g_i^+, \tag{13.42}$$

and the correspondence rules are

$$\hat{\rho} \Rightarrow \partial_\mu \hat{\Psi}(x)\, \hat{\rho}, \quad P_{\mathrm{f}}[\psi(x), \psi^+(x)] \Rightarrow \partial_\mu \psi(x)\, P_{\mathrm{f}}, \tag{13.43}$$

$$\hat{\rho} \Rightarrow \hat{\rho} \partial_\mu \hat{\Psi}(x), \quad P_{\mathrm{f}}[\psi(x), \psi^+(x)] \Rightarrow \left(P_{\mathrm{f}} \left(+\partial_\mu \frac{\overleftarrow{\delta}}{\delta \psi^+(x)} \right) \right) - P_{\mathrm{f}}\, \partial_\mu \psi(x), \tag{13.44}$$

$$\hat{\rho} \Rightarrow \partial_\mu \hat{\Psi}^\dagger(x)\, \hat{\rho}, \quad P_{\mathrm{f}}[\psi(x), \psi^+(x)] \Rightarrow \left(\left(+\partial_\mu \frac{\overrightarrow{\delta}}{\delta \psi(x)} \right) P_{\mathrm{f}} \right) - \partial_\mu \psi^+(x)\, P_{\mathrm{f}}, \tag{13.45}$$

$$\hat{\rho} \Rightarrow \hat{\rho}\, \partial_\mu \hat{\Psi}^\dagger(x), \quad P_{\mathrm{f}}[\psi(x), \psi^+(x)] \Rightarrow P_{\mathrm{f}}\, \partial_\mu \psi^+(x), \tag{13.46}$$

where, from mode expansions as in (11.149) and (11.156), the following expressions apply:

$$\partial_\mu \frac{\overrightarrow{\delta}}{\delta \psi(x)} \equiv \sum_k \partial_\mu \phi_k^*(x) \frac{\overrightarrow{\partial}}{\partial g_k}, \tag{13.47}$$

$$\partial_\mu \frac{\overleftarrow{\delta}}{\delta \psi^+(x)} \equiv \sum_k \partial_\mu \phi_k(x) \frac{\overleftarrow{\partial}}{\partial g_k^+}. \tag{13.48}$$

Note that the proof of these expressions for the fermion results does not depend on the distribution functional $P[\psi(x), \psi^+(x)]$ being equivalent to an even Grassmann distribution function $P_{\mathrm{f}}(g, g^+)$. Again, when successive operations on the distribution functional are involved, the safe way to proceed is to use this form of the correspondence rules. In applications of these rules, the field operators occur in expressions for Hamiltonians involving spatial integrals (2.79) being a typical example. Expressions involving spatial derivatives of functional derivatives, such as those arising from (13.44), are then further developed by using spatial integration by parts. This is again based on the assumption that the relevant mode functions become zero on the spatial boundary. The correspondence rules for combined boson–fermion systems are an obvious combination of the separate rules.

13.1.3 Fermion Case – Unnormalised Distribution Functional

The unnormalised fermion distribution functional has a set of correspondence rules for $B_{\mathrm{f}}[\psi(x),\psi^{+}(x)]$ that can be obtained from those above for $P_{\mathrm{f}}[\psi(x),\psi^{+}(x)]$ by introducing the relation (12.36) between the two distribution functionals. For example, when $\hat{\rho}\Rightarrow\hat{\rho}\,\hat{\Psi}(x)$, $P_{\mathrm{f}}[\psi(x),\psi^{+}(x)]\Rightarrow P_{\mathrm{f}}\,(+\overleftarrow{\delta}/\delta\psi^{+}(x)-\psi(x))$. Now

$$
\begin{aligned}
P_{\mathrm{f}}\left(\frac{\overleftarrow{\delta}}{\delta\psi^{+}(x)}-\psi(x)\right) &= \left[B_{\mathrm{f}}\exp\left(\int dx\,\psi(x)\psi^{+}(x)\right)\right]\left(\frac{\overleftarrow{\delta}}{\delta\psi^{+}(x)}-\psi(x)\right)\\
&= B_{\mathrm{f}}\left[\exp\left(\int dx\,\psi(x)\psi^{+}(x)\right)\left(\frac{\overleftarrow{\delta}}{\delta\psi^{+}(x)}\right)\right]+\left[B_{\mathrm{f}}\left(\frac{\overleftarrow{\delta}}{\delta\psi^{+}(x)}\right)\right]\exp\left(\int dx\,\psi(x)\psi^{+}(x)\right)\\
&\quad-\left[B_{\mathrm{f}}\exp\left(\int dx\,\psi(x)\psi^{+}(x)\right)\right]\psi(x)\\
&= \left[B_{\mathrm{f}}\left(\frac{\overleftarrow{\delta}}{\delta\psi^{+}(x)}\right)\right]\exp\left(\int dx\,\psi(x)\psi^{+}(x)\right),
\end{aligned}
\tag{13.49}
$$

using the product functional differentiation rule (11.164), noting that the exponential functional is even, and the result

$$
\exp\left(\int dx\,\psi(x)\psi^{+}(x)\right)\left(\frac{\overleftarrow{\delta}}{\delta\psi^{+}(x)}\right)=\exp\left(\int dx\,\psi(x)\psi^{+}(x)\right)\psi(x). \tag{13.50}
$$

Hence when $\hat{\rho}\Rightarrow\hat{\rho}\hat{\Psi}(x)$, $\quad B_{\mathrm{f}}[\psi(x),\psi^{+}(x)]\Rightarrow B_{\mathrm{f}}(\overleftarrow{\delta}/\delta\psi^{+}(x))$. The other correspondence rules are proved similarly and are

$$
\begin{aligned}
\hat{\rho}&\Rightarrow\hat{\Psi}(x)\hat{\rho}, & B_{\mathrm{f}}[\psi(x),\psi^{+}(x)]&\Rightarrow\psi(x)B_{\mathrm{f}}, && (13.51)\\
\hat{\rho}&\Rightarrow\hat{\rho}\hat{\Psi}(x), & B_{\mathrm{f}}[\psi(x),\psi^{+}(x)]&\Rightarrow B_{\mathrm{f}}\left(\frac{\overleftarrow{\delta}}{\delta\psi^{+}(x)}\right), && (13.52)\\
\hat{\rho}&\Rightarrow\hat{\Psi}^{\dagger}(x)\hat{\rho}, & B_{\mathrm{f}}[\psi(x),\psi^{+}(x)]&\Rightarrow\left(\frac{\overrightarrow{\delta}}{\delta\psi(x)}\right)B_{\mathrm{f}}, && (13.53)\\
\hat{\rho}&\Rightarrow\hat{\rho}\hat{\Psi}^{\dagger}(x), & B_{\mathrm{f}}[\psi(x),\psi^{+}(x)]&\Rightarrow B_{\mathrm{f}}\psi^{+}(x). && (13.54)
\end{aligned}
$$

Here we only state the correspondence rules that can be applied in succession. The correspondence rules involving spatial derivatives of the field operators can be established by applying ∂_{μ} to each side of (13.31)–(13.34) and are

$$\hat{\rho} \Rightarrow \partial_\mu \hat{\Psi}(x)\,\hat{\rho}, \qquad B_{\rm f}[\psi(x),\psi^+(x)] \Rightarrow \partial_\mu \psi(x)\, B_{\rm f}, \tag{13.55}$$

$$\hat{\rho} \Rightarrow \hat{\rho}\partial_\mu \hat{\Psi}(x), \qquad B_{\rm f}[\psi(x),\psi^+(x)] \Rightarrow \left[B_{\rm f} \left(\partial_\mu \frac{\overleftarrow{\delta}}{\delta\psi^+(x)} \right) \right], \tag{13.56}$$

$$\hat{\rho} \Rightarrow \partial_\mu \hat{\Psi}^\dagger(x)\,\hat{\rho}, \qquad B_{\rm f}[\psi(x),\psi^+(x)] \Rightarrow \left(\partial_\mu \frac{\overrightarrow{\delta}}{\delta\psi(x)} \right) B_{\rm f}, \tag{13.57}$$

$$\hat{\rho} \Rightarrow \hat{\rho}\,\partial_\mu \hat{\Psi}^\dagger(x), \qquad B_{\rm f}[\psi(x),\psi^+(x)] \Rightarrow B_{\rm f}\,\partial_\mu \psi^+(x). \tag{13.58}$$

As in the separate-modes case, the correspondence rules are simpler and lead to more useful functional Fokker–Planck equations.

13.2 Boson and Fermion Functional Fokker–Planck Equations

We now set out the functional Fokker–Planck equations that are equivalent to the Fokker–Planck equations treated in Section 8.7 for $P+$ distribution functions. These involved only first- and second-order derivatives with respect to the phase variables. As previously indicated, here we derive the functional Fokker–Planck equation from the ordinary Fokker–Planck equation for the distribution function because this is the simplest way to establish the form of the functional Fokker–Planck equation. In the applications to specific physical problems in Chapter 15, the functional Fokker–Planck equations will be obtained by applying the correspondence rules directly to the equation of motion for the density operator. To obtain functional Fokker–Planck equations in which functional derivatives of order higher than two are absent, approximations cutting off higher-order terms may be required. The form of the functional Fokker–Planck equations for the unnormalised distribution functions is obtained in the same way as presented here for the normalised distribution functions. Naturally, the resulting equations will differ because of the different drift vectors and diffusion matrices that apply in the unnormalised case. We also consider the generalisation to the situation where more than one pair of field operators is involved, such as for spin-1/2 fermions.

A notation using $-,+$ has been used up to now to designate phase variables, viz. $\alpha_i^- \equiv \alpha_i, \alpha_i^+ \equiv \alpha_i^+$ and $g_i^- \equiv g_i, g_i^+ \equiv g_i^+$, and also mode functions $\phi_i^-(x) \equiv \phi_i(x), \phi_i^+(x) \equiv \phi_i^*(x)$, fields $\psi_-(x) \equiv \psi(x), \psi_+(x) \equiv \psi^+(x)$, field operators $\hat{\psi}_-(x) \equiv \hat{\psi}(x), \hat{\psi}_+(x) \equiv \hat{\psi}^\dagger(x)$, matrices $F^{--}, F^{-+}, F^{+-}, F^{++}$ and C^-, C^+ etc. We will later wish to consider cases where there is more than one pair of field annihilation and creation operators, such as for spin-1/2 fermions with separate pairs of field operators for spin-up and spin-down fermions. Such operators are often designated by $+$ and $-$ (see for example Section 10.2), so, to avoid confusion, a different symbol $A = 1, 2$ will now be used to replace the $-$ and $+$ in the same positions as before. We will also use u and d for spin-up and spin-down fermions (see Section 15.2), rather than $-$ and $+$.

13.2.1 Boson Case

If we now list the phase space variables as α_i^A $(A = 1, 2;\ i = 1, \cdots, n)$, so the two sets are $\alpha^1 \equiv \{\alpha_1, \cdots, \alpha_n\} \equiv \alpha$ and $\alpha^2 \equiv \{\alpha_1^+, \cdots, \alpha_i^+, \cdots, \alpha_n^+\} \equiv \alpha^+$, and for short we write $\boldsymbol{\alpha} \equiv \{\alpha^1, \alpha^2\}$ and $\boldsymbol{\alpha}^* \equiv \{\alpha^{1*}, \alpha^{2*}\}$, then the Fokker–Planck equation (8.56) can be written in the form

$$\frac{\partial}{\partial t} P_b(\boldsymbol{\alpha}, \boldsymbol{\alpha}^*) = -\sum_{Ai} \frac{\partial}{\partial \alpha_i^A}(\mathcal{A}_i^A P_b) + \frac{1}{2} \sum_{Ai} \sum_{Bj} \frac{\partial}{\partial \alpha_i^A} \frac{\partial}{\partial \alpha_j^B} (\mathcal{D}_{ij}^{AB} P_b), \tag{13.59}$$

where, in terms of the previous notation in (8.56) and (8.57),

$$\begin{aligned} P_b(\boldsymbol{\alpha}, \boldsymbol{\alpha}^*) &= P_b(\alpha, \alpha^+, \alpha^*, \alpha^{+*}), \\ \mathcal{A}_i^1 &= C_i^-, \qquad \mathcal{A}_i^2 = C_i^+, \\ \mathcal{D}_{ij}^{11} &= F_{ij}^{--}, \qquad \mathcal{D}_{ij}^{12} = F_{ij}^{-+}, \qquad \mathcal{D}_{ij}^{21} = F_{ij}^{+-}, \qquad \mathcal{D}_{ij}^{22} = F_{ij}^{++}. \end{aligned} \tag{13.60}$$

Note that the drift vector $\mathcal{A}(\boldsymbol{\alpha})$ is a $2n \times 1$ matrix and the diffusion matrix $\mathcal{D}(\boldsymbol{\alpha})$ is a $2n \times 2n$ matrix. The symmetry properties in (8.58) show that the matrix $\mathcal{D}$ is symmetric:

$$\mathcal{D}_{ij}^{AB} = \mathcal{D}_{ji}^{BA}. \tag{13.61}$$

We now introduce a new notation for the mode functions and bosonic fields,

$$\xi_i^1(x) = \phi_i(x), \qquad \xi_i^2(x) = \phi_i^*(x), \tag{13.62}$$

$$\psi_1(x) = \psi(x), \qquad \psi_2(x) = \psi^+(x), \tag{13.63}$$

so that from (12.22) the bosonic fields $\psi_A(x)$ $(A = 1, 2)$ are now given by

$$\psi_A(x) = \sum_i \alpha_i^A \xi_i^A(x). \tag{13.64}$$

The $\xi_i^A(x)$ satisfy the usual orthogonality and completeness relationships

$$\int dx\, \xi_i^A(x)^* \xi_j^A(x) = \delta_{ij}, \tag{13.65}$$

$$\sum_i \xi_i^A(x) \xi_i^A(y)^* = \delta(x - y). \tag{13.66}$$

The mode functions may be time-dependent, but this will not be made explicit.

The expressions for the bosonic fields can be inverted, giving

$$\alpha_i^A = \int dx\, \xi_i^A(x)^* \psi_A(x), \tag{13.67}$$

so that the bosonic phase variables α_i^A can be considered as functionals of the bosonic fields $\psi_A(x)$, $\alpha_i^A = \alpha_i^A[\psi_A(x)]$. This feature also applies to any function of the α_i^A, such as the drift and diffusion terms $\mathcal{A}_i^A$ and $\mathcal{D}_{ij}^{AB}$.

In Chapter 11, we have seen that if the distribution function $P_b(\alpha^1, \alpha^2, \alpha^{1*}, \alpha^{2*})$ is equivalent to the distribution functional $P_b[\psi_1(x), \psi_2(x), \psi_1^*(x), \psi_2^*(x)]$, then from

(11.58) and (11.59) the derivatives with respect to the phase variables can be related to functional derivatives via

$$\frac{\partial}{\partial \alpha_i^A} \Rightarrow \int dx\, \xi_i^A(x) \left(\frac{\delta}{\delta \psi_A(x)}\right)_x. \tag{13.68}$$

Hence the Fokker–Planck equation (13.59) is equivalent to the functional Fokker–Planck equation

$$\begin{aligned}\frac{\partial}{\partial t} P_{\mathrm{b}}[\boldsymbol{\psi}(x), \boldsymbol{\psi}^*(x)] = &-\sum_A \int dx \left(\frac{\delta}{\delta \psi_A(x)}\right)_x (A_A(\boldsymbol{\psi}(x), x) P_{\mathrm{b}}[\boldsymbol{\psi}(x), \boldsymbol{\psi}^*(x)]) \\ &+ \frac{1}{2}\sum_A \sum_B \int\int dx\, dy \left(\frac{\delta}{\delta \psi_A(x)}\right)_x \left(\frac{\delta}{\delta \psi_{\mathrm{b}}(y)}\right)_y \\ &\times (D_{AB}(\boldsymbol{\psi}(x), x, \boldsymbol{\psi}(y), y) P_{\mathrm{b}}[\boldsymbol{\psi}(x), \boldsymbol{\psi}^*(x)]),\end{aligned} \tag{13.69}$$

where

$$A_A(\boldsymbol{\psi}(x), x) = \sum_i \xi_i^A(x) \mathcal{A}_i^A, \tag{13.70}$$

$$D_{AB}(\boldsymbol{\psi}(x), x, \boldsymbol{\psi}(y), y) = \sum_{ij} \xi_i^A(x) \mathcal{D}_{ij}^{AB} \xi_j^B(y) \tag{13.71}$$

are the drift and diffusion terms, and we have introduced the notation $\boldsymbol{\psi}(x) \equiv \{\psi_1(x), \psi_2(x)\} \equiv \{\psi(x), \psi^+(x)\}$ and $\boldsymbol{\psi}^*(x) \equiv \{\psi_1^*(x), \psi_2^*(x)\}$. Note the integrals over x, y in the non-local diffusion term and that because $\mathcal{A}$ and $\mathcal{D}$ are functions of the α_i^A, then the A_A and D_{AB} are really functionals of the $\psi_A(x)$. They may of course turn out to be ordinary functions of the ψ_A in specific applications. The symmetry of $\mathcal{D}$ leads to the condition

$$D_{AB}(\boldsymbol{\psi}(x), x, \boldsymbol{\psi}(y), y) = D_{BA}(\boldsymbol{\psi}(y), y, \boldsymbol{\psi}(x), x). \tag{13.72}$$

The last results can be inverted to give

$$\mathcal{A}_i^A = \int dx\, \xi_i^A(x)^* A_A(\boldsymbol{\psi}(x), x), \tag{13.73}$$

$$\mathcal{D}_{ij}^{AB} = \int\int dx\, dy\, \xi_i^A(x)^* D_{AB}(\boldsymbol{\psi}(x), x, \boldsymbol{\psi}(y), y) \xi_j^B(y)^*. \tag{13.74}$$

In general, the functional Fokker–Planck equations have a diffusion term with a double spatial integral. This term arises in the general situation where the two-body interaction term in the Hamiltonian contains finite-range boson–boson interactions. In cases where the zero-range approximation is applied, only a single spatial integral occurs.

13.2.2 Fermion Case

If we now list the phase space variables as g_i^A $(A = 1, 2;\ i = 1, \cdots, n)$, so the two sets are $g^1 \equiv \{g_1, \cdots, g_i, \cdots, g_n\} \equiv g$ and $g^2 \equiv \{g_1^+, \cdots, g_i^+, \cdots, g_n^+\} \equiv g^+$, and for short we write $\boldsymbol{g} \equiv \{g^1, g^2\}$, then the Fokker–Planck equation (8.91) can be written in the form

$$\frac{\partial}{\partial t} P_{\mathrm{f}}(\boldsymbol{g}) = -\sum_{Ai} (\mathcal{A}_i^A P_{\mathrm{f}}) \frac{\overleftarrow{\partial}}{\partial g_i^A} + \frac{1}{2} \sum_{Ai} \sum_{Bj} (\mathcal{D}_{ij}^{AB} P_{\mathrm{f}}) \frac{\overleftarrow{\partial}}{\partial g_j^B} \frac{\overleftarrow{\partial}}{\partial g_i^A}, \tag{13.75}$$

where, in terms of the previous notation in (8.94) and (8.84),

$$\begin{aligned} &P_{\mathrm{f}}(\boldsymbol{g}) = P_{\mathrm{f}}(g, g^+), \\ &\mathcal{A}_i^1 = C_i^-, \qquad \mathcal{A}_i^2 = C_i^+, \\ &\mathcal{D}_{ij}^{11} = -F_{ij}^{--}, \qquad \mathcal{D}_{ij}^{12} = F_{ij}^{-+}, \qquad \mathcal{D}_{ij}^{21} = -F_{ji}^{-+}, \qquad \mathcal{D}_{ij}^{22} = F_{ij}^{++}. \end{aligned} \tag{13.76}$$

The drift vector $\mathcal{A}(\boldsymbol{g})$ is a $2n \times 1$ matrix and the diffusion matrix $\mathcal{D}(\boldsymbol{g})$ is a $2n \times 2n$ matrix. The symmetry properties in (8.95) show that the matrix $\mathcal{D}$ is antisymmetric:

$$\mathcal{D}_{ij}^{AB} = -\mathcal{D}_{ji}^{BA}. \tag{13.77}$$

We now introduce a new notation for the mode functions and fermionic fields

$$\xi_i^1(x) = \phi_i(x), \qquad \xi_i^2(x) = \phi_i^*(x), \tag{13.78}$$
$$\psi_1(x) = \psi(x), \qquad \psi_2(x) = \psi^+(x), \tag{13.79}$$

so that from (12.29) the fermionic fields $\psi_A(x)$ $(A = 1, 2)$ are now given by

$$\psi_A(x) = \sum_i g_i^A \xi_i^A(x). \tag{13.80}$$

The mode functions satisfy the previous orthogonality and completeness relationships.

The expressions for the fermionic fields can be inverted, giving

$$g_i^A = \int dx\, \xi_i^A(x)^* \psi_A(x), \tag{13.81}$$

so that the fermionic phase variables g_i^A can be considered as *functionals* of the fermionic fields $\psi_A(x)$, $g_i^A = g_i^A[\psi_A(x)]$. This feature also applies to any function of the g_i^A, such as the drift and diffusion terms $\mathcal{A}_i^A$ and $\mathcal{D}_{ij}^{AB}$.

In Chapter 11, we have seen that if the distribution function $P_{\mathrm{f}}(g^1, g^2)$ is equivalent to the distribution functional $P_{\mathrm{f}}[\psi_1(x), \psi_2(x)]$, then from (11.159) and (11.160) the derivatives with respect to the phase variables can be related to functional derivatives via

$$\frac{\overleftarrow{\partial}}{\partial g_i^A} \Rightarrow \int dx\, \xi_k^A(x) \left(\frac{\overleftarrow{\delta}}{\delta \psi_A(x)} \right)_x. \tag{13.82}$$

Hence the Fokker–Planck equation (13.75) is equivalent to the functional Fokker–Planck equation

$$\frac{\partial}{\partial t}P_{\mathrm{f}}[\boldsymbol{\psi}(x)] = -\sum_{A}\int dx\,(A_A(\boldsymbol{\psi}(x),x)P_{\mathrm{f}}[\boldsymbol{\psi}(x)])\left(\frac{\overleftarrow{\delta}}{\delta\psi_A(x)}\right)_x$$
$$+\frac{1}{2}\sum_{A,B}\int dx\,dy(D_{AB}(\boldsymbol{\psi}(x),x,\boldsymbol{\psi}(y),y)P_{\mathrm{f}}[\;])\left(\frac{\overleftarrow{\delta}}{\delta\psi_B(y)}\right)_y\left(\frac{\overleftarrow{\delta}}{\delta\psi_A(x)}\right)_x, \tag{13.83}$$

where

$$A_A(\boldsymbol{\psi}(x),x) = \sum_i \xi_i^A(x)\mathcal{A}_i^A, \tag{13.84}$$

$$D_{AB}(\boldsymbol{\psi}(x),x,\boldsymbol{\psi}(y),y) = \sum_{ij}\xi_i^A(x)\mathcal{D}_{ij}^{AB}\xi_j^B(y) \tag{13.85}$$

are the drift and diffusion terms and we use the notation $\boldsymbol{\psi}(x) \equiv \{\psi_1(x),\psi_2(x)\} \equiv \{\psi(x),\psi^+(x)\}$. There are integrals over x, y in the non-local diffusion term. Note that because $\mathcal{A}$ and $\mathcal{D}$ are functions of the g_i^A, then the A_A and D_{AB} defined in the last equations are really functionals of the $\psi_A(x)$. They may of course turn out to be ordinary functions of the ψ_A in specific applications. The antisymmetry of $\mathcal{D}$ leads to the antisymmetry condition

$$D_{AB}(\boldsymbol{\psi}(x),x,\boldsymbol{\psi}(y),y) = -D_{BA}(\boldsymbol{\psi}(y),y,\boldsymbol{\psi}(x),x). \tag{13.86}$$

The last results can be inverted to give

$$\mathcal{A}_i^A = \int dx\,\xi_i^A(x)^* A_A(\boldsymbol{\psi}(x),x), \tag{13.87}$$

$$\mathcal{D}_{ij}^{AB} = \int\int dx\,dy\,\xi_i^A(x)^* D_{AB}(\boldsymbol{\psi}(x),x,\boldsymbol{\psi}(y),y)\xi_j^B(y)^*. \tag{13.88}$$

The remarks about the diffusion term at the end of the last subsection also apply here.

13.3 Generalisation to Several Fields

The functional Fokker–Planck equations treated above for bosons and fermions dealt with the situation where only one pair of field operators $\hat{\Psi}(x), \hat{\Psi}^\dagger(x)$ was involved. However, in situations where the bosons or fermions have an internal structure – as in atoms – more than one pair of field operators may be involved. One such case is that for spin-1/2 fermions, where there are two pairs of field operators, $\hat{\Psi}_u(x), \hat{\Psi}_u^\dagger(x)$ and $\hat{\Psi}_d(x), \hat{\Psi}_d^\dagger(x)$, corresponding to spin up, u, and spin down, d. In such cases the field operators will be designated $\hat{\Psi}_\alpha(x), \hat{\Psi}_\alpha^\dagger(x)$. The mode functions and field functions will in general differ for each α and will be designated

$$\xi^1_{\alpha i}(x) = \phi_{\alpha i}(x), \qquad \xi^2_{\alpha i}(x) = \phi^*_{\alpha\, i}(x), \tag{13.89}$$

$$\psi_{\alpha 1}(x) = \psi_\alpha(x), \qquad \psi_{\alpha 2}(x) = \psi^+_\alpha(x), \tag{13.90}$$

where, in terms of the phase space variables $\alpha^A_{\alpha i}$ for bosons and $g^A_{\alpha i}$ for fermions, the field functions $\psi_{\alpha A}(x)$ $(A = 1, 2)$ are

$$\psi_{\alpha A}(x) = \sum_i \alpha^A_{\alpha i}\, \xi^A_{\alpha i}(x), \quad \psi_{\alpha A}(x) = \sum_i g^A_{\alpha i}\, \xi^A_{\alpha i}(x). \tag{13.91}$$

The mode functions for a particular α are orthogonal and normalised. In terms of annihilation $(A = 1)$ and creation $(A = 2)$ operators, the field operators now become

$$\hat{\Psi}_{\alpha A}(x) = \sum_i \hat{a}^A_{\alpha i}\, \xi^A_{\alpha i}(x), \quad \hat{\Psi}_{\alpha A}(x) = \sum_i \hat{c}^A_{\alpha i}\, \xi^A_{\alpha i}(x). \tag{13.92}$$

for boson and fermion fields. The Hamiltonian may include terms corresponding to interactions between the different fields, as in (2.79) for spin-1/2 fermions.

The previous results can easily be generalised to allow for more than one pair of field operators. Thus we now write $\boldsymbol{\psi}(x) \equiv \{\psi_{\alpha 1}(x), \psi_{\alpha 2}(x)\} \equiv \{\psi_\alpha(x), \psi^+_\alpha(x)\}$ for both bosons and fermions, and also $\boldsymbol{\psi}^*(x) \equiv \{\psi^*_{\alpha 1}(x), \psi^*_{\alpha 2}(x)\}$ for bosons.

The functional Fokker–Planck equations are now of the form

$$\begin{aligned}\frac{\partial}{\partial t} P_{\mathrm{b}}[\boldsymbol{\psi}(x), \boldsymbol{\psi}^*(x)] = &-\sum_{\alpha A} \int dx \left(\frac{\delta}{\delta \psi_{\alpha A}(x)}\right)_x (A_{\alpha A}(\boldsymbol{\psi}(x), x) P_{\mathrm{b}}[\boldsymbol{\psi}(x), \boldsymbol{\psi}^*(x)]) \\ &+ \frac{1}{2} \sum_{\alpha A} \sum_{\beta B} \int\!\!\int dx\, dy \left(\frac{\delta}{\delta \psi_{\alpha A}(x)}\right)_x \left(\frac{\delta}{\delta \psi_{\beta B}(y)}\right)_y \\ &\times (D_{\alpha A\, \beta B}(\boldsymbol{\psi}(x), x, \boldsymbol{\psi}(y), y) P_{\mathrm{b}}[\boldsymbol{\psi}(x), \boldsymbol{\psi}^*(x)])\end{aligned} \tag{13.93}$$

for bosons and

$$\begin{aligned}\frac{\partial}{\partial t} P_{\mathrm{f}}[\boldsymbol{\psi}(x)] = &-\sum_{\alpha A} \int dx\, (A_{\alpha A}(\boldsymbol{\psi}(x), x) P_{\mathrm{f}}[\boldsymbol{\psi}(x)]) \left(\frac{\overleftarrow{\delta}}{\delta \psi_{\alpha A}(x)}\right)_x \\ &+ \frac{1}{2} \sum_{\alpha A, \beta B} \int dx\, dy\, (D_{\alpha A\, \beta B}(\boldsymbol{\psi}(x), x, \boldsymbol{\psi}(y), y) P_{\mathrm{f}}[\boldsymbol{\psi}(x)]) \\ &\times \left(\frac{\overleftarrow{\delta}}{\delta \psi_{\beta B}(y)}\right)_y \left(\frac{\overleftarrow{\delta}}{\delta \psi_{\alpha A}(x)}\right)_x\end{aligned} \tag{13.94}$$

for fermions. In both cases the drift vector and diffusion matrix contain more elements. The same forms apply for the case of unnormalised distribution functionals when several pairs of field operators occur – just replace $P_{\mathrm{b,f}}$ by $B_{\mathrm{b,f}}$.

14
Langevin Field Equations

In this chapter, we show how the functional Fokker–Planck equations for the phase space distribution functional are equivalent to Ito stochastic equations for stochastic fields. This may involve first truncating the Fokker–Planck equations to include only terms with at most second-order functional derivatives. The stochastic fields are defined via the expansion of the phase space field functions in terms of mode functions and then treating the expansion coefficients as stochastic variables. The derivation of the Ito equations for the stochastic fields is based on the Ito equations for stochastic expansion coefficients set out in Chapter 9. As the relationship between the Fokker–Planck equation and the Ito stochastic differential equations has now been established, we no longer need to use distinct symbols for the drift vector in the Fokker–Planck equation and the related term in the Ito stochastic differential equations (apart from in Section 14.1.3). The Ito stochastic field equations can of course also be derived directly from the functional Fokker–Planck equation through establishing the link between the classical field and noise terms in the Langevin field equations and the drift and diffusion terms in the functional Fokker–Planck equation. We show this derivation here also, and this relationship will be used directly in the applications treated in Chapter 15. The Ito stochastic field equations are the sum of a deterministic term associated with the first-order functional derivatives in the functional Fokker–Planck equation (the drift terms) and a quantum noise term associated with the second-order functional derivatives in the functional Fokker–Planck equation (the diffusion terms). The stochastic averages of products of noise field terms are obtained. For products of odd numbers of noise field terms, these averages are zero for products of even numbers of noise field terms, the results are sums of products of stochastic averages of products of diffusion matrix element terms. The latter are delta function correlated in time, but not in space. In this chapter we emphasise how the phase space distribution functionals which determine the quantum correlation functions can then be replaced by stochastic averages involving products of the stochastic fields. The relationship between the Ito stochastic field equations and the functional Fokker–Planck equation for the unnormalised distribution functions is obtained in the same way as presented here for the normalised distribution functionals. Naturally, the resulting Ito equations will differ because of the different drift vectors and diffusion matrices that apply in the unnormalised case. We also consider the generalisation to the situation where more than one pair of field operators is involved, such as for spin-1/2 fermions.

14.1 Boson Stochastic Field Equations

14.1.1 Ito Equations for Bosonic Stochastic Phase Variables

The functional Fokker–Planck equation (13.69) is equivalent to the ordinary Fokker–Planck equation (13.59),

$$\frac{\partial}{\partial t}P_{\mathrm{b}}(\boldsymbol{\alpha},\boldsymbol{\alpha}^*) = -\sum_{Ai}\frac{\partial}{\partial\alpha_i^A}(\mathcal{A}_i^A P_{\mathrm{b}}) + \frac{1}{2}\sum_{Ai}\sum_{Bj}\frac{\partial}{\partial\alpha_i^A}\frac{\partial}{\partial\alpha_j^B}(\,\mathcal{D}_{ij}^{AB}P_{\mathrm{b}}), \tag{14.1}$$

where we write $\boldsymbol{\alpha} \equiv \{\alpha^1,\alpha^2\} \equiv \{\alpha_i^A\}$ and $\boldsymbol{\alpha}^* \equiv \{\alpha^{1*},\alpha^{2*}\} \equiv \{\alpha_i^{A*}\}$. In terms of the previous notation, $p \to A,i$, $\alpha_p \to \alpha_i^A$, $A_p \to \mathcal{A}_i^A$, $D_{pq} \to \mathcal{D}_{ij}^{AB}$.

From the theory presented in Chapter 9, we can immediately write down the equivalent Ito stochastic equations for time-dependent stochastic variables $\tilde{\alpha}_i^A(t)$. As before, the procedure involves replacing the time-independent phase space variables α_i^A by time-dependent stochastic variables $\tilde{\alpha}_i^A(t)$. The Ito stochastic equations for the $\tilde{\alpha}_i^A(t)$ are such that phase space averages of functions of the α_i^A give the same result as stochastic averages of the same functions of the $\tilde{\alpha}_i^A(t)$.

The Ito equations for the stochastic expansion coefficients $\tilde{\alpha}_i^A$ can be written in several forms:

$$\begin{aligned}\delta\tilde{\alpha}_i^A(t) &= \tilde{\alpha}_i^A(t+\delta t) - \tilde{\alpha}_i^A(t)\\ &= \mathcal{A}_i^A(\tilde{\boldsymbol{\alpha}}(t))\,\delta t + \sum_{Dk}\mathcal{B}_{ik}^{AD}(\tilde{\boldsymbol{\alpha}}(t))\int_t^{t+\delta t}dt_1\,\Gamma_k^D(t_1), \end{aligned}\tag{14.2}$$

$$\frac{d}{dt}\tilde{\alpha}_i^A(t) = \mathcal{A}_i^A(\tilde{\boldsymbol{\alpha}}(t)) + \sum_{Dk}\mathcal{B}_{ik}^{AD}(\tilde{\boldsymbol{\alpha}}(t))\frac{d}{dt}w_k^D(t) \tag{14.3}$$

$$= \mathcal{A}_i^A(\tilde{\boldsymbol{\alpha}}(t)) + \sum_{Dk}\mathcal{B}_{ik}^{AD}(\tilde{\boldsymbol{\alpha}}(t))\Gamma_k^D(t_+), \tag{14.4}$$

where $\tilde{\boldsymbol{\alpha}}(t) \equiv \{\tilde{\alpha}_i^A(t)\} \equiv \{\tilde{\alpha}_k(t),\tilde{\alpha}_k^+(t)\}$ and, in the new notation, $B_a^p \to B_{ik}^{AD}$ and $\Gamma_a \to \Gamma_k^D$. The results in (9.26) have been used to relate the Ito equation to the Fokker–Planck equation.

The diffusion matrix $\mathcal{D}$ is symmetric, and as a result we can always write the diffusion matrix $\mathcal{D}$ in the Takagi factorisation form [63] (see Section 4.3)

$$\mathcal{D} = \mathcal{B}\mathcal{B}^T. \tag{14.5}$$

In detail, the matrix $\mathcal{B}$ is related to the diffusion matrix $\mathcal{D}$ as in (14.5):

$$\mathcal{D}_{ij}^{AB}(\tilde{\boldsymbol{\alpha}}(t)) = \sum_{Dk}\mathcal{B}_{ik}^{AD}(\tilde{\boldsymbol{\alpha}}(t))\mathcal{B}_{jk}^{BD}(\tilde{\boldsymbol{\alpha}}(t)). \tag{14.6}$$

The matrix elements $\mathcal{B}_{ik}^{AD}(\tilde{\boldsymbol{\alpha}}(t))$ are functions of the $\tilde{\alpha}_i^A(t)$. The quantity t_+ is to indicate that if the Ito stochastic equation is integrated from t to $t+\delta t$, the Gaussian–Markov noise term is integrated over this interval whilst the $\mathcal{A}_i^A(\tilde{\alpha}_j^C(t))$

and $\mathcal{B}_{ik}^{AD}(\tilde{\alpha}_j^C(t))$ are left at time t. The Gaussian–Markov noise terms are listed as Γ_k^D, with $D = -, +$ and $k = 1, 2, \cdots, n$ in a way that is similar to the α_i^A or $\tilde{\alpha}_i^A$. The number of D and k is the same as the number of A and i, which is $2n$.

The quantities $w_k^D(t)$ and $\Gamma_k^D(t)$ are Wiener and Gaussian–Markov stochastic variables as defined earlier in (9.11) and (9.28), but for convenience we repeat the key features in this chapter. The Gaussian–Markov quantities Γ_k^D satisfy the stochastic-averaging results

$$\begin{aligned}
\overline{\Gamma_k^D(t_1)} &= 0,\\
\overline{\{\Gamma_k^D(t_1)\Gamma_l^E(t_2)\}} &= \delta_{DE}\delta_{kl}\delta(t_1 - t_2),\\
\overline{\{\Gamma_k^D(t_1)\Gamma_l^E(t_2)\Gamma_m^F(t_3)\}} &= 0,\\
\overline{\{\Gamma_k^D(t_1)\Gamma_l^E(t_2)\Gamma_m^F(t_3)\Gamma_n^G(t_4)\}} &= \overline{\{\Gamma_k^D(t_1)\Gamma_l^E(t_2)\}}\;\overline{\{\Gamma_m^F(t_3)\Gamma_n^G(t_4)\}}\\
&\quad+\overline{\{\Gamma_k^E(t_1)\Gamma_m^F(t_3)\}}\;\overline{\{\Gamma_l^E(t_2)\Gamma_n^G(t_4)\}}\\
&\quad+\overline{\{\Gamma_k^D(t_1)\Gamma_n^G(t_4)\}}\;\overline{\{\Gamma_l^E(t_2)\Gamma_m^F(t_3)\}}\\
&\quad\cdots,
\end{aligned} \tag{14.7}$$

with stochastic averages being denoted with a bar. The stochastic average of an odd number of noise terms is always zero, whilst that for an even number is the sum of all products of stochastic averages of two noise terms. The Gaussian–Markov noise terms Γ_k^D are related to the Wiener stochastic variables w_k^D via

$$w_k^D(t) = \int_0^T dt_1\, \Gamma_k^D(t_1), \tag{14.8}$$

$$\delta w_k^D(t) = w_k^D(t + \delta t) - w_k^D(t) = \int_t^{t+\delta t} dt_1\, \Gamma_k^D(t_1), \tag{14.9}$$

$$\frac{d}{dt} w_k^D(t) = \lim_{\delta t \to 0}\left(\frac{\delta w_k^D(t)}{\delta t}\right) = \Gamma_k^D(t_+). \tag{14.10}$$

One of the rules in stochastic averaging is

$$\overline{\sum_A F_A(\tilde{\boldsymbol{\alpha}}(t))} = \sum_A \overline{F_A(\tilde{\boldsymbol{\alpha}}(t))}, \tag{14.11}$$

so the stochastic average of the sum is the sum of the stochastic averages. Also, in Ito stochastic calculus the noise terms $\Gamma_k^D(t_1)$ within the interval $t, t + \delta t$ are uncorrelated with any function of the $\tilde{\alpha}_i^C(t)$ at the earlier time t, so that the stochastic average of the product of such a function with a product of the noise terms factorises:

$$\begin{aligned}
&\overline{F(\tilde{\boldsymbol{\alpha}}(t_1))\{\Gamma_k^D(t_2)\Gamma_l^E(t_3)\Gamma_m^{\mathfrak{f}}(t_4)\cdots\Gamma_a^x(t_l)\}}\\
&= \overline{F(\tilde{\boldsymbol{\alpha}}(t_1))}\;\overline{\{\Gamma_k^D(t_2)\Gamma_l^E(t_3)\Gamma_m^{\mathfrak{f}}(t_4)\cdots\Gamma_a^x(t_l)\}}, \qquad t_1 < t_2, t_3, \cdots, t_l.
\end{aligned} \tag{14.12}$$

These key features of Ito stochastic calculus are important in deriving the properties of the noise fields in the stochastic field equations.

14.1.2 Derivation of Bosonic Ito Stochastic Field Equations

The bosonic stochastic fields $\tilde{\psi}_A(x,t)$ are defined via the same expansion as for the time-independent field functions $\psi_A(x)$ by replacing the bosonic time-independent phase space variables α_i^A by time-dependent stochastic variables $\tilde{\alpha}_i^A(t)$:

$$\tilde{\psi}_A(x,t) = \sum_i \tilde{\alpha}_i^A(t)\xi_i^A(x). \tag{14.13}$$

Note that the bosonic stochastic field is a c-number spatial function. The expansion coefficients in (14.13) are restricted to those required for expanding the particular field function $\psi_A(x)$. Also, the stochastic variations in $\tilde{\psi}_A(x,t)$ are chosen so as to be due only to stochastic fluctuations in the $\tilde{\alpha}_i^A(t)$. Although the mode functions may be time-dependent, their time variations are not stochastic in origin, so the stochastic field equations for the $\tilde{\psi}_A(x,t)$ do not allow for time variations in the mode functions. The $\tilde{\alpha}_i^A(t)$ may be considered as functionals of the stochastic field $\tilde{\psi}_A(x,t)$. The pair of stochastic fields are denoted $\tilde{\boldsymbol{\psi}}(x,t) \equiv \{\tilde{\psi}_1(x,t), \tilde{\psi}_2(x,t)\} \equiv \{\tilde{\psi}(x,t), \tilde{\psi}^+(x,t)\}$.

The change in the stochastic field is

$$\begin{aligned} \delta\tilde{\psi}_A(x,t) &= \tilde{\psi}_A(x,t+\delta t) - \tilde{\psi}_A(x,t) \\ &= \sum_i \delta\tilde{\alpha}_i^A(t)\,\xi_i^A(x). \end{aligned} \tag{14.14}$$

The Ito stochastic equation for the stochastic fields $\tilde{\psi}_A(x,t)$ can then be derived from the Ito stochastic equations for the expansion coefficients. Using (13.70), the drift term in the stochastic equation gives

$$\sum_i \mathcal{A}_i^A(\tilde{\boldsymbol{\alpha}}(t))\,\xi_i^A(x)\,\delta t = A_A(\tilde{\boldsymbol{\psi}}(x,t),x)\,\delta t, \tag{14.15}$$

which gives the drift vector A_A in the functional Fokker–Planck equation. This depends on x and is also a functional of the stochastic fields $\tilde{\boldsymbol{\psi}}(x,t)$. The diffusion term in the stochastic equation gives

$$\sum_i \sum_{Dk} \mathcal{B}_{ik}^{AD}(\tilde{\boldsymbol{\alpha}}(t))\,\xi_i^A(x) \int_t^{t+\delta t} dt_1\,\Gamma_k^D(t_1) = \sum_{Dk} \eta_k^{AD}(\tilde{\boldsymbol{\psi}}(x,t),x) \int_t^{t+\delta t} dt_1\,\Gamma_k^D(t_1), \tag{14.16}$$

where

$$\eta_k^{AD}(\tilde{\boldsymbol{\psi}}(x,t),x) = \sum_i \mathcal{B}_{ik}^{AD}(\tilde{\boldsymbol{\alpha}}(t))\,\xi_i^A(x) \tag{14.17}$$

depends on x and is a functional of the $\tilde{\psi}_A(x,t)$ via the $\tilde{\alpha}_i^A(t)$. The $\eta_k^{AD}(\tilde{\boldsymbol{\psi}}(x,t),x)$ are related via $\mathcal{B}_{ik}^{AD}(\tilde{\boldsymbol{\alpha}}(t))$ to the diffusion matrix D_{AB} in the functional Fokker–Planck equation. Using (13.71), we have

$$
\begin{aligned}
\sum_{Dk} \eta_k^{AD}(\tilde{\psi}(x,t),x)\eta_k^{BD}(\tilde{\psi}(y,t),y) &= \sum_{Dk}\sum_{i} \mathcal{B}_{ik}^{AD}\xi_i^A(x)\sum_{j}\mathcal{B}_{jk}^{BD}\xi_j^B(y) \\
&= \sum_{ij}\xi_i^A(x)\left(\sum_{Dk}\mathcal{B}_{ik}^{AD}\mathcal{B}_{jk}^{BD}\right)\xi_j^B(y) \\
&= \sum_{ij}\xi_i^A(x)\,(\mathcal{D})_{ij}^{AB}\,\xi_j^B(y) \\
&= D_{AB}(\tilde{\psi}(x,t),x,\tilde{\psi}(y,t),y), \qquad (14.18)
\end{aligned}
$$

which gives the diffusion matrix D_{AB} in the functional Fokker–Planck equation.

The stochastic field equations are then given by

$$
\begin{aligned}
\delta\tilde{\psi}_A(x,t) &= A_A(\tilde{\psi}(x,t),x)\,\delta t + \sum_{Dk}\eta_k^{AD}(\tilde{\psi}(x,t),x)\int_t^{t+\delta t} dt_1\,\Gamma_k^D(t_1) \\
&= A_A(\tilde{\psi}(x,t),x)\,\delta t + \delta\tilde{G}_A(\tilde{\psi}(x,t),\mathbf{\Gamma}(t_+)), \qquad (14.19) \\
\frac{\partial}{\partial t}\tilde{\psi}_A(x,t) &= A_A(\tilde{\psi}(x,t),x) + \sum_{Dk}\eta_k^{AD}(\tilde{\psi}(x,t),x)\,\frac{d}{dt}w_k^D(t) \\
&= A_A(\tilde{\psi}(x,t),x) + \sum_{Dk}\eta_k^{AD}(\tilde{\psi}(x,t),x)\,\Gamma_k^D(t_+) \\
&= A_A(\tilde{\psi}(x,t),x) + \frac{\partial}{\partial t}\tilde{G}_A(\tilde{\psi}(x,t),\mathbf{\Gamma}(t_+)), \qquad (14.20)
\end{aligned}
$$

where

$$
\begin{aligned}
(\eta\eta^T)^{AB} &= \sum_{Dk}\eta_k^{AD}(\tilde{\psi}(x,t),x)\eta_k^{BD}(\tilde{\psi}(y,t),y) \\
&= D_{AB}(\tilde{\psi}(x,t),x,\tilde{\psi}(y,t),y). \qquad (14.21)
\end{aligned}
$$

Here we write $\mathbf{\Gamma}(t_+) \equiv \{\Gamma_k^1(t_+),\Gamma_k^2(t_+)\}$. The first form gives the change in the stochastic field over a small time interval $t, t+\delta t$; the second is in the form of a partial differential equation. The first term in the Ito equation for the stochastic fields (14.20), $A_A(\tilde{\psi}(x,t),x)$, is the deterministic term and is obtained from the drift vector in the functional Fokker–Planck equation, and the second term, $(\partial/\partial t)\tilde{G}_A(\tilde{\psi}(x,t),\mathbf{\Gamma}(t_+))$, is the quantum noise field, whose statistical properties are obtained from the diffusion matrix, and which depends both on the stochastic fields $\tilde{\psi}(x,t)$ and on the Gaussian–Markov stochastic variables $\mathbf{\Gamma}(t_+)$.

The noise field term is

$$
\frac{\partial}{\partial t}\tilde{G}_A(\tilde{\psi}(x,t),\mathbf{\Gamma}(t_+)) = \sum_{Dk}\eta_k^{AD}(\tilde{\psi}(x,t),x)\,\frac{d}{dt}w_k^D(t) = \sum_{Dk}\eta_k^{AD}(\tilde{\psi}(x,t),x)\,\Gamma_k^D(t_+), \qquad (14.22)
$$

where the stochastic field $\eta_k^{AD}(\tilde{\psi}(x,t),x)$ is related to the diffusion matrix. The noise field term depends on x and is a functional of the stochastic fields.

14.1.3 Alternative Derivation of Bosonic Stochastic Field Equations

The Ito stochastic field equations can be derived directly from the functional Fokker–Planck equations themselves by following a similar approach to Chapter 9 and considering the equation of motion of the phase space functional average and the stochastic average for an arbitrary functional $F[\mathbf{\Psi}(x)]$ of the fields $\mathbf{\Psi}(x) \equiv \{\psi_1(x), \psi_2(x)\} \equiv \{\psi(x), \psi^+(x)\}$. The phase space functional average is given by

$$\langle F[\mathbf{\Psi}(x)]\rangle_t = \int\int D^2\psi\, D^2\psi^+ F[\mathbf{\Psi}(x)]\, P_b[\mathbf{\Psi}(x), \mathbf{\Psi}^*(x)], \tag{14.23}$$

so that correct to $O(\delta t)$ we have

$$\langle F[\mathbf{\Psi}(x)]\rangle_{t+\delta t} - \langle F[\mathbf{\Psi}(x)]\rangle_t = \int D^2\psi\, D^2\psi^+ F[\mathbf{\Psi}(x)]\, \frac{\partial}{\partial t} P_b(\boldsymbol{\alpha}, \boldsymbol{\alpha}^*)\, \delta t.$$

Substituting from the functional Fokker–Planck equation (13.69), we get, using functional integration by parts and assuming that the distribution function goes to zero fast enough on the phase space boundary,

$$\begin{aligned}
&\langle F[\mathbf{\Psi}(x)]\rangle_{t+\delta t} - \langle F[\mathbf{\Psi}(x)]\rangle_t \\
&= \int D^2\psi\, D^2\psi^+ F[\mathbf{\Psi}(x)] \left\{ -\sum_A \int dx \left(\frac{\delta}{\delta\psi_A(x)}\right)_x (A_A(\boldsymbol{\psi}(x), x) P_b[\boldsymbol{\psi}(x), \boldsymbol{\psi}^*(x)]) \right. \\
&\quad \left. + \frac{1}{2}\sum_{A,B} \int dx\, dy \left(\frac{\delta}{\delta\psi_A(x)}\right)_x \left(\frac{\delta}{\delta\psi_B(y)}\right)_y (D_{AB}(\boldsymbol{\psi}(x), x, \boldsymbol{\psi}(y), y) P_b[\boldsymbol{\psi}(x), \boldsymbol{\psi}^*(x)]) \right\} \delta t \\
&= \int D^2\psi\, D^2\psi^+ \sum_A \int dx \left(\left(\frac{\delta}{\delta\psi_A(x)}\right)_x F[\mathbf{\Psi}(x)]\right) (A_A(\boldsymbol{\psi}(x), x) P_b[\boldsymbol{\psi}(x), \boldsymbol{\psi}^*(x)]\, \delta t \\
&\quad + \int D^2\psi\, D^2\psi^+ \frac{1}{2} \sum_A \sum_B \int\int dx\, dy \left(\left(\frac{\delta}{\delta\psi_A(x)}\right)_x \left(\frac{\delta}{\delta\psi_B(y)}\right)_y F[\mathbf{\Psi}(x)]\right) \\
&\quad \times (D_{AB}(\boldsymbol{\psi}(x), x, \boldsymbol{\psi}(y), y) P_b[\boldsymbol{\psi}(x), \boldsymbol{\psi}^*(x)]\, \delta t.
\end{aligned} \tag{14.24}$$

Hence

$$\begin{aligned}
&\frac{d}{dt}\langle F[\mathbf{\Psi}(x)]\rangle_t \\
&= \left\langle \sum_A \int dx \left(\left(\frac{\delta}{\delta\psi_A(x)}\right)_x F[\mathbf{\Psi}(x)]\right) A_A(\boldsymbol{\psi}(x), x) \right\rangle \\
&\quad + \left\langle \frac{1}{2}\sum_{A,B} \int dx\, dy \left(\left(\frac{\delta}{\delta\psi_A(x)}\right)_x \left(\frac{\delta}{\delta\psi_B(y)}\right)_y F[\mathbf{\Psi}(x)]\right) D_{AB}(\boldsymbol{\psi}(x), x, \boldsymbol{\psi}(y), y) \right\rangle
\end{aligned} \tag{14.25}$$

gives the equation of motion for the phase space functional average.

We now replace the fields $\mathbf{\Psi}(x)$ by stochastic fields $\tilde{\boldsymbol{\psi}}(x,t) \equiv \{\tilde{\psi}_1(x,t), \tilde{\psi}_2(x,t)\} \equiv \{\tilde{\psi}(x,t), \tilde{\psi}^+(x,t)\}$. These stochastic fields will later be expanded in terms of mode functions, but at this stage no specific mode expansion is needed. The stochastic average of the arbitrary functional $F[\tilde{\mathbf{\Psi}}(x,t)]$ is

$$\overline{F[\tilde{\mathbf{\Psi}}(x,t)]} = \frac{1}{m}\sum_{i=1}^{m} F[\tilde{\mathbf{\Psi}}(x,t)_i], \tag{14.26}$$

where there are $i = 1, \cdots, m$ samples $\tilde{\mathbf{\Psi}}(x,t)_i$ of the stochastic fields averaged over.

The change in the stochastic functional in the interval δt, correct to second order in the field fluctuations is given by

$$\begin{aligned} & F[\tilde{\mathbf{\Psi}}(x,t+\delta t)]\text{-}F[\tilde{\mathbf{\Psi}}(x,t)] \\ &= \sum_A \int dx \left(\left(\frac{\delta}{\delta \tilde{\psi}_A(x,t)} \right)_x F[\tilde{\mathbf{\Psi}}(x,t\ )] \right) \delta \tilde{\psi}_A(x,t) \\ &+ \frac{1}{2} \sum_{A,B} \int dx\, dy \left(\left(\frac{\delta}{\delta \tilde{\psi}_A(x,t)} \right)_x \left(\frac{\delta}{\delta \tilde{\psi}_B(y,t)} \right)_y F[\tilde{\mathbf{\Psi}}(x,t)] \right) \delta \tilde{\psi}_A(x,t)\, \delta \tilde{\psi}_B(y,t), \end{aligned} \tag{14.27}$$

where we have used the Taylor expansion for a functional given in (11.66) in terms of the field fluctuations $\delta\tilde{\psi}_A(x,t)$. Expansion to second order is needed to give stochastic averages correct to $O(\delta t)$. Taking the stochastic average and using the theorems in Appendix G relating the stochastic average of a sum to the sum of the stochastic averages gives a formula for the average with the same terms as those in (14.27)

Now suppose the stochastic fields satisfy Ito stochastic field equations of the form

$$\delta \tilde{\psi}_A(x,t) = \mathcal{L}_A(\tilde{\boldsymbol{\psi}}(x,t),x)\,\delta t + \sum_{Dk} \phi_k^{AD}(\tilde{\boldsymbol{\psi}}(x,t),x) \int_t^{t+\delta t} dt_1\, \Gamma_k^D(t_1), \tag{14.28}$$

where the quantities $\mathcal{L}_A(\tilde{\boldsymbol{\psi}}(x,t),x)$ and $\phi_k^{AD}(\tilde{\boldsymbol{\psi}}(x,t),x)$ are not yet known but will be determined from the requirement that the time derivatives of the phase space functional average of the arbitrary functional F and the stochastic average are to be the same. The Γ_k^D are the usual Gaussian–Markov random noise terms ($a = 1, 2, \cdots$), whose stochastic averages are given in (9.11). The Ito stochastic field equation can also be written as

$$\frac{\partial}{\partial t}\tilde{\psi}_A(x,t) = \mathcal{L}_A(\tilde{\boldsymbol{\psi}}(x,t),x) + \sum_{Dk} \phi_k^{AD}(\tilde{\boldsymbol{\psi}}(x,t),x)\,\Gamma_k^D(t_+). \tag{14.29}$$

The non-correlation rule is now

$$\begin{aligned} & \overline{G[\tilde{\mathbf{\Psi}}(x,t_1)]\Gamma_a(t_2)\Gamma_b(t_3)\Gamma_c(t_4)\cdots\Gamma_k(t_l)} \\ &= \overline{G[\tilde{\mathbf{\Psi}}(x,t_1)]}\ \overline{\Gamma_a(t_2)\Gamma_b(t_3)\Gamma_c(t_4)\cdots\Gamma_k(t_l)}, \qquad t_1 < t_2, t_3, \cdots, t_l, \end{aligned} \tag{14.30}$$

for any quantity G that depends only on the stochastic fields at time t_1.

We can now obtain expressions for the stochastic averages. For the first-order derivative terms, we have

$$\begin{aligned}
&\overline{\sum_A \int dx \left(\left(\frac{\delta}{\delta\tilde{\psi}_A(x,t)} \right)_x F[\tilde{\mathbf{\Psi}}(x,t)] \right) \delta\tilde{\psi}_A(x,t)} \\
&= \overline{\sum_A \int dx \left(\left(\frac{\delta}{\delta\tilde{\psi}_A(x,t)} \right)_x F[\tilde{\mathbf{\Psi}}(x,t)] \right) \mathcal{L}_A(\tilde{\psi}(x,t),x)\, \delta t} \\
&+ \overline{\sum_A \int dx \left(\left(\frac{\delta}{\delta\tilde{\psi}_A(x,t)} \right)_x F[\tilde{\mathbf{\Psi}}(x,t)] \right) \sum_{Dk} \phi_k^{AD}(\tilde{\psi}(x,t),x)} \int_t^{t+\delta t} \overline{dt_1\, \Gamma_k^D(t_1)} \\
&= \overline{\sum_A \int dx \left(\left(\frac{\delta}{\delta\tilde{\psi}_A(x,t)} \right)_x F[\tilde{\mathbf{\Psi}}(x,t)] \right) \mathcal{L}_A(\tilde{\psi}(x,t),x)\, \delta t},
\end{aligned} \tag{14.31}$$

where the stochastic-average rules for sums and products have been used, the non-correlation rule (14.30) between the averages of functions of $\alpha_{p\mu}^s(t)$ at time t and the Γ at later times between t and $t+\delta t$ has been applied, and the term involving $\overline{\Gamma_k^D(t_1)}$ is equal to zero from (14.7). Note that this term is proportional to δt.

For the second-order derivative terms,

$$\begin{aligned}
&\overline{\frac{1}{2} \sum_{A,B} \int dx\, dy \left(\left(\frac{\delta}{\delta\tilde{\psi}_A(x,t)} \right)_x \left(\frac{\delta}{\delta\tilde{\psi}_B(y,t)} \right)_y F[\tilde{\mathbf{\Psi}}(x,t)] \right) \delta\tilde{\psi}_A(x,t)\, \delta\tilde{\psi}_B(y,t)} \\
&= \overline{\left\{ \begin{array}{l} \frac{1}{2} \sum_{A,B} \int dx\, dy \left(\left(\frac{\delta}{\delta\tilde{\psi}_A(x,t)} \right)_x \left(\frac{\delta}{\delta\tilde{\psi}_B(y,t)} \right)_y F[\tilde{\mathbf{\Psi}}(x,t\ )] \right) \\ \times \left[\mathcal{L}_A(\tilde{\psi}(x,t),x)\, \delta t + \sum_{Dk} \phi_k^{AD}(\tilde{\psi}(x,t),x) \int_t^{t+\delta t} dt_1\, \Gamma_k^D(t_1) \right] \\ \times \left[\mathcal{L}_B(\tilde{\psi}(y,t),y)\, \delta t + \sum_{El} \phi_l^{BE}(\tilde{\psi}(y,t),y) \int_t^{t+\delta t} dt_2\, \Gamma_l^E(t_2) \right], \end{array} \right\}}
\end{aligned} \tag{14.32}$$

and there are four terms when the factors are multiplied out. The stochastic averages for the functions of $\tilde{\psi}_A(x,t)$ and the Γ_A are uncorrelated. The resulting terms involving a single Γ have a zero stochastic average, and from (9.16) the terms with two Γ give a stochastic average proportional to δt, so that correct to order δt, the second-order derivative term is (where we write $\phi_k^{AD}(\tilde{\psi}(x,t),x) = \phi_k^{AD}$ etc.)

$$\begin{aligned}
&\overline{\frac{1}{2} \sum_{A,B} \int dx\, dy \left(\left(\frac{\delta}{\delta\tilde{\psi}_A(x,t)} \right)_x \left(\frac{\delta}{\delta\tilde{\psi}_B(y,t)} \right)_y F[\tilde{\mathbf{\Psi}}(x,t)] \right) \delta\tilde{\psi}_A(x,t)\, \delta\tilde{\psi}_B(y,t)} \\
&= \overline{\frac{1}{2} \sum_{A,B} \int dx\, dy \left(\left(\frac{\delta}{\delta\tilde{\psi}_A(x,t)} \right)_x \left(\frac{\delta}{\delta\tilde{\psi}_B(y,t)} \right)_y F[\tilde{\mathbf{\Psi}}(x,t)] \right) \left[\sum_{Dk} \phi_k^{AD} \phi_k^{BD} \right]}\, \delta t.
\end{aligned} \tag{14.33}$$

The remaining terms give stochastic averages correct to order δt^2 or higher, so that we have, correct to first order in δt,

$$\overline{F[\tilde{\mathbf{\Psi}}(x,t+\delta t)]\text{-}F[\tilde{\mathbf{\Psi}}(x,t)]}$$
$$= \overline{\sum_A \int dx \left(\left(\frac{\delta}{\delta \tilde{\psi}_A(x,t)} \right)_x F[\tilde{\mathbf{\Psi}}(x,t)] \right) \mathcal{L}_A(\tilde{\psi}(x,t),x)\, \delta t}$$
$$+ \overline{\frac{1}{2} \sum_{A,B} \int dx\, dy \left(\left(\frac{\delta}{\delta \tilde{\psi}_A(x,t)} \right)_x \left(\frac{\delta}{\delta \tilde{\psi}_B(y,t)} \right)_y F[\tilde{\mathbf{\Psi}}(x,t)] \right) \left[\sum_{Dk} \phi_k^{AD} \phi_k^{BD} \right]}\, \delta t \tag{14.34}$$

or

$$\frac{d}{dt}\overline{F[\tilde{\mathbf{\Psi}}(x,t)]}$$
$$= \overline{\sum_A \int dx \left(\left(\frac{\delta}{\delta \tilde{\psi}_A(x,t)} \right)_x F[\tilde{\mathbf{\Psi}}(x,t)] \right) \mathcal{L}_A(\tilde{\psi}(x,t),x)}$$
$$+ \overline{\frac{1}{2} \sum_{A,B} \int dx\, dy \left(\left(\frac{\delta}{\delta \tilde{\psi}_A(x,t)} \right)_x \left(\frac{\delta}{\delta \tilde{\psi}_B(y,t)} \right)_y F[\tilde{\mathbf{\Psi}}(x,t)] \right) \left[\sum_{Dk} \phi_k^{AD} \phi_k^{BD} \right]}. \tag{14.35}$$

This result is the same as that based on the phase space functional average in (14.25) if we have the following relationships between the matrices A and D in the functional Fokker–Planck equation and the matrices $\mathcal{L}$ and ϕ occurring in the Ito stochastic differential equation:

$$\mathcal{L}_A(\tilde{\psi}(x,t),x) = A_A(\tilde{\psi}(x,t),x),$$
$$\sum_{Dk} \phi_k^{AD}(\tilde{\boldsymbol{\psi}}(x,t),x)\phi_k^{BD}(\tilde{\boldsymbol{\psi}}(y,t),y) = D_{AB}(\tilde{\boldsymbol{\psi}}(x,t),x,\tilde{\boldsymbol{\psi}}(y,t),y). \tag{14.36}$$

Clearly, the term $\mathcal{L}_A$ in the Ito stochastic field equation is given by the drift term A_A in the functional Fokker–Planck equation, the same result found from the derivation based on separate modes that involves the Ito stochastic differential equation and the ordinary Fokker–Planck equation. Similarly, the terms ϕ_k^{AD} in the Ito stochastic field equation are given by the diffusion term in the functional Fokker–Planck equation, and in view of (14.18) can be taken to be the same as the quantity η_k^{AD} found from the derivation based on separate modes that involves the Ito stochastic differential equation and the ordinary Fokker–Planck equation. Note that the matrices ϕ and η are undetermined up to multiplication by an orthogonal matrix. Thus we have

$$\mathcal{L}_A = A_A,$$
$$\phi_k^{AD} = \eta_k^{AD}, \qquad (\eta\eta^T)^{AB} = D_{AB}. \tag{14.37}$$

Hence we have shown that the Ito stochastic equations for phase variables and fields are entirely equivalent.

14.1.4 Properties of Bosonic Noise Fields

To determine the properties of the noise field $(\partial/\partial t)\tilde{G}_A(\tilde{\psi}(x,t),\mathbf{\Gamma}(t_+))$ defined in (14.22), we will use the result (14.18). The quantity $[\eta(\tilde{\psi}(x,t),x)\ \eta(\tilde{\psi}(y,t),y)^T]_{AB}$ for the same time t is equal to the non-local diffusion matrix element $D_{AB}(\tilde{\psi}(x,t), x,\tilde{\psi}(y,t),y)$.

The stochastic averages of the noise field terms can now be obtained. These results follow from (14.18) and the properties (14.7), (14.11) and (14.12). For the stochastic average of each noise term,

$$\begin{aligned}
\overline{\left(\frac{\partial}{\partial t}\tilde{G}_A(\tilde{\psi}(x,t),\mathbf{\Gamma}(t_+))\right)} &= \overline{\sum_{Dk}\eta_k^{AD}(\tilde{\psi}(x,t),x)\,\Gamma_k^D(t_+)} \\
&= \sum_{Dk}\overline{\eta_k^{AD}(\tilde{\psi}(x,t),x)\,\Gamma_k^D(t_+)} \\
&= \sum_{Dk}\overline{\eta_k^{AD}(\tilde{\psi}(x,t),x)}\ \overline{\Gamma_k^D(t_+)} \\
&= 0,
\end{aligned} \tag{14.38}$$

showing that the stochastic average of each noise field is zero. For the stochastic average of the product of two noise terms, we have

$$\begin{aligned}
&\overline{\left(\frac{\partial}{\partial t}\tilde{G}_A(\tilde{\psi}(x_1,t_1),\mathbf{\Gamma}(t_{1+}))\right)\left(\frac{\partial}{\partial t}\tilde{G}_B(\tilde{\psi}(x_2,t_2),\mathbf{\Gamma}(t_{2+}))\right)} \\
&= \overline{\sum_{Dk}\eta_k^{AD}(\tilde{\psi}(x_1,t_1),x_1)\,\Gamma_k^D(t_{1+})\sum_{El}\eta_l^{BE}(\tilde{\psi}(x_2,t_2),x_2)\,\Gamma_l^E(t_{2+})} \\
&= \sum_{Dk}\sum_{El}\overline{\eta_k^{AD}(\tilde{\psi}(x_1,t_1),x_1)\eta_l^{BE}(\tilde{\psi}(x_2,t_2),x_2)\ \Gamma_k^D(t_{1+})\Gamma_l^E(t_{2+})} \\
&= \sum_{Dk}\sum_{El}\overline{\eta_k^{AD}(\tilde{\psi}(x_1,t_{1,2}),x_1)\eta_l^{BE}(\tilde{\psi}(x_2,t_{1,2}),x_2)}\overline{\Gamma_k^D(t_{1+})\Gamma_l^E(t_{2+})} \\
&= \sum_{Dk}\sum_{El}\overline{\eta_k^{AD}(\tilde{\psi}(x_1,t_{1,2}),x_1)\eta_l^{BE}(\tilde{\psi}(x_2,t_{1,2}),x_2)}\ \delta_{DE}\delta_{kl}\delta(t_1-t_2) \\
&= \overline{\sum_{Dk}\eta_k^{AD}(\tilde{\psi}(x_1,t_{1,2}),x_1)\eta_k^{BD}(\tilde{\psi}(x_2,t_{1,2}),x_2)}\ \delta(t_1-t_2) \\
&= \overline{D_{AB}(\tilde{\psi}(x_1,t_{1,2}),x_1,\tilde{\psi}(x_2,t_{1,2}),x_2)}\,\delta(t_1-t_2).
\end{aligned} \tag{14.39}$$

In going from line three to line four, it has been implicitly assumed that t_1 and t_2 are less than both t_{1+} and t_{2+}. This implies $t_1 = t_2$. If we had $t_1 < t_2$, then $\Gamma_l^E(t_{2+})$ would be uncorrelated with the remaining factors $\eta_k^{AD}(\tilde{\psi}(x_1,t_1),x_1)\eta_l^{BE}(\tilde{\psi}(x_2,t_2),x_2)$ $\Gamma_k^D(t_{1+})$ so the fourth line would have vanished, since the average of a single Gaussian–Markov noise term is zero. A similar result applies if $t_2 < t_1$. The stochastic average of the product of two noise terms is always delta function correlated in time. However, in general it is not delta function correlated in space. Instead, the spatial correlation is given by the stochastic average of the non-local diffusion term $D_{AB}(\tilde{\psi}(x_1,t),x_1,\tilde{\psi}(x_2,t),x_2)$ in the original functional Fokker–Planck equation (13.69).

Although the noise fields have some of the features of (14.7), they are not themselves Gaussian–Markov processes. The stochastic averages of products of odd numbers of noise fields are indeed zero, but although averages of products of even numbers of noise fields can be written as sums of products of stochastic averages of pairs of stochastic quantities with the same delta function time correlations as in (14.7), the pairs involved are the diffusion matrix elements $D_{AB}(\tilde{\psi}(x_1,t),x_1,\tilde{\psi}(x_2,t),x_2)$ rather than products of noise fields such as $\left((\partial/\partial t)\tilde{G}_A(\tilde{\psi}(x_1,t_1),\mathbf{\Gamma}(t_{1+}))\right)\left((\partial/\partial t)\tilde{G}_B(\tilde{\psi}(x_2,t_2),\mathbf{\Gamma}(t_{2+}))\right)$. Nevertheless, the stochastic averages of the noise field terms either are zero or are determined from stochastic averages involving only the diffusion matrix elements $D_{AB}(\tilde{\psi}(x_1,t),x_1,\tilde{\psi}(x_2,t),x_2)$. There is thus never any need to actually determine the matrices $\eta(\tilde{\psi}(x,t),x)$ such that $\eta(\tilde{\psi}(x_1,t))\eta(\tilde{\psi}(x_2,t))^T = D(\tilde{\psi}(x_1,t),x_1,\tilde{\psi}(x_2,t),x_2)$ or $D(\tilde{\psi}(x_{1,2},t),x_{1,2})\delta(x_1-x_2)$, so all the required expressions for treating the stochastic properties of the noise fields are provided in the functional Fokker–Planck equation. The average for three noise terms vanishes; for four noise terms, the result is

$$\begin{aligned}
&\overline{\left\{\begin{array}{l}\left(\frac{\partial}{\partial t}\tilde{G}_A(\tilde{\psi}(x_1,t_1),\mathbf{\Gamma}(t_{1+}))\right)\left(\frac{\partial}{\partial t}\tilde{G}_B(\tilde{\psi}(x_2,t_2),\mathbf{\Gamma}(t_{2+}))\right)\\ \times\left(\frac{\partial}{\partial t}\tilde{G}_C(\tilde{\psi}(x_3,t_3),\mathbf{\Gamma}(t_{3+}))\right)\left(\frac{\partial}{\partial t}\tilde{G}_D(\tilde{\psi}(x_4,t_4),\mathbf{\Gamma}(t_{4+}))\right)\end{array}\right\}}\\
&\quad=\overline{\left[D_{AB}(\tilde{\psi}(x_1,t_{1,2}),x_1,\tilde{\psi}(x_2,t_{1,2}),x_2)\right]\left[D_{CD}(\tilde{\psi}(x_3,t_{3,4}),x_3,\tilde{\psi}(x_4,t_{3,4}),x_4)\right]}\\
&\qquad\times\delta(t_1-t_2)\delta(t_3-t_4)\\
&\qquad+\overline{\left[D_{AC}(\tilde{\psi}(x_1,t_{1,3}),x_1,\tilde{\psi}(x_3,t_{1,3}),x_3)\right]\left[D_{BD}(\tilde{\psi}(x_2,t_{2,4}),x_2,\tilde{\psi}(x_4,t_{2,4}),x_4)\right]}\\
&\qquad\times\delta(t_1-t_3)\delta(t_2-t_4)\\
&\qquad+\overline{\left[D_{AD}(\tilde{\psi}(x_1,t_{1,4}),x_1,\tilde{\psi}(x_4,t_{1,4}),x_2)\right]\left[D_{BC}(\tilde{\psi}(x_2,t_{2,3}),x_2,\tilde{\psi}(x_3,t_{2,3}),x_3)\right]}\\
&\qquad\times\delta(t_1-t_4)\delta(t_2-t_3).
\end{aligned}\tag{14.40}$$

The result for the stochastic average of four noise field terms is not quite the same as

$$\begin{aligned}
&\overline{\left\{\left(\frac{\partial}{\partial t}\tilde{G}_A(\tilde{\psi}(x_1,t_1),\mathbf{\Gamma}(t_{1+}))\right)\left(\frac{\partial}{\partial t}\tilde{G}_B(\tilde{\psi}(x_2,t_2),\mathbf{\Gamma}(t_{2+}))\right)\right\}}\\
&\times\overline{\left\{\left(\frac{\partial}{\partial t}\tilde{G}_C(\tilde{\psi}(x_3,t_3),\mathbf{\Gamma}(t_{3+}))\right)\left(\frac{\partial}{\partial t}\tilde{G}_D(\tilde{\psi}(x_4,t_4),\mathbf{\Gamma}(t_{4+}))\right)\right\}}\\
&+\overline{\left\{\left(\frac{\partial}{\partial t}\tilde{G}_A(\tilde{\psi}(x_1,t_1),\mathbf{\Gamma}(t_{1+}))\right)\left(\frac{\partial}{\partial t}\tilde{G}_C(\tilde{\psi}(x_3,t_3),\mathbf{\Gamma}(t_{3+}))\right)\right\}}\\
&\times\overline{\left\{\left(\frac{\partial}{\partial t}\tilde{G}_B(\tilde{\psi}(x_2,t_2),\mathbf{\Gamma}(t_{2+}))\right)\left(\frac{\partial}{\partial t}\tilde{G}_D(\tilde{\psi}(x_4,t_4),\mathbf{\Gamma}(t_{4+}))\right)\right\}}\\
&+\overline{\left\{\left(\frac{\partial}{\partial t}\tilde{G}_A(\tilde{\psi}(x_1,t_1),\mathbf{\Gamma}(t_{1+}))\right)\left(\frac{\partial}{\partial t}\tilde{G}_D(\tilde{\psi}(x_4,t_4),\mathbf{\Gamma}(t_{4+}))\right)\right\}}\\
&\times\overline{\left\{\left(\frac{\partial}{\partial t}\tilde{G}_B(\tilde{\psi}(x_2,t_2),\mathbf{\Gamma}(t_{2+}))\right)\left(\frac{\partial}{\partial t}\tilde{G}_C(\tilde{\psi}(x_3,t_3),\mathbf{\Gamma}(t_{3+}))\right)\right\}},
\end{aligned}\tag{14.41}$$

because in general the stochastic average of a product of two diffusion matrix elements is not the same as the product of the stochastic averages of each element. The proofs that the averages have these forms are left as an exercise. A careful consideration of various subcases involving the time order of t_1, t_2, t_3, t_4 is required.

Classical field equations can be obtained from the Ito equations by ignoring the quantum noise term. The classical field equations are

$$\frac{\partial \psi_A^{\text{class}}(x,t)}{\partial t} = A_A(\tilde{\psi}^{\text{class}}(x,t), x) \tag{14.42}$$

for the bosonic field.

14.2 Fermion Stochastic Field Equations

14.2.1 Ito Equations for Fermionic Stochastic Phase Space Variables

The functional Fokker–Planck equation (13.83) is equivalent to the ordinary Fokker–Planck equation (13.75),

$$\frac{\partial}{\partial t} P_{\text{f}}(\boldsymbol{g}) = -\sum_{Ai} (\mathcal{A}_i^A P_{\text{f}}) \frac{\overleftarrow{\partial}}{\partial g_i^A} + \frac{1}{2} \sum_{Ai} \sum_{Bj} (\mathcal{D}_{ij}^{AB} P_{\text{f}}) \frac{\overleftarrow{\partial}}{\partial g_j^B} \frac{\overleftarrow{\partial}}{\partial g_i^A}, \tag{14.43}$$

where we write $\boldsymbol{g} \equiv \{g^1, g^2\} \equiv \{g_i^A\}$. In terms of the previous notation, $p \to A, i$, $g_p \to g_i^A$, $A_p \to \mathcal{A}_i^A$, $D_{pq} \to \mathcal{D}_{ij}^{AB}$.

From the theory presented in Chapter 9, we can immediately write down the equivalent Ito stochastic equations for time-dependent stochastic variables $\tilde{g}_i^A(t)$. As before, the procedure involves replacing the time-independent phase space variables g_i^A by time-dependent stochastic variables $\tilde{g}_i^A(t)$. The Ito stochastic equations for the $\tilde{g}_i^A(t)$ are such that phase space averages of functions of the g_i^A give the same result as stochastic averages of the same functions of the $\tilde{g}_i^A(t)$. These equations for the stochastic expansion coefficients $\tilde{g}_i^A$ can be written in several forms:

$$\begin{aligned} \delta \tilde{g}_i^A(t) &= \tilde{g}_i^A(t+\delta t) - \tilde{g}_i^A(t) \\ &= -\mathcal{A}_i^A(\tilde{\boldsymbol{g}}(t))\,\delta t + \sum_{Dk} \mathcal{B}_{ik}^{AD}(\tilde{\boldsymbol{g}}(t)) \int_t^{t+\delta t} dt_1\, \Gamma_k^D(t_1), \end{aligned} \tag{14.44}$$

$$\frac{d}{dt}\tilde{g}_i^A(t) = -\mathcal{A}_i^A(\tilde{\boldsymbol{g}}(t)) + \sum_{Dk} \mathcal{B}_{ik}^{AD}(\tilde{\boldsymbol{g}}(t)) \frac{d}{dt} w_k^D(t) \tag{14.45}$$

$$= -\mathcal{A}_i^A(\tilde{\boldsymbol{g}}(t)) + \sum_{Dk} \mathcal{B}_{ik}^{AD}(\tilde{\boldsymbol{g}}(t)) \Gamma_k^D(t_+), \tag{14.46}$$

where $\tilde{\boldsymbol{g}}(t) \equiv \{\tilde{g}_i^A(t)\} \equiv \{\tilde{g}_i(t), \tilde{g}_i^+(t)\}$ and where, in the new notation, $\mathcal{B}_A^p \to \mathcal{B}_{ik}^{AD}$, $\Gamma_A \to \Gamma_k^D$. The results in (9.51) have been used to relate the Ito equation to the Fokker–Planck equation.

The diffusion matrix $\mathcal{D}$ is antisymmetric and, as shown in Chapter 9, we can always write the diffusion matrix $\mathcal{D}$ in the factorised form (see (9.62))

$$\mathcal{D} = \mathcal{B}\mathcal{B}^T. \tag{14.47}$$

In detail, the matrix $\mathcal{B}$ is related to the diffusion matrix $\mathcal{D}$ as in (9.62):

$$\mathcal{D}_{ij}^{AB}(\tilde{\boldsymbol{g}}(t)) = \sum_{Dk} \mathcal{B}_{ik}^{AD}(\tilde{\boldsymbol{g}}(t)) \mathcal{B}_{jk}^{BD}(\tilde{\boldsymbol{g}}(t)). \tag{14.48}$$

The matrix elements $\mathcal{B}_{ik}^{AD}(\tilde{\boldsymbol{g}}(t))$ are functions of the $\tilde{g}_i^A(t)$. The quantity t_+ is to indicate that if the Ito stochastic equation is integrated from t to $t+\delta t$, the Gaussian–Markov noise term is integrated over this interval whilst the $\mathcal{A}_i^A(\tilde{g}_j^C(t))$ and $\mathcal{B}_{ik}^{AD}(\tilde{g}_j^C(t))$ are left at time t.

The quantities $w_k^D(t)$ and $\Gamma_k^D(t)$ are Wiener and Gaussian–Markov stochastic variables as defined earlier in (9.11) and (9.28). In this chapter, the key results are set out above in (14.7), (14.8), (14.11) and (14.12).

14.2.2 Derivation of Fermionic Ito Stochastic Field Equations

The fermionic stochastic fields $\tilde{\psi}_A(x,t)$ are defined via the same expansion as for the time-independent field functions $\psi_A(x)$ by replacing the fermionic time-independent phase space variables g_i^A by time-dependent stochastic variables $\tilde{g}_i^A(t)$:

$$\tilde{\psi}_A(x,t) - \sum_i \tilde{g}_i^A(t)\zeta_i^A(x). \tag{14.49}$$

Note that the fermionic stochastic fields are spatial Grassmann functions. The expansion coefficients in (14.13) are restricted to those required for expanding the particular field function $\psi_A(x)$. Also, the stochastic variations in $\tilde{\psi}_A(x,t)$ are chosen so as to be due only to stochastic fluctuations in the $\tilde{g}_i^A(t)$. Although the mode functions may be time-dependent, their time variations are not stochastic in origin, so the stochastic field equations for the $\tilde{\psi}_A(x,t)$ do not allow for time variations in the mode functions. The $\tilde{g}_i^A(t)$ may be considered as functionals of the stochastic field $\tilde{\psi}_A(x,t)$. The pair of stochastic fields are denoted $\tilde{\boldsymbol{\psi}}(x,t) \equiv \{\tilde{\psi}_1(x,t), \tilde{\psi}_2(x,t)\} \equiv \{\tilde{\psi}(x,t), \tilde{\psi}^+(x,t)\}$.

The change in the stochastic field is

$$\begin{aligned}\delta\tilde{\psi}_A(x,t) &= \tilde{\psi}_A(x,t+\delta t) - \tilde{\psi}_A(x,t)\\ &= \sum_i \delta\tilde{g}_i^A(t)\xi_i^A(x).\end{aligned} \tag{14.50}$$

The Ito stochastic equation for the stochastic fields $\tilde{\psi}_A(x,t)$ can then be derived from the Ito stochastic equations for the expansion coefficients. Using (13.84), the drift term is

$$-\sum_i \mathcal{A}_i^A(\tilde{\boldsymbol{g}}(t))\,\xi_i^A(x)\,\delta t = -A_A(\tilde{\boldsymbol{\psi}}(x,t),x)\,\delta t, \tag{14.51}$$

where A_A is the drift vector in the functional Fokker–Planck equation. This depends on x and is also a functional of the stochastic fields $\tilde{\boldsymbol{\psi}}(x,t)$. The diffusion term in the stochastic equation gives

$$\sum_i\sum_{Dk} \mathcal{B}_{ik}^{AD}(\tilde{\boldsymbol{g}}(t))\,\xi_i^A(x)\int_t^{t+\delta t} dt_1\Gamma_k^D(t_1) = \sum_{Dk}\eta_k^{AD}(\tilde{\boldsymbol{\psi}}(x,t),x)\int_t^{t+\delta t} dt_1\,\Gamma_k^D(t_1), \tag{14.52}$$

where

$$\eta_k^{AD}(\tilde{\boldsymbol{\psi}}(x,t),x) = \sum_i \mathcal{B}_{ik}^{AD}(\tilde{\boldsymbol{g}}(t))\,\xi_i^A(x) \tag{14.53}$$

depends on x and is a functional of the $\tilde{\psi}_A(x,t)$ via the $\tilde{g}_i^A(t)$. The $\eta_k^{AD}(\tilde{\boldsymbol{\psi}}(x,t),x)$ are related via $\mathcal{B}_{ik}^{AD}(\tilde{\boldsymbol{g}}(t))$ to the diffusion matrix D_{AB} in the functional Fokker–Planck equation. Using (13.85), we have

$$
\begin{aligned}
\sum_{Dk} \eta_k^{AD}(\tilde{\psi}(x,t),x))\eta_k^{BD}(\tilde{\psi}(y,t),y) &= \sum_{Dk}\sum_i \mathcal{B}_{ik}^{AD}\xi_i^A(x)\sum_j \mathcal{B}_{jk}^{BD}\xi_j^B(y) \\
&= \sum_{ij} \xi_i^A(x)\left(\sum_{Dk}\mathcal{B}_{ik}^{AD}\mathcal{B}_{jk}^{BD}\right)\xi_j^B(y) \\
&= \sum_{ij} \xi_i^A(x)\,(\mathcal{D})_{ij}^{AB}\,\xi_j^B(y) \\
&= D_{AB}(\tilde{\psi}(x,t),x,\tilde{\psi}(y,t),y),
\end{aligned} \tag{14.54}
$$

which gives the diffusion matrix D_{AB} in the functional Fokker–Planck equation.

The stochastic field equations are then given by

$$
\begin{aligned}
\delta\tilde{\psi}_A(x,t) &= -A_A(\tilde{\psi}(x,t),x)\,\delta t + \sum_{Dk}\eta_k^{AD}(\tilde{\psi}(x,t),x)\int_t^{t+\delta t} dt_1\,\Gamma_k^D(t_1) \\
&= -A_A(\tilde{\psi}(x,t),x)\,\delta t + \delta\tilde{G}_A(\tilde{\psi}(x,t),\ \Gamma(t_+)),
\end{aligned} \tag{14.55}
$$

$$
\begin{aligned}
\frac{\partial}{\partial t}\tilde{\psi}_A(x,t) &= -A_A(\tilde{\psi}(x,t),x) + \sum_{Dk}\eta_k^{AD}(\tilde{\psi}(x,t),x)\,\frac{d}{dt}w_k^D(t) \\
&= -A_A(\tilde{\psi}(x,t),x) + \sum_{Dk}\eta_k^{AD}(\tilde{\psi}(x,t),x)\,\Gamma_k^D(t_+) \\
&= -A_A(\tilde{\psi}(x,t),x) + \frac{\partial}{\partial t}\tilde{G}_A(\tilde{\psi}(x,t),\mathbf{\Gamma}(t_+)),
\end{aligned} \tag{14.56}
$$

where

$$
\begin{aligned}
(\eta\eta^T)^{AB} &= \sum_{Dk}\eta_k^{AD}(\tilde{\psi}(x,t),x)\eta_k^{BD}(\tilde{\psi}(y,t),y) \\
&= D_{AB}(\psi(x,t),x,\psi(y,t),y).
\end{aligned} \tag{14.57}
$$

Here we write $\mathbf{\Gamma}(t_+) \equiv \{\Gamma_k^1(t_+),\Gamma_k^2(t_+)\}$. The first form gives the change in the stochastic field over a small time interval $t, t+\delta t$; the second is in the form of a partial differential equation. The first term in the Ito equation for the stochastic fields (14.56), $A_A(\tilde{\psi}(x,t),x)$, is the deterministic term and is obtained from the drift vector in the functional Fokker–Planck equation, and the second term, $(\partial/\partial t)\tilde{G}_A(\tilde{\psi}(x,t),\mathbf{\Gamma}(t_+))$, is the quantum noise field, whose statistical properties are found from the diffusion matrix and which depends both on the stochastic fields $\tilde{\psi}(x,t)$ and on the Gaussian–Markov stochastic variables $\mathbf{\Gamma}(t_+)$.

The noise field term is

$$
\frac{\partial}{\partial t}\tilde{G}_A(\tilde{\psi}(x,t),\mathbf{\Gamma}(t_+)) = \sum_{Dk}\eta_k^{AD}(\tilde{\psi}(x,t),x)\,\frac{d}{dt}w_k^D(t) = \sum_{Dk}\eta_k^{AD}(\tilde{\psi}(x,t),x)\,\Gamma_k^D(t_+), \tag{14.58}
$$

where the stochastic field $\eta_k^{AD}(\tilde{\psi}(x,t),x)$ is related to the diffusion matrix. The noise field term depends on x and is a functional of the stochastic fields.

The Ito stochastic equation for Grassmann fields can be derived directly from the functional Fokker–Planck equation similarly to the above treatment for c-number boson fields. This derivation is left as an exercise.

14.2.3 Properties of Fermionic Noise Fields

To determine the properties of the noise field $(\partial/\partial t)\tilde{G}_A(\tilde{\psi}(x,t),\mathbf{\Gamma}(t_+))$ defined in (14.58), we will use the result (14.54). The quantity $[\eta(\tilde{\psi}(x,t),x)\eta(\tilde{\psi}(y,t),y)^T]_{AB}$ for the same time t is equal to the non-local diffusion matrix element $D_{AB}(\tilde{\psi}(x,t),x,\tilde{\psi}(y,t),y)$.

The stochastic averages of the noise field terms can now be obtained. These results follow from (14.54) and the properties (14.7), (14.11) and (14.12). For the stochastic average of each noise term,

$$\begin{aligned}\overline{\left(\frac{\partial}{\partial t}\tilde{G}_A(\tilde{\psi}(x,t),\mathbf{\Gamma}(t_+))\right)} &= \overline{\sum_{Dk}\eta_k^{AD}(\tilde{\psi}(x,t),x)\,\Gamma_k^D(t_+)}\\ &= \sum_{Dk}\overline{\eta_k^{AD}(\tilde{\psi}(x,t),x)\,\Gamma_k^D(t_+)}\\ &= \sum_{Dk}\overline{\eta_k^{AD}(\tilde{\psi}(x,t),x)}\;\overline{\Gamma_k^D(t_+)}\\ &= 0,\end{aligned} \tag{14.59}$$

showing that the stochastic average of each noise field is zero. For the stochastic average of the product of two noise terms, we have

$$\begin{aligned}&\overline{\left(\frac{\partial}{\partial t}\tilde{G}_A(\tilde{\psi}(x_1,t_1),\mathbf{\Gamma}(t_{1+}))\right)\left(\frac{\partial}{\partial t}\tilde{G}_B(\tilde{\psi}(x_2,t_2),\mathbf{\Gamma}(t_{2+}))\right)}\\ &= \overline{\sum_{Dk}\eta_k^{AD}(\tilde{\psi}(x_1,t_1),x_1)\Gamma_k^D(t_{1+})\sum_{El}\eta_l^{BE}(\tilde{\psi}(x_2,t_2),x_2)\,\Gamma_l^E(t_{2+})}\\ &= \sum_{Dk}\sum_{El}\overline{\eta_k^{AD}(\tilde{\psi}(x_1,t_1),x_1)\eta_l^{BE}(\tilde{\psi}(x_2,t_2),x_2)\Gamma_k^D(t_{1+})\Gamma_l^E(t_{2+})}\\ &= \sum_{Dk}\sum_{El}\overline{\eta_k^{AD}(\tilde{\psi}(x_1,t_{1,2}),x_1)\eta_l^{BE}(\tilde{\psi}(x_2,t_{1,2}),x_2)}\;\overline{\Gamma_k^D(t_{1+})\Gamma_l^E(t_{2+})}\\ &= \sum_{Dk}\sum_{El}\overline{\eta_k^{AD}(\tilde{\psi}(x_1,t_{1,2}),x_1)\eta_l^{BE}(\tilde{\psi}(x_2,t_{1,2}),x_2)}\;\delta_{DE}\delta_{kl}\delta(t_1-t_2)\\ &= \sum_{Dk}\overline{\eta_k^{AD}(\tilde{\psi}(x_1,t_{1,2}),x_1)\eta_k^{BD}(\tilde{\psi}(x_2,t_{1,2}),x_2)}\;\delta(t_1-t_2)\\ &= \overline{\sum_{Dk}\eta_k^{AD}(\tilde{\psi}(x_1,t_{1,2}),x_1)\eta_k^{BD}(\tilde{\psi}(x_2,t_{1,2}),x_2)}\;\delta(t_1-t_2)\\ &= \overline{D_{AB}(\tilde{\psi}(x_1,t_{1,2}),x_1,\tilde{\psi}(x_2,t_{1,2}),x_2)}\times\delta(t_1-t_2).\end{aligned} \tag{14.60}$$

As the $\Gamma_k^D(t_{1+})$ and $\Gamma_l^E(t_{2+})$ are c-numbers, we can commute these with the Grassmann functions in going from line two to line three. Note that in going from line three to line four it has been implicitly assumed that both t_1 and t_2 are less than both t_{1+} and t_{2+}. This implies $t_1 = t_2$. If we had $t_1 < t_2$, then $\Gamma_l^E(t_{2+})$ would be uncorrelated with the remaining factors $\overline{\eta_k^{AD}(\tilde{\psi}(x_1,t_1),x_1)\eta_l^{BE}(\tilde{\psi}(x_2,t_2),x_2)\Gamma_k^D(t_{1+})}$, so the fourth line would have been $\sum_{Dk}\sum_{El}\overline{\eta_k^{AD}(\tilde{\psi}(x_1,t_1),x_1)\eta_l^{BE}(\tilde{\psi}(x_2,t_2),x_2)\Gamma_k^D(t_{1+})}\ \overline{\Gamma_l^E(t_{2+})}$, which is zero, since the average of a single Gaussian–Markov noise term is zero. A similar result applies if $t_2 < t_1$. The stochastic average of the product of two noise terms is always delta function correlated in time. However, in general it is not delta function correlated in space. Instead, the spatial correlation is given by the stochastic average of the non-local diffusion term $D_{AB}(\tilde{\psi}(x_1,t),x_1,\tilde{\psi}(x_2,t),x_2)$ in the original functional Fokker–Planck equation (13.83).

However, although the noise fields have some of the features of (14.7), they are not themselves Gaussian–Markov processes. The stochastic averages of products of odd numbers of noise fields are indeed zero, but although averages of products of even numbers of noise fields can be written as sums of products of stochastic averages of pairs of stochastic quantities with the same delta function time correlations as in (14.7), the pairs involved are diffusion matrix elements $D_{AB}(\tilde{\psi}(x_1,t),x_1,\tilde{\psi}(x_2,t),x_2)$ rather than products of noise fields such as $\left((\partial/\partial t)\tilde{G}_A(\tilde{\psi}(x_1,t_1),\boldsymbol{\Gamma}(t_{1+}))\right)\left((\partial/\partial t)\tilde{G}_B(\tilde{\psi}(x_2,t_2),\boldsymbol{\Gamma}(t_{2+}))\right)$. Nevertheless, the stochastic averages of the noise field terms either are zero or are determined from stochastic averages involving only the diffusion matrix elements $D_{AB}(\tilde{\psi}(x_1,t),x_1,\tilde{\psi}(x_2,t),x_2)$. There is thus never any need to actually determine the matrices $\eta(\tilde{\psi}(x,t),x)$ such that $\eta(\tilde{\psi}(x_1,t))\eta(\tilde{\psi}(x_2,t))^T = D(\tilde{\psi}(x_1,t),x_1,\tilde{\psi}(x_2,t),x_2)$ or $D(\tilde{\psi}(x_{1,2},t),x_{1,2})\delta(x_1-x_2)$, so all the required expressions for treating the stochastic properties of the noise fields are provided in the functional Fokker–Planck equation.

For three noise terms the average vanishes, as already stated, and

$$
\begin{aligned}
&\overline{\left\{\begin{array}{l}\left(\frac{\partial}{\partial t}\tilde{G}_A(\tilde{\psi}(x_1,t_1),\boldsymbol{\Gamma}(t_{1+}))\right)\left(\frac{\partial}{\partial t}\tilde{G}_B(\tilde{\psi}(x_2,t_2),\boldsymbol{\Gamma}(t_{2+}))\right)\\ \times\left(\frac{\partial}{\partial t}\tilde{G}_C(\tilde{\psi}(x_3,t_3),\boldsymbol{\Gamma}(t_{3+}))\right)\left(\frac{\partial}{\partial t}\tilde{G}_D(\tilde{\psi}(x_4,t_4),\boldsymbol{\Gamma}(t_{4+}))\right)\end{array}\right\}}\\
&\quad=\overline{\left[D_{AB}(\tilde{\psi}(x_1,t_{1,2}),x_1,\tilde{\psi}(x_2,t_{1,2}),x_2)\right]\left[D_{CD}(\tilde{\psi}(x_3,t_{3,4}),x_3,\tilde{\psi}(x_4,t_{3,4}),x_4)\right]}\\
&\qquad\times\delta(t_1-t_2)\delta(t_3-t_4)\\
&\qquad+\overline{\left[D_{AC}(\tilde{\psi}(x_1,t_{1,3}),x_1,\tilde{\psi}(x_3,t_{1,3}),x_3)\right]\left[D_{BD}(\tilde{\psi}(x_2,t_{2,4}),x_2,\tilde{\psi}(x_4,t_{2,4}),x_4)\right]}\\
&\qquad\times\delta(t_1-t_3)\delta(t_2-t_4)\\
&\qquad+\overline{\left[D_{AD}(\tilde{\psi}(x_1,t_{1,4}),x_1,\tilde{\psi}(x_4,t_{1,4}),x_2)\right]\left[D_{BC}(\tilde{\psi}(x_2,t_{2,3}),x_2,\tilde{\psi}(x_3,t_{2,3}),x_3)\right]}\\
&\qquad\times\delta(t_1-t_4)\delta(t_2-t_3)
\end{aligned}
\tag{14.61}
$$

for four noise fields. The result for the stochastic average of four noise field terms is not quite the same as

$$\overline{\left\{\left(\frac{\partial}{\partial t}\tilde{G}_A(\tilde{\boldsymbol{\psi}}(x_1,t_1),\mathbf{\Gamma}(t_{1+}))\right)\left(\frac{\partial}{\partial t}\tilde{G}_B(\tilde{\boldsymbol{\psi}}(x_2,t_2),\mathbf{\Gamma}(t_{2+}))\right)\right\}}$$
$$\times\overline{\left\{\left(\frac{\partial}{\partial t}\tilde{G}_C(\tilde{\boldsymbol{\psi}}(x_3,t_3),\mathbf{\Gamma}(t_{3+}))\right)\left(\frac{\partial}{\partial t}\tilde{G}_D(\tilde{\boldsymbol{\psi}}(x_4,t_4),\mathbf{\Gamma}(t_{4+}))\right)\right\}}$$
$$+\overline{\left\{\left(\frac{\partial}{\partial t}\tilde{G}_A(\tilde{\boldsymbol{\psi}}(x_1,t_1),\mathbf{\Gamma}(t_{1+}))\right)\left(\frac{\partial}{\partial t}\tilde{G}_C(\tilde{\boldsymbol{\psi}}(x_3,t_3),\mathbf{\Gamma}(t_{3+}))\right)\right\}}$$
$$\times\overline{\left\{\left(\frac{\partial}{\partial t}\tilde{G}_B(\tilde{\boldsymbol{\psi}}(x_2,t_2),\mathbf{\Gamma}(t_{2+}))\right)\left(\frac{\partial}{\partial t}\tilde{G}_D(\tilde{\boldsymbol{\psi}}(x_4,t_4),\mathbf{\Gamma}(t_{4+}))\right)\right\}}$$
$$+\overline{\left\{\left(\frac{\partial}{\partial t}\tilde{G}_A(\tilde{\boldsymbol{\psi}}(x_1,t_1),\mathbf{\Gamma}(t_{1+}))\right)\left(\frac{\partial}{\partial t}\tilde{G}_D(\tilde{\boldsymbol{\psi}}(x_4,t_4),\mathbf{\Gamma}(t_{4+}))\right)\right\}}$$
$$\times\overline{\left\{\left(\frac{\partial}{\partial t}\tilde{G}_B(\tilde{\boldsymbol{\psi}}(x_2,t_2),\mathbf{\Gamma}(t_{2+}))\right)\left(\frac{\partial}{\partial t}\tilde{G}_C(\tilde{\boldsymbol{\psi}}(x_3,t_3),\mathbf{\Gamma}(t_{3+}))\right)\right\}}, \quad (14.62)$$

because in general the stochastic average of a product of two diffusion matrix elements is not the same as the product of the stochastic averages of each element. The proof of (14.61) is left as an exercise. Again, a careful consideration of various subcases involving the time order of t_1, t_2, t_3, t_4 is required.

As with the boson case, classical field equations can be obtained from the Ito equations by ignoring the quantum noise term. The classical field equations are

$$\frac{\partial \psi_A^{\text{class}}(x,t)}{\partial t} = -A_A(\tilde{\boldsymbol{\psi}}^{\text{class}}(x,t),x) \quad (14.63)$$

for the Grassmann fields.

14.3 Ito Field Equations – Generalisation to Several Fields

The Ito stochastic field equations treated above for bosons and fermions dealt with the situation where only one pair of field operators $\hat{\Psi}(x), \hat{\Psi}^\dagger(x)$ was involved. We now include the generalisation to the situation where more than one pair of field operators is involved, such as for spin-1/2 fermions. The stochastic fields will now be designated $\tilde{\psi}_{\alpha A}(x,t)$, following the notation in Section 13.3, and the stochastic phase space variables will be $\tilde{\alpha}_{\alpha i}^{A}$ (for bosons) and $\tilde{g}_{\alpha i}^{A}$ (for fermions). We write $\delta\tilde{\psi}_{\alpha A}(x,t) = \tilde{\psi}_{\alpha A}(x,t+\delta t) - \tilde{\psi}_{\alpha A}(x,t)$. For bosons, the Ito stochastic field equations are

$$\delta\tilde{\psi}_{\alpha A}(x,t) = A_{\alpha A}(\tilde{\boldsymbol{\psi}}(x,t),x)\,\delta t + \sum_{Dk}\eta_k^{\alpha AD}(\tilde{\boldsymbol{\psi}}(x,t),x)\int_t^{t+\delta t} dt_1\,\Gamma_k^D(t_1), \quad (14.64)$$

where $A_{\alpha A}$ is the drift term in the functional Fokker–Planck equation (13.93) and $\eta_k^{\alpha AD}$ is related to the diffusion matrix via

$$\sum_{Dk}\eta_k^{\alpha AD}(\tilde{\boldsymbol{\psi}}(x_1,t),x_1))\eta_k^{\beta BD}(\tilde{\boldsymbol{\psi}}(x_2,t),x_2) = D_{\alpha A\,\beta B}(\tilde{\boldsymbol{\psi}}(x_1,t),x_1,\tilde{\boldsymbol{\psi}}(x_2,t),x_2). \quad (14.65)$$

This is of course the standard Takagi factorisation.

For fermions, the Ito stochastic field equations are also of the form

$$\delta\tilde{\psi}_{\alpha A}(x,t) = -A_{\alpha A}(\tilde{\boldsymbol{\psi}}(x,t),x)\,\delta t + \sum_{Dk} \eta_k^{\alpha AD}(\tilde{\boldsymbol{\psi}}(x,t),x) \int_t^{t+\delta t} dt_1\,\Gamma_k^D(t_1), \quad (14.66)$$

where $A_{\alpha A}$ is the drift term in the functional Fokker–Planck equation (13.94) and $\eta_k^{\alpha AD}$ is related to the diffusion matrix as in the boson case in (14.65).

When several pairs of field operators occur, the same forms apply for the Ito stochastic field equations that apply to unnormalised distribution functionals.

14.4 Summary of Boson and Fermion Stochastic Field Equations

For convenience, we summarise here the key results derived in this section. These will be stated in terms of $P+$ distribution functionals but can easily be transferred to the unnormalised B distribution functionals, since the relationship between the functional Fokker–Planck equations and the Langevin field equations is of the same form.

14.4.1 Boson Case

The boson field operators, field functions and stochastic fields are given in terms of mode expansions as

$$\hat{\psi}_A(x) = \sum_i \hat{a}_i^A \xi_i^A(x), \quad (14.67)$$

$$\psi_A(x) = \sum_i \alpha_i^A \xi_i^A(x), \quad (14.68)$$

$$\tilde{\psi}_A(x,t) = \sum_i \tilde{\alpha}_i^A(t) \xi_i^A(x), \quad (14.69)$$

where $A = 1, 2$, $\hat{\psi}_1(x) = \hat{\psi}(x)$ and $\hat{\psi}_2(x) = \hat{\psi}^\dagger(x)$ are boson field operators, $\xi_i^1(x) = \phi_i(x)$ and $\xi_i^2(x) = \phi_i^*(x)$ are mode functions and their complex conjugates, $\hat{a}_i^1 = \hat{a}_i$ and $\hat{a}_i^2 = \hat{a}_i^\dagger$ are boson mode annihilation and creation operators, $\psi_1(x) = \psi(x)$ and $\psi_2(x) = \psi^+(x)$ are c-number field functions, $\tilde{\psi}_1(x) = \tilde{\psi}(x)$ and $\tilde{\psi}_2(x) = \tilde{\psi}^+(x)$ are c-number stochastic field functions, $\alpha_i^1 = \alpha_i$, and $\alpha_i^2 = \alpha_i^+$ are c-number phase space variables, and $\tilde{\alpha}_i^1 = \tilde{\alpha}_i$ and $\tilde{\alpha}_i^2 = \tilde{\alpha}_i^+$ are c-number stochastic phase space variables,

The boson functional Fokker–Planck equation is

$$\begin{aligned}\frac{\partial}{\partial t} P_{\mathrm{b}}[\boldsymbol{\psi}(x),\boldsymbol{\psi}^*(x)] = &-\sum_A \int dx \left(\frac{\delta}{\delta\psi_A(x)}\right)_x (A_A(\boldsymbol{\psi}(x),x) P_{\mathrm{b}}[\boldsymbol{\psi}(x),\boldsymbol{\psi}^*(x)]) \\ &+ \frac{1}{2}\sum_{A,B} \int dx\,dy \left(\frac{\delta}{\delta\psi_A(x)}\right)_x \left(\frac{\delta}{\delta\psi_B(y)}\right)_y \\ &\times (D_{AB}(\boldsymbol{\psi}(x),x,\boldsymbol{\psi}(y),y) P_{\mathrm{b}}[\boldsymbol{\psi}(x),\boldsymbol{\psi}^*(x)]).\end{aligned} \quad (14.70)$$

The corresponding Ito stochastic field equations are

$$\begin{aligned}\delta\tilde{\psi}_A(x,t) &= \tilde{\psi}_A(x,t+\delta t) - \tilde{\psi}_A(x,t) \\ &= A_A(\tilde{\boldsymbol{\psi}}(x,t),x)\,\delta t + \sum_{Dk} \eta_k^{AD}(\tilde{\boldsymbol{\psi}}(x,t),x) \int_t^{t+\delta t} dt_1\,\Gamma_k^D(t_1) \\ &= A_A(\tilde{\boldsymbol{\psi}}(x,t),x)\,\delta t + \sum_{Dk} \eta_k^{AD}(\tilde{\boldsymbol{\psi}}(x,t),x)\,\delta w_k^D, \end{aligned} \tag{14.71}$$

where

$$\sum_{Dk} \eta_k^{AD}(\tilde{\boldsymbol{\psi}}(x,t),x)\eta_k^{BD}(\tilde{\boldsymbol{\psi}}(y,t),y) = D_{AB}(\tilde{\boldsymbol{\psi}}(x,t),x,\tilde{\boldsymbol{\psi}}(y,t),y). \tag{14.72}$$

14.4.2 Fermion Case

The fermion field operators, field functions and stochastic fields are given in terms of mode expansions as

$$\hat{\psi}_A(x) = \sum_i \hat{c}_i^A \xi_i^A(x), \tag{14.73}$$

$$\psi_A(x) = \sum_i g_i^A \xi_i^A(x), \tag{14.74}$$

$$\tilde{\psi}_A(x,t) = \sum_i \tilde{g}_i^A(t) \xi_i^A(x), \tag{14.75}$$

where $A = 1, 2$, $\hat{\psi}_1(x) = \hat{\psi}(x)$ and $\hat{\psi}_2(x) = \hat{\psi}^\dagger(x)$ are fermion field operators, $\xi_i^1(x) = \phi_i(x)$ and $\xi_i^2(x) = \phi_i^*(x)$ are mode functions and their complex conjugates, $\hat{c}_i^1 = \hat{c}_i$ and $\hat{c}_i^2 = \hat{c}_i^\dagger$ are fermion mode annihilation and creation operators, $\psi_1(x) = \psi(x)$ and $\psi_2(x) = \psi^+(x)$ are Grassmann field functions, $\tilde{\psi}_1(x) = \tilde{\psi}(x)$ and $\tilde{\psi}_2(x) = \tilde{\psi}^+(x)$ are Grassmann stochastic field functions, $g_i^1 = g_i$ and $g_i^2 = g_i^+$ are Grassmann phase space variables, and $\tilde{g}_i^1 = \tilde{g}_i$ and $\tilde{g}_i^2 = \tilde{g}_i^+$ are Grassmann stochastic phase space variables,

The fermion functional Fokker–Planck equation is

$$\begin{aligned}\frac{\partial}{\partial t} P_{\mathrm{f}}[\boldsymbol{\psi}(x)] &= -\sum_A \int dx\,(A_A(\boldsymbol{\psi}(x),x) P_{\mathrm{f}}[\boldsymbol{\psi}(x)]) \left(\frac{\overleftarrow{\delta}}{\delta\psi_A(x)}\right)_x \\ &\quad + \frac{1}{2}\sum_{A,B} \int dx\,dy\,(D_{AB}(\boldsymbol{\psi}(x),x,\boldsymbol{\psi}(y),y) P_{\mathrm{f}}[\boldsymbol{\psi}(x)]) \\ &\quad \times \left(\frac{\overleftarrow{\delta}}{\delta\psi_B(y)}\right)_y \left(\frac{\overleftarrow{\delta}}{\delta\psi_A(x)}\right)_x. \end{aligned} \tag{14.76}$$

The corresponding Ito stochastic field equations are

$$\begin{aligned}\delta\tilde{\psi}_A(x,t) &= \tilde{\psi}_A(x,t+\delta t) - \tilde{\psi}_A(x,t) \\ &= -A_A(\tilde{\boldsymbol{\psi}}(x,t),x)\,\delta t + \sum_{Dk} \eta_k^{AD}(\tilde{\boldsymbol{\psi}}(x,t),x) \int_t^{t+\delta t} dt_1\,\Gamma_k^D(t_1) \\ &= -A_A(\tilde{\boldsymbol{\psi}}(x,t),x)\,\delta t + \sum_{Dk} \eta_k^{AD}(\tilde{\boldsymbol{\psi}}(x,t),x)\,\delta w_k^D, \end{aligned} \tag{14.77}$$

where

$$\sum_{Dk} \eta_k^{AD}(\tilde{\boldsymbol{\psi}}(x,t),x)\eta_k^{BD}(\tilde{\boldsymbol{\psi}}(y,t),y) = D_{AB}(\tilde{\boldsymbol{\psi}}(x,t),x,\tilde{\boldsymbol{\psi}}(y,t),y). \tag{14.78}$$

Note the sign difference in the A_A terms in the Ito stochastic field equations between the boson and fermion cases.

The above results are for the case of a single field. Cases of several fields will involve additional indices α, β etc. in the above equations. Thus $\hat{\psi}_A(x) \to \hat{\psi}_{\alpha A}(x), \xi_i^A(x) \to \xi_{\alpha i}^A(x), D_{AB} \to D_{\alpha A \beta B}$ etc.

Exercises

(14.1) Derive the Ito stochastic equation for Grassmann fields directly from the functional Fokker–Planck equation (13.83), following the same approach as for c-number fields.

(14.2) Derive the result for four noise fields in (14.40) or (14.61).

15

Application to Multi-Mode Systems

15.1 Boson Case – Trapped Bose–Einstein Condensate

15.1.1 Introduction

The physics of single-component Bose–Einstein condensates in trapping potentials is a good example of a multi-mode system with a large number of particles and will be treated in this chapter in terms of phase space distribution functionals and Ito stochastic field equations. We will develop the results both for distribution functionals of the positive P type and for those of the Wigner type. However, neither of these approaches is totally suitable for treating Bose–Einstein condensates at very low temperatures, where most of the bosons occupy only a single mode (two modes in the cases of a double-well trapping potential), with only a few bosons in other modes. The physics of this situation suggests writing the bosonic field operators as a sum of a condensate and a non-condensate term, and introducing a hybrid representation where the condensate mode(s) are treated via the Wigner representation and the non-condensate modes are treated via the positive P representation. The field functions associated with the highly occupied condensate mode(s) behave like a classical mean field satisfying Gross–Pitaevskii equations, whilst those associated with the largely unoccupied non-condensate modes embody the quantum fluctuation effects. The approach lends itself to an expansion of the Hamiltonian in inverse powers of $\sqrt{N}$, and in the weak-interaction case this expansion can be terminated after terms allowing for Bogoliubov excitations. A treatment based on the hybrid Wigner–$P+$ representation is presented elsewhere [52]. However, this chapter will be confined to the simpler case of separate representations, which nevertheless illustrate most of the main issues that arise in applying functional distribution methods to bosonic systems.

15.1.2 Field Operators

The field operators can be expanded in mode functions

$$\hat{\Psi}(\mathbf{r}) = \sum_k \hat{a}_k \phi_k(\mathbf{r}), \tag{15.1}$$

$$\hat{\Psi}^\dagger(\mathbf{r}) = \sum_k \phi_k^*(\mathbf{r}) \hat{a}_k^\dagger, \tag{15.2}$$

Phase Space Methods for Degenerate Quantum Gases. First Edition. Bryan J. Dalton, John Jeffers, and Stephen M. Barnett.
Published in 2015 by Oxford University Press.

where the mode functions are orthonormal,

$$\int d\mathbf{r}\phi_i^*(\mathbf{r})\phi_j(\mathbf{r}) = \delta_{ij}. \tag{15.3}$$

Both the mode functions and their accompanying annihilation and creation operators are time dependent in general, but for simplicity of notation the time dependence will be left implicit.

In the mode expansion, we will assume that there is a cut-off at some large mode number K (a momentum cut-off). This is to be consistent with using the zero-range approximation in the Hamiltonian. Accordingly, the completeness expression for the mode functions does not give the ordinary delta function but a restricted delta function $\delta_C(\mathbf{r}, \mathbf{s})$, which is no longer singular when $\mathbf{r} = \mathbf{s}$:

$$\sum_{k=1}^{K} \phi_k(\mathbf{r})\phi_k^*(\mathbf{s}) = \delta_C(\mathbf{r}, \mathbf{s}). \tag{15.4}$$

Thus although the annihilation and creation operators satisfy the bosonic commutation rules, the field operators satisfy modified rules, for which the non-zero results are

$$\begin{aligned}[\hat{a}_k, \hat{a}_l^\dagger] &= \delta_{kl}, \\ [\hat{\Psi}(\mathbf{r}), \hat{\Psi}^\dagger(\mathbf{s})] &= \delta_C(\mathbf{r},\mathbf{s}).\end{aligned} \tag{15.5}$$

In obtaining these rules, those for the annihilation and creation operators are treated as fundamental and those for the field operators are then derived. If the cut-off is made very large, then the restricted delta function approaches the ordinary delta function. We note that

$$\int d\mathbf{s}\, \delta_C(\mathbf{s}, \mathbf{s}) = K, \tag{15.6}$$

where K is the number of modes involved in the expansion of the field operators.

15.1.3 Hamiltonian

The full Hamiltonian in terms of field operators is given by

$$\hat{H} = \int d\mathbf{r} \left(\frac{\hbar^2}{2m} \nabla \hat{\Psi}(\mathbf{r})^\dagger \cdot \nabla \hat{\Psi}(\mathbf{r}) + \hat{\Psi}(\mathbf{r})^\dagger V \hat{\Psi}(\mathbf{r}) + \frac{g}{2} \hat{\Psi}(\mathbf{r})^\dagger \hat{\Psi}(\mathbf{r})^\dagger \hat{\Psi}(\mathbf{r}) \hat{\Psi}(\mathbf{r}) \right) \tag{15.7}$$

$$= \hat{K} + \hat{V} + \hat{U}, \tag{15.8}$$

the sum of kinetic-energy, trap potential energy and boson–boson interaction energy terms. As usual, the zero-range approximation is made, with $g = 4\pi\hbar^2 a_s/m$, where a_s is the s-wave scattering length.

15.1.4 Functional Fokker–Planck Equations and Correspondence Rules

The distribution functional $P[\boldsymbol{\psi}(\mathbf{r}),\boldsymbol{\psi}^*(\mathbf{r})]$ satisfies a functional Fokker–Planck equation of the form given in (13.69). As usual, we write $\boldsymbol{\psi}(\mathbf{r}\ ) \equiv \{\psi(\mathbf{r}),\psi^+(\mathbf{r})\}$ and $\boldsymbol{\psi}^*(\mathbf{r}) \equiv \{\psi^*(\mathbf{r}),\psi^{+*}(\mathbf{r})\}$. The functional Fokker–Planck equation is derived using the correspondence rules given in (13.9)–(13.12) for the positive P distribution functional and (13.23)–(13.26) for the Wigner functional. The field functions are

$$\psi(\mathbf{r}) = \sum_k^K \alpha_k\phi_k(\mathbf{r}), \tag{15.9}$$

$$\psi^+(\mathbf{r}) = \sum_k^K \phi_k^*(\mathbf{r})\alpha_k^+. \tag{15.10}$$

In applying the correspondence rules to the BEC problem, the following functional derivative results can be obtained as straightforward generalisations of (11.62):

$$\frac{\delta}{\delta\psi(\mathbf{s})}\psi(\mathbf{r}) = \delta_C(\mathbf{r},\mathbf{s}), \qquad \frac{\delta}{\delta\psi^+(\mathbf{s})}\psi^+(\mathbf{r}) = \delta_{C+}(\mathbf{r},\mathbf{s}) = \delta_C(\mathbf{s},\mathbf{r}), \tag{15.11}$$

$$\frac{\delta}{\delta\psi(\mathbf{s})}\psi^+(\mathbf{r}) = 0, \qquad \frac{\delta}{\delta\psi^+(\mathbf{s})}\psi(\mathbf{r}) = 0. \tag{15.12}$$

Note the reverse order of $\mathbf{r},\mathbf{s}$ in the second result, due to the third result in (11.62).

In deriving the functional Fokker–Planck equation, we make use of the product rule for functional differentiation, a spatial integration-by-parts result and the mode form for functional derivatives. The product rule for functional derivatives is

$$\begin{aligned}
&\frac{\delta}{\delta\psi(\mathbf{s})}(F[\psi(\mathbf{r}),\psi^+(\mathbf{r})]G[\psi(\mathbf{r}),\psi^+(\mathbf{r})]) \\
&\qquad = (\frac{\delta}{\delta\psi(\mathbf{s})}F[\cdots])G[\cdots] + F[\cdots](\frac{\delta}{\delta\psi(\mathbf{s})}G[\cdots]) \\
&\frac{\delta}{\delta\psi^+(\mathbf{s})}(F[\psi(\mathbf{r}),\psi^+(\mathbf{r})]G[\psi(\mathbf{r}),\psi^+(\mathbf{r})]) \\
&\qquad = (\frac{\delta}{\delta\psi^+(\mathbf{s})}F[\cdots])G[\cdots] + F[\cdots](\frac{\delta}{\delta\psi^+(\mathbf{s})}G[\cdots]),
\end{aligned} \tag{15.13}$$

where $[\cdots] \equiv [\psi(\mathbf{r}),\psi^+(\mathbf{r})]$ for short.

The standard approach to space integration gives the result

$$\int ds\,\{\partial_\mu C(\mathbf{s})\} = 0 \tag{15.14}$$

for functions $C(\mathbf{s})$ that become zero on the boundary. This then leads to a useful result involving product functions $C(\mathbf{s}) = A(\mathbf{s})B(\mathbf{s})$, enabling the spatial derivative to be applied to either $A(\mathbf{s})$ or $B(\mathbf{s})$:

$$\int ds\, \{\partial_\mu A(\mathbf{s})\} B(\mathbf{s}) = -\int ds\, A(\mathbf{s})\, \{\partial_\mu B(\mathbf{s})\}. \tag{15.15}$$

We can assume that $\psi(\mathbf{s})$ and $\psi^+(\mathbf{s})$ become zero on the boundary, since they both involve condensate mode functions or their conjugates that are localised owing to the trap potential. Also, the functional derivatives produce linear combinations of either the condensate mode functions or their conjugates, so the various $C(\mathbf{s})$ that will be involved should become zero on the boundary.

The mode-equivalent form for the functional derivatives is, from (11.54) and (11.55),

$$\begin{aligned}\left(\frac{\delta}{\delta\psi(\mathbf{r})}\right) &= \sum_k \phi_k^*(\mathbf{r})\, \frac{\partial}{\partial\alpha_k},\\ \left(\frac{\delta}{\delta\psi^+(\mathbf{r})}\right) &= \sum_k \phi_k(\mathbf{r})\, \frac{\partial}{\partial\alpha_k^+}.\end{aligned} \tag{15.16}$$

The derivation of the functional Fokker–Planck equations is set out in Appendix A.

15.1.5 Functional Fokker–Planck Equation – Positive P Case

The contributions to the functional Fokker–Planck equation may be written in the form

$$\begin{aligned}\frac{\partial}{\partial t}P[\boldsymbol{\psi}(\mathbf{r}),\, \boldsymbol{\psi}^*(\mathbf{r})] &= \left(\frac{\partial}{\partial t}P[\boldsymbol{\psi}(\mathbf{r}),\, \boldsymbol{\psi}^*(\mathbf{r})]\right)_K + \left(\frac{\partial}{\partial t}P[\boldsymbol{\psi}(\mathbf{r}), \boldsymbol{\psi}^*(\mathbf{r})]\right)_V\\ &\quad + \left(\frac{\partial}{\partial t}P[\boldsymbol{\psi}(\mathbf{r}),\, \boldsymbol{\psi}^*(\mathbf{r})]\right)_U,\end{aligned} \tag{15.17}$$

i.e. of a sum of terms from the kinetic energy, the trap potential and the boson–boson interaction. Derivations of the form for each term are given in Appendix I. As usual, we write $\boldsymbol{\psi}(\mathbf{r}) \equiv \{\psi(\mathbf{r}), \psi^+(\mathbf{r})\} \equiv \{\psi^-(\mathbf{r}), \psi^+(\mathbf{r})\}$ and $\boldsymbol{\psi}^*(\mathbf{r}) \equiv \{\psi^*(\mathbf{r}), \psi^{+*}(\mathbf{r})\} \equiv \{\psi^{-*}(\mathbf{r}), \psi^{+*}(\mathbf{r})\}$. Here and elsewhere, ∂_μ is short for $\partial/\partial s_\mu$.

The terms are, firstly, the kinetic-energy contribution,

$$\begin{aligned}\left(\frac{\partial}{\partial t}P[\boldsymbol{\psi}(\mathbf{r}),\, \boldsymbol{\psi}^*(\mathbf{r})]\right)_K &= \frac{i}{\hbar}\int ds \left\{\frac{\delta}{\delta\psi^+(\mathbf{s})}\left(\sum_\mu \frac{\hbar^2}{2m}\partial_\mu^2\psi^+(\mathbf{s})\right) P[\boldsymbol{\psi}(\mathbf{r}),\, \boldsymbol{\psi}^*(\mathbf{r})]\right.\\ &\quad \left. - \frac{\delta}{\delta\psi(\mathbf{s})}\left(\sum_\mu \frac{\hbar^2}{2m}\partial_\mu^2\psi(\mathbf{s})\right) P[\boldsymbol{\psi}(\mathbf{r}),\, \boldsymbol{\psi}^*(\mathbf{r})])\right\}.\end{aligned} \tag{15.18}$$

The trap potential contribution is

$$\begin{aligned}\left(\frac{\partial}{\partial t}P[\boldsymbol{\psi}(\mathbf{r}),\, \boldsymbol{\psi}^*(\mathbf{r})]\right)_V &= \frac{i}{\hbar}\int ds \left\{\frac{\delta}{\delta\psi(\mathbf{s})}\{V(\mathbf{s})\psi(\mathbf{s})\}P[\boldsymbol{\psi}(\mathbf{r}), \boldsymbol{\psi}^*(\mathbf{r})]\right.\\ &\quad \left. - \frac{\delta}{\delta\psi^+(\mathbf{s})}\{V(\mathbf{s})\psi^+(\mathbf{s})\}P[\boldsymbol{\psi}(\mathbf{r}), \boldsymbol{\psi}^*(\mathbf{r})]\right\}.\end{aligned} \tag{15.19}$$

The contribution from the boson–boson interaction is given by

$$\left(\frac{\partial}{\partial t}P[\boldsymbol{\psi}(\mathbf{r}),\,\boldsymbol{\psi}^*(\mathbf{r})]\right)_U = \frac{i}{\hbar}g\int ds\left\{\frac{\delta}{\delta\psi(\mathbf{s})}\{(\psi^+(\mathbf{s})\psi(\mathbf{s}))\psi(\mathbf{s})\}P[\boldsymbol{\psi}(\mathbf{r}),\boldsymbol{\psi}^*(\mathbf{r})]\right.$$
$$-\frac{\delta}{\delta\psi^+(\mathbf{s})}\{(\psi^+(\mathbf{s})\psi(\mathbf{s}))\psi^+(\mathbf{s})\}P[\boldsymbol{\psi}(\mathbf{r}),\boldsymbol{\psi}^*(\mathbf{r})]$$
$$-\frac{1}{2}\frac{\delta}{\delta\psi(\mathbf{s})}\frac{\delta}{\delta\psi(\mathbf{s})}\{\psi(\mathbf{s})\psi(\mathbf{s})\}P[\boldsymbol{\psi}(\mathbf{r}),\boldsymbol{\psi}^*(\mathbf{r})]$$
$$\left.+\frac{1}{2}\frac{\delta}{\delta\psi^+(\mathbf{s})}\frac{\delta}{\delta\psi^+(\mathbf{s})}\{\psi^+(\mathbf{s})\psi^+(\mathbf{s})\}P[\boldsymbol{\psi}(\mathbf{r}),\boldsymbol{\psi}^*(\mathbf{r})]\right\}. \tag{15.20}$$

15.1.6 Functional Fokker–Planck Equation – Wigner Case

Again the contributions to the functional Fokker–Planck equation may be written

$$\left(\frac{\partial}{\partial t}W[\boldsymbol{\psi}(\mathbf{r}),\boldsymbol{\psi}^*(\mathbf{r})]\right) = \left(\frac{\partial}{\partial t}W[\boldsymbol{\psi}(\mathbf{r}),\boldsymbol{\psi}^*(\mathbf{r})]\right)_K + \left(\frac{\partial}{\partial t}W[\boldsymbol{\psi}(\mathbf{r}),\,\boldsymbol{\psi}^*(\mathbf{r})]\right)_V$$
$$+\left(\frac{\partial}{\partial t}W[\boldsymbol{\psi}(\mathbf{r}),\,\boldsymbol{\psi}^*(\mathbf{r})]\right)_U, \tag{15.21}$$

as a sum of terms from the kinetic energy, the trap potential and the boson–boson interaction. Derivations of the form for each term are given in Appendix I. As usual, we write $\boldsymbol{\psi}(\mathbf{r}) \equiv \{\psi(\mathbf{r}),\psi^+(\mathbf{r})\} \equiv \{\psi^-(\mathbf{r}),\psi^+(\mathbf{r})\}$ and $\boldsymbol{\psi}^*(\mathbf{r}) \equiv \{\psi^*(\mathbf{r}),\psi^{+*}(\mathbf{r})\} \equiv \{\psi^{-*}(\mathbf{r}),\psi^{+*}(\mathbf{r})\}$.

The contribution to the equation from the kinetic energy is

$$\left(\frac{\partial}{\partial t}W[\boldsymbol{\psi}(\mathbf{r}),\,\boldsymbol{\psi}^*(\mathbf{r})]\right)_K = \frac{i}{\hbar}\int ds\left\{\frac{\delta}{\delta\psi^+(\mathbf{s})}\left(\sum_\mu \frac{\hbar^2}{2m}\partial_\mu^2\psi^+(\mathbf{s})\right)W[\boldsymbol{\psi}(\mathbf{r}),\boldsymbol{\psi}^*(\mathbf{r})]\right.$$
$$\left.-\frac{\delta}{\delta\psi(\mathbf{s})}\left(\sum_\mu \frac{\hbar^2}{2m}\partial_\mu^2\psi(\mathbf{s})\right)W[\boldsymbol{\psi}(\mathbf{r}),\boldsymbol{\psi}^*(\mathbf{r})])\right\}. \tag{15.22}$$

The trap potential contribution is given by

$$\left(\frac{\partial}{\partial t}W[\boldsymbol{\psi}(\mathbf{r}),\boldsymbol{\psi}^*(\mathbf{r})]\right)_V = \frac{i}{\hbar}\int ds\left\{\frac{\delta}{\delta\psi(\mathbf{s})}\{V(\mathbf{s})\psi(\mathbf{s})\}W[\boldsymbol{\psi}(\mathbf{r}),\boldsymbol{\psi}^*(\mathbf{r})]\right.$$
$$\left.-\frac{\delta}{\delta\psi^+(\mathbf{s})}\{V(\mathbf{s})\psi^+(\mathbf{s})\}W[\boldsymbol{\psi}(\mathbf{r}),\boldsymbol{\psi}^*(\mathbf{r})]\right\}. \tag{15.23}$$

The contribution from the boson–boson interaction is given by

$$\begin{aligned}\left(\frac{\partial}{\partial t}W[\boldsymbol{\psi}(\mathbf{r}),\boldsymbol{\psi}^*(\mathbf{r})]\right)_U &= \frac{i}{\hbar}\left\{g\int d\mathbf{s}\frac{\delta}{\delta\psi(\mathbf{s})}\left\{(\psi^+(\mathbf{s})\psi(\mathbf{s})-\delta_C(\mathbf{s},\mathbf{s}))\psi(\mathbf{s})\right\}W\right\}\\ &-\frac{i}{\hbar}\left\{g\int d\mathbf{s}\frac{\delta}{\delta\psi^+(\mathbf{s})}\left\{(\psi^+(\mathbf{s})\psi(\mathbf{s})-\delta_C(\mathbf{s},\mathbf{s}))\psi^+(\mathbf{s})\right\}W\right\}\\ &-\frac{i}{\hbar}\left\{\frac{g}{4}\int d\mathbf{s}\frac{\delta}{\delta\psi(\mathbf{s})}\frac{\delta}{\delta\psi(\mathbf{s})}\frac{\delta}{\delta\psi^+(\mathbf{s})}\{\psi(\mathbf{s})\}W\right\}\\ &+\frac{i}{\hbar}\left\{\frac{g}{4}\int d\mathbf{s}\frac{\delta}{\delta\psi^+(\mathbf{s})}\frac{\delta}{\delta\psi^+(\mathbf{s})}\frac{\delta}{\delta\psi(\mathbf{s})}\{\psi^+(\mathbf{s})\}W\right\},\end{aligned} \tag{15.24}$$

which involves first-order and third-order functional derivatives. The quantity $\delta_C(\mathbf{s},\mathbf{s})$ is a diagonal element of the restricted delta function.

15.1.7 Ito Equations for Positive *P* Case

The functional Fokker–Planck equation contains only first- and second-order functional derivatives. Hence the general theory from Chapters 13 and 14, can be applied to obtain Ito stochastic field equations. From (14.71) and (14.72), we have

$$\begin{aligned}\frac{\partial}{\partial t}\tilde{\psi}(\mathbf{s},t) &= -\frac{i}{\hbar}\left[-\frac{\hbar^2}{2m}\nabla^2\tilde{\psi}(\mathbf{s})+V(\mathbf{s})\tilde{\psi}(\mathbf{s})+g\{\tilde{\psi}^+(\mathbf{s})\tilde{\psi}(\mathbf{s})\}\tilde{\psi}(\mathbf{s})\right]\\ &+\sqrt{\frac{-ig}{\hbar}}\,\tilde{\psi}(\mathbf{s},t)\,\Gamma_-(t_+)\end{aligned} \tag{15.25}$$

with a similar equation for $\tilde{\psi}^+(\mathbf{s},t)$.

The stochastic averages of the noise fields are given in (14.39), where the non-zero diffusion matrix elements are

$$\begin{aligned}D_{-;-}(\tilde{\boldsymbol{\psi}}(\mathbf{s},t),\mathbf{s},\tilde{\boldsymbol{\psi}}(\mathbf{r},t),\mathbf{r}) &= -\frac{i}{\hbar}g\{\tilde{\psi}(\mathbf{s})\tilde{\psi}(\mathbf{s})\}\boldsymbol{\delta}(\mathbf{s}\text{-}\mathbf{r}),\\ D_{+;+}(\tilde{\boldsymbol{\psi}}(\mathbf{s},t),\mathbf{s},\tilde{\boldsymbol{\psi}}(\mathbf{r},t),\mathbf{r}) &= +\frac{i}{\hbar}g\{\tilde{\psi}^+(\mathbf{s})\tilde{\psi}^+(\mathbf{s})\}\boldsymbol{\delta}(\mathbf{s}\text{-}\mathbf{r}).\end{aligned} \tag{15.26}$$

This shows that the noise field terms are delta function correlated in space, as well as in time.

The classical field equation is given by

$$\frac{\partial}{\partial t}\psi_{\text{class}}(\mathbf{s},t) = -\frac{i}{\hbar}\left[-\frac{\hbar^2}{2m}\nabla^2\psi_{\text{class}}(\mathbf{s})+V(\mathbf{s})\psi_{\text{class}}(\mathbf{s})+g\{\psi^+_{\text{class}}(\mathbf{s})\psi_{\text{class}}(\mathbf{s})\}\psi_{\text{class}}(\mathbf{s})\right], \tag{15.27}$$

which is almost the same as the Gross–Pitaevskii equation.

15.1.8 Ito Equations for Wigner Case

From Chapter 12, the distribution functional can be used to determine normally ordered quantum correlation functions via phase space $P+$ distribution functional integrals of the form (12.34). A similar phase space Wigner distribution functional integral gives the symmetrically ordered quantum correlation functions. These phase space functional integrals will be replaced by stochastic averages. As an example of the use of the quantum correlation function result, consider the mean value of the number operator

$$\hat{N} = \int dr\, \hat{\Psi}(\mathbf{r})^{\dagger}\hat{\Psi}(\mathbf{r}) = \sum_{k}^{K} \hat{a}_l^{\dagger}\hat{a}_k$$
$$= \int dr \left[\left\{\hat{\Psi}(\mathbf{r})^{\dagger}\hat{\Psi}(\mathbf{r})\right\} - \frac{1}{2}\delta_{\mathrm{C}}(\mathbf{r},\mathbf{r})\right]. \tag{15.28}$$

We have from (12.34)

$$\langle\hat{N}\rangle = \int d\mathbf{r} \int D^2\psi\, D^2\psi^{+}(\psi^{+}(\mathbf{r})\psi(\mathbf{r}))P_{\mathrm{b}}^{+}[\psi(\mathbf{r}),\psi^{+}(\mathbf{r}),\psi^{*}(\mathbf{r}),\psi^{+*}(\mathbf{r})] \tag{15.29}$$
$$= \int d\mathbf{r} \int D^2\psi\, D^2\psi^{+}(\psi^{+}(\mathbf{r})\psi(\mathbf{r}) - \frac{1}{2}\delta_{\mathrm{C}}(\mathbf{r},\mathbf{r}))P_{\mathrm{b}}^{\mathrm{W}}[\psi(\mathbf{r}),\psi^{+}(\mathbf{r}),\psi^{*}(\mathbf{r}),\psi^{+*}(\mathbf{r})]$$
$$= \int d\mathbf{r} \int D^2\psi\, D^2\psi^{+}\psi^{+}(\mathbf{r})\psi(\mathbf{r})P_{\mathrm{b}}^{\mathrm{W}}[\psi(\mathbf{r}),\psi^{+}(\mathbf{r}),\psi^{*}(\mathbf{r}),\psi^{+*}(\mathbf{r})] - \frac{1}{2}K \tag{15.30}$$

for the positive P and Wigner cases, respectively. We note that the phase space functional integral of the distribution functional is unity, and for an N-boson system $\langle\hat{N}\rangle = N$. In the usual case where $N \gg K$, we then see that in both cases the field functions in the region where they are most important will scale like $\sqrt{N}$.

For the Wigner representation this result is quite significant, as it provides a justification for neglecting the third- order functional derivative term in the functional Fokker–Planck equation. This results in a functional Fokker–Planck equation that does not contain any functional derivatives of higher order than two, and hence the general theory from Chapters 13 and 14 can be applied to obtain Ito stochastic field equation. Indeed, for the Wigner representation here, there are only first-order functional derivatives. From (14.71), we have

$$\frac{\partial}{\partial t}\tilde{\psi}(\mathbf{s},t) = -\frac{i}{\hbar}\left[-\frac{\hbar^2}{2m}\nabla^2\tilde{\psi}(\mathbf{s}) + V(\mathbf{s})\tilde{\psi}(\mathbf{s}) + g\{\tilde{\psi}^{+}(\mathbf{s})\tilde{\psi}(\mathbf{s}) - \delta_{\mathrm{C}}(\mathbf{s},\mathbf{s})\}\tilde{\psi}(\mathbf{s})\right], \tag{15.31}$$

with a similar equation for $\tilde{\psi}^{+}(\mathbf{s},t)$. For the Wigner case, there is no noise field term, as all the diffusion matrix elements are zero.

The classical field equation is given by

$$\frac{\partial}{\partial t}\psi_{\mathrm{class}}(\mathbf{s},t) = -\frac{i}{\hbar}\left[-\frac{\hbar^2}{2m}\nabla^2\psi_{\mathrm{class}}(\mathbf{s}) + V(\mathbf{s})\psi_{\mathrm{class}}(\mathbf{s})\right.$$
$$\left. + g\{\psi_{\mathrm{class}}^{+}(\mathbf{s})\psi_{\mathrm{class}}(\mathbf{s}) - \delta_{\mathrm{C}}(\mathbf{s},\mathbf{s})\}\psi_{\mathrm{class}}(\mathbf{s})\right], \tag{15.32}$$

which is the same as the Gross–Pitaevskii equation in the case where there is only one mode in the expansion of the field operators.

15.1.9 Stochastic Averages for Quantum Correlation Functions

The quantum averages of symmetrically ordered products of the field operators $\{\hat{\Psi}^\dagger(\mathbf{r}_1)\cdots\hat{\Psi}^\dagger(\mathbf{r}_p)\hat{\Psi}(\mathbf{s}_q)\cdots\hat{\Psi}(\mathbf{s}_1)\}$ (Wigner representation) and normally ordered products of the field operators $\hat{\Psi}^\dagger(\mathbf{u}_1)\cdots\hat{\Psi}^\dagger(\mathbf{u}_r)\hat{\Psi}(\mathbf{v}_s)\cdots\hat{\Psi}(\mathbf{v}_1)$ are now given by stochastic averages. These replace the functional integrals involving quasi-distribution functionals given, for example, by (12.34). We have

$$\mathrm{Tr}[\hat{\rho}\hat{\Psi}^\dagger(\mathbf{u}_1)\cdots\hat{\Psi}^\dagger(\mathbf{u}_r)\hat{\Psi}(\mathbf{v}_s)\cdots\hat{\Psi}(\mathbf{v}_1)] = \overline{\psi^+(\mathbf{u}_1)\cdots\psi^+(\mathbf{u}_r)\psi(\mathbf{v}_s)\cdots\psi(\mathbf{v}_1)}$$

for the positive P case, and for the Wigner case,

$$\mathrm{Tr}[\hat{\rho}\{\hat{\Psi}^\dagger(\mathbf{r}_1)\cdots\hat{\Psi}^\dagger(\mathbf{r}_p)\hat{\Psi}(\mathbf{s}_q)\cdots\hat{\Psi}(\mathbf{s}_1)\}] = \overline{\psi^+(\mathbf{r}_1)\cdots\psi^+(\mathbf{r}_p)\psi(\mathbf{s}_q)\cdots\psi(\mathbf{s}_1)}.$$

15.2 Fermion Case – Fermions in an Optical Lattice

15.2.1 Introduction

The physics of two-component Fermi gases in trapping potentials is a good example of a multi-mode system with a large number of particles and will be treated in this chapter in terms of phase space distribution functionals and Ito stochastic field equations. If there are a large number of fermions there will of necessity be a large number of modes, so a field-based approach is suggested. We will develop the results for distribution functionals of the unnormalised B type for the case where there are both spin-up and spin-down fermions, showing that the method based on Grassmann variables can lead to equations that can be solved numerically. We show how to apply the results both to the free Fermi gas and to a Fermi gas in an optical lattice.

15.2.2 Field Operators

There are separate field operators for spin-up fermions and spin-down fermions,

$$\begin{aligned}\hat{\Psi}_{u1}(x) &= \hat{\Psi}_u(x), & \hat{\Psi}_{u2}(x) &= \hat{\Psi}_u^\dagger(x),\\ \hat{\Psi}_{d1}(x) &= \hat{\Psi}_d(x), & \hat{\Psi}_{d2}(x) &= \hat{\Psi}_d^\dagger(x).\end{aligned}$$

In terms of the notation $\alpha = u, d$ for spin-up and spin-down fermions, and with $A = 1, 2$ designating annihilation and creation operators, etc., the field operators have mode expansions in terms of fermion mode annihilation operators $\hat{c}_{\alpha i}^1$ and creation operators $\hat{c}_{\alpha i}^2$ as follows:

$$\hat{\Psi}_{\alpha A}(x) = \sum_i \hat{c}_{\alpha i}^A\,\xi_{\alpha i}^A(x), \qquad \alpha = u, d. \tag{15.33}$$

The mode functions and Grassmann field functions will in general differ for each α, and will be designated

$$\xi^1_{\alpha i}(x) = \phi_{\alpha i}(x), \qquad \xi^2_{\alpha i}(x) = \phi^*_{\alpha i}(x), \qquad \alpha = u, d, \tag{15.34}$$

$$\psi_{\alpha 1}(x) = \psi_{\alpha}(x), \qquad \psi_{\alpha 2}(x) = \psi^+_{\alpha}(x), \qquad \alpha = u, d, \tag{15.35}$$

where, in terms of Grassmann phase space variables $g^A_{\alpha i}$ and Grassmann stochastic phase space variables $\tilde{g}^A_{\alpha i}$ for fermions, the field functions $\psi_{\alpha A}(x)$ ($A = 1, 2$ and $\alpha = u, d$) and stochastic field functions are

$$\psi_{\alpha A}(x) = \sum_i g^A_{\alpha i} \xi^A_{\alpha i}(x), \tag{15.36}$$

$$\tilde{\psi}_{\alpha A}(x) = \sum_i \tilde{g}^A_{\alpha i} \xi^A_{\alpha i}(x). \tag{15.37}$$

The mode functions for a particular $\alpha = u, d$ are orthogonal and normalised:

$$\int d\mathbf{r}\, \phi^*_{\alpha i}(\mathbf{r}) \phi_{\alpha j}(\mathbf{r}) = \delta_{ij}. \tag{15.38}$$

Both the mode functions and their accompanying annihilation and creation operators are time-dependent in general, but for simplicity of notation this is usually left implicit.

In the mode expansion, we will assume that there is a momentum cut-off at some large mode number K. This is to be consistent with using the zero-range approximation in the Hamiltonian. Accordingly, the completeness expression for the mode functions does not give the ordinary delta function, but a restricted delta function $\delta_C(\mathbf{r}, \mathbf{s})$ which is no longer singular when $\mathbf{r} = \mathbf{s}$:

$$\sum_{k=1}^{K} \phi_{\alpha k}(\mathbf{r}) \phi^*_{\alpha k}(\mathbf{s}) = \delta_C(\mathbf{r}, \mathbf{s}). \tag{15.39}$$

Thus although the annihilation and creation operators satisfy the standard fermionic anticommutation rules, the field operators satisfy modified rules, for which the non-zero results are

$$\{\hat{c}_{\alpha k}, \hat{c}^\dagger_{\beta l}\} = \delta_{\alpha\beta}\delta_{kl},$$

$$\{\hat{\Psi}_{\alpha}(\mathbf{r}), \hat{\Psi}^\dagger_{\beta}(\mathbf{s})\} = \delta_{\alpha\beta}\delta_C(\mathbf{r}, \mathbf{s}). \tag{15.40}$$

In obtaining these rules, those for the annihilation and creation operators are treated as fundamental and those for the field operators are then derived. If the cut-off is made very large, then the restricted delta function approaches the ordinary delta function. We note that

$$\int d\mathbf{s}\, \delta_C(\mathbf{s}, \mathbf{s}) = K, \tag{15.41}$$

where K is the number of modes involved in the expansion of the field operators.

15.2.3 Hamiltonian

From (2.79), the full Hamiltonian in terms of field operators is given by

$$\hat{H} = \sum_{\alpha} \int d\mathbf{r} \left[\frac{\hbar^2}{2m} \nabla \hat{\Psi}_{\alpha}(\mathbf{r})^{\dagger} \cdot \nabla_{\alpha} \hat{\Psi}(\mathbf{r}) + \hat{\Psi}_{\alpha}(\mathbf{r})^{\dagger} V_{\alpha} \hat{\Psi}_{\alpha}(\mathbf{r}) \right.$$
$$\left. + \frac{g}{2} \hat{\Psi}_{\alpha}(\mathbf{r})^{\dagger} \hat{\Psi}_{-\alpha}(\mathbf{r})^{\dagger} \hat{\Psi}_{-\alpha}(\mathbf{r}) \hat{\Psi}_{\alpha}(\mathbf{r}) \right] \tag{15.42}$$
$$= \hat{K} + \hat{V} + \hat{U}, \tag{15.43}$$

the sum of kinetic-energy, trap potential energy and boson–boson interaction energy terms. Here $-u = d$ and $-d = u$. As usual, the zero-range approximation is made, with $g = 4\pi\hbar^2 a_{\mathrm{s}}/m$, where a_{s} is the s-wave scattering length. This Hamiltonian applies to spin-conserving interactions, and the trap potential may depend on the spin state.

15.2.4 Functional Fokker–Planck Equation – Unnormalised *B*

The unnormalised distribution functional $B[\boldsymbol{\psi}(\mathbf{r})]$ satisfies a functional Fokker–Planck equation of a form similar to that given in (14.76). Here, we write $\boldsymbol{\psi}(\mathbf{r}) \equiv \{\psi_u(\mathbf{r}), \psi_u^{+}(\mathbf{r}), \psi_d(\mathbf{r}), \psi_d^{+}(\mathbf{r})\}$. The functional Fokker–Planck equation is derived using the correspondence rules given in (13.51)–(13.58) for the B_{f} case. All functional derivatives have been placed on the right using the results in (11.169), (11.170) and (11.171). The derivation of the functional Fokker–Planck equations is set out in Appendix I.

The contributions to the functional Fokker–Planck equation may be written as

$$\left(\frac{\partial}{\partial t} B[\boldsymbol{\psi}(\mathbf{r})]\right) = \left(\frac{\partial}{\partial t} B[\boldsymbol{\psi}(\mathbf{r})]\right)_K + \left(\frac{\partial}{\partial t} B[\boldsymbol{\psi}(\mathbf{r})]\right)_V + \left(\frac{\partial}{\partial t} B[\boldsymbol{\psi}(\mathbf{r})]\right)_U.$$

The kinetic-energy term is given by

$$\left(\frac{\partial}{\partial t} B[\boldsymbol{\psi}(\mathbf{r})]\right)_K = \frac{i}{\hbar} \int d\mathbf{s} \left[\left\{ \left(\sum_{\mu} \frac{\hbar^2}{2m} \partial_{\mu}^2 \psi_{u1}(\mathbf{s})\, B[\boldsymbol{\psi}(\mathbf{r})] \right) \frac{\overleftarrow{\delta}}{\delta \psi_{u1}(\mathbf{s})} \right\} \right.$$
$$+ \left\{ \left(\sum_{\mu} \frac{\hbar^2}{2m} \partial_{\mu}^2 \psi_{d1}(\mathbf{s})\, B[\boldsymbol{\psi}(\mathbf{r})] \right) \frac{\overleftarrow{\delta}}{\delta \psi_{d1}(\mathbf{s})} \right\}$$
$$- \left\{ \left(\sum_{\mu} \frac{\hbar^2}{2m} \partial_{\mu}^2 \psi_{u2}(\mathbf{s})\, B[\boldsymbol{\psi}(\mathbf{r})] \right) \frac{\overleftarrow{\delta}}{\delta \psi_{u2}(\mathbf{s})} \right\}$$
$$\left. - \left\{ \left(\sum_{\mu} \frac{\hbar^2}{2m} \partial_{\mu}^2 \psi_{d2}(\mathbf{s})\, B[\boldsymbol{\psi}(\mathbf{r})] \right) \frac{\overleftarrow{\delta}}{\delta \psi_{d2}(\mathbf{s})} \right\} \right], \tag{15.44}$$

the trap potential term is

$$\left(\frac{\partial}{\partial t}B[\boldsymbol{\psi}(\mathbf{r})]\right)_V = \frac{-i}{\hbar}\int ds\left[\left\{(V_u\psi_{u1}(\mathbf{s})B[\boldsymbol{\psi}(\mathbf{r})])\frac{\overleftarrow{\delta}}{\delta\psi_{u1}(\mathbf{s})}\right\}\right.$$
$$+\left\{(V_d\psi_{d1}(\mathbf{s})B[\boldsymbol{\psi}(\mathbf{r})])\frac{\overleftarrow{\delta}}{\delta\psi_{d1}(\mathbf{s})}\right\}$$
$$-\left\{(V_u\psi_{u2}(\mathbf{s})B[\boldsymbol{\psi}(\mathbf{r})])\frac{\overleftarrow{\delta}}{\delta\psi_{u2}(\mathbf{s})}\right\}$$
$$\left.-\left\{(V_d\psi_{d2}(\mathbf{s})B[\boldsymbol{\psi}(\mathbf{r})])\frac{\overleftarrow{\delta}}{\delta\psi_{d2}(\mathbf{s})}\right\}\right] \qquad (15.45)$$

and the fermion fermion interaction term is

$$\left(\frac{\partial}{\partial t}B[\boldsymbol{\psi}(\mathbf{r})]\right)_U = \frac{i}{\hbar}\frac{g}{2}\int ds\left[\left\{\psi_{d1}(\mathbf{s})\psi_{u1}(\mathbf{s})B[\boldsymbol{\psi}(\mathbf{r})]\frac{\overleftarrow{\delta}}{\delta\psi_{d1}(\mathbf{s})}\frac{\overleftarrow{\delta}}{\delta\psi_{u1}(\mathbf{s})}\right\}\right.$$
$$+\left\{\psi_{u1}(\mathbf{s})\psi_{d1}(\mathbf{s})B[\boldsymbol{\psi}(\mathbf{r})]\frac{\overleftarrow{\delta}}{\delta\psi_{u1}(\mathbf{s})}\frac{\overleftarrow{\delta}}{\delta\psi_{d1}(\mathbf{s})}\right\}$$
$$-\left\{\psi_{d2}(\mathbf{s})\psi_{u2}(\mathbf{s})B[\boldsymbol{\psi}(\mathbf{r})]\frac{\overleftarrow{\delta}}{\delta\psi_{d2}(\mathbf{s})}\frac{\overleftarrow{\delta}}{\delta\psi_{u2}(\mathbf{s})}\right\}$$
$$\left.-\left\{\psi_{u2}(\mathbf{s})\psi_{d2}(\mathbf{s})B[\boldsymbol{\psi}(\mathbf{r})]\frac{\overleftarrow{\delta}}{\delta\psi_{u2}(\mathbf{s})}\frac{\overleftarrow{\delta}}{\delta\psi_{d2}(\mathbf{s})}\right\}\right]. \qquad (15.46)$$

We note that the functional Fokker–Planck equation involves only a single spatial integral in this case of zero-range fermion–fermion interactions.

Thus the drift vector in the functional Fokker–Planck equation is

$$[A] = \left[\begin{array}{|c|c|}\hline \mathbf{u1} & \frac{i}{\hbar}\left\{-\frac{\hbar^2}{2m}\nabla^2\psi_{u1}(\mathbf{s}) + V_u\psi_{u1}(\mathbf{s})\right\} \\ \hline \mathbf{d1} & \frac{i}{\hbar}\left\{-\frac{\hbar^2}{2m}\nabla^2\psi_{d1}(\mathbf{s}) + V_d\psi_{d1}(\mathbf{s})\right\} \\ \hline \mathbf{u2} & -\frac{i}{\hbar}\left\{-\frac{\hbar^2}{2m}\nabla^2\psi_{u2}(\mathbf{s}) + V_u\psi_{u2}(\mathbf{s})\right\} \\ \hline \mathbf{d2} & -\frac{i}{\hbar}\left\{-\frac{\hbar^2}{2m}\nabla^2\psi_{d2}(\mathbf{s}) + V_d\psi_{d2}(\mathbf{s})\right\} \\ \hline \end{array}\right] \qquad (15.47)$$

and the diffusion matrix is

$$[D] = \frac{ig}{\hbar}\left[\begin{array}{|c|c|c|c|c|}\hline & \mathbf{u1} & \mathbf{d1} & \mathbf{u2} & \mathbf{d2} \\ \hline \mathbf{u1} & 0 & \{\psi_{d1}\psi_{u1}\} & 0 & 0 \\ \hline \mathbf{d1} & \{\psi_{u1}\psi_{d1}\} & 0 & 0 & 0 \\ \hline \mathbf{u2} & 0 & 0 & 0 & -\{\psi_{d2}\psi_{u2}\} \\ \hline \mathbf{d2} & 0 & 0 & -\{\psi_{u2}\psi_{d2}\} & 0 \\ \hline \end{array}\right]. \qquad (15.48)$$

15.2.5 Ito Equations for Unnormalised Distribution Functional

The functional Fokker–Planck equation contains only first- and second-order functional derivatives. Hence the general theory from Chapter 14 can be applied to obtain Ito stochastic field equations. From (14.78), we need to factorise the diffusion matrix in the form $D = \eta\,\eta^T$. To do this, we apply a similar approach to that in Section 9.4.

We note that the diffusion matrix elements are of the form

$$D_{\alpha A\,\beta B} = \sum_{\gamma C}\sum_{\delta D} M_{\alpha A\gamma C\,;\,\beta B\,\delta D}\psi_{\gamma C}(\mathbf{s})\psi_{\delta D}(\mathbf{s}), \tag{15.49}$$

where the $M_{\alpha A\gamma C\,;\,\beta B\,\delta D}$ are all c-numbers. The rows of M are designated by $\alpha A\,\gamma C$ and the columns are $\beta B\,\delta D$. From the antisymmetry of the fermion diffusion matrix, it is straightforward to show that M is a symmetric matrix:

$$M_{\alpha A\gamma C\,;\,\beta B\,\delta D} = M_{\beta B\,\delta D\,;\,\alpha A\,\gamma C}. \tag{15.50}$$

Hence it can be factorised via Takagi factorisation into $M = KK^T$ and we have

$$M_{\alpha A\gamma C\,;\,\beta B\,\delta D} = \sum_{\xi} K_{\alpha A\gamma C\,;\,\xi} K_{\beta B\,\delta D\,;\,\xi}, \tag{15.51}$$

and thus we can write the diffusion matrix in the form

$$D_{\alpha A\,\beta B} = \sum_{\xi} \eta_{\alpha A\,;\,\xi}\,\eta_{\beta B\,;\,\xi}, \tag{15.52}$$

as is required for determining the noise term in the Ito stochastic field equations. The quantities $\eta_{\alpha A\,;\,\xi}$ and $\eta_{\beta B\,;\,\xi}$ are Grassmann fields that depend linearly on the $\psi_{\gamma C}(\mathbf{s})$. They are defined by

$$\begin{aligned}\eta_{\alpha A\,;\,\xi}(\mathbf{s}) &= \sum_{\gamma C} K_{\alpha A\gamma C\,;\,\xi}\psi_{\gamma C}(\mathbf{s}),\\ \eta_{\beta B\,;\,\xi}(\mathbf{s}) &= \sum_{\delta D} K_{\beta B\,\delta D\,;\,\xi}\psi_{\delta D}(\mathbf{s}).\end{aligned} \tag{15.53}$$

In the present case,

$[M] = \dfrac{ig}{\hbar}$

$\alpha A, \gamma C \Downarrow; \beta B, \delta D \Longrightarrow$	**u1,d1**	**d1,u1**	**u2,d2**	**d2,u2**
u1,d1	0	1	0	0
d1,u1	1	0	0	0
u2,d2	0	0	0	−1
d2,u2	0	0	−1	0

.

(15.54)

The Takagi factorisation of M can be carried out using the procedure set out in [63, p. 206] or via a first-principles determination. The result (which is easily confirmed) is given by

$$[K] = \sqrt{\frac{ig}{\hbar}}\frac{1}{\sqrt{2}}\left[\begin{array}{|l|l|l|l|l|}\hline \alpha A, \gamma C \Downarrow; \beta B, \delta D \Longrightarrow & \mathbf{u1,d1} & \mathbf{d1,u1} & \mathbf{u2,d2} & \mathbf{d2,u2} \\ \hline \mathbf{u1,d1} & 1 & i & 0 & 0 \\ \hline \mathbf{d1,u1} & 1 & -i & 0 & 0 \\ \hline \mathbf{u2,d2} & 0 & 0 & 1 & i \\ \hline \mathbf{d2,u2} & 0 & 0 & -1 & i \\ \hline \end{array}\right], \tag{15.55}$$

which gives the elements $K_{\alpha A\gamma C\,;\,\xi}$, where for the present we list the ξ as $\beta B, \delta D$.

The Ito stochastic field equations are obtained from (14.77) and are

$$\delta\tilde{\psi}_{\alpha A}(\mathbf{s},t) = -A_{\alpha A}(\tilde{\boldsymbol{\psi}}(\mathbf{s},t))\,\delta t + \sum_{\xi}\sum_{\gamma C} K_{\alpha A\gamma C\,;\,\xi}\,\tilde{\psi}_{\gamma C}(\mathbf{s})\,\delta w_{\xi}, \tag{15.56}$$

so on reading off the matrix elements we find that

$$\begin{aligned}
\delta\tilde{\psi}_{u1}(\mathbf{s},t) &= -\frac{i}{\hbar}\left\{-\frac{\hbar^2}{2m}\nabla^2\tilde{\psi}_{u1}(\mathbf{s}) + V_u\tilde{\psi}_{u1}(\mathbf{s})\right\}\delta t \\
&\quad +\sqrt{\frac{ig}{2\hbar}}\,\tilde{\psi}_{d1}(\mathbf{s})\,\delta w_{u1,d1} + \sqrt{\frac{ig}{2\hbar}}(i)\,\tilde{\psi}_{d1}(\mathbf{s})\,\delta w_{d1,u1}, \\
\delta\tilde{\psi}_{d1}(\mathbf{s},t) &= -\frac{i}{\hbar}\left\{-\frac{\hbar^2}{2m}\nabla^2\tilde{\psi}_{d1}(\mathbf{s}) + V_d\tilde{\psi}_{d1}(\mathbf{s})\right\}\delta t \\
&\quad +\sqrt{\frac{ig}{2\hbar}}\,\tilde{\psi}_{u1}(\mathbf{s})\,\delta w_{u1,d1} + \sqrt{\frac{ig}{2\hbar}}(-i)\,\tilde{\psi}_{u1}(\mathbf{s})\,\delta w_{d1,u1}, \\
\delta\tilde{\psi}_{u2}(\mathbf{s},t) &= +\frac{i}{\hbar}\left\{-\frac{\hbar^2}{2m}\nabla^2\tilde{\psi}_{u2}(\mathbf{s}) + V_u\tilde{\psi}_{u2}(\mathbf{s})\right\}\delta t \\
&\quad +\sqrt{\frac{ig}{2\hbar}}\,\tilde{\psi}_{d2}(\mathbf{s})\,\delta w_{u2,d2} + \sqrt{\frac{ig}{2\hbar}}(i)\,\tilde{\psi}_{d2}(\mathbf{s})\,\delta w_{d2,u2}, \\
\delta\tilde{\psi}_{d2}(\mathbf{s},t) &= +\frac{i}{\hbar}\left\{-\frac{\hbar^2}{2m}\nabla^2\tilde{\psi}_{d2}(\mathbf{s}) + V_d\tilde{\psi}_{d2}(\mathbf{s})\right\}\delta t \\
&\quad +\sqrt{\frac{ig}{2\hbar}}\,(-1)\tilde{\psi}_{u2}(\mathbf{s})\,\delta w_{u2,d2} + \sqrt{\frac{ig}{2\hbar}}(+i)\,\tilde{\psi}_{u2}(\mathbf{s})\,\delta w_{d2,u2}.
\end{aligned} \tag{15.57}$$

Alternatively, we can write

$$\begin{aligned}
\tilde{\psi}_{u1}(\mathbf{s},t+\delta t) &= \tilde{\psi}_{u1}(\mathbf{s},t) - \frac{i}{\hbar}\left\{-\frac{\hbar^2}{2m}\nabla^2\tilde{\psi}_{u1}(\mathbf{s},t) + V_u\tilde{\psi}_{u1}(\mathbf{s},t)\right\} \\
&\quad + \sqrt{\frac{ig}{2\hbar}}\,\tilde{\psi}_{d1}(\mathbf{s},t)\,\delta w_{u1,d1} + \sqrt{\frac{ig}{2\hbar}}(i)\,\tilde{\psi}_{d1}(\mathbf{s},t)\,\delta w_{d1,u1}, \\
\tilde{\psi}_{d1}(\mathbf{s},t+\delta t) &= \tilde{\psi}_{d1}(\mathbf{s},t) - \frac{i}{\hbar}\left\{-\frac{\hbar^2}{2m}\nabla^2\tilde{\psi}_{d1}(\mathbf{s},t) + V_d\tilde{\psi}_{d1}(\mathbf{s},t)\right\} \\
&\quad + \sqrt{\frac{ig}{2\hbar}}\,\tilde{\psi}_{u1}(\mathbf{s},t)\,\delta w_{u1,d1} + \sqrt{\frac{ig}{2\hbar}}(-i)\,\tilde{\psi}_{u1}(\mathbf{s},t)\,\delta w_{d1,u1}, \\
\tilde{\psi}_{u2}(\mathbf{s},t+\delta t) &= \tilde{\psi}_{u2}(\mathbf{s},t) + \frac{i}{\hbar}\left\{-\frac{\hbar^2}{2m}\nabla^2\tilde{\psi}_{u2}(\mathbf{s},t) + V_u\tilde{\psi}_{u2}(\mathbf{s},t)\right\} \\
&\quad + \sqrt{\frac{ig}{2\hbar}}\,\tilde{\psi}_{d2}(\mathbf{s},t)\,\delta w_{u2,d2} + \sqrt{\frac{ig}{2\hbar}}(i)\,\tilde{\psi}_{d2}(\mathbf{s},t)\,\delta w_{d2,u2}, \\
\tilde{\psi}_{d2}(\mathbf{s},t+\delta t) &= \tilde{\psi}_{d2}(\mathbf{s},t) + \frac{i}{\hbar}\left\{-\frac{\hbar^2}{2m}\nabla^2\tilde{\psi}_{d2}(\mathbf{s},t) + V_d\,\tilde{\psi}_{d2}(\mathbf{s},t)\right\} \\
&\quad + \sqrt{\frac{ig}{2\hbar}}\,(-1)\tilde{\psi}_{u2}(\mathbf{s},t)\,\delta w_{u2,d2} + \sqrt{\frac{ig}{2\hbar}}(+i)\tilde{\psi}_{u2}(\mathbf{s},t)\,\delta w_{d2,u2},
\end{aligned} \tag{15.58}$$

or, in terms of matrices,

$$\begin{bmatrix} \tilde{\psi}_{u1}(\mathbf{s},t+\delta t) \\ \tilde{\psi}_{d1}(\mathbf{s},t+\delta t) \\ \tilde{\psi}_{u2}(\mathbf{s},t+\delta t) \\ \tilde{\psi}_{d2}(\mathbf{s},t+\delta t) \end{bmatrix} = \begin{bmatrix} G^1_{u1,u1}(t) & G^1_{u1,d1}(t) & 0 & 0 \\ G^1_{d1,u1}(t) & G^1_{d1,d1}(t) & 0 & 0 \\ 0 & 0 & G^2_{u2,u2}(t) & G^2_{u2,d2}(t) \\ 0 & 0 & G^2_{d2,u2}(t) & G^2_{d2,d2}(t) \end{bmatrix} \begin{bmatrix} \tilde{\psi}_{u1}(\mathbf{s},t) \\ \tilde{\psi}_{d1}(\mathbf{s},t) \\ \tilde{\psi}_{u2}(\mathbf{s},t) \\ \tilde{\psi}_{d2}(\mathbf{s},t) \end{bmatrix}, \tag{15.59}$$

where the c-number stochastic submatrices are

$$\begin{aligned}
[G^1(t)] &= \begin{bmatrix} 1 - \frac{i}{\hbar}\left\{-\frac{\hbar^2}{2m}\nabla^2 + V_u\right\}\delta t & \sqrt{\frac{ig}{2\hbar}}\{\delta w_{u1,d1} + i\,\delta w_{d1,u1}\} \\ \sqrt{\frac{ig}{2\hbar}}\{\delta w_{u1,d1} - i\,\delta w_{d1,u1}\} & 1 - \frac{i}{\hbar}\left\{-\frac{\hbar^2}{2m}\nabla^2 + V_d\right\}\delta t \end{bmatrix}, \\
[G^2(t)] &= \begin{bmatrix} 1 + \frac{i}{\hbar}\left\{-\frac{\hbar^2}{2m}\nabla^2 + V_u\right\}\delta t & \sqrt{\frac{ig}{2\hbar}}\{\delta w_{u2,d2} + i\,\delta w_{d2,u2}\} \\ \sqrt{\frac{ig}{2\hbar}}\{-\delta w_{u2,d2} + i\,\delta w_{d2,u2}\} & 1 + \frac{i}{\hbar}\left\{-\frac{\hbar^2}{2m}\nabla^2 + V_d\right\}\delta t \end{bmatrix},
\end{aligned} \tag{15.60}$$

which shows that the Grassmann stochastic fields at time $t+\delta t$ are related to the Grassmann stochastic fields at time t via a c-number stochastic matrix, which involves the Laplacian operator. This is the same feature that applies when separate modes are treated and shows how the Grassmann stochastic fields at any time t can be related to the Grassmann stochastic fields at an initial time.

15.2.6 Case of Free Fermi Gas

In this case the trap potential is zero, and we may find a general solution to the last equation via spatial Fourier transforms. We write

$$\begin{aligned}\tilde{\psi}_{\alpha 1}(\mathbf{s},t) &= \int \frac{d\mathbf{k}}{(2\pi)^{3/2}}\,\exp(i\mathbf{k}\cdot\mathbf{s})\,\tilde{\phi}_{\alpha 1}(\mathbf{k},t)\\ \tilde{\psi}_{\alpha 2}(\mathbf{s},t) &= \int \frac{d\mathbf{k}}{(2\pi)^{3/2}}\,\exp(-i\mathbf{k}\cdot\mathbf{s})\,\tilde{\phi}_{\alpha 2}(\mathbf{k},t),\end{aligned} \tag{15.61}$$

and the last equation can be turned into stochastic equations for the Fourier transforms $\tilde{\phi}_{\alpha A}(\mathbf{k},t)$. This eliminates the Laplacian in favour of a c-number k^2. We then have

$$\begin{bmatrix}\tilde{\phi}_{u1}(\mathbf{k},t+\delta t)\\ \tilde{\phi}_{d1}(\mathbf{k},t+\delta t)\\ \tilde{\phi}_{u2}(\mathbf{k},t+\delta t)\\ \tilde{\phi}_{d2}(\mathbf{k},t+\delta t)\end{bmatrix} = \begin{bmatrix}F^1_{u1,u1}(t) & F^1_{u1,d1}(t) & 0 & 0\\ F^1_{d1,u1}(t) & F^1_{d1,d1}(t) & 0 & 0\\ 0 & 0 & F^2_{u2,u2}(t) & F^2_{u2,d2}(t)\\ 0 & 0 & F^2_{d2,u2}(t) & F^2_{d2,d2}(t)\end{bmatrix}\begin{bmatrix}\tilde{\phi}_{u1}(\mathbf{k},t)\\ \tilde{\phi}_{d1}(\mathbf{k},t)\\ \tilde{\phi}_{u2}(\mathbf{k},t)\\ \tilde{\phi}_{d2}(\mathbf{k},t)\end{bmatrix}, \tag{15.62}$$

where the submatrices are

$$\begin{aligned}[F^1(t)] &= \begin{bmatrix}1-\frac{i}{\hbar}\left\{\frac{\hbar^2}{2m}k^2\right\}\delta t & \sqrt{\frac{ig}{2\hbar}}\{\delta w_{u1,d1}+i\,\delta w_{d1,u1}\}\\ \sqrt{\frac{ig}{2\hbar}}\{\delta w_{u1,d1}-i\,\delta w_{d1,u1}\} & 1-\frac{i}{\hbar}\left\{\frac{\hbar^2}{2m}k^2\right\}\delta t\end{bmatrix},\\ [F^2(t)] &= \begin{bmatrix}1+\frac{i}{\hbar}\left\{\frac{\hbar^2}{2m}k^2\right\}\delta t & \sqrt{\frac{ig}{2\hbar}}\{\delta w_{u2,d2}+i\,\delta w_{d2,u2}\}\\ \sqrt{\frac{ig}{2\hbar}}\{-\delta w_{u2,d2}+i\,\delta w_{d2,u2}\} & 1+\frac{i}{\hbar}\left\{\frac{\hbar^2}{2m}k^2\right\}\delta t\end{bmatrix},\end{aligned} \tag{15.63}$$

which, though stochastic, no longer involve the Laplacian operator. The equations can then be solved numerically.

The four-mode situation treated in Section 10.2 for two fermions with spins u, d occupying modes with momenta $\hbar\mathbf{k}$ and $-\hbar\mathbf{k}$ can be treated here as a special case. If we introduce momentum field operators as in (2.70) as

$$\hat{\Psi}_{\alpha 1}(\mathbf{r}) = \int \frac{d\mathbf{k}}{(2\pi)^{3/2}} \exp(i\mathbf{k}\cdot\mathbf{r})\, \hat{\Phi}_{\alpha 1}(\mathbf{k}),$$
$$\hat{\Phi}_{\alpha 1}(\mathbf{k}) = \int \frac{d\mathbf{r}}{(2\pi)^{3/2}} \exp(-i\mathbf{k}\cdot\mathbf{r})\, \hat{\Psi}_{\alpha 1}(\mathbf{r}), \tag{15.64}$$

then apart from using delta function rather than box normalisation, the momentum field operators $\hat{\Phi}_{u2}(\mathbf{k}), \hat{\Phi}_{d2}(\mathbf{k})$ are the same as the mode annihilation operators $\hat{c}_1 \equiv \hat{c}_{\mathbf{k}\,(+)}, \hat{c}_2 \equiv \hat{c}_{\mathbf{k}\,(-)}$, and $\hat{\Phi}_{u2}(-\mathbf{k}), \hat{\Phi}_{d2}(-\mathbf{k})$ are the same as $\hat{c}_3 \equiv \hat{c}_{-\mathbf{k}\,(+)}, \hat{c}_4 \equiv \hat{c}_{-\mathbf{k}\,(-)}$. The initial state $|\Phi_3\rangle$ in (10.17) and the other state of interest, $|\Phi_4\rangle$, are momentum Fock states

$$|\Phi_3\rangle = \hat{\Phi}_{u2}(\mathbf{k})\hat{\Phi}_{d2}(-\mathbf{k})\,|0\rangle\,,$$
$$|\Phi_4\rangle = \hat{\Phi}_{d2}(\mathbf{k})\hat{\Phi}_{u2}(-\mathbf{k})\,|0\rangle\,, \tag{15.65}$$

and a straightforward generalisation of (12.42) and (12.43) to treat momentum Fock states shows us that the population of the initial state $|\Phi_3\rangle$ and the coherence between this state and $|\Phi_4\rangle$ are given in terms of the following phase space functional integrals:

$$P(\Phi_3) = \int D\phi^+ D\phi\, \phi_{d1}(-\mathbf{k})\phi_{u1}(\mathbf{k}) B_{\mathrm{f}}[\boldsymbol{\phi}(x)]\, \phi_{u2}(\mathbf{k})\phi_{d2}(-\mathbf{k}),$$

and

$$C(\Phi_4;\Phi_3) = \int D\phi^+ D\phi\; \phi_{d1}(-\mathbf{k})\phi_{u1}(\mathbf{k}) B_{\mathrm{f}}[\boldsymbol{\phi}(x)]\, \phi_{d2}(\mathbf{k})\phi_{u2}(-\mathbf{k})$$

where the phase space integral has been converted into a functional integral involving the fermion unnormalised distribution functional for momentum fields, and the notation in the present section has been used. These phase space functional integrals are of course equivalent to stochastic averages of products of stochastic momentum fields. Hence, from the initial conditions, the only initial stochastic average that is non-zero is that for the initial population of $|\Phi_3\rangle$, which is unity:

$$P(\Phi_3) = \overline{(\tilde{\phi}_{d1}(-\mathbf{k})\tilde{\phi}_{u1}(\mathbf{k})\tilde{\phi}_{u2}(\mathbf{k})\tilde{\phi}_{d2}(-\mathbf{k}))}_{t=0} = 1. \tag{15.66}$$

The coherence at time δt is

$$C(\Phi_4;\Phi_3) = \overline{(\tilde{\phi}_{d1}(-\mathbf{k})\tilde{\phi}_{u1}(\mathbf{k})\tilde{\phi}_{d2}(\mathbf{k})\tilde{\phi}_{u2}(-\mathbf{k}))}_{t=\delta t}. \tag{15.67}$$

From (15.62), we see that with $\chi = \sqrt{ig/2\hbar}$ and $F_k = 1 - i/\hbar\,\{(\hbar^2/2m)k^2\}\,\delta t$ we have

$$\begin{aligned}
&\overline{(\tilde{\phi}_{d1}(-\mathbf{k})\tilde{\phi}_{u1}(\mathbf{k})\tilde{\phi}_{d2}(\mathbf{k})\tilde{\phi}_{u2}(-\mathbf{k}))}_{t=\delta t}\\
&\quad = [\{\chi\{\delta w_{u1,d1} - i\delta w_{d1,u1}\}\tilde{\phi}_{u1}(-\mathbf{k})_0 + F_k\tilde{\phi}_{d1}(-\mathbf{k})_0\}\\
&\qquad \times\{F_k\tilde{\phi}_{u1}(\mathbf{k})_0 + \chi\{\delta w_{u1,d1} + i\,\delta w_{d1,u1}\}\tilde{\phi}_{d1}(\mathbf{k})_0\}\\
&\qquad \times\{\chi\{-\delta w_{u2,d2} + i\,\delta w_{d2,u2}\}\tilde{\phi}_{u2}(\mathbf{k})_0 + F_k^*\tilde{\phi}_{d2}(\mathbf{k})_0\}\\
&\qquad \times\{F_k^*\tilde{\phi}_{u2}(-\mathbf{k})_0 + \chi\{\delta w_{u2,d2} + i\,\delta w_{d2,u2}\}\tilde{\phi}_{d2}(-\mathbf{k})_0\}]_{\mathrm{StochAv}}.
\end{aligned} \tag{15.68}$$

Noting that the stochastic average factorises into products of stochastic averages of terms involving the c-number Wiener increments and products of the stochastic momentum fields at $t = 0$, and the only non-zero initial stochastic average is given by (15.66), we can then identify possible non-zero overall contributions to $C(\Phi_4; \Phi_3)$. We find that

$$C(\Phi_4; \Phi_3) = \overline{F_k F_k \chi\{-\delta w_{u2,d2} + i\,\delta w_{d2,u2}\}\chi\{+\delta w_{u2,d2} + i\,\delta w_{d2,u2}\}} \times \overline{\tilde{\phi}_{d1}(-\mathbf{k})_0 \tilde{\phi}_{u1}(\mathbf{k})_0 \tilde{\phi}_{u2}(\mathbf{k})_0 \tilde{\phi}_{u2}(\mathbf{k})_0}. \tag{15.69}$$

Since $F_k^2 = 1 + O(\delta t)$ and the stochastic averages of the products of Wiener increments are equal to δt for the same increments and zero otherwise, and using (15.66), we find that correct to $O(\delta t)$

$$\begin{aligned} C(\Phi_4; \Phi_3) &= \chi^2(-2)\,\delta t \\ &= -\frac{ig}{\hbar}\delta t, \end{aligned} \tag{15.70}$$

which, apart from the difference due to using box normalisation, is exactly the same as before in (10.44).

15.2.7 Case of Optical Lattice

In this case the trap potentials are spatially periodic and we may be able to find a solution via expanding the stochastic field functions in terms of Bloch functions $\chi_\alpha^{\mathbf{k},a}(\mathbf{r})$, where $\mathbf{k}$ ranges over the Brillouin zone and a lists the different bands. Such functions obey an eigenvalue equation of the form

$$\left\{-\frac{\hbar^2}{2m}\nabla^2 + V_\alpha\right\}\chi_\alpha^{\mathbf{k},a}(\mathbf{r}) = \hbar\omega_\alpha^{\mathbf{k},a}\chi_\alpha^{\mathbf{k},a}(\mathbf{r}). \tag{15.71}$$

This again enables the Laplacian and the trap potential to be eliminated in favour of a c-number, the Bloch energy $\hbar\omega_\alpha^{\mathbf{k},a}$.

Exercise

(15.1) For the boson Wigner distribution functional $W[\psi(\mathbf{r}), \psi^*(\mathbf{r})]$ derive the contribution to the functional Fokker–Planck equation from the boson–boson interaction given in (15.24).

16
Further Developments

Applications based on the positive P distribution have been carried out for a number of bosonic cases. Numerical simulations using the Ito stochastic equations have sometimes resulted in unstable stochastic trajectories, leading to problems in obtaining the required stochastic averages [31, 61, 86]. This has occurred in highly non-linear systems with small damping, and has led to questioning whether the distribution function in these cases drops off to zero fast enough for the critical step of discarding boundary terms in deriving the Fokker–Planck equations themselves to be valid. To try to overcome these problems, a further development of the positive P distribution, called the *gauge P distribution*, has been made. In this approach [54, 55], the density operator is still written in terms of Bargmann coherent-state projectors, but now an additional phase space variable Ω, called the *weight*, is introduced, and *drift* and *diffusion* gauges are now incorporated into the Ito stochastic equations for the enlarged set of stochastic phase variables that are equivalent to the Fokker–Planck equation. For n modes there are now $2n+1$ phase space variables, and this is the number of Gaussian–Markov noise terms Γ_p needed in the Ito equations. The diffusion gauge is related to the flexibility in choosing the matrix B that satisfies $BB^T = D$, where D is the diffusion matrix. As commented on previously, for fermion systems numerical applications based on fermion Bargmann coherent-state projectors involving Grassmann phase space variables are rare, with [14] being an example. No gauge P extensions exist so far for fermions.

Other extensions of the theory presented in this book have also been made in the bosonic case. In the present treatment it has been assumed that the mode functions are time-independent. There are situations, such as when mode functions are determined from solutions of coupled Gross–Pitaevskii equations in the case of time-dependent trapping potentials, where time-dependent modes are a more physical choice. In addition, it is often convenient to divide the modes and field operators into two groups, such the division into highly occupied condensate modes and largely unoccupied non-condensate modes that would be appropriate for studying Bose–Einstein condensates well below the transition temperature. It could then be possible to use hybrid distribution functions, such as functions of the Wigner type for the condensate modes and of the positive P type for the non-condensate modes. Developments of this type have been published in [50–53]. In a hybrid development involving time-dependent modes [53], the Fokker–Planck and Langevin equations (for both modes and fields) exhibit new features, with additional terms appearing that depend on overlap integrals of mode functions and their time derivatives.

Phase Space Methods for Degenerate Quantum Gases. First Edition. Bryan J. Dalton, John Jeffers, and Stephen M. Barnett.
Published in 2015 by Oxford University Press.

Finally, in another phase space approach for both bosons and fermions [56–58], the density operator is represented in terms of *Gaussian projectors* rather than the Bargmann coherent-state projectors previously used. These depend on a set of new phase space variables which are related to pairs of mode annihilation and creation operators ($\hat{a}_i\hat{a}_j$, $\hat{a}_i^\dagger\hat{a}_j^\dagger$, $\hat{a}_i^\dagger\hat{a}_j$ for bosons, and $\hat{c}_i\hat{c}_j$, $\hat{c}_i^\dagger\hat{c}_j^\dagger$, $\hat{c}_i^\dagger\hat{c}_j$ for fermions), as well as phase variables related to individual annihilation and creation operators ($\hat{a}_i$, $\hat{a}_i^\dagger$) in the bosonic case. An additional phase space weight variable Ω is also included. Fokker–Planck equations involving derivatives with respect to the new phase space variables are obtained and then replaced by Ito stochastic equations. For n modes there are now $n(3n+2)+1$ phase space variables for bosons, and $n(2n-1)+1$ phase space variables for fermions. The Ito equations involve these numbers of noise terms Γ_p, showing that this number increases as the square of the number of modes.

Details of these extensions beyond the positive P representation are beyond the scope of this book, and the reader should consult the articles cited and the review by He et al. [87].

Appendix A
Fermion Anticommutation Rules

We wish to establish that the fermion annihilation and creation operators satisfy the anticommutation rules

$$\{\hat{c}_i, \hat{c}_j^\dagger\} = \hat{c}_i\hat{c}_j^\dagger + \hat{c}_j^\dagger\hat{c}_i = \delta_{ij},$$
$$\{\hat{c}_i, \hat{c}_j\} = \{\hat{c}_i^\dagger, \hat{c}_j^\dagger\} = 0. \tag{A.1}$$

We first apply the expressions for the effect of these operators on the basis states,

$$\hat{c}_i|\nu_1, \cdots, \nu_i, \cdots, \nu_n\rangle = \nu_i(-1)^{\eta_i}|\nu_1, \cdots, 1-\nu_i, \cdots, \nu_n\rangle,$$
$$\hat{c}_i^\dagger|\nu_1, \cdots, \nu_i, \cdots, \nu_n\rangle = (1-\nu_i)(-1)^{\eta_i}|\nu_1, \cdots, 1-\nu_i, \cdots, \nu_n\rangle, \tag{A.2}$$

where $(-1)^{\eta_i} = +1$ or -1 according to whether there are an even or odd number of modes listed preceding the mode m_i which are occupied ($\eta_i = \sum_{j<i} \nu_j$). Here the m_i are symbols listing the modes in a suitable order, and $\nu_i = 0, 1$ is the number of fermions occupying the mode m_i.

For the anticommutation rule $\{\hat{c}_i, \hat{c}_j\} = 0$, the case with $i = j$ reduces to showing that $\hat{c}_i^2 = 0$. This is simple, since

$$\hat{c}_i^2|\nu_1, \nu_n\rangle = \hat{c}_i\nu_i(-1)^{\eta_i}|\nu_1, \cdots, 1-\nu_i, \cdots, \nu_n\rangle$$
$$= \nu_i(1-\nu_i)(-1)^{2\eta_i}|\nu_1, \cdots, 1-(1-\nu_i), \cdots, \nu_n\rangle, \tag{A.3}$$

which vanishes since either ν_i or $1-\nu_i$ is zero. If $i < j$, we have the four possible choices $(\nu_i, \nu_j) = (0,0), (0,1), (1,0), (1,1)$. The first three cases of the form $\{\hat{c}_i, \hat{c}_j\}|\nu_1, \cdots, \nu_n\rangle$ vanish trivially, as an annihilation operator acts on a ground state. The last is straightforwardly shown to vanish:

$$\begin{aligned}\{\hat{c}_i, \hat{c}_j\}|\nu_1, \cdots, 1, \cdots, 1, \cdots, \nu_n\rangle &= (\hat{c}_i\hat{c}_j + \hat{c}_j\hat{c}_i)|\nu_1, \cdots, 1, \cdots 1, \cdots, \nu_n\rangle \\ &= \hat{c}_i(-1)^{\eta_j}|\nu_1, \cdots, 1, \cdots, 0, \cdots, \nu_n\rangle \\ &\quad + \hat{c}_j(-1)^{\eta_i}|\nu_1, \cdots, 0, \cdots, 1, \cdots, \nu_n\rangle \\ &= \left[(-1)^{\eta_i+\eta_j} + (-1)^{\eta_i+\eta_j-1}\right]|\nu_1, \cdots, 0, \cdots, 0, \cdots, \nu_n\rangle,\end{aligned} \tag{A.4}$$

so that

$$\{\hat{c}_i, \hat{c}_j\}|\nu_1, \cdots, \nu_n\rangle = 0\,. \tag{A.5}$$

Phase Space Methods for Degenerate Quantum Gases. First Edition. Bryan J. Dalton, John Jeffers, and Stephen M. Barnett.
Published in 2015 by Oxford University Press.

The anticommutation rule $\{\hat{c}_i^\dagger, \hat{c}_j^\dagger\} = 0$ follows from taking the adjoint of $\{\hat{c}_i, \hat{c}_j\} = 0$.

The anticommutation rule $\{\hat{c}_i, \hat{c}_j^\dagger\} = \delta_{ij}$ with $i = j$ can be derived:

$$\begin{aligned}
\{\hat{c}_i, \hat{c}_i^\dagger\}|\nu, \cdots, \nu_i = 0, \cdots, \nu_n\rangle &= (\hat{c}_i\hat{c}_i^\dagger + \hat{c}_i^\dagger\hat{c}_i)|\nu_1, \cdots, 0, \cdots, \nu_n\rangle \\
&= (\hat{c}_i)(-1)^{\eta_i}|\nu_1, \cdots, 1, \cdots, \nu_n\rangle \\
&= (-1)^{2\eta_i}|\nu_1, \cdots, 0, \cdots, \nu_n\rangle \\
&= |\nu_1, \cdots, 0, \cdots, \nu_n\rangle, \\
\{\hat{c}_i, \hat{c}_i^\dagger\}|\nu_1, \cdots, \nu_i = 1, \cdots, \nu_n\rangle &= (\hat{c}_i\hat{c}_i^\dagger + \hat{c}_i^\dagger\hat{c}_i)|\nu_1, \cdots, 1, \cdots, \nu_n\rangle \\
&= (\hat{c}_i^\dagger)(-1)^{\eta_i}|\nu_1, \cdots, 0, \cdots, \nu_n\rangle \\
&= (-1)^{2\eta_i}|\nu_1, \cdots, 1, \cdots, \nu_n\rangle \\
&= |\nu_1, \cdots, 1, \cdots, \nu_n\rangle,
\end{aligned}$$

so that for $i = j$,

$$\{\hat{c}_i, \hat{c}_i^\dagger\}|\nu_1, \cdots, \nu_n\rangle = |\nu_1, \cdots, \nu_n\rangle.$$

If $i < j$, we have the four possible situations $(\nu_i, \nu_j) = (0,0), (0,1), (1,0), (1,1)$. The first two cases vanish, as an annihilation (or creation) operator acts on a vacuum (or one-particle) state. The others are

$$\begin{aligned}
&\{\hat{c}_i, \hat{c}_j^\dagger\}|\nu_1, \nu_i = 1, \cdots, \nu_j = 0, \cdots, \nu_n\rangle = (\hat{c}_i\hat{c}_j^\dagger + \hat{c}_j^\dagger\hat{c}_i)|\nu_1, \cdots, 1, \cdots, 0, \cdots, \nu_n\rangle \\
&\quad = (\hat{c}_i)(-1)^{\eta_j}|\nu_1, \cdots, 1, \cdots, 1, \cdots, \nu_n\rangle \\
&\qquad + (\hat{c}_j^\dagger)(-1)^{\eta_i}|\nu_1, \cdots, 0, \cdots, 0, \cdots, \nu_n\rangle \\
&\quad = \left[(-1)^{\eta_i+\eta_j} + (-1)^{\eta_i+\eta_j-1}\right]|\nu_1, \cdots, 0, \cdots, 1, \cdots, \nu_n\rangle \\
&\quad = 0, \\
&\{\hat{c}_i, \hat{c}_j^\dagger\}|\nu_1, \cdots, \nu_i = 1, \cdots, \nu_j = 1, \cdots, \nu_n\rangle = (\hat{c}_i\hat{c}_j^\dagger + \hat{c}_j^\dagger\hat{c}_i)|\nu_1, \cdots, 1, \cdots, 1, \cdots, \nu_n\rangle \\
&\quad = (\hat{c}_j^\dagger)(-1)^{\eta_i}|\nu_1, \cdots, 0, \cdots, 1, \cdots, \nu_n\rangle \\
&\quad = 0,
\end{aligned}$$

so that for $i < j$,

$$\{\hat{c}_i, \hat{c}_j^\dagger\}|\nu_1, \cdots, \nu_n\rangle = 0. \tag{A.6}$$

Note that for the $(1,0)$ case, in the second term after the $\hat{c}_i$ operation the number of occupied modes for $i < j$ will have decreased by one.

Appendix B
Markovian Master Equation

The master equation can be obtained via standard methods [4, 32] starting from the two-body form of the system–reservoir interaction

$$\widehat{V}_{\mathrm{SR}} = \hbar \sum_a \widehat{S}_a \widehat{R}_a, \tag{B.1}$$

where the $\widehat{S}_a$ are given by (2.84) and (2.85) and the reservoir operators are

$$\widehat{R}_a = \sum_{KL} \frac{1}{\hbar} \zeta_{jK\,iL}\, \widehat{A}_K^{\dagger} \widehat{A}_L, \tag{B.2}$$

$$\zeta_{jK\,iL} = \langle j(a)| \langle K(b)| \widehat{V}(a,b) |i(a)\rangle |L(b)\rangle . \tag{B.3}$$

The $\widehat{A}_K^{\dagger}$ and $\widehat{A}_L$ are creation and annihilation operators for the foreign atoms, whose single-particle states (modes) are $|K\rangle$ and $|L\rangle$, and the $\zeta_{jK\,iL}$ are two-body matrix elements for the interaction $\widehat{V}(a,b)$ between a system atom a and a reservoir atom b. The relaxation matrix elements are obtained via weak-coupling Markovian master equation theory in the form

$$\begin{aligned}\Gamma_{ab} = \frac{1}{2} \int_0^T d\tau \; \Big\{ &\left\langle \widehat{R}_a(t) \widehat{R}_b(t-\tau)^{\dagger} \right\rangle \exp[(i\omega_b - \varepsilon)\tau] \\ &+ \left\langle \widehat{R}_a(t-\tau) \widehat{R}_b(t)^{\dagger} \right\rangle \exp[(-i\omega_a - \varepsilon)\tau] \Big\}.\end{aligned} \tag{B.4}$$

The $\langle \widehat{R}_a(t)\widehat{R}_b(t-\tau)^{\dagger}\rangle, \langle \widehat{R}_a(t-\tau)\widehat{R}_b(t)^{\dagger}\rangle$ are reservoir correlation functions defined in terms of the reservoir density operator $\widehat{\sigma}_{\mathrm{R}}$ and Heisenberg picture operators $\widehat{R}_a(t)$:

$$\left\langle \widehat{R}_a(t_1) \widehat{R}_b(t_2)^{\dagger} \right\rangle = \mathrm{Tr}\{\widehat{\sigma}_{\mathrm{R}}\, \widehat{R}_a(t_1) \widehat{R}_b(t_2)^{\dagger}\}, \tag{B.5}$$

$$\widehat{R}_a(t) = \exp\{i\widehat{H}_{\mathrm{R}} t/\hbar\}\, \widehat{R}_a\, \exp\{-i\widehat{H}_{\mathrm{R}} t/\hbar\}, \tag{B.6}$$

where the Hamiltonian for the reservoir atoms is $\widehat{H}_{\mathrm{R}}$. In the Markov approximation the reservoir correlation functions decay to zero over a reservoir correlation time τ_{c} which is short compared with the system relation time $\sim 1/\Gamma$, and the master equation applies to the time where $t \gg \tau_{\mathrm{c}}$. The system transition frequencies are defined via

$$[\widehat{H}_{\mathrm{S0}}, \widehat{S}_a] = \hbar\omega_a \widehat{S}_a, \tag{B.7}$$

where $\widehat{H}_{S0}$ is the zeroth-order term in the system Hamiltonian, which in the present case will be

$$\begin{aligned}\widehat{H}_{S0} &= \sum_i \hbar\omega_i \hat{c}_i^\dagger \hat{c}_i \qquad \text{Fermions} \\ &= \sum_i \hbar\omega_i \hat{a}_i^\dagger \hat{a}_i, \qquad \text{Bosons} \\ \omega_a &= \omega_j - \omega_i = \omega_{ji}.\end{aligned} \tag{B.8}$$

In the master equation, there is an additional shift term given by

$$\left(\frac{\partial}{\partial t}\widehat{\rho}\right) = -i \sum_{ab} \Delta_{ab} \, [\widehat{S}_a \widehat{S}_b^\dagger, \widehat{\rho}], \tag{B.9}$$

whore

$$\begin{aligned}\Delta_{ab} = \frac{1}{2i} \int_0^T d\tau \, \Big\{ &\Big\langle \widehat{R}_a(t) \widehat{R}_b(t-\tau)^\dagger \Big\rangle \exp[(i\omega_b - \varepsilon)\tau] \\ &- \Big\langle \widehat{R}_a(t-\tau) \widehat{R}_b(t)^\dagger \Big\rangle \exp[(-i\omega_a - \varepsilon)\tau] \Big\},\end{aligned} \tag{B.10}$$

but we will not consider this here.

The symmetry results

$$\Gamma_{ab} = \Gamma_{ba}^*, \qquad \Delta_{ab} = \Delta_{ba}^*, \tag{B.11}$$

are easily derived from the definitions.

Appendix C
Grassmann Calculus

C.1 Double-Integral Result

The double integral $\int dh^{+}dh\ F(h,h^{+})(h-g)(h^{+}-g^{+})$ can be obtained in general, noting that the Grassmann function can always be written in the form

$$F(h,h^{+}) = F_0 + F_1 h + F_1^{+} h^{+} + F_2 hh^{+}. \tag{C.1}$$

Then

$$\begin{aligned}
&\int dh^{+}dh\ F(h,h^{+})(h-g)(h^{+}-g^{+}) \qquad\qquad (C.2)\\
&= \int dh^{+}dh\ \{F_0 + F_1 h + F_1^{+} h^{+} + F_2 hh^{+}\} \times (h-g)(h^{+}-g^{+})\\
&= F_0 \int dh^{+}dh\ (h)(h^{+}) + F_1 \int dh^{+}dh\ (-hg)(h^{+})\\
&\quad + F_1^{+} \int dh^{+}dh - (h)(-h^{+}g^{+}) + F_2 \int dh^{+}dh\ \{hh^{+}\}(-g)(-g^{+})\\
&= F_0 + F_1 g + F_1^{+} g^{+} + F_2 gg^{+}\\
&= F(g,g^{+}), \qquad\qquad (C.3)
\end{aligned}$$

as required.

C.2 Grassmann Fourier Integral

We wish to show that

$$\int dk^{+}dk\, F(k,k^{+}) \int dg^{+}dg\ \exp(ig[h^{+}-k^{+}])\exp(i[h-k]g^{+}) = F(h,h^{+}). \tag{C.4}$$

Phase Space Methods for Degenerate Quantum Gases. First Edition. Bryan J. Dalton, John Jeffers, and Stephen M. Barnett.
Published in 2015 by Oxford University Press.

Noting that $\exp(ig[h^+ - k^+]) = 1 + ig[h^+ - k^+]$ etc., we see that

$$\begin{aligned}
&\int dg^+ dg\, \exp(ig[h^+ - k^+])\, \exp(i[h-k]g^+) \\
&\quad = \int dg^+ dg\, \{1 + ig[h^+ - k^+]\}\{1 + i[h-k]g^+\} \\
&\quad = \int dg^+ \{i[h^+ - k^+]\}\{1 - ig^+[h-k]\} \\
&\quad = -\{i[h^+ - k^+]\} \int dg^+ \{1 - ig^+[h-k]\} \\
&\quad = [h-k][h^+ - k^+]\,,
\end{aligned} \tag{C.5}$$

where the standard integration and anticommutation rules for Grassmann variables and differentials have been used.

Thus we have

$$\begin{aligned}
&\int dk^+ dk\, F(k,k^+) \int dg^+ dg\, \exp(ig[h^+ - k^+])\, \exp(i[h-k]g^+) \\
&\quad = \int dk^+ dk\, F(k,k^+)\, [h-k][h^+ - k^+] \\
&\quad = F(h,h^+),
\end{aligned} \tag{C.6}$$

using (4.69). The generalisation to the multi-variable case is straightforward, noting that the $\{1 + ig_i[h_i^+ - k_i^+]\}$ etc. are even, commuting Grassmann functions.

C.3 Differentiating Multiple Grassmann Integrals of Functions of Two Sets of Grassmann Variables

To derive the rules for differentiating the multiple integral over all g_i of a function of two sets of Grassmann variables g_i and $h_j$we can write the function as a linear combination of monomials with theg_i placed to the left of the h_j. Hence we have (here and following $i_1 < i_2 < \cdots < i_p, j_1 < j_2 < \cdots < j_q$)

$$F(g_i, h_j) = \sum_{i_1 \cdots i_p, j_1 \cdots j_q} F_{p+q}^{i_1 \cdots i_p; j_1 \cdots j_q} g_{i_1} \cdots g_{i_p} h_{j_1} \cdots h_{j_q}, \tag{C.7}$$

where an ordering convention has been applied to avoid repeating terms. Each monomial may not necessarily involve the same number of g_i and h_j. Then the Grassmann integration over the g_i gives zero except for the terms in which all g_i are present, and hence

$$\int \prod_i dg_i \; F(g_i, h_j) = \sum_{j_1 \cdots j_q} F_{n+q}^{1 \cdots n; j_1 \cdots j_q} h_{j_1} \cdots h_{j_q},$$

$$\frac{\overrightarrow{\partial}}{\partial h_j}\left(\int \prod_i dg_i \; F(g_i, h_j) \right) = \sum_{j_1 \cdots j_q \neq j} (-1)^{q_j} F_{n+q}^{1 \cdots n; j_1 \cdots j_q} h_{j_1} \cdots h_{j_q},$$

$$\left(\int \prod_i dg_i \; F(g_i, h_j) \right) \frac{\overleftarrow{\partial}}{\partial h_j} = \sum_{j_1 \cdots j_q \neq j} (-1)^{q - q_j - 1} F_{n+q}^{1 \cdots n; j_1 \cdots j_q} h_{j_1} \cdots h_{j_q}, \quad \text{(C.8)}$$

where there are q_j transpositions needed to move h_j to the left of the other h_{j_m} and therefore $q - q_j - 1$ transpositions are needed to move h_j to the right of the other h_{j_m}. But, on the other hand,

$$\frac{\overrightarrow{\partial}}{\partial h_j} F(g_i, h_j) = \sum_{i_1 \cdots i_p, j_1 \cdots j_q \neq j} (-1)^{p + q_j} F_{p+q}^{i_1 \cdots i_p; j_1 \cdots j_q} g_{i_1} \cdots g_{i_p} h_{j_1} \cdots h_{j_q},$$

$$\int \prod_i dg_i \left(\frac{\overrightarrow{\partial}}{\partial h_j} F(g_i, h_j) \right) = (-1)^n \sum_{j_1 \cdots j_q \neq j} (-1)^{q_j} F_{p+q}^{1 \cdots n; j_1 \cdots j_q} h_{j_1} \cdots h_{j_q} \quad \text{(C.9)}$$

and

$$F(g_i, h_j) \frac{\overleftarrow{\partial}}{\partial h_j} = \sum_{i_1 \cdots i_p, j_1 \cdots j_q \neq j} (-1)^{q - q_j - 1} F_{p+q}^{i_1 \cdots i_p; j_1 \cdots j_q} g_{i_1} \cdots g_{i_p} h_{j_1} \cdots h_{j_q},$$

$$\int \prod_i dg_i \left(F(g_i, h_j) \frac{\overleftarrow{\partial}}{\partial h_j} \right) = \sum_{j_1 \cdots j_q \neq j} (-1)^{q - q_j - 1} F_{p+q}^{1 \cdots n; j_1 \cdots j_q} h_{j_1} \cdots h_{j_q}, \quad \text{(C.10)}$$

and hence we have the rules that

$$\frac{\overrightarrow{\partial}}{\partial h_j}\left(\int \prod_i dg_i \; F(g_i, h_j) \right) = (-1)^n \int \prod_i dg_i \left(\frac{\overrightarrow{\partial}}{\partial h_j} F(g_i, h_j) \right),$$

$$\left(\int \prod_i dg_i \; F(g_i, h_j) \right) \frac{\overleftarrow{\partial}}{\partial h_j} = \int \prod_i dg_i \left(F(g_i, h_j) \frac{\overleftarrow{\partial}}{\partial h_j} \right). \quad \text{(C.11)}$$

Thus we see that the usual c-number rule does not apply for left differentiation – an extra factor of $(-1)^n$ is involved. This factor does not occur for right differentiation. The distinction occurs because we have considered only left integration.

Appendix D
Properties of Coherent States

D.1 Fermion Coherent-State Eigenvalue Equation

The derivation of the eigenvalue equation follows from the transformation properties of the annihilation operator under the unitary displacement operator,

$$\begin{aligned}
\hat{c}_i|\mathbf{g},\mathbf{g}^*\rangle &= \hat{c}_i\hat{D}_{\mathrm{f}}(\mathbf{g},\mathbf{g}^*)|0\rangle \\
&= \hat{D}_{\mathrm{f}}(\mathbf{g},\mathbf{g}^*)\hat{D}_{\mathrm{f}}^{\dagger}(\mathbf{g},\mathbf{g}^*)\hat{c}_i\hat{D}_{\mathrm{f}}(\mathbf{g},\mathbf{g}^*)|0\rangle \\
&= \hat{D}_{\mathrm{f}}(\mathbf{g},\mathbf{g}^*)(\hat{c}_i+g_i)|0\rangle \\
&= g_i\hat{D}_{\mathrm{f}}(\mathbf{g},\mathbf{g}^*)|0\rangle \\
&= g_i|\mathbf{g},\mathbf{g}^*\rangle, \qquad \text{(D.1)}
\end{aligned}$$

using $\hat{c}_i|0\rangle = 0$ and the fact that g_i commutes with $\hat{D}_{\mathrm{f}}(\mathbf{g},\mathbf{g}^*)$. Similarly, using $\hat{c}_i^{\dagger}|1\rangle = 0$ and the fact that g_i^* commutes with $\hat{D}_{\mathrm{f}}(\mathbf{g},\mathbf{g}^*)$, we can derive

$$\hat{c}_i^{\dagger}|\mathbf{g}',\mathbf{g}'^*\rangle = \hat{c}_i^{\dagger}\hat{D}_{\mathrm{f}}(\mathbf{g},\mathbf{g}^*)|1\rangle = g_i^*|\mathbf{g}',\mathbf{g}'^*\rangle. \qquad \text{(D.2)}$$

D.2 Trace of Coherent-State Projectors

Bargmann states can help to evaluate the trace of the fermion coherent-state projector

$$\begin{aligned}
\mathrm{Tr}_{\mathrm{f}}(|\mathbf{g},\mathbf{g}^*\rangle\langle\mathbf{h},\mathbf{h}^*|) &= \exp\left[-\frac{1}{2}(\mathbf{g}^*\cdot\mathbf{g}+\mathbf{h}^*\cdot\mathbf{h})\right]\mathrm{Tr}_{\mathrm{f}}(|\mathbf{g}\rangle_{\mathrm{BB}}\langle\mathbf{h}|) \\
&= \exp\left[-\frac{1}{2}(\mathbf{g}^*\cdot\mathbf{g}+\mathbf{h}^*\cdot\mathbf{h})\right]\mathrm{Tr}_{\mathrm{f}}\left(\prod_i |g_i\rangle_{\mathrm{BB}}\langle h_i|\right). \qquad \text{(D.3)}
\end{aligned}$$

As the vacuum state commutes with Grassmann numbers, we have

$$\begin{aligned}
|g_i\rangle_{\mathrm{B}} &= (1+\hat{c}_i^{\dagger}g_i)|0_i\rangle = |0_i\rangle + |1_i\rangle g_i, \\
\langle h_i|_{\mathrm{B}} &= \langle 0_i|(1+h_i^*\hat{c}_i) = \langle 0_i| + h_i^*\langle 1_i|, \qquad \text{(D.4)}
\end{aligned}$$

so that introducing fermion Fock state basis vectors $|\nu_1,\cdots,\nu_n\rangle$ with $\nu_i = 0,1$ gives

Phase Space Methods for Degenerate Quantum Gases. First Edition. Bryan J. Dalton, John Jeffers, and Stephen M. Barnett.
Published in 2015 by Oxford University Press.

$$\begin{aligned}
\mathrm{Tr_f}\left(\prod_i |g_i\rangle_{\mathrm{B}\,\mathrm{B}}\langle h_i|\right) &= \sum_{\nu_1',\cdots,\nu_n} \langle \nu_1,\cdots,\nu_n| \left(\prod_i |g_i\rangle_{\mathrm{B}\,\mathrm{B}}\langle h_i|\right) |\nu_1,\cdots,\nu_n\rangle \\
&= \prod_i \sum_{\nu_i=0,1} \langle \nu_i|g_i\rangle_{\mathrm{B}\,\mathrm{B}}\langle h_i|\nu_i\rangle \\
&= \prod_i \{\langle 0_i|g_i\rangle_{\mathrm{B}\,\mathrm{B}}\langle h_i|0_i\rangle + \langle 1_i|g_i\rangle_{\mathrm{B}\,\mathrm{B}}\langle h_i|1_i\rangle\} \\
&= \prod_i \{1 + g_i h_i^*\} = \exp(\mathbf{g}\cdot\mathbf{h}^*).
\end{aligned} \tag{D.5}$$

Hence we have for the fermionic trace

$$\mathrm{Tr_f}\left(|\mathbf{g},\mathbf{g}^*\rangle\langle \mathbf{h},\mathbf{h}^*|\right) = \exp\left[\mathbf{g}\cdot\mathbf{h}^* - \frac{1}{2}(\mathbf{g}^*\cdot\mathbf{g} + \mathbf{h}^*\cdot\mathbf{h})\right]. \tag{D.6}$$

However, as $\langle \mathbf{h},\mathbf{h}^*|\mathbf{g},\mathbf{g}^*\rangle = \exp\left[\mathbf{h}^*\cdot\mathbf{g} - \frac{1}{2}(\mathbf{h}^*\cdot\mathbf{h} + \mathbf{g}^*\cdot\mathbf{g})\right]$ and because $\mathbf{h}^*\cdot\mathbf{g} = -\mathbf{g}\cdot\mathbf{h}^*$,we see that $\mathrm{Tr_f}(|\mathbf{g},\mathbf{g}^*\rangle\langle \mathbf{h},\mathbf{h}^*| = \langle -\mathbf{h}, -\mathbf{h}^*|\mathbf{g},\mathbf{g}^*\rangle \neq \langle \mathbf{h},\mathbf{h}^*|\mathbf{g},\mathbf{g}^*\rangle$ as it would be for bosons.

It is instructive to treat the bosonic case in a similar way:

$$\begin{aligned}
\mathrm{Tr_b}\left(|\boldsymbol{\alpha},\boldsymbol{\alpha}^*\rangle\langle \boldsymbol{\beta},\boldsymbol{\beta}^*|\right) &= \exp\left[-\frac{1}{2}(\boldsymbol{\alpha}^*\cdot\boldsymbol{\alpha} + \boldsymbol{\beta}^*\cdot\boldsymbol{\beta})\right] \mathrm{Tr_b}\left(|\boldsymbol{\alpha}\rangle_{\mathrm{B}\,\mathrm{B}}\langle\boldsymbol{\beta}|\right) \\
&= \exp\left[-\frac{1}{2}(\boldsymbol{\alpha}^*\cdot\boldsymbol{\alpha} + \boldsymbol{\beta}^*\cdot\boldsymbol{\beta})\right] \mathrm{Tr_b}\left(\prod_i |\alpha_i\rangle_{\mathrm{B}\,\mathrm{B}}\langle\beta_i|\right).
\end{aligned} \tag{D.7}$$

This time, however,

$$\begin{aligned}
|\alpha_i\rangle_{\mathrm{B}} &= \exp(\hat{a}_i^\dagger \alpha_i)|0_i\rangle = \sum_{m_i} \frac{(\alpha_i)^{m_i}}{\sqrt{m_i!}}|m_i\rangle, \\
\langle\beta_i|_{\mathrm{B}} &= \sum_{p_i} \frac{(\beta_i^*)^{p_i}}{\sqrt{p_i!}}\langle p_i|,
\end{aligned} \tag{D.8}$$

so that on introducing boson Fock state basis vectors $|\nu_1,\cdots,\nu_n\rangle$ with $\nu_i = 0,1,2,\cdots$, we have

$$\begin{aligned}
\mathrm{Tr_b}\left(\prod_i |\alpha_i\rangle_{\mathrm{B}\,\mathrm{B}}\langle\beta_i|\right) &= \sum_{\nu_1,\cdots,\nu_n} \langle \nu_1,\cdots,\nu_n| \left(\prod_i |\alpha_i\rangle_{\mathrm{B}\,\mathrm{B}}\langle\beta_i|\right) |\nu_1,\cdots,\nu_n\rangle \\
&= \prod_i \sum_{\nu_i=0,1,\cdots} \langle\nu_i|\alpha_i\rangle_{\mathrm{B}\,\mathrm{B}}\langle\beta_i|\nu_i\rangle \\
&= \prod_i \sum_{\nu_i=0,1,\cdots} \frac{(\alpha_i)^{\nu_i}}{\sqrt{\nu_i!}} \frac{(\beta_i^*)^{\nu_i}}{\sqrt{\nu_i!}} \\
&= \prod_i \exp(\alpha_i\beta_i^*) \\
&= \exp(\boldsymbol{\alpha}\cdot\boldsymbol{\beta}^*).
\end{aligned} \tag{D.9}$$

Thus the outcome has the same form as for the fermionic case even though the intermediate steps differ, and hence

$$\mathrm{Tr}_{\mathrm{b}}\left(|\boldsymbol{\alpha},\boldsymbol{\alpha}^*\rangle\langle\boldsymbol{\beta},\boldsymbol{\beta}^*|\right) = \exp\left[\boldsymbol{\alpha}\cdot\boldsymbol{\beta}^* - \frac{1}{2}(\boldsymbol{\alpha}^*\cdot\boldsymbol{\alpha} + \boldsymbol{\beta}^*\cdot\boldsymbol{\beta})\right].$$

The c-number variables commute, and hence we can conclude that $\mathrm{Tr}_{\mathrm{b}}(|\boldsymbol{\alpha}\boldsymbol{\alpha}^*\rangle\langle\boldsymbol{\beta}\boldsymbol{\beta}^*|) = \langle\boldsymbol{\beta}\boldsymbol{\beta}^*|\boldsymbol{\alpha}\boldsymbol{\alpha}^*\rangle$.

D.3 Completeness Relation for Fermion Coherent States

Consider the Grassmann operator $\hat{\Xi}$ defined by

$$\begin{aligned}\hat{\Xi} &= \int d^2\mathbf{g}\ |\mathbf{g},\mathbf{g}^*\rangle\langle\mathbf{g},\mathbf{g}^*| \\ &= \int\prod_i d^2g_i\ \hat{D}_{\mathrm{f}}(\mathbf{g},\mathbf{g}^*)|0\rangle\langle 0|\hat{D}_{\mathrm{f}}(\mathbf{g},\mathbf{g}^*)^\dagger,\end{aligned} \tag{D.10}$$

where here $d^2g_i = dg_i^*\, dg_i$, so both g_i^* and g_i are integrated over.

We then determine the matrix element of this operator between two general states $|\Phi\rangle_{\mathrm{f}}$ and $|\Theta\rangle_{\mathrm{f}}$ of the form given in (2.27), where n is the number of modes and $l \equiv \{l_1, l_2, \cdots, l_p\}$ and $m \equiv \{m_1, m_2, \cdots, m_q\}$ list two sets of occupied modes, with $0 \le p, q \le n$, and with these modes listed in a conventional order – denoted $l_1 < l_2 < \cdots < l_q$ and $m_1 < m_2 < \cdots < m_p$. In terms of this form for expanding the states, we have for $\langle g, g^*|\Phi\rangle_{\mathrm{f}}$

$$\langle\mathbf{g},\mathbf{g}^*|\Phi\rangle_{\mathrm{f}} = \sum_m D(m_1\cdots m_p)g_{m_1}^*\cdots g_{m_p}^*\exp\left(-\frac{1}{2}\mathbf{g}^*\cdot\mathbf{g}\right), \tag{D.11}$$

where $D(m_1\cdots m_p)$ is the expansion coefficients for $|\Phi\rangle_{\mathrm{f}}$ in terms of the occupied Fock state $|m_1\cdots m_p\rangle$ (see (5.66)). A related expression applies for ${}_{\mathrm{f}}\langle\Theta|g,g^*\rangle$, the expansion coefficients then being denoted as $C(l_1\cdots l_q)$. Taking the c-numbers outside the integral and using (3.39) to combine the two exponentials, we find that

$$\begin{aligned}\langle\Theta|\hat{\Xi}|\Phi\rangle_{\mathrm{f}} &= \int d^2\mathbf{g}\ {}_{\mathrm{f}}\langle\Theta|\mathbf{g},\mathbf{g}^*\rangle\langle\mathbf{g},\mathbf{g}^*|\Phi\rangle_{\mathrm{f}} \\ &= \int d^2\mathbf{g}\ \exp(-\mathbf{g}^*\cdot\mathbf{g})\sum_{\{l\}}C^*(l_1\cdots l_q)g_{l_q}\cdots g_{l_1} \\ &\quad\times\sum_{\{m\}}D(m_1\cdots m_p)g_{m_1}^*\cdots g_{m_p}^* \\ &= \sum_{\{l\},\{m\}}C^*(l_1\cdots l_q)D(m_1\cdots m_p) \\ &\quad\times\int d^2\mathbf{g}\prod_i(1-g_i^*g_i)g_{l_q}\cdots g_{l_1}g_{m_1}^*\cdots g_{m_p}^*.\end{aligned} \tag{D.12}$$

These modes are listed in a conventional order, denoted $m_1 < m_2 < \cdots < m_p$ and $l_1 < l_2 < \cdots < l_q$. If for a particular pair of sets l, m a mode is occupied, then the $g_i^* g_i$ factor for one of these occupied modes will multiply a g or g^* present in the factor $g_{l_q} \cdots g_{l_1} g_{m_1}^* \cdots g_{m_p}^*$ and give zero. Hence the product of the $(1 - g_i^* g_i)$ can be replaced by unity for modes in the sets l, m, giving

$$\langle \Theta | \hat{\Xi} | \Phi \rangle_{\mathrm{f}} = \sum_{\{l\},\{m\}} C^*(l_1 \cdots l_q) B(m_1 \cdots m_p) \times \int d^2\mathbf{g} \prod_{i \neq \{l\},\{m\}} (1 - g_i^* g_i) g_{l_q} \cdots g_{l_1} g_{m_1}^* \cdots g_{m_p}^* . \tag{D.13}$$

Then, for these unoccupied modes, the partial Grassmann integral gives

$$\int d^2 g_i \, (1 - g_i^* g_i) = - \int dg_i^* \, dg_i \, g_i^* g_i = \int dg_i^* \, g_i^* \int dg_i \, g_i = 1 . \tag{D.14}$$

Note at this point that an integral over both g_i^* and g_i is involved. Hence we have

$$\langle \Theta | \hat{\Xi} | \Phi \rangle_{\mathrm{f}} = \sum_{l,m} C^*(l_1 \cdots l_q) D(m_1 \cdots m_p) \int \prod_{k=l,m} dg_k^* \, dg_k \, g_{l_q} \cdots g_{l_1} g_{m_1}^* \cdots g_{m_p}^* , \tag{D.15}$$

where now the integration is only over modes in the l, m sets.

We can now see that there is no contribution unless the two sets of occupied modes l, m are the same. If that is not the case, then there will be no matching pairs of g_{l_i} and $g_{l_i}^*$ in the factor $g_{l_q} \cdots g_{l_2} g_{l_1} g_{m_1}^* g_{m_2}^*, \cdots, g_{m_p}^*$ and then an integral of the form $\int dg_i^*$ or $\int dg_i$ will give zero. Hence $p = q$ and $l_1 = m_1, \cdots, l_p = m_p$, and thus

$$\langle \Theta | \hat{\Xi} | \Phi \rangle_{\mathrm{f}} = \sum_{\{m\}} C^*(m_1 \cdots m_p) D(m_1 \cdots m_p) \int \prod_{k=m} dg_k^* \, dg_k \, g_{m_p} \cdots g_{m_1} g_{m_1}^* \cdots g_{m_p}^* . \tag{D.16}$$

Finally, we carry out the integrations in succession starting with $\int dg_{m_1}^* dg_{m_1}$ and find

$$\int \prod_{k=m} dg_k^* \, dg_k \, g_{m_p} \cdots g_{m_1} g_{m_1}^* \cdots g_{m_p}^* = \int \prod_{k=m, k \neq m_1} dg_k^* \, dg_k \, g_{m_p} \cdots g_{m_2} g_{m_2}^* \cdots g_{m_p}^* \times 1 = 1 \tag{D.17}$$

after all the integrals have been done. Hence we have found that

$$\langle \Theta | \hat{\Xi} | \Phi \rangle_{\mathrm{f}} = \sum_{\{m\}} C^*(m_1 \cdots m_p) D(m_1 \cdots m_p) = \langle \Theta | \Phi \rangle_{\mathrm{f}},$$

so that the operator $\hat{\Xi}$ must be the identity:

$$\int d^2\mathbf{g} \, |\mathbf{g}, \mathbf{g}^* \rangle \langle \mathbf{g}, \mathbf{g}^* | = \int d^2\mathbf{g} \, \hat{D}_{\mathrm{f}}(\mathbf{g}, \mathbf{g}^*) |0\rangle \langle 0 | \hat{D}_{\mathrm{f}}(\mathbf{g}, \mathbf{g}^*)^\dagger = \hat{1} . \tag{D.18}$$

Appendix E
Phase Space Distributions for Bosons and Fermions

E.1 Canonical Forms of Fermion Distribution Function

The canonical form (7.28) of the distribution function in terms of the *characteristic function* can also be expressed in terms of *fermion coherent states*. We substitute for the characteristic function from (7.11) to first write

$$P_{\rm c}^{\rm f}(\mathbf{g},\mathbf{g}^+) = \int d\mathbf{h}^+ d\mathbf{h}\, \exp(-i\mathbf{g}\cdot\mathbf{h}^+)\, \mathrm{Tr}\left\{\exp(i\widehat{\mathbf{c}}\cdot\mathbf{h}^+)\,\hat{\rho}\,\exp(i\mathbf{h}\cdot\widehat{\mathbf{c}}^\dagger)\right\} \exp(-i\mathbf{h}\cdot\mathbf{g}^+). \tag{E.1}$$

Inserting two completeness relationships in terms of Bargmann states from (5.57) on each side of the density operator, we have, after shifting the even Grassmann differentials $\int d^2\mathbf{k} \int d^2\mathbf{k}^{+*}$ to the left,

$$\begin{aligned} P_{\rm c}^{\rm f}(\mathbf{g},\mathbf{g}^+) &= \int d\mathbf{h}^+ d\mathbf{h}\, \exp(-i\mathbf{g}\cdot\mathbf{h}^+)\, \mathrm{Tr}\left[\int d^2\mathbf{k}\int d^2\mathbf{k}^{+*} \exp(i\mathbf{k}\cdot\mathbf{h}^+)\, \frac{|\mathbf{k}\rangle_{\rm B}\langle\mathbf{k}|_{\rm B}}{{}_{\rm B}\langle\mathbf{k}|\mathbf{k}\rangle_{\rm B}}\right. \\ &\quad \left.\times\, \hat{\rho}\frac{|\mathbf{k}^{+*}\rangle_{\rm B}\langle\mathbf{k}^{+*}|_{\rm B}}{{}_{\rm B}\langle\mathbf{k}^{+*}|\mathbf{k}^{+*}\rangle_{\rm B}} \exp(i\mathbf{h}\cdot\mathbf{k}^+)\right] (\exp - i\mathbf{h}\cdot\mathbf{g}^+), \end{aligned} \tag{E.2}$$

where we have used the eigenvalue equations for fermion annihilation and creation operators. Then, splitting $d^2\mathbf{k}d^2\mathbf{k}^{+*}$ into $d\mathbf{k}^* d\mathbf{k}\, d\mathbf{k}^+ d\mathbf{k}^{+*}$, shifting all the even exponential Grassmann functions and the even differentials $\int d\mathbf{h}^+ d\mathbf{h}$ to the right, and combining exponentials and introducing the factor $\mathrm{Tr}(|\mathbf{k}\rangle_{\rm B}\langle\mathbf{k}^{+*}|_{\rm B})$ into the numerator and denominator to produce a normalised Bargmann state projector, we find that

$$\begin{aligned} P_{\rm c}^{\rm f}(\mathbf{g},\mathbf{g}^+) &= \mathrm{Tr}\left[\int d\mathbf{k}^* d\mathbf{k} \int d\mathbf{k}^+ d\mathbf{k}^{+*} \frac{|\mathbf{k}\rangle_{\rm B}\langle\mathbf{k}^{+*}|_{\rm B}}{\mathrm{Tr}(|\mathbf{k}\rangle_{\rm B}\langle\mathbf{k}^{+*}|_{\rm B})}\, \frac{\mathrm{Tr}(|\mathbf{k}\rangle_{\rm B}\langle\mathbf{k}^{+*}|_{\rm B})}{{}_{\rm B}\langle\mathbf{k}|\mathbf{k}\rangle_{\rm B}\,{}_{\rm B}\langle\mathbf{k}^{+*}|\mathbf{k}^{+*}\rangle_{\rm B}}\right. \\ &\quad \left.\times\, {}_{\rm B}\langle\mathbf{k}|\,\hat{\rho}\,|\mathbf{k}^{+*}\rangle_{\rm B} \int d\mathbf{h}^+ d\mathbf{h}\, \exp\left[i\mathbf{h}\cdot(\mathbf{k}^+ - \mathbf{g}^+)\right] \exp\left[i(\mathbf{k}-\mathbf{g})\cdot\mathbf{h}^+\right]\right]. \end{aligned} \tag{E.3}$$

Carrying out the trace operation, which gives unity and can be performed inside the even Grassmann differentials, substituting from (5.78) and (5.52) for the trace of the Bargmann projector and the scalar product of Bargmann states, and reordering the Grassmann differentials we obtain the following, using the fact that ${}_{\mathrm{B}}\langle\mathbf{k}|\hat{\rho}|\mathbf{k}^{+*}\rangle_{\mathrm{B}}$ is an even Grassmann function of the k_i^* and k_i^{+*} and not the k_i and k_i^+:

$$\begin{aligned}
P_{\mathrm{c}}^{\mathrm{f}}(\mathbf{g},\mathbf{g}^+) &= \int d\mathbf{k}^{+*}d\mathbf{k}^*{}_{\mathrm{B}}\langle\mathbf{k}|\hat{\rho}|\mathbf{k}^{+*}\rangle_{\mathrm{B}} \int d\mathbf{k}^+d\mathbf{k}\ \exp(-\mathbf{k}^*\cdot\mathbf{k})\exp(-\mathbf{k}^+\cdot\mathbf{k}^{+*}) \\
&\quad\times\exp(\mathbf{k}\cdot\mathbf{k}^+)\int d\mathbf{h}^+d\mathbf{h}\exp\left[i\mathbf{h}\cdot(\mathbf{k}^+-\mathbf{g}^+)\right]\exp\left[i(\mathbf{k}-\mathbf{g})\cdot\mathbf{h}^+\right] \\
&= \int d\mathbf{k}^{+*}d\mathbf{k}^*{}_{\mathrm{B}}\langle\mathbf{k}|\hat{\rho}|\mathbf{k}^{+*}\rangle_{\mathrm{B}} \int d\mathbf{k}^+d\mathbf{k}\ \exp(-\mathbf{k}^*\cdot\mathbf{k})\exp(-\mathbf{k}^+\cdot\mathbf{k}^{+*}) \\
&\quad\times\exp(\mathbf{k}\cdot\mathbf{k}^+)\int d\mathbf{h}^+d\mathbf{h}\ \exp\left[i\mathbf{h}\cdot(\mathbf{g}^+-\mathbf{k}^+)\right]\exp\left[i(\mathbf{g}-\mathbf{k})\cdot\mathbf{h}^+\right],
\end{aligned} \tag{E.4}$$

where to get the last line we have replaced the integration variables h_i, h_i^+ by $-h_i, -h_i^+$ and used $d(-\mathbf{h}^+)\ d(-\mathbf{h})=(-1)^{2n}d\mathbf{h}^+d\mathbf{h} = d\mathbf{h}^+d\mathbf{h}$. We can then use the Grassmann Fourier integral result (4.72) to show that

$$\begin{aligned}
&\int d\mathbf{k}^+d\mathbf{k}\ \exp(-\mathbf{k}^*\cdot\mathbf{k})\exp(-\mathbf{k}^+\cdot\mathbf{k}^{+*})\exp(\mathbf{k}\cdot\mathbf{k}^+) \\
&\quad\times\int d\mathbf{h}^+d\mathbf{h}\ \exp\left[i\mathbf{h}\cdot(\mathbf{g}^+-\mathbf{k}^+)\right]\exp\left[i(\mathbf{g}-\mathbf{k})\cdot\mathbf{h}^+\right] \\
&\quad\quad= \exp(-\mathbf{k}^*\cdot\mathbf{g})\exp(-\mathbf{g}^+\cdot\mathbf{k}^{+*})\exp(\mathbf{g}\cdot\mathbf{g}^+) \\
&\quad\quad= \exp(\mathbf{g}\cdot\boldsymbol{k}^* + \mathbf{k}^{+*}\cdot\mathbf{g}^+ + \mathbf{g}\cdot\mathbf{g}^+),
\end{aligned} \tag{E.5}$$

giving

$$\begin{aligned}
P_{\mathrm{c}}^{\mathrm{f}}(\mathbf{g},\mathbf{g}^+) &= \int d\mathbf{k}^{+*}d\mathbf{k}^*{}_{\mathrm{B}}\langle\mathbf{k}|\hat{\rho}|\mathbf{k}^{+*}\rangle_{\mathrm{B}}\exp(\mathbf{g}\cdot\boldsymbol{k}^* + \mathbf{k}^{+*}\cdot\mathbf{g}^+ + \mathbf{g}\cdot\mathbf{g}^+) \\
&= P_{\mathrm{canon}}^{\mathrm{f}}(\mathbf{g},\mathbf{g}^+),
\end{aligned} \tag{E.6}$$

as required.

E.2 Quantum Correlation Functions

E.2.1 Boson Case – Normal Ordering

For the *bosonic* case, we can derive a relationship between the normally ordered quantum correlation functions and the normally ordered characteristic function given in (7.5). The proofs for the boson case can be carried out more simply than for the fermion case, since bosonic operators and c-numbers satisfy commutation rather than anticommutation rules and c-number differentiation traverses the trace process.

Now, as the factors $\exp(i\xi_j\hat{a}_j^\dagger)$ in $\hat{\Omega}_{\mathrm{b}}^-(\boldsymbol{\xi})$ commute for different j, we can move the jth such factor to the left and differentiate it:

$$\frac{\partial}{\partial(i\xi_j)}\hat{\Omega}_{\mathrm{b}}^{-}(\boldsymbol{\xi}) = \hat{a}_j^{\dagger}\exp(i\xi_j\hat{a}_j^{\dagger})\prod_{k\neq j}\exp(i\xi_k\hat{a}_k^{\dagger})$$
$$= \hat{a}_j^{\dagger}\hat{\Omega}_{\mathrm{b}}^{-}(\boldsymbol{\xi}) = \hat{\Omega}_{\mathrm{b}}^{-}(\boldsymbol{\xi})\hat{a}_j^{\dagger}. \tag{E.7}$$

Similarly, for the factors in $\hat{\Omega}_{\mathrm{b}}^{+}(\boldsymbol{\xi}^{+})$, we have

$$\frac{\partial}{\partial(i\xi_i^{+})}\hat{\Omega}_{\mathrm{b}}^{+}(\boldsymbol{\xi}^{+}) = \hat{\Omega}_{\mathrm{b}}^{+}(\boldsymbol{\xi}^{+})\hat{a}_i = \hat{a}_i\hat{\Omega}_{\mathrm{b}}^{+}(\boldsymbol{\xi}^{+}). \tag{E.8}$$

Hence we have

$$\frac{\partial}{\partial(i\xi_j)}\chi_{\mathrm{b}}(\boldsymbol{\xi},\boldsymbol{\xi}^{+}) = \mathrm{Tr}\left[\hat{\Omega}_{\mathrm{b}}^{+}(\boldsymbol{\xi}^{+})\hat{\rho}_{\mathrm{b}}\hat{\Omega}_{\mathrm{b}}^{-}(\boldsymbol{\xi})\hat{a}_j^{\dagger}\right],$$
$$\frac{\partial}{\partial(i\xi_i^{+})}\chi_{\mathrm{b}}(\boldsymbol{\xi},\boldsymbol{\xi}^{+}) = \mathrm{Tr}\left[\hat{a}_i\hat{\Omega}_{\mathrm{b}}^{+}(\boldsymbol{\xi}^{+})\hat{\rho}_{\mathrm{b}}\hat{\Omega}_{\mathrm{b}}^{-}(\boldsymbol{\xi})\right]. \tag{E.9}$$

Continuing in this way, we find that

$$\frac{\partial}{\partial(i\xi_{l_p})}\cdots\frac{\partial}{\partial(i\xi_{l_1})}\frac{\partial}{\partial(i\xi_{m_q}^{+})}\cdots\frac{\partial}{\partial(i\xi_{m_1}^{+})}\chi_{\mathrm{b}}(\boldsymbol{\xi},\boldsymbol{\xi}^{+})$$
$$= \mathrm{Tr}\left[\hat{\Omega}_{\mathrm{b}}^{+}(\boldsymbol{\xi}^{+})\hat{a}_{m_q}\cdots\hat{a}_{m_1}\hat{\rho}_{\mathrm{b}}\hat{a}_{l_1}^{\dagger}\cdots\hat{a}_{l_p}^{\dagger}\hat{\Omega}_{\mathrm{b}}^{-}(\boldsymbol{\xi})\right]. \tag{E.10}$$

Then, by putting all the ξ_i, ξ_i^{+} equal to zero, we get

$$\left(\frac{\partial}{\partial(i\xi_{l_p})}\cdots\frac{\partial}{\partial(i\xi_{l_1})}\frac{\partial}{\partial(i\xi_{m_q}^{+})}\cdots\frac{\partial}{\partial(i\xi_{m_1}^{+})}\chi_{\mathrm{b}}(\boldsymbol{\xi},\boldsymbol{\xi}^{+})\right)_{\boldsymbol{\xi},\boldsymbol{\xi}^{+}=0} = \mathrm{Tr}(\hat{a}_{m_q}\cdots\hat{a}_{m_1}\hat{\rho}_{\mathrm{b}}\hat{a}_{l_1}^{\dagger}\cdots\hat{a}_{l_p}^{\dagger})$$
$$= \mathrm{Tr}(\hat{\rho}_{\mathrm{b}}\hat{a}_{l_1}^{\dagger}\cdots\hat{a}_{l_p}^{\dagger}\hat{a}_{m_q}\cdots\hat{a}_{m_1})$$
$$= G_{\mathrm{b}}(l_1,\cdots,l_p;m_q,\cdots,m_1). \tag{E.11}$$

We then apply this formula to the expression for the characteristic function in terms of the distribution function in (7.20), and we can carry out c-number differentiation of the integrand inside the integrals $\int d^2\boldsymbol{\alpha}^{+}d^2\boldsymbol{\alpha}$. As

$$\frac{\partial}{\partial(i\xi_i)}\exp\left(i\boldsymbol{\alpha}\cdot\boldsymbol{\xi}^{+}\right)P_{\mathrm{b}}(\boldsymbol{\alpha},\boldsymbol{\alpha}^{+},\boldsymbol{\alpha}^{*},\boldsymbol{\alpha}^{+*})\exp(i\boldsymbol{\xi}\cdot\boldsymbol{\alpha}^{+})$$
$$= \exp\left(i\boldsymbol{\alpha}\cdot\boldsymbol{\xi}^{+}\right)P_{\mathrm{b}}(\boldsymbol{\alpha},\boldsymbol{\alpha}^{+},\boldsymbol{\alpha}^{*},\boldsymbol{\alpha}^{+*})\alpha_i^{+}\exp(i\boldsymbol{\xi}\cdot\boldsymbol{\alpha}^{+}) \tag{E.12}$$

and

$$\frac{\partial}{\partial(i\xi_i^+)}\exp\big(i\boldsymbol{\alpha}\cdot\boldsymbol{\xi}^+\big)\,P_{\rm b}(\boldsymbol{\alpha},\boldsymbol{\alpha}^+,\boldsymbol{\alpha}^*,\boldsymbol{\alpha}^{+*})\exp\big(i\boldsymbol{\xi}\cdot\boldsymbol{\alpha}^+\big)$$
$$= \exp\big(i\boldsymbol{\alpha}\cdot\boldsymbol{\xi}^+\big)\,\alpha_i P_{\rm b}(\boldsymbol{\alpha},\boldsymbol{\alpha}^+,\boldsymbol{\alpha}^*,\boldsymbol{\alpha}^{+*})\alpha_i^+\exp\big(i\boldsymbol{\xi}\cdot\boldsymbol{\alpha}^+\big), \tag{E.13}$$

we see that

$$\frac{\partial}{\partial(i\xi_i)}\chi_{\rm b}(\boldsymbol{\xi},\boldsymbol{\xi}^+) = \int d^2\boldsymbol{\alpha}^+ d^2\boldsymbol{\alpha}\exp\big(i\boldsymbol{\alpha}\cdot\boldsymbol{\xi}^+\big)\,P_{\rm b}(\boldsymbol{\alpha},\boldsymbol{\alpha}^+,\boldsymbol{\alpha}^*,\boldsymbol{\alpha}^{+*})\alpha_i^+\exp\big(i\boldsymbol{\xi}\cdot\boldsymbol{\alpha}^+\big),$$
$$\frac{\partial}{\partial(i\xi_i^+)}\chi_{\rm b}(\boldsymbol{\xi},\boldsymbol{\xi}^+) = \int d^2\boldsymbol{\alpha}^+ d^2\boldsymbol{\alpha}\exp\big(i\boldsymbol{\alpha}\cdot\boldsymbol{\xi}^+\big)\,\alpha_i P_{\rm b}(\boldsymbol{\alpha},\boldsymbol{\alpha}^+,\boldsymbol{\alpha}^*,\boldsymbol{\alpha}^{+*})\exp\big(i\boldsymbol{\xi}\cdot\boldsymbol{\alpha}^+\big). \tag{E.14}$$

Continuing in this way, we find that

$$\begin{aligned}
&\left(\frac{\partial}{\partial(i\xi_{l_p})}\cdots\frac{\partial}{\partial(i\xi_{l_1})}\frac{\partial}{\partial(i\xi_{m_q}^+)}\cdots\frac{\partial}{\partial(i\xi_{m_1}^+)}\chi_{\rm b}(\boldsymbol{\xi},\boldsymbol{\xi}^+)\right)_{\boldsymbol{\xi},\boldsymbol{\xi}^+=0}\\
&= \int d^2\boldsymbol{\alpha}^+ d^2\boldsymbol{\alpha}\;(\alpha_{m_q}\cdots\alpha_{m_1})P_{\rm b}(\boldsymbol{\alpha},\boldsymbol{\alpha}^+,\boldsymbol{\alpha}^*,\boldsymbol{\alpha}^{+*})(\alpha_{l_1}^+\cdots\alpha_{l_p}^+)\\
&= \mathrm{Tr}\left(\hat{a}_{m_q}\cdots\hat{a}_{m_1}\hat{\rho}_{\rm b}\hat{a}_{l_1}^\dagger\cdots\hat{a}_{l_p}^\dagger\right)\\
&= \mathrm{Tr}\left(\hat{\rho}_{\rm b}\hat{a}_{l_1}^\dagger\cdots\hat{a}_{l_p}^\dagger\hat{a}_{m_q}\cdots\hat{a}_{m_1}\right)\\
&= G_{\rm b}(l_1,\cdots,l_p;m_q,\cdots,m_1),
\end{aligned} \tag{E.15}$$

which is the required phase space integral. The $\alpha_{m_1}\cdots\alpha_{m_q}$ and the $\alpha_{l_p}^+\cdots\alpha_{l_1}^+$ have been rearranged in the same order as the operators.

E.2.2 Boson Case – Symmetric Ordering

We can also derive a relationship between the symmetrically ordered quantum correlation functions and the symmetrically ordered characteristic function, which is given in (7.7). After expanding the exponential, we have

$$\chi^{\rm W}(\boldsymbol{\xi},\boldsymbol{\xi}^+) = \sum_N \frac{1}{n!}\mathrm{Tr}\left[\widehat{\rho}\left(\sum_i i(\xi_i\hat{a}_i^\dagger + \hat{a}_i\xi_i^+)\right)^n\right]. \tag{E.16}$$

With $\widehat{A}(\boldsymbol{\xi}) = \sum_i i\xi_i\,\hat{a}_i^\dagger$ and $\widehat{B}(\boldsymbol{\xi}^+) = \sum_i \hat{a}_i\,i\xi_i^+$, there are $N(p,q) = (p+q)!/p!q!$ ways that the operator $\widehat{A}$ appears p times and the operator $\widehat{B}$ appears q times when we expand $(\widehat{A}+\widehat{B})^n$ (where $n = p+q$) and each order of these operators appears once. We can introduce the symbol $\{(\widehat{A})^p\,(\widehat{B})^q\}$ to denote the average of these $N(p,q)$ ordered products,

$$\{(\widehat{A})^p\,(\widehat{B})^q\} = \frac{1}{N(p,q)}\Big((\widehat{A})^p\,(\widehat{B})^q + (\widehat{A})^{p-1}\,(\widehat{B})^q\,(\widehat{A}) + \cdots + (\widehat{B})^q\,(\widehat{A})^p\Big), \tag{E.17}$$

and write

$$\chi^{\mathrm{W}}(\boldsymbol{\xi},\boldsymbol{\xi}^{+}) = \sum_{p,q}\frac{1}{p!q!}\mathrm{Tr}\left(\hat{\rho}\,\{(\widehat{A})^p\,(\widehat{B})^q\}\right). \tag{E.18}$$

In this form, it is convenient to calculate the derivatives, since $\widehat{A}(\boldsymbol{\xi})$ and $\widehat{B}(\boldsymbol{\xi}^{+})$ are functions only of ξ_i and ξ_i^{+}, respectively, so their derivatives with respect to the other variables will be zero. Then

$$\left(\frac{\partial\widehat{A}(\boldsymbol{\xi})}{\partial(i\xi_i)}\right) = \hat{a}_i^{\dagger}, \qquad \left(\frac{\partial\widehat{B}(\boldsymbol{\xi}^{+})}{\partial(i\xi_j^{+})}\right) = \hat{a}_j. \tag{E.19}$$

We see that each time that either $\widehat{A}(\boldsymbol{\xi})$ is differentiated with respect to $i\xi_i$ or $\widehat{B}(\boldsymbol{\xi}^{+})$ is differentiated with respect to $i\xi_i^{+}$ an annihilation or creation operator results, and therefore any further differentiation with respect to any $i\xi_i$ or $i\xi_j^{+}$ will give only zero. We also note that as $\boldsymbol{\xi},\boldsymbol{\xi}^{+} \longrightarrow 0$, both $\widehat{A}(\boldsymbol{\xi})$ and $\widehat{B}(\boldsymbol{\xi}^{+})$ become zero. To proceed further, we need to calculate derivatives of products of $\widehat{A}(\boldsymbol{\xi})$ and $\widehat{B}(\boldsymbol{\xi}^{+})$. This can be carried out by applying the general rule for derivatives of products of functions. Consider the term $\{(\hat{A})^p\,(\hat{B})^q\}$ (which is the average of the $N(p,q)$ ordered products where $\widehat{A}(\boldsymbol{\xi})$ appears p times and $\widehat{B}(\boldsymbol{\xi}^{+})$ appears q times). If each of the terms in $\{(\widehat{A})^p\,(\widehat{B})^q\}$ is differentiated *fewer* than p times with respect to the $i\xi_i$'s, then there will be at least one factor $\widehat{A}(\boldsymbol{\xi})$ still remaining and thus, as $\boldsymbol{\xi} \longrightarrow 0$, the result of the differentiation will be zero. Similar conclusions apply if the term in $\{(\widehat{A})^p\,(\widehat{B})^q\}$ is differentiated fewer than q times with respect to the $i\xi_j^{+}$'s. On the other hand, if each of the terms in $\{(\widehat{A})^p\,(\widehat{B})^q\}$ is differentiated *more* than p times with respect to the $i\xi_i$'s, then the result of the differentiation must be zero, because after the pth differentiation all of the $\widehat{A}(\boldsymbol{\xi})$ will have been replaced by a factor $\widehat{a}_i^{\dagger}$ and therefore further functional differentiation with respect to any $i\xi_i$ will give zero. Similar conclusions apply if the term in $\{(\widehat{A})^p\,(\widehat{B})^q\}$ is differentiated more than q times with respect to the $i\xi_j^{+}$'s. Hence only the p,q term in the last expression for $\chi^{\mathrm{W}}(\boldsymbol{\xi},\boldsymbol{\xi}^{+})$ contributes to the required result for

$$\begin{aligned}&\left(\frac{\partial}{\partial(i\xi_{l_p})}\cdots\frac{\partial}{\partial(i\xi_{l_1})}\frac{\partial}{\partial(i\xi_{m_q}^{+})}\cdots\frac{\partial}{\partial(i\xi_{m_1}^{+})}\chi_{\mathrm{b}}^{\mathrm{W}}(\boldsymbol{\xi},\boldsymbol{\xi}^{+})\right)_{\boldsymbol{\xi},\boldsymbol{\xi}^{+}\to 0}\\ &\quad = \frac{1}{p!q!}\mathrm{Tr}\left[\widehat{\rho}\left(\frac{\partial}{\partial(i\xi_{l_p})}\cdots\frac{\partial}{\partial(i\xi_{l_1})}\frac{\partial}{\partial(i\xi_{m_q}^{+})}\cdots\frac{\partial}{\partial(i\xi_{m_1}^{+})}\{(\widehat{A})^p\,(\widehat{B})^q\}\right)_{\boldsymbol{\xi},\boldsymbol{\xi}^{+}\to 0}\right].\end{aligned} \tag{E.20}$$

Consider the pth-order derivative of any term in $\{(\widehat{A})^p\,(\widehat{B})^q\}$ with respect to the $(i\xi_{l_p}),\cdots,(i\xi_{l_1})$, where the q factors $\widehat{B}(\boldsymbol{\xi}^{+})$ are just represented by dots. The result will be the sum of products of factors $\hat{a}_{l_1}^{\dagger},\cdots,\hat{a}_{l_p}^{\dagger}$ in all $p!$ orders:

$$\left(\frac{\partial}{\partial(i\xi_{l_p})}\cdots\frac{\partial}{\partial(i\xi_{l_1})}(\widehat{A}(\boldsymbol{\xi})\cdots\widehat{A}(\boldsymbol{\xi})\cdots\widehat{A}(\boldsymbol{\xi})\cdots\widehat{A}(\boldsymbol{\xi}))^{p,q}\right)_{\boldsymbol{\xi},\boldsymbol{\xi}^+\to 0}$$
$$=\sum_P(\widehat{a}^\dagger_{\mu_1}\cdots\widehat{a}^\dagger_{\mu_2}\cdots\widehat{a}^\dagger_{\mu_i}\cdots\widehat{a}^\dagger_{\mu_p}), \tag{E.21}$$

where the sum is over all permutations $P=\uparrow(\mu_1/1\mu_2/2\cdots\mu_i/i\cdots\mu_p/p)$ of $1,2,\cdots,i,\cdots p$. Considering also the qth-order derivative of the same term in $\{(\widehat{A})^p\,(\widehat{B})^q\}$ with respect to the $(i\xi^+_{m_q}),\cdots,(i\xi^+_{m_1})$, we get overall

$$\frac{\partial}{\partial(i\xi_{l_p})}\cdots\frac{\partial}{\partial(i\xi_{l_1})}\frac{\partial}{\partial(i\xi^+_{m_q})}\cdots\frac{\partial}{\partial(i\xi^+_{m_1})}$$
$$\times\left((\widehat{A}(\boldsymbol{\xi})\cdots\widehat{B}(\boldsymbol{\xi}^+)\cdots\widehat{A}(\boldsymbol{\xi})\cdots\widehat{B}(\boldsymbol{\xi}^+)\cdots\widehat{A}(\boldsymbol{\xi})\cdots\widehat{B}(\boldsymbol{\xi}^+)\cdots\widehat{A}(\boldsymbol{\xi}))^{p,q}\right)_{\boldsymbol{\xi},\boldsymbol{\xi}^+\to 0}$$
$$=\sum_{P,Q}(\hat{a}^\dagger_{\mu_1}\cdots\widehat{a}_{\lambda_q}\cdots\widehat{a}^\dagger_{\mu_2}\cdots\widehat{a}_{\lambda_2}\cdots\widehat{a}^\dagger_{\mu_p}\cdots\widehat{a}_{\lambda_1}) \tag{E.22}$$

where the second sum is over all permutations $Q=\uparrow(\lambda_q/q\cdots\lambda j/j\cdots\lambda_2/2\lambda_1/1)$ of $q,\cdots,i,\cdots,2,1$. Now, within each of the $N(p,q)=(p+q)!/p!q!$ orderings of products of $\widehat{A}$ and $\widehat{B}$ where $\widehat{A}$ appears p times and $\widehat{B}$ appears q times, there are $p!$ orderings of the $\hat{a}^\dagger$ operators and $q!$ orderings of the $\hat{a}$ operators, giving a total of $M(p,q)=N(p,q)\,p!q!=(p+q)!$ different orderings of the p operators $\hat{a}^\dagger$ and the q operators $\hat{a}$, and all possible orderings are present in view of the sum over the permutations P,Q. Taking into account the factor $1/p!q!$ we see that when the differentiation is applied to the quantity $\{(\widehat{A})^p\,(\widehat{B})^q\}$ itself, we just get $\{\hat{a}^\dagger_{l_1}\cdots\hat{a}^\dagger_{l_p}\hat{a}_{m_1}\cdots\hat{a}_{m_q}\}$. Thus

$$\left(\frac{\partial}{\partial(i\xi_{l_p})}\cdots\frac{\partial}{\partial(i\xi_{l_1})}\frac{\partial}{\partial(i\xi^+_{m_q})}\cdots\frac{\partial}{\partial(i\xi^+_{m_1})}\chi^{\mathrm{W}}_{\mathrm{b}}(\boldsymbol{\xi},\boldsymbol{\xi}^+)\right)_{\boldsymbol{\xi},\boldsymbol{\xi}^+\to 0}$$
$$=\mathrm{Tr}\left(\widehat{\rho}\,\{\widehat{a}^\dagger_{l_1}\cdots\widehat{a}^\dagger_{l_p}\widehat{a}_{m_q}\cdots\widehat{a}_{m_1}\}\right)$$
$$=G^{\mathrm{W}}_{\mathrm{b}}(l_1,\cdots,l_p;m_q,\cdots,m_1), \tag{E.23}$$

where the symmetric-ordering symbol is given by

$$\{\hat{a}^\dagger_{l_1}\cdots\hat{a}^\dagger_{l_p}\hat{a}_{m_1}\cdots\hat{a}_{m_q}\}=\frac{1}{(p+q)!}\sum_R\Re(\hat{a}^\dagger_{l_1}\cdots\hat{a}^\dagger_{l_p}\hat{a}_{m_q}\cdots\hat{a}_{m1})\,. \tag{E.24}$$

The sum over R is over all $(p+q)!$ orderings $\Re$ of the factors $\hat{a}^\dagger_{l_1}\cdots\widehat{a}^\dagger_{l_p}\hat{a}_{m_q}\cdots\hat{a}_{m_1}$.

We then apply this formula to the expression for the characteristic function in terms of the distribution function

$$\chi^{\mathrm{W}}_{\mathrm{b}}(\boldsymbol{\xi},\boldsymbol{\xi}^+)=\int d^2\boldsymbol{\alpha}\;d\boldsymbol{\alpha}\;\exp(i\boldsymbol{\alpha}\cdot\boldsymbol{\xi}^+\})\;W_{\mathrm{b}}(\boldsymbol{\alpha},\boldsymbol{\alpha}^+,\boldsymbol{\alpha}^*,\boldsymbol{\alpha}^{+*})\exp(i\boldsymbol{\xi}\cdot\boldsymbol{\alpha}^+\}), \tag{E.25}$$

and we can carry out c-number differentiation of the integrand inside the integrals $\int d^2\boldsymbol{\alpha}\;d\boldsymbol{\alpha}$. As the relationship between the symmetrically ordered characteristic function and the double-space Wigner distribution function is of the same form as in the normally ordered case, we find that

$$\left(\frac{\partial}{\partial(i\xi_{l_p})} \cdots \frac{\partial}{\partial(i\xi_{l_1})} \frac{\partial}{\partial(i\xi^+_{m_q})} \cdots \frac{\partial}{\partial(i\xi^+_{m_1})} \chi^{\mathrm{W}}_{\mathrm{b}}(\boldsymbol{\xi}, \boldsymbol{\xi}^+) \right)_{\boldsymbol{\xi},\boldsymbol{\xi}^+=0}$$
$$= \int d^2\,\boldsymbol{\alpha}^+ d^2\boldsymbol{\alpha}(\alpha_{m_q} \cdots \alpha_{m_1}) W_{\mathrm{b}}(\boldsymbol{\alpha}, \boldsymbol{\alpha}^+, \boldsymbol{\alpha}^*, \boldsymbol{\alpha}^{+*})(\alpha^+_{l_1} \cdots \alpha^+_{l_p})$$
$$= \mathrm{Tr}(\hat{\rho}_{\mathrm{b}} \{\hat{a}^\dagger_{l_1} \cdots \hat{a}^\dagger_{l_p} \hat{a}_{m_q} \cdots \hat{a}_{m_1}\})$$
$$= G^{\mathrm{W}}_{\mathrm{b}}(l_1, \cdots, l_p; m_q, \cdots, m_1)\,. \tag{E.26}$$

E.2.3 Fermion Case

For the *fermionic* case, we can derive a relationship between the quantum correlation functions and the characteristic function given in (7.11). In the fermionic case the operators and g-numbers have anticommutation properties and the Grassmann differentiation operation must traverse the trace process, so more elaborate proofs that involve dealing carefully with the order of the various factors are required for the fermion case. It is necessary to consider differentiation from both the left and the right at the same time and to include the trace process from the start. As we will see, the results apply only to quantum correlation functions with $p - q$ equal to an even integer.

Now, because the factors $\exp(\hat{c}_i h^+_i)$ in $\hat{\Omega}^+_{\mathrm{f}}(\mathbf{h}^+)$ commute for different i, we can move the ith such factor to the left and differentiate it from the left, and since the factors $\exp(ih_j\hat{c}^\dagger_j)$ in $\hat{\Omega}^-_{\mathrm{f}}(\mathbf{h})$ commute for different j, we can move the jth such factor to the right and differentiate it from the right. This gives, after writing the trace as the sum of diagonal elements with Fock states $|\nu_1, \cdots, \nu_n\rangle$,

$$\frac{\overrightarrow{\partial}}{\partial(ih^+_i)} \chi_{\mathrm{f}}(\mathbf{h}, \mathbf{h}^+) \frac{\overleftarrow{\partial}}{\partial(ih_j)}$$
$$= \frac{\overrightarrow{\partial}}{\partial(ih^+_i)} \mathrm{Tr}\left[\hat{\Omega}^+_{\mathrm{f}}(\mathbf{h}^+)\hat{\rho}_{\mathrm{f}}\hat{\Omega}^-_{\mathrm{f}}(\mathbf{h})\right] \frac{\overleftarrow{\partial}}{\partial(ih_j)}$$
$$= \frac{\overrightarrow{\partial}}{\partial(ih^+_i)} \left\{ \sum_{\{\nu\}} \langle \nu_1, \cdots, \nu_n | (1 - ih^+_i \hat{c}_i) \widehat{\Theta}^{\mathrm{f}}_{i;j} (1 - i\hat{c}^\dagger_j h_j) | \nu_1, \cdots, \nu_n \rangle \right\} \frac{\overleftarrow{\partial}}{\partial(ih_j)}\,, \tag{E.27}$$

where

$$\widehat{\Theta}^{\mathrm{f}}_{i;j} = \left(\prod_{m \neq i} \exp i\hat{c}_m h^+_m \right) \hat{\rho}_{\mathrm{f}} \left(\prod_{k \neq j} \exp ih_k \hat{c}^\dagger_k \right). \tag{E.28}$$

There are four terms associated with multiplying out the two factors $(1 - ih^+_i \hat{c}_i)$ and $(1 - i\hat{c}^\dagger_j h_j)$. Three terms associated with unity in the two factors give zero contributions after the two differentiations are carried out, since the Grassmann functions involved do not depend on at least one of h^+_i and h_j. The term remaining involves both $-ih^+_i \hat{c}_i$ and $-i\hat{c}^\dagger_j h_j$. We can commute the Grassmann variable $-ih^+_i$ with $\langle \nu_1, \cdots, \nu_n|$ and the

Grassmann variable $-ih_j$ with $|\nu_1,\cdots,\nu_n\rangle$, leading to two factors $(-1)^{\Sigma_i \nu_i}$ that cancel out for each Fock state. Carrying out the two differentiations and canceling the two -1's then leads to

$$\begin{aligned}
&\frac{\overrightarrow{\partial}}{\partial(ih_i^+)}\chi_{\mathrm{f}}(\mathbf{h},\mathbf{h}^+)\frac{\overleftarrow{\partial}}{\partial(ih_j)} \\
&\quad = \sum_{\{\nu\}}\langle \nu_1,\cdots,\nu_n|(\hat{c}_i)(1+i\hat{c}_i h_i^+)\widehat{\Theta}^f_{i;j}(1+ih_j\hat{c}_j^\dagger)(\hat{c}_j^\dagger)|\nu_1,\cdots,\nu_n\rangle \\
&\quad = \mathrm{Tr}\left[\hat{\Omega}_{\mathrm{f}}^+(\mathbf{h}^+)\hat{c}_i\hat{\rho}_{\mathrm{f}}\hat{c}_j^\dagger\hat{\Omega}_{\mathrm{f}}^-(\mathbf{h})\right],
\end{aligned} \tag{E.29}$$

where we have used $(\hat{c}_i) = (\hat{c}_i)(1+i\hat{c}_i h_i^+)$ and $(\hat{c}_j^\dagger) = (1+ih_j\hat{c}_j^\dagger)(\hat{c}_j^\dagger)$, together with the commutation of $\hat{c}_i$ with $\hat{\Omega}_{\mathrm{f}}^+(\mathbf{h}^+)$ and $\hat{c}_j^\dagger$ with $\hat{\Omega}_{\mathrm{f}}^-(\mathbf{h})$. Hence

$$\frac{\overrightarrow{\partial}}{\partial(ih_{m_1}^+)}\chi_{\mathrm{f}}(\mathbf{h},\mathbf{h}^+)\frac{\overleftarrow{\partial}}{\partial(ih_{l_1})} = \mathrm{Tr}\left[\hat{\Omega}_{\mathrm{f}}^+(\mathbf{h}^+)\hat{c}_{m_1}\hat{\rho}_{\mathrm{f}}\hat{c}_{l_1}^\dagger\hat{\Omega}_f^-(\mathbf{h})\right]. \tag{E.30}$$

Continuing, we then get

$$\begin{aligned}
&\frac{\overrightarrow{\partial}}{\partial(ih_{m_2}^+)}\frac{\overrightarrow{\partial}}{\partial(ih_{m_1}^+)}\chi_{\mathrm{f}}(\mathbf{h},\mathbf{h}^+)\frac{\overleftarrow{\partial}}{\partial(ih_{l_1})}\frac{\overleftarrow{\partial}}{\partial(ih_{l_2})} \\
&\quad = \frac{\overrightarrow{\partial}}{\partial(ih_{m_2}^+)}\mathrm{Tr}\left[\hat{\Omega}_{\mathrm{f}}^+(\mathbf{h}^+)\hat{c}_{m_1}\hat{\rho}_{\mathrm{f}}\hat{c}_{l_1}^\dagger\hat{\Omega}_{\mathrm{f}}^-(\mathbf{h})\right]\frac{\overleftarrow{\partial}}{\partial(ih_{l_2})} \\
&\quad = \mathrm{Tr}\left[\hat{\Omega}_{\mathrm{f}}^+(\mathbf{h}^+)\hat{c}_{m_2}\hat{c}_{m_1}\hat{\rho}_{\mathrm{f}}\hat{c}_{l_1}^\dagger\hat{c}_{l_2}^\dagger\hat{\Omega}_{\mathrm{f}}^-(\mathbf{h})\right],
\end{aligned} \tag{E.31}$$

as all the previous analysis still applies when we replace $\hat{\rho}_{\mathrm{f}}$ by $\hat{c}_{m_1}\hat{\rho}_{\mathrm{f}}\hat{c}_{l_1}^\dagger$. We can clearly continue in this way provided that the numbers of left and right differentiations are the same.

Suppose we have $p \geq q$. Then we can first continue to obtain the result

$$\begin{aligned}
&\frac{\overrightarrow{\partial}}{\partial(ih_{m_q}^+)}\cdots\frac{\overrightarrow{\partial}}{\partial(ih_{m_1}^+)}\chi_{\mathrm{f}}(\mathbf{h},\mathbf{h}^+)\frac{\overleftarrow{\partial}}{\partial(ih_{l_1})}\cdots\frac{\overleftarrow{\partial}}{\partial(ih_{l_q})} \\
&\quad = \mathrm{Tr}\left[\hat{\Omega}_{\mathrm{f}}^+(\mathbf{h}^+)\hat{c}_{m_q}\cdots\hat{c}_{m_1}\hat{\rho}_{\mathrm{f}}\hat{c}_{l_1}^\dagger\cdots\hat{c}_{l_q}^\dagger\hat{\Omega}_{\mathrm{f}}^-(\mathbf{h})\right].
\end{aligned} \tag{E.32}$$

However, we need to continue for a further $p-q$ right differentiations. This can be done if $p-q$ is an even number. To begin with we have

$$\begin{aligned}
&\chi_{\mathrm{f}}(\mathbf{h},\mathbf{h}^+)\frac{\overleftarrow{\partial}}{\partial(ih_j)}\frac{\overleftarrow{\partial}}{\partial(ih_i)} \\
&\quad = \left\{\sum_{\{\nu\}}\langle \nu_1,\cdots,\nu_n|\hat{\Omega}_{\mathrm{f}}^+(\mathbf{h}^+)\,\widehat{\Xi}^{\mathrm{f}}_{i;j}(1-i\hat{c}_i^\dagger h_i)(1-i\hat{c}_j^\dagger h_j)|\nu_1,\cdots,\nu_n\rangle\right\}\frac{\overleftarrow{\partial}}{\partial(ih_j)}\frac{\overleftarrow{\partial}}{\partial(ih_i)},
\end{aligned} \tag{E.33}$$

where

$$\widehat{\Xi}^{\mathrm{f}}_{i;j} = \hat{\rho}_{\mathrm{f}} \left(\prod_{k \neq i,j} \exp(ih_k \hat{c}_k^\dagger) \right). \tag{E.34}$$

A similar treatment to the above shows that only the factors $-i\hat{c}_i^\dagger h_i$ and $-i\hat{c}_j^\dagger h_j$ contribute from the four terms associated with multiplying out $(1 - i\hat{c}_i^\dagger h_i)(1 - i\hat{c}_j^\dagger h_j)$. The Grassmann variables $(-ih_i)$ and $(-ih_j)$ commute with $|\nu_1, \nu_2, \cdots, \nu_n\rangle$, producing two canceling factors $(-1)^{\Sigma_i \nu_i}$, but there is an extra -1 factor that occurs because the $-ih_i$ anticommutes with the $\hat{c}_j^\dagger$. Carrying out the two right derivatives, cancelling two of the -1 factors and using the remaining -1 to change $(\hat{c}_i^\dagger)(\hat{c}_j^\dagger)$ into $(\hat{c}_j^\dagger)(\hat{c}_i^\dagger)$, we have, finally,

$$\begin{aligned}
&\chi_{\mathrm{f}}(\mathbf{h}, \mathbf{h}^+) \frac{\overleftarrow{\partial}}{\partial(ih_j)} \frac{\overleftarrow{\partial}}{\partial(ih_i)} \\
&= \left\{ \sum_{\{\nu\}} \langle \nu_1, \cdots, \nu_n | \hat{\Omega}_{\mathrm{f}}^+(\mathbf{h}^+) \, \widehat{\Xi}^{\mathrm{f}}_{i;j} (1 + ih_i \hat{c}_i^\dagger)(1 + ih_j \hat{c}_j^\dagger)(\hat{c}_j^\dagger)(\hat{c}_i^\dagger) | \nu_1, \cdots, \nu_n \rangle \right\} \\
&= \mathrm{Tr}\left[\hat{\Omega}_{\mathrm{f}}^+(\mathbf{h}^+) \hat{\rho}_{\mathrm{f}} \hat{c}_j^\dagger \hat{c}_i^\dagger \hat{\Omega}_{\mathrm{f}}^-(\mathbf{h}) \right],
\end{aligned} \tag{E.35}$$

where we note that the $\hat{c}_j^\dagger \hat{c}_i^\dagger$ are in the same order as the right differentiations.

The same analysis applies when $\hat{\rho}_{\mathrm{f}}$ is replaced by $\hat{c}_{m_q} \cdots \hat{c}_{m_1} \hat{\rho}_{\mathrm{f}} \hat{c}_{l_1}^\dagger \cdots \hat{c}_{l_q}^\dagger$. Hence, provided $p - q$ is even, we can continue the right differentiations to give

$$\begin{aligned}
&\frac{\overrightarrow{\partial}}{\partial(ih_{m_q}^+)} \cdots \frac{\overrightarrow{\partial}}{\partial(ih_{m_1}^+)} \chi_{\mathrm{f}}(\mathbf{h}, \mathbf{h}^+) \frac{\overleftarrow{\partial}}{\partial(ih_{l_1})} \cdots \frac{\overleftarrow{\partial}}{\partial(ih_{l_p})} \\
&\quad = \mathrm{Tr}\left[\hat{\Omega}_{\mathrm{f}}^+(h^+) \hat{c}_{m_q} \cdots \hat{c}_{m_1} \hat{\rho}_{\mathrm{f}} \hat{c}_{l_1}^\dagger \cdots \hat{c}_{l_p}^\dagger \hat{\Omega}_{\mathrm{f}}^-(h) \right]
\end{aligned} \tag{E.36}$$

for $p \geq q$. A similar analysis applies to the case $p \leq q$ with $q - p$ even. Hence the result applies to all cases provided $p - q$ is an even integer $0, \pm 2, \pm 4, \ldots$.

Thus, by replacing h and h^+ by zero after the Grassmann differentiation process, we have

$$\begin{aligned}
&\left(\frac{\overrightarrow{\partial}}{\partial(ih_{m_q}^+)} \cdots \frac{\overrightarrow{\partial}}{\partial(ih_{m_1}^+)} \chi_{\mathrm{f}}(\mathbf{h}, \mathbf{h}^+) \frac{\overleftarrow{\partial}}{\partial(ih_{l_1})} \cdots \frac{\overleftarrow{\partial}}{\partial(ih_{l_p})} \right)_{\mathbf{h}, \mathbf{h}^+ = 0} \\
&\quad = \mathrm{Tr}(\hat{c}_{m_q} \cdots \hat{c}_{m_1} \hat{\rho}_{\mathrm{f}} \hat{c}_{l_1}^\dagger \cdots \hat{c}_{l_p}^\dagger) \\
&\quad = G_{\mathrm{f}}(l_1, \cdots, l_p; m_q, \cdots, m_1), \qquad p - q = 0, \pm 2, \pm 4, \cdots .
\end{aligned} \tag{E.37}$$

The similarity of the result to the boson case is clear.

We then apply this formula to the expression in (7.23) for the characteristic function in terms of the distribution function We first differentiate the last expression for the characteristic function with respect to h_i and h_i^+ by using the results in (4.77) and (4.78), which show that we can differentiate under the Grassmann integral. Using $\exp(ig_ih_i^+) = (1 - i\{h_i^+g_i\})$ and $\exp(ih_jg_j^+) = (1 - i\{g_j^+h_j\})$, we get the following after carrying out the two differentiations via the product rules (4.28) and then using $(g_i) = (g_i)(1 + i\{g_ih_i^+\})$ and $(g_j^+) = (1 + i\{h_jg_j^+\})(g_j^+)$:

$$\frac{\overrightarrow{\partial}}{\partial(ih_i^+)}\chi_{\mathrm{f}}(\mathbf{h},\mathbf{h}^+)\frac{\overleftarrow{\partial}}{\partial(ih_j)} = \frac{\overrightarrow{\partial}}{\partial(ih_i^+)}\left(\int d\mathbf{g}^+ d\mathbf{g}\ (1 - i\{h_i^+g_i\})F_{i;j}^{\mathrm{f}}(1 - i\{g_j^+h_j\})\right)\frac{\overleftarrow{\partial}}{\partial(ih_j)}$$
$$= \left(\int d\mathbf{g}^+ d\mathbf{g}\ \exp(i\mathbf{g}\cdot\mathbf{h}^+)\ g_iP_{\mathrm{f}}(\mathbf{g},\mathbf{g}^+)g_j^+ \exp(i\{\mathbf{h}\cdot\mathbf{g}^+)\right), \tag{E.38}$$

where $F_{i;j}^{\mathrm{f}}$ is an even Grassmann function of $\mathbf{g},\mathbf{g}^+$ and $\mathbf{h},\mathbf{h}^+$ except for h_i^+ and h_j:

$$F_{i;j}^{\mathrm{f}} = \prod_{m\neq i}\exp i\{g_mh_m^+\}P_{\mathrm{f}}(\mathbf{g},\mathbf{g}^+)\prod_{k\neq j}\exp i\{h_kg_k^+\}. \tag{E.39}$$

Applying this process twice, we get

$$\frac{\overrightarrow{\partial}}{\partial(ih_{m_2}^+)}\frac{\overrightarrow{\partial}}{\partial(ih_{m_1}^+)}\chi_{\mathrm{f}}(\mathbf{h},\mathbf{h}^+)\frac{\overleftarrow{\partial}}{\partial(ih_{l_1})}\frac{\overleftarrow{\partial}}{\partial(ih_{l_2})}$$
$$= \int d\mathbf{g}^+ d\mathbf{g}\ \exp(i\mathbf{g}\cdot\mathbf{h}^+)\ g_{m_2}g_{m_1}P_{\mathrm{f}}(\mathbf{g},\mathbf{g}^+)g_{l_1}^+g_{l_2}^+\exp(i\mathbf{h}\cdot\mathbf{g}^+), \tag{E.40}$$

as all the previous analysis still applies when we replace $P_{\mathrm{f}}(\mathbf{g},\mathbf{g}^+)$ by $g_{m_1}P_{\mathrm{f}}(\mathbf{g},\mathbf{g}^+)g_{l_1}^+$. Suppose $p \geq q$; then, continuing in this way, we have

$$\frac{\overrightarrow{\partial}}{\partial(ih_{m_q}^+)}\cdots\frac{\overrightarrow{\partial}}{\partial(ih_{m_2}^+)}\frac{\overrightarrow{\partial}}{\partial(ih_{m_1}^+)}\chi_{\mathrm{f}}(h,h^+)\frac{\overleftarrow{\partial}}{\partial(ih_{l_1})}\frac{\overleftarrow{\partial}}{\partial(ih_{l_2})}\cdots\frac{\overleftarrow{\partial}}{\partial(ih_{l_q})}$$
$$= \left(\int\prod_i dg_i^+dg_i\prod_j\exp i\{g_jh_j^+\}\ g_{m_q}\cdots g_{m_2}g_{m_1}P_{\mathrm{f}}(g,g^+)g_{l_1}^+g_{l_2}^+\cdots g_{l_q}^+\prod_k\exp i\{h_kg_k^+\}\right). \tag{E.41}$$

However, we wish to continue differentiation from the right for a further $p - q$ times. This gives extra factors $(-g_{l_{q+1}}^+)$, $(-g_{l_{q+2}}^+)$, etc., which give the required product $g_{l_{q+1}}^+g_{l_{q+2}}^+\cdots$ times a factor $(-1)^{p-q}$. As we are restricted to the case where $p - q$ is an even integer, this factor equals $+1$, so it can be ignored. Similar considerations apply when $p \leq q$ with $q - p$ an even integer, so that in all cases

$$
\begin{aligned}
&\left(\frac{\overrightarrow{\partial}}{\partial(ih^+_{m_q})}\cdots\frac{\overrightarrow{\partial}}{\partial(ih^+_{m_1})}\chi_{\mathrm{f}}(\mathbf{h},\mathbf{h}^+)\frac{\overleftarrow{\partial}}{\partial(ih_{l_1})}\cdots\frac{\overleftarrow{\partial}}{\partial(ih_{l_p})}\right)_{\mathbf{h},\mathbf{h}^+=0} \\
&\quad = \int d\mathbf{g}^+ d\mathbf{g}\ (g_{m_q}\cdots g_{m_1})P_{\mathrm{f}}(\mathbf{g},\mathbf{g}^+)(g^+_{l_1}\cdots g^+_{l_p}) \\
&\quad = \mathrm{Tr}\left(\hat{c}_{m_q}\cdots\hat{c}_{m_1}\hat{\rho}_{\mathrm{f}}\hat{c}^\dagger_{l_1}\cdots\hat{c}^\dagger_{l_p}\right) \\
&\quad = G_{\mathrm{f}}(l_1,\cdots,l_p;m_q,\cdots,,m_1), \qquad p-q=0,\pm2,\pm4,\cdots,
\end{aligned}
\tag{E.42}
$$

which is the required phase space integral.

E.3 Normal, Symmetric and Antinormal Distribution Functions

Distribution functions for a single boson mode in a single phase space are well known in quantum optics. These include cases for dealing with normally ordered quantum correlation functions (Glauber–Sudarshan P functions), symmetrically ordered correlation functions (Wigner W functions) and antinormally ordered correlation functions (Husumi Q functions). These are well described in other textbooks and papers, for example in the papers of Cahill and Glauber [11, 12] and in Chapter 4 of *Methods in Theoretical Quantum Optics* by Barnett and Radmore [4]. The extension to multi-mode systems is straightforward. Fokker–Planck equations can always be derived, and Langevin stochastic equations often follow.

A similar approach involving complex pairs of Grassmann variables g, g^* can be introduced to treat distribution functions for a single fermion mode in a single phase space. The extension to multi-mode systems is straightforward. Details can be found in the paper of Cahill and Glauber [10].

Appendix F
Fokker–Planck Equations

F.1 Correspondence Rules

F.1.1 Grassmann and Operator Formulae

In the Liouville–von Neumann equation, the density operator $\hat{\rho}$ may be replaced by $\hat{\Xi}\hat{\rho}$ or $\hat{\rho}\hat{\Xi}$ or $\hat{\Xi}\hat{\rho}\hat{\Gamma}$, where $\hat{\Xi}$ and $\hat{\Gamma}$ are boson or fermion operators. There will be a consequent change in the characteristic function and hence in the distribution function. These changes are expressed as correspondence rules.

The definition of the characteristic function $\chi(\boldsymbol{\xi},\boldsymbol{\xi}^+,\mathbf{h},\mathbf{h}^+)$ is

$$\begin{aligned}
\chi(\boldsymbol{\xi},\boldsymbol{\xi}^+,\mathbf{h},\mathbf{h}^+) &= \mathrm{Tr}\left[\hat{\Omega}_{\mathrm{b}}^+(\boldsymbol{\xi}^+)\hat{\Omega}_{\mathrm{f}}^+(\mathbf{h}^+)\hat{\rho}\hat{\Omega}_{\mathrm{f}}^-(\mathbf{h})\hat{\Omega}_{\mathrm{b}}^-(\boldsymbol{\xi})\right], \\
\hat{\Omega}_{\mathrm{b}}^+(\xi^+) &= \exp(i\hat{\mathbf{a}}\cdot\boldsymbol{\xi}^+) = \prod_i \exp(i\hat{a}_i\xi_i^+), \\
\hat{\Omega}_{\mathrm{b}}^-(\boldsymbol{\xi}) &= \left(\hat{\Omega}_{\mathrm{b}}^+(-\boldsymbol{\xi}^*)\right)^\dagger, \\
\hat{\Omega}_{\mathrm{f}}^+(\mathbf{h}^+) &= \exp(i\hat{\mathbf{c}}\cdot\mathbf{h}) = \prod_i \exp(i\hat{c}_i h_i^+) = \prod_i (1+i\hat{c}_i h_i^+), \\
\hat{\Omega}_{\mathrm{f}}^-(\mathbf{h}) &= \left(\hat{\Omega}_{\mathrm{f}}^+(-\mathbf{h}^*)\right)^\dagger,
\end{aligned} \tag{F.1}$$

where $\boldsymbol{\xi} \equiv \{\xi_1,\xi_2,\cdots,\xi_p\}$ and $\boldsymbol{\xi}^+ \equiv \{\xi_1^+,\xi_2^+,\cdots,\xi_p^+\}$ are two sets of c-numbers associated with the p bosonic modes, and $\mathbf{h} \equiv \{h_1,h_2,\cdots,h_n\}$ and $\mathbf{h}^+ \equiv \{h_1^+,h_2^+,\cdots,h_n^+\}$ are two sets of Grassmann numbers associated with the n fermionic modes.

The characteristic function and the distribution function $P(\boldsymbol{\alpha},\boldsymbol{\alpha}^+,\boldsymbol{\alpha}^*,\boldsymbol{\alpha}^{+*},\mathbf{g},\mathbf{g}^+)$ are related via Grassmann and c-number integrals:

$$\begin{aligned}
\chi(\boldsymbol{\xi},\boldsymbol{\xi}^+,\mathbf{h},\mathbf{h}^+) = &\int d\mathbf{g}^+ d\mathbf{g}\, d^2\boldsymbol{\alpha}^+ d^2\boldsymbol{\alpha} \exp(i\boldsymbol{\alpha}\cdot\boldsymbol{\xi}^+)\exp(i\mathbf{g}\cdot\mathbf{h}^+) \\
&\times P(\boldsymbol{\alpha},\boldsymbol{\alpha}^+,\boldsymbol{\alpha}^*,\boldsymbol{\alpha}^{+*},\mathbf{g},\mathbf{g}^+)\exp(i\mathbf{h}\cdot\mathbf{g}^+)\exp(i\boldsymbol{\xi}\cdot\boldsymbol{\alpha}^+),
\end{aligned} \tag{F.2}$$

where $\boldsymbol{\alpha} \equiv \{\alpha_1,\alpha_2,\cdots,\ \alpha_p\}$ and $\boldsymbol{\alpha}^+ \equiv \{\alpha_1^+$ and $\alpha_2^+,\cdots,\alpha_p^+\}$ are also two sets of c-numbers associated with the p bosonic modes, and $\mathbf{g} \equiv \{g_1,g_2,\cdots,g_n\}$ and $\mathbf{g}^+ \equiv \{g_1^+,g_2^+,\cdots,g_n^+\}$ are also two sets of Grassmann numbers associated with the n fermionic modes.

The correspondence rules for fermion annihilation and creation operators may be obtained by starting with the canonical form of the density operator for the combined boson–fermion system. There will be a consequent change in the Bargmann state projectors and hence in the distribution function. These changes are expressed as correspondence rules.

The density operator is related to the canonical form of the distribution function as follows:

$$\begin{aligned}\hat{\rho} &= \int d\mathbf{g}^{+}d\mathbf{g}\int d^2\boldsymbol{\alpha}^{+}d^2\boldsymbol{\alpha}\ P_{\text{canon}}(\boldsymbol{\alpha},\boldsymbol{\alpha}^{+},\boldsymbol{\alpha}^{*},\boldsymbol{\alpha}^{+*},\mathbf{g},\mathbf{g}^{+})\hat{\Lambda}(\mathbf{g},\mathbf{g}^{+},\boldsymbol{\alpha},\boldsymbol{\alpha}^{+})\\ &= \int d\mathbf{g}^{+}d\mathbf{g}\int d^2\boldsymbol{\alpha}^{+}d^2\boldsymbol{\alpha}\ \hat{\Lambda}(\mathbf{g},\mathbf{g}^{+},\boldsymbol{\alpha},\boldsymbol{\alpha}^{+})P_{\text{canon}}(\boldsymbol{\alpha},\boldsymbol{\alpha}^{+},\boldsymbol{\alpha}^{*},\boldsymbol{\alpha}^{+*},\mathbf{g},\mathbf{g}^{+}),\end{aligned} \tag{F.3}$$

where

$$\hat{\Lambda}(\mathbf{g},\mathbf{g}^{+},\boldsymbol{\alpha},\boldsymbol{\alpha}^{+}) = \hat{\Lambda}_{\text{f}}(\mathbf{g},\mathbf{g}^{+})\hat{\Lambda}_{\text{b}}(\boldsymbol{\alpha},\boldsymbol{\alpha}^{+}) \tag{F.4}$$

is a product of normalised projectors, with $\hat{\Lambda}_{\text{f}}(\mathbf{g},\mathbf{g}^{+})$ and $\hat{\Lambda}_{\text{b}}(\boldsymbol{\alpha},\boldsymbol{\alpha}^{+})$ being the usual fermion and boson Bargmann state projectors:

$$\begin{aligned}\hat{\Lambda}_{\text{f}}(\mathbf{g},\mathbf{g}^{+}) &= \frac{|\mathbf{g}\rangle_{\text{BB}}\langle\mathbf{g}^{+*}|}{\text{Tr}(|\mathbf{g}\rangle_{\text{BB}}\langle\mathbf{g}^{+*}|)},\\ \hat{\Lambda}_{\text{b}}(\boldsymbol{\alpha},\boldsymbol{\alpha}^{+}) &= \frac{|\boldsymbol{\alpha}\rangle_{\text{BB}}\langle\boldsymbol{\alpha}^{+*}|}{\text{Tr}(|\boldsymbol{\alpha}\rangle_{\text{BB}}\langle\boldsymbol{\alpha}^{+*}|)}.\end{aligned} \tag{F.5}$$

The two forms for the density operator apply because $\hat{\Lambda}(\mathbf{g},\mathbf{g}^{+},\boldsymbol{\alpha},\boldsymbol{\alpha}^{+})$ is an even Grassmann operator and therefore commutes with $P_{\text{canon}}(\boldsymbol{\alpha},\boldsymbol{\alpha}^{+},\ \boldsymbol{\alpha}^{*},\boldsymbol{\alpha}^{+*},\mathbf{g},\mathbf{g}^{+})$.

The derivation of the correspondence rules involves using standard quantum rules such as

$$\begin{aligned}\hat{\Xi}(\exp\hat{S}) &= (\exp\hat{S})\left\{\hat{\Xi} - [\hat{S},\hat{\Xi}] + \frac{1}{2!}[\hat{S},[\hat{S},\hat{\Xi}]] - \frac{1}{3!}[\hat{S},[\hat{S},[\hat{S},\hat{\Xi}]]] + \cdots\right\},\\ (\exp\hat{S})\hat{\Xi} &= \left\{\hat{\Xi} + [\hat{S},\hat{\Xi}] + \frac{1}{2!}[\hat{S},[\hat{S},\hat{\Xi}]] + \frac{1}{3!}[\hat{S},[\hat{S},[\hat{S},\hat{\Xi}]]] + \cdots\right\}(\exp\hat{S}),\end{aligned} \tag{F.6}$$

involving operators $\hat{\Xi}$ and $\hat{S}$ to change the order of exponentials and other operators for the boson terms. In addition, the commutation of boson operators with fermion operators is used extensively. Care must be taken when Grassmann differentiation of a trace of Grassmann operators is involved. The results for fermion operators make use of the anticommutation features of fermion operators and Grassmann variables and the rules for Grassmann left and right differentiation. It is important to move the Grassmann variable being differentiated to the left of the expression to implement left differentiation, and to the extreme right for carrying out right differentiation. These movements involve the use of anticommutation rules for pairs of Grassmann variables or a Grassmann variable and a fermion operator, plus of course the anticommutation rules in (2.18) for pairs of fermion operators. Grassmann variables and fermion operators commute with c-numbers and boson operators.

The derivation of the correspondence rules involves considering differentiating the product of two functions $e(\mathbf{g})$ and $f(\mathbf{g})$, each of which is either even or odd. Results for differentiating products of functions that are neither even nor odd can be obtained by first expressing each function as the sum of even and odd components and then using the linearity property of differentiation. The product rules are

$$\frac{\overrightarrow{\partial}}{\partial g_i}[e(\mathbf{g})f(\mathbf{g})] = \left(\frac{\overrightarrow{\partial}}{\partial g_i}e(\mathbf{g})\right) f(\mathbf{g}) + \sigma(e)e(\mathbf{g}) \left(\frac{\overrightarrow{\partial}}{\partial g_i}f(\mathbf{g})\right), \tag{F.7}$$

$$[f(\mathbf{g})e(\mathbf{g})]\frac{\overleftarrow{\partial}}{\partial g_i} = f(\mathbf{g}) \left(e(\mathbf{g})\frac{\overleftarrow{\partial}}{\partial g_i}\right) + \sigma(e) \left(f(\mathbf{g})\frac{\overleftarrow{\partial}}{\partial g_i}\right) e(\mathbf{g}), \tag{F.8}$$

where $\sigma(e), \sigma(f) = +1$ or -1 depending on whether e or f is even or odd.

The integration-by-parts result for Grassmann functions can be derived by applying the product differentiation rules and noting that the full phase space Grassmann integral of a Grassmann derivative is zero. We have

$$\int d^2\mathbf{g} \left(\frac{\overrightarrow{\partial}}{\partial g_i}e(\mathbf{g})\right) f(\mathbf{g}) = -\sigma(e) \int d^2\mathbf{g}\; e(\mathbf{g}) \left(\frac{\overrightarrow{\partial}}{\partial g_i}f(\mathbf{g})\right), \tag{F.9}$$

$$\int d^2\mathbf{g} \left(\frac{\overrightarrow{\partial}}{\partial g_i^+}e(\mathbf{g})\right) f(\mathbf{g}) = -\sigma(e) \int d^2\mathbf{g}\; e(\mathbf{g}) \left(\frac{\overrightarrow{\partial}}{\partial g_i^+}f(\mathbf{g})\right), \tag{F.10}$$

$$\int d^2\mathbf{g} \left(f(\mathbf{g})\frac{\overleftarrow{\partial}}{\partial g_i}\right) e(\mathbf{g}) = -\sigma(e) \int d^2\mathbf{g}\; f(\mathbf{g}) \left(e(\mathbf{g})\frac{\overleftarrow{\partial}}{\partial g_i}\right), \tag{F.11}$$

$$\int d^2\mathbf{g} \left(f(\mathbf{g})\frac{\overleftarrow{\partial}}{\partial g_i^+}\right) e(\mathbf{g}) = -\sigma(e) \int d^2\mathbf{g}\; f(\mathbf{g}) \left(e(\mathbf{g})\frac{\overleftarrow{\partial}}{\partial g_i^+}\right). \tag{F.12}$$

Some other rules needed are those for differentiating the Grassmann integral over all g_i, g_i^+ of a function $F(\mathbf{g}, \mathbf{g}^+, \mathbf{h}, \mathbf{h}^+)$ of two sets of Grassmann variables $\{g_i, g_i^+\}$ and $\{h_j, h_j^+\}$. The Grassmann integral will be a function of the h_j, h_j^+. The rule is that differentiation from either direction after integration gives the same result as differentiating first and then integrating. Thus we have

$$\frac{\overrightarrow{\partial}}{\partial h_j}\left(\int d\mathbf{g}^+ d\mathbf{g}\; F(\mathbf{g}, \mathbf{g}^+, \mathbf{h}, \mathbf{h}^+)\right) = \int d\mathbf{g}^+ d\mathbf{g} \left(\frac{\overrightarrow{\partial}}{\partial h_j}F(\mathbf{g}, \mathbf{g}^+, \mathbf{h}, \mathbf{h}^+)\right), \tag{F.13}$$

$$\frac{\overrightarrow{\partial}}{\partial h_j^+}\left(\int d\mathbf{g}^+ d\mathbf{g}\; F(\mathbf{g}, \mathbf{g}^+, \mathbf{h}, \mathbf{h}^+)\right) = \int d\mathbf{g}^+ d\mathbf{g} \left(\frac{\overrightarrow{\partial}}{\partial h_j^+}F(\mathbf{g}, \mathbf{g}^+, \mathbf{h}, \mathbf{h}^+)\right), \tag{F.14}$$

$$\left(\int d\mathbf{g}^+ d\mathbf{g}\; F(\mathbf{g}, \mathbf{g}^+, \mathbf{h}, \mathbf{h}^+)\right)\frac{\overleftarrow{\partial}}{\partial h_j} = \int d\mathbf{g}^+ d\mathbf{g} \left(F(\mathbf{g}, \mathbf{g}^+, \mathbf{h}, \mathbf{h}^+)\frac{\overleftarrow{\partial}}{\partial h_j}\right), \tag{F.15}$$

$$\left(\int d\mathbf{g}^+ d\mathbf{g}\; F(\mathbf{g}, \mathbf{g}^+, \mathbf{h}, \mathbf{h}^+)\right)\frac{\overleftarrow{\partial}}{\partial h_j^+} = \int d\mathbf{g}^+ d\mathbf{g} \left(F(\mathbf{g}, \mathbf{g}^+, \mathbf{h}, \mathbf{h}^+)\frac{\overleftarrow{\partial}}{\partial h_j^+}\right). \tag{F.16}$$

In addition, certain rules involving the traces of Grassmann operators are needed. These include the cyclic rules

$$\mathrm{Tr}[\hat{\Omega}_{\mathrm{f}}(h)\hat{\Delta}_{\mathrm{f}}(k)] = \mathrm{Tr}[\hat{\Delta}_{\mathrm{f}}(k)\hat{\Omega}_{\mathrm{f}}(-h)] \tag{F.17}$$

$$= \mathrm{Tr}[\hat{\Delta}_{\mathrm{f}}(-k)\hat{\Omega}_{\mathrm{f}}(h)]\,, \tag{F.18}$$

which apply when the two Grassmann operators are either both even or both odd. The rules for differentiating or multiplying the traces of Grassmann operators are also needed. For differentiation, these are

$$\begin{aligned}
\frac{\overrightarrow{\partial}}{\partial g_i}\frac{\overrightarrow{\partial}}{\partial g_j}\left(\mathrm{Tr}(\hat{\Omega}_{\mathrm{f}}(g))\right) &= \mathrm{Tr}\left(\frac{\overrightarrow{\partial}}{\partial g_i}\frac{\overrightarrow{\partial}}{\partial g_j}\hat{\Omega}_{\mathrm{f}}(g)\right), \\
\frac{\overrightarrow{\partial}}{\partial g_i}\left(\mathrm{Tr}(\hat{\Omega}_{\mathrm{f}}(g))\right)\frac{\overleftarrow{\partial}}{\partial g_j} &= \mathrm{Tr}\left(\frac{\overrightarrow{\partial}}{\partial g_i}\hat{\Omega}_{\mathrm{f}}(g)\frac{\overleftarrow{\partial}}{\partial g_j}\right), \\
\left(\mathrm{Tr}(\hat{\Omega}_{\mathrm{f}}(g))\right)\frac{\overleftarrow{\partial}}{\partial g_j}\frac{\overleftarrow{\partial}}{\partial g_i} &= \mathrm{Tr}\left(\hat{\Omega}_{\mathrm{f}}(g)\frac{\overleftarrow{\partial}}{\partial g_j}\frac{\overleftarrow{\partial}}{\partial g_i}\right),
\end{aligned} \tag{F.19}$$

with analogous results involving multiplication. The rules require two differentiation or multiplication processes to occur – no simple rules exist for just a single process.

F.1.2 Boson Case – Canonical-Density-Operator Approach

Let us introduce, for simplicity of notation, the fermionic operator

$$\hat{\Pi}_{\mathrm{f}} = \int d\mathbf{g}^{+}d\mathbf{g}\,P(\boldsymbol{\alpha},\boldsymbol{\alpha}^{+},\boldsymbol{\alpha}^{*},\boldsymbol{\alpha}^{+*},\,\mathbf{g},\,\mathbf{g}^{+})\hat{\Lambda}_{\mathrm{f}}(\mathbf{g},\,\mathbf{g}^{+})\,, \tag{F.20}$$

which clearly commutes with $\hat{\Lambda}_{\mathrm{b}}(\boldsymbol{\alpha},\boldsymbol{\alpha}^{+})$. Hence we can write the density operator in two alternative forms as

$$\hat{\rho} = \int d^2\boldsymbol{\alpha}^{+}d^2\boldsymbol{\alpha}\hat{\Pi}_{\mathrm{f}}\hat{\Lambda}_{\mathrm{b}}(\boldsymbol{\alpha},\boldsymbol{\alpha}^{+}) = \int d^2\boldsymbol{\alpha}^{+}d^2\boldsymbol{\alpha}\;\hat{\Lambda}_{\mathrm{b}}(\boldsymbol{\alpha},\boldsymbol{\alpha}^{+})\hat{\Pi}_{\mathrm{f}}\,. \tag{F.21}$$

Note that $\hat{\Pi}_{\mathrm{f}}$ does not involve any bosonic operators.

Certain results involving the bosonic projector can be obtained from Chapter 5, which give $\hat{a}_i|\alpha\rangle_{\mathrm{B}} = \alpha_i|\alpha\rangle_{\mathrm{B}}$, ${}_{\mathrm{B}}\langle\alpha^{+*}|\hat{a}_i^{\dagger} = {}_{\mathrm{B}}\langle\alpha^{+*}|\alpha_i^{+}$, $\hat{a}_i^{\dagger}|\alpha\rangle_{\mathrm{B}} = (\partial/\partial\alpha_i)\,|\alpha\rangle_{\mathrm{B}}$ and ${}_{\mathrm{B}}\langle\alpha^{+*}|\hat{a}_i = \left(\partial/\partial\alpha_i^{+}\right){}_{\mathrm{B}}\langle\alpha^{+*}|$. We use the fact that $\mathrm{Tr}(|\alpha\rangle_{\mathrm{BB}}\langle\alpha^{+*}|) = \exp\left(\boldsymbol{\alpha}\cdot\boldsymbol{\alpha}^{+}\right)$ depends only on the α_i and α_i^{+}, and that both $|\alpha\rangle_{\mathrm{B}}$ and ${}_{\mathrm{B}}\langle\alpha^{+*}|$ commute with any c-number. Moreover, ${}_{\mathrm{B}}\langle\boldsymbol{\alpha}^{+*}|$ depends only on the α_i^{+} and $|\boldsymbol{\alpha}\rangle_{\mathrm{B}}$ depends only on the α_i, and it follows that $\hat{\Lambda}_{\mathrm{b}}(\boldsymbol{\alpha},\boldsymbol{\alpha}^{+})$ is an analytic function of the α_i and α_i^{+}. It therefore can be differentiated with respect to the α_i and α_i^{+}, and differentiation with respect to either α_{ix} or $i\alpha_{iy}$ and either α_{ix}^{+} or $i\alpha_{iy}^{+}$ yields the same result in the two cases. We find that

$$\begin{aligned}
\hat{a}_i \hat{\Lambda}_{\rm b}(\boldsymbol{\alpha}, \boldsymbol{\alpha}^+) &= \alpha_i \hat{\Lambda}_{\rm b}(\boldsymbol{\alpha}, \boldsymbol{\alpha}^+) \\
\hat{\Lambda}_{\rm b}(\boldsymbol{\alpha}, \boldsymbol{\alpha}^+)\hat{a}_i^\dagger &= \alpha_i^+ \hat{\Lambda}_{\rm b}(\boldsymbol{\alpha}, \boldsymbol{\alpha}^+), \\
\hat{a}_i^\dagger \hat{\Lambda}_{\rm b}(\boldsymbol{\alpha}, \boldsymbol{\alpha}^+) &= \left(\frac{\partial}{\partial \alpha_i} + \alpha_i^+\right) \hat{\Lambda}_{\rm b}(\boldsymbol{\alpha}, \boldsymbol{\alpha}^+) \\
\hat{\Lambda}_{\rm b}(\boldsymbol{\alpha}, \boldsymbol{\alpha}^+)\hat{a}_i &= \left(\frac{\partial}{\partial \alpha_i^+} + \alpha_i\right) \hat{\Lambda}_{\rm b}(\boldsymbol{\alpha}, \boldsymbol{\alpha}^+),
\end{aligned} \tag{F.22}$$

where the standard product differentiation rules for complex functions have been used. By way of illustration, we obtain the third of these as follows:

$$\begin{aligned}
\hat{a}_i^\dagger \hat{\Lambda}_{\rm b}(\boldsymbol{\alpha}, \boldsymbol{\alpha}^+) &= \left(\frac{\partial}{\partial \alpha_i} |\boldsymbol{\alpha}\rangle_{\rm B}\right) {}_{\rm B}\langle \alpha^{+*}| \exp(-\boldsymbol{\alpha}\cdot\boldsymbol{\alpha}^+) \\
&= \frac{\partial}{\partial \alpha_i} \hat{\Lambda}_{\rm b}(\boldsymbol{\alpha}, \boldsymbol{\alpha}^+) - |\boldsymbol{\alpha}\rangle_{\rm B} \frac{\partial}{\partial \alpha_i} \left\{ {}_{\rm B}\langle \boldsymbol{\alpha}^{+*}| \exp(-\boldsymbol{\alpha}\cdot\boldsymbol{\alpha}^+) \right\} \\
&= \frac{\partial}{\partial \alpha_i} \hat{\Lambda}_{\rm b}(\boldsymbol{\alpha}, \boldsymbol{\alpha}^+) - |\boldsymbol{\alpha}\rangle_{\rm BB}\langle \boldsymbol{\alpha}^{+*}| \left(-\alpha_i^+\right) \exp(-\boldsymbol{\alpha}\cdot\boldsymbol{\alpha}^+) \\
&= \left(\frac{\partial}{\partial \alpha_i} + \alpha_i^+\right) \hat{\Lambda}_{\rm b}(\boldsymbol{\alpha}, \boldsymbol{\alpha}^+).
\end{aligned} \tag{F.23}$$

We can now apply these results to determine the changes to the distribution function. For example, if $\hat{\rho}$ is replaced by $\hat{a}_i^\dagger \hat{\rho}$, then

$$\begin{aligned}
\hat{a}_i^\dagger \hat{\rho} &= \int d^2\boldsymbol{\alpha}^+ d^2\boldsymbol{\alpha} \left(\left(\frac{\partial}{\partial \alpha_i} + \alpha_i^+\right) \hat{\Lambda}_{\rm b}(\boldsymbol{\alpha}, \boldsymbol{\alpha}^+) \right) \hat{\Pi}_{\rm f} \\
&= \int d^2\boldsymbol{\alpha}^+ d^2\boldsymbol{\alpha}\, \hat{\Lambda}_{\rm b}(\boldsymbol{\alpha}, \boldsymbol{\alpha}^+) \left[\left(-\frac{\partial}{\partial \alpha_i} + \alpha_i^+\right) \hat{\Pi}_{\rm f} \right] \\
&= \int d\mathbf{g}^+ d\mathbf{g}\, d^2\boldsymbol{\alpha}^+ d^2\boldsymbol{\alpha} \left[\left(-\frac{\partial}{\partial \alpha_i} + \alpha_i^+\right) P(\boldsymbol{\alpha}, \boldsymbol{\alpha}^+, \boldsymbol{\alpha}^*, \boldsymbol{\alpha}^{+*}, \mathbf{g}, \mathbf{g}^+) \right] \hat{\Lambda}(\mathbf{g}, \mathbf{g}^+, \boldsymbol{\alpha}, \boldsymbol{\alpha}^+),
\end{aligned} \tag{F.24}$$

where we have used integration by parts, assuming that the non-analytic function $P(\boldsymbol{\alpha}, \boldsymbol{\alpha}^+, \boldsymbol{\alpha}^*, \boldsymbol{\alpha}^{+*}, \mathbf{g}, \mathbf{g}^+)$ goes to zero on the boundary of the $\boldsymbol{\alpha}, \boldsymbol{\alpha}^+$ phase space. In addition, as $P(\boldsymbol{\alpha}, \boldsymbol{\alpha}^+, \boldsymbol{\alpha}^*, \boldsymbol{\alpha}^{+*}, \mathbf{g}, \mathbf{g}^+)$ and hence $\hat{\Pi}_{\rm f}$ are not analytic functions, differentiation with respect to α_{ix} or $i\alpha_{iy}$ gives different results and hence the interpretation of differentiation with respect to α_i is taken to mean that one of these choices has been made. Thus, as $P(\boldsymbol{\alpha}, \boldsymbol{\alpha}^+, \boldsymbol{\alpha}^*, \boldsymbol{\alpha}^{+*}, \mathbf{g}, \mathbf{g}^+)$ is a valid distribution function, we see that

$$\hat{\rho} \Rightarrow \hat{a}_i^\dagger \hat{\rho} \qquad P(\boldsymbol{\alpha}, \boldsymbol{\alpha}^+, \boldsymbol{\alpha}^*, \boldsymbol{\alpha}^{+*}, \mathbf{g}, \mathbf{g}^+) \Rightarrow \left(-\frac{\partial}{\partial \alpha_i} + \alpha_i^+\right) P, \tag{F.25}$$

where

$$\frac{\partial}{\partial \alpha_i} \Rightarrow \frac{\partial}{\partial \alpha_{ix}} \quad \text{or} \quad \frac{\partial}{\partial (i\alpha_{iy})}. \tag{F.26}$$

The other three cases are proved similarly. To summarise the results,

$$\begin{aligned}
\hat{\rho} &\Rightarrow \hat{a}_i\hat{\rho}, & P(\boldsymbol{\alpha},\boldsymbol{\alpha}^+,\boldsymbol{\alpha}^*,\boldsymbol{\alpha}^{+*},\mathbf{g},\mathbf{g}^+) &\Rightarrow \alpha_i P, \\
\hat{\rho} &\Rightarrow \hat{\rho}\hat{a}_i, & P(\boldsymbol{\alpha},\boldsymbol{\alpha}^+,\boldsymbol{\alpha}^*,\boldsymbol{\alpha}^{+*},\mathbf{g},\mathbf{g}^+) &\Rightarrow \left(-\frac{\partial}{\partial\alpha_i^+}+\alpha_i\right)P, \\
\hat{\rho} &\Rightarrow \hat{a}_i^\dagger\hat{\rho}, & P(\boldsymbol{\alpha},\boldsymbol{\alpha}^+,\boldsymbol{\alpha}^*,\boldsymbol{\alpha}^{+*},\mathbf{g},\mathbf{g}^+) &\Rightarrow \left(-\frac{\partial}{\partial\alpha_i}+\alpha_i^+\right)P, \\
\hat{\rho} &\Rightarrow \hat{\rho}\hat{a}_i^\dagger, & P(\boldsymbol{\alpha},\boldsymbol{\alpha}^+,\boldsymbol{\alpha}^*,\boldsymbol{\alpha}^{+*},\mathbf{g},\mathbf{g}^+) &\Rightarrow \alpha_i^+ P,
\end{aligned} \tag{F.27}$$

where

$$\begin{aligned}
\frac{\partial}{\partial\alpha_i} &\Rightarrow \frac{\partial}{\partial\alpha_{ix}} \quad \text{or} \quad \frac{\partial}{\partial(i\alpha_{iy})}, \\
\frac{\partial}{\partial\alpha_i^+} &\Rightarrow \frac{\partial}{\partial\alpha_{ix}^+} \quad \text{or} \quad \frac{\partial}{\partial(i\alpha_{iy}^+)}.
\end{aligned} \tag{F.28}$$

F.1.3 Boson Case – Characteristic-Function Approach

The characteristic function may be written as

$$\chi(\boldsymbol{\xi},\boldsymbol{\xi}^+,\mathbf{h},\mathbf{h}^+) = \mathrm{Tr}\left[\hat{\Omega}_{\mathrm{b}}^+(\boldsymbol{\xi}^+)\hat{\Theta}(\mathbf{h},\mathbf{h}^+)\hat{\Omega}_{\mathrm{b}}^-(\boldsymbol{\xi})\right], \tag{F.29}$$

where we have incorporated the density operator and the Grassmann fermion operators into the single operator

$$\hat{\Theta} = \hat{\Omega}_{\mathrm{f}}^+(\mathbf{h}^+)\hat{\rho}\hat{\Omega}_{\mathrm{f}}^-(\mathbf{h}) = \hat{\Theta}(\mathbf{h},\mathbf{h}^+). \tag{F.30}$$

We shall normally leave the $\mathbf{h},\mathbf{h}^+$ dependence implicit. Note that $\hat{\Theta}(\mathbf{h},\mathbf{h}^+)$ is a Grassmann operator that involves the full Bose–Fermi density operator and fermion annihilation and creation operators.

Some bosonic identities will be useful:

$$\hat{\Omega}_{\mathrm{b}}^+(\boldsymbol{\xi}^+)\hat{a}_i = \frac{\partial}{\partial(i\xi_i^+)}\hat{\Omega}_{\mathrm{b}}^+(\boldsymbol{\xi}^+) = \hat{a}_i\hat{\Omega}_{\mathrm{b}}^+(\boldsymbol{\xi}^+), \tag{F.31}$$

$$\hat{\Omega}_{\mathrm{b}}^-(\boldsymbol{\xi})\hat{a}_i^\dagger = \frac{\partial}{\partial(i\xi_i)}\hat{\Omega}_{\mathrm{b}}^-(\boldsymbol{\xi}) = \hat{a}_i^\dagger\hat{\Omega}_{\mathrm{b}}^-(\boldsymbol{\xi}). \tag{F.32}$$

These are used, in particular, to determine how the characteristic function changes when $\hat{\rho}$ is replaced by $\hat{a}_i\hat{\rho}$ etc. For example, if $\hat{\rho}$ is replaced by $\hat{a}_i^\dagger\hat{\rho}$, then $\hat{\Omega}_{\mathrm{b}}^+(\boldsymbol{\xi}^+)\hat{\Theta}(\mathbf{h},\mathbf{h}^+)\ \times\hat{\Omega}_{\mathrm{b}}^-(\boldsymbol{\xi})$ becomes $\hat{\Omega}_{\mathrm{b}}^+(\boldsymbol{\xi}^+)\hat{a}_i^\dagger\hat{\Theta}(\mathbf{h},\mathbf{h}^+)\hat{\Omega}_{\mathrm{b}}^-(\boldsymbol{\xi})$. Then, denoting that with $\hat{S} = i\hat{\mathbf{a}}\cdot\boldsymbol{\xi}^+$, we have $[\hat{S},\hat{a}_i^\dagger] = i\xi_i^+$, $[\hat{S},[\hat{S},\hat{a}_i^\dagger]] = 0,\cdots$, we find

$$\begin{aligned}
\hat{\Omega}_{\mathrm{b}}^+(\boldsymbol{\xi}^+)\hat{a}_i^\dagger &= (\hat{a}_i^\dagger + i\xi_i^+)\hat{\Omega}_{\mathrm{b}}^+(\boldsymbol{\xi}^+) \\
\Rightarrow \mathrm{Tr}\left[\hat{\Omega}_{\mathrm{b}}^+(\boldsymbol{\xi}^+)\hat{a}_i^\dagger\hat{\Theta}(\mathbf{h},\mathbf{h}^+)\hat{\Omega}_{\mathrm{b}}^-(\hat{\Omega}_{\mathrm{b}}^+(\boldsymbol{\xi}^+))\right] &= \mathrm{Tr}\left[\hat{\Omega}_{\mathrm{b}}^+(\boldsymbol{\xi}^+)\hat{\Theta}(\mathbf{h},\mathbf{h}^+)\hat{\Omega}_{\mathrm{b}}^-(\boldsymbol{\xi})(\hat{a}_i^\dagger + i\xi_i^+)\right] \\
&= \left(\frac{\partial}{\partial(i\xi_i)} + i\xi_i^+\right)\chi(\boldsymbol{\xi},\boldsymbol{\xi}^+,\mathbf{h},\mathbf{h}^+),
\end{aligned} \tag{F.33}$$

where we have made use of the cyclic property of the trace. The other cases may be derived in a similar fashion.

Summarising the results, we have the correspondence rules:

$$
\begin{aligned}
\hat{\rho} &\Rightarrow \hat{a}_i\hat{\rho}, & \chi &\Rightarrow \frac{\partial}{\partial(i\xi_i^+)}\chi, \\
\hat{\rho} &\Rightarrow \hat{\rho}\hat{a}_i, & \chi &\Rightarrow \left(\frac{\partial}{\partial(i\xi_i^+)} + i\xi_i\right)\chi, \\
\hat{\rho} &\Rightarrow \hat{a}_i^\dagger\hat{\rho}, & \chi &\Rightarrow \left(\frac{\partial}{\partial(i\xi_i)} + i\xi_i^+\right)\chi, \\
\hat{\rho} &\Rightarrow \hat{\rho}\hat{a}_i^\dagger, & \chi &\Rightarrow \frac{\partial}{\partial(i\xi_i)}\chi.
\end{aligned}
\tag{F.34}
$$

Our next task is to translate these into correspondence rules for the distribution function. We recall that the characteristic and distribution functions are related via

$$
\chi(\boldsymbol{\xi}, \boldsymbol{\xi}^+, \mathbf{h}, \mathbf{h}^+) = \int d^2\boldsymbol{\alpha}^+ d^2\boldsymbol{\alpha}\, \exp(i\boldsymbol{\alpha}\cdot\boldsymbol{\xi}^+)\,\Theta\exp(i\boldsymbol{\xi}\cdot\boldsymbol{\alpha}^+), \tag{F.35}
$$

where we have introduced the simplifying notation

$$
\Theta = \int d\mathbf{g}^+ d\mathbf{g}\, \exp(i\{\mathbf{g}\cdot\mathbf{h}^+\})\, P(\boldsymbol{\alpha}, \boldsymbol{\alpha}^+, \boldsymbol{\xi}^*, \boldsymbol{\xi}^{+*}\mathbf{g}, \mathbf{g}^+)\exp(i\mathbf{h}\cdot\mathbf{g}^+), \tag{F.36}
$$

which is, of course, a Grassmann function $\Theta(\boldsymbol{\alpha}, \boldsymbol{\alpha}^+, \boldsymbol{\xi}^*, \boldsymbol{\xi}^{+*}, \mathbf{h}, \mathbf{h}^+)$ of $\mathbf{h}, \mathbf{h}^+$, although this will be left implicit. As an example of the procedure, if $\hat{\rho}$ is replaced by $\hat{a}_i^\dagger\hat{\rho}$ then the characteristic function becomes

$$
\begin{aligned}
\chi(\boldsymbol{\xi}, \boldsymbol{\xi}^+, \mathbf{h}, \mathbf{h}^+) &\Rightarrow \left(\frac{\partial}{\partial(i\xi_i)} + i\xi_i^+\right)\chi(\boldsymbol{\xi}, \boldsymbol{\xi}^+, \mathbf{h}, \mathbf{h}^+) \\
&= \left(\frac{\partial}{\partial(i\xi_i)} + i\xi_i^+\right)\int d^2\boldsymbol{\alpha}^+ d^2\boldsymbol{\alpha}\, \exp(i\boldsymbol{\alpha}\cdot\boldsymbol{\xi}^+)\,\Theta\exp(i\boldsymbol{\xi}\cdot\boldsymbol{\alpha}^+) \\
&= \int d^2\boldsymbol{\alpha}^+ d^2\boldsymbol{\alpha}\;\, \exp(i\boldsymbol{\alpha}\cdot\boldsymbol{\xi}^+)\,(\Theta\alpha_i^+)\exp(i\boldsymbol{\xi}\cdot\boldsymbol{\alpha}^+) \\
&\quad + \int d^2\boldsymbol{\alpha}^+ d^2\boldsymbol{\alpha}\left[\frac{\partial}{\partial\alpha_i}\exp(i\boldsymbol{\alpha}\cdot\boldsymbol{\xi}^+)\right]\Theta\exp(i\boldsymbol{\xi}\cdot\boldsymbol{\alpha}^+).
\end{aligned}
\tag{F.37}
$$

We note that $\exp(i\boldsymbol{\alpha}\cdot\boldsymbol{\xi}^+)$ is an analytic function of α_i, and it follows that

$$
\frac{\partial}{\partial\alpha_i}\exp(i\boldsymbol{\alpha}\cdot\boldsymbol{\xi}^+) = \frac{\partial}{\partial\alpha_{ix}}\exp(i\boldsymbol{\alpha}\cdot\boldsymbol{\xi}^+) = \frac{\partial}{\partial(i\alpha_{iy})}\exp(i\boldsymbol{\alpha}\cdot\boldsymbol{\xi}^+). \tag{F.38}
$$

Applying these to (F.37) and performing integration by parts, we recover the associated transformation:

$$
\Theta \Rightarrow \left(\alpha_i^+ - \frac{\partial}{\partial\alpha_{ix}}\right)\Theta \equiv \left(\alpha_i^+ - \frac{\partial}{\partial(i\alpha_{iy})}\right)\Theta\,. \tag{F.39}
$$

Note that the function $\Theta(\boldsymbol{\alpha}, \boldsymbol{\alpha}^+, \boldsymbol{\alpha}^*, \boldsymbol{\alpha}^{+*})$ is not an analytic function of α_i, but it is still differentiable with respect to α_{ix} and α_{iy}. Hence only derivatives with respect to these variables may be used in performing the integration-by-parts step, which also relies on $P(\boldsymbol{\alpha}, \boldsymbol{\alpha}^+, \boldsymbol{\alpha}^*, \boldsymbol{\alpha}^{+*}, \mathbf{g}, \mathbf{g}^+)$ and hence $\Theta(\boldsymbol{\alpha}, \boldsymbol{\alpha}^+, \boldsymbol{\alpha}^*, \boldsymbol{\alpha}^{+*})$ going to zero fast enough on the boundaries of the α_i planes to allow boundary terms to be ignored. It follows that the change to the distribution function is as follows:

$$\hat{\rho} \Rightarrow \hat{a}_i^\dagger \hat{\rho},$$
$$P(\boldsymbol{\alpha}, \boldsymbol{\alpha}^+, \boldsymbol{\alpha}^*, \boldsymbol{\alpha}^{+*}, \mathbf{g}, \mathbf{g}^+) \Rightarrow \left(\alpha_i^+ - \frac{\partial}{\partial \alpha_{ix}}\right) P \equiv \left(\alpha_i^+ - \frac{\partial}{\partial (i\alpha_{iy})}\right) P. \quad \text{(F.40)}$$

The other cases may be derived in a similar fashion to give the correspondence rules:

$$\begin{aligned}
&\hat{\rho} \Rightarrow \hat{a}_i \hat{\rho}, && P \Rightarrow \alpha_i P, \\
&\hat{\rho} \Rightarrow \hat{\rho} \hat{a}_i, && P \Rightarrow \left(\alpha_i - \frac{\partial}{\partial \alpha_{ix}^+}\right) P \equiv \left(\alpha_i - \frac{\partial}{\partial (i\alpha_{iy}^+)}\right) P, \\
&\hat{\rho} \Rightarrow \hat{a}_i^\dagger \hat{\rho}, && P \Rightarrow \left(\alpha_i^+ - \frac{\partial}{\partial \alpha_{ix}}\right) P \equiv \left(\alpha_i^+ - \frac{\partial}{\partial (i\alpha_{iy})}\right) P, \\
&\hat{\rho} \Rightarrow \hat{\rho} \hat{a}_i^\dagger, && P \Rightarrow \alpha_i^+ P.
\end{aligned} \quad \text{(F.41)}$$

The symmetrically ordered case can be obtained from the normally ordered case, since the two characteristic functions are related as in (7.10):

$$\chi_b^W(\boldsymbol{\xi}, \boldsymbol{\xi}^+) = \exp\left(-\frac{1}{2}\boldsymbol{\xi} \cdot \boldsymbol{\xi}^+\right) \chi_b(\boldsymbol{\xi}, \boldsymbol{\xi}^+). \quad \text{(F.42)}$$

We can use this relationship together with the results for the normally ordered case to determine the change in the symmetrically ordered characteristic function and hence in the corresponding distribution function. For example, if $\hat{\rho}$ is replaced by $\hat{a}_i \hat{\rho}$, we see that

$$\begin{aligned}
\chi_b = \exp\left(\frac{1}{2}\boldsymbol{\xi} \cdot \boldsymbol{\xi}^+\right) \chi^W \Rightarrow& \frac{\partial}{\partial (i\xi_i^+)} \left\{\exp\left(\frac{1}{2}\boldsymbol{\xi} \cdot \boldsymbol{\xi}^+\right) \chi^W\right\} \\
&= \exp\left(\frac{1}{2}\boldsymbol{\xi} \cdot \boldsymbol{\xi}^+\right) \left(-\frac{1}{2} i\xi_i + \frac{\partial}{\partial (i\xi_i^+)}\right) \chi^W.
\end{aligned} \quad \text{(F.43)}$$

It follows, therefore, that pre-multiplying the density operator by $\hat{a}_i$ corresponds to the transformation

$$\chi^W \Rightarrow \left(\frac{\partial}{\partial (i\xi_i^+)} - \frac{1}{2} i\xi_i\right) \chi^W. \quad \text{(F.44)}$$

The other cases are obtained in a similar manner to give

$$\begin{aligned}
\hat{\rho} &\Rightarrow \hat{a}_i\hat{\rho}, & \chi^{\mathrm{W}} &\Rightarrow \left(\frac{\partial}{\partial(i\xi_i^+)} - \frac{1}{2}i\xi_i\right)\chi^{\mathrm{W}}, \\
\hat{\rho} &\Rightarrow \hat{\rho}\hat{a}_i, & \chi^{\mathrm{W}} &\Rightarrow \left(\frac{\partial}{\partial(i\xi_i^+)} + \frac{1}{2}i\xi_i\right)\chi^{\mathrm{W}}, \\
\hat{\rho} &\Rightarrow \hat{a}_i^\dagger\hat{\rho}, & \chi^{\mathrm{W}} &\Rightarrow \left(\frac{\partial}{\partial(i\xi_i)} + \frac{1}{2}i\xi_i^+\right)\chi^{\mathrm{W}}, \\
\hat{\rho} &\Rightarrow \hat{\rho}\hat{a}_i^\dagger, & \chi^{\mathrm{W}} &\Rightarrow \left(\frac{\partial}{\partial(i\xi_i)} - \frac{1}{2}i\xi_i^+\right)\chi^{\mathrm{W}}.
\end{aligned} \tag{F.45}$$

It is straightforward to apply these to the relationship between the symmetrically ordered characteristic function and the double-space Wigner distribution function to derive the correspondence rules for the Wigner distribution function. We find

$$\begin{aligned}
\hat{\rho} &\Rightarrow \hat{a}_i\hat{\rho}, & W &\Rightarrow \left(\alpha_i + \frac{1}{2}\frac{\partial}{\partial\alpha_{ix}^+}\right)W \equiv \left(\alpha_i + \frac{1}{2}\frac{\partial}{\partial(i\alpha_{iy}^+)}\right)W, \\
\hat{\rho} &\Rightarrow \hat{\rho}\hat{a}_i, & W &\Rightarrow \left(\alpha_i - \frac{1}{2}\frac{\partial}{\partial\alpha_{ix}^+}\right)W \equiv \left(\alpha_i - \frac{1}{2}\frac{\partial}{\partial(i\alpha_{iy}^+)}\right)W, \\
\hat{\rho} &\Rightarrow \hat{a}_i^\dagger\hat{\rho}, & W &\Rightarrow \left(\alpha_i^+ - \frac{1}{2}\frac{\partial}{\partial\alpha_{ix}}\right)W \equiv \left(\alpha_i^+ - \frac{1}{2}\frac{\partial}{\partial(i\alpha_{iy})}\right)W, \\
\hat{\rho} &\Rightarrow \hat{\rho}\hat{a}_i^\dagger, & W &\Rightarrow \left(\alpha_i^+ + \frac{1}{2}\frac{\partial}{\partial\alpha_{ix}}\right)W \equiv \left(\alpha_i^+ + \frac{1}{2}\frac{\partial}{\partial(i\alpha_{iy})}\right)W.
\end{aligned} \tag{F.46}$$

F.1.4 Fermion Case – Density Operator Approach

Let us begin, in order to condense the expressions that arise, by introducing the operator

$$\hat{\Pi}_{\mathrm{b}} = \int d^2\boldsymbol{\alpha}^+ d^2\boldsymbol{\alpha}\; P(\boldsymbol{\alpha},\boldsymbol{\alpha}^+,\boldsymbol{\alpha}^*,\boldsymbol{\alpha}^{+*},\;\; \mathbf{g},\mathbf{g}^+)\hat{\Lambda}_{\mathrm{b}}(\boldsymbol{\alpha},\boldsymbol{\alpha}^+). \tag{F.47}$$

We note that this is both a bosonic operator and also an even Grassmann function and that, because it is even, $\hat{\Pi}_{\mathrm{b}}$ commutes with $\hat{\Lambda}_{\mathrm{f}}(\mathbf{g},\mathbf{g}^+)$. It follows that we can write the density operator in two alternative forms,

$$\hat{\rho} = \int d\mathbf{g}^+ d\mathbf{g}\, \hat{\Pi}_{\mathrm{b}}\hat{\Lambda}_{\mathrm{f}}(\mathbf{g},\mathbf{g}^+) = \int d\mathbf{g}^+ d\mathbf{g}\, \hat{\Lambda}_{\mathrm{f}}(\mathbf{g},\mathbf{g}^+)\hat{\Pi}_{\mathrm{b}}. \tag{F.48}$$

Note that $\hat{\Pi}_{\mathrm{b}}(\mathbf{g},\mathbf{g}^+)$ does not involve any fermionic operators.

Certain results involving the fermion projector can be obtained from Chapter 5, which give $\hat{c}_i|\mathbf{g}\rangle_{\mathrm{B}} = g_i|\mathbf{g}\rangle_{\mathrm{B}}$, ${}_{\mathrm{B}}\langle\mathbf{g}^{+*}|\hat{c}_i^\dagger = {}_{\mathrm{B}}\langle\mathbf{g}^{+*}|g_i^+$, $\hat{c}_i^\dagger|\mathbf{g}\rangle_{\mathrm{B}} = \left(-\overrightarrow{\partial}/\partial g_i\right)|\mathbf{g}\rangle_{\mathrm{B}}$ and ${}_{\mathrm{B}}\langle\mathbf{g}^{+*}|\hat{c}_i = {}_{\mathrm{B}}\langle\mathbf{g}^{+*}|\left(-\overleftarrow{\partial}/\partial g_i^+\right)$. We use the fact that $\mathrm{Tr}(|\mathbf{g}\rangle_{\mathrm{BB}}\langle\mathbf{g}^{+*}|) = \exp(\mathbf{g}\cdot\mathbf{g}^+)$ is an even Grassmann function, and that both $|\mathbf{g}\rangle_{\mathrm{B}}$ and ${}_{\mathrm{B}}\langle\mathbf{g}^{+*}|$ will commute with any Grassmann number, as they just involve factors like $\hat{c}_i^\dagger g_i$ or $\hat{c}_i g_i^+$ where two anticommuting quantities are involved. Also, ${}_{\mathrm{B}}\langle\mathbf{g}^{+*}|$ depends only on the g_i^+ and $|\mathbf{g}\rangle_{\mathrm{B}}$ depends only on the g_i. These are neither even nor odd functions, however, although $\hat{\Lambda}_{\mathrm{f}}(\mathbf{g},\mathbf{g}^+)$ is an even Grassmann operator. We find that

$$\begin{aligned}
\hat{c}_i\hat{\Lambda}_{\mathrm{f}}(\mathbf{g},\mathbf{g}^+) &= g_i\frac{|g\rangle_{\mathrm{BB}}\langle g^{+*}|}{\mathrm{Tr}(|g\rangle_{\mathrm{BB}}\langle g^{+*}|)} = g_i\hat{\Lambda}_{\mathrm{f}}(\mathbf{g},\mathbf{g}^+),\\
\hat{\Lambda}_{\mathrm{f}}(\mathbf{g},\mathbf{g}^+)\hat{c}_i^\dagger &= \frac{|g\rangle_{\mathrm{BB}}\langle g^{+*}|}{\mathrm{Tr}(|g\rangle_{\mathrm{BB}}\langle g^{+*}|)}g_i^+ = \hat{\Lambda}_{\mathrm{f}}(\mathbf{g},\mathbf{g}^+)g_i^+,\\
\hat{c}_i^\dagger\hat{\Lambda}_{\mathrm{f}}(\mathbf{g},\mathbf{g}^+) &= \left(-\frac{\overrightarrow{\partial}}{\partial g_i} - g_i^+\right)\hat{\Lambda}_{\mathrm{f}}(\mathbf{g},\mathbf{g}^+),\\
\hat{\Lambda}_{\mathrm{f}}(\mathbf{g},\mathbf{g}^+)\hat{c}_i &= \hat{\Lambda}_{\mathrm{f}}(\mathbf{g},\mathbf{g}^+)\left(-\frac{\overleftarrow{\partial}}{\partial g_i^+} - g_i\right),
\end{aligned} \tag{F.49}$$

where the product differentiation rules (4.28) have been used. By way of illustration, we obtain the third of these as follows:

$$\begin{aligned}
\hat{c}_i^\dagger\hat{\Lambda}_{\mathrm{f}}(\mathbf{g},\mathbf{g}^+) &= \left(-\frac{\overrightarrow{\partial}}{\partial g_i}|\mathbf{g}\rangle_{\mathrm{B}}\right){}_{\mathrm{B}}\langle\mathbf{g}^{+*}|\exp(-\mathbf{g}\cdot\mathbf{g}^+)\\
&= -\frac{\overrightarrow{\partial}}{\partial g_i}\hat{\Lambda}_{\mathrm{f}}(\mathbf{g},\mathbf{g}^+) + |\mathbf{g}\rangle_{\mathrm{B}}\frac{\overrightarrow{\partial}}{\partial g_i}\{{}_{\mathrm{B}}\langle\mathbf{g}|\exp(-\mathbf{g}\cdot\mathbf{g}^+)\}\\
&= -\frac{\overrightarrow{\partial}}{\partial g_i}\hat{\Lambda}_{\mathrm{f}}(\mathbf{g},\mathbf{g}^+) + |\mathbf{g}\rangle_{\mathrm{BB}}\langle\mathbf{g}|(-g_i^+)\exp(-\mathbf{g}\cdot\mathbf{g}^+)\\
&= \left(-\frac{\overrightarrow{\partial}}{\partial g_i} - g_i^+\right)\hat{\Lambda}_{\mathrm{f}}(\mathbf{g},\mathbf{g}^+).
\end{aligned} \tag{F.50}$$

All of the above results can also be written in another form, with the Grassmann number differentiation or multiplication applied from the other direction. The Grassmann Bargmann projector $\hat{\Lambda}_{\mathrm{f}}(\mathbf{g},\mathbf{g}^+)$ is an even Grassmann operator and behaves like an even Grassmann function, as if it is expanded out, there are always an even number of anticommuting quantities in every term. This is because either the Grassmann numbers appear in pairs, as in the $\exp(-\mathbf{g}\cdot\mathbf{g}^+)$ factor, or they appear in a pair with a fermion annihilation or creation operator, as in the expressions for $|\mathbf{g}\rangle_{\mathrm{B}}$ or $\langle\mathbf{g}^{+*}|_{\mathrm{B}}$. In regard to multiplication by a Grassmann number g_i or g_i^+, there are always an even number of anticommutations involved in taking the number from one side of $\hat{\Lambda}_{\mathrm{f}}(\mathbf{g},\mathbf{g}^+)$. Also, as mentioned before, the proof of the result (4.21) that left and right differentiation of an even Grassmann function are related by a factor -1 can be extended to the case of an even Grassmann operator because, in the movement of the Grassmann number to the left or right for differentiation, it would not matter if fermion annihilation or creation operators replaced some of the Grassmann numbers, since they would still anticommute with the Grassmann number being differentiated. Hence we *also* have

$$\begin{aligned}
\hat{c}_i\hat{\Lambda}_{\mathrm{f}}(\mathbf{g},\mathbf{g}^+) &= \hat{\Lambda}_{\mathrm{f}}(\mathbf{g},\mathbf{g}^+)g_i,\\
\hat{\Lambda}_{\mathrm{f}}(\mathbf{g},\mathbf{g}^+)\hat{c}_i^\dagger &= g_i^+\hat{\Lambda}_{\mathrm{f}}(\mathbf{g},\mathbf{g}^+),\\
\hat{c}_i^\dagger\hat{\Lambda}_{\mathrm{f}}(\mathbf{g},\mathbf{g}^+) &= \hat{\Lambda}_{\mathrm{f}}(\mathbf{g},\mathbf{g}^+)\left(+\frac{\overleftarrow{\partial}}{\partial g_i}-g_i^+\right),\\
\hat{\Lambda}_{\mathrm{f}}(\mathbf{g},\mathbf{g}^+)\hat{c}_i &= \left(+\frac{\overrightarrow{\partial}}{\partial g_i^+}-g_i\right)\hat{\Lambda}_f(\mathbf{g},\mathbf{g}^+),
\end{aligned} \tag{F.51}$$

where the sign of the derivative changes but not that of the Grassmann number.

We can apply these results to determine the changes to the distribution function. For example, if $\hat{\rho}$ is replaced by $\hat{c}_i^\dagger\hat{\rho}$, then

$$\begin{aligned}
\hat{c}_i^\dagger\hat{\rho} &= \int d\mathbf{g}^+d\mathbf{g}\left[\left(\left(-\frac{\overrightarrow{\partial}}{\partial g_i}-g_i^+\right)\hat{\Lambda}_{\mathrm{f}}(\mathbf{g},\mathbf{g}^+)\right)\hat{\Pi}_{\mathrm{b}}\right.\\
&= \int d\mathbf{g}^+d\mathbf{g}\,\hat{\Lambda}_{\mathrm{f}}(\mathbf{g},\mathbf{g}^+)\left[\left(+\frac{\overrightarrow{\partial}}{\partial g_i}-g_i^+\right)\hat{\Pi}_{\mathrm{b}}\right]\\
&= \int d\mathbf{g}^+d\mathbf{g}\;d^2\boldsymbol{\alpha}^+d^2\boldsymbol{\alpha}\;\hat{\Lambda}(\mathbf{g},\mathbf{g}^+,\boldsymbol{\alpha},\boldsymbol{\alpha}^+)\left[\left(+\frac{\overrightarrow{\partial}}{\partial g_i}-g_i^+\right)P(\boldsymbol{\alpha},\boldsymbol{\alpha}^+,\boldsymbol{\alpha}^*,\boldsymbol{\alpha}^{+*},\,\mathbf{g},\mathbf{g}^+)\right],
\end{aligned} \tag{F.52}$$

where we have used the third of the integration-by-parts rules (F.9), noting that $\hat{\Lambda}_{\mathrm{f}}(\mathbf{g},\mathbf{g}^+)$ is an even Grassmann operator.

As $P(\boldsymbol{\alpha},\boldsymbol{\alpha}^+,\boldsymbol{\alpha}^*,\boldsymbol{\alpha}^{+*},\,\mathbf{g},\mathbf{g}^+)$ is a valid distribution function, we see that

$$\hat{\rho}\Rightarrow\hat{c}_i^\dagger\hat{\rho},\qquad P(\boldsymbol{\alpha},\boldsymbol{\alpha}^+,\boldsymbol{\alpha}^*,\boldsymbol{\alpha}^{+*},\mathbf{g},\mathbf{g}^+)\Rightarrow\left(+\frac{\overrightarrow{\partial}}{\partial g_i}-g_i^+\right)P=P\left(-\frac{\overleftarrow{\partial}}{\partial g_i}-g_i^+\right), \tag{F.53}$$

where the last step is based on (4.21) and depends on $P(\boldsymbol{\alpha},\boldsymbol{\alpha}^+,\boldsymbol{\alpha}^*,\boldsymbol{\alpha}^{+*},\,\mathbf{g},\mathbf{g}^+)$ being an even Grassmann function. The other cases may be treated in a similar manner. The overall results are

$$\begin{aligned}
\hat{\rho}&\Rightarrow\hat{c}_i\hat{\rho}, && P(\boldsymbol{\alpha},\boldsymbol{\alpha}^+,\boldsymbol{\alpha}^*,\boldsymbol{\alpha}^{+*},\mathbf{g},\mathbf{g}^+)\Rightarrow g_iP=Pg_i,\\
\hat{\rho}&\Rightarrow\hat{\rho}\hat{c}_i, && P(\boldsymbol{\alpha},\boldsymbol{\alpha}^+,\boldsymbol{\alpha}^*,\boldsymbol{\alpha}^{+*},\mathbf{g},\mathbf{g}^+)\Rightarrow P\left(+\frac{\overleftarrow{\partial}}{\partial g_i^+}-g_i\right)=\left(-\frac{\overrightarrow{\partial}}{\partial g_i^+}-g_i\right)P,\\
\hat{\rho}&\Rightarrow\hat{c}_i^\dagger\hat{\rho}, && P(\boldsymbol{\alpha},\boldsymbol{\alpha}^+,\boldsymbol{\alpha}^*,\boldsymbol{\alpha}^{+*},\mathbf{g},\mathbf{g}^+)\Rightarrow\left(+\frac{\overrightarrow{\partial}}{\partial g_i}-g_i^+\right)P=P\left(-\frac{\overleftarrow{\partial}}{\partial g_i}-g_i^+\right),\\
\hat{\rho}&\Rightarrow\hat{\rho}\hat{c}_i^\dagger, && P(\boldsymbol{\alpha},\boldsymbol{\alpha}^+,\boldsymbol{\alpha}^*,\boldsymbol{\alpha}^{+*},\mathbf{g},\mathbf{g}^+)\Rightarrow Pg_i^+=g_i^+P.
\end{aligned} \tag{F.54}$$

The symmetry between the first forms for the results is to be noted, with a related symmetry applying to the second forms.

F.1.5 Fermion Case – Characteristic-Function Method

The characteristic function is related to the distribution function via Grassmann and c-number integrals:

$$\chi(\boldsymbol{\xi},\boldsymbol{\xi}^{+},\mathbf{h},\mathbf{h}^{+}) = \int d\mathbf{g}^{+}d\mathbf{g}\int d^{2}\boldsymbol{\alpha}^{+}d^{2}\boldsymbol{\alpha}\exp(i\boldsymbol{\alpha}\cdot\boldsymbol{\xi}^{+})\exp(i\mathbf{g}\cdot\mathbf{h}^{+})$$
$$\times P(\boldsymbol{\alpha},\boldsymbol{\alpha}^{+},\boldsymbol{\alpha}^{*},\boldsymbol{\alpha}^{+*},\mathbf{g},\mathbf{g}^{+})\exp(i\mathbf{h}\cdot\mathbf{g}^{+})\exp(i\boldsymbol{\xi}\cdot\boldsymbol{\alpha}^{+})\ . \quad (F.55)$$

One way of justifying the correspondence rules is to *assume* their effect on the distribution function and *derive* their effect on the characteristic function, given that the latter is required to be related to the distribution function via the phase space integral form (7.51). We can then return to the basic definition of the characteristic function (7.50) and *confirm* that the new characteristic function does correspond to the characteristic function that would apply when the density operator $\hat{\rho}$ is replaced by its product with a fermion annihilation or creation operator. However, the process of multiplying or differentiating a trace with Grassmann variables is involved, and to carry out these processes inside the trace requires there to be two Grassmann processes involved (see (5.137)). Hence the correspondence rules can only be confirmed in this way for cases where the density operator $\hat{\rho}$ is replaced by its product with *two* fermion annihilation and/or creation operators. Unfortunately, though, there are a total of 12 different replacements that might be considered, corresponding to pre-multiplication by two operators, post-multiplication by two operators or sandwiching the density operator between two operators. In terms of applying the correspondence rules in physical situations, we note that kinetic-energy and one-body interaction terms in the Hamiltonian always involve pairs of fermion operators $\hat{c}_i^{\dagger}\hat{c}_j$, two-body interaction terms involve two pairs $\hat{c}_i^{\dagger}\hat{c}_j^{\dagger}\hat{c}_k\hat{c}_l$, terms in master equations involve forms such as $[\hat{c}_i^{\dagger}\hat{c}_j,\hat{\rho}]$, $[\hat{c}_i^{\dagger}\hat{c}_j^{\dagger}\hat{c}_k\hat{c}_l,\hat{\rho}]$, $[\hat{c}_i^{\dagger},\hat{\rho}\hat{c}_j]$, $[\hat{c}_i,\hat{\rho}\hat{c}_j^{\dagger}]$ and so on. Even in the case of combined boson–fermion systems where pairs of fermions may be created or destroyed, the terms in the Hamiltonian will involve pairs of annihilation or creation operators. All these require the density operator $\hat{\rho}$ to replaced in succession by a product of the density operator with two fermion annihilation or creation operators. Hence, in physical situations it is not necessary to consider the replacement of the density operator by its product with *only one* fermion annihilation or creation operator.

A second approach involving the characteristic function is to consider the effect of multiplying the density operator by fermion annihilation or creation operators within the definition (7.11) of the characteristic function and then use its relationship to the distribution function to determine the correspondence rule. However, the same problems as described in the last paragraph also occur, and this approach also requires consideration of the effect of pairs of fermion operators on the density operator – as we will now illustrate.

As an example of the second approach to deriving the fermion correspondence rules for pairs of fermion operators from the characteristic function, consider the case when $\hat{\rho} \Rightarrow \hat{c}_i^\dagger \hat{\rho} \hat{c}_j$. Because pair cases are complicated, we will ignore the boson operators and variables. In this case

$$\chi(\mathbf{h}, \mathbf{h}^+) \to \mathrm{Tr}\left[\exp(i\hat{\mathbf{c}} \cdot \mathbf{h}^+)\,(\hat{c}_i^\dagger \hat{\rho} \hat{c}_j)\exp\Big(i\{\mathbf{h} \cdot \hat{\mathbf{c}}^\dagger\Big)\right]. \tag{F.56}$$

It is straightforward to show, using the anticommutation rules, that

$$\begin{aligned}\exp(i\hat{\mathbf{c}} \cdot \mathbf{h}^+)\,\hat{c}_i^\dagger &= (\hat{c}_i^\dagger - ih_i^+)\exp(i\hat{\mathbf{c}} \cdot \mathbf{h}^+), \\ \hat{c}_j \exp\Big(i\{\mathbf{h} \cdot \hat{\mathbf{c}}^\dagger\Big) &= \exp\Big(i\mathbf{h} \cdot \hat{\mathbf{c}}^\dagger\Big)\,(\hat{c}_j - ih_j),\end{aligned} \tag{F.57}$$

so that

$$\chi(\mathbf{h}, \mathbf{h}^+) \to \mathrm{Tr}\left[(\hat{c}_i^\dagger - ih_i^+)\exp\;(i\hat{\mathbf{c}} \cdot \mathbf{h}^+)\,\hat{\rho}\exp\Big(i\mathbf{h} \cdot \hat{\mathbf{c}}^\dagger\Big)\,(\hat{c}_j - ih_j)\right]. \tag{F.58}$$

This is the sum of four terms

$$\begin{aligned}\chi(\mathbf{h}, \mathbf{h}^+)_1^\# &= \mathrm{Tr}\left[(\hat{c}_i^\dagger)\exp\,(i\hat{\mathbf{c}} \cdot \mathbf{h}^+)\,\hat{\rho}\;\exp\Big(i\mathbf{h} \cdot \hat{\mathbf{c}}^\dagger\Big)\,(\hat{c}_j)\right] \\ \chi(\mathbf{h}, \mathbf{h}^+)_2^\# &= \mathrm{Tr}\left[(-ih_i^+)\exp\,(i\hat{\mathbf{c}} \cdot \mathbf{h}^+)\,\hat{\rho}\;\exp\Big(i\mathbf{h} \cdot \hat{\mathbf{c}}^\dagger\Big)\,\hat{c}_j\right], \\ \chi(\mathbf{h}, \mathbf{h}^+)_3^\# &= \mathrm{Tr}\left[\hat{c}_i^\dagger\exp\,(i\hat{\mathbf{c}} \cdot \mathbf{h}^+)\,\hat{\rho}\;\exp\Big(i\mathbf{h} \cdot \hat{\mathbf{c}}^\dagger\Big)\,(-ih_j)\right], \\ \chi(\mathbf{h}, \mathbf{h}^+)_4^\# &= \mathrm{Tr}\left[(-ih_i^+)\exp\,(i\hat{\mathbf{c}} \cdot \mathbf{h}^+)\,\hat{\rho}\;\exp\Big(i\mathbf{h} \cdot \hat{\mathbf{c}}^\dagger\Big)\,(-ih_j)\right].\end{aligned} \tag{F.59}$$

Note that we are not yet in a position to take Grassmann variables outside the trace.

Using the cyclic properties of the trace and the fermion anticommutation rule, we have

$$\begin{aligned}\chi(\mathbf{h}, \mathbf{h}^+)_1^\# &= \mathrm{Tr}\left[\exp(i\hat{\mathbf{c}} \cdot \mathbf{h}^+)\,\hat{\rho}\;\exp\Big(i\mathbf{h} \cdot \hat{\mathbf{c}}^\dagger\Big)\,(\delta_{ij} - \hat{c}_i^\dagger \hat{c}_j)\right] \\ &= \delta_{ij}\chi(\mathbf{h}, \mathbf{h}^+) - \mathrm{Tr}\left[\hat{c}_j\exp(i\hat{\mathbf{c}} \cdot \mathbf{h}^+)\,\hat{\rho}\;\exp\Big(i\mathbf{h} \cdot \hat{\mathbf{c}}^\dagger\Big)\,\hat{c}_i^\dagger\right].\end{aligned} \tag{F.60}$$

It is then straightforward to show that

$$\begin{aligned}-\frac{\overrightarrow{\partial}}{\partial(ih_j^+)}\exp(i\hat{\mathbf{c}} \cdot \mathbf{h}^+) &= \hat{c}_j\exp(i\hat{\mathbf{c}} \cdot \mathbf{h}^+), \\ \exp\Big(i\mathbf{h} \cdot \hat{\mathbf{c}}^\dagger\Big)\left(-\frac{\overleftarrow{\partial}}{\partial(ih_i)}\right) &= \exp\Big(i\mathbf{h} \cdot \hat{\mathbf{c}}^\dagger\Big)\,\hat{c}_i^\dagger,\end{aligned} \tag{F.61}$$

so that

$$\begin{aligned}\chi(\mathbf{h},\mathbf{h}^+)_1^{\#} &= \delta_{ij}\chi(\mathbf{h},\mathbf{h}^+) \\ &\quad - \mathrm{Tr}\left\{\left[\left(-\frac{\overrightarrow{\partial}}{\partial(ih_j^+)}\right)\exp\left(i\hat{\mathbf{c}}\cdot\mathbf{h}^+\right)\right]\hat{\rho}\left[\exp\left(i\mathbf{h}\cdot\hat{\mathbf{c}}^\dagger\right)\left(-\frac{\overleftarrow{\partial}}{\partial(ih_i)}\right)\right]\right\} \\ &= \delta_{ij}\chi(\mathbf{h},\mathbf{h}^+) - \frac{\overrightarrow{\partial}}{\partial(ih_j^+)}\mathrm{Tr}\left[\exp\left(i\hat{\mathbf{c}}\cdot\mathbf{h}^+\right)\hat{\rho}\,\exp\left(i\mathbf{h}\cdot\hat{\mathbf{c}}^\dagger\right)\right]\frac{\overleftarrow{\partial}}{\partial(ih_i)} \\ &= \delta_{ij}\chi(\mathbf{h},\mathbf{h}^+) - \frac{\overrightarrow{\partial}}{\partial(ih_j^+)}\chi(\mathbf{h},\mathbf{h}^+)\frac{\overleftarrow{\partial}}{\partial(ih_i)}\,, \end{aligned} \tag{F.62}$$

where we have used the result for double Grassmann differentiation of a trace.

Using a similar approach, we can show that

$$\begin{aligned}\chi(\mathbf{h},\mathbf{h}^+)_2^{\#} &= \frac{\overrightarrow{\partial}}{\partial(ih_j^+)}\chi(\mathbf{h},\mathbf{h}^+)(+ih_i^+), \\ \chi(\mathbf{h},\mathbf{h}^+)_3^{\#} &= (+ih_j)\chi(\mathbf{h},\mathbf{h}^+)\frac{\overleftarrow{\partial}}{\partial(ih_i)}, \\ \chi(\mathbf{h},\mathbf{h}^+)_4^{\#} &= (+ih_i^+)\chi(\mathbf{h},\mathbf{h}^+)(+ih_j)\,. \end{aligned} \tag{F.63}$$

Collecting these terms together, we have

$$\begin{aligned}\chi(\mathbf{h},\mathbf{h}^+) \rightarrow \delta_{ij}\chi(\mathbf{h},\mathbf{h}^+) &- \frac{\overrightarrow{\partial}}{\partial(ih_j^+)}\chi(\mathbf{h},\mathbf{h}^+)\left(\frac{\overleftarrow{\partial}}{\partial(ih_i)}\right) + \frac{\overrightarrow{\partial}}{\partial(ih_j^+)}\chi(\mathbf{h},\mathbf{h}^+)(+ih_i^+) \\ &+ (+ih_j)\chi(\mathbf{h},\mathbf{h}^+)\frac{\overleftarrow{\partial}}{\partial(ih_i)} + (+ih_i^+)\chi(\mathbf{h},\mathbf{h}^+)(+ih_j), \end{aligned} \tag{F.64}$$

showing that the overall effect is to take the original characteristic function and apply to it an even number of Grassmann derivatives or multipliers.

We now express the characteristic function in terms of the distribution function,

$$\chi(\mathbf{h},\mathbf{h}^+) = \int d\mathbf{g}^+ d\mathbf{g}\,\exp\left(i\mathbf{g}\cdot\mathbf{h}^+\right)P(\mathbf{g},\mathbf{g}^+)\exp\left(i\{\mathbf{h}\cdot\mathbf{g}^+\right), \tag{F.65}$$

and carry out the required operations on it. We note that

$$\begin{aligned}
\frac{\overrightarrow{\partial}}{\partial(ih_j^+)} \exp\bigl(i\mathbf{g}\cdot\mathbf{h}^+\bigr) &= -g_j \exp\bigl(i\mathbf{g}\cdot\mathbf{h}^+\bigr),\\
\exp\bigl(i\mathbf{h}\cdot\mathbf{g}^+\bigr)\frac{\overleftarrow{\partial}}{\partial(ih_i)} &= \exp\bigl(i\mathbf{h}\cdot\mathbf{g}^+\bigr)\,(-g_i^+),\\
\frac{\overrightarrow{\partial}}{\partial g_i} \exp\bigl(i\mathbf{g}\cdot\mathbf{h}^+\bigr) &= ih_i^+ \exp\bigl(i\mathbf{g}\cdot\mathbf{h}^+\bigr),\\
\exp\bigl(i\mathbf{h}\cdot\mathbf{g}^+\bigr)\frac{\overleftarrow{\partial}}{\partial g_j^+} &= \exp\bigl(i\mathbf{h}\cdot\mathbf{g}^+\bigr)\,(ih_j).
\end{aligned} \tag{F.66}$$

On performing the Grassmann differentiations, we then find

$$\begin{aligned}
\chi(\mathbf{h},\mathbf{h}^+) \to &\int d\mathbf{g}^+ d\mathbf{g}\, \exp\bigl(i\mathbf{g}\cdot\mathbf{h}^+\bigr)\,\delta_{ij}P(\mathbf{g},\mathbf{g}^+)\exp\bigl(i\mathbf{h}\cdot\mathbf{g}^+\bigr)\,\chi(\mathbf{h},\mathbf{h}^+)\\
&-\int d\mathbf{g}^+ d\mathbf{g}\,(-g_j)\exp\bigl(i\mathbf{g}\cdot\mathbf{h}^+\bigr)\,P(\mathbf{g},\mathbf{g}^+)\exp\bigl(i\mathbf{h}\cdot\mathbf{g}^+\bigr)\,(-g_i^+)\\
&+\int d\mathbf{g}^+ d\mathbf{g}\,(-g_j)\exp\bigl(i\mathbf{g}\cdot\mathbf{h}^+\bigr)\,P(\mathbf{g},\mathbf{g}^+)\exp\bigl(i\mathbf{h}\cdot\mathbf{g}^+\bigr)\,(+ih_i^+)\\
&+\int d\mathbf{g}^+ d\mathbf{g}\,(+ih_j)\exp\bigl(i\mathbf{g}\cdot\mathbf{h}^+\bigr)\,P(\mathbf{g},\mathbf{g}^+)\exp\bigl(i\mathbf{h}\cdot\mathbf{g}^+\bigr)\,(-g_i^+)\\
&+\int d\mathbf{g}^+ d\mathbf{g}\,(+ih_i^+)\exp\bigl(i\{\mathbf{g}\cdot\mathbf{h}^+\bigr)\,P(\mathbf{g},\mathbf{g}^+)\exp\bigl(i\mathbf{h}\cdot\mathbf{g}^+\bigr)\,(+ih_j)\,.
\end{aligned} \tag{F.67}$$

So, after moving the Grassmann variables, we get

$$\begin{aligned}
\chi(\mathbf{h},\mathbf{h}^+) \to &\int d\mathbf{g}^+ d\mathbf{g}\, \exp\bigl(i\mathbf{g}\cdot\mathbf{h}^+\bigr)\,\delta_{ij}P(\mathbf{g},\mathbf{g}^+)\exp\bigl(i\mathbf{h}\cdot\mathbf{g}^+\bigr)\,\chi(\mathbf{h},\mathbf{h}^+)\\
&+\int d\mathbf{g}^+ d\mathbf{g}\, \exp\bigl(i\mathbf{g}\cdot\mathbf{h}^+\bigr)\,g_i^+P(\mathbf{g},\mathbf{g}^+)g_j\exp\bigl(i\mathbf{h}\cdot\mathbf{g}^+\bigr)\\
&+\int d\mathbf{g}^+ d\mathbf{g}\,(+ih_i^+)\exp\bigl(i\mathbf{g}\cdot\mathbf{h}^+\bigr)\,P(\mathbf{g},\mathbf{g}^+)g_j\exp\bigl(i\mathbf{h}\cdot\mathbf{g}^+\bigr)\\
&+\int d\mathbf{g}^+ d\mathbf{g}\, \exp\bigl(i\mathbf{g}\cdot\mathbf{h}^+\bigr)\,g_i^+P(\mathbf{g},\mathbf{g}^+)\exp\bigl(i\mathbf{h}\cdot\mathbf{g}^+\bigr)\,(+ih_j)\\
&+\int d\mathbf{g}^+ d\mathbf{g}\,(+ih_i^+)\exp\bigl(i\mathbf{g}\cdot\mathbf{h}^+\bigr)\,P(\mathbf{g},\mathbf{g}^+)\exp\bigl(i\{\mathbf{h}\cdot\mathbf{g}^+\bigr)\,(+ih_j)
\end{aligned} \tag{F.68}$$

and, using the results in (F.66), we then have

$$\begin{aligned}\chi(\mathbf{h},\mathbf{h}^+) \to &\int d\mathbf{g}^+ d\mathbf{g}\, \exp(i\mathbf{g}\cdot\mathbf{h}^+)\, \delta_{ij} P(\mathbf{g},\mathbf{g}^+) \exp(i\mathbf{h}\cdot\mathbf{g}^+)\, \chi(\mathbf{h},\mathbf{h}^+)\\
&+ \int d\mathbf{g}^+ d\mathbf{g}\, \exp(i\mathbf{g}\cdot\mathbf{h}^+)\, g_i^+ P(\mathbf{g},\mathbf{g}^+) g_j) \exp(i\mathbf{h}\cdot\mathbf{g}^+)\\
&+ \int d\mathbf{g}^+ d\mathbf{g} \left[\frac{\overrightarrow{\partial}}{\partial g_i} \exp(i\mathbf{g}\cdot\mathbf{h}^+)\right] P(\mathbf{g},\mathbf{g}^+) g_j \exp(i\mathbf{h}\cdot\mathbf{g}^+)\\
&+ \int d\mathbf{g}^+ d\mathbf{g}\, \exp(i\mathbf{g}\cdot\mathbf{h}^+)\, g_i^+ P(\mathbf{g},\mathbf{g}^+) \left[\exp(i\mathbf{h}\cdot\mathbf{g}^+) \frac{\overleftarrow{\partial}}{\partial g_j^+}\right]\\
&+ \int d\mathbf{g}^+ d\mathbf{g} \left[\frac{\overrightarrow{\partial}}{\partial g_i} \exp(i\{\mathbf{g}\cdot\mathbf{h}^+)\right] P(\mathbf{g},\mathbf{g}^+) \left[\exp(i\mathbf{h}\cdot\mathbf{g}^+) \frac{\overleftarrow{\partial}}{\partial g_j^+}\right]. \qquad \text{(F.69)}\end{aligned}$$

Next we use Grassmann integration by parts to move the derivatives onto the distribution function to give

$$\begin{aligned}\chi(\mathbf{h},\mathbf{h}^+) \to &\int d\mathbf{g}^+ d\mathbf{g}\, \exp(i\mathbf{g}\cdot\mathbf{h}^+)\, \delta_{ij} P(\mathbf{g},\mathbf{g}^+) \exp(i\mathbf{h}\cdot\mathbf{g}^+)\\
&+ \int d\mathbf{g}^+ d\mathbf{g}\, \exp(i\mathbf{g}\cdot\mathbf{h}^+)\, g_i^+ P(\mathbf{g},\mathbf{g}^+) g_j \exp(i\mathbf{h}\cdot\mathbf{g}^+)\\
&+ \int d\mathbf{g}^+ d\mathbf{g}\, \exp(i\mathbf{g}\cdot\mathbf{h}^+) \left[-\frac{\overrightarrow{\partial}}{\partial g_i} P(\mathbf{g},\mathbf{g}^+) g_j\right] \exp(i\mathbf{h}\cdot\mathbf{g}^+)\\
&+ \int d\mathbf{g}^+ d\mathbf{g}\, \exp(i\mathbf{g}\cdot\mathbf{h}^+) \left[-g_i^+ P(\mathbf{g},\mathbf{g}^+) \frac{\overleftarrow{\partial}}{\partial g_j^+}\right] \exp(i\mathbf{h}\cdot\mathbf{g}^+)\\
&+ \int d\mathbf{g}^+ d\mathbf{g}\, \exp(i\mathbf{g}\cdot\mathbf{h}^+) \left[\frac{\overrightarrow{\partial}}{\partial g_i} P(\mathbf{g},\mathbf{g}^+) \frac{\overleftarrow{\partial}}{\partial g_j^+}\right] \exp(i\mathbf{h}\cdot\mathbf{g}^+). \qquad \text{(F.70)}\end{aligned}$$

Finally, we can simplify this by making use of the following:

$$\begin{aligned}\left(\frac{\overrightarrow{\partial}}{\partial g_i} - g_i^+\right) P(\mathbf{g},\mathbf{g}^+) \left(\frac{\overleftarrow{\partial}}{\partial g_j^+} - g_j\right) &= \frac{\overrightarrow{\partial}}{\partial g_i} \left[P(\mathbf{g},\mathbf{g}^+) \frac{\overleftarrow{\partial}}{\partial g_j^+}\right] - \frac{\overrightarrow{\partial}}{\partial g_i} \left[P(\mathbf{g},\mathbf{g}^+) g_j\right]\\
&\quad - g_i^+ \left[P(\mathbf{g},\mathbf{g}^+) \frac{\overleftarrow{\partial}}{\partial g_j^+}\right] + g_i^+ P(\mathbf{g},\mathbf{g}^+) g_j\\
&= \frac{\overrightarrow{\partial}}{\partial g_i} P(\mathbf{g},\mathbf{g}^+) \frac{\overleftarrow{\partial}}{\partial g_j^+} - \frac{\overrightarrow{\partial}}{\partial g_i} \left[P(\mathbf{g},\mathbf{g}^+) g_j\right]\\
&\quad - g_i^+ P(\mathbf{g},\mathbf{g}^+) \frac{\overleftarrow{\partial}}{\partial g_j^+} + \delta_{ij} P(\mathbf{g},\mathbf{g}^+)\\
&\quad + g_i^+ P(\mathbf{g},\mathbf{g}^+) g_j\,. \qquad \text{(F.71)}\end{aligned}$$

In this way, we arrive at

$$\chi(\mathbf{h},\mathbf{h}^{+}) \rightarrow \int d\mathbf{g}^{+}d\mathbf{g}\,\exp(i\mathbf{g}\cdot\mathbf{h}^{+}) \\ \times\left(\frac{\overrightarrow{\partial}}{\partial g_i}-g_i^{+}\right)P(\mathbf{g},\mathbf{g}^{+})\left(\frac{\overleftarrow{\partial}}{\partial g_j^{+}}-g_j\right)\exp(i\mathbf{h}\cdot\mathbf{g}^{+}), \qquad \text{(F.72)}$$

which we recognise as the correct correspondence rule for $\hat{\rho}\rightarrow\hat{c}_i^{\dagger}\hat{\rho}\hat{c}_j$.

F.1.6 Boson Case – Canonical-Distribution Rules

To obtain the bosonic canonical correspondence rules from the standard correspondence rules, we note that

$$\frac{\partial}{\partial\alpha_i}=\frac{1}{2}\left(\frac{\partial}{\partial\gamma_i}+\frac{\partial}{\partial\delta_i}\right),\qquad \frac{\partial}{\partial\alpha_i^{+}}=\frac{1}{2}\left(\frac{\partial}{\partial\gamma_i^{*}}-\frac{\partial}{\partial\delta_i^{*}}\right),$$
$$\frac{\partial}{\partial\alpha_i^{*}}=\frac{1}{2}\left(\frac{\partial}{\partial\gamma_i^{*}}+\frac{\partial}{\partial\delta_i^{*}}\right),\qquad \frac{\partial}{\partial\alpha_i^{+*}}=\frac{1}{2}\left(\frac{\partial}{\partial\gamma_i}-\frac{\partial}{\partial\delta_i}\right) \qquad \text{(F.73)}$$

and that

$$\left(\frac{\partial}{\partial\delta_i^{*}}+\delta_i\right)\exp\left(-\sum_i\delta_i\delta_i^{*}\right)=0=\left(\frac{\partial}{\partial\delta_i}+\delta_i^{*}\right)\exp\left(-\sum_i\delta_i\delta_i^{*}\right). \qquad \text{(F.74)}$$

Using (8.8), we see that

$$\alpha_i+2\frac{\partial}{\partial\alpha_i^{*}}=\left(\frac{\partial}{\partial\gamma_i^{*}}+\gamma_i\right)+\left(\frac{\partial}{\partial\delta_i^{*}}+\delta_i\right),$$
$$\left(\alpha_i-\frac{\partial}{\partial\alpha_i^{+}}\right)+2\frac{\partial}{\partial\alpha_i^{*}}=\gamma_i+\left(\frac{\partial}{\partial\delta_i^{*}}+\delta_i\right),$$
$$\left(\alpha_i^{+}-\frac{\partial}{\partial\alpha_i}\right)+2\frac{\partial}{\partial\alpha_i^{+*}}=\gamma_i^{*}-\left(\frac{\partial}{\partial\delta_i}+\delta_i^{*}\right),$$
$$\alpha_i^{+}+2\frac{\partial}{\partial\alpha_i^{+*}}=\left(\frac{\partial}{\partial\gamma_i}+\gamma_i^{*}\right)-\left(\frac{\partial}{\partial\delta_i}+\delta_i^{*}\right). \qquad \text{(F.75)}$$

When these are applied to the canonical form (8.15) for the distribution function, the correspondence rules (8.18) follow. The effect of the operators involving δ_i, δ_i^{*} gives zero.

F.2 Successive Correspondence Rules

We wish to consider situations where $\hat{\rho}$ is replaced by its product with several boson or fermion operators. To proceed further it is convenient to consider a general operator $\hat{\sigma}$, where $\hat{\sigma}=\hat{\rho},\,\hat{c}_i\hat{\rho},\,\hat{\rho}\hat{c}_i,\,\hat{c}_i^{\dagger}\hat{\rho},\,\hat{\rho}\hat{c}_i^{\dagger},\hat{c}_i\,\hat{\rho}\hat{c}_j^{\dagger},\hat{c}_j^{\dagger}\hat{c}_i\,\hat{\rho},\hat{\rho}\hat{c}_j^{\dagger}\hat{c}_i,\cdots$ as appropriate. In general, $\hat{\sigma}$ involves cases where $\hat{\rho}$ is replaced by successive products. As in the single-operator case, we can determine the outcome via the use of canonical forms for the density operator or using the characteristic-function approach. We can define a general characteristic function $\kappa(\boldsymbol{\xi},\boldsymbol{\xi}^+,\mathbf{h},\mathbf{h}^+)$ via

$$\kappa(\boldsymbol{\xi},\boldsymbol{\xi}^+,\mathbf{h},\mathbf{h}^+) = \mathrm{Tr}\left[\hat{\Omega}_{\mathrm{b}}^+(\boldsymbol{\xi}^+)\,\hat{\Omega}_{\mathrm{f}}^+(\mathbf{h}^+)\,\hat{\sigma}\,\hat{\Omega}_{\mathrm{f}}^-(\mathbf{h})\hat{\Omega}_{\mathrm{b}}^-(\boldsymbol{\xi})\right], \tag{F.76}$$

which is related to a general distribution function $S(\boldsymbol{\alpha},\boldsymbol{\alpha}^+,\boldsymbol{\alpha}^*,\boldsymbol{\alpha}^{+*},\mathbf{g},\mathbf{g}^+)$ via

$$\begin{aligned}\kappa(\boldsymbol{\xi},\boldsymbol{\xi}^+,\mathbf{h},\mathbf{h}^+) &= \int d\mathbf{g}^+ d\mathbf{g} \int d^2\boldsymbol{\alpha}^+ d^2\boldsymbol{\alpha}\, \exp(i\boldsymbol{\alpha}\cdot\boldsymbol{\xi}^+)\exp(i\mathbf{g}\cdot\mathbf{h}^+) \\ &\quad\times S(\boldsymbol{\alpha},\boldsymbol{\alpha}^+,\boldsymbol{\alpha}^*,\boldsymbol{\alpha}^{+*},\mathbf{g},\mathbf{g}^+)\exp(i\mathbf{h}\cdot\mathbf{g}^+)\exp(i\boldsymbol{\xi}\cdot\boldsymbol{\alpha}^+).\end{aligned} \tag{F.77}$$

The effect of the successive multiplications on the characteristic function can be determined via their effect on the factors $\hat{\Omega}_{\mathrm{b}}^+(\boldsymbol{\xi}^+)\,\hat{\Omega}_{\mathrm{f}}^+(\mathbf{h}^+)$ and $\hat{\Omega}_{\mathrm{f}}^-(\mathbf{h})\,\hat{\Omega}_{\mathrm{b}}^-(\boldsymbol{\xi})$ followed by taking the trace (fermion operators must be taken in pairs) and then working out the effect on the distribution function using (F.77).

The general operator $\hat{\sigma}$ will have a canonical representation in the standard form

$$\begin{aligned}\hat{\sigma} &= \int d\mathbf{g}^+ d\mathbf{g} \int\!\!\int d^2\boldsymbol{\alpha}^+ d^2\boldsymbol{\alpha}\, S_{\mathrm{canon}}(\boldsymbol{\alpha},\boldsymbol{\alpha}^+\boldsymbol{\alpha}^*,\boldsymbol{\alpha}^{+*},\mathbf{g},\mathbf{g}^+)\,\widehat{\Lambda}(\mathbf{g},\mathbf{g}^+,\boldsymbol{\alpha},\boldsymbol{\alpha}^+) \\ &= \int d\mathbf{g}^+ d\mathbf{g} \int\!\!\int d^2\boldsymbol{\alpha}^+ d^2\boldsymbol{\alpha}\, \widehat{\Lambda}(\mathbf{g},\mathbf{g}^+,\boldsymbol{\alpha},\boldsymbol{\alpha}^+) S_{\mathrm{canon}}(\boldsymbol{\alpha},\boldsymbol{\alpha}^+\boldsymbol{\alpha}^*,\boldsymbol{\alpha}^{+*},\mathbf{g},\mathbf{g}^+),\end{aligned} \tag{F.78}$$

where

$$\begin{aligned}\widehat{\Lambda}(\mathbf{g},\mathbf{g}^+,\boldsymbol{\alpha},\boldsymbol{\alpha}^+) &= \frac{{}_{\mathrm{B}}|\mathbf{g}\rangle\langle\mathbf{g}^{+*}|_{\mathrm{B}}}{\mathrm{Tr}({}_{\mathrm{B}}|\mathbf{g}\rangle\langle\mathbf{g}^{+*}|_{\mathrm{B}})}\,\frac{{}_{\mathrm{B}}|\boldsymbol{\alpha}\rangle\langle\boldsymbol{\alpha}^{+*}|_{\mathrm{B}}}{Tr({}_{\mathrm{B}}|\boldsymbol{\alpha}\rangle_{\mathrm{B}}\,{}_{\mathrm{B}}\langle\boldsymbol{\alpha}^{+*}|)} \\ &= \widehat{\Lambda}_{\mathrm{f}}(\mathbf{g},\mathbf{g}^+)\widehat{\Lambda}_{\mathrm{b}}(\boldsymbol{\alpha},\boldsymbol{\alpha}^+)\end{aligned} \tag{F.79}$$

are normalised projectors, and $\widehat{\Lambda}_{\mathrm{f}}(\mathbf{g},\mathbf{g}^+)$ and $\widehat{\Lambda}_{\mathrm{b}}(\boldsymbol{\alpha},\boldsymbol{\alpha}^+)$ are the usual fermionic and bosonic Bargmann projectors. The effect of successive multiplications with annihilation and creation operators can be determined by first applying the process to the Bargmann projectors using (8.29) and (8.33). These operators depend analytically on the $\boldsymbol{\alpha},\boldsymbol{\alpha}^+$ and are even Grassmann functions of the $\mathbf{g},\mathbf{g}^+$. The use of integration by parts then enables the effect on the canonical distribution function to be determined.

For the more general result in which an existing product $\hat{\sigma}$ of the density operator $\hat{\rho}$ with boson and fermion operators is replaced by $\hat{\sigma}$ times a *single* fermion annihilation or creation operator, the existing distribution function $S(\boldsymbol{\alpha},\boldsymbol{\alpha}^+,\boldsymbol{\alpha}^*,\boldsymbol{\alpha}^{+*},\mathbf{g},\mathbf{g}^+)$ is replaced as follows:

$$\hat{\sigma} \Rightarrow \hat{a}_i\,\hat{\sigma}, \qquad S(\boldsymbol{\alpha},\boldsymbol{\alpha}^+,\boldsymbol{\alpha}^*,\boldsymbol{\alpha}^{+*},\mathbf{g},\mathbf{g}^+) \Rightarrow \alpha_i S, \tag{F.80}$$

$$\hat{\sigma} \Rightarrow \hat{\sigma}\hat{a}_i, \qquad S(\boldsymbol{\alpha},\boldsymbol{\alpha}^+,\boldsymbol{\alpha}^*,\boldsymbol{\alpha}^{+*},\mathbf{g},\mathbf{g}^+) \Rightarrow \left(\alpha_i - \frac{\partial}{\partial\alpha_i^+}\right)S, \tag{F.81}$$

$$\hat{\sigma} \Rightarrow \hat{a}_i^\dagger\hat{\sigma}, \qquad S(\boldsymbol{\alpha},\boldsymbol{\alpha}^+,\boldsymbol{\alpha}^*,\boldsymbol{\alpha}^{+*},\mathbf{g},\mathbf{g}^+) \Rightarrow \left(\alpha_i^+ - \frac{\partial}{\partial\alpha_i}\right)S, \tag{F.82}$$

$$\hat{\sigma} \Rightarrow \hat{\sigma}\hat{a}_i^\dagger, \qquad S(\boldsymbol{\alpha}, \boldsymbol{\alpha}^+, \boldsymbol{\alpha}^*, \boldsymbol{\alpha}^{+*}, \mathbf{g}, \mathbf{g}^+) \Rightarrow \alpha_i^+ S, \tag{F.83}$$

$$\hat{\sigma} \Rightarrow \hat{c}_i\hat{\sigma}, \qquad S(\boldsymbol{\alpha}, \boldsymbol{\alpha}^+, \boldsymbol{\alpha}^*, \boldsymbol{\alpha}^{+*}, \mathbf{g}, \mathbf{g}^+) \Rightarrow g_i S = e(S) S g_i, \tag{F.84}$$

$$\begin{aligned} \hat{\sigma} \Rightarrow \hat{\sigma}\hat{c}_i, \qquad S(\boldsymbol{\alpha}, \boldsymbol{\alpha}^+, \boldsymbol{\alpha}^*, \boldsymbol{\alpha}^{+*}, \mathbf{g}, \mathbf{g}^+) &\Rightarrow S\left(+\frac{\overleftarrow{\partial}}{\partial g_i^+} - g_i\right) \\ &= e(S)\left(-\frac{\overrightarrow{\partial}}{\partial g_i^+} - g_i\right) S, \end{aligned} \tag{F.85}$$

$$\begin{aligned} \hat{\sigma} \Rightarrow \hat{c}_i^\dagger\hat{\sigma}, \qquad S(\boldsymbol{\alpha}, \boldsymbol{\alpha}^+, \boldsymbol{\alpha}^*, \boldsymbol{\alpha}^{+*}, \mathbf{g}, \mathbf{g}^+) &\Rightarrow \left(+\frac{\overrightarrow{\partial}}{\partial g_i} - g_i^+\right) S \\ &= e(S) S\left(-\frac{\overleftarrow{\partial}}{\partial g_i} - g_i^+\right), \end{aligned} \tag{F.86}$$

$$\hat{\sigma} \Rightarrow \hat{\sigma}\hat{c}_i^\dagger, \qquad S(\boldsymbol{\alpha}, \boldsymbol{\alpha}^+, \boldsymbol{\alpha}^*, \boldsymbol{\alpha}^{+*}, \mathbf{g}, \mathbf{g}^+) \Rightarrow S g_i^+ = e(S)\, g_i^+ S, \tag{F.87}$$

where $e(S) = +1$ if $S(\boldsymbol{\alpha}, \boldsymbol{\alpha}^+, \boldsymbol{\alpha}^*, \boldsymbol{\alpha}^{+*}, \mathbf{g}, \mathbf{g}^+)$ is an even Grassmann function and $e(S) = -1$ if $S(\boldsymbol{\alpha}, \boldsymbol{\alpha}^+, \boldsymbol{\alpha}^*, \boldsymbol{\alpha}^{+*}, \mathbf{g}, \mathbf{g}^+)$ is odd.

The safest way to proceed for fermion operators is always to use the result which does not depend on whether $S(\boldsymbol{\alpha}, \boldsymbol{\alpha}^+, \boldsymbol{\alpha}^*, \boldsymbol{\alpha}^{+*}, \mathbf{g}, \mathbf{g}^+)$ is even or odd:

$$\hat{\sigma} \Rightarrow \hat{c}_i\hat{\sigma}, \qquad S(\boldsymbol{\alpha}, \boldsymbol{\alpha}^+, \boldsymbol{\alpha}^*, \boldsymbol{\alpha}^{+*}, \mathbf{g}, \mathbf{g}^+) \Rightarrow g_i S, \tag{F.88}$$

$$\hat{\sigma} \Rightarrow \hat{\sigma}\hat{c}_i, \qquad S(\boldsymbol{\alpha}, \boldsymbol{\alpha}^+, \boldsymbol{\alpha}^*, \boldsymbol{\alpha}^{+*}, \mathbf{g}, \mathbf{g}^+) \Rightarrow S\left(+\frac{\overleftarrow{\partial}}{\partial g_i^+} - g_i\right), \tag{F.89}$$

$$\hat{\sigma} \Rightarrow \hat{c}_i^\dagger\hat{\sigma}, \qquad S(\boldsymbol{\alpha}, \boldsymbol{\alpha}^+, \boldsymbol{\alpha}^*, \boldsymbol{\alpha}^{+*}, \mathbf{g}, \mathbf{g}^+) \Rightarrow \left(+\frac{\overrightarrow{\partial}}{\partial g_i} - g_i^+\right) S, \tag{F.90}$$

$$\hat{\sigma} \Rightarrow \hat{\sigma}\hat{c}_i^\dagger, \qquad S(\boldsymbol{\alpha}, \boldsymbol{\alpha}^+, \boldsymbol{\alpha}^*, \boldsymbol{\alpha}^{+*}, \mathbf{g}, \mathbf{g}^+) \Rightarrow S g_i^+. \tag{F.91}$$

For bosonic systems, a similar analysis gives

$$\hat{\sigma} \Rightarrow \hat{a}_i\hat{\sigma}, \qquad S(\boldsymbol{\alpha}, \boldsymbol{\alpha}^+, \boldsymbol{\alpha}^*, \boldsymbol{\alpha}^{+*}, \mathbf{g}, \mathbf{g}^+) \Rightarrow \alpha_i S, \tag{F.92}$$

$$\hat{\sigma} \Rightarrow \hat{\sigma}\hat{a}_i, \qquad S(\boldsymbol{\alpha}, \boldsymbol{\alpha}^+, \boldsymbol{\alpha}^*, \boldsymbol{\alpha}^{+*}, \mathbf{g}, \mathbf{g}^+) \Rightarrow \left(-\frac{\partial}{\partial \alpha_i^+} + \alpha_i\right) S, \tag{F.93}$$

$$\hat{\sigma} \Rightarrow \hat{a}_i^\dagger\hat{\sigma}, \qquad S(\boldsymbol{\alpha}, \boldsymbol{\alpha}^+, \boldsymbol{\alpha}^*, \boldsymbol{\alpha}^{+*}, \mathbf{g}, \mathbf{g}^+) \Rightarrow \left(-\frac{\partial}{\partial \alpha_i} + \alpha_i^+\right) S, \tag{F.94}$$

$$\hat{\sigma} \Rightarrow \hat{\sigma}\hat{a}_i^\dagger, \qquad S(\boldsymbol{\alpha}, \boldsymbol{\alpha}^+, \boldsymbol{\alpha}^*, \boldsymbol{\alpha}^{+*}, \mathbf{g}, \mathbf{g}^+) \Rightarrow \alpha_i^+ S. \tag{F.95}$$

There are many options in applying the boson correspondence rules, owing to the analytic features of Bargmann projectors or characteristic functions.

Appendix G
Langevin Equations

G.1 Stochastic Averages

G.1.1 Basic Concepts

Consider first a c-number variable x for which there is defined a phase space quasi-distribution function $P(x,t)$ such that at time t the average value of some function $F(x)$ of the c-number variable is given by

$$\langle F(x)\rangle_t = \int dx\; F(x)P(x,t)\,. \tag{G.1}$$

For simplicity, the notation here will refer only to a single variable x, but the notation and treatment are easily extended to the case where x is a multi-component variable $x \equiv \{x_1, \cdots, x_n\}$. This type of average has been applied to treating quantum correlation functions for bosonic systems, with $P_{\rm b}(\boldsymbol{\alpha}, \boldsymbol{\alpha}^+, \boldsymbol{\alpha}^*, \boldsymbol{\alpha}^{+*})$ a time-dependent positive-P-type quasi-distribution function in a double phase space $\boldsymbol{\alpha}, \boldsymbol{\alpha}^+$, with the integration over $d^2\boldsymbol{\alpha}\; d^2\boldsymbol{\alpha}^+ \equiv d\boldsymbol{\alpha}_x\; d\boldsymbol{\alpha}_y\; d\boldsymbol{\alpha}_x^+\; d\boldsymbol{\alpha}_y^+$ and where the function of interest is of the form $(\alpha_{m_q}\cdots\alpha_{m_1})(\alpha_{l_1}^+\cdots\alpha_{l_p}^+)$.

The stochastic average is essentially a replacement for the phase space average. At each time t, we consider an ensemble of representative values for the c-number variable x. We denote this stochastic version of x at time t as $\widetilde{x}(t)$ and list the m members of the ensemble as $\widetilde{x}_i(t)$, with $i = 1, 2, \cdots, m$. Then the stochastic average of $F(x)$ is denoted $\overline{F(\widetilde{x}(t))}$ and is defined by

$$\overline{F(\widetilde{x}(t))} = \frac{1}{m}\sum_{i=1}^{m} F(\widetilde{x}_i(t))\,. \tag{G.2}$$

This process involves determining the value of the function at time t for the sample member $\widetilde{x}_i(t)$ and then averaging over all members of the ensemble.

Clearly, if the stochastic average and the phase space average are to be the same, then the approach used to construct the ensemble of $\widetilde{x}_i(t)$ must be related to the approach used to determine the quasi-distribution function, and this must generate the same averages for any choice of F. Since the quasi-distribution function $P(x,t)$ is determined as a function of time from a Fokker–Planck equation, the equation determining the stochastic version of the phase variable $\widetilde{x}(t)$ must be determined from a related equation. Such an equation is the stochastic differential equation. For the case of Fokker–Planck equations involving only derivatives up to second order, the

Phase Space Methods for Degenerate Quantum Gases. First Edition. Bryan J. Dalton, John Jeffers, and Stephen M. Barnett.
Published in 2015 by Oxford University Press.

form of the stochastic differential equation for the $\widetilde{x}(t)$ is determined from the drift and diffusion matrices in the Fokker–Planck equation. The stochastic differential equation will also involve some random quantity $\Gamma(t)$ with properties suitable for generating an ensemble of $\widetilde{x}_i(t)$ that is equivalent to the quasi-distribution function. Two forms of the stochastic differential equation are known, one called the Stratonovich form and the other the Ito form, and it is the latter which we use here. The stochastic averaging approach can be used analytically in certain situations, but perhaps the main reason for replacing the phase space average by the stochastic-average is that numerical calculations of averages usually require much less computer memory and computer time, since the stochastic method uses a suitably representative sample of phase space points rather than the quasi-distribution function over much more of phase space.

The ideas of averaging can be extended to sums of functions $F_a(x)$, and

$$\begin{aligned}\left\langle \sum_a F_a(x) \right\rangle_t &= \sum_a \langle F_a(x) \rangle_t, \\ \overline{\sum_a F_a(\widetilde{x}(t))} &= \sum_a \overline{F_a(\widetilde{x}(t))}\,, \end{aligned} \tag{G.3}$$

so the average of the sum is the sum of the averages.

The average of a product $F(x)G(x)$ is also important, and

$$\begin{aligned}\langle F(x)G(x) \rangle_t &= \int dx \ F(x)G(x)P(x,t), \\ \overline{F(\widetilde{x}(t))G(\widetilde{x}(t))} &= \frac{1}{N}\sum_{i=1}^{N} F(\widetilde{x}_i(t))G(\widetilde{x}_i(t))\,. \end{aligned} \tag{G.4}$$

Combining this result with that for a sum, we find that the average of the sum of products is the sum of the average of each product:

$$\begin{aligned}\left\langle \sum_a (F_a(x)G_a(x)) \right\rangle_t &= \sum_a \langle F_a(x)G_a(x) \rangle_t, \\ \overline{\sum_a (F_a(\widetilde{x}(t))G_a(\widetilde{x}(t)))} &= \sum_a \overline{(F_a(\widetilde{x}(t))G_a(\widetilde{x}(t)))}\,. \end{aligned} \tag{G.5}$$

The case where the quantities F, G are uncorrelated is when the average of the product equals the product of the averages, i.e.

$$\begin{aligned}\langle F(x)G(x) \rangle_t &= \langle F(x) \rangle_t \langle G(x) \rangle_t, \\ \overline{F(\widetilde{x}(t))G(\widetilde{x}(t))} &= \overline{F(\widetilde{x}(t))}\ \overline{G(\widetilde{y}(t))}, \end{aligned} \tag{G.6}$$

and obviously some special feature of the quasi-distribution function and the functions is required for this to occur. One such situation is when the phase space involves two variables $x \to x, y$, the distribution function factorises as $P(x,t) \to p(x,t)q(y,t)$ and the functions are $F(x)$ and $G(y)$. In the phase version of the average $\langle F(x)G(y)\rangle_t$, the factorisation of the double-phase-space integral gives the uncorrelated result. In the stochastic version of the average, a double ensemble of $\widetilde{x}_i(t)$ and $\widetilde{y}_j(t)$ can be used, with $m_x m_y$ members overall. The stochastic average is then

$$\begin{aligned}
\overline{F(\widetilde{x}(t))G(\widetilde{y}(t))} &= \frac{1}{m_x m_y}\sum_{i=1}^{m_x}\sum_{j=1}^{m_y} F(\widetilde{x}_i(t))G(\widetilde{y}_j(t)) \\
&= \frac{1}{m_x}\sum_{i=1}^{m_x} F(\widetilde{x}_i(t))\frac{1}{m_y}\sum_{j=1}^{m_y} G(\widetilde{y}_j(t)) \\
&= \overline{F(\widetilde{x}(t))}\;\overline{G(\widetilde{y}(t))},
\end{aligned} \tag{G.7}$$

as required. In the case where each pair of functions in a product are uncorrelated, we have the result that

$$\begin{aligned}
\left\langle \sum_a (F_a(x)G_a(x)) \right\rangle_t &= \sum_a \langle F_a(x)\rangle_t \langle G_a(x)\rangle_t, \\
\overline{\sum_a (F_a(\widetilde{x}(t))G_a(\widetilde{x}(t)))} &= \sum_a \overline{F_a(\widetilde{x}(t))}\;\overline{G_a(\widetilde{x}(t))}.
\end{aligned} \tag{G.8}$$

The previous concepts can be extended to the case where the variable is a Grassmann number g. The concept of a quasi-distribution function $P(g,t)$ applies, with the average value at time t of a Grassmann function $F(g)$ being given by a Grassmann phase space integral:

$$\langle F(g)\rangle_t = \int dg\; F(g)P(g,t). \tag{G.9}$$

Again, for simplicity the notation here will refer only to a single variable g, but the notation and treatment are easily extended to the case where $\mathbf{g}$ is a multi-component variable $\mathbf{g} \equiv \{g_1,\ldots,g_n\}$. This type of average has been applied to treating quantum correlation functions for fermionic systems, with $P_{\mathrm{f}}(\mathbf{g},\mathbf{g}^+)$ being a time-dependent positive-P-type quasi-distribution function in a double phase space g, g^+, with the integration over $d\mathbf{g}^+ d\mathbf{g}$ and where the Grassmann function of interest is of the form $(g_{m_q}\cdots g_{m_1})(g_{l_1}^+\cdots g_{l_p}^+)$. In this case the distribution function is initially sandwiched between the two factors, and thus $\langle (g_{m_q}\cdots g_{m_1})(g_{l_1}^+\cdots g_{l_p}^+)\rangle_t = \int dg^+ dg\; (g_{m_q}\cdots g_{m_1})P_{\mathrm{f}}(\mathbf{g},\mathbf{g}^+,t)(g_{l_1}^+\cdots g_{l_p}^+)$. However, because P_{f} is even, the result is the same as

$$\int d\mathbf{g}^+ d\mathbf{g}(g_{m_q}\cdots g_{m_1})\;(g_{l_1}^+\cdots g_{l_p}^+)P_{\mathrm{f}}(\mathbf{g},\mathbf{g}^+,t).$$

The stochastic average is essentially a replacement for the phase space average. At each time t, we consider an ensemble of representative values for the Grassmann variable g. We denote this stochastic version of g at time t as $\widetilde{g}(t)$ and list the m

members of the ensemble as $\widetilde{g}_i(t)$, with $i = 1, 2, \cdots, m$. Then the stochastic average of $F(g)$ is denoted $\overline{F(\widetilde{g}(t))}$ and is defined by

$$\overline{F(\widetilde{g}(t))} = \frac{1}{m}\sum_{i=1}^{m} F(\widetilde{g}_i(t)). \tag{G.10}$$

Again, this process involves determining the Grassmann function at time t for the sample member $\widetilde{g}_i(t)$ and then averaging over all members of the ensemble.

In order that the stochastic and phase space averages coincide as a function of time for all $F(g)$, similar requirements to those in the c-number case apply. Thus the Fokker–Planck equation for the Grassmann quasi-distribution function will be related to a stochastic differential equation for the $\widetilde{g}(t)$. However, because of the anticommuting feature of Grassmann variables and the lack of any numerical realisation for Grassmann numbers, there will be differences from the c-number cases.

G.1.2 Gaussian–Markov Stochastic Process

Of particular importance in the specific forms of stochastic equations used in phase space theories are the Gaussian–Markov random noise terms $\Gamma_a(t)$, which have the properties

$$\begin{aligned}
\overline{\Gamma_a(t_1)} &= 0, \\
\overline{\Gamma_a(t_1)\Gamma_b(t_2)} &= \delta_{ab}\delta(t_1 - t_2), \\
\overline{\Gamma_a(t_1)\Gamma_b(t_2)\Gamma_c(t_3)} &= 0, \\
\overline{\Gamma_a(t_1)\Gamma_b(t_2)\Gamma_c(t_3)\Gamma_d(t_4)} &= \overline{\Gamma_a(t_1)\Gamma_b(t_2)}\,\overline{\Gamma_c(t_3)\Gamma_d(t_4)} + \overline{\Gamma_a(t_1)\Gamma_c(t_3)}\,\overline{\Gamma_b(t_2)\Gamma_d(t_4)} \\
&\quad + \overline{\Gamma_a(t_1)\Gamma_d(t_4)}\,\overline{\Gamma_b(t_2)\Gamma_c(t_3)}, \\
&\quad \cdots,
\end{aligned} \tag{G.11}$$

so that the stochastic averages of products of odd numbers of Γ are zero and the stochastic averages of products of even numbers of Γ are the sums of products of stochastic averages of pairs of Γ.

These functions are related to *Wiener stochastic functions*, defined by

$$\widetilde{w}_a(t) = \int_{t_0}^{t} dt_1\ \Gamma_a(t_1)\,. \tag{G.12}$$

Hence the Wiener stochastic functions satisfy the Ito stochastic differential equation

$$\delta\widetilde{w}_a(t) = \widetilde{w}_a(t + \delta t) - \widetilde{w}_a(t) = \int_{t}^{t+\delta t} dt_1\ \Gamma_a(t_1)\,. \tag{G.13}$$

This is a particular case of the integral form of the stochastic differential equation in (9.26), but with $\mathcal{A}^p = 0$ and $\mathcal{B}^p_a = \delta_{ap}$. The quantity $\delta\widetilde{w}_a(t)$ is known as the *Wiener increment.*

If we now consider the w_a as phase space variables with a quasi-distribution function $P(w, t)$ with $w \equiv \{w_1, w_2, \cdots, w_a, \cdots w_{2n}\}$, it then follows from the Fokker–Planck equation in (9.26) with $A_a = 0$ and, $F_{ab} = \sum_p \mathcal{B}_a^p \mathcal{B}_a^p = \delta_{ab}$ that the Fokker–Planck equation must be

$$\frac{\partial}{\partial t} P(w,t) = \frac{1}{2} \sum_a \frac{\partial^2}{\partial w_a^2} (P(w,t)) \,. \tag{G.14}$$

This is a Fokker–Planck equation with no drift terms, and equal diffusion terms for all components.

The solution to the Fokker–Planck equation for the Wiener distribution involves *Gaussian* functions. If the initial condition at time t_0 is

$$P(w,t_0) = \delta(w - w_0), \tag{G.15}$$

then

$$P(w,t) = [2\pi(t - t_0)]^{-n/2} \exp[-(w - w_0)^2/2(t - t_0)] \,. \tag{G.16}$$

This solution gives the following phase space averages:

$$\begin{aligned} \langle (w_a - w_{a0}) \rangle &= 0, \\ \langle (w_a - w_{a0})(w_b - w_{b0}) \rangle &= \delta_{ab}(t - t_0) \,. \end{aligned} \tag{G.17}$$

These results describe Brownian motion allowing for diffusion processes only.

G.2 Fluctuations

G.2.1 Boson Correlation Functions

From (9.4) and with

$$\begin{aligned} \frac{\partial}{\partial \alpha_{rx}} (\alpha_x^p + i\alpha_y^p)(\alpha_x^q + i\alpha_y^q) &= \delta_{rp}(\alpha_x^q + i\alpha_y^q) + (\alpha_x^p + i\alpha_y^p)\delta_{rq}, \\ \frac{\partial}{\partial \alpha_{ry}} (\alpha_x^p + i\alpha_y^p)(\alpha_x^q + i\alpha_y^q) &= i\{\delta_{rp}(\alpha_x^q + i\alpha_y^q) + (\alpha_x^p + i\alpha_y^p)\delta_{rq}\}, \\ \frac{\partial}{\partial \alpha_{rx}} \frac{\partial}{\partial \alpha_{sx}} (\alpha_x^p + i\alpha_y^p)(\alpha_x^q + i\alpha_y^q) &= \delta_{sp}\delta_{rq} + \delta_{rp}\delta_{sq}, \\ \frac{\partial}{\partial \alpha_{ry}} \frac{\partial}{\partial \alpha_{sx}} (\alpha_x^p + i\alpha_y^p)(\alpha_x^q + i\alpha_y^q) &= i\{\delta_{sp}\delta_{rq} + \delta_{rp}\delta_{sq}\}, \\ \frac{\partial}{\partial \alpha_{rx}} \frac{\partial}{\partial \alpha_{sy}} (\alpha_x^p + i\alpha_y^p)(\alpha_x^q + i\alpha_y^q) &= i\{\delta_{sp}\delta_{rq} + \delta_{rp}\delta_{sq}\}, \\ \frac{\partial}{\partial \alpha_{ry}} \frac{\partial}{\partial \alpha_{sy}} (\alpha_x^p + i\alpha_y^p)(\alpha_x^q + i\alpha_y^q) &= i^2\{\delta_{sp}\delta_{rq} + \delta_{rp}\delta_{sq}\}, \end{aligned} \tag{G.18}$$

we find for the time derivative of the phase space average

$$\begin{aligned}\frac{d}{dt}\langle\alpha_p\alpha_q\rangle_t &= \langle(\alpha_x^q + i\alpha_y^q)(A_x^p + iA_y^p) + (\alpha_x^p + i\alpha_y^p)(A_x^q + iA_y^q)\rangle \\ &\quad + \frac{1}{2}\langle(D_{xx}^{qp} + D_{xx}^{pq})\rangle + \frac{1}{2}i\langle(D_{yx}^{qp} + D_{yx}^{pq})\rangle \\ &\quad + \frac{1}{2}i\langle(D_{xy}^{qp} + D_{xy}^{pq})\rangle + \frac{1}{2}i^2\langle(D_{yy}^{qp} + D_{yy}^{pq})\rangle \\ &= \langle(\alpha_p A_q + A_p\alpha_q)\rangle + \langle F_{pq}\rangle, \end{aligned} \tag{G.19}$$

where (8.71), (8.76) and (8.67) have all been used.

Details for the time derivative of the correlation function $\langle\widehat{a}_j^\dagger\widehat{a}_i\rangle_t$ are

$$\begin{aligned}\frac{d}{dt}\left\langle\widehat{a}_j^\dagger\widehat{a}_i\right\rangle_t &= -\frac{1}{i\hbar}\sum_k k_{jk}^*\left\langle\alpha_i\alpha_k^+\right\rangle - \frac{1}{i\hbar}\sum_{mkl}\xi_{jm\,kl}^*\left\langle\alpha_i\alpha_k^+\alpha_l^+\alpha_m\right\rangle \\ &\quad + \sum_{mkl}(\{\Gamma_{mk\,jl}^* - \Gamma_{lj\,km}^*\}\left\langle\alpha_i\alpha_m^+\alpha_k\alpha_l^+\right\rangle - \Gamma_{lj\,km}^*\,\delta_{lk}\left\langle\alpha_i\alpha_m^+\right\rangle) \\ &\quad + \frac{1}{i\hbar}\sum_k k_{ik}\left\langle\alpha_k\alpha_j^+\right\rangle + \frac{1}{i\hbar}\sum_{mkl}\xi_{im\,kl}\left\langle\alpha_m^+\alpha_l\alpha_k\alpha_j^+\right\rangle \\ &\quad + \sum_{mkl}(\{\Gamma_{mk\,il} - \Gamma_{li\,km}\}\left\langle\alpha_l\alpha_k^+\alpha_m\alpha_j^+\right\rangle - \Gamma_{li\,km}\delta_{lk}\left\langle\alpha_m\alpha_j^+\right\rangle) \\ &\quad + 2\sum_{kl}\{\Gamma_{il\,jk}\}\left\langle\alpha_l^+\alpha_k\right\rangle \\ &= -\frac{1}{i\hbar}\sum_k k_{jk}^*\left\langle\alpha_i\alpha_k^+\right\rangle - \frac{1}{i\hbar}\sum_{mkl}\xi_{jm\,kl}^*\left\langle\alpha_i\alpha_m\alpha_k^+\alpha_l^+\right\rangle \\ &\quad + \sum_{mkl}(\{\Gamma_{mk\,jl}^* - \Gamma_{lj\,km}^*\}\left\langle\alpha_i\alpha_k\alpha_l^+\alpha_m^+\right\rangle - \Gamma_{lj\,km}^*\,\delta_{lk}\left\langle\alpha_i\alpha_m^+\right\rangle) \\ &\quad + \frac{1}{i\hbar}\sum_k k_{ik}\left\langle\alpha_k\alpha_j^+\right\rangle + \frac{1}{i\hbar}\sum_{mkl}\xi_{im\,kl}\left\langle\alpha_l\alpha_k\alpha_m^+\alpha_j^+\right\rangle \\ &\quad + \sum_{mkl}(\{\Gamma_{mk\,il} - \Gamma_{li\,km}\}\left\langle\alpha_l\alpha_m\alpha_j^+\alpha_k^+\right\rangle - \Gamma_{li\,km}\,\delta_{lk}\left\langle\alpha_m\alpha_j^+\right\rangle) \\ &\quad + 2\sum_{kl}\{\Gamma_{il\,jk}\}\left\langle\alpha_k\alpha_l^+\right\rangle. \end{aligned} \tag{G.20}$$

G.2.2 Fermion Correlation Functions

Details for the time derivative of the correlation function $\langle\widehat{c}_j^\dagger\widehat{c}_i\rangle_t$ are

$$\begin{aligned}
\frac{d}{dt}\left\langle \hat{c}_j^\dagger \hat{c}_i \right\rangle_t = & +\frac{1}{i\hbar}\sum_k h_{jk}^* \langle g_i g_k^+ \rangle - \frac{1}{i\hbar}\sum_{mkl} \nu_{jm\,kl}^* \langle g_i g_k^+ g_l^+ g_m \rangle \\
& + \sum_{mkl} \left(\{\Gamma_{mk\,jl}^* - \Gamma_{lj\,km}^*\} \langle g_i g_m^+ g_k g_l^+ \rangle + \Gamma_{lj\,km}^* \delta_{lk} \langle g_i g_m^+ \rangle \right) \\
& - \frac{1}{i\hbar}\sum_k h_{ik} \langle g_k g_j^+ \rangle + \frac{1}{i\hbar}\sum_{mkl} \nu_{im\,kl} \langle g_m^+ g_l g_k g_j^+ \rangle \\
& + \sum_{mkl} \left(\{\Gamma_{mk\,il} - \Gamma_{li\,km}\} \langle g_l g_k^+ g_m g_j^+ \rangle + \Gamma_{li\,km} \delta_{lk} \langle g_m g_j^+ \rangle \right) \\
& + 2\sum_{kl} \{\Gamma_{jk\,il}\} \langle g_l g_k^+ \rangle \\
= & +\frac{1}{i\hbar}\sum_k h_{jk}^* \langle g_i g_k^+ \rangle - \frac{1}{i\hbar}\sum_{mkl} \nu_{jm\,kl}^* \langle g_i g_m g_k^+ g_l^+ \rangle \\
& + \sum_{mkl} \left(\{\Gamma_{mk\,jl}^* - \Gamma_{lj\,km}^*\} \langle g_k g_i g_m^+ g_l^+ \rangle + \Gamma_{lj\,km}^* \delta_{lk} \langle g_i g_m^+ \rangle \right) \\
& - \frac{1}{i\hbar}\sum_k h_{ik} \langle g_k g_j^+ \rangle + \frac{1}{i\hbar}\sum_{mkl} \nu_{im\,kl} \langle g_l g_k g_m^+ g_j^+ \rangle \\
& + \sum_{mkl} \left(\{\Gamma_{mk\,il} - \Gamma_{li\,km}\} \langle g_m g_l g_k^+ g_j^+ \rangle + \Gamma_{li\,km} \delta_{lk} \langle g_m g_j^+ \rangle \right) \\
& + 2\sum_{kl} \{\Gamma_{jk\,il}\} \langle g_l g_k^+ \rangle . \qquad \text{(G.21)}
\end{aligned}$$

Appendix H
Functional Calculus for Restricted Boson and Fermion Fields

H.1 General Features

It is necessary also to consider functionals involving c-number and Grassmann field functions $\psi^K(x)$ which are still based on an expansion in terms of orthonormal mode functions $\phi_k(x)$ but where there is some restriction on the modes that are included. Such functions will be referred to as *restricted functions*. Examples include the fields $\psi_{\mathrm{C}}(\mathbf{r}), \psi_{\mathrm{C}}^{+}(\mathbf{r}), \psi_{\mathrm{NC}}(\mathbf{r}), \psi_{\mathrm{NC}}^{+}(\mathbf{r})$ used for condensate and non-condensate modes in the theory of Bose condensates [52], where even the combined condensate and non-condensate modes are subject to a restriction, in that modes associated with a momentum greater than a cut-off value are excluded. The definitions of functionals are similar, and again the main processes are functional differentiation and integration. Many of the results are similar to those presented in Chapter 11, but now a new restricted Dirac delta function appears in the theory.

H.2 Functionals for Restricted C-Number Fields

H.2.1 Restricted Functions

The expansion for the field function is

$$\psi^K(x) = \sum_k^K \beta_k \phi_k(x), \tag{H.1}$$

where the specific restricted mode expansion for the *restricted set* K is signified by the symbol K. Other restricted sets involving different modes will be designated L, M etc., with expansion coefficients γ_k, δ_k etc.

Orthonormality conditions still apply to all modes, i.e.

$$\int dx\ \phi_k^*(x)\phi_l(x) = \delta_{kl}, \tag{H.2}$$

and this gives the well-known result for the expansion coefficients

$$\beta_k = \int dx\ \phi_k^*(x)\psi^K(x). \tag{H.3}$$

However, the completeness relationship is now

$$\sum_{k}^{K} \phi_k(y)\phi_k^*(x) = \delta_K(y,x), \tag{H.4}$$

which defines the *restricted delta function* $\delta_K(y,x)$ for the set K. This is a function of two variables x and y, and does not depend on $y - x$.

The restricted delta functions have some interesting properties:

$$\begin{aligned}\int dx\, dy\; \phi_l^*(y)\delta_K(y,x)\phi_m(x) &= \delta_{l,m} \qquad (l, m \in K),\\ &= 0 \qquad (l \notin K, m \notin K),\end{aligned} \tag{H.5}$$

$$\begin{aligned}\int dx\; \delta_K(y,x)\delta_L(x,z) &= \int dx \sum_{k}^{K} \phi_k(y)\phi_k^*(x) \sum_{l}^{L} \phi_l(x)\phi_l^*(z)\\ &= \sum_{k}^{K} \phi_k(y)\delta_{k,l} \sum_{l}^{L} \phi_l^*(z)\\ &= \delta_{K,L}\delta_K(y,z)\end{aligned} \tag{H.6}$$

and

$$\int dx\; \delta_K(x,x) = N_K, \tag{H.7}$$

where N_K is the number of mode functions in the set K. Unlike the normal delta function, the restricted delta functions are non-singular and can be treated as standard c-number functions within expressions.

H.2.2 Functionals

As with general functions, a *functional* $F[\psi^K(x)]$ of restricted functions $\psi^K(x)$ maps the *c-number function* $\psi^K(x)$ onto a *c-number* that depends on *all* the values of $\psi^K(x)$ over its entire range.

The *restricted function* $\psi^K(y)$ can be expressed as a functional $F_y[\psi^K(x)]$ of the restricted function $\psi^K(x)$. In terms of the restricted delta function, we have

$$\begin{aligned}\psi^K(y) &= \int dx\; \delta_K(y,x)\psi^K(x)\\ &= F_y[\psi^K(x)],\end{aligned} \tag{H.8}$$

showing how $\psi^K(y)$ can still be written as a functional $F_y[\psi^K(x)]$ of $\psi^K(x)$, but now (H.8) applies, which involves the restricted delta function $\delta_K(y,x)$ as a kernel, rather than (11.18), which involved the normal delta function and applied to functions $\psi(x)$ with unrestricted mode expansions.

The *spatial derivative* $\nabla_y \psi^K(y)$ of the restricted function $\psi^K(y)$ can also be expressed as a functional $F_{\nabla y}[\psi^K(x)]$ of $\psi^K(x)$. Using (H.8), we have

$$\begin{aligned}\nabla_y \psi^K(y) &= \int dx\ \nabla_y \delta_K(y,x)\psi^K(x) \\ &= F_{\nabla y}[\psi^K(x)],\end{aligned} \tag{H.9}$$

which now involves $\nabla_y \delta_K(y,x)$ as a kernel. We can confirm the validity of (H.9), by substituting for $\psi^K(x)$ from (H.1), which gives

$$\begin{aligned}\int dx\ \nabla_y \delta_K(y,x)\psi^K(x) &= \sum_k^K \beta_k \int dx\ \nabla_y \delta_K(y,x)\phi_k(x) \\ &= \sum_k^K \beta_k \sum_l^K \int dx\ \nabla_y \phi_l(y)\phi_l^*(x)\phi_k(x) \\ &= \sum_k^K \beta_k \nabla_y \phi_k(y) \\ &= \nabla_y \psi^K(y),\end{aligned}$$

as required.

As the value of the function at any point in the range for x is determined uniquely by the expansion coefficients $\{\beta_k\}$, the functional $F[\psi^K(x)]$ must therefore also just depend on the c-number expansion coefficients, and hence may also be viewed as a *function* $g(\beta_1, \beta_2, \cdots, \beta_k, \cdots, \beta_n)$ of the *expansion coefficients*, a useful equivalence when functional differentiation and integration are considered:

$$F[\psi^K(x)] \equiv g(\beta_1, \beta_2, \cdots, \beta_k, \cdots, \beta_n). \tag{H.10}$$

H.2.3 Related Restricted Sets

We may also consider restricted functions based on the *conjugate modes*. This set will be referred to as K^* or K^+. Thus the previous equations become

$$\psi^{K+}(x) = \sum_k^K \phi_k^*(x)\beta_k^+, \tag{H.11}$$

$$\beta_k^+ = \int dx\ \phi_k(x)\psi^{K+}(x), \tag{H.12}$$

$$\delta_{K+}(y,x) = \sum_k^K \phi_k^*(y)\phi_k(x), \tag{H.13}$$

where the last equation defines the restricted delta function for the K^+ case. We note that the restricted delta function $\delta_{K+}(y,x)$ is related to the previous one via

$$\delta_{K+}(y,x) = \delta_K(x,y). \tag{H.14}$$

We can again write the restricted function $\psi^{K+}(y)$ as a functional $F_y[\psi^{K+}(x)]$ via

$$\begin{aligned}\psi^{K+}(y) &= \int dx\ \delta_{K+}(y,x)\psi^{K+}(x) \\ &= \int dx\ \delta_K(x,y)\psi^{K+}(y) \\ &= F_y[\psi^{K+}(x)]. \end{aligned} \tag{H.15}$$

Similarly, the *spatial derivative* $\nabla_y\psi^{K+}(y)$ of the restricted function is also a functional $F_{\nabla_y}[\psi^{K+}(x)]$, given by

$$\begin{aligned}\nabla_y\psi^{K+}(y) &= \int dx\ \nabla_y\delta_{K+}(y,x)\psi^{K+}(x) \\ &= \int dx\ \nabla_y\delta_K(x,y)\psi^{K+}(x) \\ &= F_{\nabla_y}[\psi^{K+}(x)]. \end{aligned} \tag{H.16}$$

Note that, considered as a function of y, the restricted delta function $\delta_K(y,x)$ is a member of the set K of restricted functions $\psi^K(y)$ (the expansion coefficients are $\phi_k^*(x)$). On the other hand, considered as a function of x, the restricted delta function $\delta_K(y,x)$ is a member of the *conjugate set* K^+ of mode functions $\phi_k^*(x)$ (the expansion coefficients are $\phi_k(y)$).

As the value of the function at any point in the range for x is determined uniquely by the expansion coefficients $\{\beta_k^+\}$, the functional $F[\psi^{K+}(x)]$ must therefore also just depend on the c-number expansion coefficients, and hence may also be viewed as a *function* $g^+(\beta_1^+,\beta_2^+,\cdots,\beta_k^+,\cdots,\beta_n^+)$ of the expansion coefficients, a useful equivalence when functional differentiation and integration are considered:

$$F[\psi^{K+}(x)] \equiv g^+(\beta_1^+,\beta_2^+,\cdots,\beta_k^+,\cdots,\beta_n^+). \tag{H.17}$$

A second related restricted set is the *complementary set* $\overline{K}$, which includes all of the other orthonormal mode functions not included in the set K.

Clearly, *any* function can be expanded in terms of modes in the restricted sets K and $\overline{K}$. Thus we now have

$$\psi(x) = \sum_k^K \gamma_k\phi_k(x) + \sum_k^{\overline{K}} \gamma_k\phi_k(x), \tag{H.18}$$

$$\gamma_k = \int dx\ \phi_k^*(x)\psi(x), \qquad k \in K, \overline{K}, \tag{H.19}$$

$$\delta_{\overline{K}}(y,x) = \sum_k^{\overline{K}} \phi_k(y)\phi_k^*(x) = \sum_{L\neq K} \delta_L(y,x), \tag{H.20}$$

and now the *full Dirac delta function* is

$$\delta(y,x) = \sum_k^K \phi_k(y)\phi_k^*(x) + \sum_k^{\overline{K}} \phi_k(y)\phi_k^*(x). \tag{H.21}$$

The general function $\psi(y)$ may be written as a functional $F_y[\psi(x)]$ of $\psi(x)$ involving the full delta function:

$$\begin{aligned}\psi(y) &= \int dx\ \delta(y,x)\psi(x) \\ &= F_y[\psi(x)].\end{aligned} \tag{H.22}$$

Applying (H.6), we obtain the interesting result

$$\int dx\ \delta_K(y,x)\delta_{\overline{K}}(x,z) = 0. \tag{H.23}$$

Note that the full delta function is still written as a function of x and y. Because the total set of functions is still restricted, it will have a narrow though finite width and can be treated like a normal function.

H.3 Functional Differentiation for Restricted C-Number Fields

H.3.1 Definition of Functional Derivative

The *functional derivative* $\delta F[\psi^K(x)]/\delta\psi^K(x)$ is defined by

$$F[\psi^K(x) + \delta\psi^K(x)] \approx F[\psi^K(x)] + \int dx\ \delta\psi^K(x)\left(\frac{\delta F[\psi^K(x)]}{\delta\psi^K(x)}\right)_x, \tag{H.24}$$

where $\delta\psi^K(x)$ is a small change in $\psi^K(x)$. Since, as in (H.10), the functional is equivalent to a function of the expansion coefficients β_k, the only meaningful change to $\psi^K(x)$ would be associated with changes $\delta\beta_k$ in these expansion coefficients, and thus $\delta\psi^K(x)$ will be within the restricted function space K. In this equation, the left side is a functional of $\psi^K(x) + \delta\psi^K(x)$ and the first term on the right side is a functional of $\psi^K(x)$. The second term on the right side is a functional of $\delta\psi^K(x)$ and thus the functional derivative must be a function of x, hence the subscript x. In most situations this subscript will be left understood. Thus the functional derivative will be defined in terms of changes to the restricted function of the form

$$\delta\psi^K(x) = \sum_k^K \delta\beta_k\ \phi_k(x). \tag{H.25}$$

As the functional derivative is just a function of x, we can expand it in terms of *all* of the *conjugate* modes (these also form a full basis set of orthogonal functions):

$$\left(\frac{\delta F[\psi^K(x)]}{\delta\psi^K(x)}\right)_x = \sum_l \eta_l\phi_l^*(x).$$

Then we have

$$\int dx\ \delta\psi^K(x)\left(\frac{\delta F[\psi^K(x)]}{\delta\psi^K(x)}\right)_x = \sum_l \eta_l \int dx\ \delta\psi^K(x)\phi_l^*(x)$$
$$= \sum_k^K \eta_k\ \delta\beta_k,$$

as the contributions from modes l not in the set K^+ will be zero using orthogonality. This shows that any contribution to the functional derivative from modes $\phi_l^*(x)$ outside the set K^+ cannot contribute to $F[\psi^K(x) + \delta\psi^K(x)] - F[\psi^K(x)]$, and hence can be arbitrarily set to zero in determining the functional derivative with respect to restricted functions $\psi^K(x)$ in the set K. Thus we have

$$\left(\frac{\delta F[\psi^K(x)]}{\delta\psi^K(x)}\right)_x = \sum_k^K \eta_k\phi_k^*(x), \tag{H.26}$$

showing that the functional derivative is a function in the set $K^+ \equiv K^*$.

Noting that the function $\delta_K(x,y)$ is within the restricted function space, we may obtain a useful expression for the functional derivative by applying (H.8) for a function in the set K^*. Since $\delta_{K^+}(y,x) = \delta_K(x,y)$, this shows that the functional derivative may be obtained by choosing $\delta\psi^K(x) = \epsilon\delta_K(x,y)$ for small ϵ in the definition (H.24):

$$\left(\frac{\delta F[\psi^K(x)]}{\delta\psi^K(x)}\right)_y = \int dx\ \delta_{K^+}(y,x)\left(\frac{\delta F[\psi^K(x)]}{\delta\psi^K(x)}\right)_x$$
$$= \lim_{\epsilon\to 0}\left(\frac{F[\psi^K(x) + \epsilon\delta_K(x,y)] - F[\psi^K(x)]}{\epsilon}\right).$$

To confirm that the right side of the last equation does in fact give $\left(\delta F[\psi^K(x)]/\delta\psi^K(x)\right)_y$, we substitute from (H.26):

$$\int dx\ \delta_k(x,y)\left(\frac{\delta F[\psi^K(x)]}{\delta\psi^K(x)}\right)_x = \sum_k^K \eta_k \int dx\ \delta_K(x,y)\phi_k^*(x)$$
$$= \sum_k^K \eta_k \sum_l^K \phi_l^*(y)\int dx\ \phi_l(x)\phi_k^*(x)$$
$$= \sum_k^K \eta_k\phi_k^*(y)$$
$$= \left(\frac{\delta F[\psi^K(x)]}{\delta\psi^K(x)}\right)_y,$$

as required. Thus we have the useful expression for the functional derivative of a restricted function

$$\left(\frac{\delta F[\psi^K(x)]}{\delta\psi^K(x)}\right)_y = \lim_{\epsilon\to 0}\left(\frac{F[\psi^K(x)+\epsilon\delta_K(x,y)]-F[\psi^K(x)]}{\epsilon}\right). \tag{H.27}$$

We may also have functionals $F[\psi^K(x),\psi^L(x)]$ that involve two functions $\psi^K(x)$, $\psi^L(x)$ in two different restricted sets K, L. The straightforward generalisation of the useful result (H.27) is

$$\left(\frac{\delta F[\psi^K(x),\psi^L(x)]}{\delta\psi^K(x)}\right)_y = \lim_{\epsilon\to 0}\left(\frac{F[\psi^K(x)+\epsilon\delta_K(x,y),\psi^L(x)]-F[\psi^K(x),\psi^L(x)]}{\epsilon}\right), \tag{H.28}$$

$$\left(\frac{\delta F[\psi^K(x),\psi^L(x)]}{\delta\psi^L(x)}\right)_y = \lim_{\epsilon\to 0}\left(\frac{F[\psi^K(x),\psi^L(x)+\epsilon\delta_L(x,y)]-F[\psi^K(x),\psi^L(x)]}{\epsilon}\right). \tag{H.29}$$

Similar results apply for the functional derivative $\delta F[\psi^{K+}(x)]/\delta\psi^{K+}(x)$ with respect to the restricted function $\psi^{K+}(x)$ in the set $K^+\equiv K^*$, which is defined by

$$F[\psi^{K+}(x)+\delta\psi^{K+}(x)] \approx F[\psi^{K+}(x)] + \int dx\;\delta\psi^{K+}(x)\left(\frac{\delta F[\psi^{K+}(x)]}{\delta\psi^{K+}(x)}\right)_x, \tag{H.30}$$

where $\delta\psi^{K+}(x)$ is a small change in $\psi^{K+}(x)$. The function $\delta\psi^{K+}(x)$ is associated with changes $\delta\beta_k^+$ in these expansion coefficients, and thus $\delta\psi^{K+}(x)$ will be within the restricted function space K^+. We then have

$$\left(\frac{\delta F[\psi^{K+}(x)]}{\delta\psi^{K+}(x)}\right)_x = \sum_k^K \eta_k^+\phi_k(x), \tag{H.31}$$

showing that the functional derivative is a function in the set K. Also, the function $\delta_{K+}(x,y)$ is within the restricted function space, so we may obtain a useful expression for the functional derivative as

$$\left(\frac{\delta F[\psi^{K+}(x)]}{\delta\psi^{K+}(x)}\right)_y = \lim_{\epsilon\to 0}\left(\frac{F[\psi^{K+}(x)+\epsilon\delta_{K+}(x,y)]-F[\psi^{K+}(x)]}{\epsilon}\right) \tag{H.32}$$

$$= \lim_{\epsilon\to 0}\left(\frac{F[\psi^{K+}(x)+\epsilon\delta_K(y,x)]-F[\psi^{K+}(x)]}{\epsilon}\right). \tag{H.33}$$

H.3.2 Examples of Restricted Functional Derivatives

To obtain the functional derivative of the function $\psi^K(y)$ with respect to $\psi^K(x)$, we note that this derivative exists as a function of x in the restricted set $K^+\equiv K^*$ since $\psi^K(y)$ is also a functional: $\psi^K(y)=F_y[\psi^K(x)]$. We can thus use the expression (H.27):

$$
\begin{aligned}
\left(\frac{\delta}{\delta\psi^K(x)}\psi^K(y)\right)_x &= \lim_{\epsilon\to 0}\left(\frac{F_y[\psi^K(u)+\epsilon\delta_K(u,x)]-F_y[\psi^K(u)]}{\epsilon}\right) \\
&= \lim_{\epsilon\to 0}\left(\frac{\int du\ \delta_K(y,u)\{\psi^K(u)+\epsilon\delta_K(u,x)\}-\int du\ \delta_K(y,u)\psi^K(u)}{\epsilon}\right) \\
&= \lim_{\epsilon\to 0}\left(\int du\ \delta_K(y,u)\delta_K(u,x)\right) \\
&= \delta_K(y,x), \qquad\qquad (H.34)
\end{aligned}
$$

where (H.6) has been used. As noted before, considered as a function of x, the derivative of $\psi^K(y)$ with respect to $\psi^K(x)$ is in the restricted set K^*, but is in the set K considered as a function of y. This result is the modification of (11.27) for restricted functions.

A further result can be derived for when the functional derivative is with respect to $\psi^L(x)$ in a different restricted set, L. Applying (H.29), we get

$$
\begin{aligned}
\left(\frac{\delta}{\delta\psi^L(x)}\psi^K(y)\right)_x &= \lim_{\epsilon\to 0}\left(\frac{F_y[\psi^K(u),\psi^L(u)+\epsilon\delta_L(u,x)]-F_y[\psi^K(u),\psi^L(u)]}{\epsilon}\right) \\
&= \lim_{\epsilon\to 0}\left(\frac{\int du\ \delta_L(y,u)\psi^K(u)-\int du\ \delta_L(y,u)\psi^K(u)}{\epsilon}\right) \\
&= 0, \qquad\qquad (H.35)
\end{aligned}
$$

as the functional $\psi^K(y)=F_y[\psi^K(x),\psi^L(x)]$ does not involve $\psi^L(x)$ at all.

For the functional derivative of the *spatial derivative* $\nabla_y\psi^K(y)$ of the function $\psi^K(y)$ with respect to $\psi^K(x)$, we note that this derivative exists as a function of x in the restricted set $K^+\equiv K^*$, since $\nabla_y\psi^K(y)$ is also a functional: $\nabla_y\psi^K(y) = F_{\nabla y}[\psi^K(x)]$. We can thus use the expression (H.27):

$$
\begin{aligned}
&\left(\frac{\delta}{\delta\psi^K(x)}\nabla_y\psi^K(y)\right)_x \\
&= \lim_{\epsilon\to 0}\left(\frac{F_{\nabla y}[\psi^K(u)+\epsilon\delta_K(u,x)]-F_{\nabla y}[\psi^K(u)]}{\epsilon}\right) \\
&= \lim_{\epsilon\to 0}\left(\frac{\int du\ \nabla_y\delta_K(y,u)\{\psi^K(u)+\epsilon\delta_K(u,x)\}-\int du\ \nabla_y\delta_K(y,u)\psi^K(u)}{\epsilon}\right) \\
&= \lim_{\epsilon\to 0}\left(\int du\ \nabla_y\delta_K(y,u)\delta_K(u,x)\right) \\
&= \nabla_y\left(\int du\ \delta_K(y,u)\delta_K(u,x)\right) \\
&= \nabla_y\delta_K(y,x), \qquad\qquad (H.36)
\end{aligned}
$$

showing that the functional derivative involves a spatial derivative of the restricted delta function with respect to y. As pointed out previously, considered as a function of

x, the functional derivative is in the set K^+. Note also that we see that the functional-derivative and spatial-derivative processes can be carried out in either order:

$$\begin{aligned}\left(\frac{\delta}{\delta\psi^K(x)}\nabla_y\psi^K(y)\right)_x &= \nabla_y\left(\frac{\delta}{\delta\psi^K(x)}\psi^K(y)\right)_x \\ &= \nabla_y\delta_K(y,x). \end{aligned} \tag{H.37}$$

We also can obtain similar results for the function $\psi^{K+}(y)$, which is in the restricted set $K^+ \equiv K^*$ and can be written as a functional $F_y[\psi^{K+}(x)] \equiv \psi^{K+}(y)$. Thus

$$\begin{aligned}&\left(\frac{\delta}{\delta\psi^{K+}(x)}\psi^{K+}(y)\right)_x \\ &= \lim_{\epsilon\to 0}\left(\frac{F_y[\psi^{K+}(u)+\epsilon\delta_{K+}(u,x)] - F_y[\psi^{K+}(u)]}{\epsilon}\right) \\ &= \lim_{\epsilon\to 0}\left(\frac{\int du\ \delta_{K^+}(y,u)\{\psi^{K+}(u)+\epsilon\delta_{K+}(u,x)\} - \int du\ \delta_{K+}(y,u)\psi^{K+}(u)}{\epsilon}\right) \\ &= \lim_{\epsilon\to 0}\left(\int du\ \delta_{K+}(y,u)\delta_{K+}(u,x)\right) \\ &= \delta_{K+}(y,x) \\ &= \delta_K(x,y). \end{aligned} \tag{H.38}$$

For the spatial derivative $\nabla_y\psi^{K+}(y)$ of the function $\psi^{K+}(y)$, we have

$$\begin{aligned}&\left(\frac{\delta}{\delta\psi^{K+}(x)}\nabla_y\psi^{K+}(y)\right)_x \\ &= \lim_{\epsilon\to 0}\left(\frac{F_{\nabla y}[\psi^{K+}(u)+\epsilon\delta_{K+}(u,x)] - F_{\nabla y}[\psi^{K+}(u)]}{\epsilon}\right) \\ &= \lim_{\epsilon\to 0}\left(\frac{\int du\ \nabla_y\delta_{K+}(y,u)\{\psi^{K+}(u)+\epsilon\delta_{K+}(u,x)\} - \int du\ \nabla_y\delta_{K+}(y,u)\psi^{K+}(u)}{\epsilon}\right) \\ &= \lim_{\epsilon\to 0}\left(\int du\ \nabla_y\delta_{K+}(y,u)\delta_{K+}(u,x)\right) \\ &= \left(\int du\ \nabla_y\delta_K(u,y)\delta_k(x,u)\right) \\ &= \nabla_y\delta_K(x,y) \\ &= \nabla_y\delta_{K+}(y,x), \end{aligned} \tag{H.39}$$

as expected. Note also that

$$\begin{aligned}\left(\frac{\delta}{\delta\psi^{K+}(x)}\nabla_y\psi^{K+}(y)\right)_x &= \nabla_y\left(\frac{\delta}{\delta\psi^{K+}(x)}\psi^{K+}(y)\right)_x \\ &= \nabla_y\delta_{K+}(y,x). \end{aligned} \tag{H.40}$$

H.3.3 Restricted Functional Derivatives and Mode Functions

We can obtain an expression for the functional derivative $\left(\delta F[\psi^K(x)]/\delta\psi^K(x)\right)_x$ with respect to a restricted function $\psi^K(x)$ in terms of the ordinary derivatives of the function in (H.10) that is equivalent to the functional. Substituting from (H.25), we see that

$$F[\psi^K(x) + \delta\psi^K(x)] - F[\psi^K(x)] \approx \int dx\ \delta\psi^K(x) \left(\frac{\delta F[\psi^K(x)]}{\delta\psi^K(x)}\right)_x$$

$$\approx \sum_k^K \delta\beta_k \int dx\ \phi_k(x) \left(\frac{\delta F[\psi(x)]}{\delta\psi(x)}\right)_x,$$

but the left side is the same as

$$g(\beta_1 + \delta\beta_1, \cdots, \beta_k + \delta\beta_k, \cdots) - g(\beta_1, \cdots, \beta_k, \cdots) \approx \sum_k^K \delta\beta_k \frac{\partial g(\beta_1, \cdots, \beta_k, \cdots)}{\partial\beta_k}.$$

Equating the coefficients of the independent $\delta\alpha_k$ and then using the completeness relationship in (H.13) gives the key result

$$\left(\frac{\delta F[\psi^K(x)]}{\delta\psi^K(x)}\right)_x = \sum_k^K \phi_k^*(x) \frac{\partial g(\beta_1, \cdots, \beta_k, \cdots)}{\partial\beta_k}. \tag{H.41}$$

This relates the functional derivative to the mode functions and to the ordinary partial derivatives of the function $g(\beta_1, \cdots, \beta_k, \cdots, \beta_n)$ that was equivalent to the original functional $F[\psi^K(x)]$. Again, we see that the result is a function of x. Note that the functional derivative involves an expansion in terms of the conjugate mode functions $\phi_k^*(x)$ rather than the original modes $\phi_k(x)$. The last result can be put in the form of a useful *operator identity*

$$\left(\frac{\delta}{\delta\psi^K(x)}\right)_x = \sum_k^K \phi_k^*(x) \frac{\partial}{\partial\beta_k}, \tag{H.42}$$

where it is understood that the left side operates on an arbitrary functional $F[\psi^K(x)]$ of the restricted function $\psi^K(x)$ and the right side operates on the equivalent function $g(\beta_1, \cdots, \beta_k, \cdots)$.

We can obtain a similar expression for the functional derivative $\left(\delta F[\psi^{K+}(x)]/\delta\psi^{K+}(x)\right)_x$ with respect to a restricted function $\psi^{K+}(x)$ in the set K^+ in terms of the ordinary derivatives of the function (H.17) that is equivalent to the functional:

$$\left(\frac{\delta F[\psi^{K+}(x)]}{\delta\psi^{K+}(x)}\right)_x = \sum_k^K \phi_k(x) \frac{\partial g^+(\beta_1^+, \cdots, \beta_k^+, \cdots)}{\partial\beta_k^+}. \tag{H.43}$$

The last result can be put in the form of a useful operator identity

$$\left(\frac{\delta}{\delta\psi^{K+}(x)}\right)_x = \sum_k^K \phi_k(x)\frac{\partial}{\partial\beta_k^+}, \tag{H.44}$$

where it is understood that the left side operates on an arbitrary functional $F[\psi^{K+}(x)]$ of the restricted function $\psi^{K+}(x)$ and the right side operates on the equivalent function $g^+(\beta_1^+,\cdots,\beta_k^+,\cdots)$.

The *spatial derivative* of a *functional derivative* can be found from

$$\nabla_x\left(\frac{\delta F[\psi^K(x)]}{\delta\psi^K(x)}\right)_x = \sum_k^K\{\nabla_x\phi_k^*(x)\}\frac{\partial g(\beta_1,\cdots,\beta_k,\cdots)}{\partial\beta_k}, \tag{H.45}$$

$$\nabla_x\left(\frac{\delta F[\psi^{K+}(x)]}{\delta\psi^{K+}(x)}\right)_x = \sum_k^K\{\nabla_x\phi_k(x)\}\frac{\partial g^+(\beta_1^+,\cdots,\beta_k^+,\cdots)}{\partial\beta_k^+} \tag{H.46}$$

for the two cases of functionals of $\psi^K(x)$ or $\psi^{K+}(x)$. Clearly, the spatial derivative acts *only* on either the $\phi_k^*(x)$ or the $\phi_k(x)$. These results can also be put in the form of *operator identities*

$$\nabla_x\left(\frac{\delta}{\delta\psi^K(x)}\right)_x = \sum_k^K\{\nabla_x\phi_k^*(x)\}\frac{\partial}{\partial\beta_k}, \tag{H.47}$$

$$\nabla_x\left(\frac{\delta}{\delta\psi^{K+}(x)}\right)_x = \sum_k^K\{\nabla_x\phi_k(x)\}\frac{\partial}{\partial\beta_k^+}, \tag{H.48}$$

where it is understood that the left side operates on an arbitrary functional $F[\psi^K(x)]$ or $F[\psi^{K+}(x)]$ of the restricted function $\psi^K(x)$ or $\psi^{K+}(x)$, respectively, and the right side operates on the equivalent function $g(\beta_1,\cdots,\beta_k,\cdots)$ or $g^+(\beta_1^+,\cdots,\beta_k^+,\cdots)$. These operator forms are useful in deriving results for applying functional derivatives in succession.

As an example of applying these operator identities, consider the case of the functionals $F_y[\psi^K(x)] \equiv \psi^K(y) = \sum_k \beta_k\phi_k(y)$ and $F_y[\psi^{K+}(x)] \equiv \psi^{K+}(y) = \sum_k \phi_k^*(y)\beta_k^+$. As in these cases

$$\frac{\partial g(\beta_1,\cdots,\beta_k,\cdots)}{\partial\beta_k} = \phi_k(y), \qquad \frac{\partial g^+(\beta_1^+,\cdots,\beta_k^+,\cdots)}{\partial\beta_k^+} = \phi_k^*(y),$$

we have

$$\left(\frac{\delta\psi^K(y)}{\delta\psi^K(x)}\right)_x = \sum_k^K \phi_k^*(x)\phi_k(y) = \delta_K(y,x),$$

$$\left(\frac{\delta\psi^{K+}(y)}{\delta\psi^{K+}(x)}\right)_x = \sum_k^K \phi_k(x)\phi_k^*(y) = \delta_{K+}(y,x)$$

for the functional derivatives as before, and

$$\nabla_x \left(\frac{\delta \psi^K(y)}{\delta \psi^K(x)} \right)_x = \sum_k^K \{\nabla_x \phi_k^*(x)\} \phi_k(y) = \nabla_x \delta_K(y,x), \tag{H.49}$$

$$\nabla_x \left(\frac{\delta \psi^{K+}(y)}{\delta \psi^{K+}(x)} \right)_x = \sum_k^K \{\nabla_x \phi_k(x)\} \phi_k^*(y) = \nabla_x \delta_{K+}(y,x) \tag{H.50}$$

for the spatial derivatives of the functional derivatives.

For the spatial-derivative functionals, $F_{\nabla y}[\psi^K(x)] \equiv \nabla_y \psi^K(y) = \sum_k \beta_k \nabla_y \phi_k(y)$ and $F_{\nabla y}[\psi^{K+}(x)] \equiv \nabla_y \psi^{K+}(y) = \sum_k \nabla_y \phi_k^*(y) \beta_k^+$. As in these cases

$$\frac{\partial g(\beta_1, \cdots, \beta_k, \cdots)}{\partial \beta_k} = \nabla_y \phi_k(y), \qquad \frac{\partial g^+(\beta_1^+, \cdots, \beta_k^+, \cdots)}{\partial \beta_k^+} = \nabla_y \phi_k^*(y),$$

we have

$$\left(\frac{\delta \nabla_y \psi^K(y)}{\delta \psi^K(x)} \right)_x = \sum_k^K \phi_k^*(x) \nabla_y \phi_k(y) = \nabla_y \delta_K(y,x),$$

$$\left(\frac{\delta \nabla_y \psi^{K+}(y)}{\delta \psi^{K+}(x)} \right)_x = \sum_k^K \phi_k(x) \nabla_y \phi_k^*(y) = \nabla_y \delta_{K+}(y,x),$$

which are the same results as before. Note the distinction between $\left(\delta \nabla_y \psi^K(y)/\delta \psi^K(x)\right)_x$ and $\nabla_x \left(\frac{\delta \psi^K(y)}{/} \delta \psi^K(x)\right)_x$ – the first being the functional derivative of the y spatial derivative $\nabla_y \psi^K(y)$ with respect to $\psi^K(x)$, and the second being the x spatial derivative of the functional derivative of $\psi^K(y)$ with respect to $\psi^K(x)$.

H.4 Functional Integration for Restricted C-Number Functions

H.4.1 Definition of Functional Integral

Functional integration involving restricted functions $\psi^K(x)$ is defined via

$$\int D\psi \; F[\psi^K(x)] \\ = \lim_{n\to\infty} \lim_{\epsilon\to 0} \int \cdots \int d\psi_1^K \cdots d\psi_n^K w(\psi_1^K, \cdots, \psi_n^K) F(\psi_1^K, \cdots, \psi_i^K, \cdots, \psi_n^K), \tag{H.51}$$

where the average values ψ_i^K in the interval Δx_i are defined via

$$\psi_i^K = \frac{1}{\Delta x_i} \int_{\Delta x_i} dx \; \psi^K(x) \tag{H.52}$$

and only take on values allowed by restricting $\psi^K(x)$ to the restricted function space. The weight function depends on such average values. Similar considerations apply for the conjugate restricted functions $\phi^{K^+}(x)$.

Apart from where stated otherwise, the same general rules (the product rule for functional differentiation, functional integration by parts etc.) also apply to functionals based on restricted functions. These rules can be derived in a straightforward way, and will not be presented here.

H.5 Functionals for Restricted Grassmann Fields

The theory for Grassmann fields follows lines similar to that for c-number fields. As restricted fields are used mainly for boson cases, the details for fermion cases will be not be covered.

Appendix I
Applications to Multi-Mode Systems

I.1 Bose Condensate – Derivation of Functional Fokker–Planck Equations

I.1.1 Positive P Case

In this section, we denote the distribution functional $P[\psi(\mathbf{r}), \psi^+(\mathbf{r}), \psi^*(\mathbf{r}), \psi^{+*}(\mathbf{r})]$ as $P[\psi, \psi^+]$ for short, leaving the dependence on $\psi^*(\mathbf{r}), \psi^{+*}(\mathbf{r})$ implicit. The Hamiltonian is given in (15.7). The correspondence rules are given in (13.9)–(13.12) and (13.17)–(13.20). We will assume the modes are restricted by a cut-off K.

P+ Kinetic-Energy Terms. If $\widehat{\rho} \to \widehat{T}\widehat{\rho}$ then

$$
\begin{aligned}
P[\psi(\mathbf{r}), \psi^+(\mathbf{r})] &\to \frac{\hbar^2}{2m} \sum_{\mu} \int ds \left\{ \left(\partial_\mu \psi^+(\mathbf{s}) - \partial_\mu \frac{\delta}{\delta\psi(\mathbf{s})} \right) (\partial_\mu \psi(\mathbf{s})) \right\} P[\psi, \psi^+] \\
&= \frac{\hbar^2}{2m} \sum_{\mu} \int ds \left\{ (\partial_\mu \psi^+(\mathbf{s})) (\partial_\mu \psi(\mathbf{s})) \right\} P[\psi, \psi^+] \\
&\quad + \frac{\hbar^2}{2m} \sum_{\mu} \int ds \left\{ \left(-\partial_\mu \frac{\delta}{\delta\psi(\mathbf{s})} \right) (\partial_\mu \psi(\mathbf{s})) \right\} P[\psi, \psi^+]
\end{aligned}
\tag{I.1}
$$

and if $\widehat{\rho} \to \widehat{\rho}\widehat{T}$ then

$$
\begin{aligned}
P[\psi(\mathbf{r}), \psi^+(\mathbf{r})] &\to \frac{\hbar^2}{2m} \sum_{\mu} \int ds \left\{ \left(\partial_\mu \psi(\mathbf{s}) - \partial_\mu \frac{\delta}{\delta\psi^+(\mathbf{s})} \right) (\partial_\mu \psi^+(\mathbf{s})) \right\} P[\psi, \psi^+] \\
&= \frac{\hbar^2}{2m} \sum_{\mu} \int ds \left\{ (\partial_\mu \psi(\mathbf{s})) (\partial_\mu \psi^+(\mathbf{s})) \right\} P[\psi, \psi^+] \\
&\quad + \frac{\hbar^2}{2m} \sum_{\mu} \int ds \left\{ \left(-\partial_\mu \frac{\delta}{\delta\psi^+(\mathbf{s})} \right) (\partial_\mu \psi^+(\mathbf{s})) \right\} P[\psi, \psi^+],
\end{aligned}
\tag{I.2}
$$

so for $\widehat{\rho} \to (-i/\hbar)[\widehat{T}, \widehat{\rho}]$ the terms involving $(\partial_\mu \psi(\mathbf{s}))\,(\partial_\mu \psi^+(\mathbf{s}))$ cancel and we are left with

$$P[\psi(\mathbf{r}),\psi^{+}(\mathbf{r})] \rightarrow -\frac{i}{\hbar}\frac{\hbar^2}{2m}\sum_{\mu}\int d\mathbf{s}\left\{\left(-\partial_{\mu}\frac{\delta}{\delta\psi(\mathbf{s})}\right)(\partial_{\mu}\psi(\mathbf{s})) - \left(-\partial_{\mu}\frac{\delta}{\delta\psi^{+}(\mathbf{s})}\right)(\partial_{\mu}\psi^{+}(\mathbf{s}))\right\}P[\psi,\psi^{+}]. \quad (I.3)$$

We can then use spatial integration by parts to simplify this result. Using the mode expressions given in (13.21) and (13.22) for the spatial derivatives of functional derivatives and the field function, and noting that the distribution functional is equivalent to a distribution function $p(\alpha_k, \alpha_k^{+}, \alpha_k^{*}, \alpha_k^{+*})$, the first term becomes

$$\begin{aligned}
&\int d\mathbf{s}\left\{\left(\partial_{\mu}\frac{\delta}{\delta\psi(\mathbf{s})}\right)(\partial_{\mu}\psi(\mathbf{s}))\right\}P[\psi,\psi^{+}] \\
&= \int d\mathbf{s}\sum_{k=1}^{K}\{\partial_{\mu}\phi_k^{*}(\mathbf{s})\}\frac{\partial}{\partial\alpha_k}\sum_{l=1}^{K}\alpha_l\{\partial_{\mu}\phi_l(\mathbf{s})\}p(\alpha_k,\alpha_k^{+}) \\
&= -\int d\mathbf{s}\sum_{k=1}^{K}\{\phi_k^{*}(\mathbf{s})\}\frac{\partial}{\partial\alpha_k}\sum_{l=1}^{K}\alpha_l\{\partial_{\mu}^{2}\phi_l(\mathbf{s})\}p(\alpha_k,\alpha_k^{+}) \\
&= -\int d\mathbf{s}\left\{\left(\frac{\delta}{\delta\psi(\mathbf{s})}\right)(\partial_{\mu}^{2}\psi(\mathbf{s}))\right\}P[\psi,\psi^{+}].
\end{aligned} \quad (I.4)$$

We can assume that the $\psi(\mathbf{s})$ and $\psi^{+}(\mathbf{s})$ become zero on the boundary, since they both involve mode functions or their conjugates that are localised owing to the trap potential. A similar treatment simplifies the second term, and if $\widehat{\rho} \rightarrow (-i/\hbar)[\widehat{T},\widehat{\rho}]$ then the kinetic-energy term in the Fokker–Planck equation is

$$\begin{aligned}
P[\psi(\mathbf{r}),\psi^{+}(\mathbf{r})] \rightarrow & \frac{-i}{\hbar}\left\{-\int d\mathbf{s}\,\frac{\delta}{\delta\psi^{+}(\mathbf{s})}\left(\sum_{\mu}\frac{\hbar^2}{2m}\partial_{\mu}^{2}\psi^{+}(\mathbf{s})\right)P[\psi,\psi^{+}]\right\} \\
&+\frac{-i}{\hbar}\left\{+\int d\mathbf{s}\left\{\frac{\delta}{\delta\psi(\mathbf{s})}\left(\sum_{\mu}\frac{\hbar^2}{2m}\partial_{\mu}^{2}\psi(\mathbf{s})\right)P[\psi,\psi^{+}])\right\}\right\},
\end{aligned} \quad (I.5)$$

which involves only first-order functional derivatives.

P+ Trap Potential Terms. If $\widehat{\rho} \rightarrow \widehat{V}\widehat{\rho}$ then

$$\begin{aligned}
P[\psi(\mathbf{r}),\psi^{+}(\mathbf{r})] &\rightarrow \int d\mathbf{s}\left\{\left(\psi^{+}(\mathbf{s}) - \frac{\delta}{\delta\psi(\mathbf{s})}\right)V(\mathbf{s})\,(\psi(\mathbf{s}))\right\}P[\psi,\psi^{+}] \\
&= \int d\mathbf{s}\,\{\psi^{+}(\mathbf{s})V(\mathbf{s})\psi(\mathbf{s})\}\,P[\psi,\psi^{+}] \\
&\quad + \int d\mathbf{s}\left\{\left(-\frac{\delta}{\delta\psi(\mathbf{s})}\right)V(\mathbf{s})\psi(\mathbf{s})\right\}P[\psi,\psi^{+}]
\end{aligned} \quad (I.6)$$

and if $\widehat{\rho} \rightarrow \widehat{\rho}\widehat{V}$ then

$$\begin{aligned}P[\psi(\mathbf{r}),\psi^{+}(\mathbf{r})] &\rightarrow \int d\mathbf{s}\left(\psi(\mathbf{s})-\frac{\delta}{\delta\psi^{+}(\mathbf{s})}\right)V(\mathbf{s})\left(\psi^{+}(\mathbf{s})\right)P[\psi,\psi^{+}] \\ &= \int d\mathbf{s}\left(\psi(\mathbf{s})\right)V(\mathbf{s})\left(\psi^{+}(\mathbf{s})\right)P[\psi,\psi^{+}] \\ &\quad+\int d\mathbf{s}\left(-\frac{\delta}{\delta\psi^{+}(\mathbf{s})}\right)V(\mathbf{s})\left(\psi^{+}(\mathbf{s})\right)P[\psi,\psi^{+}],\end{aligned} \tag{I.7}$$

so for $\widehat{\rho} \rightarrow (-i/\hbar)[\widehat{V},\widehat{\rho}]$ the terms involving $\{\psi^{+}(\mathbf{s})V(\mathbf{s})\psi(\mathbf{s})\}$ cancel and then the trap potential-energy term in the Fokker–Planck equation is

$$\begin{aligned}P[\psi(\mathbf{r}),\psi^{+}(\mathbf{r})] \rightarrow &\frac{-i}{\hbar}\left\{-\int d\mathbf{s}\left(\frac{\delta}{\delta\psi(\mathbf{s})}\right)\{V(\mathbf{s})\psi(\mathbf{s})\}P[\psi,\psi^{+}]\right\} \\ &-\frac{i}{\hbar}\left\{+\int d\mathbf{s}\left(\frac{\delta}{\delta\psi^{+}(\mathbf{s})}\right)\{V(\mathbf{s})\left(\psi^{+}(\mathbf{s})\right)\}P[\psi,\psi^{+}]\right\},\end{aligned} \tag{I.8}$$

which involves only first-order functional derivatives.

P+ Boson–Boson Interaction Terms. If $\widehat{\rho} \rightarrow \widehat{U}\widehat{\rho}$ then

$$\begin{aligned}P[\psi(\mathbf{r}),\psi^{+}(\mathbf{r})] &\rightarrow \frac{g}{2}\int d\mathbf{s}\left(\psi^{+}(\mathbf{s})-\frac{\delta}{\delta\psi(\mathbf{s})}\right)\left(\psi^{+}(\mathbf{s})-\frac{\delta}{\delta\psi(\mathbf{s})}\right)\left(\psi(\mathbf{s})\right)\left(\psi(\mathbf{s})\right)P[\psi,\psi^{+}] \\ &= \frac{g}{2}\int d\mathbf{s}\left\{\psi^{+}(\mathbf{s})\psi^{+}(\mathbf{s})\psi(\mathbf{s})\psi(\mathbf{s})\right\}P[\psi,\psi^{+}] \\ &\quad+\frac{g}{2}\int d\mathbf{s}\left\{-\psi^{+}(\mathbf{s})\frac{\delta}{\delta\psi(\mathbf{s})}\psi(\mathbf{s})\psi(\mathbf{s})\right\}P[\psi,\psi^{+}] \\ &\quad+\frac{g}{2}\int d\mathbf{s}\left\{-\frac{\delta}{\delta\psi(\mathbf{s})}\psi^{+}(\mathbf{s})\psi(\mathbf{s})\psi(\mathbf{s})\right\}P[\psi,\psi^{+}] \\ &\quad+\frac{g}{2}\int d\mathbf{s}\left\{\frac{\delta}{\delta\psi(\mathbf{s})}\frac{\delta}{\delta\psi(\mathbf{s})}\psi(\mathbf{s})\psi(\mathbf{s})\right\}P[\psi,\psi^{+}]\end{aligned} \tag{I.9}$$

and if $\widehat{\rho} \rightarrow \widehat{\rho}\widehat{U}$ then

$$\begin{aligned}&P[\psi(\mathbf{r}),\psi^{+}(\mathbf{r})] \\ &\rightarrow \frac{g}{2}\int d\mathbf{s}\left(\psi(\mathbf{s})-\frac{\delta}{\delta\psi^{+}(\mathbf{s})}\right)\left(\psi(\mathbf{s})-\frac{\delta}{\delta\psi^{+}(\mathbf{s})}\right)\left(\psi^{+}(\mathbf{s})\right)\left(\psi^{+}(\mathbf{s})\right)P[\psi,\psi^{+}] \\ &= \frac{g}{2}\int d\mathbf{s}\left\{\psi(\mathbf{s})\psi(\mathbf{s})\psi^{+}(\mathbf{s})\psi^{+}(\mathbf{s})\right\}P[\psi,\psi^{+}] \\ &\quad+\frac{g}{2}\int d\mathbf{s}\left\{-\psi(\mathbf{s})\frac{\delta}{\delta\psi^{+}(\mathbf{s})}\psi^{+}(\mathbf{s})\psi^{+}(\mathbf{s})\right\}P[\psi,\psi^{+}] \\ &\quad+\frac{g}{2}\int d\mathbf{s}\left\{-\frac{\delta}{\delta\psi^{+}(\mathbf{s})}\psi(\mathbf{s})\psi^{+}(\mathbf{s})\psi^{+}(\mathbf{s})\right\}P[\psi,\psi^{+}] \\ &\quad+\frac{g}{2}\int d\mathbf{s}\left\{\frac{\delta}{\delta\psi^{+}(\mathbf{s})}\frac{\delta}{\delta\psi^{+}(\mathbf{s})}\psi^{+}(\mathbf{s})\psi^{+}(\mathbf{s})\right\}P[\psi,\psi^{+}],\end{aligned} \tag{I.10}$$

so for $\widehat{\rho} \rightarrow (-i/\hbar)[\widehat{U}, \widehat{\rho}]$ the terms involving $\{\psi(\mathbf{s})\psi(\mathbf{s})\psi^+(\mathbf{s})\psi^+(\mathbf{s})\}$ cancel and then we are left with

$$\begin{aligned}
P[\psi(\mathbf{r}), \psi^+(\mathbf{r})] \rightarrow & \frac{-i}{\hbar}\left\{\frac{g}{2}\int d\mathbf{s}\ \frac{\delta}{\delta\psi(\mathbf{s})}\frac{\delta}{\delta\psi(\mathbf{s})}\{\psi(\mathbf{s})\psi(\mathbf{s})\}P[\psi, \psi^+]\right\} \\
& -\frac{i}{\hbar}\left\{-\frac{g}{2}\int d\mathbf{s}\ \frac{\delta}{\delta\psi^+(\mathbf{s})}\frac{\delta}{\delta\psi^+(\mathbf{s})}\{\psi^+(\mathbf{s})\psi^+(\mathbf{s})\}P[\psi, \psi^+]\right\} \\
& -\frac{i}{\hbar}\left\{-\frac{g}{2}\int d\mathbf{s}\ \frac{\delta}{\delta\psi(\mathbf{s})}\{\psi^+(\mathbf{s})\psi(\mathbf{s})\psi(\mathbf{s})\}P[\psi, \psi^+]\right\} \\
& -\frac{i}{\hbar}\left\{+\frac{g}{2}\int d\mathbf{s}\ \frac{\delta}{\delta\psi^+(\mathbf{s})}\{\psi(\mathbf{s})\psi^+(\mathbf{s})\psi^+(\mathbf{s})\}P[\psi, \psi^+]\right\} \\
& -\frac{i}{\hbar}\left\{-\frac{g}{2}\int d\mathbf{s}\ \left\{\psi^+(\mathbf{s})\frac{\delta}{\delta\psi(\mathbf{s})}\psi(\mathbf{s})\psi(\mathbf{s})\right\}P[\psi, \psi^+]\right\} \\
& -\frac{i}{\hbar}\left\{+\frac{g}{2}\int d\mathbf{s}\ \left\{\psi(\mathbf{s})\frac{\delta}{\delta\psi^+(\mathbf{s})}\psi^+(\mathbf{s})\psi^+(\mathbf{s})\right\}P[\psi, \psi^+]\right\}. \qquad (I.11)
\end{aligned}$$

The last two terms can be simplified using the product rule (15.13) for functional derivatives, together with $((\delta/\delta\psi(\mathbf{s}))\psi^+(\mathbf{s})) = ((\delta/\delta\psi^+(\mathbf{s}))\psi(\mathbf{s})) = 0$ from (15.12). We have

$$\begin{aligned}
\left\{\psi^+(\mathbf{s})\frac{\delta}{\delta\psi(\mathbf{s})}\psi(\mathbf{s})\psi(\mathbf{s})\right\}P &= \left(\frac{\delta}{\delta\psi(\mathbf{s})}\left\{\psi^+(\mathbf{s})\psi(\mathbf{s})\psi(\mathbf{s})\right\}P\right) \\
&\quad -\left(\frac{\delta}{\delta\psi(\mathbf{s})}\psi^+(\mathbf{s})\right)\{\psi(\mathbf{s})\psi(\mathbf{s})\}P \\
&= \left(\frac{\delta}{\delta\psi(\mathbf{s})}\left\{\psi^+(\mathbf{s})\psi(\mathbf{s})\psi(\mathbf{s})\right\}P\right), \qquad (I.12) \\
\left\{\psi(\mathbf{s})\frac{\delta}{\delta\psi^+(\mathbf{s})}\psi^+(\mathbf{s})\psi^+(\mathbf{s})\right\}P &= \left(\frac{\delta}{\delta\psi^+(\mathbf{s})}\left\{\psi(\mathbf{s})\psi^+(\mathbf{s})\psi^+(\mathbf{s})\right\}P\right), \qquad (I.13)
\end{aligned}$$

so for $\widehat{\rho} \rightarrow (-i/\hbar)[\widehat{U}, \widehat{\rho}]$ the boson–boson interaction energy term in the Fokker–Planck equation is

$$\begin{aligned}
P[\psi(\mathbf{r}), \psi^+(\mathbf{r})] \rightarrow & \frac{-i}{\hbar}\left\{-g\int d\mathbf{s}\ \frac{\delta}{\delta\psi(\mathbf{s})}\{\psi^+(\mathbf{s})\psi(\mathbf{s})\psi(\mathbf{s})\}P[\psi, \psi^+]\right\} \\
& -\frac{i}{\hbar}\left\{+g\int d\mathbf{s}\ \frac{\delta}{\delta\psi^+(\mathbf{s})}\{\psi(\mathbf{s})\psi^+(\mathbf{s})\psi^+(\mathbf{s})\}P[\psi, \psi^+]\right\} \\
\rightarrow & \frac{-i}{\hbar}\left\{\frac{g}{2}\int d\mathbf{s}\ \frac{\delta}{\delta\psi(\mathbf{s})}\frac{\delta}{\delta\psi(\mathbf{s})}\{\psi(\mathbf{s})\psi(\mathbf{s})\}P[\psi, \psi^+]\right\} \\
& -\frac{i}{\hbar}\left\{-\frac{g}{2}\int d\mathbf{s}\ \frac{\delta}{\delta\psi^+(\mathbf{s})}\frac{\delta}{\delta\psi^+(\mathbf{s})}\{\psi^+(\mathbf{s})\psi^+(\mathbf{s})\}P[\psi, \psi^+]\right\}, \qquad (I.14)
\end{aligned}$$

which involves both first-order and second-order functional derivatives.

I.1.2 Wigner Case

In this section, we denote the distribution functional $P[\psi(\mathbf{r}), \psi^+(\mathbf{r}), \psi^*(\mathbf{r}), \psi^{+*}(\mathbf{r})]$ as $W[\psi, \psi^+]$ for short, leaving the dependence on $\psi^*(\mathbf{r}), \psi^{+*}(\mathbf{r})$ implicit.

Wigner Kinetic-Energy Terms. If $\widehat{\rho} \to \widehat{T}\widehat{\rho}$ then

$$
\begin{aligned}
&W[\psi(\mathbf{r}), \psi^+(\mathbf{r})] \\
&\to \frac{\hbar^2}{2m}\sum_\mu \int ds \left\{ \left(\partial_\mu \psi^+(\mathbf{s}) - \frac{1}{2}\partial_\mu \frac{\delta}{\delta\psi(\mathbf{s})}\right) \left(\partial_\mu \psi(\mathbf{s}) + \frac{1}{2}\partial_\mu \frac{\delta}{\delta\psi^+(\mathbf{s})}\right) \right\} W[\psi, \psi^+] \\
&= \frac{\hbar^2}{2m}\sum_\mu \int ds \left\{ (\partial_\mu \psi^+(\mathbf{s}))\, (\partial_\mu \psi(\mathbf{s})) \right\} W[\psi, \psi^+] \\
&+ \frac{\hbar^2}{2m}\sum_\mu \int ds \left\{ (\partial_\mu \psi^+(\mathbf{s})) \left(\frac{1}{2}\partial_\mu \frac{\delta}{\delta\psi^+(\mathbf{s})}\right) \right\} W[\psi, \psi^+] \\
&+ \frac{\hbar^2}{2m}\sum_\mu \int ds \left\{ \left(-\frac{1}{2}\partial_\mu \frac{\delta}{\delta\psi(\mathbf{s})}\right) (\partial_\mu \psi(\mathbf{s})) \right\} W[\psi, \psi^+] \\
&+ \frac{\hbar^2}{2m}\sum_\mu \int ds \left\{ \left(-\frac{1}{2}\partial_\mu \frac{\delta}{\delta\psi(\mathbf{s})}\right) \left(+\frac{1}{2}\partial_\mu \frac{\delta}{\delta\psi^+(\mathbf{s})}\right) \right\} W[\psi, \psi^+] \qquad \text{(I.15)}
\end{aligned}
$$

and if $\widehat{\rho} \to \widehat{\rho}\widehat{T}$ then

$$
\begin{aligned}
&W[\psi(\mathbf{r}), \psi^+(\mathbf{r})] \\
&\to \frac{\hbar^2}{2m}\sum_\mu \int ds \left\{ \left(\partial_\mu \psi(\mathbf{s}) - \frac{1}{2}\partial_\mu \frac{\delta}{\delta\psi^+(\mathbf{s})}\right) \left(\partial_\mu \psi^+(\mathbf{s}) + \frac{1}{2}\partial_\mu \frac{\delta}{\delta\psi(\mathbf{s})}\right) \right\} W[\psi, \psi^+] \\
&= \frac{\hbar^2}{2m}\sum_\mu \int ds \left\{ (\partial_\mu \psi(\mathbf{s}))\, (\partial_\mu \psi^+(\mathbf{s})) \right\} W[\psi, \psi^+] \\
&+ \frac{\hbar^2}{2m}\sum_\mu \int ds \left\{ (\partial_\mu \psi(\mathbf{s})) \left(\frac{1}{2}\partial_\mu \frac{\delta}{\delta\psi(\mathbf{s})}\right) \right\} W[\psi, \psi^+] \\
&+ \frac{\hbar^2}{2m}\sum_\mu \int ds \left\{ \left(-\frac{1}{2}\partial_\mu \frac{\delta}{\delta\psi^+(\mathbf{s})}\right) (\partial_\mu \psi^+(\mathbf{s})) \right\} W[\psi, \psi^+] \\
&+ \frac{\hbar^2}{2m}\sum_\mu \int ds \left\{ \left(-\frac{1}{2}\partial_\mu \frac{\delta}{\delta\psi^+(\mathbf{s})}\right) \left(+\frac{1}{2}\partial_\mu \frac{\delta}{\delta\psi(\mathbf{s})}\right) \right\} W[\psi, \psi^+]. \qquad \text{(I.16)}
\end{aligned}
$$

For $\widehat{\rho} \to (-i/\hbar)[\widehat{T}, \widehat{\rho}]$, the terms involving $\{(\partial_\mu \psi^+(\mathbf{s}))\,(\partial_\mu \psi(\mathbf{s}))\}$ cancel. We can treat the terms that involve $\int ds \left\{ \left(-\frac{1}{2}\partial_\mu(\delta/\delta\psi(\mathbf{s}))\right) (\partial_\mu \psi(\mathbf{s})) \right\} W\psi, \psi^+]$ and also those involving $\int ds \left\{ \left(-\frac{1}{2}\partial_\mu(\delta/\delta\psi^+(\mathbf{s}))\right) (\partial_\mu \psi^+(\mathbf{s})) \right\} W[\psi, \psi^+]$ via the same spatial integration-by-parts approach as for the positive P case. The fourth terms, involving two spatial derivatives of functional derivatives, can be shown to be equal and thus will cancel. For these terms, we introduce mode expansions and find that

$$\int ds \left\{ \left(\partial_\mu \frac{\delta}{\delta\psi(\mathbf{s})} \right) \left(\partial_\mu \frac{\delta}{\delta\psi^+(\mathbf{s})} \right) \right\} W[\psi, \psi^+]$$
$$= \int ds \sum_k^K \{\partial_\mu \phi_k^*(\mathbf{s})\} \frac{\partial}{\partial\alpha_k} \sum_l^K \{\partial_\mu \phi_l(\mathbf{s})\} \frac{\partial}{\partial\alpha_l^+} w(\alpha_k, \alpha_k^+)$$
$$= \int ds \sum_l^K \{\partial_\mu \phi_l(\mathbf{s})\} \frac{\partial}{\partial\alpha_l^+} \sum_k^K \{\partial_\mu \phi_k^*(\mathbf{s})\} \frac{\partial}{\partial\alpha_k} w(\alpha_k, \alpha_k^+)$$
$$= \int ds \left\{ \left(\partial_\mu \frac{\delta}{\delta\psi^+(\mathbf{s})} \right) \left(\partial_\mu \frac{\delta}{\delta\psi(\mathbf{s})} \right) \right\} W[\psi, \psi^+]. \tag{I.17}$$

For the third terms, the spatial derivative of the functional derivative can be removed via spatial integration by parts and applied to the spatial function. For example,

$$\int ds \left\{ (\partial_\mu \psi^+(\mathbf{s})) \left(\partial_\mu \frac{\delta}{\delta\psi^+(\mathbf{s})} \right) \right\} W[\psi, \psi^+]$$
$$= \int ds \sum_k^K \{\partial_\mu \phi_k^*(\mathbf{s})\} \alpha_k^+ \sum_l^K \{\partial_\mu \phi_l(\mathbf{s})\} \frac{\partial}{\partial\alpha_l^+} w(\alpha_k, \alpha_k^+)$$
$$= -\int ds \sum_k^K \{\partial_\mu^2 \phi_k^*(\mathbf{s})\} \alpha_k^+ \sum_l^K \{\phi_l(\mathbf{s})\} \frac{\partial}{\partial\alpha_l^+} w(\alpha_k, \alpha_k^+)$$
$$= -\int ds \left\{ (\partial_\mu^2 \psi^+(\mathbf{s})) \left(\frac{\delta}{\delta\psi^+(\mathbf{s})} \right) \right\} W[\psi, \psi^+]. \tag{I.18}$$

Applying the product rule then enables the functional differentiation to apply to the product of $(\partial_\mu^2 \psi^+(\mathbf{s}))$ with the distribution functional:

$$(\partial_\mu^2 \psi^+(\mathbf{s})) \left(\frac{\delta}{\delta\psi^+(\mathbf{s})} W[\psi, \psi^+] \right)$$
$$= \left(\frac{\delta}{\delta\psi^+(\mathbf{s})} (\partial_\mu^2 \psi^+(\mathbf{s}))\, W[\psi, \psi^+] \right) - \left(\frac{\delta}{\delta\psi^+(\mathbf{s})} (\partial_\mu^2 \psi^+(\mathbf{s})) \right) W[\psi, \psi^+]$$
$$= \left(\frac{\delta}{\delta\psi^+(\mathbf{s})} (\partial_\mu^2 \psi^+(\mathbf{s}))\, W[\psi, \psi^+] \right) - \left(\partial_\mu^2 \frac{\delta}{\delta\psi^+(\mathbf{s})} (\psi^+(\mathbf{s})) \right) W[\psi, \psi^+]$$
$$= \left(\frac{\delta}{\delta\psi^+(\mathbf{s})} (\partial_\mu^2 \psi^+(\mathbf{s}))\, W[\psi, \psi^+] \right) - (\partial_\mu^2 \delta_C(\mathbf{s}, \mathbf{s}))\, W[\psi, \psi^+], \tag{I.19}$$

where the functional derivative of $\psi^+(\mathbf{s})$ has been evaluated using (15.12) and involves the restricted delta function (15.4). Thus

$$\int ds \left\{ (\partial_\mu \psi^+(\mathbf{s})) \left(\partial_\mu \frac{\delta}{\delta\psi^+(\mathbf{s})} \right) \right\} W[\psi, \psi^+]$$
$$= -\int ds \left\{ \frac{\delta}{\delta\psi^+(\mathbf{s})} (\partial_\mu^2 \psi^+(\mathbf{s})) \right\} W[\psi, \psi^+] + \int ds \{\partial_\mu^2 \delta_C(\mathbf{s}, \mathbf{s})\} W[\psi, \psi^+]. \tag{I.20}$$

A similar treatment shows that

$$\int ds \left\{ (\partial_\mu \psi(\mathbf{s})) \left(\partial_\mu \frac{\delta}{\delta\psi(\mathbf{s})} \right) \right\} W[\psi, \psi^+]$$
$$= -\int ds \left\{ \frac{\delta}{\delta\psi(\mathbf{s})} \left(\partial_\mu^2 \psi(\mathbf{s}) \right) \right\} W[\psi, \psi^+] + \int ds \left\{ \partial_\mu^2 \delta_C(\mathbf{s}, \mathbf{s}) \right\} W[\psi, \psi^+], \qquad \text{(I.21)}$$

so the terms involving $\{\partial_\mu^2 \delta_C(\mathbf{s}, \mathbf{s})\}$ cancel out.

For $\widehat{\rho} \to (-i/\hbar)[\widehat{T}, \widehat{\rho}]$, the kinetic-energy term in the functional Fokker–Planck equation then is

$$W[\psi(\mathbf{r}), \psi^+(\mathbf{r})] \to \frac{-i}{\hbar} \left\{ -\int ds \frac{\delta}{\delta\psi^+(\mathbf{s})} \left(\sum_\mu \frac{\hbar^2}{2m} \partial_\mu^2 \psi^+(\mathbf{s}) \right) W[\psi, \psi^+] \right\}$$
$$- \frac{i}{\hbar} \left\{ + \int ds \frac{\delta}{\delta\psi(\mathbf{s})} \left(\sum_\mu \frac{\hbar^2}{2m} \partial_\mu^2 \psi(\mathbf{s}) \right) W[\psi, \psi^+]) \right\}. \qquad \text{(I.22)}$$

The first-order functional derivative terms combine to remove the factors $\frac{1}{2}$. Thus only a first-order functional derivative term occurs. The result is the same as for the positive P case.

Wigner Trap Potential Terms. If $\widehat{\rho} \to \widehat{V}\widehat{\rho}$ then

$$W[\psi(\mathbf{r}), \psi^+(\mathbf{r})] \to \int ds \left\{ \left(\psi^+(\mathbf{s}) - \frac{1}{2}\frac{\delta}{\delta\psi(\mathbf{s})} \right) V(\mathbf{s}) \left(\psi(\mathbf{s}) + \frac{1}{2}\frac{\delta}{\delta\psi^+(\mathbf{s})} \right) \right\} W[\psi, \psi^+]$$
$$= \int ds \left\{ \psi^+(\mathbf{s}) V(\mathbf{s}) \psi(\mathbf{s}) \right\} W[\psi, \psi^+]$$
$$+ \int ds \left\{ \psi^+(\mathbf{s}) V(\mathbf{s}) \left(\frac{1}{2}\frac{\delta}{\delta\psi^+(\mathbf{s})} \right) \right\} W[\psi, \psi^+]$$
$$+ \int ds \left\{ \left(-\frac{1}{2}\frac{\delta}{\delta\psi(\mathbf{s})} \right) V(\mathbf{s}) \psi(\mathbf{s}) \right\} W[\psi, \psi^+]$$
$$+ \int ds \left\{ \left(-\frac{1}{2}\frac{\delta}{\delta\psi(\mathbf{s})} \right) V(\mathbf{s}) \left(\frac{1}{2}\frac{\delta}{\delta\psi^+(\mathbf{s})} \right) \right\} W[\psi, \psi^+], \qquad \text{(I.23)}$$

and for $\widehat{\rho} \to \widehat{\rho}\widehat{V}$ we have

$$W[\psi(\mathbf{r}), \psi^+(\mathbf{r})] \to \int ds \left(\psi(\mathbf{s}) - \frac{1}{2}\frac{\delta}{\delta\psi^+(\mathbf{s})} \right) V(\mathbf{s}) \left(\psi^+(\mathbf{s}) + \frac{1}{2}\frac{\delta}{\delta\psi(\mathbf{s})} \right) W[\psi, \psi^+]$$
$$= \int ds \, (\psi(\mathbf{s})) \, V(\mathbf{s}) \, (\psi^+(\mathbf{s})) \, W[\psi, \psi^+]$$
$$+ \int ds \, (\psi(\mathbf{s})) \, V(\mathbf{s}) \left(\frac{1}{2}\frac{\delta}{\delta\psi(\mathbf{s})} \right) W[\psi, \psi^+]$$
$$+ \int ds \left(-\frac{1}{2}\frac{\delta}{\delta\psi^+(\mathbf{s})} \right) V(\mathbf{s}) \, (\psi^+(\mathbf{s})) \, W[\psi, \psi^+]$$
$$+ \int ds \left(-\frac{1}{2}\frac{\delta}{\delta\psi^+(\mathbf{s})} \right) V(\mathbf{s}) \left(\frac{1}{2}\frac{\delta}{\delta\psi(\mathbf{s})} \right) W[\psi, \psi^+]. \qquad \text{(I.24)}$$

For $\widehat{\rho} \rightarrow (-i/\hbar)[\widehat{T}, \widehat{\rho}]$, the terms involving $\{\psi^+(\mathbf{s})V(\mathbf{s})\psi(\mathbf{s})\}$ cancel, and it is straightforward to show that the terms involving two functional derivatives are equal and also cancel. The treatment of the terms involving a single functional derivative is essentially the same as for the positive P case, and the trap potential-energy term in the functional Fokker–Planck equation then is

$$W[\psi,\psi^+] \rightarrow \frac{-i}{\hbar}\left\{-\int d\mathbf{s}\left\{\frac{\delta}{\delta\psi(\mathbf{s})}\{V(\mathbf{s})\psi(\mathbf{s})\}\right\}W[\psi,\psi^+]\right\}$$
$$-\frac{i}{\hbar}\left\{+\int d\mathbf{s}\left\{\frac{\delta}{\delta\psi^+(\mathbf{s})}\{V(\mathbf{s})\psi^+(\mathbf{s})\}\right\}W[\psi,\psi^+]\right\}. \qquad (I.25)$$

Wigner Boson–Boson Interaction Terms. If $\widehat{\rho} \rightarrow \widehat{U}\widehat{\rho}$ then

$$W[\psi,\psi^+] \rightarrow \frac{g}{2}\int d\mathbf{s}\left(\psi^+(\mathbf{s}) - \frac{1}{2}\frac{\delta}{\delta\psi(\mathbf{s})}\right)\left(\psi^+(\mathbf{s}) - \frac{1}{2}\frac{\delta}{\delta\psi(\mathbf{s})}\right)\left(\psi(\mathbf{s}) + \frac{1}{2}\frac{\delta}{\delta\psi^+(\mathbf{s})}\right)$$
$$\times\left(\psi(\mathbf{s}) + \frac{1}{2}\frac{\delta}{\delta\psi^+(\mathbf{s})}\right)W[\psi,\psi^+], \qquad (I.26)$$

and if $\widehat{\rho} \rightarrow \widehat{\rho}\widehat{U}$ then

$$W[\psi(\mathbf{r}),\psi^+(\mathbf{r})] \rightarrow \frac{g}{2}\int d\mathbf{s}\left(\psi(\mathbf{s}) - \frac{1}{2}\frac{\delta}{\delta\psi^+(\mathbf{s})}\right)\left(\psi(\mathbf{s}) - \frac{1}{2}\frac{\delta}{\delta\psi^+(\mathbf{s})}\right)\left(\psi^+(\mathbf{s}) + \frac{1}{2}\frac{\delta}{\delta\psi(\mathbf{s})}\right)$$
$$\times\left(\psi^+(\mathbf{s}) + \frac{1}{2}\frac{\delta}{\delta\psi(\mathbf{s})}\right)W[\psi,\psi^+]. \qquad (I.27)$$

There are a total of 32 terms to consider. These can all be developed using the product rules for functional differentiation in (15.13) together with the rules $(\delta/\delta\psi(\mathbf{s}))\psi^+(\mathbf{s}) = (\delta/\delta\psi^+(\mathbf{s}))\psi(\mathbf{s}) = 0$ and $(\delta/\delta\psi(\mathbf{s}))\psi(\mathbf{s}) = (\delta/\delta\psi^+(\mathbf{s}))\psi^+(\mathbf{s}) = \delta_C(\mathbf{s},\mathbf{s})$ from (15.12). As with the previous terms, the overall strategy is to move all the functional derivatives to the left – as required for functional Fokker–Planck equations. It turns out that all of the even-order functional derivative terms cancel, as do the terms with no functional derivative. Unlike the positive P case, not all of the terms involving $\delta_C(\mathbf{s},\mathbf{s})$ cancel out. The derivation is left as an exercise.

For $\widehat{\rho} \rightarrow (-i/\hbar)[\widehat{U}, \widehat{\rho}]$, the boson–boson interaction energy term in the Fokker–Planck equation is

$$W[\psi(\mathbf{r}),\psi^+(\mathbf{r})] \rightarrow \frac{-i}{\hbar}\left\{-g\int d\mathbf{s}\,\frac{\delta}{\delta\psi(\mathbf{s})}\{(\psi^+(\mathbf{s})\psi(\mathbf{s}) - \delta_C(\mathbf{s},\mathbf{s}))\psi(\mathbf{s})\}W[\psi,\psi^+]\right\}$$
$$-\frac{i}{\hbar}\left\{+g\int d\mathbf{s}\,\frac{\delta}{\delta\psi^+(\mathbf{s})}\{(\psi^+(\mathbf{s})\psi(\mathbf{s}) - \delta_C(\mathbf{s},\mathbf{s}))\psi^+(\mathbf{s})\}W[\psi,\psi^+]\right\}$$
$$-\frac{i}{\hbar}\left\{+g\int d\mathbf{s}\,\frac{1}{4}\frac{\delta}{\delta\psi(\mathbf{s})}\frac{\delta}{\delta\psi(\mathbf{s})}\frac{\delta}{\delta\psi^+(\mathbf{s})}\{\psi(\mathbf{s})\}W[\psi,\psi^+]\right\}$$
$$-\frac{i}{\hbar}\left\{-g\int d\mathbf{s}\,\frac{1}{4}\frac{\delta}{\delta\psi^+(\mathbf{s})}\frac{\delta}{\delta\psi^+(\mathbf{s})}\frac{\delta}{\delta\psi(\mathbf{s})}\{\psi^+(\mathbf{s})\}W[\psi,\psi^+]\right\}, \qquad (I.28)$$

which involves first-order and third-order functional derivatives. The appearance of the term $\delta_C(\mathbf{s},\mathbf{s})$ may be surprising at first, but we recall that $\psi^+(\mathbf{s})\psi(\mathbf{s})$ represents a symmetrically ordered number density, so $\psi^+(\mathbf{s})\psi(\mathbf{s}) - \delta_C(\mathbf{s},\mathbf{s})$ corresponds to the true atom number density.

I.2 Fermi Gas – Derivation of Functional Fokker–Planck Equations

I.2.1 Unnormalised B Case

In this section, we denote the distribution functional $B[\psi_u(\mathbf{r}), \psi_u^+(\mathbf{r}), \psi_d(\mathbf{r}), \psi_d^+(\mathbf{r})]$ as $B[\boldsymbol{\psi}(\mathbf{r})]$ for short. The Hamiltonian is given in (15.42). The correspondence rules are given in (13.51)–(13.58). We will assume the modes are restricted by a cut-off K.

B+ Kinetic-Energy Terms. If $\widehat{\rho} \to \widehat{T}\widehat{\rho}$ then

$$B[\boldsymbol{\psi}(\mathbf{r})] \to \frac{\hbar^2}{2m}\sum_{\alpha}\sum_{\mu}\int d\mathbf{s}\left\{\left(\partial_\mu \frac{\overrightarrow{\delta}}{\delta\psi_{\alpha 1}(\mathbf{s})}\right)(\partial_\mu\psi_{\alpha 1}(\mathbf{s}))\right\}B[\boldsymbol{\psi}(\mathbf{s})]$$

and if $\widehat{\rho} \to \widehat{\rho}\widehat{T}$ then

$$B[\boldsymbol{\psi}(\mathbf{r})] \to \frac{\hbar^2}{2m}\sum_{\alpha}\sum_{\mu}\int d\mathbf{s}\; B[\boldsymbol{\psi}(\mathbf{s})]\left\{(\partial_\mu\psi_{\alpha 2}(\mathbf{s}))\left(\partial_\mu \frac{\overleftarrow{\delta}}{\delta\psi_{\alpha 2}(\mathbf{s})}\right)\right\},$$

so for $\widehat{\rho} \to (-i/\hbar)[\widehat{T}, \widehat{\rho}]$ we have

$$\begin{aligned} B[\boldsymbol{\psi}(\mathbf{r})] \to &-\frac{i}{\hbar}\left\{\frac{\hbar^2}{2m}\sum_{\alpha}\sum_{\mu}\int d\mathbf{s}\left\{\left(\partial_\mu \frac{\overrightarrow{\delta}}{\delta\psi_{\alpha 1}(\mathbf{s})}\right)(\partial_\mu\psi_{\alpha 1}(\mathbf{s}))\right\}B[\boldsymbol{\psi}(\mathbf{s})]\right\} \\ &-\frac{i}{\hbar}\left\{-\frac{\hbar^2}{2m}\sum_{\alpha}\sum_{\mu}\int d\mathbf{s}\; B[\boldsymbol{\psi}(\mathbf{s})]\left\{(\partial_\mu\psi_{\alpha 2}(\mathbf{s}))\left(\partial_\mu \frac{\overleftarrow{\delta}}{\delta\psi_{\alpha 2}(\mathbf{s})}\right)\right\}\right\}. \end{aligned} \tag{I.29}$$

Using the mode expansions in (11.149) and (11.156) for the functional derivatives, we have for the first term

$$\begin{aligned} &\int d\mathbf{s}\left\{\left(\partial_\mu \frac{\overrightarrow{\delta}}{\delta\psi_{\alpha 1}(\mathbf{s})}\right)(\partial_\mu\psi_{\alpha 1}(\mathbf{s}))\right\}B[\boldsymbol{\psi}(\mathbf{s})] \\ &= \sum_i\sum_j\int d\mathbf{s}\left\{\left(\partial_\mu\xi_{\alpha i}^2(\mathbf{s})\frac{\overrightarrow{\partial}}{\partial g_{\alpha i}^1}\right)(\partial_\mu\xi_{\alpha j}^1(\mathbf{s})\,g_{\alpha j}^1)\right\}B(\mathbf{g}) \\ &= -\sum_i\sum_j\int d\mathbf{s}\left\{\left(\xi_{\alpha i}^2(\mathbf{s})\frac{\overrightarrow{\partial}}{\partial g_{\alpha i}^1}\right)(\partial_\mu^2\xi_{\alpha j}^1(\mathbf{s})\,g_{\alpha j}^1)\right\}B(\mathbf{g}) \\ &= -\int d\mathbf{s}\left\{\left(\frac{\overrightarrow{\delta}}{\delta\psi_{\alpha 1}(\mathbf{s})}\right)(\partial_\mu^2\psi_{\alpha 1}(\mathbf{s}))\right\}B[\boldsymbol{\psi}(\mathbf{s})] \\ &= -\int d\mathbf{s}\; B[\boldsymbol{\psi}(\mathbf{s})]\left\{\partial_\mu^2\psi_{\alpha 1}(\mathbf{s})\left(\frac{\overleftarrow{\delta}}{\delta\psi_{\alpha 1}(\mathbf{s})}\right)\right\}, \end{aligned} \tag{I.30}$$

where we have used spatial integration by parts and then (11.169) to reverse the functional derivative, noting that $(\partial_\mu^2 \psi_{\alpha 1}(\mathbf{s}))\, B[\boldsymbol{\psi}(\mathbf{s})]$ is an odd Grassmann function. Similarly,

$$\int d\mathbf{s}\; B[\boldsymbol{\psi}(\mathbf{s})] \left\{ (\partial_\mu \psi_{\alpha 2}(\mathbf{s})) \left(\partial_\mu \frac{\overleftarrow{\delta}}{\delta \psi_{\alpha 2}(\mathbf{s})} \right) \right\}$$
$$= -\int d\mathbf{s}\; B[\boldsymbol{\psi}(\mathbf{s})] \left\{ (\partial_\mu^2 \psi_{\alpha 2}(\mathbf{s})) \left(\frac{\overleftarrow{\delta}}{\delta \psi_{\alpha 2}(\mathbf{s})} \right) \right\}, \tag{I.31}$$

though here no reversal of the functional derivative is needed. Combining these results for $\widehat{\rho} \to (-i/\hbar)[\widehat{T}, \widehat{\rho}]$ gives the kinetic-energy term in the functional Fokker–Planck equation as

$$\left(\frac{\partial}{\partial t} B[\boldsymbol{\psi}(\mathbf{s})] \right)_K = \frac{-i}{\hbar} \left\{ -\sum_\alpha \int d\mathbf{s} \left\{ \left(\sum_\mu \frac{\hbar^2}{2m} \partial_\mu^2 \psi_{\alpha 1}(\mathbf{s}) B[\boldsymbol{\psi}(\mathbf{s})] \right) \frac{\overleftarrow{\delta}}{\delta \psi_{\alpha 1}(\mathbf{s})} \right\} \right\}$$
$$- \frac{i}{\hbar} \left\{ +\sum_\alpha \int d\mathbf{s} \left\{ \left(\sum_\mu \frac{\hbar^2}{2m} \partial_\mu^2 \psi_{\alpha 2}(\mathbf{s}) B[\boldsymbol{\psi}(\mathbf{s})] \right) \frac{\overleftarrow{\delta}}{\delta \psi_{\alpha 2}(\mathbf{s})} \right\} \right\}. \tag{I.32}$$

B+ Potential-Energy Terms. If $\widehat{\rho} \to \widehat{V}\widehat{\rho}$ then

$$B[\boldsymbol{\psi}(\mathbf{r})] \to \sum_\alpha \int d\mathbf{s} \left\{ \left(\frac{\overrightarrow{\delta}}{\delta \psi_{\alpha 1}(\mathbf{s})} \right) V_\alpha(\mathbf{s})\, (\psi_{\alpha 1}(\mathbf{s})) \right\} B[\boldsymbol{\psi}(\mathbf{s})]$$
$$= \sum_\alpha \int d\mathbf{s}\, \{ V_\alpha(\mathbf{s})\, (\psi_{\alpha 1}(\mathbf{s})) \}\, B[\boldsymbol{\psi}(\mathbf{s})] \left(\frac{\overleftarrow{\delta}}{\delta \psi_{\alpha 1}(\mathbf{s})} \right), \tag{I.33}$$

where the functional derivative in the first term has been reversed, and we have used the fact that $\{V_\alpha(\mathbf{s})\, (\psi_{\alpha 1}(\mathbf{s}))\} B[\boldsymbol{\psi}(\mathbf{s})]$ is an odd Grassmann function.

If $\widehat{\rho} \to \widehat{\rho}\widehat{V}$ then

$$B[\boldsymbol{\psi}(\mathbf{r})] \to \sum_\alpha \int d\mathbf{s}\; B[\boldsymbol{\psi}(\mathbf{s})] \left\{ (\psi_{\alpha 2}(\mathbf{s}))\, V_\alpha(\mathbf{s}) \left(\frac{\overleftarrow{\delta}}{\delta \psi_{\alpha 2}(\mathbf{s})} \right) \right\},$$

so for $\widehat{\rho} \to (-i/\hbar)[\widehat{V}, \widehat{\rho}]$ the potential-energy term in the functional Fokker–Planck equation becomes

$$B[\boldsymbol{\psi}(\mathbf{r})]_V \to -\frac{i}{\hbar} \left\{ +\sum_\alpha \int d\mathbf{s}\, (V_\alpha(\mathbf{s})\, \psi_{\alpha 1}(\mathbf{s}) B[\boldsymbol{\psi}(\mathbf{s})]) \left(\frac{\overleftarrow{\delta}}{\delta \psi_{\alpha 1}(\mathbf{s})} \right) \right\}$$
$$- \frac{i}{\hbar} \left\{ -\sum_\alpha \int d\mathbf{s}\, (V_\alpha(\mathbf{s})\, \psi_{\alpha 2}(\mathbf{s}) B[\boldsymbol{\psi}(\mathbf{s})]) \left(\frac{\overleftarrow{\delta}}{\delta \psi_{\alpha 2}(\mathbf{s})} \right) \right\}. \tag{I.34}$$

B+ Interaction Energy Terms. If $\widehat{\rho} \rightarrow \widehat{U}\widehat{\rho}$ then

$$\begin{aligned}
B[\boldsymbol{\psi}(\mathbf{s})] &\rightarrow \frac{g}{2}\sum_{\alpha}\int d\mathbf{s}\left\{\left(+\frac{\overrightarrow{\delta}}{\delta\psi_{\alpha 1}(\mathbf{s})}\right)\left(+\frac{\overrightarrow{\delta}}{\delta\psi_{-\alpha 1}(\mathbf{s})}\right)\psi_{-\alpha 1}(\mathbf{s})\psi_{\alpha 1}(\mathbf{s})\right\}B[\boldsymbol{\psi}(\mathbf{s})] \\
&= -\frac{g}{2}\sum_{\alpha}\int d\mathbf{s}\left\{\left(+\frac{\overrightarrow{\delta}}{\delta\psi_{\alpha 1}(\mathbf{s})}\right)\left\{\psi_{-\alpha 1}(\mathbf{s})\psi_{\alpha 1}(\mathbf{s})B[\boldsymbol{\psi}(\mathbf{s})]\left(+\frac{\overleftarrow{\delta}}{\delta\psi_{-\alpha 1}(\mathbf{s})}\right)\right\}\right\} \\
&= -\frac{g}{2}\sum_{\alpha}\int d\mathbf{s}\left\{\left\{\psi_{-\alpha 1}(\mathbf{s})\psi_{\alpha 1}(\mathbf{s})B[\boldsymbol{\psi}(\mathbf{s})]\left(+\frac{\overleftarrow{\delta}}{\delta\psi_{-\alpha 1}(\mathbf{s})}\right)\right\}\left(+\frac{\overleftarrow{\delta}}{\delta\psi_{\alpha 1}(\mathbf{s})}\right)\right\},
\end{aligned} \tag{I.35}$$

where the second line is obtained using (11.169), noting that $\psi_{-\alpha 1}(\mathbf{s})\psi_{\alpha 1}(\mathbf{s})B[\boldsymbol{\psi}(\mathbf{s})]$ is an even Grassmann function, and the third line is obtained from the same equation, but noting that $\psi_{-\alpha 1}(\mathbf{s})\psi_{\alpha 1}(\mathbf{s})B[\boldsymbol{\psi}(\mathbf{s})] + \overleftarrow{\delta}/\delta\psi_{-\alpha 1}(\mathbf{s})$ is an odd Grassmann function.

If $\widehat{\rho} \rightarrow \widehat{\rho}\widehat{U}$ then

$$\begin{aligned}
B[\boldsymbol{\psi}(\mathbf{s})] &\rightarrow \frac{g}{2}\sum_{\alpha}\int d\mathbf{s}\; B[\boldsymbol{\psi}(\mathbf{s})]\left\{\psi_{\alpha 2}(\mathbf{s})\psi_{-\alpha 2}(\mathbf{s})\left(+\frac{\overleftarrow{\delta}}{\delta\psi_{-\alpha 2}(\mathbf{s})}\right)\left(+\frac{\overleftarrow{\delta}}{\delta\psi_{\alpha 2}(\mathbf{s})}\right)\right\} \\
&= -\frac{g}{2}\sum_{\alpha}\int d\mathbf{s}\; B[\boldsymbol{\psi}(\mathbf{s})]\left\{\psi_{-\alpha 2}(\mathbf{s})\psi_{\alpha 2}(\mathbf{s})\left(+\frac{\overleftarrow{\delta}}{\delta\psi_{-\alpha 2}(\mathbf{s})}\right)\left(+\frac{\overleftarrow{\delta}}{\delta\psi_{\alpha 2}(\mathbf{s})}\right)\right\},
\end{aligned} \tag{I.36}$$

where we have reversed the order of the fields.

Hence, if $\widehat{\rho} \rightarrow (-i/\hbar)[\widehat{U},\widehat{\rho}]$, then the interaction energy term in the functional Fokker–Planck equation becomes

$$\begin{aligned}
B[\boldsymbol{\psi}(\mathbf{r})]_U &\rightarrow \frac{i}{\hbar}\frac{g}{2}\sum_{\alpha}\int d\mathbf{s}\left\{\psi_{-\alpha 1}(\mathbf{s})\psi_{\alpha 1}(\mathbf{s})B[\boldsymbol{\psi}(\mathbf{s})]\left(+\frac{\overleftarrow{\delta}}{\delta\psi_{-\alpha 1}(\mathbf{s})}\right)\left(+\frac{\overleftarrow{\delta}}{\delta\psi_{\alpha 1}(\mathbf{s})}\right)\right\} \\
&= -\frac{i}{\hbar}\frac{g}{2}\sum_{\alpha}\int d\mathbf{s}\left\{\psi_{-\alpha 2}(\mathbf{s})\psi_{\alpha 2}(\mathbf{s})B[\boldsymbol{\psi}(\mathbf{s})]\left(+\frac{\overleftarrow{\delta}}{\delta\psi_{-\alpha 2}(\mathbf{s})}\right)\left(+\frac{\overleftarrow{\delta}}{\delta\psi_{\alpha 2}(\mathbf{s})}\right)\right\}.
\end{aligned} \tag{I.37}$$

References

[1] R. J. Glauber, *Phys. Rev.* **131**, 2766 (1963).
[2] R. J. Glauber, *Phys. Rev.* **130**, 2529 (1963).
[3] W. H. Louisell, *Quantum Statistical Properties of Radiation* (Wiley, New York, 1973).
[4] S. M. Barnett and P. M. Radmore, *Methods in Theoretical Quantum Optics* (Clarendon Press, Oxford, 1997).
[5] Z. Ficek and S. Swain, *Quantum Interference and Coherence: Theory and Experiments* (Springer, Berlin, 2004).
[6] J. Zinn-Justin, *Quantum Field Theory and Critical Phenomena* (Clarendon Press, Oxford, 2002).
[7] R. J. Rivers, *Path Integral Methods in Quantum Field Theory* (Cambridge University Press, Cambridge, 1987).
[8] J.-P. Blaizot and G. Ripka, *Quantum Theory of Finite Systems* (MIT Press, Cambridge, MA, 1986).
[9] F. A. Berezin, *The Method of Second Quantization* (Academic Press, New York, 1966).
[10] K. E. Cahill and R. J. Glauber, *Phys. Rev. A* **59**, 1538 (1999).
[11] K. E. Cahill and R. J. Glauber, *Phys. Rev.* **177**, 1857 (1969).
[12] K. E. Cahill and R. J. Glauber, *Phys. Rev.* **177**, 1882 (1969).
[13] A. Gatti, H. Wiedemann, L. Lugiato, I. Marzoli, G.-L. Oppo and S. M. Barnett, *Phys. Rev. A* **56**, 877 (1997).
[14] L. Plimak, M. J. Collett and M. K. Olsen, *Phys. Rev. A* **64**, 063409 (2001).
[15] L. Plimak and S. Stenholm, *Ann. Phys.* **324**, 600 (2009).
[16] S. Shresta, C. Anastopoulos, A. Dragulescu and B. L. Hu, *Phys. Rev. A* **71**, 022109 (2005).
[17] C. Anastopoulos and B. L. Hu, *Phys. Rev. A* **62**, 033801 (2000).
[18] T. Tyc, B. Hamilton, B. C. Sanders and W. D. Oliver, *Found. Phys.* **37**, 1519 (2007).
[19] J. Tempere and J. T. Devreese, *Phys. Rev. A* **72**, 063601 (2005).
[20] V. V. Shcherbakov, *Theor. Math. Phys.* **96**, 23 (1993).
[21] V. V. Shcherbakov, *Theor. Math. Phys.* **96**, 182 (1993).
[22] V. V. Shcherbakov, *Theor. Math. Phys.* **96**, 1323 (1993).
[23] R. J. Glauber, in *Quantum Optics and Electronics*, eds C. DeWitt, A. Blandin and C. Cohen-Tannoudji (Gordon & Breach, New York, 1965), p65.
[24] H. Risken, *The Fokker–Planck Equation*, 2nd edition (Springer, Berlin, 1989).
[25] H. A. Kramers, *Physica* **7**, 284 (1940).
[26] J. E. Moyal, *J. Roy. Stat. Soc. (London) B* **11**, 150 (1949).

[27] A. J. Leggett, *Rev. Mod. Phys.* **73**, 307 (2001).
[28] R. Bach and K. Rzazewskii, *Phys. Rev. A* **70**, 063622 (2004).
[29] B. J. Dalton and S. Ghanbari, *J. Mod. Opt.* **59**, 287 (2012); *J. Mod. Opt.* **60**, 602 (2013).
[30] D. T. Pegg and S. M. Barnett, *Phys. Rev. A* **39**, 1665 (1989).
[31] C. W. Gardiner and P. Zoller, *Quantum Noise*, 3rd edition (Springer, Berlin, 2004).
[32] B. J. Dalton, Z. Ficek and S. Swain, *J. Mod. Opt.* **46**, 379 (1999).
[33] T. Matsubara, *Prog. Theor. Phys.* **14**, 351 (1955).
[34] M. Kaku, *Quantum Field Theory* (Oxford University Press, Oxford, 1993), p52.
[35] R. Courant, *Differential and Integral Calculus, Vols I and II* (Blackie & Son, London, 1935).
[36] E. T. Copson, *An Introduction to the Theory of Functions of a Complex Variable* (Oxford University Press, Oxford, 1935).
[37] G. Stephenson and P. M. Radmore, *Advanced Mathematical Methods for Engineering and Science Students* (Cambridge University Press, Cambridge, 1990).
[38] P. D. Drummond and C. W. Gardiner, *J. Phys. A: Math. Gen.* **13**, 2353 (1980).
[39] P. A. M. Dirac, *The Principles of Quantum Mechanics*, 4th edition (Oxford University Press, Oxford, 1958).
[40] G. J. Milburn, *J. Phys. A: Math. Gen.* **17**, 737 (1984).
[41] R. Loudon, *Phys. Rev. A* **58**, 4904 (1998).
[42] R. A. Campos, B. E. A. Saleh and M. C. Teich, *Phys. Rev. A* **40**, 1371 (1989).
[43] C. K. Hong, Z. Y. Ou and L. Mandel, *Phys. Rev. Lett.* **59**, 2044 (1987).
[44] R. Loudon, *The Quantum Theory of Light*, 3rd edition (Oxford University Press, Oxford, 2000).
[45] R. Loudon and P. L. Knight, *J. Mod. Opt.* **34**, 709 (1997).
[46] H. Umezawa, H. Masumoto and M. Tachiki, *Thermofield Dynamics and Condensed States* (North-Holland, Amsterdam, 1982).
[47] L. Pitaevskii and S. Stringari, *Bose–Einstein Condensation* (Oxford University Press, Oxford, 2003).
[48] C. J. Pethick and H. Smith, *Bose–Einstein Condensation in Dilute Bose Gases* (Cambridge University Press, Cambridge, 2002).
[49] F. A. M. de Oliveira, *Phys. Rev. A* **45**, 5104 (1992).
[50] B. J. Dalton, *J. Phys. C: Conf. Ser.* **67**, 012059 (2007).
[51] S. E. Hoffmann, J. F. Corney and P. D. Drummond, *Phys. Rev. A* **78**, 013622 (2008).
[52] B. J. Dalton, *Ann. Phys.* **326**, 668 (2011).
[53] B. J. Dalton, *Ann. Phys.* **327**, 2432 (2012).
[54] P. Deuar and P. D. Drummond, *Phys. Rev. A* **66**, 033812 (2002).
[55] P. D. Drummond and P. Deuar, *J. Opt. B: Semiclass. Opt.* **5**, S281 (2003).
[56] J. F. Corney and P. D. Drummond, *Phys. Rev. A* **68**, 063822 (2003).
[57] J. F. Corney and P. D. Drummond, *Phys. Rev. B* **73**, 125112 (2006).
[58] J. F. Corney and P. D. Drummond, *J. Phys. A: Math. Gen.* **39**, 269 (2006).

[59] I. S. Gradsteyn and I. M. Ryzhik, *Tables of Integrals, Series and Products* (Academic Press, New York, 1965), p4.
[60] R. Schack and A. Schenzle, *Phys. Rev. A* **44**, 682 (1991).
[61] A. Gilchrist, C. W. Gardiner and P. D. Drummond, *Phys. Rev. A* **55**, 3014 (1997).
[62] T. Takagi, *Japan. J. Math.* **1**, 83 (1925).
[63] R. A. Horn and C. R. Johnson, *Matrix Analysis* (Cambridge University Press, Cambridge, 1985).
[64] The authors are grateful to Dr C. Gilson, Department of Applied Mathematics, University of Glasgow, for the first derivation of a solution to this problem.
[65] M. A. Kasevich, *Science* **298**, 1363 (2002).
[66] A. D. Cronin, J. Schmiedmayer and D. E. Pritchard, *Rev. Mod. Phys.* **81**, 1051 (2009).
[67] C. Gross, *J. Phys. B: At. Mol. Opt. Phys.* **45**, 103001 (2012).
[68] A. Sinatra and Y. Castin, *Eur. Phys. J. D* **8**, 319 (2000).
[69] M. Egorov, R. P. Anderson, V. Ivannikov, B. Opanchuk, P. Drummond, B. V. Hall and A. I. Sidorov, *Phys. Rev. A* **84**, 021605 (2011).
[70] B. Opanchuk, M. Egorov, S. Hoffmann, A. I. Sidorov and P. D. Drummond, *Europhys. Lett.* **97**, 50003 (2012).
[71] E. T. Jaynes and F. W. Cummings, *Proc. IEEE* **51**, 89 (1963).
[72] B. R. Mollow, *Phys. Rev.* **188**, 1969 (1969).
[73] C. Cohen-Tannoudji and S. Reynaud, *J. Phys. B: At. Mol. Opt. Phys.* **10**, 345 (1976).
[74] F. Scuda, C. R. Stroud Jr. and M. Hercher, *J. Phys. B: At. Mol. Opt. Phys.* **7**, L198 (1974).
[75] F. Y. Wu, R. E. Grove and S. Ezekiel, *Phys. Rev. Lett.* **35**, 1426 (1975).
[76] W. Hartig, W. Rasmussen, R. Schieder and H. Walther, *Z. Phys. A* **278**, 205 (1976).
[77] C. C. Gerry and P. L. Knight, *Introductory Quantum Optics* (Cambridge University Press, Cambridge, 2005).
[78] G. Rempe, H. Walther and N. Klein, *Phys. Rev. Lett.* **58**, 353 (1987).
[79] M. Brune, F. Schmidt-Kaler, A. Maaili, J. Dreyer, E. Hagley, J. M. Raimond and S. Haroche, *Phys. Rev. Lett.* **76**, 1800 (1996).
[80] S. Stenholm, *Opt. Commun.* **36**, 75 (1981).
[81] J. Eiselt and H. Risken, *Phys. Rev. A* **43**, 346 (1991).
[82] B. J. Dalton, B. M. Garraway, J. Jeffers and S. M. Barnett, *Ann. Phys.* **334**, 100 (2013).
[83] S. Stenholm, *Phys. Rep.* **6**, 1 (1973).
[84] R. P. Feynman and A. R. Hibbs, *Quantum Mechanics and Path Integrals* (McGraw-Hill, New York, 1965).
[85] J. Zinn-Justin, *Path Integrals in Quantum Mechanics* (Oxford University Press, Oxford, 2005).
[86] A. M. Smith and C. W. Gardiner, *Phys. Rev. A* **39**, 3511 (1989).
[87] Q.-Y. He, M. D. Reid, B. Opanchuk, R. Polkinghorne, L. E. C. Rosales-Zarate and P. D. Drummond, *Front. Phys.* **7**, 16 (2012).

Index

The manufacturer's authorised representative in the EU for product
safety is Oxford University Press España S.A. of El Parque Empresarial
San Fernando de Henares, Avenida de Castilla, 2 - 28830 Madrid
(www.oup.es/en or product.safety@oup.com). OUP España S.A. also acts
as importer into Spain of products made by the manufacturer.
Printed and bound by CPI Group (UK) Ltd, Croydon, CR0 4YY
14/09/2026
02216307-0020